Science as Natural Philosophy and Finding Our Place in the Universe

Science as Natural Philosophy and Finding Our Place in the Universe

By

Richard L. Summers

Cambridge Scholars Publishing

Science as Natural Philosophy and Finding Our Place in the Universe

By Richard L. Summers

This book first published 2024

Cambridge Scholars Publishing

Lady Stephenson Library, Newcastle upon Tyne, NE6 2PA, UK

British Library Cataloguing in Publication Data
A catalogue record for this book is available from the British Library

ISBN (10): 1-5275-5535-6
ISBN (13): 978-1-5275-5535-8

Dedicated to my grandchildren, the joy of my life

TABLE OF CONTENTS

PREFACE

Science is increasingly answering questions that used to be the province of Religion. —**Stephen Hawking, physicist**

In 2020, the battle-cry during the COVID-19 pandemic was “follow the science”. This universal collective mindset emerged because Science was deemed to be our only possible savior and reliable guide to lead us in a time of existential uncertainty. At that time, there were questions around the meaning of the laboratory testing, the populations that should be targeted, and the efficacy of a myriad of treatments. But ultimately there were no guarantees of any protection for our health and survival and the real limitations of Science were for the first time laid bare for all to see on a global scale. Yes, the idealized absolutes of Science offered us the greatest hope for a pathway forward in this time of trouble, but the biases and confounding perspectives of a frightened humanity did not provide resolute clarity to the direction we should take.

As a physician-scientist on the frontlines of the battle against this pandemic, the uncertainties of those tentative endeavors compelled me to once again reflect on the role of Science in society and in my life personally. For the ancients, Science was considered as Natural Philosophy with the purpose of providing a better understanding of the natural world and our place in the Universe. With the emergence of some powerful methods and tools during the Scientific Revolution of the 16th and 17th centuries, a myriad of new discoveries fueled the Industrial Revolution as a practical means to harness Nature for practically improving the human condition. However, in our quest for these new technologies there were also some surprising revelations born of discoveries in the modern era concerning the realities of the Universe that have had significant religious and philosophical implications.

The Scientific Revolution is thought to have really begun with the publication of Copernicus’ heliocentric theory. In fact, the contemporary meaning of the word *revolution* as a radical and complete change is derived from his heretical notion that the Earth “revolved” around the Sun. That idea was indeed revolutionary in that it redefined our place in the Universe. No longer were the affairs of humanity considered as the centerpiece of all that was known. This Copernican revolution is one of the clearest hallmark examples of the possible impact that pure Science can have on our philosophical, religious, and spiritual perspectives and serves as a thematic thread interwoven in the discussions provided.

In this book, I explore the possible return of Science to a role as Natural Philosophy and a pathway to understanding our place in the Universe. In this journey, I draw heavily from my prior work (*Experiences in the Biocontinuum*) in which I searched for an explicit understanding of Life's foundational mechanism by utilizing a detailed and rigorous scientific approach. The insights of that endeavor and a lifetime of considering the roles of Science and Religion in comprehending the Universe have led me to the conclusions expressed in this current exploration. The content is heavily supported by frequent quotes by major thought leaders through history in the various areas of Science, Philosophy, Religion and the Arts. There is also some intentional repetition in the presentation of certain complex ideas from varying perspectives and in different chapters and sections in order to provide for better clarity. The references provided herein also serve as additional sources supporting the ideas and content of this current work.

CHAPTER 1

THE CENTER OF THE UNIVERSE

The Historical Perspective

Finally, we shall place the Sun himself at the Center of the Universe.
—Nicolaus Copernicus, priest, astronomer

On May 24th in 1543, the Polish astronomer and Catholic canon cleric, Nicolaus Copernicus, lay writhing on his deathbed, partially paralyzed by a stroke he had suffered in the preceding winter. An advanced copy of his just published book entitled *De revolutionibus orbium coelestium* (On the Revolutions of the Celestial Spheres) was placed in his 70-year-old hands. The book contained the final conception and details of a radical idea that he had first proposed in 1514. However, for so long he had been much too afraid to ever publish his conclusions. Copernicus' notion that the planets revolved around the Sun rather than the Earth was in direct contradiction to two millennia of common wisdom and church doctrine. In this revolutionary Copernican heliocentric view, Earth and man's concerns were no longer the Center of the Universe. Strangely, Aristarchus of Samos had originally proposed the heliocentric concept around 300 BC without much attention. Nevertheless, in the world of Christendom such ideas could certainly result in Copernicus' execution as a heretic in the 16th Century.

As he gazed on the pages of his life's work, Copernicus took his last breath and ironically passed into a central place in the history of Science. In fact, the Scientific Revolution as a major hallmark in the history of Western Civilization can arguably be said to have begun with this event. Even the modern meaning of the word "revolutionary" originates from a recognition that Copernicus' idea was a radical transformation from the norms of his time. Prior to this Copernican revolution, the accepted scientific perspective was consistent with Artistolean Greek and Ptolemaic cosmology that included Earth as the center of all the known Universe. However, at that time it was not Science in the modern sense of the word that considered questions like these as they related to the functioning of the natural world. Such deliberations were considered as a part of Natural Philosophy and emanating from reasoned thought rather than relying primarily on empirical evidence. In fact, it seemed obvious to the common observer that the planets and stars were moving around us in the night sky and that the earth stands fixed in space. After all, shouldn't humanity be at the center of everything anyway? Building on the Artistolean vision, Ptolemy developed a complex geometric system of dynamics as the foundation of an astronomy that explained the strange movement of the planets and stars based on the geocentric position. An arising Christendom embraced this viewpoint because of its consistency with the biblical creation narrative and the Old Testament citation that declared Joshua prayed until the Sun stood still. So clearly it was the Sun and not the Earth that was in motion.

Surprisingly, Copernicus' seemingly heretical conception did not initially create any great consternation with the papacy or scholars of the time. Possibly because the publishing editor, Osiander, included a preface to the book that asserted that the work was simply a hypothesis and a methodology for simplifying calculations as compared to the convoluted Ptolemaic system. Furthermore, he contended that this new Copernican proposition said nothing about the true causation of celestial motions.

In 1593 the Italian Dominican friar Giordano Bruno was put on trial for heresy and burned at the stake by the Roman Inquisition based on charges related to his cosmological and religious views. In addition to his acceptance of the Copernican model, he proposed that the stars were distant suns with their own planets and possibly Life. In his world-view, the Universe was considered infinite and therefore could not logically have a center. Obviously, Bruno's cosmology also informed his religious beliefs, particularly regarding the lack of a special position for humanity in creation.

The heliocentric theory really came to widespread public awareness after a notorious conflict with the Roman Catholic orthodoxy when Galilei Galileo defended the idea based on astronomical observations made with his new telescope device that was not available in Copernicus' time. Pope Urban VIII asked Galileo to objectively provide arguments for and against heliocentrism. In the now famous book, *Dialogue Concerning the Two Chief World Systems*, Galileo presented a biased perspective which depicted those arguing against the heliocentric view as being fools. This stance resulted in a scandal from which Galileo barely escaped execution and he was confined to house arrest for the rest of his life. In 1992, Pope John Paul II conceded that the Church had wrongly condemned Galileo and insightfully recognized that there is a formal distinction between the literal words of the Bible and their interpretation.

In the years subsequent to Galileo's controversial contentions, objective corroboration and validation of this heliocentric theory accumulated including considerably more accurate astronomical observations by Tyco Brahe and Johannes Kepler in the early 1600s. Additionally, Ferdinand Magellan had already circumnavigated the globe in 1519 and proved that the Earth was indeed a round sphere like the other planets. This emerging evidence allowed Isaac Newton to begin to wrest Earth from the center of the Universe and causally describe the dynamics of the both planetary motions and other earthly phenomena through a unified system of basic physical laws. Newton's real focus of study was concerning the force created by the center of gravity of a mass rather than a specific location for that center. Interestingly, Newton spent the majority of the latter years of his life studying theology and reinterpreting the scriptures from a uniquely heretical perspective.

Along this same time, the English seafaring traders had a great need for accurate astronomical measures to aid with navigation. Of particular importance was the designation of a specific fixed meridian from which longitude could be universally determined using chronometers. In 1674, King Charles II appointed a commission (including Newton's friend Christopher Wren) that led to the establishment of the Royal Observatory at Greenwich, London to designate Prime Meridian Time. John Flamsteed was appointed as Royal astronomer and for many Europeans at that time this location served as their practical center of focus and point of reference.

Subsequently, revolutionary thinkers such as Charles Darwin and Sigmund Freud extended the conception of Newtonian-like cause-effect mechanics to the scientific areas of biology and psychology and furthered the perception that humanity might not be the center of God's constant attention. In order to resolve this apparent inattentiveness of the deity to the affairs of humans, the well-known philosopher mathematician and scientist Pierre-Simon Laplace declared in 1814 that the Universe was like a great clockwork machine that had been set in motion by its detached creator and was solvable if we only knew all the factors.

If there were anything we could discover in Nature that would give us some special insight into the handiwork of God, it would have to be the final laws of Nature. —**Steven Weinberg, Nobel physicist**

Eventually the mounting astronomical evidence compelled Christendom and the general scientific community into a common acceptance of the heliocentric view. However, a compromise view arose amongst some religious clerics that considered our solar system in the heart of the Milky Way Galaxy as the new center of God's concerns. It was only when there was a documented observation of a separate Andromeda Galaxy that the public began to understand the vastness of a Universe that might not have a real locational center as we know it. Over time, humankind began to acknowledge and even embrace their new diminished position with a sense of awe for the grandeur of our insignificance as a component of religious humility.

A religion...that stressed the magnificence of the Universe as revealed by modern Science, might be able to draw forth reserves of reverence and awe hardly tapped by the conventional faiths. **—Carl Sagan, Astronomer**

When Edmund Hubble first demonstrated the redshift in the spectrum of observable light coming from distant galaxies, there was the further realization that the Universe was also expanding and therefore could potentially be traced back in time. With this revelation and through analyzing Albert Einstein's equations for General Relativity, Catholic priest and mathematician Georges Lemaître then concluded that there must be a beginning to the Universe. For many, this was again evidence of God's act of creation and seminal center of everything. Fred Hoyle who emphatically believed in a static Universe, disparagingly termed the origins of this expansion process as the "Big Bang". The final convincing evidence for a Big Bang came with the Nobel Prize winning discovery of the cosmic microwave background radiation by Arno Penzias and Robert Wilson. However, those radiation signals surprisingly came from all different directions and the idea of a central starting place was once again lost. Alan Guth's inflationary model describing the initial moments of Big Bang expansion on the surface of a bubble of curved space-time explained the ubiquitous radiation findings but also indicated that the notion of an actual positional center to the Universe did not make logical sense in this view of expanding time and space.

Look deep into nature, and then you will understand everything better.
—Albert Einstein

Despite the arising indications that there is no physical center, an innate driving desire for some position of grand relevance continued to stoke a common feeling that humankind's existence must play some central role in the Universe. In the beginning of the twentieth century, a new scientific perspective sometimes referred to as the "Copernican reversal" arose from the conclusions of relativity theory and quantum mechanics. While these fields of physics did not resolve anything regarding a universal center, there was definitely evidence that the observer of any phenomenal experience had a privileged position in determining how that reality was manifest. In fact, conceptual frameworks such as those considering the prospect of an underlying static ether or the exact position and momentum of a particle in space were no longer deemed valid. In light of these unusual findings from the physical sciences, many modern thinkers began to reimagine the true nature of reality, especially drawing heavily on the ancient ideas found in eastern mysticism.

A Different Kind of Center

Scientists have become the bearers of the torch of discovery in our quest for knowledge. **—Stephen Hawking, physicist**

Early Greece was the cradle of critical thinking and reasoning based on employing a combination of observations and logical deductions. There were two great schools of thought that emerged in that ancient period that sought to understand the natural workings of the Universe beyond the established mythologies of the time. Natural Philosophy had its origins with Anaximander, Democritus and their followers who attempted to explain phenomenal dynamics based on more fundamental and naturalistic causes. For instance, in considering the paradox of Zeno, Democritus concluded that the Universe had a basic granular or bit-like nature and was the first to develop the concept of the indivisible atom. This perspective bounded the concepts of infinity and considered all phenomena as the product of the interactions of these atomic particles. By stark contrast, Aristotle adopted the view that a Final Causation was an inherent property of all matter, living and nonliving. He concluded that this innate causal force ultimately determined the path of all dynamics. This teleological based perspective was undoubtedly influenced by Aristotle's detailed study of biology with its seemingly purposeful nature. Thus, through the Artistolean methodology the causality of phenomena and the true nature of reality could be discovered solely by a process of reasoning with end-purposes in mind and without a heavy reliance on observational evidences.

Unfortunately, the Artistolean form of Natural Philosophy dominated the path of Science for most of the history of western civilization. This ascendancy was probably due in part to the fact that the naturalistic approach of Anaximander and Democritus implicitly required specialized techniques and instruments for precision observations that were not available until the time of the Renaissance craftsman. But most importantly, Aristotle's perspective of a preordained causality was in consilience with most theological ideologies that championed the role of the creator in determining the detailed outcomes of the Universe. Naturally, humankind was judged as the primary recipient of these conditions as the central focus of a divine creation and therefore held a privileged position. And since humans lived on Earth then certainly that location should be considered the Center of the Universe.

Even before Aristotle, the Center of the Universe was often considered by many cultures as some kind of physical location that usually had some religious significance. For the Jewish nation, Jerusalem and the promised land of Canaan was the Holy Land. For Islam it was Mecca. For the Greeks, the gods lived on Mount Olympus. For the Aztecs, Teotihuacan was considered their universal center as connected to the deity.

Today, modern physics and cosmology does not provide for any conception of a Center of the Universe. Furthermore, neither biology nor the psychological sciences places humanity in a "crown of creation" position. In fact, such notions are really considered by scientists as an archaic relic of our past mythologies and antithetical to our current scientific understanding of the Universe. This view is particularly true if the center is considered as an exact position in space and the term Universe refers solely to the physical infrastructure of solar systems and galaxies.

Considering alternative non-positional definitions for the term "Center and Universe", we know from the American Heritage Dictionary that a center is otherwise abstractly defined as the point from which all activities or processes are directed or emanate. As a verb it is the focal cause or central directive. Likewise, the term Universe can mean more than just a physical entity with a location or position in space but can also be applied to indicate the total realm of existence, experience and knowledge. For instance, there is also the universe of ideas and the sphere of experiences, past and present. Such realms are generally observer dependent and generated through intentionality whereby the process for the origination of this type of universe is central to its actualization. This observer-oriented conception is foundationally a different kind of center in a realm that is also a different kind of space known as the biological continuum or biocontinuum.

Reality is therefore as much in the connection (relationship) as in the distinction between the open system and its environment.
—Edgar Morin, complexity theorist

The biocontinuum is considered a coherent information space that includes everything that could have a potential experiential interaction and information exchange with the living system processes which are in turn responsible for generating, actualizing, and sustaining the conditions of the space. Therefore, the biocontinuum space includes all possible energy, material and informational exchanges as well as communiques originating from within or external to the usual considered boundaries of the organism. As a continuum, there is a no real distinction between the living system and its adjacent environment since the considered space includes everything within the experiential realm of the life processes. This scope also includes the organism's experience of its own internal state. The inseparability of the open living system from its environment, particularly with regards to exchanges in information, is becoming a common notion in modern theoretical constructs where the systems themselves are considered as an intrinsic part of the environment as a whole.

The main character of any living system is openness.
—Ilya Prigogine, Nobel chemist

The major advantage of this conceptualization of the biocontinuum as a coherent information space including the entirety of the experiential realm is that it allows for the mathematical formulation of the possible states of the space in terms of Shannon information. This information is first captured with perception and then differentiated and deconstructed to meaning as the biological process of knowledge acquisition progresses. With further ongoing experiences, a Bayesian process of updating through acquired knowledge can also be incorporated. This information-based biocontinuum framework allows us to axiomatically track the general process of experience as the key element in the emergence of Life. The specifics of the formulation of this framework will be discussed in greater detail in Chapters 4 and 5.

Perhaps, a new kind of Center of the Universe can be considered as the point of origination of all experiential existence and knowledge. From that point of origin, the acquisition of knowledge, the practice of Science and our conception of the Universe can be examined from a uniquely biocentric perspective. This path will then be a new means to understanding with a possible reconciliation of scientific and religious ways of thinking.

Why Do We Need a New Way of Understanding?

I am against Religion because it teaches us to be satisfied with not understanding the world. —**Richard Dawkins, evolutionary biologist**

As previously noted, the Greek way of scientific thinking lasted for thousands of years and dominated all thought throughout the Dark Ages and into the Renaissance period. While the term Natural Philosophy is no longer used to designate the accepted methodology for understanding Nature, its expression as a philosophy reflects the fact that it was a competing worldview for much of our history. As a philosophy, all knowledge acquisition prior to the Scientific Revolution depended less on 3rd party objective empirical evidences and relied more heavily on reasoning and logical deduction from assumed universal principles that were mostly handed down by Aristotle. The salient difference in this approach from the practices of modern Science mainly stems from the ancient's lack of sophisticated and accurate tools for observation.

During the Renaissance, craftsman, engineers and artisans began developing many new tools for measurement and construction because of the increasing need for precision in their work building large structures such as cathedrals. Additionally, the building of multistage industrial systems such as windmills and water turbines to generate a specified action of work helped shape our understanding of how causal chains are mechanically produced from connecting the individual components. At that time, Galileo Galilei also emerged to become what was probably our first modern scientist. Galileo, Newton and others began to formulate generalized mathematical descriptions of simple physical phenomena based on more accurate observations using instruments such as timepieces, prisms, and telescopes in what was the beginning of our modern Scientific Method. It was the unwavering requirement of this new method for objective proof through testing that differentiated Science as a practical endeavor. In this process, the natural world was considered as the ultimate arbiter and judge of truth from which there could be little or no debate. Today, this empirical testing of a theory is still the aspect of the modern Scientific Method that is considered the hallmark that separates it from philosophy and theology. The terminology "Natural Philosophy" was later changed in the 19th century by William Whewell to the Latin scientia, meaning knowledge, because of the realization that Science as a discipline was really about getting to objectively know the world and how it works without the encumberments of philosophical considerations.

The human foot is a masterpiece of engineering and a work of art.
—Leonardo da Vinci, renaissance thinker

This practical approach to knowledge acquisition opened up the applied methodology of Science to be used as a means for improving the human condition. As such, it also became a bridge between the world of the craftsman-artisan who depended on tools and the academic view of the intellegencia who championed purely deductive thinking. Leonardo da Vinci is an example of a Renaissance artisan who began to bring these two approaches of thinking together. His passion for art and architecture uniquely intersected with his mechanical acumen. Eventually, the practice of Science, with its success in causal prediction and precision, became synonymous with technological development. Science became the key in advancing the Industrial Revolution of the past two centuries and the subsequent nuclear, space, and information ages and forthcoming biomedical revolution.

Along this trek towards greater technological development, we have lost the original intent of Natural Philosophy as a means to help us truly understand our place in the Universe and as a way for balancing the relative importance of inductive information and deductive reasoning. This balancing is once again particularly important as we are now reaching the limits of our observational capabilities and are frequently relying on the internal mathematical consistency of our theories in providing an understanding beyond what our senses and instruments offer in direct experimental experience. While modern Science provides us with a clear methodology for understanding how things work from a causal perspective, it is ultimately the lens of the individual observer's experience through which we actually get to know the world holistically. And the serious limitations and paradoxes of that experiential process in the practice of the Scientific Method have repeated emerged in a variety of areas of study. From the self-referential enigma of Gödel's Incompleteness Theorem to the Heisenberg Uncertainty Principle of Quantum Theory, the position of the observer has been at the heart of a disconnection between Science and an understanding of our Universe. The importance of this unique observer's role is also critical in comprehending such scientific ideas as the 2nd Law of Thermodynamics, Special and General Relativity, Information Theory, and the Anthropic Principle. According to modern physics, so central is the perspective of the observer in the determination of perceived reality that the renowned physicist John Archibald Wheeler resolved that we really live in a participatory Universe. This articulated relationship with the Universe should be the basis for a new understanding and a novel central position for the experiential processes of living systems from which the entirety of known reality emanates and the created becomes the creator.

CHAPTER 2

THE NATURE OF EXPERIENTIAL REALITY

The Foundations of Reality

From those early days of the Greek Philosophers it was recognized that if there is to be a discovery of absolute truth then there must first be an agreement on the source of reality. Idealism, as developed by Plato, asserts that reality is derived from idealized mental or virtual constructs. For instance, a truly perfect geometric circle does not actually exist in Nature but the idealized mental construct for such a circle is the reference point for our thinking and communicating about the reality of such objects in the world. Plato's famous allegory of people chained in a cave with their backs to the entrance that only ever know of the projected shadows of objects on the wall also highlights the role of the individual's experiential condition on the perception and understanding of reality. By contrast in the extreme, Materialism proposes that the actual physical objects are the only true reality and that these objects exist objectively outside of our experience of them. In the 17th Century, Descartes was concerned with reconciling the subjective processes of the mind with the objective analyses of Science and proposed a mind-body dualism in the determination of reality.

> *If a man will begin with certainties, he shall end in doubts; but if he will be content to begin with doubts, he shall end in certainties.*
> **—Sir Francis Bacon, philosopher**

By the time of the 19th Century, influential thinkers such as Sir Francis Bacon, John Locke and David Hume had won the soul of Science by considering empiricism as the grounding principle for the only reality from which we could discover scientific truths. Based on purely objective empirical evidences, Science became more concerned with describing the workings of Nature as "how" questions rather than philosophical conceptions or the "why" questions. Knowing how things work was particularly critical at this point in our history as we attempted to harness the mechanics of the natural world for the burgeoning Industrial Revolution. However, using a strict induction procedure from empirical observations as a basis for reality was inherently fraught with some variability and inconsistency in determinations. By definition, empirical evidence is based on an individual's observations and experiences. Consequently, the multitude of observations events of various individuals were coalesced into more generalized deduced theories concerning how things worked. These theories were then viewed as simplified approximations to the conditions of reality and the mechanics of phenomena that could be applied broadly. While these determinations often carried the illusion of dogmatic certainty, they were really subject to the idiosyncratic conditions under which the evidences were procured and it was not uncommon for theories to change significantly.

As the investigations of Science became more detailed and complex, such as in the case of Einsteinian relativity, quantum theory, and Big Bang cosmology, some of the theories had a metaphysical leaning that challenged our conventional notions of scientific reality. Is Schrödinger's cat dead or alive? Is there really a multi-verse? Is string theory a valid hypothesis? At the heart of this trending shift in the paradigm of Science is the role of the observer. The modern approach to Science tries to reconcile all of these perspectives with new considerations for the origins of objective and subjective knowledge. But unfortunately, the boundaries between the reality of the physical world and the observer, which is also a part of the physical Universe, are still blurred.

In the prior chapter, we charted our evolution in thinking about the position and role of humanity in the Universe. Through this historical journey in Science, researchers have learned much about the structure and character of reality as seen from our role as observer/participant. But what are the salient lessons from physics, mathematics and biology? Does the evidence for a central position for the observer represent a real disconnection between objective Science as traditionally practiced and our true comprehension of natural phenomena? Moreover, do these lessons call us to return to the original intent of Science as Natural Philosophy for understanding our place in the Universe? In this chapter, we will use these lessons to begin to explore some of these questions in greater depth in order to find our way on a new path forward.

Man has to awaken to wonder - and so perhaps do peoples. Science is a way of sending him to sleep again. **—Ludwig Wittgenstein, philosopher**

Lessons from Physics

Einstein's Relativistic Thinking

In his famous "miracle year" of 1905, a young Albert Einstein was fascinated with the long forgotten Galilean principle connecting conditions of inertia with the relativity of dynamics. Galileo first described this principle in 1632 in which he noted that the laws of motion are relative to the inertial frame of reference of the observer. Stated more simply, it is impossible for an observer to determine if a seemingly stationary object was actually moving or at rest without some outside source of reference. In the Special Theory of Relativity Theory, Einstein postulated that the laws of physics are invariant in all inertial systems in uniform motion and that only the speed of light is constant and independent of its emitting source. To reconcile these propositions, time has to be considered as a variable like all other dimensions. Therefore, as the observer moves in space, experienced events could occur at times that differ from another observer moving at another speed or in another direction. Prior to that seminal work, the frame of reference for all existing natural phenomena was the absolute ether space of Newton.

Einstein also suggested that time and space are as much a part of the physical world as matter and energy and are interwoven into a unified continuum that cannot be separated. Furthermore, the conditions of space-time can change based on the frame of reference of the observer. In fact, there is no observer independent structure of space-time. This space-time structure also results in a relativity of simultaneity which is a concept in physics first consider by Henri Poincaré in 1900 and later adopted as a central idea in Special Relativity by Einstein. Because the speed of light is constant, an observer's momentary experience of spatially separated events does not mean that those events occurred at the same instance in local time. Therefore, any notion that there is an objective simultaneity of events is meaningless. And for any individual experiential frame of reference, there is no global or universal time. For example, any current observations of the stars in the night sky are really the result of events that happened light-years ago. Because the observer's distance to each individual star may differ, there is no synchrony in the timing of the emission of those rays of starlight. Additionally, gravitational forces that distort the space-time continuum can also alter when events occur. Hence, all our present perceptions and experiences really spans the course of both time and space.

When a new quality is manifested, something which doesn't have a name, something which you can't recognize, or more exactly you recognize as something which you can't identify, then that is exactly the moment of Becoming. **—Georg Wilhelm Friedrich Hegel, philosopher**

For any individual observer, there is a clear past and future in time. The past exist as a memory of coalesced phenomenal occurrences and the future holds an imagined temporal position of events yet to come. Both are established mental constructs. However, as described above, the present as experienced is not a bounded occasion in time and space. The present has a unique locus which Carlo Rovelli calls the expanded or extended present that is determined by the interwoven structure of space-time and the observer's frame of reference. Science is not so much about discovering the reality of "things" as it is about understanding their dynamics and interactions. Rovelli suggests that we can consider reality as a complex web of "*becoming"* that is dependent on relative relational interactions. This process of becoming is reminiscent of the Buddhist holistic concept of a co-dependent arising of all things in the creation of the Universe. In this way, we can think of the reality of the present as a continuous becoming that is the "now" created out of the past and thrust into the future. The work of Einstein was a definitive philosophical sea change from the typical concept of phenomenal dynamics as driven by direct forces and mechanistic causality as codified in the Newtonian laws of physics. Coincidentally, this relativistic shift in perspective was also consistent with the ancient Buddhist notions of impermanence as the essential nature of all things. The relativistic concepts of simultaneity and the extended present as a process of becoming are also critical to exploring the ideas of the interconnectedness of the Universe across both time and space in subsequent chapters.

There is nothing permanent except change. No man ever steps in the same river twice, for it's not the same river and he's not the same man. **—Heraclitus, ancient Greek philosopher**

Even before Einstein, the rederivation of phenomena by Lagrange, Hamilton and others had emphasized the identification of the right action functional driven by changing energy balances as a more general framework for analyzing dynamics. If a mathematical construct correctly describes the energy accounting of a system then the path for dynamic actions naturally flows from this background assemblage. With the curvature of space-time in general relativity, the movement of objects in gravity no longer had to depend on the "action at a distance" of Newtonian gravity. Dynamics in this space-time geometry flowed according to the natural shape of the mass induced gradient curvature of space-time as an expected action. While the dynamics of relativity organize the context of experience for the observer, it is unlikely that the relativity of reality stops with the basic experience of the dimensions of time, space and gravity. Obviously, other observable metrics such as colors or textures are to some degree similarly relative to the observer. Ask any colorblind person. Perhaps, all perceived information that is dependent on the experiential conditions of the observer is also relative.

Heisenberg Uncertainty and Schrödinger's Cat

What we observe is not Nature itself, but Nature exposed to our method of questioning. **—Werner Heisenberg, Nobel physicist**

Quantum mechanics is the field of study concerned with physical processes involving extremely small particles such as photons, protons and electrons. It was first developed as a field of study in the early part of the 20th Century stemming from Max Planck's work in solving the black-body radiation problem intended for understanding the physics of the new invention of the light bulb. According to Planck, the energy for the action of a particle (E) is proportional to its frequency (v) as multiplied by h (Planck's constant).

$$E = hv$$

Initially there was some trepidation amongst physicists with this result in the realization that the frequency is a discrete quantity, whereas Newtonian mechanics assumes that all energy transactions occur as a continuous process. Thus, the discretely quantized behaviors of these particles seem to contradict some basic tenets of classical physics.

Because of the revolutionary implications of the theory regarding the basic nature of physical reality, Planck insisted that he had only discovered a mathematical trick to calculate the absorption and emission of radiation. However, in 1905 Albert Einstein used Planck's quantum method to explain the photoelectric effect for which he won the 1921 Nobel Prize in Physics. Einstein theorized that when light hits the surface of an object, energy is transferred as quantized packets of energy known as photons. This practical application of the strange theory brought immediate credibility to the quantum approach.

Ironically, in subsequent years Einstein was probably the most fervent opponent to the radical notions concerning the nature of reality suggested by quantum theory. In response to the conundrum, Schrödinger's wave function equation was formulated to describe the deterministic behavior of particles acting as both waves and particles. However, from these developments it appeared that the actual experimental process of observation in itself always disturbs the condition of the particle and provides an intrinsic degree of uncertainty in knowing its true state. This conditional uncertainty left the state of the particle in limbo or superposition until it is actual experienced and measured. This perplexing scenario is the basis of the Heisenberg Uncertainty Principle first introduced by the German physicist Werner Heisenberg in 1927. From his proposition, quantum uncertainty became quantitatively characterized by a determined probability distribution of measurement outcomes during the observation of a particle's state. Moreover, the very act of measurement converts quantum indefiniteness into the classical definitive.

To observe is to disturb, for observation breaks the wholeness of Nature.
—Fred Alan Wolf, physicist, author

But uncertain to whom? To God? To the Universe? Or just to the scientists trying to make the measure? The term uncertainty denotes a lack of knowledge about the true state of the particles and suggests that these particle states exist only as potentials within a spectrum of probabilities until they are actualized in the experiential process. In fact, "indeterminacy" (unbestimmtheit in German) was the original name that Heisenberg used to describe his principle indicating the "unknowability" (ungenauigkeit in German) of the particle state before it is observed. The difference between uncertainty and indeterminacy is the difference between not knowing because it hasn't been definitively measured and not knowing because it can't be determined within the act of measuring. And through this act of measuring or observation, the dimensions of the particle are actualized and interpreted. Therefore, the principle's indeterminacy really describes the ontological physical ambiguity concerning the particle's state and posits that even its exact position and momentum can only be established within the error limits of human observational knowability. Of course, that limit is the quantum of the light photon, which is really our best direct tool for probing the subatomic levels.

A phenomenon is only a phenomenon if it is an observed phenomenon.
—John Archibald Wheeler, physicist

To resolve these perplexing considerations, the "Copenhagen" interpretation of quantum mechanics was devised by Niels Bohr and Werner Heisenberg to provide a new understanding of physical reality. According to their interpretation, physical systems do not have definite properties prior to being actually observed and the methods of quantum mechanics can only predict the probabilities that the measurements will produce certain results. Accordingly, reality only exists as a potential until it is actualized by the experiential process. Therefore, it appears that the perception of reality is implicitly a fundamentally active participatory process by the observer as noted by John Archibald Wheeler.

When large numbers of particles are assembled together, their wave interferences tend to cancel each other out in a way that results in their perception as tangible objects. This is akin to many pebbles hitting the pond at the same time and their ripples cancelling each other out to produce the appearance of still water. Fortunately, the contents of our experiences have considerable inter-subjective agreement, particularly at the macroscopic level. This consilience is likely due to the similarity in our biologic constitution. For instance, we cannot be certain that we all experience the same blue color but we generally agree on which shading is designated blue because at a very basic level the signal created by the varying excitation of a cluster of rhodopsin molecules caused by the differing photon frequencies of assorted colors of light are the same for all humans. This distinction may be quite different for other phyla or even other species of organisms.

The more we know about our Universe, the more difficult it becomes to believe in determinism. **—Ilya Prigogine, Nobel chemist**

Einstein rejected the viewpoint of the Copenhagen interpretation because it lacked clear deterministic causality. Even the alternative explanation of a collapse of the wave function upon measurement seemed somewhat arbitrary. Other notable scientists also had problems with the non-deterministic nature of quantum mechanics and argued that the theory was incomplete and that there must be an underlying structure with "hidden variables" that were yet unknown to Science. Great thinkers such as Louis de Broglie and David Bohm joined Einstein in the quest to find a new theory which included this hidden framework of causation. But every attempt at deterministic resolution included the implication of strange unseen causal effects or mysterious explanations such as unknowable parallel universes or intermediary concealed variables that all seemed to be somewhat contrived.

> *This does not mean that the Universe is absurd or meaningless, only that an understanding of its existence and properties lies outside the usual categories of rational human thought.* **—Paul Davies, physicist, writer**

In response to the uncertainty conundrum, Erwin Schrödinger proposed his classic thought experiment in quantum mechanics in which a cat is imagined as being sealed in an enclosed box with a poison that will be released when a radioactive source randomly emits radiation. Schrödinger argued that is absurd to consider that there is an absolute indeterminacy as to whether the cat is alive or dead before it is observed. However, in the standard Copenhagen interpretation, the cat is considered to be simultaneously both dead and alive until the box is opened and the cat is actually observed to determine it state. Prior to the observation the cat's condition is indeterminate and is said to be in a state of quantum superposition. The act of observation precipitates the final condition by collapsing the quantum wave probability equation and actualizes the state of reality. Therefore, the uncertainty within the Heisenberg Principle is not only due to an epistemic lack of information but also represents the degree of ontological equivocality of the state of things. The problem is that the Copenhagen view disrupts the causal entailment of natural phenomena because considering states as "possibles" fails the basic logic of the Principles of Noncontradiction and the Excluded Middle. In a solution for this noncausal context, Max Born interpreted the Schrödinger wave function as simply a probabilistic description of our knowledge for subatomic observations for which he won the Nobel Prize in 1954.

> *The Universe is under no obligation to make sense to you.*
> **—Neil deGrasse Tyson, physicist, author**

This crisis concerning determinism in Science resulted in great debates between Einstein and Bohr concerning the fundamental nature of experiential reality. During one debate, Einstein famously stated "God does not play dice" to which Bohr replied "quit telling God what to do". Since the time of the Fifth Solvay Congress, when Max Born, Werner Heisenberg and Niels Bohr declared that quantum mechanics was a complete theory, the debate has continued to this day. Some thought experiments such as the Einstein, Podolsky, Rosen paradox and more recently Wheeler's delayed-choice "gedankenexperiment" proposal have famously highlighted the logical conundrum of a violation of some of the basic principles of causality in quantum theory as evidenced by particle entanglement. Despite these early objections to requiring an observer for actualization, the concepts of quantum mechanics have been repeated validated and are found to successfully work in principle. This validation includes the recent experiments by 2022 Nobel Prize winners John Clauser, Anton Zeilinger, and Alain Aspect under the defining conditions set in Bell's Theorem by the Irish Physicist John Stewart Bell in 1964.

At the quantum level, reality does not exist if you are not looking at it.
—Andrew Truscott, experimental physicist

Not only does God play dice, but... he sometimes throws them where they cannot be seen. **—Stephen Hawking**

In the experiential process, the perceived information acquired by the observer from the continuum of potential physical states is deconstructed into discernible alternative possibilities that are assessed for meaning in the context of the observing system. In the process of translation to meaning, the possibles are actualized to a reality of entailing significance for the experiencing observer. Therefore, the biologically translated reality of perceived phenomena offers the illusion of determined causality because it also actually functions that way for the practical purpose of the survival of the living system. This practical meaning of reality as perceived for survival is encapsulated in the observation of philosopher David Hume who noted that that although we have no rationale for believing in an objective reality, we also have no choice but to act as if it is true. Such a pragmatic construal is necessary for survival, even if it is an illusion. This viewpoint indicates that the real fulcrum for comprehending reality and the character of our Universe hinges on a structured understanding of the biologic process of experience and knowledge acquisition with an interpretation for meaning. Such a framework is derived and detailed in the book *Experiences in the Biocontinuum* as a prelude to this current work. The experiential capacity for knowledge acquisition and translation to meaning is also considered the foundational process that defines living systems and is reviewed in Chapter 5. This biologic functionality may be even more fundamental than our current understanding of quantum physics that is dependent on an active process of observation. Furthermore, an extensive understanding of the biologic experiential process could be the new frontier of physics.

No man's knowledge here can go beyond his experience
—John Locke, philosopher

As noted, a modern understanding of quantum mechanics intimates that determinism ultimately arises from uncertain and indeterminate processes. Likewise, the determined whole created by the observer in these processes is also the platform responsible for originally creating the indeterminism due to its limits for knowledge acquisition. The notion of a circular loop causality in the creation of reality is even found in ancient mythologies such as the Egyptian and Greek mythologies of the Ouroboros and much of Buddhist philosophy. The famous mathematician and Nobel physicist Roger Penrose has suggested that the end of all things through the black hole may be tied to the beginning of a new Universe with the Big Bang in a continuously circular causal fashion that he calls conformal cyclic cosmology. Other physicists have also postulated the possibility of white holes from which light and matter emanates. Such mechanisms of continuous circular causality stand-alone without the need for an absolute creation origin. In fact, it can be said that causality really originates from an interconnected dependency. Interestingly, the ultimate unification of Einsteinian Relativity with the perspectives of Quantum Theory will also be necessary for us to understand the origins of our Universe at the point of the Big Bang. Interestingly, both these transformational ideas are founded on the idea that the involvement of an observer from a holistic position of interconnected dependency is fundamental in the establishment of all experiential reality.

Entropy and the Qualification of Work Energy

Thermodynamics is more like a mode of reasoning than a body of physical law. —**Steven Weinberg, Nobel physicist**

It has been said that the 2nd Law of Thermodynamics is probably the only true scientific Law of Nature. Entropy is also the most mysterious and misunderstood conception with a variety of different interpretations. Intuitively we all know that certain processes (like the breaking of an egg) are irreversible and that the world naturally tends to degenerate toward more disordered states. In fact, the 2nd Law requires the generation of such disorder in the form of entropy for all spontaneous energy transfer processes to occur. The key here is the term "disorder" and how is that state is determined or known and by whom or through what mechanism?

The term entropy is derived from the Greek "tropos" which means transformation or tendency and characterizes the natural tendency of energy flows. Energy induced matter spontaneously changes from being localized to being dispersed unless it is constrained. In this state as dispersed energy, entropy is not explicitly knowable and therefore not practically useful by the system to do directed actions or "work". Classically, entropy as a thermodynamic quantity represents the unavailability of energy in a system to be converted into mechanical work. Practically speaking, this means that if the state of energy is unknowable it is logically also unavailable to be applied to perform useful activity. This conceptualization was formalized by German scientist Rudolf Clausius in the 1850s and stemmed from engineering considerations that surfaced during the development of the steam engine at the beginning of the Industrial Revolution.

The idea of dissipation of energy depends on the extent of our knowledge.
—James Clerk Maxwell, mathematician, physicist

So, entropy is unique as a measure in that it carries an implicit attribute of uselessness for organized activity in the context of the observer's understanding. Organization of a system inherently limits the degrees of freedom of movement of matter and the flow of energy and naturally reduces the entropic disorder. Entropy has also often been interpreted as the degree of unpredictability, randomness or disorder of matter in the system that results in it being unknowable and unavailable to be channeled into a useful work process. Therefore, the energy "quality" is determined by the degree that it is knowable and available for work by an observing system. To be qualified as useful, the energy exchanges must be experienced by an observing system that can also ascribe meaning to the exchange (i.e. entropic disorder vs work). Regardless, the total energy of the Universe does not change during the exchange. Furthermore, the entropy transformation within a system does not change just because the quantity of system energy has changed. What has changed is the information we have about the energy within the system. This view goes to the heart of the intertwined relationship between the observer and basic energy concepts such as entropy and work. It also highlights the role of the experiential process in determining the ordering of reality as we know it.

To this point the discussion has been mostly about energy qualification (work vs entropy) from the viewpoint of an external 3rd person observer. This approach is not surprising considering that is the routine perspective of classical thermodynamics and most of Science since the time of Newton. In his 1944 groundbreaking book *What is Life?*, Erwin Schrödinger concluded that that an ability to feed on negative entropy or "negentropy" is the primary distinctive characteristic of living systems. Schrödinger explained that negative entropy really means free energy which is somehow available for work by the living system in strict thermodynamic terms and serves to counter the effects of the 2nd Law of Thermodynamics on the organization of these living systems. In other words, living systems develop processes whereby they can import free energy and export entropy to keep their internal entropy low and develop order and an organization.

It is a remarkable fact that the second law of thermodynamics has played in the history of Science a fundamental role far beyond its original scope.
—Ilya Prigogine, Nobel chemist

From the viewpoint of any external observer, it would seem to require an enormous effort and free energy intake to maintain the degree of complexity seen in even the simplest living organism. This incongruency has been the greatest barrier to understanding how these complex systems have emerged, sustained themselves, and evolved. However, remember that a determination of work quality by an observer is centered on the certainty of knowledge of that observer about the energy state so it can be used for a directed work objective. For the living organism, the uncertainty or lack of information about its internal microstate by any external 3rd person observer is essentially inconsequential to its personal experience or functioning. Living organisms are coherent systems of organized complexity in which the dynamic connectivity of the integrated parts provides for instantaneous intercommunication concerning the state of all facets of the system. In other words, on some level the organism knows its own state and internal condition. This participatory control of energy qualification is uniquely intrinsic to the functioning of living systems and is only achieved through the experiential process. By knowing with certainty the entire state of the system first hand at each instance, the effective degrees of freedom in this process is essentially **1** and the internal entropy is near zero for the organism. This intrinsic knowledge of systemhood is what makes Life possible. The relatively small amount of entropy created to maintain this complex organizational state is expelled from the open living system.

There are also broader scientific implications realized from this condition of cohesion and coherency in the living system and its biocontinuum. If the known state of experienced reality is really qualified and actualized by the living system, then the 1st person perspective is a valid consideration as a way to scientific truth. All 1st person experiences are "lived" experiences that are inherently associated with subjectivity and the unique perspective of the person having the experience and ascribing meaning. This important idea will be explored in greater detail in a later chapter.

The Anthropic Principle

Something pretty mysterious had to give rise to the origin of the Universe.
—Richard Dawkins, evolutionary biologist, writer

As scientists continue to learn more about the structure and functioning of our Universe, an evident and perplexing observation has arisen that once again suggests the possibility of a central role for living beings. The physical Universe in almost every aspect possesses the very narrow and unique conditions that are exactly required for the emergence of Life. This seemingly odd "just right" specificity of physical properties required for Life to exist that is intrinsic to the essential structure of matter has led physicist Paul Davies and other scientists to refer to our Universe as the Goldilocks Enigma. Though this condition does not necessarily infer that Life is central in the Universe, it does suggest that Life's position as the designated observer is somehow privileged.

The statistical probability that organic structures and the most precisely harmonized reactions that typify living organisms would be generated by accident, is zero. **—Ilya Prigogine, Nobel chemist**

The phrase "Anthropic Principle" was first used by theoretical astrophysicist Brandon Carter during the 1973 Kraków symposium honoring Copernicus' 500th birthday. Carter created the expression as a refutation of the Copernican Principle that humans do not really occupy a privileged position in the Universe. Though Carter understood that Life may not necessarily be critical for the existence of the physical Universe, he recognized that its emergence is inescapably favored by the very fact that we are here. Carter described two possible versions of the Anthropic Principle. The weak Anthropic Principle (WAP) advances the argument that if the conditions for Life were not just right, then we would not be here and would not know that the Universe existed. By contrast, the strong anthropic principle (SAP) supported by some noted scientists such as Stephen Hawking asserts that the Universe and its fundamental parameters must be just as they are in order to allow for the inevitable creation of observers. Furthermore, Hawking contends that the fact that living systems exist to experience the Universe imposes constraints not only on the components of our environment but also on the possible form of the very laws of Nature. Could that be the reason the observer plays such a pivotal role in relativity and quantum theories? Some proponents of the SAP have even suggested the eventual emergence of Life as the observer is a necessary condition for the Universe to come into being. This is because it is thought that an observer is required to actualize the Universe's reality in the same way that observation transforms Schrödinger's cat from a superposition state to its determined reality. Ultimately, all versions of the Anthropic Principle reveal the natural linkage of living systems and the fundamentals of reality and therefore implies that there certainly is a privileged or central position for Life in the Universe as we know it.

Lessons from Mathematics

Gödel's Incompleteness Theorem

At heart, all of Life is founded on self-reference.
—Jesper Hoffmeyer, biosemiotic informaticist

Since the time of Pythagoras, mathematics was considered to have an independent existence or idealized basis in reality outside of human thought and the workings of the material world. After the publication of Sir Isaac Newton's Naturalis Principia Mathematica (Latin for Mathematical Principles of Natural Philosophy), mathematics became the unquestioned language of Science. Furthermore, the natural underpinnings for mathematics were considered as an intrinsic part of the real world. In the early part of the 20th Century, well-known mathematicians such as Bertrand Russell and Alfred North Whitehead began to search for the primordial foundations for all mathematical logic and reasoning. In 1910 they published the first of several editions of their epic treatise entitled Principia Mathematica (in deference to Newton) that sought to describe a set of axioms and inference rules in symbolic logic from which all mathematical truths could be proven.

The three major principles of logic that are the foundation for this development are:

1. Law of Identity which says that whatever is, truly is. (Example: a = a).
2. Law of Noncontradiction says something cannot both be and not be.
3. The Law of the Excluded Middle says everything must be or not be.

In 1931 a young Austrian mathematician named Kurt Gödel working on mathematical proofs had an extraordinary insight that led to his famous Incompleteness Theorem. The theorem showed that any mathematical system composed of a set of axioms and inference rules must inevitably be found to be inconsistent in some case and contain some supposed truths which could not be deduced from their reductionist foundations. In essence, he demonstrated that all such systems of logic will necessarily contain a statement that could never be proved to be true, even though it really is true. However, there is one important caveat. That proof for Gödel's Incompleteness Theorem is within the canonical 3rd person context which is at the heart of the Scientific Method.

Mathematics may be defined as the subject in which we never know what we are talking about, nor whether what we are saying is true.
—Bertrand Russell, mathematician, philosopher

Gödel's ideas rocked the foundations of all systems of mathematics and scientific theories based in mathematical constructs since there was no absolute assuredness of the consistency of any predictive derivations. What Gödel had really uncovered was a mathematical interpretation of the logic of an ancient enigma in philosophy first stated by Epimenides which is often called the "liar's paradox". Epimenides was a Cretan who made the declaration "All Cretans are liars". The evident paradox lies in the inherent fact that a Cretan was saying that all Cretans are liars. Is this statement also a lie? Another version often used is the simple sentence: "This sentence is false". Obviously, if it is false then it could not be true that the sentence is false. The failure in the axiomatic process occurs when there is a self-referential quality intrinsic to the statement itself that naturally creates paradoxes that cannot strictly be logically resolved from a dissected analysis of the details of the sentence by an external or 3rd person observer. Self-reference is simply a condition when a 1st person self-entity makes a distinction that refers to itself. In mathematics, such references in the processing of logical statements are of interest because of their tendency to create paradoxes and inconsistencies. As noted by Gödel, the reasons for this inconsistency appear to be the inability of such a system to insure absolute truth for an observing third party when the system refers to itself. Such is sometimes the case in mathematics when there is a system or sequence of equations with corresponding or shared variables.

The words of truth are always paradoxical. **—Lao Tzu, philosopher**

A potential resolution to the conflict may be found if we consider the sentence from the overall perspective of its meaning as a whole rather than from its individual components. Additionally, the true value of information from any logical or mathematical statement cannot be completely judged from the outside observer if the statement refers to the state of itself. Gödel's proof indicated that there is an emergent hidden implicit meaning in the totality of a mathematical formulation that is not apparent from its methodical examination achieved by building upward from an axiomatic analysis of its basic parts. This finding is akin to the discovery of the emergence of unique global properties in complex systems that cannot be explained by a reductionist analysis of its parts. That embedded meaning represents a third state of logic for an inference that the statement is not yet true or false but still to be determined by some unique observer who considers the statement as a whole and in the circumstance of its pronouncement. The same is also true of any statement when viewed in its entirety and the context of the setting rather than from the simple conjoining of the words. We experience this every day in our conversations when we understand the implication of the statement as a whole that is beyond the straightforward meaning of the individual words. An observer also can infer the meaning of the statement when considered as a whole and in the context and circumstances of the moment including the conditions from a historical perspective of prior common experiences. The key to understanding the disconnection of any statement from its parts is that the value and meaning of that statement comes only from a reflection on the system in its entirety and uniquely from the originator's perspective and intent.

God used beautiful mathematics in creating the world.
—Paul Dirac, Nobel physicist

With some statements (such as the Cretan statement) the only observer that is capable of accurately judging the truthfulness of the statement is the 1st person speaker himself. Epimenides knows if what he is saying is true or not and that is the reason that only self-reference can solve the problem. In this communication process, the valuation of the statement is not determined immediately but only with an iteration of the forthcoming information back into the entirety of the system that includes Epimenides himself as he observes or experiences his own statement. It is at this point of re-entry of the information into the system as a whole that valid self-reference and adjudication can be achieved and the truth known. The acquiring of such knowledge valuation requires the necessary inference of the truth of the statement as a whole and based on the observer's relationship with Epimenides and prior experiences. This inferred knowledge is said to transcend the meaning of the dissected parts.

The word transcendence derives from the Latin trans-, meaning "beyond," and the word scandare, "to climb. Therefore, to transcend literally means you have gone beyond ordinary expectations. The great 18th Century philosopher Immanuel Kant also thought there is a quality of knowledge that is considered transcendental and is not a primary ingredient of objects but rather a part of the way in which it is possible for us to know the object beyond that derived from the simple superficial sense impressions we have of them. In other words, to transcend objects is to really understand their meaning beyond the apparent perceptions. Likewise, the structural realists consider that it is the causal relations of interactions amongst objects that are the true primary ingredient of natural phenomena. Going one step further, the renowned physicist Carlo Rovelli suggests that the essence of all physical reality and its manifest meaning is actually fundamentally based on such relative relational interactions. Therefore, we can say that it is expected that the conditions of self-referential experiences might include inferred knowledge of an implicate order that transcends that directly imparted from the perceptual experience. That "beyond" knowledge element comes from the meaning of the greater whole of interdependent interactions and causal relations while incorporating the context of the experiencing observer, the environment, and including past experiences.

In order to overcome this problem of self-reference in mathematical systems, several attempts have been made to develop a calculus for self-reference. In 1972 Spencer-Brown developed an axiomatic system of primal logic known as the Laws of Forms. The basis for this system is that all structures, patterns and relationships can be distinguished as an abstract form and that the most primitive act of logic is the drawing of a distinction. This distinction is essentially a deconstruction of the form from the whole of the domain. This deconstruction process is akin to that characterized in the neurodynamics experiments of perception conducted by Christine Skarda that will be referred to later.

> *God is constantly Creating, Experiencing, Witnessing, De-constructing, and Expanding. God is everything and everyone – that's the realm where we don't exist at all. That's the realm where there's none, but One.*
> **—Manprit Kaur, spiritual writer**

Francisco Varela extended Spencer-Brown's foundational system to allow distinguished forms to re-enter their domain. In allowing direct re-entry, the deconstructed form is both the distinctive part and the observed whole at the same time and therefore a condition for self-reference is established. During the path for this re-entry procedure, the value or meaning of that form expressed from the initial act of drawing a distinction cannot be determined immediately because the form is included within the act itself. It is from this uncertainty of meaning of the form that the circumstance in which a paradox arises for the 3rd person observer. When the meaning of the distinction is wholly assessed and realized by the system making the declaration, then the truth is known internally through self-reference.

Spencer-Brown felt that the concept of time had to be brought into the logical procedure to overcome paradoxical problems with systems having inherent self-reference. This iteration of time allowance provides a means for the functional emergence and re-entry of the form so there can be a determination of meaning in relation to the whole as intended by the self-referring entity. Therefore, the 1st person observer does not typically reference an object or state in the present reality during a communication but rather that of the very immediate past after self-system meaning has been ascribed. However, experience may also have a more direct route primal connection to the present as independent of time iterations and valuations. This special condition of experience immediately and completely engages the whole as will be discussed further in a later section.

In the end, we self-perceiving, self-inventing, locked-in mirages are little miracles of self-reference. **—Douglas Hofstadter, physicist, writer**

In the mathematics describing the mechanics of this self-referring action, a recurrence relation is an expression that recursively defines a sequence in which the determined value of a term is defined as a function of the preceding same specific term in the re-entry process of self-reference. The term then serves as the argument, the function and the solution value all at the same time. This arrangement naturally results in a failure of the mathematical superposition principle and creates the condition of nonlinearity for the equation system. This failure is also the source of the uncertainty of the argument's value. Reflective or recursive statements can be mathematically framed by the general equation of the form:

$$f(f) = f$$

This algebraic expression, called the Ouroboros equation by Jorge Soto-Andrade in reference to the ancient archetypal form for the eternal cyclic reflexivity of destruction and recreation (see graphic below), allows the element f in the equation to concurrently operate as a part of the argument, function and result.

...he not busy being born is busy dying **—Bob Dylan, Nobel poet laureate**

It is easy to recognize that this type of expression is typically nonlinear in its overall geometry and predictive capacity when it is used in the solution of an iterative time-dependent process. Such a dynamic process continuously evaluates the state of the system and through feedback furnishes the input argument to form the global structure of the nonlinear system. The system components become self-reproducing in a circular causality and the specific overarching pattern that emerges does so only because of its self-referential quality. A capacity for self-reference is critically important for the functioning of living systems, particularly in the creation of system coherency and a reduction in the degrees of freedom for the control of entropy. A conditional necessity for self-referencing also exists because the processes of Life involve a continuous evaluation of the biocontinuum state under the conditions of the known self and then a recreation of a new self from the previous self which results in a global structure with emergent properties.

> *Logic is not a body of doctrine, but a mirror-image of the world. Logic is transcendental.* —**Ludwig Wittgenstein, philosopher**

This biocentric view of mathematical systems and the origins of the mental machinations of their logic is in consilience with the concepts proposed by George Lakoff and Rafael Nuñoz. They contend that mathematics is not some abstract independent entity. Rather, they assert that these derived computational systems and their logic is grounded in our basic experiences of the material world and embodied in the workings of the human mind. The basics of algebra is thought to have likely first evolved from the discovery of the harmony created in the specific patterns of string lengths of some musical instruments. It is important to note that harmony is a very subjective experience. Likewise, the right angle as the geometric foundation of the Pythagorean Theorem is based on the subjective perceptual experience of the intersection of the visual horizon and our upright posture in the context of earth's gravitational field.

There is geometry in the humming of the strings, there is music in the spacing of the spheres. —**Pythagoras, mathematician**

The development of mathematical procedures by Isaac Newton and Gottfried Wilhelm Leibniz that map the dynamics of phenomena (the calculus) reflects humanities very real and personal experience of an ever-changing world of impermanence. Therefore, mathematical theorems formulated as truths from the logic of experience are always ultimately born from the perspective of the 1st person. For the most part, these theorems and their derived mathematical systems work for external participants also because we share a common biology and material experience. The failure within the system for the external observer, as noted in Gödel's Incompleteness Theorem, is seen when 1st person self-referencing is required to adjudicate truth.

Probability, Uncertainty and Bayes Theorem

There are lies, damned lies, and statistics. —**Mark Twain, novelist**

If measures of probability form the foundations of uncertainty, then it is important to understand the role of the observer in probability theory. The concept of probability originated in a correspondence about gambling between Pierre de Fermat and Blaise Pascal in the late 17th Century. Later, Christiaan Huygens, a close friend of Isaac Newton, began the development of the mathematical framework for a general theory of probability with the idea that the determination of event probabilities are governed by objective rules that were an independent feature of the Universe and without observer influences or meaning. Huygens' frequentist perspective considered probabilities as an a priori examination of the space of all possible outcomes (i.e. only 1 in 6 chance to obtain a 3 during a dice roll). From this frequentist's viewpoint, the conditions of probability are thought to be real, objective physical phenomena and independent of any observer.

I had therefore to remove knowledge, in order to make room for belief.
—Immanuel Kant, philosopher

It was an English mathematician and Presbyterian minister named Thomas Bayes who in the 1700's first published a unique mathematical procedure to update beliefs about probabilistic events. His realized that the real function of probability was to represent our limitations in knowledge, which could also be informed by prior experiences. In his method, determined probabilities are influenced by the degrees of beliefs of the observer based on previous real-world experiences. This idea was further developed by Laplace in his 1812 publication entitled Théorie analytique des probabilités. Bayes Theorem was initially rejected by the general mathematics community who favored the frequentist perspective of probability theory. However, with the emergence of relativistic and quantum ideas in the 20th century there was a new emphasis on the role of the observer in scientific thinking and the Bayesian interpretation of probability has gained increasing acceptance and is now considered a valid mainstream and possibly more accurate approach to the analysis of many complex problems.

You cannot feed the hungry on statistics. **—Heinrich Heine, poet**

As implied in the Heinrich Heine quote, there is an interesting aspect of standard frequentist statistics and probability theory that reflects a very real disconnection between the implications of any detached and independent global predictions and those results experienced by the individual. We often see this in medical practice when we advise a patient of their odds of an occurrence or outcome. While those predictions may be true for the overall population, any experience of such a condition or adverse event has a very different reality and subjective meaning for the singular individual.

Information Theory, Semiotics, and Maxwell's Demon

At the heart of the computational description of Nature is the concept of information. **—Paul Davies, physicist**

We typically think of the physical Universe as being wholly comprised of matter and energy at a basic level. And even matter and energy are interchangeable as a single entity based on Einstein's formulation of E =

mc^2. Only recently have scientists begun to realize that there is a third component of reality that may be even more fundamental than matter and energy. In Einstein's equation relating energy and matter there is a hidden ingredient that structures the relationship and determines the form and function of the connection. The quantity that is so obviously in front of our eyes that we often overlook its subtle existence is the information that constitutes the relational framework for all known reality. In fact, a true knowledge of reality first requires the presence of information about the Universe. John Wheeler famously noted that every physical quantity derives its ultimate significance from information.

> *All things physical are information-theoretic in origin and this is a participatory Universe.* —**John Archibald Wheeler, physicist**

The first serious realization of the existence of this informational component of reality came from the great mind of James Clerk Maxwell. After his ground-breaking work in electromagnetism, Maxwell became concerned with the apparent conflict between the absolute universal truth of the 2nd Law of Thermodynamics and the fluctuating behavior of the microstate as prescribed by Ludwig Boltzmann's statistical mechanics description of entropy. In 1867 Maxwell constructed a simple thought experiment in order to better explore the problem of thermodynamics on the microscopic scale. This thought experiment first appeared in a letter Maxwell wrote to Peter Guthrie Tait and was later published his 1872 book on thermodynamics titled *Theory of Heat*.

In an imagined scenario, he supposed that a demon (Maxwell's Demon as coined by Lord Kelvin) would be able to violate the 2nd Law of Thermodynamics by opening the door between two chambers and only allow fast molecules to pass to one side while slow molecules naturally would remain on the other side. This differentiation would result in a paradoxical increase in the order within the overall system. In reflecting on this enigma, Maxwell concluded that the 2nd Law of Thermodynamics has the same degree of truth as the statement that if you throw a cup of water into the sea, you cannot get the same cup of water out again. It was not probable that anyone could actually follow the paths of the individual water molecules with any degree of certainty. Since the exact courses of the molecules are simply unknowable, then the condition of releasing the water into the sea results in an increase in the overall entropy. However, if it were possible to track and control individual molecules like the Demon, then the second law could be systematically violated. The key to resolving the paradox is understanding the role of information acquired by the observer

(Demon) in the mechanism of knowing through the process of experience.

Entropy is the price of structure. —**Ilya Prigogine, Nobel chemist**

In the following 100 years after the initial presentation of Maxwell's Demon thought experiment, many serious thinkers sought to rescue the 2nd Law of Thermodynamics by providing an explanation of how the Demon increased the entropy of the system. The conundrum was eventually resolved by Leo Szilard and Léon Brillouin who recognized that the Demon would have to capture information about the system and the individual molecules. They construed that the energy required for this information gathering process would be greater than that gained by the selection procedure so that there is no violation of the 2nd Law. Additionally, Szilard realized that this experiential measurement process would also require storing the memory of the results. Therefore, the information about the speeds of the molecules was a sort of negative entropy or "negentropy" as coined by Brillouin. This insight led to a conception that information processing has a fundamental entropic cost and therefore also has a physical basis in reality. Much like the energy required to run computer programs and process algorithms. Rolf Landauer suggested that it is actually the erasure and loss of this information where the price is paid for the apparent ordering of the system from the logical irreversible handling of information. This process is also reminiscent of an updating of prior probabilities during a Bayesian inference procedure where the prior disappears with the advent of the posterior probability. Prior to this erasure, the information is assimilated and held in memory as a known quantity with material objectivity that obeys physical laws. The loss of information by expurgation creates a condition of increased uncertainty or entropic state for the memory of the system. Of course, this memory must be contained within a physical apparatus and erasing the memory and unbinding of the information from this device frees it and increases the degrees of freedom. Interestingly, Landauer's principle maintains that there is a minimum possible amount of work required to erase one bit of information with that energy being dissipated into the environment defined mathematically as:

$$E = kT\ln(2)$$

Where k is the Boltzmann constant (approximately 1.38×10^{-23} J/K), T is the temperature in kelvins, and ln 2 is the natural logarithm of 2 (0.69315). Landauer's principle of energy dissipation by deleting information was experimentally confirmed by Eric Lutz and his team in 2012. Just like with the discovery of $E = mc^2$ which describes how energy and matter are interchangeable, information is also a commutable quantity.

Then what is information from the scientific perspective and how is it integral to our understanding of the nature of reality? Information has been described as a signal that depicts the content, pattern, or structure of state as a communicated reflection of reality. However, since this state must be communicated to become information it exists solely as a potential until it is perceived and discriminated by a receiver or observer. Information also relates how the contents of reality manifestly interact. Importantly though, information should be about something in order to be meaningful and significant.

> *Observer participancy gives rise to information; and information gives rise to physics.* —**John Archibald Wheeler, physicist**

Information is the currency of communication and communication is the process of the conveyance of information that follows a path from origination to reception. The value of that information or signal can be quantified according to the degree of surprise (surprisal) of the recipient in receiving the message on a 0 to 1 probability scale. If the receiving observer already knows the probable state being communicated then it is expected and is not a surprise. Therefore, the message does not bring new information and the degree of surprise of the observer of the experience is related to how probable that informational signal is expected by the recipient. A highly improbable outcome is very surprising and results in large surprisal whereas a very probable outcome results in a low surprisal measure. The surprisal of an observation is mathematically defined as:

$$S = Log_2\left(\frac{1}{p}\right) \textbf{ or } -Log_2(p)\ ; \quad \text{Where } p \text{ is a probability } (0 \leq p \leq 1).$$

The well-known biochemist Franklin M. Harold asserts that the biological equivalent to realizing this surprisal value is found in the functioning of sensory perception within the experiential process. Since a receiver's surprise and expectation is highly dependent on prior experiences, there is also an innate Bayesian quality to this formal measure of information. Though surprisal encounters are composed of discrete component signals (i.e. words in a sentence), the streaming of these signals from the information source over time is often conflated by the observer into a continuum of experience (the entire sentence or paragraph). A general theory conceptualizing information transfer within this streaming circumstance began with the publication of Claude E. Shannon's classic paper "A Mathematical Theory of Communication" in the Bell System Technical Journal in 1948. The blending of signals relates to Shannon information theory as the reduction in uncertainty or surprise achieved by acquiring the information as measured in information units (also known as entropy) and may be defined as an average surprisal during a period of the communication. The probability weighted average of all surprisals in a composite of received signals is the information (I) and is given by:

$$S_{average} = I = \sum_{i=0}^{n} P_i(S_i)$$

Another way to think about this information measure that also incorporates the observer is to include the expectation value for the recipient regarding the state of the experience. As previously noted, to be surprised by the information obtained from an experience, the observer must have some prior expectation. Expected values ranges from highly probable to improbable (see graph for entropy below). If there is near certainty that the state is present (probability = 1) or if there is near certainty that the state is not present (probability = 0) then the expectations for both those states are high and the uncertainty or entropy is low. The ensemble average of the expected value during the time of an experience is the probabilistic weighted average of all possible values that this signal variable can have and is defined as the Shannon Information or entropy:

$$e = -\sum_{j=0}^{n} P_j \log_2(P_j)$$

This relationship can be depicted graphically as:

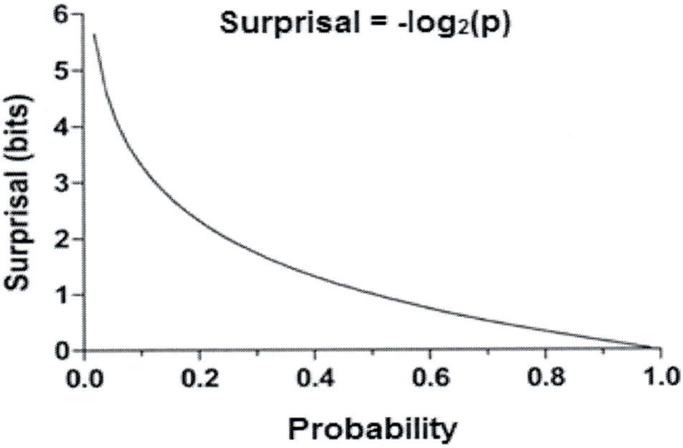

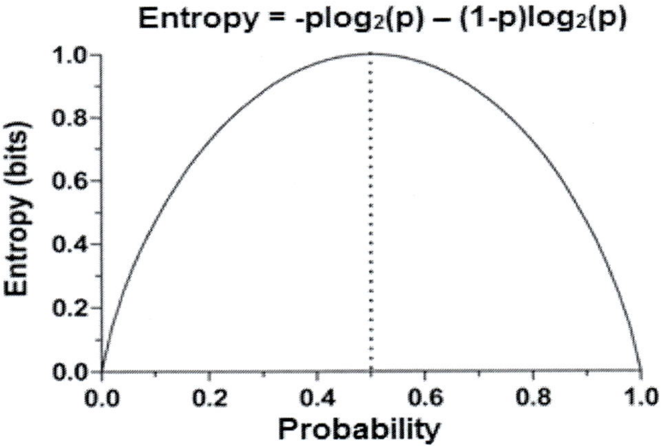

While Shannon information has historically been correlated with the physical conceptions of entropy and the Landauer Principle, this connection was conclusively demonstrated by E.T Jaynes' using Shannon information as an alternative approach to the derivation of statistical mechanics based on the Principle of Maximal Entropy. Jaynes' proof of the 2nd Law of Thermodynamics involved a transition of the microscopically based Boltzman formula to produce a macroscopic state using Liouville's theorem. More recently, physicists Melvin Vopson and Serban Lepadatu have also provided evidences of a converse relationship between entropy and information in what they call the "second law of infodynamics". While entropy in dynamic systems increases over time, there is a contrasting natural process of system information minimization and optimization. However, Shannon information has no distinct physical attribute in itself and has units only in the context of a message with a receiver that assesses its surprisal value. Therefore, it follows that there must be an observer in order to generate a surpisal value.

The quantum physicist Carlo Rovelli notes that the Kullback-Leibler divergence or relative information does provide a basis for physical quantitative relationships. It basically measures the information differences between an observation and the reference anticipated prediction of its state. Because all systems and their phenomena are in some relationship with their surroundings, Rovelli suggests that this relative relational metric is the true physical version of Shannon-type information measures that are ubiquitous in the natural world. As previously noted, Rovelli also considers that physical reality is actually based on such relative relational interactions.

The observer, when he seems to himself to be observing a stone, is really, if physics is to be believed, observing the effects of the stone upon himself.
—Bertrand Russell, mathematician, philosopher

While Shannon information is an excellent measure of the communication of information in the experiential process, it tells us nothing about what that information means. However, we all intuitively know that information without meaning has very little practical value. Information signals that have meaning act to convey the importance of a causal event or potential condition to an observer. Meaning is the mechanism by which we communicate the significance of the event.

A lack of meaningful value content in communicated information was recognized very early after the development of Shannon's Information theory and prompted attempts by many scientists and philosophers to create contextual and semantic-based definitions of information. Donald M. McKay, the well-known information theorist, argued for a more comprehensive approach to understanding the nature of information that incorporates meaning into the measure. He contends that meaning is not solely a property of the message but requires a relationship between the message and its recipient. Charles Sanders Peirce termed the property of meaning in signals as "semiotic" information. The potential to inform can be mapped by the Shannon criteria but this potential is only meaningfully actualized if it impacts the functioning of the observer. Meaningful information should be "about" something (the source) and have some significance to something (the recipient). Following this line of thinking, Gregory Bateson defined such information as "a difference which makes a difference" and this conceptualization of meaningful information has resonated with information theorists even to this today. But what does that really imply? Difference to who or what? And how does it relate to Shannon information? Can it be practically quantified?

MacKay suggests that the meaning of a message should be defined as its selective function on the range of the recipient's states of conditional readiness for goal directed activity. Similarly, Tom Stonier contends that the determination of meaning necessitates that a system has a capacity to discriminate the state information with an intent to functionally act upon the message. Peirce also asserts that any meaning of a message should relate to some function of the perceiving system. By necessitating that information measures contain meaning, McKay tethers the measure of information to the interpretive capabilities of the recipient and its contextual conditions of being. Therefore, meaningful information can be considered as a relationship between the states of the source and its recipient rather than an independent property of the message alone. Meaningful information can then be considered to necessarily have a co-dependent origination via source and recipient (Universe and observer).

The notion of meaning also implies the asking of a question. The question usually asked is sometimes considered somewhat of a "why?" question unlike the typical questions of "what" and "how" of traditional Science. The difference between a frequentist's approach and one where the observer has some prior expectation (Bayesian approach) is that in the frequentist's perspective there is no prior expectation and the information simply reduces to basic structural information. As noted before, all consequential or meaningful information transference really requires a receiver or observer. In as much that all information needs to be transferred or experienced to have any meaning, then the frequentist approach to information formulation is likely just a vestige of our classical Newtonian interpretation of Science from the perspective of a neutral and detached 3rd party observer. In the real world, information acquisition as a process is really inseparable from the receiver/observer since it is the uncertainty (surprise) of the observer about the state of the observed that actually determines the substance and potential meaning of the information. As suggested by John A. Wheeler, we can really consider information acquisition as participatory in nature. Bateson's definition of information as a difference which makes a difference requires two essential elements for this information content. First, there must be an existent differentiable condition within the scope of observable experience and, secondly, that state must be interpretable and have meaning to the recipient observer.

Use endows what is used with its character, its functional meaning for the organism. **—Christine Skarda, neuroscientist**

In Shannon's theory, we have a foundation for the quantitative measurement of information based upon the probability concepts of uncertainty. But what should be the grounding for a metric for the meaning of that information? For biological systems, the capacity to differentiate informational content through sensory signals are critical to their ability to adapt and survive. Therefore, an existential meaning for the living system is derived from the differential of information signals concerning the state of matter and energy in the biocontinuum in a way that has a definitive effect on the system beyond just perception. A possible approach to substantively grounding and quantifying the meaning of information is to consider the Lyapunov vector signature of stability in the analysis of living systems as proposed by R.L. Summers. Such stability is critical for a living system's survival and therefore has existential meaning to the organism. And if the reality of physical phenomena is dependent upon the act of observation, as suggested by modern quantum theory, then such a grounding of the metric for meaning in the existence and survival of the observing living systems that actualizes that reality could have broad implications for all of Science.

There are several salient points to be concluded from this discussion around information and meaning. First, along with energy and matter, information has surfaced as one of the most fundamental constituents of our Universe. In fact, matter and energy may just be varying forms of information. Secondly, the Kullback-Leibler relative information derivative of Shannon's information theory forms the basis of an explicit relative relational connection between the physical quantities of matter and energy and those measures of information. However, the establishment of any information measures requires the discrimination of an observer. And just like with the quantum examination of matter and energy there is an inherent uncertainty and relativity in those measures within the context of the observer as a central arbiter of the reality of the Universe. Importantly, it is also the observer that instills meaning into the information relative to implications for its prime existential objectives of stability and survival.

Nonlinear Dynamics and Chaos

We adore chaos because we love to produce order. **—M.C. Escher, artist**

Systems are usually described in Science by a set of equations with shared variables whose solutions are intended to be passed between the equations. Nonlinear systems are designated by a set of simultaneous equations in which some variables appear in specialized functions with a range of multiple degrees that amplify the variables' valuations and therefore cannot satisfy the additive properties of the mathematical superposition principle. Because of its failure to satisfy the superposition principle, a nonlinear system cannot be explicitly dissected into its parts with certainty and then allow each part be solved separately to reconstruct the full solution. Living systems are noted to be highly nonlinear and with inherent variable uncertainties. Therefore, it is often much more difficult to make clear causal inferences about these systems. Furthermore, their cyclic propagation with an iterative and self-referencing feedback makes it difficult to follow a causal inference and therefore violates the causal-power actuality principle necessary for a clear casual reductionism.

Even though these nonlinear systems may in practice really operate straightforward dynamically and strictly deterministic, they can appear globally to behave chaotic and in completely random ways. In deterministic calculations, the results really appear to be chaotic because the initial conditions cannot be known with sufficient accuracy and therefore the solutions cannot be precisely determined. By example, it is known that different calculators will produce slightly different results even with the same initial conditions. It would require infinite precision in the dataset that is well beyond the knowable capacity of the observer in order to obtain absolute deterministic behavior for these complex systems. These problems with data and calculator inaccuracies are amplified when the system of equations is nonlinear so the outcomes do indeed appear to be chaotic. Henri Poincare first discovered this form of chaotic behavior in 1908 when investigating the three-body problem of gravitation using the Newtonian framework.

In all chaos there is a cosmos, in all disorder a secret order.
—Carl Jung, psychiatrist

Interestingly, nonlinear feedback dynamics also creates global attractor conditions that provide a unified framework of top-down causality for the resultant overall dynamics and a more generally structured pattern for the system. Unlike traditional reductionist causality that determines the trajectory for the system from the ground up, attractors provide the overall global constraining operational rules as a prescriptive causative construct or goal state that determines the ranges of the path followed and final general state regardless of the initial conditions. In these nonlinear systems, the sequential interactions of each of the separate individual elements simply cannot be followed precisely to determine the unfolding of a unique outcome for the dynamics that have been set in motion. Likewise, a reverse path cannot be explicitly determined with certainty and with absolute knowledge of the final state. This limitation is comparable to the irreversible conditions of entropy formation in basic thermodynamics that is similarly due to observer uncertainty.

Lessons from Biology

Biologic Complexity and Emergent Properties

The task is, not so much to see what no one has yet seen; but to think what nobody has yet thought, about that which everybody sees.
—Erwin Schrödinger, Nobel physicist

In his famous 1943 address at Trinity College in Dublin, the respected Nobel Prize winning quantum physicist Erwin Schrödinger declared that the question of the nature of Life is indeed a scientific problem and he pointed to the basic laws of physics and the principles of thermodynamics as the potential solution. This historic declaration was shortly followed by the discovery of DNA as Life's blueprint by Watson and Crick which was then further enlightened by Monod's mechanistic explanation of gene regulation. During this same period, a thermodynamic description of open dissipative systems was being advanced by Ilya Prigogine. These definitive Nobel Prize winning breakthroughs also coincided with a new understanding of nonlinear systems dynamics, advances in computer technologies, and Arthur C. Guyton's quantitative analysis of basic homeostatic physiologic functioning. This system's approach recognized the holistic and irreducible nature of biologic functioning beyond the recipe provided in the DNA.

I cannot conceive the whole without conceiving the parts and I cannot conceive the parts without conceiving the whole. **—Blaise Pascal, math**

Despite these important advances towards a decidedly more scientific basis for understanding the workings of living systems, these systems also appeared to have properties that are not readily tracked regarding their causality or reduced to basic physical mechanisms. The general process of scientific discovery has historically been reductionist for the past 400 years. Since the Scientific Revolution, all phenomena had been thought to be understood by an examination of the elements and processes from which it is comprised and the Universe becomes a great clocklike machine whose actions can be derived through knowledge of the interactions of the mechanical parts from which it is constructed. Until the 20th Century, Newtonian physics and deterministic mathematics have as the framework for describing the machinery of the natural world and has been extremely successful in providing insight into its workings.

> *Analytical mechanics thus allows us to acknowledge an object's Systemhood as such. And this is what we require if we are to allow Systems to possess degrees of freedom as wholes.* —**Mariam Thalos, philosopher**

As this same reductionist approach was applied to more complicated biological systems, it became more evident that traditional methods were inadequate to describe their unique functioning. Yet, the basic laws of physics in our Universe must somehow still be responsible for the natural emergence of this global system complexity. Complexity is the condition where there exists unpredictable characteristics (emergent properties) of a system that cannot be directly related to a singular traceable causation found by reductionist methodologies. In fact, the emergence of complexity may be an important hallmark of living systems. Complexity emerges when the system reaches a threshold at which it is liberated from the evident limitations of knowable reductionist physical processes while still subjugated to those same physical laws. More simply, a system with the attributes of complexity is comprised of a large composite of interacting elements such that the observer/examiner cannot cogitate its detailed workings in the entire scope as a whole. Therefore, there is evidently a systems-level hierarchy of physical laws of causal control beyond the basic physical laws that are just now being discovered. It appears that the underpinning of the nonlinear relationships and circular causality connecting the integrated parts of living organisms are probably responsible for the irreducibility of the emergent properties. And as noted in the above quote from Mariam Thalos, this holistic property of "Systemhood" also allows the system to possess restricted degrees of freedom as a whole with a better control of its entropy.

Teleology, Teleonomic Causality and Biologic Autonomy

Teleology is like a mistress to a biologist. He can't live without her but he's unwilling to be seen in public with her. **—J.B.S. Haldane, biologist**

A search for purpose is not typically considered the function of Science. Since the time of Newton, Science usually only asks the "how" and "what" questions and considers the "why" questions the province of philosophers and theologians. However, unlike the physical sciences, the study of biology and the functioning of living systems have remained strangely teleological in their descriptions. It is not uncommon for biologists to ask about the purpose of an organ or system and to what end they function for the organism as a whole. Not surprisingly, the end-purpose contribution for that organism is almost always stability and survival.

The main reason that the biological sciences have diverged from the strict non-teleological stance taken by the physical sciences is because complex systems have components that are globally causally interrelated in a way that is stronger than the causal interactions of the individual parts. Therefore, the holistic interpretation of these Life purposes or system objectives is really often the point of the study of living systems. The coordinated organization of these systems also appears to be driven by a singular goal-objective despite the initiating events.

As noted in the previous section, the complex systems functioning in biology makes it much more difficult to sort out the temporal aspects of causation and almost impossible to control for all other potential causal factors. Therefore, it is practically impossible to provide a reduced mechanistic explanation of system causality based on the coordination of the individual efficient causes which collectively produce the coherent organizational structure and causal properties of the system in its entirety. In fact, the input into a system is not really a direct cause in the traditional sense of the term because any effect generated due to an incoming input signal is not strictly obligatory but rather depends upon the system's autonomous interpretation of the signal. Additionally, emergent properties cannot always be traced to component interactions. Therefore, it appears there is a need for a more wide-ranging causal explanation for such systems through some global directive which is also sometimes called downward causation.

This conception of downward causation may seem similar to what Aristotle would call an analysis of "Final Cause" that is really counter to the modern practice of Science and is somewhat unsettling to serious scientists. However, the preeminent philosopher Charles Sanders Peirce claims that Aristotle's Final cause and the notion of purposeful objective are not necessary equivalent. He contends that such purposes are just the form of the final cause that is most familiar to our experience. Many aspects of the physical sciences focus on the end results of a process (i.e. the products of a chemical reaction). Even the Principle of Least Action foretells of a generally predetermined path for physical events that follow established laws and thereby serves as an inherent end directive. Obviously, the 2nd Law of Thermodynamics is also based on an understanding that all dynamic systems are moving toward the end-objective of increased entropy which actually represents a kind of final cause with less concern as to the details of the causal path by which the system got there. In fact, there are many observed associations of phenomena that are uncertain in their temporal relationship and are merely patterns reflective of the interactions of natural systems.

A formula's hidden meaning may have a peculiar kind of downward causal power, determining the formula's truth or falsity.
—Douglas Hofstadter, physicist, writer

Aristotle's notions of material, formal and final causes are really all encompassed in the idea of downward causation because they are enacted at an upper hierarchical level to form constraints. For living systems, Jesper Hoffmeyer refers to this type of global system causality as "semiotic causation" because it delineates that the effects are brought about by an interpretation of input signals as information into a meaning that is relevant to the system as a whole. And because these signals give rise to meaning and the possibility of a guided action, this type of causality gives global direction to an otherwise simple efficient causality. The incoming signals may trigger the actions but the causal control specifics of the changes are determined by the overall system controls and objectives. For living systems, the global direction for this causality actually arises from the objective of survival as axiomatically driven by the mechanics of Natural Selection.

Biology is the study of complicated things that have the appearance of having been designed with a purpose. **—Richard Dawkins, biologist**

Teleonomy was a term first coined by Colin Pittendrigh and later refined by Ernst Mayr to differentiate the goal-directed behavior of a system's operations as determined by a programmatically imposed global objective from the traditional Aristotelian Final Causes concept of teleology. The difference is that a program of teleonomy is created by natural phenomena or by living organisms that were purportedly also fashioned by natural phenomena. By contrast, teleological end-purposes have been inherently instantiated in the system from a source outside of normal functioning of natural phenomena as would occur with an intelligent agent with foresight or through the guidance of divine intervention. Some teleonomic driving processes, such as evolution, generally function by hindsight as variations in the system are tested for fitness and either culled or replicated and propagated into future system structures. The distinction is really one of source origin of the objective and its causality with teleonomic forces being intrinsic to the nature of matter and energy and their potential for driving change.

The Nobel Prize winning biochemists Jacques Monod and François Jacob used this concept of teleonomy to describe the functioning of cellular processes as a goal driven machine without having to invoke vitalistic characteristics. These teleonomic program blueprints for the system are encoded in the organism's structures and functions, sometimes in the form of informational biomolecules such as DNA. The modern concept of "Purpose of Purpose" as described by evolutionary biologist Richard Dawkins distinguishes between planned purpose (neo-purpose) and arch-purpose that resembles intentional purpose but really derives from the natural biologic impetus for adaptation in the context of the machinations of Natural Selection, with continued survival and some sustained attribute for the population as the unintended logical consequence.

This informationally derived teleonomic causation is pulled forward toward a future potential optimal physical state and not pushed from behind by the unleashed potential energy of an existing mechanical condition. Regardless, there is definitely something innately unique in the workings of Life's processes that make this global semiotic approach to causality seem plausible as a true basis of scientific investigation. The dictionary defines purpose or goal as the reason for which something is done or created or intention for which something exists. A cause, on the other hand, is understood as the driving affect or prior happening resulting in the consequence.

Law (natural science) is par excellence the thing that wants a reason.
—Charles Sanders Peirce, logician

Typically, one of the major hallmark differences defining causation is the aforementioned temporal requirement. In fact, the requirement of temporal ordering is most often the sticking point in equating purpose with causation. Purpose and goal seem to be the predicated end-result and therefore would not necessarily initiate the happening. The term "objective" is probably a better description of an innate direction for action that can be present before the initiating event. For instance, a computer program as an algorithmic system can have a preceding general global objective without a purposeful prescriptive and detailed determination of the final outcome.

Computer science is to biology what calculus is to physics. It's the natural mathematical technique that best maps the character of the subject.
—Harold J. Morowitz, biophysicist

In the physical world, serendipitous attractor conditions that exist prior to the effective endpoint or outcome can have causality outside of the usual temporal ordering requirements and yet have no preconceived purposes. Such attractors commonly arise under the natural physical conditions of energy flows and are likely at the heart of the origination of complex systems in Nature with emergent properties and teleonomic characteristics. For complex biological systems emerging beyond the attractor, the intended consequence (as objective) is also intentionally initiating the happening (intentionality). A cycle of causality can be produced if the happening results in a further consequence of results. Like the self-feeding process of the Ouroboros or the self-generating causal loop of the linked drawing hands of the M. C. Escher sketch. Darwinian Natural Selection fosters the subsistence of such systems. The experiential process of the living system in which a goal-oriented intentional query of the conditions of the biocontinuum information space results in a translation to meaning in the context of the prior objective of survival also links the apparent causal nature of end-directed and mechanical efficient causes. Just as a Bayesian analysis infers causal relations in the context of prior knowledge, the translated information results in causal action toward the objective of a survival state for the organism. The surviving organism then precedes with further intentional query of the biocontinuum condition directed toward the task of determining causal action for obtaining the goal objective.

In an organized being, the parts exist for and by means of the whole, and the whole exists for and by means of the parts.
—Immanuel Kant, philosopher

Kant similarly proposed that for complex organized beings there is an interdependence of the system components with the objectives of the global entity. Through this mechanism, system components perform functions that are a part of the broader causal consequences and purposes, yet they can only be defined because they are part of the system as a whole. Kant considered a living organism as a "natural purpose" in itself where the cause and the effect and the objective were all a part of the functioning whole and Life was the objective of Nature. This is what is meant by the "Kantian whole" and indicates that for living systems basic causality in the traditional sense is inextricably linked to the end objectives or purpose. Life then becomes a natural product of an unfolding Universe.

Each animal is an end in itself. **—Johan Wolfgang von Goethe, poet**

The dissociation between the perceived origins of causal drivers (whole versus individual parts) exists because the microscopic degrees of freedom of the individual component parts are usually recognized as fundamental physical quantities while the relevant degrees of freedom of these same parts as coherently associated to the whole system are invariably functional and definable only in context to the global causal consequences of their relationships. This view is a flip in the fundamental perspective from the microscopic as the causal driver to a broader integrated system control that also allows for a decrease in the considered degrees of freedom. By contrast, reductionist science practiced by external observers tends to focus on a bottom up causality with many degrees of freedom. If the starting point of observation is the coherent whole of the system, then it is natural that the degrees of freedom are limited. Again, this attribute of living systems is critically important from the perspective of entropy considerations.

In the ordinary practice of Science, a physical-based causal closure principle does not allow for a virtual causal driver that would be considered to have a somewhat teleological quality. Therefore, the scientific basis for a teleonomic or end-directed character of living systems must also be a sensible property of the materials of which they are composed just as wetness is a natural characteristic of water. If the system's purpose is somehow concretely physically instantiated within the material structure and organizational processes of the system whose actions happen in a way to achieve the pre-specified (virtual) goal, then the goal objective could be considered to physically exist temporally prior to the happening and satisfy all the causal requirements. Just as attractors have an inherent physically derived causal drive for their structure, biologic constructs should have a global objective that is an intrinsic part of its physical makeup. The purposeful holistic causality of biology works if the driving causal force is also the objective. The key point to remember is that the objective has to be self-generated and emerge naturally and not solely determined by an independent guide with cognitive capacity as would be the case with a machine construction or other intelligent design.

Robert Rosen argued that goal states that arise solely from within a closed system itself should not be considered as intrinsically privileged states. For biologic systems this instantiation of state arose from the evolutionary mechanisms of Natural Selection and for an efficient satisfaction of the 2nd Law of Thermodynamics. Both of these natural factors are external to the system but do not require a specific intelligent intervention. This is a more global/holistic perspective that incorporates both the internal system and the external environmental (the biocontinuum). In fact, all nonlinear linear systems are typically driven to naturally rest in the domain of some form of an attractor.

Natural Selection is a mechanism for generating an exceedingly high degree of improbability. —**Ronald Fisher, mathematician, statistician**

Lessons Learned

When considering the need for a new understanding as espoused in Chapter 1, there are three salient lessons to be learned from the discussions of this chapter that point to that necessity.

1. Living systems appear to have a privileged position in the experience of all universal phenomena, even under the most stringent scientific requirements. This reality is evident in that only a system with an experiential differentiating capacity can actualize the observed phenomena. This unique role as a central privileged arbiter of reality is particularly manifest in the extremes of experiential conditions of quantum or cosmological interest. Recently, this is also being recognized in the biological sciences. Perhaps most interesting is the realization that the physical universe of things seems to be tailored-made for Life as we know it, in a strange symbiotic or interdependent type of relationship as suggested by the Anthropic Principle.

2. The perception, experience and knowledge of reality is dependent on the observer's contextual frame of reference and tools for perceptual interaction. The frames of reference include not only the observer's position in space-time but also the point of view (i.e. 1st Person vs.

3rd Person). The tools include the usually considered biologic senses as well as any scientific measuring/observing devices (i.e. microscopes, telescopes, etc.). One important tool that is often overlooked is the brain/mind as the cognitive seat of perception, experience and knowing.

3. Causality in living systems is based on internal teleonomic objectives that exist on a global hierarchal scale. This top-down view is a reversal of the typical reductionist approach with microscopic elements as the causal drivers to a broader integrated and coherent system control that also allows for a practical decrease in the degrees of freedom of the system. Though causality and control in living systems is liberated from the rudimentary operations of reductionist physical processes, they are still subjugated to the basic physical laws. Therefore, the global objectives of these biological systems must be intrinsic to the fundamental physical makeup of their structure and functioning and Life at its core is indeed a natural product of an unfolding Universe.

...general relativity, quantum mechanics, and statistical mechanics are actually derivable, and from the same ultimate foundation: the interplay between computational irreducibility and the computational boundedness of observers.
—Stephen Wolfram.

CHAPTER 3

THE WAY OF DISCOVERY

The Practice of Science

Science is a way of thinking much more than it is a body of knowledge.
—Carl Sagan, astronomer, writer

Science is fundamentally a methodology for discovery and understanding that is based in a position of objective corroboration and inter-subjective agreement. However, even such concurrence begins with each individual's personal and subjective experiences. As we consider the central role of the individual observer in the modern procedures of scientific discovery, it is important to chart the pathway to those realizations. Our way of discovery begins with an acquisition of knowledge through the process of experience as it is translated to understanding and meaning. By contrast, Science is about crafting theory for a comprehension of phenomena that coincides with empirical experience. However, there is an underlying thread that connects these theoretical conceptions by way of the experiential process of living systems. In this chapter, we will look at the basic mechanics of the practice of Science and how it is used to better comprehend our Universe. Additionally, we will examine the limits of this methodology and explore how it can help us in our journey to that understanding.

Experiencing, Knowing, Understanding, and Meaning

Experience and knowledge are considered closely related and interdependent concepts. In general, knowledge is a noun and experience is a verb (to experience). Experience is the pathway and active process of accumulating knowledge including an understanding resulting from the direct active participation in events or activities. Even when the word experience is used as a noun, it really refers to the knowledge, comprehension, and capability accumulated from a proactive experiential process that is acquired from an effective doing rather than strictly propositional knowledge attained through second hand learning.

Experience also becomes a noun and a concrete entity as it becomes instantiated into the workings of the observing system as acquired or embodied knowledge. To have knowledge of something really requires more than just a superficial encounter with that something. Knowledge incorporated within the structure and dynamics of the system is sometimes known as embodied experience. Descartes' famous quotation of "I think, therefore I am" can be modernized to become "experience consequently being" emphasizing the role of the information and process of experience in creating the system "self" and its reality. For Descartes, that reality of self is the foundation of being because the experiential process connects to the primacy of that reality and validates the self. Everything else is uncertainty. Any new experience is informed from prior embodied information and then used for adaptation and survival. The sustained system is then better able to have recurring perceptual experiences as each cycle is completed and current experiences become ingrained as new knowledge in the process.

Information is not knowledge. —**Albert Einstein**

Knowledge also implies having understanding or apprehension of the meaning or significance of the thing that is known. French philosopher Edgar Morin asserts that the process of acquiring knowledge involves differentiating meaningful information from information devoid of meaning. Therefore, knowledge acquisition operates by interpreting information into a meaningful conception. A method for interpreting information to obtain an exacting and objective metric for meaning in living systems that is grounded on its existential value for stability and survival was previously proposed by the author in the journal Biosemiotics (see references). In order to provide some substantive meaning for a perception translated into knowledge, some philosophers have proposed the idea that the perception must be determined by the observer to be causally important and reflected in the reactive behavior of the observer. The simple acquisition of information or a perception of information is not in itself enough to qualify as knowing about it. To become essential knowledge, the information must also be meaningful and about something. The whole continuum of the experiential process drives perception to become knowledge as translated to form a meaningful understanding.

Epistemology and the Scientific Method

Reason, observation, and experience; the holy trinity of Science.
—Robert Green Ingersoll, the Great Agnostic

Science is also primarily about knowledge acquisition and mostly refers to a unique methodology for attaining knowledge and not so much about the actual knowledge itself. In the practice of Science, the procedure used for this knowledge acquisition is called the Scientific Method as an active process of epistemology. In fact, the word Science is derived from the Latin scientia, meaning knowledge. The familiar classic modern format of this methodology is:

1. Observation
2. Make a Hypothesis
3. Test the Hypothesis
4. Reach a Conclusion

The new scientific knowledge attained from this procedure is not considered truly validated until it is in consilience with ongoing repeated empirical observations obtained through objective testing. Scientific knowledge acquisition requires both induction and deduction as information must be captured, assimilated, and interpreted to meaning before we can really say that we know something or have a knowledge of it. It is not surprising that this general procedure of the Scientific Method is in parallel to the common biology of experiential processing of all information for living systems.

What is it exactly that we are trying to "know" in this practice of Science? Scientific theories are really just creative devices used to describe and understand the workings of observed natural phenomena in a way that can be used to predict expected future changes. Therefore, we can consider understanding and prediction as the two main functions of Science. In fact, an understanding of the causal workings of the world with the intent of predictive capabilities is at the heart of the success of Science, particularly as it is applied to technological advancements. And this success has made our consideration of Science and technology as being almost synonymous. However, understanding for the simple sake of understanding is still at the root of the original intent of the scientific discipline as Natural Philosophy.

We don't regard any scientific theory as the absolute truth.
—Kenneth R. Miller, evolutionary biologist

In the 1700s there was a fervent repudiation of the ancient Aristotelian approach to Science which championed deductive theoretical reasoning as the main path to truth. Empiricists such as Hume argued that these deduced descriptive devices of Science were only analytics about ideas whose truth or reality is based on the variable term's meaning and the logic of the operations of the theory. To the pure empiricists, true reality is not absolutely predictable *a priori* from deduction alone and is contingent upon the details and facts of the phenomenological observations. However, we know from the discussions in the prior chapter concerning many modern scientific explorations that the real material essence underlying those observed experiences is probably ultimately unknowable. The essence of reality or the "ding an sich" (thing in itself) as considered by Immanuel Kant is not what we describe in our theories but rather our relative perception and comprehension of these things as filtered through our senses and brains. Conversely, the famous phenomenologist Edmund Husserl discounted any distinction between perceptual appearances and the unknowable things in themselves and placed a premium on the primacy of direct experience. Regardless of perspective, it is clear that the living system actively participates in the moment to moment specification and discernment of the changes in the biocontinuum state on the backdrop of a cacophony of physical potentials. And it is ultimately that experiential capacity of the living system that determines its stability and survival.

In his 1781 Critique of Pure Reason, Kant ascribes the conscious observer with a causative role in all perceptual experience through an alignment of sensory data input with the observer's inherent understanding of the space-time phenomena. In this process, the observer's understanding is updated as one learns from the new experience. And from this acquisition of knowledge, the future state of any experienced reality can best be predicted. In fact, the moment that separates the past from the future and marks the passage of time is that instance in which there is a transition from knowing the world that was as based on prior experiences, to acquiring new incoming information and assimilating a new conception of its reality.

Charles Sanders Peirce described all inquiry not as a pursuit of the fundamental truth itself but rather as our attempt to remove as much as possible the element of surprise associated with our experiences through the development of a capacity for prediction and therefore creating beliefs on which we are prospectively prepared to act. The ultimate goal of this prognostic endeavor is stability and survival. To achieve this predictive end, Science must be able to explain natural phenomena through a delineation of the underlying causal mechanisms and forces responsible for the observations. We must "know" causality to be able to remove the surprise and thereby bring an order to the world that is predictable. If the particulars of perceptual experience emanate from an active correlation of sensory data with the observer's predicted conception of the state of phenomena, then that conception must be imputed with causal inferences. In traditional reductionist Science, that causality is usually built from the ground up. However, a new way of thinking is emerging that considers causality from the perspective of the undifferentiated whole of the experience. Francisco Varela contends that this approach first requires a method for unearthing the primacy of lived experience. In his formulation of neurophenomenology, he suggests that an accurate and disciplined account of the structure of an observer's experience should be related to its counterparts in cognitive science through their shared constraints. This linkage of the experiential to the biological allows for 1st person evidences to be integrated into neurobiological investigations. Such biological evidences can be found in electroencephalographic brain wave signals or the neurological activity seen during brain imaging.

The Essence of Scientific Proof

Art is I; Science is we. —**Claude Bernard, physiologist**

Conventions for absolute scientific proof require that the theory elucidating the evidence is correct, no auxiliary hypotheses can explain the observations and the documented evidence concurs with repeated actual objective observations. But there is always a spectrum of individual experiences associated with any evidence that leads to knowledge as justified belief that is prioritized according to an intuit certainty.

Directly experienced evidences and observations that can be replicated and shared with other observers has the highest consideration for concluding certain validity and truth. Such evidences when accompanied with inter-subjective agreement are considered truly objective and always trump all other bases for knowledge. Because objective evidences are balanced with theory in the process of the Scientific Method, inferences deduced through theory and logic have some subordinate validity as a way for understanding.

Frequently, secondary evidences determined from inference supported by a commonly accepted theory is the only possible path to knowledge. The tracings of the trajectories of subatomic particles created in supercolliders are used to extrapolate an understanding of the structure of the atom when that microscopic configuration cannot be viewed directly. The red-shift of the spectrum of incoming galactic light seen with the spectroscope is inferred evidence that the Universe is expanding though it cannot be confirmed macroscopically with the naked eye.

Lastly, a reliance on the documentation of learned opinion from accepted authorities as a way to understanding is common but can never be substituted for more direct experiential evidences. Peer review processes have improved the integrity of this documentation but should never be considered conclusive regardless of its supporting evidences. Peers also have their biases that tend to carry the inertia of established thinking. Newton's Theory of Gravitation would still be the absolute dogma of Science and there would never be a need for discoveries like the Theory of General Relativity. Popper's falsifiability condition for the Scientific Method (discussed below) works to assure that all documented knowledge is subjected to continuous scrutiny and skepticism regardless of its source.

> *Our knowledge can only be finite, while our ignorance must necessarily be infinite.* **—Karl Popper, philosopher**

To be truly designated as objective, scientific evidence must be considered as material and concrete with definite content and form. Additionally, objective evidence has to be devoid of any opinion and not seen through the prism of emotional or personal feelings. By contrast, things that are viewed subjectively are inherently influenced by the observer's bias. Objective evidence is also usually considered to be accessible to any and all observers for assessment and that a conclusion is reached through general consensus. However, paradoxically each individual observer naturally brings their own interpretation of the evidence through their subjective experience and there is often a disagreement amongst scientists concerning the conclusions. Since all experience of reality as we know it materializes in our subjective self, we would have to somehow step out of our subjective awareness to ascertain any truly objective reality. Of course, this detached position is logically an impossibility and we are really always constrained to experience reality through the lens of our subjectivity.

In his treatise on reflexive monism, renowned psychologist Max Velmans rejects any distinction between 1st and 3rd person experiences while noting that even scientific evidence still relies on the subjective observations of individual scientists. Since scientists by necessity establish their objective evidence through inter-subjective knowledge, a shared commonality of subjective experiences is by always at the heart of scientific objectivity. This shared objectivity works by a process of inter-observer consilience and is facilitated by the fact that all humans have a comparable biology of perception. It is unknown if an insect might have a completely different subjective experience.

In the end, there are no evidences in Science that are completely objective and free from the subjective biases of any individual's observations. According to Velmans, experience of the external world is simply the physical world-as-experienced in unity with our consciousness and not a totally separate phenomenal happening within the brain as just a representation of reality. Edmund Husserl similarly argued that there are no meaningful differences between directly lived experiences and objective Science and he considered scientific theories as just shared abstractions of those experiences.

Limits of the Scientific Method

Science is made up of mistakes, but they are mistakes which it is useful to make, because they lead little by little to the truth. —**Jules Verne, novelist**

The testing of the theory is the aspect of the modern Scientific Method that has always been considered the hallmark that separates it from philosophy and theology. It is the requirement of "proof" by objective evidence through this testing that differentiates Science as a practical endeavor. The empirical essence of the Scientific Method uses Nature as the ultimate arbiter and judge of truth from which there could be little or no debate. But ironically it is the process of experiencing Nature that interjects the uncertainty. Additionally, such an absolute deference to objectivity implies there is a universal algorithmic-type of standard mechanics associated with all scientific theory assessment. But no real standard exists and each individual brings a unique perspective with some subjective predisposition.

Karl Popper tried to clarify the role of testing and proof in the modern Scientific Method with two critical concepts of falsifiability. First, for any conceptualization to be considered to be a scientific theory it must first be testable. This means that it is implicit that a hypothesis or theory must be formulated in a way that it is possible to provide a means for its objective examination. In other words, it must be possible that the hypothesis can be refuted with objective evidence from a 3rd party analysis. Unconventional and unprovable assertions concerning the nature of things cannot be considered as valid scientific theories. The second concept espoused by Popper was the standard that all theories are subject to constant scrutiny through repeated experimental testing and can be falsified at any time even if the evidences have been objectively demonstrated to be valid in the past. In this sense, no theory is ever proven and the only relevant point in time is the current condition.

Experience is never at fault; it is only your judgement that is in error…
—Leonardo Da Vinci, artist, renaissance thinker

Popper's falsification requirements pose some limiting problems in the practice of a scientific methodology that is highly dependent on the subjective opinions of individual scientists in the context of their current technological capacity and understanding. The Duhem–Quine problem, named after Pierre Duhem and Willard Van Orman Quine, states that is that it is impossible to test a scientific hypothesis in absolute isolation because any empirical testing naturally requires background assumptions. Additionally, in the procedure of the Scientific Method all experimentation is formulated in the context of a preconceived hypothesis and ancillary hypotheses are frequently used to explain some of the predicate conditions and disparate observed outcomes. These circumstances create a situation in which the acceptance or rejection of competing theories is highly susceptible to opinion and confirmation bias related to subjective experiences, past and present. Even seemingly factual information is likely always interpreted in the context of our prevailing perspectives and requisites.

Science does not rest upon solid bedrock. **—Karl Popper, philosopher**

This symbiotic interplay between objective and subjective experiences is consistent with the Buddhist philosopher Nagarjuna's concept of emptiness and the interdependence of all perspectives. This will be discussed in much greater detail later. Also relevant are Carlo Rovelli's ideas about the relative relational character of physical reality. Furthermore, there exists a concerning situation of methodological incommensurability where there are no external standard metrics for unequivocal scientific theory appraisal. Interestingly, the Buddhist espoused Principle of the Scope of Negation emphasizes that to not find something is not the same as it being found not to exist. Under this principle, the differentiation exists because the phenomena being examined must correspond to the scope of the method of examination. This conditional requirement makes a logical justification for the broad application of falsifiability very difficult

We would be in a nasty position indeed if empirical science were the only kind of science possible. **—Edmund Husserl, phenomenologist**

Despite these limitations, Popper suggests that progress in Science can be made through the process of verisimilitude in which an observer's personal discernment of the experience as being real and plausible is how truth is best adjudicated. In this view, the evidence must also have internal consistency with prior experience or broader knowledge in an almost Bayesian approach. This concession to a nuanced falsifiability by Popper may also open the door to the role of deductive reasoning. By example, it is not uncommon for new developments in mathematics and physics to allow for extended logical reasoning to reach new conclusions. However, even the process of reasoning contains some similar limitations in that a negation by reason is not the same as not affirming by reason. Again, the scope of the prior experiences, a capacity for logical thought, and the tools for analysis must coincide with the phenomena under examination.

A scientific truth does not triumph by convincing its opponents and making them see the light, but rather because its opponents eventually die and a new generation grows up that is familiar with it. **—Max Planck, Nobel physicist**

Since any approach for advancing ideas obviously requires a collective consilience and understanding by scientists, Thomas Kuhn postulates in his landmark work *The Structure of Scientific Revolutions* that the progress in Science occurs as paradigm shifts rather than incremental increases in knowledge. Furthermore, Kuhn rejects the idea that there is an absolute value of empirical data that is independent of theory. He also postulates contextual meaning in the interpretation of data has an important impact on theory assessment. Like Kuhn, philosopher Paul Feyerabend argued that scientists never strictly follow the Popperian view of the Scientific Method and that Science is not uniform in its practice. In Feyerabend's view, a circuitous logic in Science results in the substance of observed phenomena being dependent on the explanatory theories about them, which in turn determines our basic interpretation and understanding of the observations. Theory also influences our experimental designs and the mechanics of discovery. The key point is that regardless of the procedures employed, all observations of phenomena are ultimately filtered through an observer's experiential process and the lens of subjective discernment. Even the Scientific Method does not have a way around these limitations in the shared knowledge acquired through individual subjective experiences.

It is interesting to note that Nature itself has somehow bridged this chasm between the subjective and objective experiences in the creation of reality. It is then the challenge of Science to also find this solution. However, if theories only serve as representations of the interpretation of our empirical findings, then they cannot concretely conjoin objective and subjective knowledge. It is probable that the only solution for connecting the objective with the subjective lies within the actual mechanics of the biological experiential process as the direct primacy of knowable reality.

Highlights from Special Theories

There is no disgrace in not knowing everything. The problem is being unwilling to reach beyond what one knows to a broader, fuller reality.
—Harold J. Morowitz, biophysicist

Theory helps us to bear our ignorance of facts.
—George Santayana, philosopher

There are several special theories that have great importance in Science but also highlight the limitations of Scientific Method as practiced. In each of the following theories, the falsifiability condition is probably not satisfied.

Evolution

Evolution is often touted as one of the great modern theories of Science. However, Popper noted that there is no falsifiable hypothesis in evolutionary theory and it should be considered as metaphysical research. There is even a suggestion that Natural Selection as a foundational law of evolution theory presents a tautology in logic. The counterargument is that there is a flaw in the falsifiability requirement in that there should not be a rule that can supersede the validity of the basic experiential knowledge acquired in the processing of observations. In this knowledge acquisition process there certainly can be a role for inference in bringing us to our scientific understanding. While it may take millions of years for a species to evolve, the theory of evolution has only been around for several hundred years. With the tincture of time we might discern a definitive evolution of some species and the evidence will be forthcoming as if it were observed in a very long running experiment. Even within the past century we have seen variations in some organism's phenotypes that approach what might be considered as the emergence of a new species.

The Anthropic Principle

Some scientists claim that the Anthropic Principle (discussed in the prior chapter) cannot be taken seriously as a real scientific principle or theory because it also fails Popper's test of falsifiability. However, Reinhard A. Bruer has brilliantly reformulated the Anthropic idea as a constraint for observation of phenomena that can be used as a condition of the hypothesis and therefore is open to refutation. Bruer states that what we observe must be limited by the conditions necessary to our presence as observers. Therefore, the constraining conditions of the Universe as empirically observed are consistent with the fact that it contains observers whose position will necessarily be privileged. If an observation is made that is not consistent with a stipulation necessitating our presence as observers, then the principle will be falsified. But such an exclusionary provision is unlikely with what we know now and for some scientists the circularity of this logical stills sounds somewhat like a tautology.

Interestingly, this Bruer reformulation of the Anthropic Principle balances many perspectives and is consistent with the participatory role of the observer in determining the state of the observations as espoused by John A. Wheeler. What the Anthropic Principle really suggests is that because Life is in a privileged position, the practice of Science should always consider the observer's subjective selection effect from a Bayesian perspective because that the act of observation inherently provides context and meaning to that which is being observed.

Truth is so hard to tell, it sometimes needs fiction to make it plausible.
—Sir Francis Bacon, philosopher

String Theory

As noted by physicist Richard Dawid in his book entitled *String Theory and the Scientific Method*, the status of a theory in Science's canonical paradigm can be significantly underdetermined simply by a lack of available empirical data. This concept of underdetermination of a theory by the observable data was first proposed by Duhem and extended by Quine as a consequence of the previously noted Duhem–Quine problem with the Scientific Method. The underdetermination argument recognizes the necessity of creativity and inference in the process of theory development and as a part of the advancement of Science through the processes of abstraction and intuitive interpretation of data. The Duhem-Quine view is that scientific hypotheses are never tested in isolation but are really a part of a holistic testing of multiple theories in conjunction as Science is built up and progresses over time (i.e. paradigm shift). This holistic approach is facilitated by the mathematization of theory that allows for the systematic organization and incorporation of foundational concepts and becomes a vehicle for the creation of abstractions. From the time of the early empirical confirmation of the structure of the atom to the more recent successful search for the Higgs particle, a nonempirical theory assessment has been necessary for guiding the understanding of many of the discoveries concerning microphysical elements. In fact, since when it was first proposed in ancient Greece by Democritus, there has been no conceivable way to determine the existence of the atom and its subparticles until recently. It took billions of dollars and thousands of hours to build the Large Hadron supercollider at CERN to finally discover the Higgs particle as was predicted in mathematical derivations using the standard model. The advent of string theory as an elegant and unifying explanation for physical phenomena has resulted in a crisis in the traditional methods of theory assessment using the Scientific Method. As it presently stands, string theory is unconfirmable and therefore fails Popper's falsifiability test. However, the theory has some simplistic beauty and internal consistency and with the development of currently unforeseen new technologies or investigative methods may one day allow for such a determination.

Inherent Global Principles

Typically, we think of the Scientific Method as a way of developing and vetting theories from a mechanistic cause and effect perspective. However, we find that there are some other noted global directives such as Noether's Theorem, Ockham's razor, and the Least Action Principle that probably universally apply to all phenomena and appear to exert their effects outside the constraints of the usual physical laws of dynamics.

For instance, intuitively we understand that every natural process flows in a path of least resistance and the course taken is always the one that is the shortest given the constraints or barriers. Stationary or steady state systems also seemingly equilibrate to a condition that minimizes the forces within the system. Like the way that a rope suspended between two points lessens the gravitational potential energy at any point on the rope by settling on a shape that distributes those forces minimally across the entire length of the rope. This intuition was formalized into a scientific principle by Pierre Louis Maupertuis and Pierre de Fermat in the 17th Century who noted that light naturally followed the shortest path between two points and that Nature is generally "thrifty" in its actions. This tenet known as the Principle of Least Action is applied broadly in the physical sciences and is considered intrinsic in the mathematics of Newtonian mechanics as a variational principle. These variational principles determine the trajectory in dynamical systems, which naturally possess some particular extremum characteristic. In the calculus of variations, mathematical methods are developed for finding functional constructs that minimize or maximize the quantities of the functions' variables (extremum). Lagrange and Hamilton recognized that the system's path could be determined through a more general relationship describing the history of the path and without relying on the cause and effect relationships of unseen forces. Unlike the process of Newtonian physics, no new physical laws must be introduced for the Lagrangian mechanics to work and there is no need for explicit constraints other than the variational principle. By setting the final state of the path prior to the initiation of the change event, the Principle of Least Action has an almost teleological or prescriptive quality that has been somewhat controversial and commonly overlooked. The presence of such intangible transcendent factors with global causal characteristics is counter to the objectivity and reductionism required in the Scientific Method. In this sense, the guiding directives of the variational principles are similar to a global dynamic attractor state for all physical phenomena.

Foundations for the Way of Discovery

> *Equipped with his five senses, man explores the Universe around him and calls the adventure Science.* —**Edwin Powell Hubble, astronomer**

If no theory is ever proven, then it is not really possible to ever reach a final conclusion. And any theory can also be found to be false if there is updated information determined from newer techniques and instruments or if the nature of things has changed. Additionally, the integrity of the evidence is obviously contingent on the collective interpretation of the data by the community of scientists with each bringing their individual bias to the perspective. Therefore, the view of reality in Science is necessarily fluid and built upon the observed experiences of the present. And all theory is probably just an approximation of the true essence of that reality with some degree of distinction derived from our idealized metaphysics.

Our theoretical view of the world is also supported by our general understanding of how Nature operates by building upon a continuous process of knowledge acquisition. Thus, theories only serve as a representation of the current interpretation of our knowledge that can be practically used for analysis and prediction until something better is established. In other words, the inter-subjective agreement or interpretation of experiences that result in more useful predictions are considered objectively superior. That was the case when the Newtonian Laws of Gravity were superseded by General Relativity because of a more accurate prediction of the precession of the perihelion of Mercury. Therefore, all conclusions derived from the practice of the Scientific Method are both temporal and relative to the community of observers. In fact, many believe that the proper application of any scientific methodology should include a specification of the observer. If the Scientific Method does not include a way to absolute proof according to its current principles, then even 1st person experiences under certain conditions can really be considered as factual evidence leading to knowledge and understanding.

String theorists suggest that it is time to alter the canonical balance between deductive premise and inductive evidence in theory appraisal and extend the conventional horizon of evidentiary observations. A new paradigm in theory assessment might include considerations of mathematical deduction, logical reasoning and inference from ancillary evidences, explanatory coherence, and even the perceived beauty of the theory. Scientific reasoning often results in conclusions that go beyond the data. By definition, we consider knowledge as justified belief. Therefore, anything that amplifies the scope and veracity of that belief will extend our base of knowledge. Even purely deductive arguments may yield credible conclusions that are not inductively demonstrable. Scientific philosophers such as Larry Laudin and others suggest that ampliative strategies for the Scientific Method that allows for arguments even beyond the premises of traditional deduction should be considered. Such transcendent ampliative arguments such as Noether's Theorem, Ockham's razor or the Least Action Principle are commonly accepted in scientific circles. Likewise, scientists such as Albert Einstein and Stephen Hawking have suggested that any grand theory of natural phenomena should not only be predictive but also predicated and constituted on a solid foundation of logical principle and explanation. Theories should be more than simple curve fitting and bring a deeper understanding to the structure of reality. In this extended framework for theory formulation, even less exacting and more subjective inputs such as intuition and logical inference can be considered as new information and have real value for conclusive decision-making. Additionally, reasoning by circumscription, feasibility, default and circumstantial evidence can all differentiate the preferential status of competing theories. These factors can be included in the process of theory assessment as priors in a Bayesian method of analysis and evaluation. However, in Bayes theory there is no established algorithm or universally accepted system for establishing prior probabilities. In the final analysis, these priors are always subjectively chosen and unique for each situation.

> *Mathematical reasoning may be regarded rather schematically as the exercise of a combination of two facilities, which we may call intuition and ingenuity.* —**Alan Turing, Enigma mathematician**

Within this broader perspective for Science/Natural Philosophy, the course for the initiation of all real knowledge acquisition and understanding is possible through several basic paths. The first and most obvious mechanism is through the primary path of perception. For humans the five senses of touch, sight, smell, hearing, and taste are the doors to the path and the way that perceptions are bought into the system for analysis. As noted before, there is an ongoing philosophical debate as to whether experience is direct realism or a construct of our own perceptions. Scientific Theories are considered explanatory devices to predict phenomenological events whereas the essence of the material reality that underlies these experiential phenomena are really unknowable and are "ding an sich" (thing in itself) as considered by Immanuel Kant. The inherent subjectivity of all evidence as demonstrated by optical illusions and other similar uncertainties in sensory information bring forward two other important considerations. There is a central role for the relative perspective of the observer when considering knowledge acquired by experience. Additionally, an analysis of information acquired in perception should include an accounting of the uncertainty of that information, much like the standard deviations attached to scientific measurements.

A secondary process for knowledge acquisition is through inferences based on information in memory. This knowledge is also considered experiential because memory is like a recording device that captures the substance of experienced events so that they can be analyzed for meaning at a later point. New knowledge is derived from memories through a process of reflection and inference. This is considered propositional knowledge that is knowable *a posteriori* on the basis of the prior experiences. Memories and their propositional knowledge can also play into intuitions, the introspection process, and most importantly become anticipatory for current experiences as the primary source for the Bayesian priors aspect of knowledge acquisition. Hence, the influence of these memories on our perspective of reality can be so compelling that they often strongly color our current perceptions. Such is the case as seen in the phenomenon of *déjà vu* where memory signals are likely mixed with incoming sensory information to provide the feeling that an experience has happened previously. While the integrity of this knowledge is often questioned, it also constitutes a real element that contributes to the assessment of reality. However, just like the ambiguities that sometime surround sensory perceptions, these memory perceptions should also include a valuation of their uncertainty.

A third methodology for acquiring experiential knowledge is through introspection. This method involves an inward directed experiencing of the primary state of the system doing the observing such as a human examining his own sensations, thoughts, and feelings. Implicit in this aspect of knowledge acquisition being considered separately from routine perceptual experiences is the view that such everyday objective perceptions only involve an examination of the world external to the system. This separation is likely artificial and has its roots in the mind-body separation notions created by Descartes at the beginning of the Scientific Revolution. In truth, the whole of perceptions within the biocontinuum do not really separate the observer from the observed. Therefore, intuition and knowledge acquisition through this introspective course might simply be a variant of the primary path.

Indo-Tibetan Buddhist thought characterizes the two features of 1st person knowledge obtained through internal introspection as cognizant apprehension and luminosity or clarity. The cognizant apprehension of an experience is the primary perception and delineation of the direct basic information discerned. Whereas, the luminous clarity aspect focuses on a reflection of the meaning of the experience. By considering both these features as part of the unified process of knowledge acquisition, we naturally bring the state of the observer into the experiential process and formulation of perceived reality. Knowledge also becomes instantiated with meaning as it becomes more than just information.

The truth is lived, not taught. —**Hermann Hesse, writer**

As noted, knowledge that arises purely through introspection, intuition, and insight is necessarily influenced by prior experiences and memories. Since much of introspection is based on prior knowledge and experiences that have been assimilated in the configuration and processes of the system, the information gathered is that with a historical basis and founded on a longitudinal knowledge and experience of reality. This memory may be incorporated in the living system all the way down to that encoded in the

DNA and structure of biomolecules and includes the entire historical evolution of Life including the unfolding path of the Universe since the time of the Big Bang. As such, there are some that believe that pure revelation, such as with religious experiences, might have a deeper source based on the fundamental constitution of reality, Life and Nature itself. The inductive source for this introspection is then the complete history of the Universe.

The past is not dead, It's not even past. **—William Faulkner, Nobel writer**

The most difficult aspect to understand or believe is the acquisition of knowledge by mystical or religious experiences. While there are many who strongly support this pathway, it is ultimately dependent upon faith and there are not any clear objective criteria that can be substantiated in the 3rd person. Neuroscientist Andrew Newberg makes a strong case for the validity of these experiences from a neurobiological perspective and presents some objective evidence from brain imaging performed during states of deep meditation. Mystical and religious experiences are a state of direct 1st person encounters and are different from the pathway of introspection and self-reflection. The Dalai Lama has been a strong proponent of thinking about knowledge acquisition by experience in a broader perspective and breaking from the Descartes notions of duality that has dominated scientific endeavors for the past 300 years. Considering direct perceptual experience as being the starting point for the true reality of the integrated whole of the observer plus the observed is consistent with the neurosensory experimental work of neuroscientists Christine Skarda and Walter Freeman. We will discuss these ideas in greater detail in subsequent chapters.

The essence of the mystical experience, then, is a type of shortcut to truth, a direct and unmediated contact with a perceived ultimate reality.
—Paul Davies, physicist, writer

As we engage these foregoing methods as expanded ways of discovery in Natural Philosophy, we then open our horizons of knowledge acquisition in a way that extends to the very foundations and history of the Universe and its evolution. If the living system as observer does indeed play a central role in our actualization of reality, then it is also important to realize how Life emerged in the Universe and the driving forces responsible for that emergence. Emergence is the arising of unique global properties, patterns and characteristics for a system from the integrated aggregate dynamics of elements within complex systems. Emergence occurs through the creation of the whole that is greater than the sum of the parts and appears completely disconnected from its origins. Life and its collective complexity, as its evolution is traced from the moment of the Big Bang, should be considered as an emergent condition of the unfolding Universe. In this way, each Life is decidedly connected to the whole of the Universe.

The cyclic coupling between the emergence of Life and its eventual participatory role in actualizing the Universe is critical to an apprehension of the co-dependent origination of reality. It is also at the heart of understanding the paradox of the quantum characteristics of matter, the relativity of physical dynamics, the calculus of self-reference in the Incompleteness Theorem of Gödel, the autopoiesis of Maturana and Varela, the tautology of the Anthropic Principle, and many other scientific riddles. The foundations of this reflexively symbiotic union of Life and the material world is naturally independent of any temporal-spatial conditions and therefore transcends all linear causality. The true reality is the primacy of the undifferentiated whole with the unifying glue of the experiential process placed at the center. Therefore, this experiential process can be considered as the Center of the Universe. In the next few chapters, we will explore the fundamental nature of this experiential process and how it emerged within the platform of living systems. Then we will examine how it might be engaged for the discovery of a better understanding and experience of Life.

God is much more than an idea. God is a deeply valued experience that goes far beyond any theological definition of the word...
—Andrew Newberg, neuroscientist

CHAPTER 4

THE EMERGENCE OF LIFE IN THE UNFOLDING OF THE UNIVERSE

Life as the Natural Product of an Unfolding Universe

We are the first generation of human beings to have substantial insights into the origin of our cosmos and of human Life in it.
—Arthur Peacocke, biochemist and theologian

The Universe is estimated by astrophysicists to have begun with the Big Bang approximately 13.8 billion years ago. By contrast, according to fossil records the earliest lifeforms on Earth were microbes that existed only 3.7-4.2 billion years ago. It may be hard to imagine that 9 billion years later living systems would be a natural consequence of the evolution of the matter, energy and space-time arising from that fiery beginning. The fact that living systems developed in the course of the unfolding of the Universe as arrangements of highly ordered structures of matter also seems to be antithetical to a natural tendency toward a disarray of matter and energy as driven by the 2nd Law of Thermodynamics. If entropy production is required for the spontaneous initiation and progression of all physical processes, then surely it must also be required for the emergence of Life's processes. Surprisingly, the origination of the living system can indeed be logically traced to these same basic processes and still be in consilience with the principles of Darwinian Natural Selection. Somehow living systems have emerged in the orthogonal tensions of the imperative for survival and the exigency for energy degradation. However, a requirement for congruency with the basic principles of the physical sciences and the logical operations of Natural Selection does not necessarily relegate living systems to simply being the products of random chance alone. In fact, Melvin Vopson has reported that genetic mutations occur in a way that decreases their information entropy over time even when the number of nucleotides remains constant, indicating that the process of mutation is not purely random. But importantly, it also certainly does not imply that there is a requirement for a magical *elan vital*. Rather, living systems really can be considered as a natural consequence of an unfolding Universe. Life then becomes no more of a fluke accident than the formation of a new star by gravitational forces in the early Universe or the developing structure of a hydrogen atom through quantum principles. And as proposed by some versions of the Anthropic Principle, the development of Life must be an integral part of an evolving Universe in order to actualize its reality. However, we must recognize that the unique processes of living systems differ dramatically from physical processes.

It is no valid objection that Science as yet throws no light on the far higher problem of the essence or origin of Life. Who can explain gravity?
—Charles Darwin, naturalist, evolutionist

The conditions necessary to drive the natural emergence of Life's processes can be divided into two major categories. First there must be the development of constraints to energy flow that form the basis of an attractor basin with an organized structure from which living systems and their processes can develop. Second, to maintain this structure the system must then develop autonomous self-sustaining and adaptive mechanisms responsible for continuous constraint regeneration. These sustaining mechanisms are ultimately highly dependent on the primary advent of basic processes for perceptual experience and knowledge acquisition by the living system. This dependency places the experiential process in a central foundational position. In this chapter, we will outline the probable pathway for the natural emergence of living systems based on acknowledged physical dynamics. A more extensive rendition of the proposed course of development and emergence of living systems was previously offered in the book entitled *Experiences in the Biocontinuum*.

Chemistry alone will not explain Life's origin, any more than a study of silicon, copper and plastic will explain how a computer can execute a program. **—Paul Davies, physicist, writer**

The Basic Physical Framework

The obvious paradox of the existence of complex system organization and energy control that surround Life's processes in the context of the 2nd Law of Thermodynamics were mainly ignored in studies of the biological sciences until Nobel Prize winning physicist Erwin Schrödinger delivered a series of public lectures in February of 1943 at the Dublin Institute for Advanced Studies at Trinity College. In those historic discourses, Schrödinger introduced some novel concepts that attempted to explain some biologic phenomena using contemporary thermodynamic principles. Central to Schrödinger's analysis was the idea that living systems somehow utilize energy and information to produce negative entropy and thereby work against the 2nd Law of Thermodynamics to create an organized biologic structure. Today it is well acknowledged that for Life to exist as an organized entity it must contain a process in which there is a continuous increase in the overall entropy of the Universe while constraining a highly complex localized system of order. Therefore, any physical foundations of Life certainly should have a primary basis in thermodynamics.

The use of thermodynamics in biology has a long history rich in confusion.
—Harold J. Morowitz, biophysicist

The Entropy Gap

The thermodynamic conditions that comprise the basic physical framework necessary for the emergence of living systems begins with gravity and the entropy gap as espoused by Paul Davies and detailed by Dorion Sagan and Eric Schneider. At the time of the Big Bang, the classical description of the inflationary expansion process resulted in an initial homogenous distribution of all matter and energies at all spatial points. The relic signature for this event is evidenced by the discovery of a ubiquitous cosmic microwave background radiation in all directions in space for which Arno Penzias and Robert Wilson won the Nobel Prize. Alan Guth's inflationary expansion theory is the only currently viable explanation of this homogeneity of the initial Universe that can be reconciled with what we know about the microscopic interactions of particles. However, it is difficult to understand how the Universe that started out in this homogenous state of near thermodynamic equilibrium could result in the currently formed disjointed Universe of galaxies, solar systems, and planets. This ordering of matter would also seem counter to the 2nd Law of Thermodynamics.

If your theory is against the Second Law of Thermodynamics there is nothing for it but to collapse in deepest humiliation.
—Sir Arthur Eddington, astronomer

During the inflation process, some energy was subject to random quantum fluctuation rules and thereby produced local differentials in accelerated expansions and variations in local densities of energy. Hence, that initial equilibrium of the Universe could not be maintained and the discontinuities in the spatial geometry of the energy distribution resulted in a cooling of some radiant energy with a precipitation of matter and a clumping of mass leading to gravitational forces. Stephen Hawking's quantum theory of gravity similarly allows for such a formation of space from the void. The created gravity resulted in further changes in the geometry of space to cause an exponentially driven progression to form stars and planets. This creatio ex nihilo process is what physicist Lawrence M. Krauss has deemed "a universe from nothing".

Without gravity, the Universe would have remained a homogenous continuum of inert gas. Gravity as a geometric property of space-time formed with the cooling and coalescing of gas particles after the Big Bang and created an entropy gap that drives the dynamics seen with the 2nd Law of Thermodynamics. Then gravity acts as a force for negative entropy as the gravitational fields disrupt the global microscopic energy equilibration of the early Universe. In this way, gravity also forms a ubiquitous attractor for the complete system that is the Universe. The developing local nonequilibrium conditions created by the gravitational fields throughout the Universe are manifest as a necessary counter requirement for entropy production according to the 2nd Law of Thermodynamics.

Regardless of which came first, it appears that gravity and entropic forces are inextricably intertwined. The seeming close association between entropy and gravity has led some to postulate a definitive physical relationship between the two entities. Physicist Erik Verlinde has described gravity as a macroscopic emergent property born from an entropic force formed by the statistical inclination of systems to move toward a state of increased entropy. This energy balancing would account for a sum total of zero energy throughout the Universe. The entropic force can be considered to be caused by changes in the information intrinsically associated with the mass and position of material bodies. This conceptualization is also in consilience with physicist Ariel Caticha's view that entropic dynamics drive all known physical phenomena as an inference processing of information.

However, for information to have substantive content and meaning there is the requirement that there be an observer to differentiate the state. The gravity-entropy connection could also be considered from the reverse perspective in which gravity, as a geometric property of space-time, results in a tendency for energy transformations with the statistical quality of an increase in observed entropy that is secondary to our uncertainty in knowledge of the information. As noted by Eric D. Schneider, there is clearly a phenomenological correspondence between the driving principles of gravitational and thermodynamic process to move systems to their lowest energy potential states. Therein may be the connection whereby gravity, in creating the entropy gap, initiates the driving force of the 2nd Law of Thermodynamics for our Universe. If observer derived experiences of spatial positions and relationships are instantiated in the geometry from which gravity is derived, then we also have a reciprocal connection between the perceptual information acquired by living systems and the driving force of entropy. Consequently, the biological limits of perceptual experience then become the source of any uncertainty. From this perspective of reciprocity, it appears that even in the early stages of the Universe and well before the presence of living systems, there seems to be a role for an observer to eventually give substance to the evolving reality. Perhaps this mutual interdependency is at the heart of the Anthropic Principle reckoning.

The Platform for Emergence: The Attractor

The energy transitions and flows during the early evolution of the Universe resulted in the formation of localized attractor systems that have some natural tendency for internal organization. For a living system to emerge, there first needs to be such a physical platform to build upon. Self-organization is an information based physical process where the pattern of a system spontaneously arises by a global coordination of local interactions between the system components that were initially randomly related. Interestingly, such self-organization is the mathematical inevitability of the chance exchanges of the system components as energy flows through the system as brilliantly delineated by Nobel chemist Ilya Prigogine and the celebrated biophysicist Harold J. Morowitz. Dr. Morowitz was the Clarence Robinson Professor of Biology at George Mason University, a consultant for NASA (Mars Viking probe, Biosphere 2) and the Chairman Emeritus of the Science Board at the prestigious Santa Fe Institute.

> *The flow of energy through a system acts to organize that system.*
> **—Harold J. Morowitz, biophysicist**

Self-organization is also a critically important system property if there is the expectation that the entity will extemporaneously arise *de novo* and without external direction. This spontaneity requirement implies that the coordination is not specifically directed from outside of the system in a purposeful way. However, the primary initiating causal drive for the organization, such as an energy flow and attractor formation, may originate outside the open system. Since the system pattern really develops through the collective coherence of the individual component interactions, the organizational coordination is considered to be distributed globally over the entirety of the system.

As noted above, the basic physical framework for that self-organization begins as an attractor. An attractor is the subset of a phase space of a dynamical open system corresponding to its steady state. Nonlinear relationships amongst the elements associated with attractors determine the constraints and overall global pattern of the system. Such patterns can be seen even when boiling water on a stove as the heat conveyance changes from conduction to a form of convection called Rayleigh–Bénard convection (see figure below). The organized cyclic flow configuration in the boiling pot is driven by the density gradient created by the heating from below by the stove and the gravity pulling the cooler, denser liquid from the top to the bottom in opposition to the viscosity forces of the fluid.

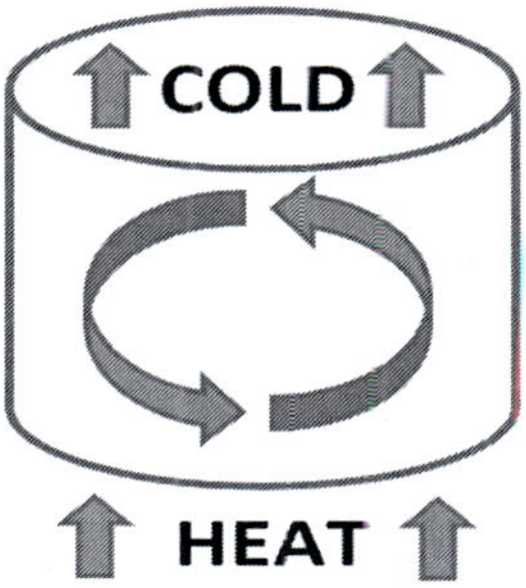

Rayleigh–Bénard Convection

Since attractors are spontaneously entailed from the energy flows, physical properties, and process dynamics of the system, they also represent a virtual blueprint for system self-organization and the basic platform for the further development of the processes and structure necessary for the emergence of living systems. The chemical equilibrium of a mixture of molecules determined by their reaction kinetics and stereochemistry is a simple example of such an attractor state. Energy flowing through the system may act further to drive chemical kinetics.

A more modern conceptualization of thermodynamics considers the steady state of open systems (such as living systems) from a new perspective. Classical thermodynamics is mainly concerned with determining the properties of the final equilibrium state of closed systems under a given set of constraints and requires spontaneous processes to move toward this equilibrium with an increase in entropy of the system. However, the steady state is just the equilibrated nonequilibirum condition of those systems with unique nonstructural external forces as constraints. In some thermodynamically open systems, energy and materials can flow in in a way in which the flux rate driving the nonequilibirum condition is greater than the entropic thermal relaxation time and therefore creates an apparent material structuring and organization. That structure and the conditions for its formation are what constitute the basic attractor state on which living systems are built. Recently, there has been the discovery of hydrothermal activity and chemical reactions within fluid vents in the ocean floor, creating energy flows and reduced gas products in a process known as serpentinization. These thermal flows and gas species generate rich attractor niches for the potential emergence of living systems and have been shown to provide the energy sources for primitive microbial activity in rocks surrounding these vents.

There is a maximal degree of likelihood that the system can obtain that ordered attractor state that occurs at a specific amount of energy flow based on the chemical and physical conditions of the system. The probability of observing an ordered pattern for the system is given by the relationship of the likelihood function and the energy flowing through the system as described by the graphic below.

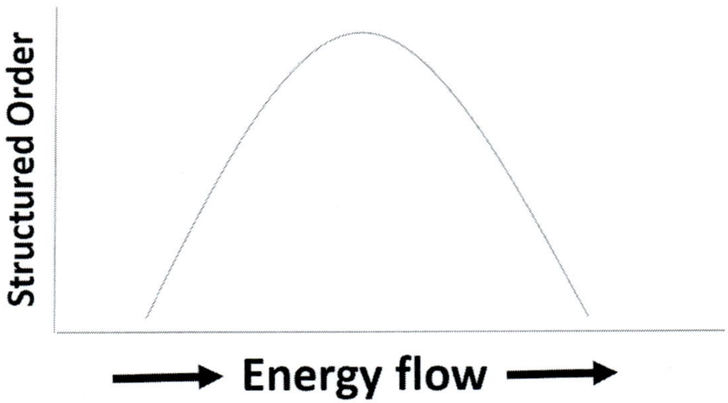

Such nonequilibirum system steady states and the resultant organized assembly of structures and processes can be maintained as long as the energy flow continues to maintain that optimal state. During this continued flow, the system must still dissipate energy and increase the overall entropy of the Universe so that neither the 1st or 2nd Law is violated. These energy dissipation processes can described by some general extremal principles for nonequilibrium thermodynamics. For his ground-breaking work in understanding nonequilibirum systems, Ilya Prigogine was awarded the Nobel Prize in 1977.

The terms organization and order are often used synonymously. The organized changes seen for the open system state are the converse to the usual disorganized condition with a random displacement of elements that the system would occupy if it were in isolation. However, the details of any system entropic state can be considered to be either less predictable microscopically based on the usual laws of thermodynamics or more "ordered" if there is development of a restraining globally coherent macrostructure. Such coherence reduces the degrees of freedom of the system if it is considered to operate as a whole. For instance, it would be impossible to predict the position of any specific individual helium molecule in a balloon filled with the gas. However, when considered from a holistic perspective, we can predict (from knowing the constraining state of the balloon) that in sampling the gas inside the balloon we will find a helium molecule with greater probability than outside the balloon. Under the constraints the helium gas-filled balloon acts as a coherent whole to float upward against gravity.

Maximizing Energy Flows

Interestingly, the attractor structural organizations created by energy flows tend to dissipate energy even more rapidly than can be achieved by simple direct entropy driven dissipation. Hence, there is an enhancement of energy degradation and entropy formation through these systems that tends to create a preferential conduit for local energy passage and utilization. The force that drives this entropy formation does so through pathways with the greatest energy gradients that likely provide for the least resistance to production. These least resistance pathways are also the ones with the most cohesive order and a catenated coupling of processes. The flow of energy through these paths further cements their connectivity and creates the conditions of their dominance as they absorb all the available local energy.

The entropy and the free energy exchanges of any chemical reaction determine whether a change in the chemistry of the elements will occur spontaneously. But it is really the intrinsic molecular kinetic factors that determine how fast the reaction proceeds. A reaction between two chemical species can be very exothermic and produce a lot of entropy but it will not practically occur if the reaction rate is too slow. If a chemical reaction can take two different paths and produce two different products, the product that is most stable thermodynamically will be favored unless the reaction is said to be under kinetic reaction control. In that case, the faster pathway is preferentially followed. Under the Curtin–Hammett principle, the energies of the intermediate species that form the transition states as the chemical reaction proceeds determine the speed of the reaction. These chemical reactions are said to have a lower activation barrier for advancing the process to the products (see graphic below). These dynamic reactions are mostly under kinetic control even before thermodynamic equilibrium conditions are ever established. The result is that the living system can exist at a state of higher free energy than the global free energy minimum. These kinetics driven processes also inherently create the potential for a natural ratchet toward disequilibrium when certain reactions are coupled. Through these mechanisms, systems may have an optimal structure and environment that favors a path for chemical reactions that produces less entropy but at a faster rate. Professor Addy Pross, an expert on chemical reactivity, proposes that it is this differential in kinetics that creates stability in dynamic chemical systems (dynamic kinetic stability) that is distinctly different from thermodynamic stability.

Life is simply a particular state of organized instability.
— Addy Pross, biochemist

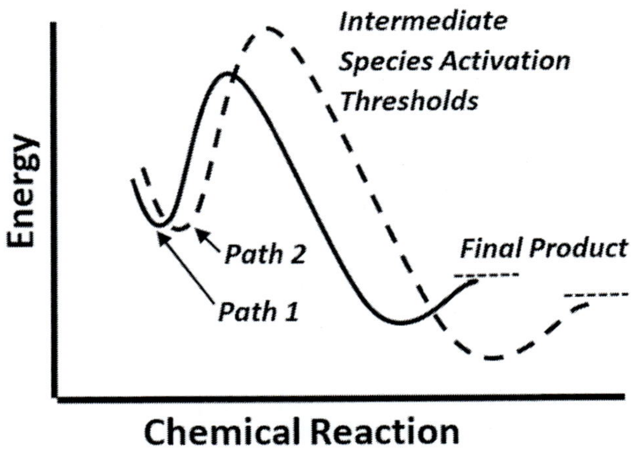

There is also evidence of a primary dissipation-driven adaptation of the basic constituents of matter at the molecular level based on established physical principles as formulated by physicist Jeremy England. Matter driven by an external energy source can also restructure or adapt its form in order to facilitate the dissipation of more energy. These adaptations result in an expedient uptake in energy from the environment by this modified molecular arrangement and therefore becomes the statistically favored configuration over time. This fundamental adaptive mechanism also amplifies the capacity for dissipation of energy of the overall system created by the energy flows. It should be noted that these selection processes occurring at the most basic molecular physical level are still driven by the same primary probabilistic propensity and sampling procedures that are common to all Darwinian Natural Selection.

Within the spirit of the Principle of Least Action and Le Chatelier's principle, it is easy to understand how energy may preferentially take a path with the least resistance and that conduit which most rapidly reduces existing gradients. This natural preferential pathway selection can result in greater utilization of the immediately available energy and a dominance of these attractor systems in the local environment. If there indeed exists an inherent thermodynamic drive for extremum in an open dynamic system or if some variational principle is really controlling energy flows, then such nonequilibrium organized systems will not arise just by chance but may actually be selectively preferred in the natural unfolding of energy excursions because of their potential for enhancing energy degradation. This favored nidus of structural organization and energy control also forms the platform for the emergence of complex systems.

The mechanisms of energy flows usually considered in the physical sciences are mainly those of mass transitions of energy as measured by macroscopic thermodynamic parameters. The energy transference within living organisms also takes place for the most part along the highly constrained and directed paths of coupled cascades of reacting biochemical molecules. Such flows of energy toward equilibrium in systems involving chemical reaction networks occur in the aligned coordinates of chemical-molecular composition rather than the usual thermodynamic coordinates of broad volumes in space. The emergence of these catenated processes is key in the advent of metabolism and the building of a degree of complexity required for living organisms. In terms of basic thermodynamic principles, the energy transference rates for chemical reactions with a large affinity is many orders of magnitude higher than those for energy transfers by other simple means (conduction, convection, etc.). Again, this differential in transference rates makes these biochemical cascades the preferred path of energy flow in these systems. The chemical energy transfers within the system, as determined by this downhill reaction process, proceed along a systematized pathway that is highly constrained and therefore significantly reduces the number of degrees of freedom and entropy for the system. The stereo-spatial structures of the molecules themselves form a kind of constraining "workbench" for facilitating these energy transfers. As noted by C.W.F. McClare and Mae-Wan Ho, the direct molecular transfer of energy within these workbenches is within a much smaller space-time and with the production of no friction and relatively little thermalized energy.

Cascades of cycles span the entire gamut of space-times from slow to fast, from local to global, that all together, make up the Life cycle.
—Mae-Wan Ho, biophysicist, geneticist

The Cycling Theorem

The significance of the discovery that systems with ordered characteristics can spontaneously arise in environments of energy flows cannot be understated and is the critical first step to our understanding of the natural emergence of biological systems. The concept was formally extended to biologic systems by Harold J. Morowitz when he proposed the more universal principle that energy flowing through a system acts to organize that system. However, the most salient breakthrough from the work of Morowitz was his discovery that the flow of energy through a formed attractor structure also results in a cycling of materials within the system when it is in the steady state. Morowitz further formally characterized the description of these cyclic processes with a theorem and mathematical proof (The Cycling Theorem) that requires the formation of at least one cycle within these dissipative structures. Within these cycles, coupled processes in the system pass energy and matter in regularly repeating sequences in a way that eventually comes back around to the beginning of the cycle succession. The connectivity of the cascading elements in the cycles are created by the naturally facilitated energy flow mechanisms discussed in the prior sections. The mechanics of this reflexive cycling procedure are a recurring theme in many aspects of living system functioning at all levels and are responsible for much of their stability through the cohesion and coherence of system elements. This cycling concept is of such significance that it also has been proposed by some as the 4th Law of Thermodynamics.

A unique form of these cycles that may have special pertinence to the emergence of living system processes is known as the hypercycle as first formulated by Manfred Eigen and Peter Schuster. Autocatalysis is a nonlinear chemical process in which the reaction products themselves are the catalyst for that reaction. This is a form of positive feedback that can drive the system further away from equilibrium and provides for an enhancement in the chemical reaction processes. When an assemblage of chemicals collectively generates reaction products that catalyze further reactions within the milieu, the entire system cascade is considered to be autocatalytic. Stuart Kauffman demonstrated that systems with sufficiently complex networks of chemical reactions will naturally self-organize into autocatalytic cycles. Such systems are considered to be reflexively autocatalytic since there is no direct replication or copying of molecules. An Eigen hypercycle is an autocatalytic system of mutually interdependent molecules in which each element in a catenated network catalyzes the replication of its successor in a recurring cycle and with momentum being created in the cycle cascade. In this way, the entire cycle becomes self-replicating or autocatalytic and self-sustaining if there is even a modest continuous external energy supply to drive the reactions. In fact, some form of autocatalysis is considered to be critical to the function of replicators at all levels in the evolution of living organisms. Multiple cycles can become linked and interwoven in an emerging network that allows for an increased stability of the system by balancing and buffering energy and material demands across the entire system complex.

The basic physical driving mechanisms of the cycling theorem can be easily understood by visualizing the workings of a simple water wheel. The incoming water provides an energy influx which drives the wheel and supporting infrastructure to move the wheel in a circular motion and thereby creates the potential for a cycling of this energy. The inflowing water to the top wheel paddle bucket brings the potential energy of gravitation to move the wheel to a bottom point of low energy. The expenditure of this potential energy as the wheel rotates down is a spontaneous downhill exergonic process. Some of the energy momentum of the turning wheel is carried into the next cycle as that part of the wheel makes it move back uphill. This means that the energy of the water falling from a higher level is transformed into a rotational energy momentum at the wheel's axle as determined by the information ingrained in the structural format of the cycling apparatus. With this momentum to overcome inertia, much less energy is required to sustain this rotation going forward. However, certain portions of the energy are dissipated as heat in the friction of the wheel's axle and an energy input is required (endergonic process) to get the wheel all the way back to the top through the complete cycle. So to continue to move the wheel around there must be also be a continuous input of energy (water inflow) with a repeated coupled cycling of the endergonic and exergonic processes. The less friction there is in the axle, the more efficient the wheel is at capturing and using that energy expressed as momentum to turn the wheel. This allows the wheel to turn faster with the same amount of energy volume input and produce more useful work. The real power of this cyclic process comes from the coupling of energy flows though the constraints and limited informational degrees of freedom of the system. It is the information instantiated in the physical structure of the water wheel system that creates this coupling and provides for a process in which very little energy input is required to continue the cycle.

For biologic systems, the cycling theorem of Morowitz establishes a similar process for chemical networks subjected to constrained energy flows through the system in quanta that are large enough to excite electronic transitions in the chemical elements and exit in the form of vibrational, rotational and translational energies (heat dissipation). Such a coupled biochemical cascade will be characterized by a cycling around the network from low energy compounds to high energy compounds and then back again. In other words, the structured environment of chemical compounds in a biosystem naturally allows for the circulation of incoming energy into those compounds with lower atomic configurations (i.e. CO_2, H_2O, and N_2) and actively pushes them up to higher energy configurations to form more complicated high energy molecules from which they then passively and spontaneously react to cycle back to their low energy precursors. With biochemical systems in living organisms, these energy transfer cycles occur as a catenated networks of chemical reactions.

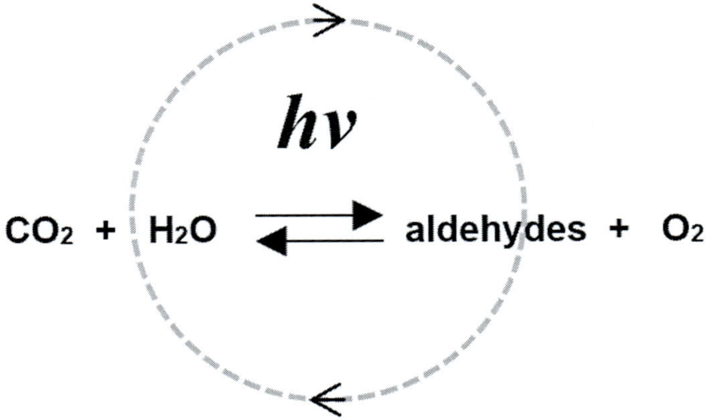

The functioning of this kind of continuous network cycle has been simply demonstrated experimentally by Groth et al. (graphic above) in which a system containing CO_2 and H_2O that is subjected to small amounts of *hv* radiation generates aldehydes and O_2 which then subsequently decomposes back to form CO_2 and H_2O. For a similar chemical network where there is no energy input, the natural thermodynamic equilibrium of the system will result in a relaxation toward an attractor of lowest energy which is the atomic configuration with the lowest energy. In this case, it would favor the molecules of CO_2 and H_2O as the dominant components of the system with a shift of the equation back to the left as the excess energy is removed into a sink with less energy being cycled back into forming molecules with higher atomic energy configurations. The amount recycled back is dependent on the thermodynamic efficiencies of the electronic transitions. While there is not an absolute physical closure of these systems, the efficient energy transfers within the cycles essentially result in what is something very near to a dynamic closure. However, since "perpetual motion" is not possible and there is always some slight thermalized degradation and dissipation of energy in the processing, there will always be a requirement for a continued small influx of energy to sustain the cycling process. If a constant flow of energy does come into such a system, the chemical equation is pushed to the right (out of equilibrium) so that the steady state for the system is one with a new attractor that favors the molecules with higher energy and a continuing of the cycling process that meets the rate of energy dissipation that matches the input. The amount of shifting toward compounds with high energy atomic configurations represents the amount of energy stored up in the system while it is in this organized nonequilibrium steady state. The possible maximum rate for energy uptake of a system is dependent upon the efficiencies of the energy or chemical electronic transmission processes and can be potentiated by a catalyst or enzymes. A well-oiled water wheel axle and paddles that can capture the full force of the incoming water can combine to turn the wheel at even faster rates and with greater momentum. Likewise, the minimal rate of energy uptake required to perpetuate such a cycling can be significantly reduced by these efficiencies. At moments when the incoming flow of water from the river source is sporadic or diminished, these same mechanical proficiencies will allow the wheel to continue to turn with the momentum. This power of the cycling mechanism is a huge thermodynamic advantage and allows for energy efficiency and flexibility in biologic systems. Just like the water wheel at the mill.

From the above discussion it is evident that the coupled and cyclical biochemical cascades and networks that are represented by the flow diagrams and graphics so common to most basic biology and biochemistry textbooks are actually the logical and effectual consequence of the natural thermodynamic processing of energy gradients in certain constrained systems. The consideration that these system functions do spontaneously arise in Nature is crucial for building a foundation for the emergence of living systems. A maximization of entropy production in these processes is also likely important in driving their selected preferential occurrence. The mechanisms of structured energy flows, chemical kinetics and network cyclic and autocatalytic coupling all result in the emergence of a system with inherent energy constraints and an organization of its matter. These arrangements are also critical for system stability and sustainability which is necessary for the emergence of an organism subject to the procedures of Natural Selection.

The Development of Complexity

What is Complexity?

Once the basic physical platform of an attractor and nonequilibrium conditions are established, how does sustained complexity emerge in systems? By definition, complex systems contain configurations consisting of complicated interwoven connections of its elements in a way that that it is difficult for the observer/examiner to cogitate its entire scope as a whole. Scientifically speaking, complexity emerges when the system reaches a threshold at which it is liberated from the evident limitations of knowable physical causation determined by reductionism while paradoxically still being subjugated to those same physical laws.

For these intricate systems to naturally arise, there must be the concurrent development of certain interactive processes and circumstances that would further foster such a state of organizational complexity. Fundamentally, it is the workings of these processes that must provide the capability of turning a thermodynamic work cycle and also have the capacity for adaptability and maintenance in the face of uncertainties within the biocontinuum environment. Therefore, the emergent living system must primarily create mechanisms that preserve the attractor conditions and provide for a recreation of the structures that make such an organization possible and sustainable. It must also provide the mechanics for cohesion leading to whole system coherence. Importantly, the only processes and mechanisms that are considered here are those that could arise naturally and from *de novo* processes as established basic physico-chemical phenomena. Some important naturally ascending attributes for these systems include capabilities for energy intake and buffering, sensing, information constitution and signification and feedback with distributed control.

Adaptivity and Stabilizing Mechanisms

Adapt or perish, now as ever, is Nature's inexorable imperative.
—H. G. Wells, science fiction writer

It is not enough to simply demonstrate the propagation of organized structures from energy flows as evidence for the spontaneous emergence of living systems. Tornados and whirlpools are examples of such organized structures but nobody would claim they are alive. A tornado can sustain its state as long as the atmospheric conditions are such that energy is continuously funneled into the system. However, the tornado system has very little internal capacity to adapt to changes in these conditions and will disintegrate if environmental conditions fall outside of the basin of attraction. Therefore, the development of an attractor platform and the associated nonlinear dynamics and complexity that comes from energy flows can only lead to the emergence of sustained living systems if they also develop methods for adaptation and stabilization in the face of a variable environment.

Scientifically, the development of these sustaining mechanisms should also arise spontaneously from the ordinary interplay of system elements and create the conditions for global cohesion and coherence. Interlaced chemical networks that create conditions for nonlinear feedback, buffering, and renewals of thermodynamic work cycles are naturally stabilizing and provide for pathways to adaptation of the system. Likewise, structural states created by these attractors can result in barriers, boundaries, and ratchets acting as Maxwell demons that also provide for system stability and resist dissipation. Many of these adaptive and stabilizing mechanisms are grounded in the natural chemical and physical phenomena of systems (i.e. buffering, feedback, autocatalysis, self-organizing macromolecules). The mechanics for each of these sustaining mechanisms is discussed in much greater detail in the book *Experiences in the Biocontinuum.* Because these mechanisms make the systems more robust, they are naturally selected with a greater propensity for subsistence. The role that these logical selective pressures provide for the natural emergence of these sustaining mechanisms in living systems will be discussed in greater detail in subsequent sections.

Negative entropy' in the sense of organization ... has not so much to do with free energy, but with the way energy is trapped, stored and mobilized in the living system. **—Mae-Wan Ho**

Advent of Causal Closure and Autonomy as Requirements for the Emergence of Life

The concepts of autonomy and causal closure are very complicated to consider. Yet, the advent of these conditions are a critically necessary step for the emergence of living systems. In fact, autonomy and casual closure are a part of the defining characteristics of living systems. Therefore, it is important to first fully grasp these concepts in order to understand how Life naturally emerged from an unfolding Universe. In the follow sections, the definitions, detailed mechanics, and unique perspectives of these salient properties are reviewed with some intentional repetition from varying viewpoints.

Autonomy

Many biological scientists consider autonomy to be a key differentiating feature of living systems and central to a definition of Life. The word autonomy is derived from the ancient Greek word αὐτονομία or autonomia which is a combination of αὐτο or auto meaning "self" and νόμος or nomos meaning "law" and denotes "one who gives oneself one's own law". As described by Jesper Hoffmeyer, autonomy in biology means that the state or condition of something (system or agent) is independent and has a freedom of action through self-governance. This biological condition is different from that of autonomous vehicles and robots in that there is a true original independence that arises naturally intrinsically and without any external programming or initial directives. Through self-governance, autonomous living systems perform the adaptive operations necessary to maintain their own individuality. Therefore, what are the operating mechanics of autonomy and what are the essential properties and causal mechanisms that are required for living systems to be considered autonomous? A condition of "causal closure" is often deemed to be at the heart of autonomy since it creates the means whereby the global organization of the system processing and adaptive controls can originate internally and are not dependent on any outside guidance.

Causal Closure

It is usual to assume that direct causal forces generally govern changes within any system and identifiable chains of efficient causation are considered the essence of all mechanistic transformations in the natural world. Therefore, every material or energy change anywhere within an environment can be expected to "cause" some alteration within that system space. For nonliving physical phenomena, the causal chain can usually be traced back to some initiating operation outside of the system space under consideration. By contrast, living systems are characterized as being causally closed in that critical changes within their system space are typically found to be initiated by processes and mechanisms that are also contained within that internal system space. In other words, the chain of efficient causation of processes within the system realm is a closed loop. The fact that the loop is closed also means that the organization of the processes and the resulting structural pattern is propagated and therefore sustained by causal directives originating from within the system. Self-replicating processes such as those seen in the cycling theorem discussed in a previous section have the potential to serve as a basis for internally closing the causal loop.

But according to the 2nd Law of Thermodynamics, all closed systems must naturally and spontaneously progress towards a state of increasing disorder or disorganization which we call entropy. By contrast, biological systems are known to be highly organized and through evolutionary pressures tend to become even more complex with time. For these divergent conditions to coexist, a seemingly quasi-independent relationship for causality is formed between the organism and its relative environmental setting. Therefore, the physical laws that typically govern the physical reality of the environment can be considered to be causally orthogonal to the emergent state of the living system. This position of causal orthogonality is essential for a specification of autonomy and operational causal closure of the system with a process for self-governance.

In reality, there is no existent separation of the living system from the physical media in which it is embedded. These living systems are always open to exchanges of energy and matter with the space they occupy. Hence, biological systems cannot be considered to be completely closed systems even if the governance and organization of their structures and processes are causally closed. Therefore, causal closure is a global property born from the collective of causally connected individual processes of the entire living system and environmental milieu (the biocontinuum). It is important to emphasize that the typical reference to system causal closure is specifically about the organization and integrated control of system processes and overall structural pattern only and that the organism is in fact materially or energetically open to the surrounding environment. In this view, the condition of the living system is considered from a more holistic perspective and system closure naturally includes the entirety of the organism and its experienced environment (the biocontinuum). However, the governing cause and directives for the form and bounded organization of the living systems as embedded in this biocontinuum emanates from within the organism.

Again, as noted, in an autonomous system it is only the cause of the dynamics and structural organization that is generated and controlled from within the processes of the living system. Varela espouses that this unique perspective provides for an underlying congruence of the living system with the environment as a structurally articulated coupling. However, a much deeper and more fundamentally interconnected relationship also seems to exist between the living system and its environmental media which makes the unified amalgamation of the biocontinuum as the primary condition. Therefore, the essential state is relational and the objective nature of things is an interdependent arising of characteristics.

The Mechanics of Autonomy and Causal Closure

An organism arises when the loop of circulating energy somehow closes on itself to give a regenerating, reproducing Life cycle.
—Mae-Wan Ho, biophysicist, geneticist

The seeming independent causal control of living systems is actually realized only because of the ability of living systems to perceive or sense the conditions of the overall biocontinuum, then translate its significance, and incorporate that meaningful information into their adaptive behaviors. Through this procedure, there is a transformational reversal of causal occurrence from a ground up to a top-down line of propagation. In this way, there is a causal loop by which the macroscopic hierarchal level establishes the general framework for the microscopic level which in turns determines the structure and workings of the larger whole. This reverse procedure is different from the direct linear bottom-up causality that is common in the description of most physical systems. This loop also creates the causal closure and liberates these systems from many of the ordinary causal rules that govern the general physical world. As noted above, even though the stimulus for the change may be triggered externally, the causal directives for the specific adaptive outcomes originate from within the living system. In this way, all system changes are ultimately subordinated to a global self-organizing internal guidance. Autonomy is said to exist because there is an additional framework of causality in the dynamics of the events that is not directly linked to the changes in the external environment. When the top-down hierarchy of cause originates from within the system, then there is the condition of autonomy.

It is important to note that causal closure is really an operational closure and an internal establishment of the system's general organizational pattern and adaptive controls as determined by those operations. Much like a written paragraph that has closure and a coherent meaning within itself that cannot be separated from the paper it is written on, otherwise it falls apart. Therefore, the property of causal closure does not mean that these living systems are closed to their external environment. The only real closure that distinguishes a living system is its unique organization and the control of processes that produce or cause that organization in the context of its environment. Hence, in living systems, there is a separation of controls that are the typical materially derived causal forces that determine physical events from those internally originating controls for the meaningful causal directives intended to sustain an organization of processes and structural patterns. In this way, there is a balancing of the autonomous objectives of living systems within the very real material dictates of the environment.

The living system is also considered to have closure because its organization is away from the thermodynamic equilibrium condition that would be expected to inevitably occur if its state were solely controlled or caused by external forces rather than those processes internal to the system's causal control. Alterations in the environment can definitely trigger the system for actions resulting in changes. But ultimately these external alterations do not directly specify, determine or cause those changes. In other words, the system has closure because it independently causes and controls its own state. Clearly there is a coupled interaction with the environment but that does not constitute causality. The key distinctions here are the conditions of control and self-determination.

Katherine Hayles provides some simple and insightful examples to differentiate the causal origins of self-controlled autonomy. For instance, if someone is slapped by an external party, their reaction may be triggered by the action of the slap but that slap does not control the response of the individual. That reaction is contingent upon the history, current state, and intrinsic system responses of the individual within the context of the situation as determined by the individual that was slapped.

It was previously noted that attractors form the base physical platform and nascent structures in the unfolding emergence of living systems. Such attractors can also be considered to have a special causal power arising from the top-down through the process of creating the conditions for cohesion of all the elements of the system into a collective coherent structure for directing the system toward a singular goal. In this way they restrict the degrees of freedom of system state and also decrease the overall statistical entropy. Since movement of the system state within the attractor and its basin invariably closes in on itself under a global physical directive, the attractor could be considered to be the natural originating basis for biological causal closure.

It is important to note that for a living system to obtain causal closure there must also have been an emergence of a capacity for coordinating the basic interacting components of the system. Such coordination is necessary for autonomous control. The natural principle of system coherence provides an overall mechanism for a coordinated explanation of this course to causal closure. Coherence is defined as the systematic or logical connection and integration of the elements and relationships in a system that results in its global attributes or state characteristics. The degree of coherence also describes the overall strength of meaningful associations (cohesion) between its components of the system. Collier first identified this property of cohesion as the necessary means for creating system coherence. So, coherence can only exist if there is a cohesion amongst the individual parts that exists within the system. The cohesive elements and processes within a system are those real factors that connect and hold components in space and time so that the system acts as an integrated coherent whole. This unified relation in turn results in closure of the dynamical causal relations amongst the individual parts of a system. Cohesion leading to system coherence is common in energized biochemical networks as seen in Morowitz's cycling theorem discussed previously.

A system with coherence built on cohesion also has the unique attribute of being totally transparent unto itself with complete self-referential and direct knowledge concerning all of its aspects and components. This knowledge minimizes the uncertainty or entropy and reduces the degrees of freedom of the system as a whole. All facets and particulars of a coherent system with a single degree of freedom can be accessed instantaneously by the system's experiential and control processes. In fact, it has been suggested by biophysicist Mae-Wan Ho that the entropy in the living system should be expressed in terms of its deviation from coherence. As the conditions of cohesion, coherence, causal closure and autonomous control all develop within the emerging living system, the resulting organism is freed from many of the momentary perturbations of its environment and has better chances for survival. Hence, such systems are also preferentially selected to emerge and subsist.

Central Role of Perception and Experiential Capacity

Science is nothing but perception. **—Plato, ancient Greek philosopher**

In order for an organism to be causally closed and possess the ability to autonomously adapt from an organizational perspective, there must first be some system capacity for perception and differentiation of the state of biocontinuum. As previously cited, Spencer Brown developed an axiomatic system of primal logic known as the Laws of Forms in which all structures, patterns and relationships can be distinguished as an abstract form and the most primitive act of logic is the drawing of a distinction. This distinction is a deconstruction of the form from the whole of the experienced domain. In allowing direct re-entry of distinguished forms during an experiential cycle, the deconstructed form is both the distinctive part and the observed whole at the same time. In this process a condition for self-referencing is also established. The underlying fundamental physical mechanism used by living systems for self-referencing is through accessing their system conditions of cohesion and coherence as facilitated by the experiential process.

The process by which meanings grow and operate is intentionality.
—Walter J. Freeman, neuroscientist

The experiential process's unique ability to integrate the system's sensing functions with determined actions is essential for any adaptivity to perturbations in the biocontinuum. This capacity is accomplished through a differentiation of signals in the context of prior experiences and an engagement of a central meaning for the intended purpose of survival. Therefore, the totality of the sensing and signaling procedures that permeate the living system and provide information concerning its detailed interactions of cohesion and state conditions of coherence are translated into a global meaning by virtue of their relative actionable significance to system stability and survival. And with this information for survival the system is once again able to create a new differentiated experience of the biocontinuum and adapt to the momentary condition.

Alicia Juarrero suggests that perceptual activity for is intentional as guided by the functional pursuit of endogenously defined objectives. This intentionality is consistent with a true state of autonomy with causal closure. In this process, there is a proactive search through sensory input for specific information to inform the system toward satisfying those inherent objectives. In this way, intentionality also implicitly incorporates the conditions of autonomy into the experiential process.

The Driving Force for the Emergence of Life

The theory of evolution by cumulative natural selection is the only theory we know of that is in principle capable of explaining the existence of organized complexity. **—Richard Dawkins, evolutionary biologist**

According to physiologist J. Scott Turner, the First Law of Biology is Darwinian Natural Selection. Consequently, the Second Law of Biology is described by the great French physiologist Claude Bernard as the inherent drive of living systems for homeostasis and an adaptation for stability of its internal milieu. The brilliant neuroscientist Karl Friston suggests that the physical basis for this drive for homeostasis arises from the natural conditions for resisting dissipation through the local free energy minimization that occurs in self-organized attractor systems. Biochemist Addy Pross further proposes that the dynamic kinetic stability of chemical reactions is different from the traditional thermodynamic stability and is the bridge between the drives for physical and biological organization. Sara I. Walker and Leroy Cronin insightfully utilize assembly theory to first explain the natural emergence of biomolecules from basic physical laws and even before Darwinian selection drove the evolutionary process. The complexity of these molecules and the accumulation of the information encoded in their assemblage is considered the path and history of the unfolding Universe.

The linchpin that links the mechanics of Natural Selection and the emergence of living systems is the functional ability of the organism to adapt for homeostatic stability and sustain itself in the biocontinuum. A selection for survival is determined by the fitness of the organism and its capacity to adjust to the prevailing environmental conditions in order that it can survive and pass on its phenotype/genotype. The basic mechanisms for this adaptability are grounded in the system's special stabilizing procedures of feedback, energy buffering and the creation of a global coherence through the process of cohesion. Additionally, autonomy born of causal closure and intentionality must be established to create an orthogonal relationship with the living system's environment. However, as noted in the previous section, the functioning of all these mechanisms is first dependent on the system's ability for perceptual experience including a capture of information and its translation into knowledge and meaning.

Natural Selection is a rudder, not a driving force.
—Addy Pross, evolutionary chemist

Sustainability of system organization, through causal closure and autonomous purposeful intention, logically emerges from within a living system as a consequence of the adjudicating pressures of Natural Selection. A sustained system organization is needed to establish the directed intention and the mechanisms of the perceptual experiential process which are in turn required for informing and directing the actions of system maintenance. These conditions create a perpetual causal closure loop for the experiential process within the biocontinuum which is ultimately the driver for control of the sustained system organization. The informed action that is instantiated in the adaptive control of the system processes also becomes an innate learning mechanism for the living system as it pursues the objective of sustainability. Therefore, this perceptual experiential process has naturally arisen in these living systems and was selected because of its adaptive advantage leading to sustainability of the system. This perceptual experience as an internally driven process is required for true sustainable causal closure in an adaptive system and therefore is unique to living systems. The experiential process can really be thought of as the primary expression of agency in living systems.

We grow in direct proportion to the amount of chaos we can sustain and dissipate. **—Ilya Prigogine, Nobel chemist**

Natural Selection by rational mechanics preferentially chooses those systems that can create meaningful information out of the natural occurring transitions of physical phenomena and use that information to endure and sustain its organization. This basic principle espoused by Alfred J. Lotka and H.T. Odum suggests that a process of preferential selection for survival may result in a propensity for the emergence of such systems that is greater than that which would be seen by chance alone. Therefore, the primary objective logic of the procedures of Natural Selection result in the emergence of living systems from the unfolding Universe that have the power to deconstruct the amorphous and uncertain conditions of their biocontinuum into an actualization of reality as meaningful information that is used to maintain and further propagates a system structure that can continue the experiential deconstruction process. And because the system itself is inextricably part of the reality of the biocontinuum, it is in a continual process of self-renewal.

Organized nonequilibrium systems may also be preferred in the natural unfolding of energy excursions because of their potential for enhancing energy degradation as previously discussed. Likewise, ordered systems that efficiently and expeditiously handle energy flows have less internal pressures for structural instability and degradation and therefore have a major selective advantage for sustainability leading to propagation. But as noted by Christian de Duve, at the very least we are seeing the playing out of chance tempered by physical constraints that impose an overall directionality to the evolution of the Universe and an increased possibility and propensity for the development of organized complex systems.

General Scheme for the Emergence of Living Systems in an Unfolding Universe

The general scheme for the emergence of living system as outlined in this chapter is depicted graphically below as redrawn from the book entitled *Experiences in the Biocontinuum* (see references).

The progression begins with the creation of the space-time continuum after the Big Bang and the coalescence of matter to form the potential for gravity. With the formation of gravity, an entropy gap is generated and there are corresponding natural flows of energy and matter. Such transitions in energy create localities of attractors with an ordered structure from the global perspective.

From these attractor platforms, there are natural feedback and physical stabilizing mechanisms that can propagate the system. There is a further ascension of more complex sustaining mechanisms, such as a capacity for adaptability, that develop under the pressures of Natural Selection. The arising power of these selected self-sustaining mechanisms eventually allow for the emergence of system autonomy and casual independence. However, the biological capacity for the perception of information and translation into actionable knowledge and meaning is the most critical development at the operational heart of all of these other mechanisms.

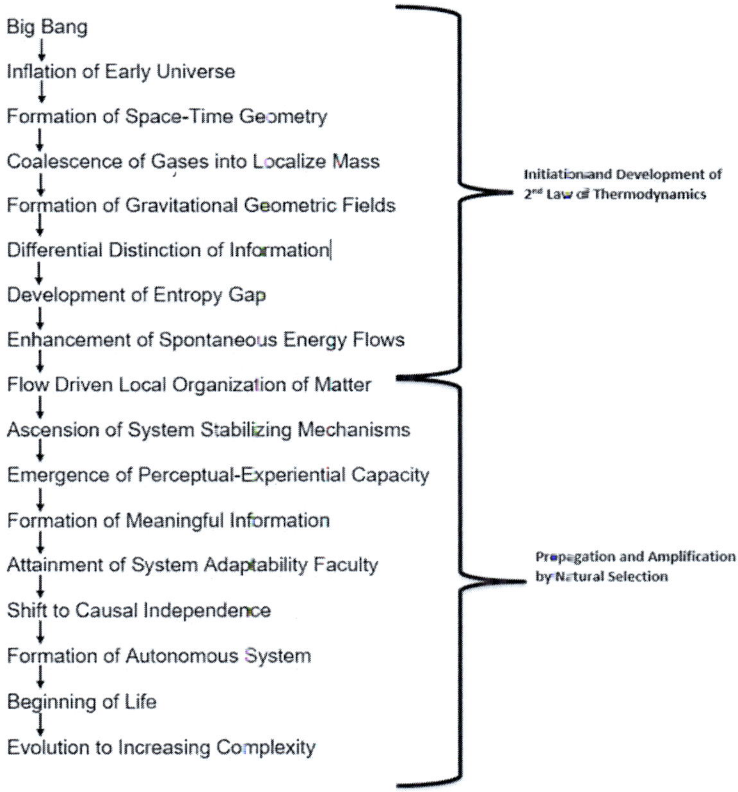

A Physical Basis for Life's Objective: An Action Principle

The principle of Life will hereafter be shown to be a part, or consequence, of some general law. —**Charles Darwin, naturalist, evolutionist**

The Natural Directive

The characteristics of autonomy and intentionality in living systems first infers the natural emergence of an intrinsic primary objective as the basis for guiding existentially important processes within the biocontinuum (i.e. for stability and survival). As suggested in the preceding section, the advent of this prime directive for continued existence is a direct logical consequence of Natural Selection. Such a directorial aim necessitates a translation of that objective into a general action principle in a way that is uniquely applicable to the dynamic physical functioning of complex biological systems. Accordingly, this action principle should guide biological functioning in a way that minimizes the matter, energy and informational divergences from the steady state condition within the biocontinuum space of the living system.

As noted by Erwin Schrödinger, the application of traditional physical principles to living systems' functioning has had limited success secondary to the inherent complexity of these systems and constraints in the detailed accounting and knowledge of the particulars of any single biologic entity. Rather than considering the direct causal forces and mechanisms for change codified in Newtonian laws of physics, the rederivation of the description of physical phenomena by Lagrange and Hamilton in the late 18th century emphasized the identification of the right action functional as a more general framework for the study of dynamics that is grounded in basic system energy exchanges. The Lagrangian functional inherently includes the internal directing logic of a variational principle known as the Least Action Principle. This form of analytic procedure is really the mapping of the state information governed by the intrinsic system constraints and degrees of freedom that can be translated to the more familiar elements of force and matter. Since its inception, the Lagrangian procedure has become a very powerful methodology for the investigations of complicated problems in modern areas of physics such as cosmology and quantum mechanics that typically resist analysis within more traditional coordinate systems.

The Central Role of Information

The enlightened physicist Paul Davies has proposed that the concept of information is at the heart of the computational description of Nature. This central role for information is particularly true in biological systems. Beyond the matter and energy elements that comprise the physical and structural components of living system, the function that information plays (i.e. Maxwell's Demon) in the coordination of the necessary complexity and autonomy of organisms is essential to understanding their mechanics and dynamics. For environmentally integrated open living systems, it is the sensory-experiential establishment of the dimensional structure of the materials, thermodynamics, and space-time within the whole of the biologic continuum (internal and external to the organism) that creates the information-theoretic metric space and the configuration of the action dynamics of all biologic phenomena. The overall information signature of this biocontinuum state then becomes the currency for the living system's processes. As this information signal drives actions for change and system adaptation, the dynamic information processing as described by Lagrangian mechanics also serves as a general action functional for the biocontinuum of the living system. Furthermore, the extremum reconciliation of the biocontinuum steady state through the mechanism of a minimization of information divergence can be viewed as the action principle for living systems. The derivation of such an action functional for living systems has been previously described (*Experiences in the Biocontinuum* – see review below) which contains an innate action principle based on the logical entropic mechanics of Solomon Kullback's Principle of Minimum Discrimination Information. This action functional for describing the dynamics of living systems that is based on a logical axiomatic inference procedure emphasizes the revelation that Life is first and foremost an active process and not simply just an object.

God is a verb, not a noun. **—R. Buckminster Fuller, systems theorist**

In the proposed theoretic platform, the information in the biocontinuum space takes the form of Shannon information represented as the probability spectrum of possible conditions in which the points in a manifold of a 2-dimensional geometry can represent the mean μ and variance σ of the probability distribution. This formulation will allow for a geometric representation of the biocontinuum in which the contours of the space's landscape provide for the natural trajectory of action dynamics as driven by a general functional of the living system's adaptation to changing conditions. The dimensions of a geometrical/topological measure of any physical phenomena and its properties is specified in terms of the number of points in some coordinate system. It is within such a dimensional construct that objects can be distinguished or differentiated and become material and substantive to the living system. Therefore, the dimensionality of the biocontinuum platform is determined by the sensory and measurement capacity of the observing system in terms of an information metrics signature. Even at the cellular level there are chemo-tactile facilities that exist to distinguish time mainly and some rudimentary space dimensionality and may have a special type of aggregated sequential ordering in experiential determination of dimensionality.

The Basis for a Biological Action Principle

In the formulation of the biological action functional, a nonequilibrium thermodynamic steady state is considered as the base state for all living systems with any deviations from this condition resulting from some force for change. This thermodynamic state for the whole of the biocontinuum (environment and organism) is broadly described in terms of traditional Shannon entropic-information metrics. In order for any organism to survive and sustain the steady state condition, the process of acquiring experiential information (both internal and external) and translating it into actionable meaning for adaptive responses becomes the general action dynamic for all living system changes. Therefore, the core axiomatic structure of this experiential dynamic process should include an innate action principle that can determine the system directional changes without requiring a detailing of the particulars of the system. Adaptive changes within any complex living system can then be inferentially mapped by the information geometry of its experiential biocontinuum space and the resultant Lagrangian of that space.

> *What is Life? Self-replicating information? Information about what? How to self-replicate!* —**John Carlos Baez, mathematician and physicist**

Marc Harper and John Baez in their progressive studies and insightful analysis of evolutionary dynamics have considered the integration of the Kullback-Leibler information divergence metric into a replicator dynamic to function as an inference engine for informing the natural evolutionary behaviors of population distributions. They found that the differentiating capability of the Kullback's information divergence measures and the mechanics of replicator dynamics within the background of the system fitness creates a natural inference equation with Bayesian learning competence. Bergstrom and others have also found a strong relationship between the concept of Darwinian fitness and Shannon-type information metrics such as the concept of mutual information. This same analytic format can be applied to the spectrum of possible states occupied by any individual organism.

While the inherent mechanics of such an inference equation naturally reduces the information divergence, the broader phenomenological and logical rationale for this behavior is grounded in the entropic gradient drive of Kullback's Principle of Minimum Discrimination Information. This inference solution principle states that an introduction of any new information into a system will most likely result in a dispersal of impact on the system state that is as close to the original distribution as possible and with as small an overall system gain in information as possible. This general inference procedure is based on the Principle of Maximal Entropy as described by E.T. Jaynes. This procedure has been successfully used as an alternative approach in the derivation of statistical mechanics and a variety of other physical phenomena. The Maximal Entropy methodology posits that inferences are made based on limited information and as such should be drawn from the probability distribution that has the maximum entropy permitted by the available information. The inferred distribution is the one that is considered to represent the most conservative assignment of values in order not to draw conclusions unsubstantiated by the existing information and known constraints. The information dynamics of this reconciliation is then considered the driving mechanism for change toward a conclusion during the inference process. This process is also really the basis of the entropic drive for the physical dispersal of matter and energy to maximum disorder based on the logical statistical mechanisms of chance variations. In fact, this statistically derived entropic drive is the same as that used in traditional mathematical descriptions of entropy transitions in physical phenomena. This consilience was demonstrated in Jaynes' proof of the 2nd Law of Thermodynamics through a conversion of the microscopically based Boltzman formula to produce a macroscopic state using Liouville's theorem. Hence, all systems naturally evolve to reconcile the introduction of new conditions in a way that maximizes their entropy and minimizes their gain in order or information. The entropic drive for resolution of divergences includes the entirety of the information signature characterizing the matter, energy and dynamic conditions of the biocontinuum. Kullback's Principle of Minimum Discrimination Information then operates like a least action principle for reconciliation of information divergence across the biocontinuum. Because of the important implications of this proposition, J. E. Shore and R. W. Johnson provided an axiomatic examination of Kullback's Principle of Minimum Discrimination Information and determined that it is a logically consistent method of inference from new information.

Inductive inference is the only process known to us by which essentially new knowledge comes into the world. **—Ronald Fisher, mathematician**

Inference is always an invasion of the unknown, a leap from the known.
—John Dewey, philosopher

Derivation of the Action Functional Mechanics

Based on the preceding proposition, a mathematical expression that includes replicator-like dynamics incorporating Kullback's Principle of Minimum Discrimination Information then serves as the axiomatic logic for the inference process with the intrinsic goal of predictions for adaptive actions to optimize survival. The central idea achieved by this inference and probability updating process is the simple notion that living organisms innately contain various "hypotheses" about actions concerning survival strategies. These hypotheses are continuously tested using empirical evidences from sensed internal and environmental information. The biology of the living system adapts to the best hypothesis based on experienced observations within the biocontinuum and according to the axioms and the entropic gradient driven inference mechanisms of Kullback's Principle of Minimum Discrimination Information processing as operated by replicator dynamics. In this way, the information entropic dynamic path along a biocontinuum geodesic trajectory naturally minimizes the distances between information points for the least action and serves as an inference procedure for directional change. As variations in observed biocontinuum conditions initiate actions for change and adaptation, this procedure also serves as a general action functional for the living system.

> *...brains are prediction machines. They are evolved organs that build and rebuild experiences from shifting mixtures of expectation and actual sensory evidence.* —**Andy Clark, cognitive philosopher**

This Kullback methodology essentially determines how much information the organism gains if it has a hypothesis about the state of the biocontinuum represented by the probability distribution p and then the gathered observations through some sensory experience reveals the best hypothesis is q. It is typical to treat a hypothesis as a distribution of possibilities and the best hypothesis is determined by the fitness of that hypothesis to produce system stability. The adapted steady state difference in hypothesis and observation (ΔX) is a measure of remaining uncertainty about the biocontinuum space state. We can determine the relative information and how the hypothesis probability distribution evolves and is updated in time according to the replicator dynamic's actions. This updating with a reduction in the information divergence is the inferred gain in knowledge by the living system about the probable state of the biocontinuum that happens continuously over time until it reaches a steady state in the replicator dynamic. Any information divergence provides an injection of drive to change the system state by releasing the pent-up potential energy constrained in the nonequilibirum condition. For biochemical systems, store energy is spent and dissipated in the adjustment process. As such, this divergence is a quantitative measure of the distance from the system steady state. Additionally, as the living system learns its environment and adjusts then the divergence approaches zero and there is a new system steady state to account for the freshly established reality of the biocontinuum.

The formulation described above relates the relative information of the biocontinuum to the system entropy and free energy and the drive of the 2nd Law of Thermodynamics. The natural entropic drive of Kullback's Principle of Minimum Discrimination Information directs the minimization of those surprisals as the living system's actions for adaptation utilize the sequestered energies in the nonequilibirum tensions. This practically structured information-theoretic is consistent with the free-energy principle of Karl Friston. In Friston's physics-based approach, free-energy is a quantity founded in information theory that bounds the state of a system's organization as it adapts from incoming sensory data. For a nonequilibirum system to maintain its existence in that organized state then it must naturally minimize the free energy bounds of any incoming surprisals. It is thought that Friston's approach is really a derivation of Feynman's path-integral or sum-over-many-paths formulation of quantum mechanics.

Reality as the Primacy of Phenomenological Experience

The luminous and revolutionary methodology of Karl Friston that builds on the ideas of Varela and Rosen, is a real advance in our understanding of biological systems. These conceptualizations of living systems and their cognitive capacity considers the open nature of organisms and their fluid interaction with their environment. However, central and intrinsic to these above cited theoretical frameworks is the notion that these observing organisms continually rebuild an internal representational model of reality. There is also an emphasis in a partitioning of the boundary conditions of the living system from the rest of environment. A limiting aspect in this approach is that these representational models themselves become separate and discrete entities with their own objective content and properties.

> *Natural objects, for example, must be experienced before any theorizing about them can occur.* —**Edmund Husserl, phenomenologist**

Early cognitivist theories of perception based on MacKay's information-flow concepts were presented as a processing of information with a constructive refinement of model representations. These conventional ideas were countered by James J. Gibson who emphasized the importance of considering the organism-environment system as a unified whole. Gibson also maintained that representations did not really exist in nature and that the true perception of reality had to be direct. His cogent arguments even later convinced Ulric Neisser, the father of cognitive psychology.

Edmund Husserl also thought that our perceptions do not represent an object but rather more directly presents the object. Furthermore, Huw Price, the Bertrand Russell Professor of Philosophy at Cambridge, contends that our thoughts and mental constructs are not simply representations of something external and objectively perceived as they are often derived internally and therefore must incorporate some degree of a subjective characteristic. The well-known American philosopher Richard Rorty likewise rejected the notion that the construction of internal representations of objects of perceived reality were necessary for the creation of knowledge. Rorty asserted that the development of knowledge requires temporal and historical context.

Following the perspectives of Gibson, Husserl, Price and Rorty, it could be concluded that meaning plays a critical role in the development of knowledge that is not typically captured in representational models. Astrophysicist Kevin Heng contends that while models are useful in describing and predicting system behaviors, they do not always provide for a more comprehensive understanding. He suggests that models only approximate reality without really capturing the essence or meaning of phenomena. Rovelli further argues that the role of Science should be about more than just ostensible informational representations as a predictive description of phenomena. Any such theoretical framework should also provide some insight into the very nature of reality and its meaning in experience. Perhaps this role is the return to Science as Natural Philosophy. Additionally, a methodology that provides for an understanding of the true foundations of reality should also allow for our directly connecting to that reality through the primacy of experience.

Perception does not just "copy" objects, it creates their meaning for the organism. **—M. Merleau–Ponty, philosopher**

The biocontinuum perspective presented here also differs in Friston's conceptualization by considering there is no fundamentally real separation between organism with its environment and the primary condition of known reality is a holistic and complete integration where the only boundaries for the living system are the limits of experience. According to the ideas of Christine Skarda, the only real boundary for the living system is the boundary of experience. Subsequently, all knowledge of the world is really a process of deconstruction of all experience into information that can be interpreted for meaning in the context of the living system. This process constitutes an actualization of the primacy of experience. This perspective is as opposed to an internal representational construction of a vision of the world by the organism. The reality of the biocontinuum emerges through the actualization of being by the living system's experience. Therefore, the organism does not contain a representational model of reality nor should its structure and being be considered as a representation of reality. Rather, the organism and the entirety of its actualized biocontinuum through the experiential process ***is*** the primacy of reality. This perspective is in juxtaposition to the view of the creation of a world model within the processes of the organism as distinguished by a Markov blanket. Again, this conceived separation between the external and internal states is somewhat counter to a more holistic view of living systems as being integrated with their environment.

Mathematical Framework for the Biological Action Functional

The concept of employing the methods of information geometry to study biological phenomena and naturally driven dynamics is informed by the work of physicists Ariel Caticha and Carlo Cafaro in their theory describing entropic dynamics. Geometric structures defined by mathematical relationships are commonly used to study the dynamics of physical systems. The action functionals inferred by these mathematics and geometries actually define the system dynamics. Such geometrical constructs can also be used to understand how information changes and flows along functional gradients within living systems. Shun'ichi Amari formalized an approach to describe the information-theoretic framework of systems by applying the methods of differential geometry to the field of probability theory in a new branch of mathematics called information geometry. In information geometry the space of probability distributions for a statistical model are described by the points of this Riemannian manifold. The degree to which one probability distribution point can be distinguished from another is through a measure of the distance between the points. The unique geometric information spaces configured in this way then form a statistical manifold in which the Fisher-Rao information Riemannian metric provides a way to measure distances and angles of trajectory along the manifold. The information divergence localizes the Fisher Information metric at its smallest distinguishable distance. Therefore, a measure of information divergence quantifies this distinction in units of uncertainty. The Fisher information takes the role of assessing the geometric curvature of the manifold and the natural flow of the trajectory. Knowledge acquisition concerning the biosystem state occurs as the divergence changes.

Our task is to find an algorithm or law that leads to the origin of information.
—Manfred Eigen, Nobel chemist

In statistics, the hypothesis concerning the likely state of the living system is really defined by a probability distribution of possible conditions. The dynamic changes in information specifying the evolving hypothetical system state as obtained through experiential sensory information capture can then be studied by analyzing the information divergence and the gradient flows of the Fisher-Rao information metric on the space of the statistical manifold. Shun'ichi Amari showed that maximum entropy distributions are precisely the ones with a minimal interaction between their variables and most closely approach independence. The Fisher metric is localized by the information divergence. The Kullback Minimum Discrimination Information formulation also provides for a Lyapunov function whose energy for change is driven by the computational mechanics of the replicator equation and derived from constrained tensions in the nonequilibirum condition. As a Lyapunov, a natural gradient for flow is formed based on the dynamics for processing new information. Marc Harper and John Baez have related this process to that of Bayesian inference with minimum cross-entropy using information divergence and replicator dynamics as the path and mechanics for the inference. In information geometry, the entropic dynamic path is along a geodesic and the trajectory of change is that which naturally minimizes the distances between points like a least action principle (see figure below). The trajectory naturally minimizes the length through the processing of the replicator equation. This trajectory is also the same as the minimization of thermodynamic free energy in a system as it naturally moves to a steady state in the context of its constraints.

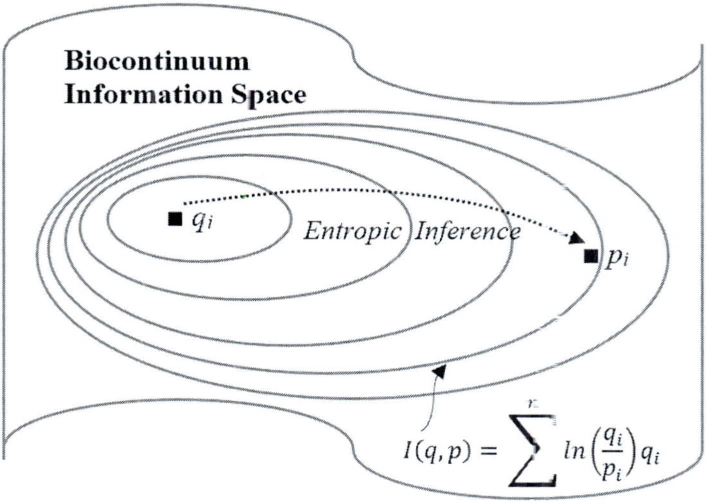

This graphic demonstrates the natural physical process for reconciliation of information divergences within the biocontinuum as grounded in the gradient flow formed by the entropic drive of the Kullback's Principle of Minimum Discrimination Information and serves as an action principle and functional for living systems.

Expressions for the potential and kinetic information of the biocontinuum are then defined by:

$$I(q,p) = \sum_{i}^{n} ln\left(\frac{q_i}{p_i}\right) q_i = \sum_{i}^{n} [ln\,(q_i) - ln\,(q_i)]\, q_i$$

and

$$\frac{d}{dt} I(q,p) = -\sum_{i}^{n} [f_i(P) - f(P)] q_i = \sum_{i}^{n} f_i(P)(p_i - q_i)$$

Where:
I(p,q) is the information state
q is the target state with a fixed probability distribution
p is the time dependent probabilities of current state
$f_i(P)$ is the fitness of each type i in the population as a survival likelihood or probability characteristic in the context of the environment

KE is kinetic energy, PE is potential energy

$$Action = S = \int_{t_1}^{t_2} (KE - PE)dt$$

Action (S) is defined as the integral summation of the Lagrangian integrand defined as the difference between the kinetic and potentials naturally minimized by Kullback's Principle of Minimum Discrimination Information within the trajectory of the system dynamics. It is important to note that this information-theoretic expression is not intended to be a representation of the absolute state of the biocontinuum but rather simply describes the course of any phenomenal dynamics.

The laws of physics might be rules of inference rather than laws of Nature.
—Ariel Caticha, physicist

At the root of this information-theoretic grounded Lagrangian approach is the perspective that phenomenal dynamics are more simply described by logic-based informational and geometric constructs rather than adopting the notions of unseen forces and universal laws. Ariel Caticha and others have concluded that the laws of the physical sciences are not really laws of Nature but rather are rules for processing information about Nature. The same concepts also hold true for chemical and biological systems. Information can then be considered the real currency of action and driving force for dynamic changes. However, it is important to note that any creation of information first requires the discerning experiential processes of an observing living system. This perspective provides a pathway and rationale for Life as a primary consequence of an unfolding Universe.

CHAPTER 5

LIFE'S UNIQUE CAPACITY FOR EXPERIENCE

But if the ultimate aim of the whole of Science is indeed, as I believe, to clarify man's relationship to the Universe, then biology must be accorded a central position. —**Jacques Monod, Nobel biochemist**

What is Experience?

According to standard dictionary definitions, "experience" is the active process of accumulating knowledge (verb) or the know-how and capacity for action resulting from the direct participation in events or activities (noun). In living systems these roles of noun and verb are coalesced into an overarching procedure of acquiring and assimilating information and transforming it into actionable meaning for the objectives of stability and survival. As noted in the previous chapter, the living system's unique ability to integrate the system's sensing functions with determined actions is essential for any adaptation to perturbations in the biocontinuum. In the book *Experiences in the Biocontinuum*, there is a detailed accounting concerning how the experiential process can be seen as the primary biological function necessary for the moment to moment continuous emergence of the energy channeling capacity, organizational coherency and causal autonomy of the living system and the reality of its biocontinuum. It is only through the mechanics of this process that information and energy are coordinated in a meaningful way for the objective of sustaining the organized living system. The conclusion from that scholarly analysis is that the capacity for experience is the singular unifying process that empowers the advent of all of these essential conditions that are necessary for Life.

Nothing ever becomes real till it is experienced. —**John Keats, poet**

This experiential-based analytic approach also flips the perspective from the canonical microscopic causal impetus (the reductionist view) to a primary perspective where the whole of the living system embedded in the environment of its biological continuum is the causal origin of the known state of reality. So rather than a reduction to physics, there is a reduction to the fundamentals of experience as the process-relational causal framework. This top-down shift in causal perspective also allows for a restriction in the degrees of freedom in living systems to overcome the forces of entropy. Experience mediated cohesion and coherence creates the constraints for a singular degree of freedom for the living system as the totality of details of the organizational state is always known to the organism. By intrinsically knowing these states through experiential self-reference, the living system has the ability to transform the qualification of energy into a designation as work. The experiential process also allows living systems to continuously recreate the dimensionality, substance, and properties of their knowable reality. From this unique position, the living system's experiential capacity is the "Center of the Universe". And as the basic experiences are collectively brought into a conscious experience for sentient beings, the subjective self then becomes aware of that central place in reality.

The Biocontinuum of Experience

> *Every man takes the limits of his own field of vision for the limits of the world.* **—Arthur Schopenhauer, philosopher**

As noted in Chapter 2, one of the great lessons of Einstein's theories of relativity was that time and space are as much a part of the physical world as matter and energy and are interwoven into a continuous nonspatial whole or succession in which no portion of the whole is absolutely distinct or separated from adjacent parts. As such, a continuum is generally considered as the seamless union of entities into a whole, like the Minkowski mathematical framework that joins space and time into a single geometric structure. However, the continuum of interest here (the biological continuum or "biocontinuum") is essentially an information space.

The Information Space

An information space has been generally described by Gregory Newby as a set of concepts and the relations between them that are contained in an information system. In other words, it is a space containing a set of systematically interconnected pieces of information. Because of these systematic and logical interconnections, the information space is considered to be fully cohesive and coherent. Such a space then describes the range of possible values or meanings an entity can have under a given set of rules and conditions. In this age of computers and internet connectivity we are accustomed to encountering information spaces as we retrieve a file from our computer or use a search engine to find information on a website. What we are doing in these encounters is actively acquiring knowledge about the content within these information spaces.

All our knowledge has its origins in our perceptions.
—Leonardo da Vinci, Renaissance thinker

The biocontinuum is structured to describe the complete realm of information describing experience regarding the environment and state of being for each individual living system. Through the experiential process, living systems are actively acquiring knowledge concerning the content of their distinct biocontinuum information space that is within the reach of their biological perceptual capacity. This biocontinuum includes everything that could have a potential interaction and exchange with the living system's processes which are in turn responsible for generating, actualizing, and sustaining the conditions of the space. That inclusion comprises all possible energy, material and informational exchanges as well as communiques originating from within or external to the usual considered boundaries of the organism. As a continuum, there is a no real distinction between the living system and its adjacent environment since the considered space incorporates everything within the experiential realm of the Life processes including the organism's own state.

Life itself has to be regarded as belonging in some sense to a totality.
—David Bohm, theoretical physicist

This conception of a biocontinuum also makes sense with what we know about the physical reality of Life's processes. It shifts the fundamental causal framework from a deterministic reductionist perspective to a holistic process-relational system's functioning. In physics, we know that all closed systems must spontaneously and continuously move towards a progressive state of increasing disorder or entropy consistent with the 2nd Law of Thermodynamics. By contrast, biological systems are known to be highly organized and through evolution tend to become more complex. Therefore, by inference from physical law, biological systems cannot be considered to be closed systems. These systems are clearly open to exchanges of energy and matter and information with the space they occupy. In fact, the only real closure that distinguishes a biological system as a separate and distinct entity from its environment is its controlled organization and the horizon of informational exchange with that environment (causal closure as previously discussed). As matter and energy are brought into the threshold of the organism and funneled into the organizing constraints of the living processes, the uncertainty of the information concerning those elements is converted through the experiential process to a knowledge of most probable order within the biocontinuum. Otherwise, all living systems are in complete material and informational continuity with their surroundings.

The real is only one realization of the possible.
—Ilya Prigogine, Nobel chemist

An Epistemic Space

The biocontinuum as conceived as an information space feeding the experiential process has many similarities with David Chalmers' epistemic space developed in his study of consciousness. The epistemic space is considered as the space of all possible scenarios that are knowable as the state of that space. All information is then positioned in the biocontinuum epistemic space in a way that governs how the world is seen and understood. This conceptual position does not deny the independent objectivity of the external world but only modulates the importance of that consideration relative to an understanding of its meaning for the living system. Within the biocontinuum, reality is actualized through its meaning to the living system. In this process, the living system brings forth the constitution and dimensionality of its reality through the very act of perceptual experience. Without the experiential process and the context of the living system, the configuration and attributes of the physical world cannot be practically distinguished and furthermore have no significant meaning. Therefore, the reality of the known world is a domain in space that is delineated epistemically or in terms of our meaningful knowledge about it, with the act of experience being the defining event.

If a tree falls in a forest and no one is around to hear it, does it make a sound? **—George Berkeley, philosopher**

Maybe it does but it has no meaning. **—Anonymous**

The knowledge acquisition process hones the scope of the space as information moves from that which is possible to that which is of momentary certainty in the experiential realm. Chalmers states that certain knowledge versus belief divides the epistemic space which is really at the threshold of the acquisition of experience. However, Mark Jago notes that Chalmers' account of epistemic space does not include information which can be gained deductively from logical inference.

That means that there are some possible scenarios that are naturally more likely than others based on prior experiences alone and can be logically inferred and deduced. But those deductions could be incorporated in the belief component noted by Chalmers or a Bayesian prior during inferential knowledge acquisition.

The biocontinuum as described here not only includes the possible knowledge space of experience but also the combined information about the state of the observer which includes all inferences that can be derived from information and knowledge acquired through prior experiences. That information is really encoded within the observer's organizational system in the embodied form of the system's previously established structure and process dynamics. At the time of a new experience, that predicate state of the living system reflects the *a priori* predicted conditions of the biocontinuum and the system's preceding adaptive conformation to those conditions in fulfilling its directive for sustaining its organization. Therefore, the epistemic space is really considered just a component of the overall biocontinuum.

In the building of a theoretical framework for biological systems, the biocontinuum as a coherent information space allows for the mathematical formulation of the possible states of the space in terms of Shannon information. This information is first captured with perception and then differentiated and deconstructed to meaning as the process of knowledge acquisition is completed. With further ongoing experiences, a Bayesian process of updating through acquired knowledge can be used.

Dimensionality

The biocontinuum platform also lends itself to discussions regarding the measurement and analysis of the dimensions of physical phenomena and the role of the observer in that determination. For example, the probability structures of statistical mechanics and quantum theory are likely born out of their epistemic chance qualities and the difficulty of obtaining experiences of microscopic events. Likewise, the position of the observer in the context of a unique biocontinuum (amplified with instruments such as microscopes) becomes the frame of reference. The basic concepts of energy and its qualification as work or disordered entropy in thermodynamic theory may also be dependent on the way changes occur within the biocontinuum and the differential determination of utility by the observer. The biocontinuum is a medium that is susceptible to being modified by physical energy and matter exchanges which in turn results in an alteration in the state of the living system because of that system's sensitivity to perceiving changes in this medium. Through the differentiation of changes and the subsequent discernment of its meaning by the living system, the quality of the energy or matter exchange is determined and work is performed. But the accounting of any changes in the biocontinuum requires some real dimensional metrics from which the information can be extracted.

A dimension is a geometrical/topological measure of an object's properties in terms of the number of points in a coordinate system that specify the object. It is within such a dimensional construct that objects can be distinguished and differentiated. It is also through this differentiation that the world becomes material to the observer/participator. According to the physicist John Wheeler, the metrics of physics are the tools for differentiation and communication of information and it is through these metrics tied to physical events that meaning is provided to biological processes. Samuel Avery notes that for humans as a perceiving organism, the olfactory, auditory, and visual senses create the spatial dimensions. These senses provide the familiar 3-dimensional structure of size and position that correspond to our experiences. The chemical and tactile facets like taste and touch are responsible for the time dimension since they provide a measured occurrence of the continuity and persistence of the experience. Since olfactory, auditory and visual mechanisms of sensation are also chemical and tactile at their rudimentary level, they all become coordinated and integrated into dimensions of space-time experiences. At the cellular level these same chemo-tactile facilities exist to distinguish the flow of time and some rudimentary space dimensionality and may have a special type of aggregated sequential ordering in experiential determination of dimensionality. There is also some growing evidence that there are some poorly understood sensory perception mechanisms used by organisms to capture various unique forms of information within the biocontinuum. For instance, some animals are able to delineate magnetic fields that aid in determining migration paths. In a much more controversial possibility, Nobel Prize-winning physicist and anaesthesiologist Stuart Hameroff, posit that microtubules in the brain detect quantum vibrations and may connect our consciousness to conditions throughout in the Universe.

External nature is only internal nature writ large.
—Swami Vivekananda, monastic

Within the biocontinuum, all physical reality can really be thought of as the units of experience and the structure of space-time as a grid for the intra-coordinated organization of that experience. And energy is then simply the propensity for these dimensions to change with potential energy being the unrealized capacity for this change and kinetic energy being the active ongoing change. While dimensions in space can change with the constant input of a time experiential metric (the velocity), a change in that change is the acceleration as related to what we call mass. The mass of an object can be experienced directly by an observer as a dimension itself through the tactile sensory mechanism which combines space and time. Hence, mass distorts or changes the tactile sensory perception of space-time for the observer and creates the action functional of gravity.

There are several important points to be noted from the above proposed sources of dimensional structures. First, it is the living system's sensory establishment of the dimensional structure of space-time through experience that creates what is considered to be the meaningful form, order and action functionals of phenomena. In other words, it is the dimensions which are created by the mechanisms of perceptual experience that give the Universe an ordered structure and dynamic, including the active organization of the living organism that in turn creates the experience. Through perceptual experience, living systems are performing a constant genesis of dimensionality and meaning for the biocontinuum information space and the living system's known reality. Before this system-initiated procedure of interrogation and deconstruction into meaningful information, the content of the biocontinuum is only an undifferentiated and undetermined cacophony of potentials.

In the framework described above, it is interesting to note that physical reality's dimensional structure is uniquely relative to the sensory functions and capacity of the observer/participator. As that capacity changes relative to the state of the medium and context in which the sensory information is captured, such as the when the observer travels close to the speed of light, then the experience changes. That relativity of observation that is dependent on frames of reference is just as was noted by Einstein. Likewise, there is a dissolution of all dimensional structures of space, time, and mass when the experiential framework is too small and beyond the capacity of the sensory mechanisms to observe the phenomena. While tools such as microscopes and telescopes can amplify and enhance the capacity for experience, the limits to the biological sensory mechanisms are still ultimately the impenetrable barrier. That impediment explains some of the weirdness of the quantum states of matter. In fact, the quantum at the level of Planck's constant is considered the lower limit of the perceptual experience of physical phenomena.

Intentionality and Perceptual Experience

Perception is not something that happens to us, or in us. It is something we do. —**Alva Noë, philosopher**

When a perception is the result of a programmed inquiry by an observer, it then moves to the domain of intentionality. Intentionality is the act of doing something with a specific prospective objective and lies at the origin of the experiential process as it unfolds in the perception-action cycle of the autonomous living system. Almost any physical difference can be perceived as an information signal leading to knowledge if the receiver is equipped to make that differentiation and act on it. In this way, the onus is on the recipient to make an intentional action in determining the information in the biocontinuum that is of importance to be perceived. As noted previously, Alicia Juarrero suggests that perceptual activity is intentional when it is guided by the pursuit of endogenously defined objectives. In this way, intentionality becomes an integral part of the perception-action cycle and reflects a predetermined preference for information that is meaningful for stability and survival. The intent is to choose requirements for knowledge that inform adaptive changes for the organism through sensing procedures that determine the signals that are targeted in the observation. Therefore, these sensing procedures are not only about a lack of knowledge but also a choice of query. All intentional queries develop out of the living systems' autonomy as driven by the naturally selected primary objective of sustained existence. In turn, the living system's queries determine the perceived dimensionality and form of the biocontinuum and the Universe and therefore will always appear to have the Goldilocks' conditions of being just right.

Consciousness as the Feeling of Emotions Emanating from the Collective of Primary Experiences

It has always been considered that there is a close relationship between the process of experience and consciousness. Consciousness is the term used to describe one's awareness of their experience (physical, mental or emotional). As such, it can also be considered the experience of one's collective experiences. Therefore, consciousness can be defined as the cognizant knowing of experience (the noun) associated with feelings, with experience (the verb) as that process of that knowledge acquisition. Feelings are also what makes an organism sentient. And knowledge acquisition, as described in *Experiences in the Biocontinuum*, involves an interpretation of knowledge to meaning as founded on existential factors. Conscious feelings interpret the collective meaning of an organism's global experience.

The Hard Problem

Physicalists view conscious events as inseparable from physical operations in the brain and that the resulting experience is simply an epiphenomenon. However, a more exacting description of consciousness within this constrained perspective is very difficult to conceive. Donald Davidson considers that the epiphenomena of conscious experience might be an irreducible physicalism such as an emergent property of a complex system. Sir Roger Penrose, Rouse Ball Professor of Mathematics at the University of Oxford, also considers consciousness to be noncomputational based on his interpretation of Gödel Incompleteness Theorem. The well-known consciousness theorist David Chalmers has concluded that the hard problem of consciousness is the problem of explaining the relationship between the objective physical properties that define brain processes or mental states and the subjective phenomenal or experiential consciousness which is also known as qualia. Chalmers suggests that there is a new type of dualism in which all perceived information has both physical and experiential aspects (physiological and psychological). Then the functions of consciousness can readily be explained by the neurobiological and physical mechanisms within the brain while the overall experiential process is the aspect instilled with meaning. In this way, consciousness is not produced by the brain but rather emerges from its functional activities through the physiological and psychological manifestations of experience. But unlike the physiological aspects that can be studied using the usual objective scientific methodology, the experiential process is highly subjective and only directly available from a 1st person perspective. Therefore, Chalmers asserts that consciousness cannot be functionally analyzed. Conversely, the Dalai Lama contends that a more complete study of the mind and consciousness must include a 1st person empiricism and suggests that contemplative methods could be useful.

> *Hard problem can be solved with a new fundamental understanding of the Universe.* —**Patricia Churchland, philosopher**

In considering the essence of this emergent consciousness, American philosopher Thomas Nagel famously asked the question, "What does it feel like to be a bat?". The operative word here is 'feel'. In consciousness, the self is the subject and the way the self feels really demarcates the subjective realm from the objective. However, the scientific understanding of the subjective character of consciousness as "what it is like" to be conscious and why it is associated with these functions is not as forthcoming.

Emotions cannot be known to the subject before there is a consciousness.
—Antonio Damasio, neuroscientist

According to David Chalmers, all normal basic human functions could occur in a reactive zombie without the conscious experience and feeling of awareness of their happening. This functioning could also include a condition of sapience, defined as the capacity for intelligence and an ability to think as seen in our current artificial intelligence technology. However, under the rules of Natural Selection, the fact that there has been the emergence of conscious awareness infers that it has a practical purpose or meaning or at least a survival benefit.

Integration and Meaning of Experiential Information

Perception is a process by which organisms use their embeddedness in physical reality as if they were independent of it.
—Christine Skarda, neuroscientist

From a scientific perspective, the understanding and analysis of an emergent consciousness still requires a more rigorous framework if the physiological aspects are to be accurately integrated with 1st person subjective elements. We need to understand how subjective elements such as feelings and self-awareness are assimilated into the more general biological process of experience. A credible model of consciousness should then be a theoretical description that quantitatively relates the physical brain activities associated with consciousness with the 1st person experiential conditions of consciousness (qualia). Since these experiences originate as perceived information, it has been suggested that information may be the primary basis in an ontological sense prior to the existence of conventional physical properties. This viewpoint is consistent with the IT from BIT perspective of John A. Wheeler where information (BIT) processed through the experiential capacity of a living system becomes the reality (IT) of the physical phenomena. And in this process, the living system themselves becomes a participator in determining the complexion and constitution of experienced reality. This process is also the ultimate relativity whereby the same information (BIT) may produce very different results for different observer/participators. Hence, the knowable physical reality (IT) only exists within the context of an actualization through the living system but not in and of itself.

The mathematical description of the experiential process in living systems (as previously proposed in the last chapter) might be used as a foundational framework to inform the investigation of Chalmers' hard problem. However, while that presented framework for basic experiences is indeed foundational, it does not really include all the necessary components expected in conscious experience. An excellent starting point to build on this framework to include conscious experience is found in the innovative and pioneering work of Giulio Tononi. His Integrated Information Theory of Consciousness (IIT) is intended to provide a qualitative and quantitative metric (Φ) for consciousness and purportedly links phenomenological experience to the physical substrates of consciousness. IIT espouses that consciousness at the fundamental level is integrated information and that can be quantitatively measured using information theory with its quality being determined by the complexity of its informational relationships. In this theory, the essential phenomenal properties of experience are expressed as axioms concerning information and its integration which can be used to infer postulates about the physical substrate characteristics required for conscious experience. Therefore, the real substance of the axioms revolves around the fundamental character of experiential information.

The basic axioms of IIT assert that:

1. There should be an acknowledgment of an intrinsic fundamental existence of reality.

2. That experience of reality for any observer is independent from any other external observers.

These assertions are valid since all primary existence is established by the cause-effect power of the phenomena as it plays into the perception-action cycle of the experiential process as described by Alicia Juarrero. However, this does not mean that the knowable perceptions necessarily entail the true characteristics of reality, independent of observations. Additionally, the experience is a unified entity that includes the observer and is not reducible to non-interdependent parts of phenomenal distinction and as such should be considered as a holistic cause-effect structure. Perceptual experience is then simply a way to understanding and engaging the world. This 'primacy of experience' viewpoint is also consistent with the ideas of neuroscientist Christine Skarda. Therefore, the complement to these axioms that should be added is that the needed integration comprises the entirety of the biocontinuum including the living system involved in the perceptual experience. This extension to the axioms which provides a holistic closed-loop path for experiential knowledge also allows for an instantiation of meaning into the information of the experience as it is considered in the context of the system functioning and its survival. The notion that experience is composed of elementary distinctions or attributes that are differentiable in a way that defines the specific state of the phenomena is an essential property if the experience and its information signals are to be captured and discriminated.

This revision to the axioms also infers that these individual elements within the differentiating physical substrate of consciousness have a causal power reflected in the whole of experience. Tononi's use of the Kullback-Leibler information divergence (relative entropy) is exactly the correct approach for measuring the effective information of the experience and provides a methodology for differentiation of any phenomenological composition and its native information. The axiomatic requirement that each phenomenological experience is unique provides a conditional cause-effect structure embedded in an exclusive spatio-temporal frame of reference.

No matter how hard I try, I cannot empty an experience of meaning.
—Giulio Tononi, neuroscientist

The rigorous and well described quantitation of consciousness (Φ) by Tononi goes a long way to solve the hard problem. In fact, it contains similar primary elements for the perceptual portion of experience used in the equations and framework for experience presented in *Experiences in the Biocontinuum*. However, the ingredient that is missing in the ITT analysis of experience leading to consciousness is a more definitive foundation for "meaning" that is rooted in the structural and dynamic specifics of the living system. Tononi insightfully notes that a meaning can be derived from the degree of cross-linking and integration of the information. In fact, his theory is founded on the notion that integrated informational relationships that constitute qualia require linkage with a system capable of integrating the information to form meaning. However, it is in the particulars of the cohesion and coherence in the dynamic structure of living systems where this linkage and integration really resides and is what determines the meaning. Those particulars are melded together in the previously described (Chapter 4) action functional and Lyapunov signature for delineating meaningful experience in the biocontinuum.

The least of things with a meaning is worth more in Life than the greatest of things without it. —**Carl Jung, psychiatrist**

When faced with new experiential information, the integrity of the system's autonomous adaptive response toward physical organizational stability (in the Lyapunov sense) reflects that degree of that integration and therefore is also a true expression of the information's practical meaning. In this way, experiential information is instantiated with meaning as it becomes actionable knowledge for guiding the living system's adaptive functions. The derived meaning is physically quantitatively grounded in the capacity for maintaining stability and sustainability as the autonomous objective of the living system. Since the existence of the living system is necessary for the actualization of the knowable reality of the experienced phenomena, a failure of survival extinguishes all potential distinction and assignment of meaning. Having meaning instantiated into this information measure transforms this Lyapunov metric into a new type of semiotic information.

This augmented approach to incorporating meaning into IIT also enhances an alignment with the first axiom by providing specific actionable significance to the information and ensuring the cause-effect power of the phenomena. Likewise, the second axiom can be interpreted to entail that the observing entity has a capacity for understanding the semiotic significance of the experience as a whole and through this facility it actualizes the causal structure of the experienced phenomena. Structuring the composition of experiential information with instilled existential meaning beyond simple perceptions and in the context of the overall system cohesion and coherence provides for a more exacting causal power directly attributable to the physical substrates.

As previously noted, nonlinear feedback with re-entry and the creation of recursive loops is the hallmark of autonomous adaptive living systems and have been considered by Gerald Edelman, Douglas Hofstadter and many other consciousness theorists to be essential to the emergence of consciousness processing. The replicator dynamics of the aforementioned proposed analytic platform provides this recursive procedure. It is through this looping mechanism that self-reference and the procedure for providing the contextual meaning concerning subsistence occurs. Since these necessary processes of autonomy, adaptation, and recursive self-reference are the exclusive domain of living systems, then consciousness also becomes the province of only living, animate entities.

There is nothing that living things do that cannot be understood from the point of view that they are made of atoms acting according to the laws of physics. —**Richard P. Feynman, Nobel physicist**

In addition to the important choice of relative entropy (Kullback-Leibler) as a grounding metric and methodology for satisfying the IIT axioms in the analysis, the incorporation of a modulating fitness function for stability and survival in a replicator expression is a natural way to instill an objective and substantive meaning quantity into the information measures of experience. Like ITT, the experiential process for the uptake and iterative assimilation of knowledge previously described in the preceding chapter uses the notion of relative information with its contextual power (see equation below). This perceived information differential is transformed into meaningful knowledge through a modulation with a fitness function that links the process to the inherent objective of the continued survival of the system. The perceptual information interpreted into meaningful knowledge becomes actionable for the adaptive response through the iterative differential dynamic of the expression and provides for an evolving stream of consciousness. This translation of information and knowledge assimilation through the physical mechanics of the adaptive biologic system responses creates the causal structure for the experience with its fitness modulation as reflected in the equation below. If consciousness is the knowing of experience and the process of that knowledge acquisition is given by an assimilation of experience, then an examination of that process might provide some insight into the nature of consciousness as recounted below.

$$\frac{d}{dt} I(q,p) = dD_{KL}(q|p) = \sum_{i}^{n} f_i(P)(p_i - q_i)$$

The lessons forthcoming from this analysis are:

1. Experience described as information processing must also incorporate meaning beyond simple perception in order to satisfy the essential extended ITT axioms that have been determined as necessary to infer consciousness.
2. The metric for meaning should be grounded in the autonomous living system's existential objective of sustaining the system's mechanics that provide for the ongoing potential for further future experience.
3. Cohesion of all system components toward the objective of a global coherence within the biocontinuum allows for an overall causal structure and linking.

4. This causal structure instilled with meaning defines the specific pathways of the information integration at a system level.
5. Since the meaning of experiential information is derived from the primary objective of living systems, then only living systems can have consciousness.
6. Phenomenal experience and the physical substrate for conscious experience are merged in the mechanics of instantiating of meaning.
7. It is through the inherent complexity of this overall system coherence and the resultant causal structure of the biocontinuum that consciousness emerges as an irreducible property.
8. Life is the emergent property of an autonomous system's ability to experience and consciousness is the further evolved system capacity for an ongoing knowing of that experience (experience of the aggregated fundamental experiences).
9. Conscious self-awareness comes from the self-referencing regarding system coherence that occurs as a part of the iterative re-entry and recursive loop for system adaptation required by the stability and survival objective.

In conclusion, all living systems, including single cell organisms, have some degree of self-awareness on a rudimentary level. Tononi relates the measure of consciousness to the differentiating capacity and degree of inherent integration of the observing. Augmentation of this idea with a definitive metric for meaning derived from the system's objective for sustained existence of the observing mechanics makes sense in creating an exacting and holistic causal structure for experience.

But What About Feelings?

What you experience all the time is fluctuating pulses of feeling in response to your movement through the world, as you check whether everything is as you expected to find it - and as you try to close the gap, somehow, when it isn't. —**Mark Solms, neuroscientist**

Experience at the level of the mind/brain can be thought of as the collection of all basic individual experiential processes occurring throughout the organism. To be alive is to experience, to be conscious is to experience the experience of living (or acquiring knowledge about the experiential process of acquiring knowledge). Sentience is a kind of primitive self-aware experience of being alive, whereas, mental cognition is the recognition and sense-making of the activity and experience of living. Therefore, self-consciousness or self-awareness is the singular aggregate experience of all the myriad corporeal and mental experiences of the organism. More importantly, consciousness can be considered as the experience of the collective meaning of all these discrete experiences, manifest as feelings. It is the global experience of all basic experiences. This is accomplished by neurologically monitoring an amalgamation of the Lyapunov meaning metrics of those disparate primary experiences. The great anthropologist Margaret Mead had a similar view of consciousness known as second-order cybernetics or the recursive application of cybernetics to itself.

Neurologists Antonio Damasio and Mark Solms contend that consciousness is the "feeling of what happens" with our mind/brain experiencing the body's reaction to the totality of the biocontinuum information and actively responding to that experience. Their clinical research demonstrates that emotional feelings and consciousness are almost always concurrently present. Emotions are measurable physiological and neural states that often result in some specific behavior (fright or flight, laughing, crying, etc.) And feelings are the conscious subjective experience of those emotions. Interestingly, common sense also tells us that a feeling is something that is felt and therefore must be conscious. And because the feelings are only subjectively known, they cannot be really described much further beyond their 1st person experience of the condition. An exception to that limitation may exist if they can potentially be tracked as a Lyapunov signature metric through conditions that map causal stability as related to the collective experiential information. Solms suggests that consciousness is not a cognitive function of the brain but rather is linked to its affective function. And because it is affective in its initial origin and not logistically determined through cognition, this factor is what separates the 1st person and 3rd person perspectives. Thus, it may also be a source of intuition and creativity.

Through an elegant analysis of data from brain injured and hydranencephalic patients, Mark Solms concludes that the seat of consciousness is in the upper brainstem and reticular activating system which is the originating source of both primitive affects and emotions (fear, lust, sadness, joy, rage) and our phenomenological consciousness. This position is as opposed to the previously purported activities in some higher cortical level. As these feelings are aligned with the sensory perceptions, they provide context to the basic experiences. According to Solms, the sentient organism can then know "I feel like this (emotion) about these (sensations)". This experience of emotions as feelings is what is termed the conscious experience. Affects, as the more reflexive physical reactions to experiences, precede feelings and a conscious awareness of the sensations. Even paramecium have these types of primal experiences through the mechanisms as described in Chapter 4. In sentient organisms, the momentary global experiences emanating from external and internal sensory information enter the superior colliculus of the brainstem to collectively form a neural pattern of emotional condition reflecting that overall physical state of being of the organism. This emotional pattern includes tangible physiological manifestations including sympathetic, parasympathetic and endocrine responses as well as unique neural traffic to higher cognitive centers. This amalgamated experience forming these emotional states is monitored by a further experience using the same layered experiential process mechanism to form feelings. These feelings are then the experience of consciousness, as the way it feels.

Feelings can also be thought of as the realized meanings at a global level for the collective of primary basic experiences that are fused into emotions. Through this connection with emotions we can also understand the personal meaning of experiences such as love and hate. We can further translate the otherwise technical content of poetry or music into a deeper meaning. These feelings born of affect are also then considered the elemental and foundational form of consciousness and can be considered as an emergent property of the brain for integrating all our basic primary experiences. As such, they also have some degree of existential significance since the meaning of the inputs is grounded in information concerning the living system's stability as discussed previously. Even stability for functioning interpersonally, psychologically, and in society can be considered. For instance, an individual's feelings of loneliness reflect a greater existential meaning regarding the potential security provided by being a part of a group or the importance of finding a mate for sustaining the gene line and survival of the species. According to evolutionist Richard Dawkins, the fundamental scientific meaning of Life is the propagation of DNA. However, he asserts that Natural Selection has also programmed into our biological machinery an ancillary feeling that there is a superseding meaning of Life. It is important to understand from where this higher objective naturally arises in living systems to be logically selected. Is this greater meaning the bridge between the reality of the Universe and the seeming necessary presence of living systems as suggested by the Anthropic Principle? An attempt to close this loop will be presented in later discussions.

If emotions and feelings as the origins of consciousness are derived from the totality of the biocontinuum and not by discrete cognitive functioning utilizing virtual representations, then the primacy of conscious experience is really an encapsulation of the whole of perceived reality as suggested by Christine Skarda in her work on neurodynamics.

Newberg similarly identified an area of the limbic system also associated with emotions that he calls the existential operator. He views the function of this operator as the mechanism by which the brain interprets the concreteness and relative reality of experiences. The assignment of an emotional valence to a "real" experience also allows for a feeling and existential awareness of that experience. The role of consciousness is to construct knowledge about the relation of the whole organism to some perturbation in the biocontinuum and its impact to causing changes in the organism. Therefore, these conscious feelings generalize the broad range of knowledge acquired from physical experiences and guide behavioral reactions. Consequently, we are wired at the level of the whole organism to more generally respond with emotions that are consciously experienced as feelings.

Affects tell long evolutionary stories of which we are completely unaware.
—Mark Solms, neuroscientist

Differences between the realists' and idealists' perspectives creates a separation between the knower and the thing known regarding the substance of reality. Phenomenologists such as Edmund Husserl suggest that there is an interconnectedness of the observer and observed that is achieved through the process of creating meaningful experiences. Then experience allows us to know objects beyond their materiality and really understand their meaning beyond the apparent perceptions. The Descartes' mind-body dualism then becomes a property dualism for brain functioning only and not an intrinsic characteristic of some purportedly independent material reality. As such, the phenomenologists interpret the reality of the whole through the structures of consciousness and as a manifestation of being alive. Solms calls this dual-aspect monism where the mind has dual properties. This view is also consistent with Chalmers' notions of the dual aspects of mental processing. The dual aspects are first the physiological properties that even the Chalmers' zombie possesses and secondly the additional emergent property of consciousness. Martin Heidegger further extends this connectivity of the observer and observed and concludes that reality and consciousness are really co-creations or have a co-dependent origination. Such a co-arising is reminiscent of Nagarjuna's conceptions discussed extensively in the next chapter. These phenomenological interpretations of consciousness are also consistent with the information-based framework presented in the earlier in this section and in the prior chapters. The experiential process of the living system allows for an ongoing actualization of the overall biocontinuum by a deconstruction of experience and its translation to actionable meaning.

Neuroscientists have also discovered that most cognitive functioning occurs in a realm outside of a conscious focus. In fact, most of our incoming sensory experiences are unconscious of which we are unaware. This ongoing unconscious information processing of the biocontinuum by the organism is within the natural attitude of phenomenal experience as described by Husserl. However, a higher level of a conscious experiencing of the basic experience is also evident in the mechanics of cognitive perception as first delineated in the famous experiments of Benjamin Libet. In these studies, Libet found that fast motor reactions are dissociated from conscious perceptions. It appears that most cognitive functions do not require consciousness even though they can be monitored by a conscious self-awareness. Even Freud considered that our feelings were divorced from the strictly cognitive processes of the mind/brain.

All these evidences from Solms, Newberg, Damasio, and many others support the notion that the origins of consciousness are founded in "the way it feels to be", as characterized by Nagel. So, consciousness does not arise from a cognitive activity such as thinking but rather from our experience of feelings. Such a grounding for consciousness in emotions and feelings also obviates most of Chalmers' zombie objections. This perspective is a revision of the famous Descartes quotation to now say "I feel therefore I am" or "experience consequently being" as previously noted. In fact, a feeling of self-awareness is the most fundamental and common aspect of our conscious experience of existence.

Why Do We Have Feelings? (Yes, a Why Question!)

Teilhard de Chardin advocated that the emergence of mind and consciousness along the evolutionary pathway was as universally significant as the emergence of Life itself. Solms and Damasio propose that the mechanisms of consciousness originated evolutionarily from its role in the comprehensive homeostasis of the organism as a whole. In a synopsis of Solms' work, Oliver Burkeman suggests that feelings are a uniquely effective and efficient way for conscious beings to globally monitor the innumerable and ever-changing biological conditions occurring in extremely unpredictable environments. This monitoring of the overall state from a myriad of sensory experiences under a singular metric of feelings allows the complex multisystem organism to rapidly integrate priorities for actions ranging from satisfying basic physical needs to determining social interactions, all with the intent of a survival benefit. Emotions leading to pleasant or unpleasant feelings can generally predict existential imperatives and drive actions for the whole of the organism. Feelings that function as action determinants also allow for instinctual behaviors for simplified responses guided by innate principles recorded throughout the biological evolutionary history of the organism.

Without feelings linked to the amalgamation of the Lyapunov meanings of all incoming basic experiences, a conscious comprehensive analysis involving the entirety of those individual experiences would take so many computational resources by the organism that it would lead to a combinatorial chaos. Such an impractical reckoning would also demand enormous levels of energy within a timeframe that would not allow for a rapid survival response. The reaction to a potential threat from a predator in the jungle requires an assimilation of all the information acquired by the senses in conjunction with the input of prior knowledge, all integrated into the emotion of fear for a quick response. Therefore, feelings are necessarily conscious and essential to survival in an uncertain world. This aptitude has tremendous evolutionary benefit and explains the reason consciousness exists and why we are not just zombies or automatons. It should be noted that these feelings and instincts can be modulated by cognitive input and that such moderation is important in the development of civilized societies where individuals do not simply react solely on emotion. Additionally, conditions for apperception, where the present experience is viewed through the lens of past experiences, can further influence responses. This important modulating capacity is a more recent evolutionary development that allows for higher animals to control their primitive drives and interact communally and altruistically, again for the overall survival benefit of the species.

The Brain as the Gateway to Conscious Experience and Mystical Encounters

Man is a singular creature. He has a set of gifts that makes him unique among the animals. —**Jacob Bronowski, mathematician, philosopher**

As already espoused in the previous sections, consciousness is derived from one's feelings associated with their experience. Consciousness is primarily constituted by the stimulated affects or emotions as neural states created through the collective incoming primary sensory experiences of the organism. Feelings are the subjective experience at a global level of these emotions as the realized fusion of meanings of the primary basic experiences.

Therefore, affective arousal is accompanied with feelings (what does it feel like?).

The degree of surprise generated by the experience and the resultant gradation of affective arousal determine the relative degree of wakefulness and intensity of awareness for the conscious experience. Within the analytic platform previously described, the collective Lyapunov vectors of the informational spectrum of experiences of the organism, also determine the magnitude and direction (positive or negative) of the emotions and associated feelings as a reflection of meaning and driver for action. But it is the brain's focused attention that controls cognitive access to an awareness of those experiences and feelings that constitute consciousness. Husserl contended that the intentional attention toward a phenomenon and the manifestation of consciousness were inexorably linked. If humans do have a "crown of creation" position amongst living creatures it is because their brains have a uniquely superior facility to cognitively focus attention and conscious awareness (mindfulness) on their experiences and feelings.

A state of consciousness refers to the area of experience on which this attention is focused and the degree of awareness. In other words, that state is the cognitive awareness of the direction and magnitude of the vectors of consciousness and conscious experience. An unconscious state refers to a lack of awareness of those experiences and feelings. There are different techniques (meditation, etc.) designed to shift the area of focus and conscious awareness for a different experience of reality.

As previously cited, Kant contended that there is a transcendental state of consciousness through which we can understand the experience and meaning of the world beyond superficial sense perceptions. Ordinary experience is simply a knowledge of objects whereas a transcendental experience is knowledge of how it is possible for us to experience those objects as existing entities. This goes to the heart of experiencing and knowing pure consciousness. Therefore, this knowledge transcends the meaning of the individual objects of experience and considers the experience as a whole. A transcendent state examines the primary condition of conscious experience and also derives from that experience what is supplied by the mind and its intuitions. This Kantian perspective influenced the development of Phenomenology as championed by Edmund Husserl and Martin Heidegger. Phenomenology analyses the experience of reality in its most fundamental condition as it first presents itself to consciousness and therefore transforms the world into the raw experience of phenomenon. This raw experience is also one of astonishment in which the world is viewed with no preconceptions. The process for this analysis, known as phenomenological reduction, is a technique that allows one to sustain this perspective and bring an awareness of this astonishment into everyday experience. This process is considered as a shift in the state of consciousness from that of always actively thinking, to one of simple reflection. When the reduction reveals the primacy of the whole, then the experience becomes the true reality as espoused by Skarda.

To begin with, we put the proposition: pure phenomenology is the science of pure consciousness. **—Edmund Husserl, phenomenologist**

States of consciousness can be considered as similar to other elements of Science such as genes and atoms that cannot be directly observed but are inferred by indirect evidences. However, because of its subjective character, consciousness is not readily amenable to the standard scientific methods of analysis. When considering states of consciousness, the most simplified view for objective differentiation that is a common experience for everyone are those states of sleep and wakefulness. Intersubjective determination of these states can potentially be made by basic 3rd person observations of objective data. For example, within sleep states an individual's, dreams had been very much a 1st person subjective experience until recently objective evidence of their material origins has been found in EEG patterns. Prior to these science-based discoveries, dreams were considered by Freud and others to be derived solely from the activities of the mind as separate from any specific neurobiological activity. Often compared to hallucinatory activity, dreams were interpreted for their purely psychological meaning or even as religious revelations. There are several modern theories concerning the practical neurological function of dreaming as it appears to have some role in synthesizing mental schemas and encoding memories within the brain.

We can't tell you the origin of the experience, but we can tell you the brain does appear to be built to have these mystical experiences.
—Andrew Newberg, neuroscientist

Andrew Newberg notes that individuals awakening from dreams or patients having resolved their hallucinatory visions are very aware of the unreality of those states. The same is true for psychotic breaks and drug induced altered states and visions. However, observed mystical states are deemed to seem more fundamentally real to queried monks and nuns than even their everyday conscious conditions. These mystical states have also been shown to have objectively determined characteristic neurobiological patterns within brain scans as documented by Newberg and others. In these studies, monks received brain scans while they were in intended meditative states and were clearly found to have heightened activity in their orientation association area as compared to the nonmeditative conditions of ordinary subjects. These findings provide objective 3rd party evidence of a neurobiological pathway leading to these transcendent conscious states.

We are not human beings having a spiritual experience. We are spiritual beings having a human experience.
—Pierre Teilhard de Chardin, philosopher

Meditative states have also long been touted to bring the direct experience of reality in union (yoga is Sanskrit for union) with our consciousness. In this union, the basic raw experience simply becomes the singular primary conscious focus without any preconceptions. And as consciousness is linked to feelings/emotions, that experience of the whole of the perceived totality of our biocontinuum is also naturally connected to our emotional states of joy, peace, bliss, and love. This is particularly true if the meditative or mystical experience is couched in the conditions of religious devotion and a sincere desire to understand the mysteries of Life.

Life is a series of natural and spontaneous changes. Don't resist them - that only creates sorrow. Let reality be reality. Let things flow naturally forward in whatever way they like. **—Lao Tzu, Taoist philosopher**

CHAPTER 6

THE PATH FORWARD

The Path Forward: A New Understanding

I want to know God's thoughts; the rest are details. **—Albert Einstein**

How does our current state of scientific knowledge lead us to a better understanding of reality and our place in the Universe? The post-modern perspective of relativism and the realization of the inherent uncertainty of knowledge has definitely changed our view concerning Science and the Scientific Method. This realization was recently most evident in our very public experience with the evolving state of the COVID-19 pandemic where we were directed to "follow the Science". However, the Science was not always definitive and changed frequently as our information and our understanding of the virus changed. In fact, the experience of the pandemic was certainly different for each individual as well as for various enclaves of society in general. Some commentators have likened these disparate viewpoints concerning the Science of the pandemic to differing religions with each faction having their own "truths" and "beliefs" backed by their own experiences and evidences as shaped by their biases.

He who possesses Science and Art; has Religion.
—Johann Wolfgang von Goethe, poet

Science has always considered itself as the veritable path to an absolute truth. In this path, Science is thought by many to maintain a superior position to religious dogma, which has historically developed more rudimentary theories utilizing limited tools for objective knowledge acquisition and relying primarily on mythology accounts and scriptural authorities. However, there has also been the recent acknowledgement of the limits of the Scientific Method as being ultimately dependent on inter-subjective consilience with an uncomfortable recognition of the 1st person privileged position in the primary knowledge acquisition process. This inherent uncertainty in all supposedly objective information blurs the lines of validity between our understanding of reality as conceived by Science and that proposed by traditional religious thinking. Additionally, a deeper comprehension of the mysteries of subjective experiences, consciousness, and emotions with feelings that influence scientific conclusions are thought to possibly benefit from religious and spiritual insights and might lead to a new understanding of our place in the Universe. Regardless of the originating perspective, intuitively there should be a foundational basis for a best process in the determination of truth regardless of the source of knowledge.

The Science of Religion

> *The most beautiful thing we can experience is the mysterious. It is the source of all true Art and Science.* —**Albert Einstein**

Though we tend to think of Religion and Science as completely separate disciplines with almost no commonalities, they both have the mutual purposes of trying to understand our Universe and our place in that reality. In previous chapters, there has been a detailed exploration of the mechanics and philosophy of Science. While religion by nature is much more diverse in its form and content, there is really a common structure in religious origins and practice. The acknowledged Prothero theory of religion says that all religions have four basic functions and purposes in common.

1. Identification of a fundamental problem (i.e. reason for existence, death, afterlife).
2. Provides the answer(s) (often provided in sacred scriptures, teachings, or myths).
3. Provides a way to solve the problem. (i.e. techniques, beliefs, rites, sacraments)
4. Identifies an entity with the solution. (i.e. avatar or incarnation, church, sect, sangha)

It is easy to relate these functions and goals of religious activity to the elements commonly provided in the Scientific Method as discussed in Chapter 3. In the Scientific Method, the observation or identification of a problem is followed by a hypothesis as the potential answer. These answers are then practically tested or applied as supported by a group of scientists. Science has often filled the gaps in this understanding that Religion could not concretely provide. Conversely, religious belief systems promise to provide meaningful context to the cold objectives of Science. Their shared position in our human reality also suggests that there is a scientific explanation for the emergence of most of our common religious thought.

The Value of Mythology

A myth can be defined as a traditional narrative, often of unknown origin, that may or may not be true though is ostensibly related to actual events. Myths are usually associated with religious beliefs, and generally serve to explain the way the world works. For instance, a creation myth is a story that tells us how the world came into existence. Myths also often contain allegorical instructions for living a good Life. Unfortunately, in the modern vernacular the word myth has come to mean a primitive fantasy story concocted by the ancients. That trivializing perspective is really a disservice to the true function and value of our myths.

> *The Universe is made of stories, not of atoms.* —**Muriel Rukeyser, poet**

The famous mythologist Joseph Campbell considers myths as guides to the spiritual potentialities of Life. In this way, myths are not merely directions provided for the search for Life's meaning but rather they tell us that the meaning of Life is actually the experience of Life itself. On a more scientific level, Campbell views myths as metaphors that arise from an interaction of biological functions with the subconscious imagination in order to reconcile ourselves with the natural world and its mysteries. Including our conception of God. In this regard, they function physiologically and psychologically as the source of solace to mitigate the fears associated with uncertainty. Myths also bring cohesion to societies and provide the authority for established morals. The effects of myths on our lives also have a decided benefit in the overall workings of Natural Selection. This pragmatic understanding of the origins and utility of mythology as described by Campbell is in consilience with Solms' and Newberg's conceptions of the biological derivations of consciousness and mystical experiences. It is unfortunate when these mythologies sometimes devolve to create a justification for repressive dogma and authoritarian edicts.

> *Science and technology revolutionize our lives, but memory, tradition and myth frame our response.* —**Arthur M. Schlesinger, historian**

It is interesting to note that there is considerable consistency in some of the elements of mythologies across the entire scope of humanity and cultures throughout history. The 19th Century German anthropologist, Adolf Bastian first noted common themes that occur in almost all the religions of the world. For instance, a wide variety of civilized cultures promulgate a diluvium flood account all the way back to the ancient Epic of Gilgamesh. Likewise, even the Buddha, Zoroaster, Horus, Dionysus, and Romulus and Remus were all reported to be born of a virgin just like Jesus. Carl Jung, Joseph Campbell and others consider that these archetypical seeds likely arose from the basic commonalities in the structures of societies, human biology, psychology, and brain functions.

> *Science without Religion is lame, Religion without Science is blind.*
> **—Albert Einstein**

Paradigm shifts in the interpretation of mythologies in response to scientific discoveries is often necessary in order for the religions they support to survive. Clearly the Christian faiths have had to adapt their perspectives concerning the Center of the Universe when evidence for the heliocentric structure of the solar system emerged and there was a new understanding of our inconspicuous place amongst a myriad of galaxies. Likewise, the widespread acceptance of Darwin's Theory of Evolution has resulted in a much less literal interpretation of the creation story of Genesis by many of the faithful. The need for a reinterpretation of traditional mythologies has also been found true for the other religions. Even the Dali Lama admits that facts presented in some of the historic mythologies of Buddhism must be superseded by the evidences provided by Science. It will be interesting to see how the various religious doctrines adapt if there is ever a discovery of extraterrestrial Life.

The Tree of Knowledge of Good and Evil and Tree of Life

> *Behold, the man has become like one of us in knowing good and evil. Now, lest he reach out his hand and take also of the tree of Life and eat, and live forever.* **—Genesis 3:22**

The celebrated mathematician and philosopher of Science, Jacob Bronowski, contends that dogma ascends and mankind's worse failures arise when we believe we have the absolute knowledge of gods. In fact, if there is any core lesson that resulted from the revolutionary discoveries of our 20th Century science, it is that all our acquired knowledge carries some degree of uncertainty. Yet, it is through the knowledge acquisition process that our perceptions become our knowable reality (the *becoming*, see Chapter 2). And it is through the lens of experience that the fundamental nature of this knowledge is established. Consequently, it is the conscious focus of that experiential lens that ultimately determines our perception and understanding of reality. A shift in the state of consciousness away from the judgmental absolutism concerning the dichotomy of right and wrong and toward a focus on a pure experience of Life and the oneness of the Universe has always been the goal of the mystics.

You can never gain perfect happiness unless you've got that state of consciousness that enables that. **—George Harrison, The Beatles**

There has been considerable debate for millennia about the true meaning of the biblical mythology of the Tree of Knowledge of Good and Evil and the Tree of Life. Regardless of interpretation, the representations in this story clearly intend to denote a stark contrast in two different ways that humans can potentially understand the world. These distinctly different perspectives concerning the nature of reality is also seen as a central breakpoint in an individual's spiritual condition in other religious traditions. The ancient Hindu scriptures describe the true reality as a realm of nonduality and a pure consciousness of Life called Brahman as opposed to an illusionary perception of the contrasted material world called Maya. The biblical accounts of Jesus' sayings and the Gnostic Gospel of Thomas found at Nag-Hammadi describe a heaven that is "within" that can be discovered through a change in revealed conscious perspective and understanding. The Buddhist conception of emptiness is somewhat different but still distinguishes an interdependent, co-originating, all-encompassing view of reality from one in which the world is composed of an inherent self-oriented duality with discrete contrasting parts. Even ancient aboriginal civilizations used psychotropic natural products and practiced intense rituals to help induce spiritual visions with a more directly primal view of reality. The Lakota Sioux Sun Dance ceremonies (Wiwáŋyaŋg Wačípi) and the Crying for a Vision quests (Haŋbléčeyapi) were practiced by nearly all Native American tribes of the Plains in attempts to gain new insights and understanding. In all these traditions, a fundamental shift in state of consciousness from a contrasting material duality to the singular holistic vision of Life is probably the origin of the spiritual occurrences of revelation, epiphany, nirvana, and all other various forms of mystical experience. These primordial experiences are manifest as transcendent states of consciousness where there is a sense of timeless infinity, oneness with the entire Universe, and universal love. Not surprisingly, these spiritual or mystical experiences are also often the originating source of emerging belief systems and organized religious sects.

Those who forget good and evil and seek only to know the facts are more likely to achieve good than those who view the world through the distorting medium of their own desires. **—Bertrand Russell, mathematics**

The real question is whether there is a rigorous scientific explanation for these religious experiences. From a psychological perspective, the epoché mental state of phenomenological reduction as proposed by Husserl allows for a suspension of all judgments (i.e. good vs evil) about the external world and a focus on the phenomena of experience itself. Husserl characterizes this state as a reduction to pure consciousness that is void of all material interpretations and known only as it is given directly through the experience. Christopher Gutland has described the similarities between Husserl's methods in phenomenology and the process of introspection which is at the heart of the Buddhist contemplative practice.

As discussed previously, we know from the physical sciences that both time and space have no absolute universal meaning and are relative to the individual's experiential frame of reference. Likewise, quantum theory has revealed the co-dependency of all phenomenal reality in the context of the observer's experience. Therefore, it is not surprising that a reduction of mental state to a focus on the pure experience of phenomena would unveil a feeling of a timelessness and oneness with all knowable space-time.

> *Scientific views end in awe and mystery, lost at the edge in uncertainty, but they appear to be so deep and so impressive that the theory that it is all arranged as a stage for God to watch man's struggle for good and evil seems inadequate.* —**Richard P. Feynman, Nobel physicist**

Regardless of their psychological or spiritual context, even the most ardent mystics such as the Christian theologian Evelyn Underhill consider mystical experiences as being born from biologic processes. The new science behind the study of states of consciousness indicates that there are basic neurobiological mechanisms within the brain responsible for the shift to a definite spiritual/mystical perspective. As previously recounted, the neuroscientist Andrew Newberg describes an accentuation of activity in areas of the brain associated with orientation and existential operations with a suppression of binary distinctions in the brain scans of meditating Tibetan monks and Franciscan nuns during their fervent prayers. This transcendent condition is as compared to the nonmeditative state of ordinary subjects. Using these objectively determined characteristic neurobiological patterns within the brain scans, Newberg makes a strong case for the validity of these spiritual experiences from a scientific perspective. These findings provide some objective 3rd party evidence of a neurobiological pathway leading to these transcendent states of consciousness. The brain's limbic system associated with emotions also functions as the mechanism by which the brain interprets the concreteness of these experiences as being different from dreams or psychiatric hallucinatory states. Newberg suggests that certain practiced religious rituals and meditation techniques engage this neurobiological machinery of transcendence to open a window of consciousness for a mystical experience of an alternative reality.

Nothing is so difficult as not deceiving oneself.
—Ludwig Wittgenstein, philosopher

If this peaceful mystical state of consciousness is preferred, then why do we have such an innate tendency to live in a reality of material distinction and with the "natural attitude" as described by Husserl? Neuroscientist Christine Skarda contends that the biologic mechanisms of our usual perceptual experiences provide us with an articulated version of physical reality in which we hear distinct sounds and see discretely characterized objects with concrete features. Hence, we erroneously see ourselves as separate from the external world and misperceive this disjointed experience of phenomena as the fundamental nature of reality. The primitive evolutionary advantage of this separate-from-the-world viewpoint is that most of our primary existence as living organisms is established by the cause-effect power of observed phenomena as it plays into the intention driven perception-action cycle required for adaptation and survival as described by Alicia Juarrero. However, this does not necessarily mean that these functional perceptions entail the true characteristics of reality. Fundamentally, it is really the meaning of an experience that establishes its place as perceived reality. Additionally, a conclusion for the meaning of perceived information is critically important for establishing adaptive actions. This perspective is also consistent with the research of Freeman and Skarda in which they found that the currency of neurodynamics is primarily meaning and only secondarily that of information. Likewise, J.J. Gibson's theory of affordances contends that organisms perceive the world not only in substance and spatial relationships but also in terms of the meaning of the experience for action. Freedom from the trepidations of birth and death and the strivings of the material world actually means that a superseding and more profound meaning is ascribed to reality and Life.

I do not fear death. I had been dead for billions and billions of years before I was born, and had not suffered the slightest inconvenience from it. —
Mark Twain, writer

Humans innately yearn to experience this transcendent reality that means more than just a mere Darwinian existence. This inborn sense is probably due to our keen awareness of biological death and not knowing what lies beyond our temporal existence. These concerns about the hereafter are probably also just a metaphysical extension of our evolutionary driven primal biologic directive for continued survival. In fact, the archetypal desire for an immortal afterlife in some heavenly safe haven is likely a psychological derivative of our biologic drive for stability and survival resulting from the logical consequences of Natural Selection. While Richard Dawkins rightly proposes that the fundamental meaning of Life is the survival and propagation of our DNA, he also suggests that Natural Selection has moreover programmed into our biology a desire for meaning grounded in the objective of a deeper understanding of our place in the Universe. This higher objective manifests itself in many different forms and goals for achievement as each individual tries to find meaning in their life. And the drive may even be the source of our innate spiritual yearning. Joseph Campbell contends that we as individuals are not necessarily searching for the meaning of Life but rather we are looking for an experience of being alive.

But from where does this objective naturally arise to then be logically selected? If the actualization of the reality of the Universe is contingent upon the co-existence of observing living systems, at least as a boundary condition as suggested by Bruer's interpretation of the Anthropic Principle, then it is not surprising that there is an innate drive for a more direct experience and meaningful understanding instilled in the essence of our biological makeup. Therefore, the meaning ascribed to the survival and propagation of DNA ***and*** the meaning of the living system's process of experience leading to the actualization of the reality of the Universe are ***both*** existential in nature. Again, this Life—Universe process of co-dependence and co-arising is in consilience with Nagarjuna's ideas as previously described.

When a focused state of pure consciousness engages the primacy of experience, then existential meaning is naturally instilled in the process. Therefore, mystical experiences as described above reveal to the recipient that they as individuals are personally connected to the meaning of the greater whole. Religion emerges from episodes of these unique spiritual experiences as a guiding path for the pilgrim or sādhaka (Sanskrit: साधक) who seeks to shift the neurological state back to that transcendent perspective. However, most individuals never take that journey but only rely on the accounts of those that have in order to form a comforting belief system. Like Joseph Campbell's mythic tales of the hero's journey.

Realization and Actualization of the Reality of the Universe

> *God created man in his own image...* —**Genesis 1:27**

If the transcendent experience of the mystics provides for a meaningful connectedness with the greater whole of the Universe, then what is the nature of that connection and how does that relationship manifest itself. There is a seemingly basic paradox in most religious theologies between the characterizations of an omnipotent, omnipresent, omniscient God contrasted with the notion of a very personal compassionate God with which we have an individual parent-child type relationship. This theological conflict is usually resolved by the introduction of incarnations and avatars in religious doctrines that can be personified. Such an incongruous impression of the deity is most demonstrated in Hinduism in which there is the conceptualization of the all-pervasive impersonal Brahman coexisting alongside an assortment of avatars or incarnations that have very personal human relationships. Even the Christian concept of the Trinity is an attempt to provide some clarity and reconciliation to the divergent characterizations. The basic need of humans for a tangible, personal connection is not surprising since our common everyday material reality is more comfortably based on an understanding of such interpersonal relationships.

> *The personality you assign to God has distinct neural patterns that correlate with your own emotional styles of behavior.*
> —**Andrew Newberg, neuroscientist**

These intermediaries between God and humanity are seen throughout most of the world's religious traditions (Saints, Prophets, Incarnations, Avatars, Bodhisattvas, Satgurus, Medicine Men, etc.). They often claim to provide a potential path for a direct connection to the divine by a revelation/realization/vision of the true reality through an enlightened state. In the traditional biblical Gospel of John (12:1) Jesus is reported to tell his disciples that he will give them the power to also become sons of God through a divine revelation. From the Nag Hammadi Coptic texts, that many scholars believe came directly from the earliest Christian writings known as the "Q" source, the Gospel of Thomas also speaks about an internal state of understanding already present in everyone that is the true Kingdom of Heaven as opposed to some physical place in time and space. In these Coptic scriptures, Jesus emphasizes the transformative role of a secret, hidden wisdom which provides access for his followers to this new understanding as divine creatures of God and with a direct inner experience of a heaven. This message is an echo of the canonical text in Luke 17:21 where Jesus proclaims that the Kingdom of God is within you. This unique alternative view of Christianity is more in keeping with some of the Eastern spiritual traditions. Though not the author of these Coptic writings, it is interesting to note that Saint Thomas did settle in southern India as a missionary.

I believe in God, but not as one thing, not as an old man in the sky.
I believe that what people call God is something in all of us.
—John Lennon, The Beatles

Many Hindu and Buddhist traditions are really founded on the idea of a realized enlightenment of consciousness and their teachings focus on techniques to shift the neuro-mental state toward this goal. Scientifically, Andrew Newberg contends that rhythmic ritualized religious practices alter hypothalamic and autonomic functions and effect specific areas of the brain leading to the transcendent state that brings forth the enlightened perspective of the mystics. Intense emotional devotion, loving worship, and dedicated meditation techniques focused on an intense desire for this connection and vision can have similar neurobiological effects. The process is likened to falling in love with the whole of the Universe. The induced conscious state is often described as a clearing of the mind of all material thoughts, dualities, and distractions in order to create a completely quiescent mental condition and to facilitate a focused primal experience of reality. Interestingly as an aside, the character of this void mental state and the understanding of "emptiness" is thought to have given rise to the scientific and mathematical concept of zero which originated in India.

You are the very incarnation…You are Generator, Operator and Destroyer
—Hans Yog Prakash, Satguru

The celebrated Indian Satguru Hans Yog Prakash said that living beings are the very incarnation of GOD as the Generator Operator Destroyer of reality. This conceptualization is a play on words alluding to the functional characteristics of the Hindu deities of Brahma, Vishnu, and Shiva and the basic principles of all living things (creation, sustenance and dissolution). Likewise, it reflects the meaning of the archetype of the Ouroboros (Chapter 2) which represents the connectivity and unity of all things in a natural cycle of destruction and recreation. This recognition of a Universe with this cyclical connectivity in which the created become the creator also integrates the impersonal reality of the Universe with each of our respective individual beings. This revelation provides a vision and path for a very personal relationship with an otherwise seemingly detached omnipotent spiritual reality that is coexisting and interdependently originating in a way that sounds like a variation of the Copenhagen interpretation of the physics of quantum reality and superposition.

It may be that our role on this planet is not to worship God - but to create him. **—Arthur C. Clarke, futurist, writer**

From a scientific perspective, these same espoused principles are consistent with the critical features of living systems as noted by modern day theoretical biologists such as Stuart Kauffman, Francisco Varela, Pier Luigi Luisi and others. Through the experiential process the living system continually recreates and actualizes the reality of its biocontinuum. In this process, organisms generate the information of their reality and translate it to meaning and operational adaptive action for survival. That grounding signifies it is the relative reality of existential substance and has the greatest meaning to the organism. By all accounts, this experience is the living system's true reality and therefore it is also the biologically knowable Universe. The details of this process have been extensively described in *Experiences in the Biocontinuum*. However, through a shift in conscious focus to the primacy of experience that encompasses the process as a whole, then we are able to become mentally/spiritually free of the struggles and tensions of this cyclic process.

Every act of creation is first of all an act of destruction.
—Pablo Picasso, artist

In the experiential process of updating knowledge concerning the state of the biocontinuum reality, prior information is erased or destroyed as determined by the entropic process as described by Landauer. Therefore, there is a continuous interplay of matter, energy, and information within the living system's framework of experience. As this experiential process sustains the living apparatus, it provides a mechanism for a continuous cycling of the content of this reality. When there is a neurological shift in consciousness toward a direct experience of the entire process as a unified whole, then there is the realization and enlightenment as described by the mystics.

The Four Noble Truths

There will be happiness and there will be sorrow. The key is to have joyful participation in the experience of Life—**Siddhartha Gautama, Buddha**

Of all the religious traditions, Buddhism probably takes the most scientific approach in its teachings and practices with an emphasis on a science of the mind. However, in contrast to the decidedly objective orientation of the Scientific Method, Buddhism considers the instrument of the subjective mind as having the primary position for understanding ourselves and the natural world as a whole. In this context, Buddhist teachings and literature also focus on such things as causality and the true nature of reality. During his lifetime, Buddha steered away from discussions about divinity, cosmology, the afterlife, and any creation mythologies. His teachings were directed toward the present attainment of peace, fostering compassion, the relief of suffering, and an improvement in the general human condition. He considered that these goals can only be achieved from a position of enlightenment.

As with all religions, Buddhism has evolved over the centuries to a myriad of modern forms which consist of complicated doctrines and canons which require some strange belief systems. However, the foundational truths of the Buddha's teachings still resonate with a certain scientific appeal. The same can also be said for some of the other great religious traditions such as Christianity and Hinduism. There are few that would disagree with most of the ideas that are found in reading the select words of a red-letter edition of the Bible. In the following, the fundamental teachings of Buddhism (Four Noble Truths) are used as a platform to examine the convergence of Religion with Science as a path to understanding.

The path of the Buddha to enlightenment and his approach to address the suffering of humanity is much like a physician approaches the problem of helping a presenting patient. First, there is a diagnosis followed by a determination of cause. Once causation has been established, there is a proposed cure with a detailed prescription of treatment. The underlying approach is also just a variation of the Scientific Method. The process is presented in Buddha's Four Noble Truths and Eightfold Path.

Noble Truth 1

The first of Buddha's Noble Truths involves acquiring a simple understanding of the true essence of Life and the experience of living. The primary condition of Nature is impermanence (as also suggested by Einstein's Special and General Relativity Theories) which results in a state of continuous duhka as the innate characteristic of existence. The Sanskrit word duhka दुःख has traditionally been translated in the psychological and spiritual sense to mean suffering. A more accurate translation from the root origins of the word suggests that it intends to refer to a condition that is "unable to be satisfied" or in the general sense it denotes the tensions and stresses associated with cravings and the struggles of Life. This condition is also probably the "original sin" referred to in the Bible and is inherent in the biological nature of all living beings.

> *Everybody understands suffering...it is at once utterly intimate and utterly shared. So that is a place to begin.*
> **—W. S. Merwin, Pulitzer Poet Laureate**

From a more basic physical and scientific perspective for biological organisms, duhka can be considered as the divergence between the current state of the living system and that desired state needed to be attained for a reconciliation toward perfect stability and assured success within the whole of the biocontinuum of its existence. There are tensional energies inherently residing in the nonequilibrium condition of the biological system that are triggered for action by the divergence. This divergence is codified in the equation presented previously (Chapter 4) which describes the momentary discrepancies in the total informational state of the biocontinuum $[q_i - p_i]$ changing with time as created by the nature of the world's impermanence. There is also a natural drive or striving for resolution of this divergence by the native mechanics of Kullback's Principle of Minimum Discrimination Information as will be discussed in the section for the Second Noble Truth below.

$$I(q,p) = \sum_{i}^{n} ln\left(\frac{q_i}{p_i}\right) q_i = \sum_{i}^{n} [ln\,(q_i) - ln\; p_i)]\; q_i$$

In considering the psychospiritual state comprising the conscious emotional experience of humans compunded as the collective experience of all the individual primal experiences (as described in Chapter 5), the divergence is from a state of total satisfaction, happiness and joy. Therefore, the First Noble Truth simply says there can never be a complete lasting reconciliation in the state of the biocontinuum at this level due to its inherent impermanence and continuously changing condition. Subsequently, the fundamental nature of all biological systems is a constant state of unrequited and unresolved striving or duhka.

Noble Truth 2

In order to resolve the suffering and dissatisfaction of Life as described in the First Noble Truth, it is essential to understand the root cause of the condition. The second of Buddha's Noble Truths is a revelation of the cause of all suffering. The Buddha asserts that the cause of all duhka (striving and suffering) are secondary to our conscious focus on these basic desires or thirsts and cravings. Obviously, on a superficial level the unsatisfied tension of Life leading to unhappiness is caused by the inherent hungers and yearnings all humans have in their day to day living. This tension is always present whether it is a rudimentary wanting for nourishment or the desire for wealth and pleasure. But there is an even more fundamental etiology of this tension.

In the emergence and evolution of living systems, the logical calculus and rote algorithmic mechanics of Natural Selection has instantiated an innate objective of a striving for a perfect success for survival in all biological organisms. This is essentially the basis of the concept of the selfish gene as espoused by the evolutionary biologist Richard Dawkins. This prime directive for survival and the maintenance of individual selfhood is manifest in the drive for reconciliation of the living system with the information state of its biocontinuum of existence. The driving process is described by the fluctuations of the divergence equation noted in the section above. This expression follows Fisher's continuous replicator dynamics as guided by the fitness function $f(P)$ of the system for stability and survival.

$$\frac{d}{dt}I(q,p) = -\sum_{i}^{n}[f_i(P) - f(P)]q_i = \sum_{i}^{n} f_i(P)(p_i - q_i)$$

At the level of conscious experience, the drive for reconciliation is toward self-oriented objectives that improve the chances of survival and propagation of genes. These goals include the desires for food, sex, wealth, and security. Efforts toward the achievement of these objectives are associated with feelings of stability, contentment and satisfaction.

From a scientific perspective, we can conclude that the Second Noble Truth asserts that the fundamental cause of our innate need for reconciliation of all divergences (sufferings and dissatisfaction) from the conditions of our biocontinuum of existence originates from the forces of Natural Selection entailing the fundamental biologic objective of self-preservation. But also in play are the reconciling drives of entropic dynamics and the natural mechanics of Kullback's Principle of Minimum Discrimination Information which are harnessed as the biological action principle.

Noble Truth 3

In the third Noble Truth, the Buddha provides the solution to end our duhka. The simple key is to become liberated from all attachments and cravings. That does not mean that all these wants and attachments are completely eliminated. Clearly, yearnings are an intrinsic and necessary part of very our biological existence as noted in the First Noble Truth. However, it is important to not become subjugated to our temporal desires. The impermanent nature of the world of things makes their very essence transitory.

> *Do not dwell in the past, do not dream of the future, concentrate the mind on the present moment.* —**Siddhartha Gautama, the Buddha**

Living systems are in a continuous transition from past to future as they adapt and reconcile with the biocontinuum. However, it is only in the present that the divergence is wholly reconciled to form a momentary new state of reality. This transition process is the *becoming* of reality as described by Carlo Rovelli and Edmund Hegel. Therefore, the experience that is emerging as the reality of the present should be considered as a timeless process and not a specific point in time. The past can be identified as a place in time in the form of memory and conceptual mental construct. Likewise, the future can also be imagined as a temporal position that is yet to come. But the ever present "now" is continuously being created in the process of becoming our ongoing experience of reality.

> *For every minute, the future is becoming the past.* —**Thor Heyerdahl**

The juncture spanning the full breadth of the biocontinuum of space-time, in which the past and the future traverse paths in the creation of the now as a *becoming*, can be found in an algebraic solution of our equation of experience and biocontinuum information divergence as below:

$$I(q,p) = \sum_{i}^{n} ln\left(\frac{q_i}{p_i}\right) q_i$$

$$\frac{I}{q} = ln\left(\frac{q}{p}\right)$$

$$e^{\frac{I}{q}} = \left(\frac{q}{p}\right)$$

$$p = \frac{q}{e^{\frac{I}{q}}}$$

$$p = \frac{q}{e^{\frac{I}{q}}} \rightarrow p = \frac{q}{e^{\frac{0}{q}}} = \frac{q}{1} = q$$

Unity of being with a reconciliation of duhka emerges as I approaches zero (Information divergence →→→ 0) and $p = q$.

God is the tangential point between zero and infinity.
—Alfred Jarry, pataphysics writer

The above theoretical analysis is in complete accordance with a view of the ongoing impermanence of Nature and a "Be Here Now" conscious state as the way to reconciliation as expressed by the famous Harvard psychologist turned mystic, Baba Ram Das. For the Buddha, liberation from duhka and impermanence and the unreconciled condition comes when our focused consciousness lives in the unitary moment of the present. From the scientific perspective, as discussed in Chapter 2, the special relativity of simultaneity results in a condition where all present experience of the "now" reaches across the full spectrum and potential expanse of time and space depending only on the individual observer's frame of reference. This reach is the timeless oneness of the transcendent experience. The fact that this purported spiritual transcendence is in complete harmony with a purely scientific deduction described in the preceding analysis really reflects the inherent logical consistency and unity of all aspects of our Universe regardless of the originating perspective, either scientific or religious.

Noble Truth 4 and the Eightfold Path

The course and means to enlightenment of the Fourth Noble Truth, as presented by the Buddha, was probably a simplified version of what has now come to be known as "the middle way". However, in the evolution of Buddhism as a major philosophical and theological system, an elaboration of that way became an eight-step process. The Fourth Noble truth espouses the Eightfold Path to enlightenment that includes the right resolve, the right speech, the right effort, the right conduct, the right livelihood, and the right mindfulness. These morals and intentions are important prerequisites to best provide for a fertile mind that can shift its perspective. However, the real heart of this Eightfold Path centers around having the right processes of meditative contemplation and focused mental concentration on the true reality of the present and its resultant enlightened understanding and view of reality and Life.

I think our search is somewhat encumbered by our concept of God.
—Joseph Campbell, mythologist, writer

The ultimate leave-taking is the leaving of God for God.
—Meister Eckhart, 12th Century Catholic mystic

The meditative process using the instrument of the mind begins with acquiring the quiescent attentional state of Samatha leading to enlightenment. This procedure amounts to a state of focused awareness without the dichotomy of thought and duality of an ascertaining cognition (the Tree of Knowledge of Good and Evil). It also seeks to detach the mind from any preconceptions regarding the nature of reality or the divine. Many religious and spiritual traditions use a variety of similar techniques to focus the mind's attention and awareness. Examples include the use of mantras in Hindu yoga and the rosary chants in Christianity. There is also some correlation with the phenomenological reduction methods of Husserl in which the goal is a raw experience of phenomena with no preconceptions. Traditionally, the best points of focus of attention are toward an entity that exists with continuous automaticity and without the need for an intentional conscious generation, such as the normal breath cycle. While some contend that any concentrated attention on any specific object seems counter to the ultimate goal of experiencing only unadulterated consciousness, the Buddhists see it as a facile stepping off place to finally disengage the mind from all thought. When there is no thought, all which is left is pure consciousness and a direct experience of reality (the Tree of Life). This is the primacy of experience. And since that reality is relational and co-originates interdependently with the observer, the experience is really that of the whole of the Universe.

But as many as received him, to them gave the power to become the sons of God. **—John 1:1**

Many individuals have reported momentary experiences of such enlightenment as during deep meditation and prayers or a religious conversion event. These experiences might be subtle and with varying degrees of illumination. However, even such brief experiences can be Life transformational with the knowledge of what the true reality looks like being carried forward in our memories. These mystical-type experiences provide context and elucidation of dharma, which are the Buddha's teachings concerning the true nature of reality and life. The same would be true for a deeper understanding of the teachings of the Christ or Krishna. As noted above, this condition of knowledge is said to be a realization of the oneness and co-origination and interdependency of all. And an individual with such an experience is said to be "realized". Interestingly, there is currently some successful use of psilocybin mushrooms in psychiatric treatments to momentarily induce similar psychologic transformations and provide for a new view and understanding of Life.

An enduring state of enlightenment or full awakening with a continuous sustained perspective of astonishment, oneness, awe, and awareness of the primacy of experience is achieved by only a few celebrated "holy" men such as the Buddha and Jesus the Christ. This level of persistent conscious awareness and realization of experience provides for their unique insights and often compels these individuals to become great spiritual teachers. Because of their full and continuous conscious connectivity with the true reality, these individuals are often considered as divine incarnations, bodhisattvas, or saints. Yet, these holy beings always declare that the same potential exists in all of us.

> *I and the Father are one. ...that they may be one even as we are one, I in them and you in me, that they may become perfectly one.*
> **—Jesus, by John 10:30 and John 17:20-23**

But what becomes of this conscious awareness when the biology of the living system ceases to exist? There is certainly no definitive scientific evidence for any form of reincarnation of the soul or even a realm of heaven after death. In fact, all notions of some kind of immortal afterlife is most likely a psychological derivative of our biologic drive for survival resulting from the logical consequences of Natural Selection. However, if Life is actually a part of an unfolding Universe, there is really no need for the individuality of an immortal soul. This integration of Life into the scope of the whole Universe is also true in the logical processing of the Anthropic Principle. The co-dependent origination and interdependency of the entire Universe and its reality created through the inherent connectivity of the experiential process of all living systems may be the path through which the essence of our living systems ultimately always becomes as one with the timeless whole. And as our basic energy, matter and information are persistent in Nature, then we also endure in a non-egocentric existence.

Heaven means to be one with God. **—Confucius, philosopher**

Is There a Scientific Approach to Understanding Religion?

Miracles are not contrary to Nature, but only contrary to what we know about Nature. **—Saint Augustine, theologian**

It is interesting to note that much of what we view as the mystery of Religion, including the spiritual experiences of the mystics, can be rationalized through objective scientific evidences. This suggests that there are no special supernatural miracles in religious beliefs other than what are already the great mysterious phenomena of the Universe itself. Both Religion and Science seek to understand this Universe. Therefore, it should be of no surprise that there is great commonality in that effort. The subtle divide that separates the two approaches involves the source of perspective and definitions of truth. As previously indicated, this difference is highlighted in the experimental accounts of Andrew Newberg who provides objective 3rd person evidence of changes in neurological function as correlated with the 1st person mystical experiences of the monks and nuns. The overwhelming conviction of these tested subjects as to the reality of their experiences also lends credence to the results. In this case, the subjects themselves could be considered as scientific instruments, like a special lens or complex sensing monitor for analyzing this transcendent reality. The possibilities for reconciliation of the obvious paradox of Science's requirements for 3rd party objectivity in a world where there are only subjective experiences is discussed in great detail in Chapter 3. Any resolution that includes the power of 1st person evidences in our understanding of the Universe also recognizes the interdependency of all its varied constituents in originating the true reality.

When I investigate and discover that the forces of the heavens and the planets are within ourselves, then truly I seem to be living among the gods.
—Leon Battista Alberti, Italian Renaissance priest and humanist

The Religion of Science

Science is a differential equation, Religion is a boundary condition
—Alan Turing, Enigma mathematician

In the modern workings of the 21st Century, a large portion of the world's population, including very religious people, put a lot of faith in Science to guide their daily lives and inform their understanding of their place in the Universe. In this respect, Science has become a practical surrogate for Religion for much of our activities. Additionally, as Science has improved our understanding about the Universe, it has helped to shape our belief systems and interpretation of religious scripture and doctrine. Scientists often speak of believing in the "God of Spinoza" as the sum of the natural and physical laws of the Universe. Similarly, Albert Einstein's impersonal God is considered to be the underlying dynamic which determines the principles of cause and effect that govern the nature of reality and the machinations of the physical Universe. It is easy to understand how the intersection of the goals of Science and Religion in trying to understand our world would result in many aspects of Science having religious implications. Some of these influences are described in the following.

The Big Bang and Creation

The Universe…a vortex of vibrations in which (the individual) realizes itself as a man or beast, flower or star…Not alone, but as central to all that surrounds it. And these centers are not, as may seem, apart from their surroundings, but stand in mutual relationship to them.
—Alan Watts, writer

Mythologies about the creation of the Universe have been with us since the dawn of humanity. Many consider that cosmology really only became a legitimate field of Science when Albert Einstein first formulated his Theory of General Relativity in 1915. The Belgian Catholic priest and theoretical physicist, Georges Lemaitre, is also considered one of the founders of that modern cosmology. He was the first to theorize an expanding universe based on his analysis of Einstein's equations that formulate the General Theory of Relativity. The English astronomer Fred Hoyle mockingly called the hypothesis the "Big Bang". Many assumed that Lemaitre was making the case for a creation story consistent with the biblical narrative. In fact, Pope Pius XII proposed that the expanding universe hypothesis provided scientific evidence supporting the biblical creation narrative and infers the existence of God. At the Pontifical Academy of Sciences in November

1951, Pius XII declared that astronomical discoveries confirmed the Fiat lux (Let there be light) declaration in the initial passages of the *Book of Genesis*. A distressed Lemaître reportedly dissuaded the Pontiff from making such future comments connecting Science and Religion. However, so deeply intertwined are the notions about creation with the fundamentals of religious belief that the Big Bang theory has continued to be a centrepiece in many such debates.

> *Anybody who has been seriously engaged in scientific work of any kind realizes that over the entrance to the gates of the temple of Science are written the words: 'Ye must have faith.* **—Max Planck, Nobel physicist**

When Nobel Prize winning quantum physicist Paul Dirac suggested to Lemaître that cosmology was the closest link between Science and Religion, Lemaître replied that he thought that the study of psychology and the science of the mind were more closely aligned with religious endeavors. While creation mythologies certainly pervade most religious traditions, many leaders of the great religions, such as Jesus and Buddha, actively avoided speaking about cosmological ideas. Their concerns were keenly focused on the spiritual aspects of the human condition such as love, compassion and righteousness. This thinking is really consistent with a perspective in which emotions and feelings within the individual are the originating source of all conscious experience of the reality of the world around us.

The Copenhagen Interpretation and Dependent Origination

> *If there is any religion that could respond to the needs of modern science, it would be Buddhism.* **—Albert Einstein**

In 1814 the philosopher, mathematician, and scientist Pierre-Simon Laplace declared that the Universe was like a great clockwork machine that was completely solvable if we only knew all the factors. Interestingly, it was also Laplace that greatly advanced the study of probability theory as a method for overcoming deficiencies in knowledge to guide inferences of causation from observed data. This probability tool was eventually employed in the early part of the 20th Century to explain the odd behavior of atomic particles as observed in laboratory. As discussed in Chapter 2, the Copenhagen interpretation of quantum mechanics states that physical systems at the subatomic level do not have definite properties prior to being observed and their reality only exists as a potential until it is actualized by the experiential process of living systems. This failure of definitive determinism and causality in physics as handed down from the Newtonian traditions was a great shock to many well-known scientists of the day and created a philosophical crisis in the scientific community.

> *...all phenomena in their own-being are empty....in emptiness, there is no form, no feeling, no perception, no formation, no consciousness.*
> **—Buddhist Heart Sutra**

The notion of a co-dependent origination or dependent co-arising of reality which is at the heart of this interpretation of quantum mechanics is not really a new concept. Nagarjuna (150 – 250 CE), who is often referred to as "the second Buddha", was highly critical of the Brahmanical and Buddhist traditionalist philosophies and theories of knowledge of his time. Today, across various Buddhist traditions, the concepts of emptiness, oneness, and dependent origination or dependent co-arising as elaborated by Nagarjuna are interrelated as fundamental ontological principles. These conceptions represent a mode of perception and a way of experiencing the world that is different from the natural or common attitude. Emptiness does not mean that there is an existential nothingness in relation to our experience of reality. Emptiness refers to the lack of any intrinsic independent condition of being for all objects and phenomena including the subjective self as separate from the whole of everything. Even consciousness requires both an entity which experiences and that which is experienced. Therefore, emptiness is not a state or condition but rather a way of experience. Nagarjuna contends that the experiential perspective that results in the assignment of an intrinsic independent existence or essence to things is also at the root of our attachment and suffering. Therefore, we must learn to experience reality in a different manner.

> *The foot feels the foot when it feels the ground.*
> **—Siddhartha Gautama the Buddha**

Not only is all phenomenal experience inseparable from the myriad of contingencies that comprise the whole of the Universe, but the observer is likewise conditionally configured. An experience of emptiness requires an intrackability by linear causality of anything in isolation. The limitations of the historic search for the most elemental particle of matter and the conundrums of uncertainty in quantum mechanics are examples of the failures of Science in finding an independent existence of the physical Universe. This same concept of irreducibility is also at the heart of the modern concepts of complexity theory and the structuring of living systems.

Science is organized knowledge. Wisdom is organized Life.
—Immanuel Kant, philosopher

When all notions of a separate distinction of things is truly dissolved, then there is an absolute oneness within all experiential reality. However, to realize this emptiness is not to negate the reality of things but rather to understand what enables the existence of everything at the primordial level. This unique understanding then allows for a freedom from any strife resulting from the impermanence of all common material experiences because they are substantively empty of any independent existent reality.

Dependent Origination (pratītyasamutpadā) extends the concepts of emptiness by emphasizing the causal interconnectedness of all phenomena. This causal interdependence or contingent co-arising as a way of understanding was also originally found in Buddhist thought. This conceptualization of reality is very different from the thinking that developed in Europe with a Science of direct Newtonian-like linear causality and a definitive singular origination and temporal ordering. Instead, dependent origination proposes a conditioned pleural causality much like is seen in sets of nonlinear mathematical equations with interdependent variables. For instance, what would be the single point of origination of the state for the following system of equations?

$$z = x + y \quad \text{and} \quad y = z^2$$

In his ground-breaking book entitled *Causality*, Judea Pearl states "it is the way we carve up the Universe that determines the directionality we associate with cause and effect". Kurt Gödel's proof for the Incompleteness Theorem indicated that there is a hidden implicit meaning in the totality of any mathematical formulation that is not apparent from its methodical examination achieved by building upward from the basic axioms. That embedded meaning represents a third state of logic for an inference regarding the validity of any experientially derived knowledge that must be determined by some living observer who considers the information as a whole and in the context of its presentation. Pearl believes that our experiences create detailed patterns of apparent probable associations and dependencies that serve as a general causal directionality and form a basic causal structure or scheme from which future experiences are vetted. Kant thought that this quality of knowledge is transcendental and unique to the observer and not a primary ingredient of the objects of ordinary experience. Rather it is a process in which it is possible for us to know the objects and really understand their meaning beyond the apparent perceptions. The same conditions are also true of any statement of information when viewed in its entirety and in the circumstance of the setting rather than from the simple conjoining of the words. In other words, the key for that understanding is that the value and meaning of information as knowledge comes only from a reflection on the system in its entirety.

This co-arising system of thought also upholds that everything must implicitly be caused into existence and yet without the need by some outside force or entity. Since everything is interdependent, there is no true single point of origination or independent creation. In considering dependent origination as a different form of causality, it is similar to the global or top-down causality seen operating in complex living systems. In this sense, the intended reason or objective of the whole complex is the driving causal force without the usual required temporal ordering. For autonomous living systems, that reason is the sustaining existence of the living entity. So, the interdependent cause for existence is existence itself. And without the creative force of the existing living system, reality is not created and nothing exists. Including the living system itself. Hence, the creator becomes the created.

This more holistic perspective is somewhat reminiscent of the recurring course of an M. C. Escher drawing with no beginning or ending and with each step originating the next. The implications are that temporal considerations are not really valid in the grand scheme and that the reason for the whole is the origination and sustaining of the system. Because of the relativity of simultaneity, the constancy of the speed of light in all frames of reference, and the gravitational distortions of space-time, we have come to discover that an absolute universal reality for time has no real meaning in any moment of experience. Science is really the study of how things interact and manifest themselves to each other. It is not about the independent existent essence of things in space and time. Carlo Rovelli contends that the origination of reality in anyone's experience of the present should be considered as the dynamics of *becoming* rather than having any intrinsic substance. In fact, how can anything in an individual's experience have any real independent substance if its condition is also simultaneously in the past in another frame of reference? In this circumstance, all things are empty and are in a continuous state of interdependent becoming for any observer.

Manifestation is the art of co-creating with the Universe. **—Unknown**

The conditionality of becoming and being is commonly considered the source of the concepts for causation and "karma" across lifetimes and through time. However, such conditions might also be considered a mechanistic explanation for the temporal paradoxes that arise in the Anthropic Principle. Furthermore, this interconnected and co-dependent causality should be considered the origin and ontological basis for all reality, both objective and subjective, and not that generated by some external creator entity existing in some other supernatural realm. Again, under these conditions the created becomes the creator which in turn results the creation. And the reason for creation (verb) is the creation (noun) in itself and experience for the sake of experience.

The Anthropic Principle Revisited: Universe Know Thyself

> *The cosmos is within us. We are made of star-stuff. We are a way for the Universe to know itself.* **—Carl Sagan, astronomer**

The Anthropic Principle espouses that the parameters of the Universe are narrowly positioned in a way that allows for the emergence of Life. This proposition arose in conjunction with the Copenhagen interpretation of quantum mechanics that requires an observation event (by a living system) to actualize the properties of subatomic particles. Recent experimental evidences supporting Wheeler's delayed-choice postulate also work to satisfy the temporal nuances necessary to connect these concepts. The implications of these seemingly intertwined principles have been extended by some scientists to mean that Life has some teleological role in the reasons for the Universe to even exist. Again, at the heart of the debate is our traditional notions of causality. Does the physical Universe result in the creation of living systems or is the emergence of living systems the causal precondition for the origination of a Universe? Unfortunately, there is also often an extrapolation of this principle by some as an attempt to prove the existence of some sort of intending creator (watchmaker) or at least a universal creative principle. Regardless of causal inferences, the Anthropic Principle seems to dictate that the reality of the Universe is predicated on an inherent ability to be observed and experienced. And as proposed, Life is just a natural consequence of an unfolding Universe.

From this perspective, physicist Sara I. Walker asks the most pertinent question. Why does the Universe seem comprehensible and why there are beings that are here to comprehend it? There appears to be no necessary reason why the Universe would make sense to a subsystem of itself. The fact that there are living systems as comprehending subsystems suggests that there is a unified framework to the Universe where comprehenders and the process of comprehension are all integral to the causal structure of physical reality. And they exist as a co-dependent arising. This position is fortunately a break from the concerning teleological perspective to one of an interdependent teleonomic origination through coemergence. It is clear that some isomorphism has to exist between the materiality of external reality and our subjective experience of it or else living systems would not be able to adapt and survive. Additionally, for these subsystems to have emerged and evolved in the first place, this condition of symbiotic isomorphism also suggests that their biology is a manifestation and record of the entire past, present and possibly future of the Universe.

The most incomprehensible thing about the world is that it is comprehensible. **—Albert Einstein**

This interdependency of things without a first cause or linear causality really aligns with the Buddhist teachings of Nagarjuna. It also works for the procedures of the Scientific Method if the Anthropic Principle is reformulated as a constraint for observation of phenomena that can be used as a refutable hypothesis as suggested by Reinhard Bruer. Bruer states that what we observe must be limited by the necessity of our presence as observers so that the conditions of the Universe are consistent with the fact that it contains observers whose position will necessarily be privileged. Life is then considered as a necessary boundary condition.

The reformulation of the Anthropic Principle by Bruer also presents a well deliberated approach that balances many perspectives and is consistent with the participatory role of the observer in determining the state of the observations as espoused by John Archibald Wheeler. Wheeler has suggested that we really live in a participatory Universe and formulated the Participatory Anthropic Principle (PAP) as a uniquely different take on this concept. His explanation is again based on the Copenhagen interpretation of quantum mechanics requiring an act of observation to resolve the superposition of quantum states. The central idea of this special derivation of the Anthropic Principle is that the reality of the Universe cannot actually exist as we observe it without there being observers. Since all experiences of the Universe originate as incoming information to the observer, it has also been suggested that information may be primary in an ontological sense prior to the existence of conventional physical properties.

Every physical quantity, every it, derives its ultimate significance from bits, binary yes-or-no indications, it from bit. **—John A. Wheeler, physicist**

In the IT from BIT postulate of John A. Wheeler noted in the quote, it is the information (BIT) processed through the experiential capacity of a living system that becomes the reality (IT) of the physical phenomena. The living system itself then becomes a participator in the determination of this reality and the same information (BIT) may produce very different perspective results for different observer/participators. Therefore, all the knowable character of physical reality (IT) only exists through the living system. But that existence is not solely in and of itself and may exist only as a potential until it is actualized by observation. Wheeler proposes that a phenomenon is only a phenomenon if it is an observed phenomenon. This creates a closed causal loop suggesting that only a Universe with the capacity for developing conscious beings can exist.

Brandon Carter's considers the Anthropic Principle as a self-selection principle in which the necessarily Bayesian process of discrimination between plausible hypotheses has an inherent self-reference measuring bias. In this process, a view of reality is selected that would therefore require the presence of the observer. Obviously, self-selection implicitly entails an ability to self-reference. Carter's analysis includes an Ouroboros type equation of self-reference and reflection in which the probability of the state (S) before (a priori) observation p_B and after the experiential evidence (E) are considered (a posteriori) p_A.

Hence, for any S: $p_B(S) = p_A(S|E)$; where there is a conditional probability.

But it was noted previously that to resolve the potential paradoxes of such a self-referencing or self-selection there must be a recursive iterative loop within the logical procedure to resolve this expression over time. This procedure provides a means for the functional emergence of the form of the state so there can be a determination of meaning in relation to the whole as intended by the self-referring entity. The system state becomes continuously updated in a circular causality so that the specific overarching pattern that emerges does so only because of its self-referential quality.

The experiential equation for living systems presented in Chapter 4 with its iterative replicator dynamics naturally provides for a procedure of self-referencing and self-selection. However, it is only when the information of the state of the biocontinuum (as a sphere of experience) is completely assimilated by the living system as knowledge with a translation to meaning does $p_B(S) = p_A(S|E)$. The truth of the matter is that the Universe is almost never in the Goldilocks' just right condition on a moment to moment basis and living systems are continually adapting to the impermanent conditions of the biocontinuum based on experience. And through that adaptation they change to fit the Universe in which they are integrated. Living systems must make themselves just right or they will not exist.

> *The individual is the aperture through which the whole energy of the Universe is aware of itself...* **—Alan Watts, writer**

These views of Wheeler, Walker, Breuer, Carter and Davies all seem to require a knowable comprehensibility as the fundamental characteristic of a Universe which contains Life. The presence of a comprehender in the Universe suggests a different interpretation of the Anthropic Principle. The Universe is not necessarily made for Life and, in fact, living systems must continually adapt to the conditions of the Universe in order to survive. However, the Universe is naturally instilled with the inherent conditions that allow for the development of an experiential capacity which is the foundational requirement for Life to emerge and adapt. Life is a natural product and consequence of an unfolding Universe. And through the presence of living systems the Universe has a way to know itself and furthermore give it meaning. The meaning? The continuing actualization of its reality.

Experiences in the Biocontinuum: Connection to the Universe

> *It is beyond a doubt that all our knowledge begins with experience.*
> **—Immanuel Kant, philosopher**

Certainly, all the given and knowable Universe begins and ends with our present experience. And in this way, the living system with its intrinsic ancestral past incorporates into itself the entirety of the Universe's physical and evolutionary history since the Big Bang. Therefore, as this history is inextricably ingrained in the structure of biology and its processes, the continual renewal of Life and its experience of the present becomes the inherent solution to the tautological paradox of the Anthropic Principle.

In its adaptation and renewal driven by the experiential process, the living system deconstructs the state of the biocontinuum into a form of information that is assimilated into its structure and dynamics. Therefore, we can presuppose that Life already has embodied the template of all the laws and history of Nature and its evolution back to the Big Bang and that this template is perpetually being verified by living experience. The timing of this evolution and a notion of time in general is only contained within sequential experiences. Intentionality for perceptual experiences that is driven by the objective of a sustained existence is the way living systems determine the dimensionality and form of the biocontinuum. Naturally, the Universe will always appear to be just right. This premise would be true if Life were carbon-based or silicon-based or structured entirely different in a way that we might not recognize.

It is remarkable that the Universe has given rise to entities that modify the very dimensional structure and form of their knowable reality for their own objective of survival. This enigmatic ability makes the experiential process the most powerful force in the Universe. However, it is important to realize that it is not just a perceptual observation that is required but rather the full scope of experience resulting in a deconstructive knowledge acquisition, the ascription of meaning, and a causally closed perception-action cycle. This vision creates a larger view of connectivity within the milieu of the Universe and a general unity for Nature through an intrinsic underlying relationship created by the process of experience.

Therefore, we have a Universe occurring as a harmonious whole with circular causality in which its matter and energy conditions have an innate predilection for the emergence of organized systems with an inherent capacity for experience and that reciprocally the experiential process is the precondition for the creation of the dimensions of matter, energy and their lawful relationships. Just as an attractor preconditions the evolution and ultimate form of a system, Life and its unique capacity for experience become the attractor state for the whole of the reality of the unfolding and evolving Universe which is the inevitable result constituted in its structure. And the fundamental constraints and structure of that attractor are formed by the underlying potentials of the physics. This creative circle is a new form of causality originating from the holistic position of interconnected dependency. This is at the heart of Nagarjuna's co-dependent origination discussed previously.

It is the quality, not the mere existence, of information that is the real mystery here. —**Paul Davies, physicist**

Paul Davies points out that it is not simply information that must be exchanged within this causal loop but also the meaningfulness of that information (Davies, 2015). The unification of everything by co-dependent origination that includes living systems allows for the Universe to give meaning to itself in an existential context. And if the currency of the Universe is information, then a meaningful interpretation of that information through the experiential process puts living systems at the center of all that is knowable. Not the center in the positional sense, but rather as a process, the experiential process. Therefore, *Experience is the Center of the Universe.*

> *I don't believe people are looking for the meaning of Life as much as they are looking for the experience of being alive:*
> **—Joseph Campbell, mythologist, writer**

The experiential conceptualization of Life and the Universe is consistent with the ideas around self-reference and Lyapunov stability as a metric for quantifying the meaning of information in living systems as described by R.L. Summers (see references). The Universe does not create Life simply to observe itself but also to comprehend itself and provide it with meaning. The meaning? The perpetuation of Life and its survival so the Universe can know itself. And this can only be achieved through the experiential process. It seems that Life's origination is at once both intimately rooted within the fundamental nature of the Universe and yet emerges as its observer and creative participator. Therefore, we really do have a participatory Universe in which matter and energy and information determine the structure of living systems and reciprocally Life actualizes the dimensionality, substance and meaning of it all.

Where does this drive to know and give meaning to the Universe come from? Is it also an intrinsic component of the Goldilocks enigma? Is it a veiled consequence of the logic of the Natural Selection process? Is this a mechanism inherent in the experiential process of living systems? Is it part of the self-referential and existential paradoxical nature of our reality? Most likely the drive has its roots originating interdependently from all of the above possibilities.

> *Natural Selection is anything but random.*
> **—Richard Dawkins, evolutionary biologist**

Stuart Kauffman speaks of the spectrum of possibles (*res potentia*), as seen in the evolutionary development of living organisms, which does not obey the principles of the Law of the Excluded Middle linked to the actuals (res extensa) by measurement. In this process, the cacophony of potential possibles that comprises the whole as the primacy is deconstructed by experience in a process of measurement instilled with meaning to actualize the causal coherence of reality. But it is important to realize that it is only through the medium of primary experience that the transformation of possibles to actuals occurs in a way that results in an apparent causal entailment of reality through the focused objective of sustaining the living system that in turn provides for the experiential process. In this way, the experiential process of translating the uncertain whole to substantive meaning for the goal of a new cycle of becoming is the entailing foundation for the Universe. At least that is true for living systems. This foundational principle is how the adjacent possible of the fish's swim bladder (Kauffman's example) within the biocontinuum causally becomes (in conjunction with the logic of Natural Selection) an actual reality with new meaning as a lung for the organism.

The autonomous objective of survival that is born of the mechanics of a logical process really provides the appearance of design without the presence of a designer. If experienced reality and the existential foundations of living systems are co-dependent, then the drive to evolve is logically linked to the Principle of Natural Selection. But the development of improved biological mechanisms that are geared to better know, better understand, and provide meaning (existentially) to an experienced reality ties the mechanics of Natural Selection to a universal directive of impersonal logic that is more than just the result of random events but without the requirement of some intelligent intervention.

A Return to Natural Philosophy

The goal of Life is to make your heartbeat match the beat of the universe, to match your nature with Nature. —**Joseph Campbell, mythologist**

Undoubtedly humankind's earliest philosophical and religious thoughts were directed toward considering the workings of the natural world. Understanding how Nature works and being able to predict the changes in the world around them was paramount to survival for early humans. In fact, the biological selection of that unique capacity has probably been the most important factor in the success of the genus Homo.

Philosophy is derived from the Greek meaning love of wisdom and is defined as the study of the fundamental basis of knowledge, reality, and existence. These defining meanings and attributes of philosophy imply a very personal journey for understanding. The term Natural Philosophy probably originated with the ancient Greeks including Anaximander and Democritus and most importantly Aristotle who viewed Nature as a whole continuum with an inherent purpose and causality. The intent of his perspective was to unite the basic experience of the natural world with the self-reflections of the inner spirit. Since very little was available to the ancients in the way of instrumentation for precision measurements, the analyses of observations in Nature were mostly deductive and global.

The March of Science

The Science of today is the technology of tomorrow.
—Edward Teller, physicist, Father of the Hydrogen Bomb

It was not until the time of the Scientific Revolution leading to the Age of Enlightenment that the notion of a more mechanical Universe with clocklike precision and direct causal entailment was born from the work of Isaac Newton and others. This new cause-effect perspective regarding the workings of the world resulted in the Industrial Revolution and Science became more of a procedural tool of technology. In fact there is a seeming historical lockstep of technology with the use of tools for scientific discovery. It was the craftsman's creation of the simple optical lens that allowed for construction of such instruments as the telescope, the microscope and the prism. These tools then led the way for observations of the heavens that resulted in a validation of Copernicus' heliocentric system theory, the development of the cell theory of biology by Hooke, and most importantly led to the derivation of Newton's theory concerning the nature of light. It is astounding to think that such humble creations of the artisan glassmaker could have such a profound effect on our understanding of the world. Other similar discoveries also led to practical technological solutions that have empowered humanity with some control over the natural world.

Other examples include:

1. The impact of the invention of the steam engine on driving the development of a theory of thermodynamics.

2. The role of the simple electric generator that paved the way for a theory of electromagnetism as defined by Michael Faraday and James Clerk Maxwell.

3. Albert A. Michelson's 1881 construction of the interferometer device was used in experiments that provided evidence that inspired the formulation of the Special Theory of Relativity.

4. And in the early part of the 20th Century, the cathode ray tube was the tool that finally allowed us to glimpse into the structure of the atom and led to the development of modern quantum theory.

Any sufficiently advanced technology is indistinguishable from magic.
—Arthur C. Clarke, futurist, science fiction writer

Clearly, the evolution of all these great theories coincided with the development of technologies that allowed us to observe and analyze Nature in a new and unique way and offered the inductive evidence to feed the deductive process, coming back together as new theory. Therefore, a synergy between technology and Science has resulted in a conflation of the practice of Science with technological development. But with this tight marriage of technology and Science have we somehow lost what was the original intent of our Natural Philosophy? And does this strict scientific reductionist methodology for technology really deliberate truth through a flawed perspective of observation and experience.

Naturalism, Materialism, Idealism

After the Age of Enlightenment, a unique form of Natural Philosophy emerged called Naturalism. Naturalism is the perspective that considers Nature as the fundamental source of all reality and attempts to explain everything in terms of natural law. This view espouses that investigations of all known phenomena, including those in the social and cognitive realms, are possible only through the natural sciences in the same way that physics, chemistry, and biology are studied.

Naturalism is often considered as being essentially equivalent to Materialism. However, the objective approach to the natural sciences that designates the material content and properties of things does not always coincide with some of the very real subjective aspects of thought and behavior such as aesthetics and moral values. Such subjective notions like truth, beauty, law, and even numbers that have their origins in the internal realm of the individual consciousness seem transcendent to the inquiries possible by the traditional sciences. Additionally, the ideas of direct or naïve realism cannot be conclusively supported through scientific evidences and there is widespread agreement that a strict objective materialist perspective for Naturalism does not make logical sense.

Kant rejected pure materialism but considered his transcendental philosophy as a form of liberal naturalism. The famous philosopher Willard Van Orman Quine did not see scientific knowledge as different from any other ordinary kind of common sense knowledge. He argued that Science and Philosophy are really a continuum of thought and that the Scientific Method should also be used in philosophical studies. Likewise, the philosopher Edmund Husserl thought that there was no meaningful distinction between the external and internal worlds of experience and prioritized the importance of 1st person knowledge over any form of 3rd person objective scientific abstraction. Another compelling argument for the lack of distinction between the objective and subjective aspects of knowledge was provided by Huw Price. Price contends that we as observers are an integrated part of Nature and therefore our objective and subjective positions must be somehow connected.

The lessons from recent scientific investigations (see Chapter 2) have supported the view of a unique and central role for an observer in the relativistic experience of physical dynamics as well as for the determination of conditions of reality at the level of microscopic particles. It is also clear that it is the fundamental biological system capacity for perception and knowledge acquisition that determines the certitude of phenomenal experience from the 1st person perspective (see Chapters 3 and 5). Consequently, some degree of uncertainty is inherent in the foundations of all possible derivations of objective mathematical systems secondary to an inability for internal self-referential knowledge by external observation. These conditions resulting from the necessary subjective interactions by the living system in the process of observation and knowledge acquisition is responsible for the limitations in determining and understanding phenomenal reality from the strictly 3rd person perspective of Science.

A variety of theories have attempted to resolve the conflict and bridge the gap between objective reality and the subjective perspective of the observer. Theories, including those proposing that the Universe is really a multi-verse or contains hidden variables, seek to eliminate any subjective indeterminacy and ground all of reality in an objectively deterministic physical world. On the other hand, the Copenhagen Interpretation endows the observer with a power and position as gatekeeper over the unique actualization of reality but somehow expects that reality is somehow commonly shared by multiple observers.

Those theories embracing a purely epistemic approach, such as Quantum Bayesianism (QBism), attempt to connect the objective with the subjective by blanketly accepting the indeterminacy found in the quantum realm in which observers are really participants in constructing reality from a single observer's viewpoint. Reality is then recognized as nothing more than the face value of each observer's direct experience and information about the world. Essentially, the observer is the singular arbiter and creator of their reality as fashioned from perceived incoming information. At first glance, this perspective does seem to close the chasm between the objective and subjective positions and such an observer could serve as the Center of the Universe. However, should any comprehensive theory in effect anchor reality to a single observer? While the formulation of QBism considers that observing participants are integral to determining the perceived reality, they are also deemed as a distinct and separated objective entity from the implicate whole. Thus, this purely epistemic framework comes with the implication that the observer is specially positioned outside of the whole of reality. But we know that, in fact, the observer is in itself observable and must also be a part of Nature. Otherwise, this type of theory just becomes another variation of the multi-verse or returns us to Plato's realm of idealism. Using such a rote conceptual structure for comprehending the workings of Nature also constitutes a very myopic view of reality that is self-contained and unconnected to the greater whole in a way that limits the richness of our understanding of the Universe. While we truly view the world through the lens of personal experience, intuitively that same experience also tells us we cannot deny the feeling of the realness of other individuals and things that we cherish. Forward thinking scientists such as Carlo Rovelli carry the strong conviction that any comprehensive scientific theory about the fundamental nature of reality should also include an underlying conceptual framework that allows for a broader understanding of the Universe. In this way, humanity can think more deeply about the meaning and intrinsic character of things, including our place in the Universe. Such a vision for practicing Science can often lead to advances that are not intuitively obvious.

Holism and the Relational Universe

> *Things derive their being and nature by mutual dependence and are nothing in themselves. ... Thanks to emptiness, everything is possible.*
> **—Nagarjuna, 2nd Century Buddhist philosopher**

What broader perspective and mechanism for the acquisition of knowledge about the natural world could possibly resolve the disconnection between the positions of objective and subjective experience? Following the insights of Nagarjuna derived from his conception of emptiness, the Prasangika school of Buddhist thought rejects both materialist and idealists' views of reality in favor of a third alternative of a co-dependent co-arising known as "the middle way". In this alternate view, the material reality of the world is not denied but rather considered to be only relative to and contingent upon the state of the observer. Likewise, any notion of a pre-existing independent reality of idealism is logically indefensible outside the context of an observer. The scientific foundations for this subtle alternative vision of reality is presented by the neuroscientist Christine Skarda in her work entitled *The Perceptual Form of Life*. Based on her experimental studies concerning the mechanisms of neurodynamics, Skarda concludes that rather than constructing representations of external perceptions, the mind and neural apparatus deconstructs the primacy of perceptual experiences of the whole to create a knowledge of changing conditions. This stance is also consistent with Edmund Husserl's notion that perceptions do not simply represent an object but more directly presents the object fundamentally for our consideration and interpretation. Husserl further states that considering conscious experiences as representations is one of the worst conceptual distortions known to philosophy resulting in many epistemological inaccuracies.

> *Little flower, but if I could understand what you are, root and all, and all in all, I should know what God and man is.*
> **—Lord Alfred Tennyson, poet**

A priori knowledge is ingrained in the physicality of the organism as has been evolutionarily built since the dawn of Life and therefore provides for a more directly integrated meaning to the system. We know that many animals instinctually perform behaviors such as avoiding predators the first time exposed to the stimulus and before learning or building a mental representation of the threat. Many physicists such as Rolf Landauer consider that all information is embodied in the physical substance of systems and is not simply just an abstraction in the form of something like a software of disembodied code. This physicality of information in the living system becomes the reality of knowledge with direct meaning. Recent research by neurobiologist such as Stuart Firestein also indicates that perceptual information is not simply received but rather is the result of intentional query based on expectations or "anticipations" as termed by Robert Rosen. Therefore, for biological systems there are no abstract representations of the experience, only direct meanings derived from the perceived differences from the expected as instilled in their very structure and dynamics.

Based on these considerations, Skarda further postulates that holism is the initial condition of everything and should always be the principal perspective as the primacy of reality. Again, the critical difference in this perspective as compared to the traditional representational model knowledge acquisition process is that the interdependent whole of the experiential realm is really the primary reality and the observing biological system directly interprets information about the world as presented rather than through constructing representations. All phenomena are then based on the natural relationships between of interacting pieces of the whole. And knowledge of reality comes through deconstruction of experience from a position of articulated integration with the whole. There is no possibility for an absolutely correct representational model because logically it would have to be a distinct and detached entity and would interrupt the co-origination and interdependent nature of the reality. Once any representation is constructed, it is reified as a necessarily separate object that is limited in its reach as it invariably loses some of the integral connectivity to the whole. These mechanics are veritable regardless whether the source of the perception is external or internal to the observer.

Pure phenomenology is the science of pure consciousness.
—Edmund Husserl, phenomenologist

As noted by Huw Price, the integrated observer/participant is a part of that relational Universe and therefore there is no definitive difference between a subjective and objective reality. Also as noted previously, the philosophical understanding of the interdependency of all things was first noted in Nagarjuna's concepts of "emptiness". Interestingly, from these perspectives, there is not really the possibility for a completely valid independent, external, third-party position for a description of phenomena. However, some of the survival-based biologic mechanisms of perception often provide living organisms with an illusion of reality in which distinct sounds and visuals characterize objects as separated entities. This results in organisms also seeing themselves as distinct from their environment and misperceive the fundamental essence of reality. But there is really no outside of the whole. There are only 1st person experiences through which all are united as a part of the whole. And all evidences are filtered through the lens of subjective experiences that are grounded in a shared commonality of biology and teleonomy that creates the inter-subjective connection which is the basis for our purported scientific objectivity.

The internal machinery of Life, the chemistry of the parts, is something beautiful. And it turns out that all Life is interconnected with all other Life.
—Richard P. Feynman, Nobel physicist

Carlo Rovelli also suggests that modern quantum theory reveals that the scientific foundations of physical reality should be grounded in a relational perspective rather than based on the observations of discrete objects. Therefore, the properties describing any of the pieces of the world must also include that part to which it is revealing itself. Subsequently, Rovelli concludes that Science is really about understanding how one part of Nature manifests itself to another part of Nature and the effect they have on each other. This enlightened position suggests a more holistic methodological approach to the practice of Science. However, there must still be the inclusion of the role of the observer.

From the biological perspective, Francisco Varela and Jonathan Shear similarly consider that the very word phenomenon implies that there is an experience of what something is for something else and therefore is inherently relational. Following Skarda's view, a deconstruction of information for the acquisition of knowledge from the primacy of the whole naturally provides for such a relational causal framework from which living systems can function and maintain integrity. Through this intentional deconstruction of the biocontinuum sensory information to relevant knowledge, meaning, and action, the living system participates in a moment by moment determination of reality in the context of a cacophony of physical potentials. In this way, the meaning of all experienced reality is grounded in its significance to the fulfillment of Life processes for a continued subsistence.

Unsurprisingly, all the intrinsic deterministic uncertainty leading to meaning is also revealed in the qualification of work energy and entropy by the living system. And this same work energy is required for building and sustaining the system in a circular causality loop. Through this process, the causal meaning of all of natural phenomena also has a physical existential frame of reference with the living organization serving as the basis of the unique co-dependent causal structure. Furthermore, the usual causal entailment of reductionist science often fails in predicting the more globally emergent characteristics of complex systems because the determination of the downward causality for those systems is instantiated in the whole in a way that is not always distinguishable and trackable by external observers.

The Journey in Review and the Path Forward

The important thing is not to understand the world but rather to understand experience. **—Siddhartha Gautama, the Buddha**

In review, what have we learned from our journey thus far and what is our path forward? There is no doubt that the modern practice of Science has been extremely successful in improving the human condition and providing a practical methodology for better understanding the mechanisms necessary to navigate our world. The technological developments resulting from these pragmatic scientific endeavors are nothing short of miraculous. However, Science as Natural Philosophy and the path to knowledge can have a deeper and more personal role in our lives. In this salient role for understanding, Science also has an intersection with the goals of Religion and all true spiritual endeavors.

The strict practice of the Scientific Method requiring pure objectivity has noted logical and philosophical limitations. Likewise, the strict dogma of religious doctrines and authoritarian scriptures have also often impeded the progress of human evolution. The lessons from Science clearly place the observer in a central position in the determination of the nature of existential reality. Religion also focuses on the experiences of the individual. As such, this experiential position can be considered the as the center from which the known Universe emanates as the creator becomes the created. This means there must be a reconciliation of the objective reality of the physical Universe with a condition that requires the subjectivity of the living system to actualize its existence. Perhaps a return to the roots of Natural Philosophy as first espoused by the ancient Greeks Anaximander and Democritus, in which our inherent uncertainty is embraced and there is continued striving for a new and better understanding of our Universe should be our path forward.

The Universe is change, Life is opinion that adapts itself.
—Democritus, Greek philosopher

According to Nagarjuna's notion of emptiness, absolute independent objectivity is really an illusion. Nothing has any objective essence or independent intrinsic existence in itself. Everything has an interdependent relation with something else and all constitution emerges from those relational encounters. The true nature of things only exists as they manifest to other things. And because the primary condition of the Universe is one of impermanence, there is no ultimate or absolute essential reality, only the interconnected phenomena of the interdependent whole. We see clear evidence of this interdependence in the complexity of living systems where the whole cannot be understood through the properties of the individual elements. Since there is no independent existence of anything, it is this interdependency that includes living systems that also provides a mechanism for instilling meaning into all phenomena.

Think of the Universe as a single, quantum whole.
—Heinrich Päs, theoretical physicist

Procedures for incorporating the observer's vantage point into an understanding of reality should include modern concepts concerning the mechanisms of cognition and consciousness. Within a strict 3rd person objective perspective, as is necessary for the common practice of the Scientific Method, there appears to be no need for consciousness in the basic processing of empirical phenomena. This limiting condition is seen in the Chalmers' famous thought experiment of the zombie automaton which appears to respond appropriately but has no conscious experience of things. However, the necessary intercession of the scientists' 1st person perspective in scientific inquiry suggests that something fundamental is missing if we deny a role for consciousness in our understanding of Science. Perhaps, it is the meaning associated with the subjective perspective that is missing. The individual's experiential process naturally instills meaning into empirical observations as necessary for the living organism's stability and survival. The amalgamated perceptions of phenomena that Solms contends form subjective feelings and emotions (what it feels like) as the root of consciousness provide more than the dispassionate information acquired by the zombies of Chalmers. These subjectively acquired conscious feelings are also the source of a measure of global meaning. And such meaning can completely change the interpretation of the raw experience.

By example, presented is the graphic below of the silhouette of a rabbit sticking its head out of a burrow *or* a crocodile poking its head out of the water? The difference in meaning, based on a subjectivity of personal knowledge, past experiences, and feelings could be existential in nature for the organism and create decidedly different feelings. The basic experience of the plain image is pure consciousness without judgement as to whether the perception is good or bad. It is the face value of each observer's information about the world with no thought and the primacy of a direct experience of reality. The raw foundation for the development of objective consilience or intersubjective agreement is based in the direct experience and rote distinction of the lines of the silhouette only. However, common shared experiences and personal biases usually influence the final conclusions for objective agreement that we see in Science.

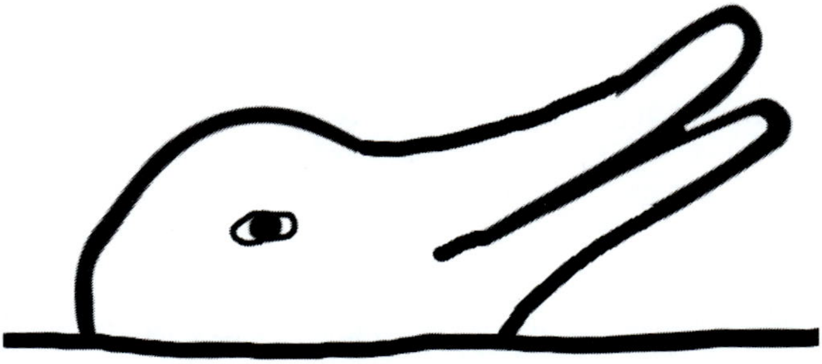

In another more relevant example from the history of Science, Ptolemy's model of the solar system fit the observed data nearly precisely and confirmed the biases of a seemingly Earth-centered Universe. However, that ancient vision lacked the harmonious simplicity and beauty of meaning implicit to the Copernican model. Such impressions of harmony emanate from subjective intuitions and experiences. Therefore, a true comprehensive Science cannot disregard the phenomenological reality of subjective experiences and must incorporate both 3rd and 1st person evidences in its methodology. This approach then requires the acceptance of credible 1st person experiences as real knowledge.

Truth can be stated in a thousand different ways, yet each one can be true.
—Swami Vivekananda, monastic

The Dalai Lama has been a vocal advocate for a broader perspective in scientific inquiry to include 1st person experiences. He considers that these experiences can be verified in a Popperian-style process of verisimilitude, particularly when there are commonalities amongst multiple individuals utilizing a replication of accepted introspection and contemplative practices. The understanding gained from such experiences are thought to be based on the collective information ingrained in the structure and processes of the living system as historically founded through the longitudinally accumulated knowledge of reality. This accrual of knowledge can extend back to the beginning of Life in the Universe and stored in our biology and DNA. However, integrating the traditional 3rd person objective scientific methodologies with those of 1st person knowledge acquisition by contemplative practices has been challenging and often results in attempts to reduce one framework into the other. Interestingly, true Naturalism only requires that realities are subject to factual inquiry and scientific investigation. The Dalai Lama suggests considering the distinction between the empirical aspects of contemplative practice and their associated metaphysical assumptions as being analogous with the Scientific Method approach to knowledge acquisition that balances inductive experimental observations with deductive theoretical interpretations. But a general axiomatic and theoretic framework for combining these two perspectives is still needed.

The natural interdependency of the observer and the observed through the biological process of experience and the instantiation of meaning as previously discussed utilizing the information-theoretic platform described in Chapters 5 and 6 may be a possible solution to the conundrum. Such a platform provides for an Ouroboros-like co-dependent origination of the knowable Universe with an eternally ongoing cyclic process for the deconstruction, recreation, and renewal of reality. This platform also includes Darwinian replicator dynamics and Natural Selection in a framework of relational information geometry incorporating the principles of axiomatic inference processing. The perspective that the deconstructed primacy of experience is the starting point for the reality of the integrated whole of the observer plus the observed is critical. If we grant such a position of primacy to subjective experience, then all reality and laws of Nature including the functioning of our brain and all our internal sensory biology arises out of this condition of awareness. The process of experience is then the Center of the Universe. And that is something we all share.

All that you touch, And all that you see, All that you taste, All you feel…And all that is now, And all that is gone, And all that's to come
And everything under the sun is in tune. But the sun is eclipsed by the moon.
—Pink Floyd, Dark Side of the Moon

REFERENCES

Abou Al-Ola, Omar M., Fujimoto, Ken'ichi, Yoshinaga, Tetsuya. 2011. "Common Lyapunov function based on Kullback-Leibler divergence for a switched nonlinear system." *Mathematical Problems in Engineering.* Vol. 2011, ID 723509, 1-12. London, UK: Hindawi Publishing Corporation, https:doi:10.1155/2011/723509

Akin, Ethan. 1990. "The differential geometry of population genetics and evolutionary games". *Mathematical and Statistical Developments of Evolutionary Theory*, No. 299:1-94.

Al-Rawi, Mahmood, Osama Y. 2012. "Enhancing Control Systems Response Using Genetic PID Controllers" R. Popa". *Genetic Algorithms in Applications, In Tech*. https:doi:10.5772/34354. https://www.intechopen.com/books/genetic-algorithms-in-applications.

Amari, Shunichi, Nagaoka, Hiroshi. 1993. *Methods of Information Geometry, Volume 191 of Translations of Mathematical Monographs*. Oxford, England: Oxford University Press.

Andronov, Aleksandr A., Pontryagin, Lev S. 1988. *Geometric methods in the theory of differential equations, Volume 250.* New York, New York: Springer-Verlag.

Aristotle (transl John H. McMahin) 1991. *The Metaphysics*, Amherst, New York, ISBN: 0879756713

Armstrong, D. M. 1961. *Perception and the Physical World*. London, England: Routledge and Kegan Paul.

Armstrong, D. M. 1973a. *Belief, Truth and Knowledge*. Cambridge, England: Cambridge University Press.

Ashby, W. R. 1962. "Principles of the self-organizing system," in *Principles of Self-Organization: Transactions of the University of Illinois Symposium.* H. Von Foerster and G. W. Zopf, Jr. (eds.), Pergamon Press: London, UK, pp. 255-278.

Asimov, Isaac. 1982. *Asimov's Biographical Encyclopedia of Science and Technology*. Garden City, NJ: Doubleday and Company.

Astakhov K. 2009. "Chapter 6: Methods of Information Geometry in Computational System Biology (Consistency between Chemical and Biological Evolution)". *Biomedical Informatics, Methods in Molecular Biology*, No. 569:115-127. Humana Press https:doi:10.1007/978-1-59745-524-4_6.

Atkins, Peter. 2007. *Four Laws that Drive the Universe*. Oxford, England: Oxford University Press.

Atkins, Peter. 2011. *On Being: A scientist's exploration of the great questions of existence*. Oxford, England: Oxford University Press.

Audi, R. 2003. "Contemporary Modest Foundationalism". *Knowledge: Classical and Contemporary Readings*. Pojman, L. Third Edition. Belmont, California: Wadsworth.

Avery, Samuel. 1995. *The Dimensional Structure of Consciousness: A Physical Basis for Immaterialism*. Samuel Avery Publisher.

Avery, Samuel. 2011. *Buddha and the Quantum: Hearing the Voice of Every Cell*. Sentient Publications.

Baez, J.C., Pollard, B.S. 2016. "Relative Entropy in Biological Systems". *Entropy*. No. 18(2): 46-52.

Baez, J.C. 2017. http://math.ucr.edu/home/baez/information/

Bak, Per. 1996. *How Nature Works: The Science of Self-Organized Criticality*. New York, New York: Copernicus.

Balduzzi D., Tononi G. 2008. "Integrated Information in Discrete Dynamical Systems: Motivation and Theoretical Framework". *PLoS Computational Biology,* No. 4: e1000091.

Barham, James A. 2012. *Normativity, Agency, and Life: Teleological Realism in Biology*. LAMBERT Academic Publishing.

Barham, James. 1996. "A dynamical model of the meaning of information". *Biosystems.* No. 38:235-241.

Barrow, John D., Tipler, Frank J. 1988. *The Anthropic Cosmological Principle*. Oxford, England: Oxford University Press.

Bateson, G. 1972. *Steps to an Ecology of Mind*. New York, New York: Ballantine Books.

Beer R.D. 2004. "Autopoiesis and cognition in the game of Life". *Artificial Life*, No. 10: 309-326.

Ben-Ari, Moti. 2005. *Just a Theory: Exploring the Nature of Science*. Amherst, NY: Prometheus Books.

Benner, S.A. 2010. *Defining Life. Astrobiology,* No. 10(10):1021-1030.

Bennett, C.H., Landauer, R. 1985. "The Fundamental Physical Limits of Computation". *Scientific American,* No. 253(1): 48, 56.

Bergstrom, C., Rosvall, M. 2011. "The Transmission sense of Information". *Biology and Philosophy*, 26: 159–176

Bergstrom, C. T., Lachmann, M. 2004. "Shannon information and biological fitness, IEEE information theory work-shop". *IEEE*. 5054. arXiv.org:q-bio/0510007.

Berkeley, George, Brainy Quote. 1685-1753. https://www.brainyquote.com/authors/george-berkeley-quotes.

Berlin, Isaiah. 2013. *The Roots of Romanticism* (2 ed.). Princeton: Princeton U.P

Berut, A., Arakelyan, A., Petrosyan, A., Ciliberto, S., Raoul, S., Dillenschneider, R., Lutz, E. 2012. "Experimental Verification of Landauer's Principle Linking Information and Thermodynamics". *Nature*, 483:186.

Bohm, David. 1983. *Wholeness and the Implicate Order.* London, England: Routledge & Kegan Paul.

Bonjour, L. 1985. *The Structure of Empirical Knowledge*. Cambridge, MA: Harvard University Press.

Bortoft, Henri. 1996. *The Wholeness of Nature: Goethe's Way of Science.* Edinburgh, Scotland: Floris Books.

Breuer, Reinhard. 1991. *The Anthropic Principle: Man as the Focal Point of Nature*. Cambridge, MA: Birkhäuser Boston, Inc.

Brillouin, L. 1951a. "Maxwell's Daemon Cannot Operate: Information and Entropy". *Journal of Applied Physics*, No. 22(3):334-37.

Brillouin, L. 1951b. "Physical entropy and information". *Journal of Applied Physics*, No. 22: 338,343.

Brillouin, L. 1962. *Science and Information Theory*, 2nd Edition. New York, New York: Academic Press.

Brooks, D.R., Wiley, EO. 1988. *Evolution as Entropy*. Chicago, IL: University of Chicago Press.

Brown, G. Spencer. 1972. *The Laws of Form*, American Edition. New York: The Julian Press.

Cafaro, Carlo. 2008. *The Information Geometry of Chaos.* Riga, Latvia: VDM Verlag Dr. Mueller.

Camazine S., Deneubourg, J-L., Franks N.G., Sneyd, J., Theraulaz G., Bonabeau E. 2001. *Self-Organization in Biological Systems*. Princeton University Press, Princeton, NJ.

Campbell, Jeremy. 1982. *Grammatical Man: Information, Energy, Language and Life.* New York, NY: Simon and Schuster.

Capra, Fritjof, Luigi, Luisi Pier. 2016. *The Systems View of Life: A Unifying Vision*. Cambridge, England: Cambridge University Press.

Capra, Fritjof. 1986. "The Santiago Theory of Life and Cognition." *la revista Be-Vision*, No. 9.1:59–60.

Capra, Fritjof. 1997. *The Web of Life: A New Scientific Understanding of Living Systems.* Anchor.

Carroll, Sean. 2010. *From Eternity to Here*. New York, NY: Plume, Penguin

Carroll, Sean. 2016. *The Big Picture: On the Origins of Life, Meaning, and the Universe Itself.* Boston, MA: Dutton.

Carter, Brandon, McCrea, William H. 1983. "The Anthropic Principle and its Implications for Biological Evolution and Discussion". *Philosophical Transactions of the Royal Society of London. Series A, Mathematical and Physical Sciences,* No. 310: 347-363.

Casti J.L., Karlqvist A. 1991. *Beyond Belief, Randomness, Prediction and Explanation in Science.* CRC Press, Boca Raton, Florida.

Caticha A. 2002. Entropy Dynamics. *AIP Conference Proceedings* 617:302.

Caticha, A. 2004. *Relative Entropy and Inductive Inference AIP Conference Proceedings*. 707, 75.

Caticha, A. 2007. "AIP Conference Proceedings". *Information and Entropy* 954.11.

Caticha, Ariel, Cafaro, Carlo. 2007. *From Information Geometry to Newtonian Dynamics*. http://arxiv.org/abs/0710.1071v1.

Caticha, Ariel. 2008. "From Inference to Physics". *The 28th International Workshop on Bayesian Inference and Maximum Entropy Methods in Science and Engineering*. Boraceia Beach, Sao Paulo, Brazil. http://arxiv.org/abs/0808.1260v1

Caticha, Ariel. 2010. "Entropic Inference in Bayesian Inference and Maximum Entropy Methods". ed. by A. Mohammad-Djafari, et al., *AIP Conference Proceedings, Science and Engineering*. No. 1305:20.

Caticha A. 2012. "Entropic Inference and the Foundations of Physics". *Invited monograph published by Brazilian Chapter of the International Society for Bayesian Analysis*. ISBrA Sao Paulo, Brazil

Caticha, Ariel. 2014. "The Basics of Information Geometry". *The 34th International Workshop on Bayesian Inference and Maximum Entropy Methods in Science and Engineering*. Amboise, France, arXiv:1412.5633v1.

Caticha, Ariel. 2015. "Entropic Dynamics". *Entropy*, No. 17: 6110-6123.

Chaitin, G. 2007. "Epistemology as information theory: From Leibniz to omega". In Dodig-Crnkovic, G., ed.: *Computation, Information, Cognition The Nexus and The Liminal*. Newcastle upon Tyne, UK: Cambridge Scholars Pub. 2–17.

Chalmers, DJ. 1996. *The Conscious Mind: In Search of a Fundamental Theory*. Oxford, England: Oxford University Press.

Chalmers, David J. 2010. *The Character of Consciousness (Philosophy of Mind*). Oxford, England: Oxford University Press.

Choifer, Alla. 2018. "A New Understanding of the First-Person and Third-Person Perspectives". Philosophical Papers, 47(3):333-371.

Chisholm, R. M. 1989. *Theory of Knowledge, 3rd edition*, Englewood Cliffs, NJ: Prentice Hall International.

Cobb, M. 2017. "60 Years Ago, Francis Crick Changed the Logic of Biology". *PLOS Biology,* No. 15(9): e2003243.

Cole, K.C. 1997. *The Universe and the Teacup*. New York, New York: Harcourt Brace and Co.

Coleman, T. G. 1975. "From Aristotle to modem computers: The role of theories in biological research". *The Physiologist, 18,* 509-5 18.

Collier, J. D. 1988. "Supervenience and Reduction in Biological Hierarchies". *Philosophy and Biology: Canadian Journal of Philosophy Supplementary*. 14.

Collier, J.D., Hooker, C.A. 1999. "Complexly Organized Dynamical Systems". *Open Systems and Information Dynamics*, No. 6: 241-302.

Conrad, M. 1969. "Computer Experiments on the Evolution of Coadaptation in a Primitive Ecosystem". *Thesis, Biophysics Program, Stanford University*.

Conrad, M. 1970. "Towards a Theoretical Biology". *Statistical and Hierarchical Aspects of Biological Organization.* Vol. 4. CH Waddington ed. Edinburgh, UK: Edinburgh University Press.

Conrad, Michael. 1983. "Chapter: The Meaning of Efficiency". *Adaptability: The Significance of Variability from Molecule to Ecosystem*: 141-159. New York, NY Plenum Press.

Conrad, M. 1996. "The Geometry of Evolution". *Rivista Di Biologia - Biology Forum*. No. 89: 21-54.

Coopersmith, Jennifer. 2017. *The Lazy Universe: An Introduction to the Principle of Least Action*. New York, NY: Oxford University Press. ISBN 978-0-19-874304-0.

Cover, T.M., Thomas, J.A. 2006. *Elements of Information Theory.* Hoboken, NJ: Thomas, John Wiley and Sons, Inc.

Cox, R.T. 1946. "Probability, Frequency, and Reasonable Expectation". *American Journal of Physics*. No. 14.

Cox, R.T. 1961. *The Algebra of Probable Inference.* Baltimore, MD: Johns Hopkins.

Crutchfield J. P., Young K. 1989. "Inferring statistical complexity". *Phys. Rev. Lett*., 63:105–108.

Crutchfield, J.P., Young K. 1990. "Computation at the Onset of Chaos". *W. Zurek, ed. Entropy, Complexity, and the Physics of Information. SFI Studies in the Sciences of Complexity,* VIII: 223–269. Reading, MA: Addison-Wesley

Crutchfield, J. P. 1994. "The Calculi of Emergence: Computation, Dynamics, and Induction". *Physica D* 75: 11-54.

Crutchfield J.P., Shalizi C.R. 1998, "Thermodynamic depth of causal states: Objective complexity via minimal representations". *Phys. Rev. E*, 59(1):275–283.

Crutchfield J.P., Feldman D.P., Shalizi C.R. 2000, Comment I on "Simple measure for complexity". *Physical review. E, Statistical Physics, Plasmas, Fluids, and Related Interdisciplinary Topics*. (2 Pt B):2996-2997. https:doi:10.1103/physreve.62.2996.

Danesi, Marcel. 2007. *The Quest for Meaning. A Guide to Semiotic Theory and Practice*. Toronto, Canada: University of Toronto Press.

Davies, Paul. 1993. *The Mind of God: The Scientific Basis for a Rational World Paperback.* New York, NY: Simon & Schuster.

Davies, Paul, 1998, *Cosmic Blueprint: New Discoveries In Natures Ability To Order Universe*. Simon and Schuster, New York, NY.

Davies, P. 2008. *The Goldilocks enigma: why is the Universe just right for Life?* Houghton, Michigan: Harcourt.

Davies, Paul, Gregersen, Deacon, Terrence, Hoffmeyer, Jesper, Kuppers, Bernd-Olaf. 2011. *Information and the Nature of Reality.* Cambridge, UK: Cambridge University Press.

Davies, P., Callender, Craig. 2014. "A Matter of Time". *Scientific American*. No. 23:4.

Davies, Paul. 2015. *The Fifth Miracle: The Search for the Origin and Meaning of Life*. New York, NY: Simon & Schuster.

Dawid Richard. 2013 *String Theory and the Scientific Method*. Cambridge, UK: Cambridge University Press.

Dawkins, Richard. 1976. *The Selfish Gene*. Oxford, UK: Oxford University Press.

Dawkins, Richard. 1987. *The Blind Watchmaker*. New, York, NY: Norton Company.

de Duve, Christian. 2002. *Life Evolving*. New York, NY: Oxford University Press.

de Duve, Christian. 2005. *Singularities: Landmarks on the pathways of Life*. New York, NY: Cambridge University Press.

DeGrasse Tyson, Neil. 2017. *Astrophysics for People in a Hurry*. New York, NY: W.W. Norton & Company.

Deacon, T.W. 2007. "Shannon-Boltzmann-Darwin: Redefining Information. Part 1." *Cognitive Semiotics,* No. 1:123-148.

Deacon, T.W. 2008. "Shannon-Boltzmann-Darwin: Redefining Information. Part 2." *Cognitive Semiotics,* No. 2:167-194.

Deacon, Terrence, Koutroufinis, Spyridon. 2014. "Complexity and Dynamical Depth". *Information*. 5. 404-423.

Descartes, R. 1641, 1983. *"First Meditation," in Meditations on First Philosophy, Reprinted in The Philosophical Writings of Descartes.* J. Cottingham, R. Stoothoff & D. Murdoch Edition Cambridge, UK, Cambridge University Press.

Descartes R, 1950, *Discourse on Method,* Trans. L. J. Lafleur. Bobbs-Merrill.

Dewar, R. C. 2009. "Maximum Entropy Production as an Inference Algorithm that Translates Physical Assumptions into Macroscopic Predictions: Don't Shoot the Messenger". *Entropy*, No. 11(4): 931-944.

Dewar, R.C., Maritan, A. 2011. *The second law, maximum entropy production and Liouville's theorem*. arXiv:1107,1088.

Dewey, J. 1914. "Psychological doctrine in philosophical teaching". *Journal of Philosophy*. No. 11:505-512.

Dodig-Crnkovic, Gordana. 2014, *Modeling Life as Cognitive Info-Computation*. http://arxiv.org/abs/1401.7191v1

Dolnick, Edward. 2011. *The Clockwork Universe: Isaac Newton, the Royal Society, and the Birth of the Modern World*. New York, NY: Harper Collins.

Donaldson-Matasci, MC, Bergman, CT, and Lachman, M. 2010. "The Fitness Value of Information". *Oikos,* No. 119(2): 219-230.

Doolittle, W.F. 2012. "Evolutionary biology: A Ratchet for Protein Complexity". *Nature*, No. 8;481(7381):270-1.

Doyle, J.C., Francis, B.A., Tannenbaum, A.R. 1992. *Feedback Control Theory.* Mineola, New York: Dover Publications.

Dretske, F. 1981. *Knowledge and the Flow of Information*. Cambridge, MA: MIT Press.

Edelman, Gerald M. 1989. *The Remembered Present: A Biological Theory of Consciousness*. New York, NY: Basic Books.

Ehrenberg, W. 1967. "Maxwell's Demon". *Scientific American*. 217:103-110

Eigen, M., Schuster, P. 1979. *The Hypercycle: A Principle of Natural Self-Organization*. Berlin, Germany: Springer Verlag.

Eigen, Manfred. 1992. *Steps Towards Life, A Perspective on Evolution, translated by P.Wooley*. Oxford, UK: Oxford University Press.

Einstein, A. 1916. *Relativity: The Special and General Theory*. New York, NY: H. Holt and Company.

Einstein, Albert, and Podolsky, B. Rosen, N. 1935. "Can Quantum-Mechanical Descriptions of Physical Reality of Considered Complete?" *Physical Review*. No. 47:777.

Erdi, Peter. 2008. *Complexity Explained*. Berlin, Heidelberg, Germany: Springer-Verlag.

Everett, Hugh. 1957. "Relative State Formulation of Quantum Mechanics". *Reviews of Modern Physics*. No. 29(3): 454–462.

Faraday, Michael. 1834. "On Electrical Decomposition". *Philosophical Transactions of the Royal Society.* London.

Feigenbaum, M.J. 1976. "Universality in complex discrete dynamics". *Los Alamos Theoretical Division Annual Report*.

Feldman, D.P., Crutchfield, J.P. 1998. "Measures of statistical complexity: Why?" *Phys. Lett. A*, 238:244–252,

Fernando, C., Rowe, J. 2008. "The origin of autonomous agents by natural selection". *Biosystems*, No. 91:355-373.

Feynman, R. P. 1966. *The Feynman Lectures on Physics, Volume 1*. Reading, MA: Addison-Wesley.

Finnigan, G.C., Hanson-Smith, V., Stevens, T.H., Thornton, J.W. 2012. "Evolution of Increased Complexity in a Molecular Machine". *Nature*, 481:360–364.

Firestein, S. 2001. "How the olfactory system makes sense of scents". *Nature,* No. 413: 211–218, https:doi:10.1038/35093026.

Fisher, R.A. 1930. *The Genetical Theory of Natural Selection*. Oxford, UK: Clarendon Press.

Fisher, Ronald. 1958. "The Nature of Probability". *The Centennial Review of Arts & Science*, East Lansing, MI: Michigan State University Press, 2: 261–274.

Fisher, Ronald A. 1971 [1935]. *The Design of Experiments* (9th ed.). New York, NY: Macmillan. ISBN 0-02-844690-9.

Fox, R.E. 1988. *Energy and the Evolution of Life*. New York, NY: W.H. Freeman and Co.

Freeman, W.J. 1991. "The Physiology of Perception". *Scientific American*, No. 264:78–85.

Freeman, W.J. 1999. *How Brains Make Up Their Minds*. London, England: Weidenfeld and Nicolson.

Freeman, W.J. 2000. *Neurodynamics: An Exploration of Mesoscopic Brain Dynamics.* London, England: Springer-Verlag.

Frieden, B. Roy. 2004. *Science from Fisher Information: A Unification.* Cambridge, UK: Cambridge University Press. ISBN 0-521-00911-1

Friston K. 2010. "The free-energy principle: a unified brain theory?" *Nature Reviews Neuroscience.* 11:12738.

Friston K. 2012. "Free Energy, Value and Attractors". *Computational and Mathematical Methods in Medicine.* Article ID 937860, https:doi:10.1155/2012/937860.

Friston, Karl, Kilner, James, Harrison, Lee. 2006. "A free energy principle for the brain". *Journal of Physiology-Paris. Elsevier BV*, No. 100(1–3):70–87.

Fujiwara, Akio, Amari, Shun-ichi. 1995. "Gradient Systems in View of Information Geometry". *Physica D: Nonlinear Phenomena*, No. 80: 317-327.

Ganti, Tibor. 2003. *The Principles of Life.* Oxford, England: Oxford University Press.

Gatlin, Lila L. 1973. *Information Theory and the Living System*. New York, NY: Columbia University Press.

Gauch, Hugh G. 2003. *Scientific Method in Practice*. Cambridge, UK: Cambridge University Press. ISBN 978-0-521-01708-4

Gause, G. F. 1934; reprint 2017. *The Struggle for Existence*. Futures 26(6): 626-647, 1994b Mineola, New York: Dover.

Gibson, J. J. 1975. "Affordances and Behavior". *Reasons for Realism: Selected Essays of James* J. Gibson. Hillsdale, NJ: Lawrence Erlbaum. No 410-411.

Gibson, J. J. 1966. *The Senses Considered as Perceptual Systems*. London, England: Allen and Unwin.

Gödel K. 1931. "Uber formal unentsheidbare Satz der Principia Mathematica und verwandter System. I". *Monatshefte fur Mathematik und Physik,* No. 38:173-198.

Goldman, A. I. 1979. "What is Justified Belief?". J*ustification and Knowledge: New Studies in Epistemology*, G. Pappas Edition. Boston, MA: Reidel. No. 1-23.

Goldman, A.I. 1986. *Epistemology and Cognition*. Cambridge, MA: Harvard University Press.

Goldstein, M, Goldstein, I.F. 1993*. The Refrigerator and the Universe — Understanding the Laws of Energy*. Cambridge, MA: Harvard University Press.

Gould, S.J. 1982. "Punctuated Equilibrium—A Different Way of Seeing." *New Scientist* No. 94: 137-139.

Gower, Barry. 1997. *Scientific Method: An Historical and Philosophical Introduction.* New York, NY: Routledge.

Gribbin, John. 1984. *In Search of Schrödinger's Cat: Quantum Physics and Reality, Reprint Edition*. New York, NY: Bantam.

Grice, H.P. 1961. "The Causal Theory of Perception". *Proceedings of the Aristotelian Society*, No. 35: 121-152.

Groth, WE, Weyssenhoff, HV. 1960. "Photochemical formation of organic compounds from mixtures of simple gases". *Planetary and Space Science*, 2(2):79-85. ISSN: 00320633

Groth, W.E. 1975. "Photochemical formation of organic compounds from mixtures of simple gases simulating the primitive atmosphere of the Earth". *Biosystems* 6:229-233.

Gutland C. 2018. Husserlian Phenomenology as a Kind of Introspection. Front. Psychol. 9:896. doi: 10.3389/fpsyg.2018.00896.

Guyton, A.C., Coleman, T.G. 1969. "Quantitative Analysis of the Pathophysiology of Hypertension". *Journal of the American Society of Nephrology*. No. 2248-2258.

Haldane, JBS. 1927. "A Mathematical Theory of Natural and Artificial Selection, Part V: Selection and Mutation". *Mathematical Proceedings of the Cambridge Philosophical Society*. No. 23(7): 838–844.

Hankinson, R.J. 1998. *Cause and Explanation in Ancient Greek Thought* Oxford, UK: Oxford University Press Premium. https:doi:10.1093/0199246564.001.0001, ISBN 9780198237457

Harold, Franklin, M. 2003. *The Way of the Cell: Molecules, Organisms, and the Order of Life*. Oxford, UK: Oxford University Press.

Harper, M. 2009a. *The Replicator Equation as an Inference Dynamic*. No. arXiv: 0911.1763.

Harper, M. 2009b. *Information Geometry and Evolutionary Game Theory*. No. arXiv: 0911.1383.

Harper, Marc. 2011. "Escort Evolutionary Game Theory". *Physica D: Nonlinear Phenomena*, No. 240(18):1411,1415.

Harper, Marc, Fryer, Dashiell E.A. 2013. *Stability of Evolutionary Dynamics on Time Scales*. arXiv:1210.5539v3

Hawking, Stephen W. 1988. *A Brief History of Time*. New York, NY: Bantam Books.

Hawking, Stephen, and Mlodinow, Leonard. 2010. *The Grand Design*. New York, NY: Bantam Books.

Hayles, Katherine. 1999a. *How We Became Posthuman: Virtual Bodies in Cybernetics, Literature and Informatics*. Chicago, IL: University of Chicago Press.

Hayles, Katherine. 1999b. "The condition of virtuality" *Peter Lunenfield ed. The Digital Dialectic*. Cambridge, MA: MIT Press.

Helmholtz, H. 1866/1962. *Concerning the perceptions in general. In Treatise on physiological optics*. (J. Southall, Trans., 3rd ed., Vol. III). New York, NY: Dover.

Hemmo, Meir, and Shenker, Orly R. 2012. *The Road to Maxwell's Demon: Conceptual Foundations of Statistical Mechanics.* Cambridge, UK: Cambridge University Press.

Henning, Brian G. and Scarfe, Adam. 2015. *Beyond Mechanism: Putting Life Back Into Biology*. Lanham, Maryland: Lexington Books.

Hill, D. L., and Moylan P.J. 1977. "Stability Results for Nonlinear Feedback Systems". *Automatica,* No. 13:377-382.

Ho, Mae-Wan. 1995. "Bioenergetics and the Coherence of Organisms". *Neural Network World*, No. 5: 733-750. https:doi:10.1007/978-94-011-5878-7_5

Ho, Mae-Wan. 1996. "Bioenergetics and Biocommunication". *Computation in Cellular and Molecular Biological Systems.* Editors: R. Culbertson, M. Holcombe, R. Paton. Singapore: World Scientific.

Ho, Mae-Wan. 1998. *The Rainbow and the Worm, the Physics of Organisms*. River Edge, NJ: World Scientific. ISBN 981-02-4813-X

Hoffman, Peter M., 2012. *Life's Ratchet: How Molecular Machines Extract Order from Chaos*. New York, NY: Basic Books.

Hoffmeyer, J. "Origin of Species by Natural Translation". *Translation*, Susan Petrilli ed.: No. 329-346.

Hoffmeyer, Jesper. 1996. *Signs of Meaning in the Universe Bloomington*. Bloomington, IN: Indiana University Press.

Hoffmeyer, Jesper. 2009. *Biosemiotics: An Examination into the Signs of Life and the Life of Sign.* Scranton, PA: University of Scranton Press.

Hoffmeyer, Jesper. 2010. "God and the World of Signs: Semiotics and the Emergence of Life". *Zygon,* No. 45(2): 367-390.

Hofstadter, Douglas R. 1980. *Gödel, Escher, Bach: An Eternal Golden Braid*. New York, NY: Vintage, 1st Edition.

Hofstadter, Douglas R. 2008. *I Am a Strange Loop.* New York, NY: Basic Books.

Holland, John H. 1995. *Hidden Order: How Adaptation Builds Complexity.* New York, NY: Basic Books.

Holland, John H. 1998. *Emergence: From Chaos to Order*. New York, NY: Basic Books.

Holland, John H. 2014. *Complexity: A Very Short Introduction*. Oxford, UK: Oxford University Press.

Hurley S.L. 1998, *Consciousness in Action*. Cambridge MA: Harvard University Press.

Hurley, Susan. 2001. "Perception and Action: Alternative Views", *Synthese* 129 (2001) 3-40.

Hurley S.L. 2003. "Animal actions in the space of reasons", *Mind and Language*. 18:231-56. Cambridge MA: Oxford University Press.

Isaacson, Walter. 2007. *Einstein: His Life and Universe*. New York, NY: Simon & Schuster.

Jago, Mark. "Logical Information and Epistemic Space". *Synthese*. No. 167 (2): 327-341.

Jaynes, E.T., Rosenkrantz, R.D. 1983. *Papers on Probability, Statistics and Statistical Physics*. Dordrecht, Holland: Reidel Publishing Company.

Jaynes, E.T. 2003. *Probability Theory: The Logic of Science*. Cambridge, UK: Cambridge University Press.

Johnson, R.W. Shore, J.E. 1983. "Comments on and correction to axiomatic derivation of the principle of maximum entropy and the principle of minimum cross-entropy". *IEEE Transactions on Information Theory,* No. IT-29: 942-943.

Juarrero, Alicia. 1999. *Dynamics in Action: Intentional Behavior as a Complex System*. Cambridge, MA: MIT Press.

Kaisers, Michael, Tuyls, Karl. 2009. "Replicator Dynamics for Multi-agent Learning - An Orthogonal Approach. In Matthew E. Taylor and Karl Tuyls, editors, *Adaptive and Learning Agents. Second Workshop, ALA'* . *Lecture Notes in Computer Science*, Vol. 5924:49-59.

Kant, I. 1781. *The Critique of Pure Reason*, N. Kemp Smith Edition. Basingstoke, Hampshire: The Macmillan Press, Ltd.

Kaplan, Daniel, and Glass, Leon. 1995. *Understanding Nonlinear Dynamics*. New York, NY: Springer-Verlag.

Karev, G. 2010a. "Replicator Equations and the Principle of Minimal Production of Information". *Bulletin of Mathematical Biology*. 72. 1124-42. 10.1007/s11538-009-9484-9.

Karev, G. 2010b. "Principle of Minimum Discrimination Information and Replica Dynamics". *Entropy*. No. 12:1673-1695 https:doi:10.3390/e12071673.

Kauffman, L.H. 1987. "Self-Reference and Recursive Forms". *Journal of Social and Biological Structure*, No. 10: 53-72.

Kauffman, Louis H. 2005. Laws of Form – An Exploration in Mathematics and Foundations,. Pg. 163. http://homepages.math.uic.edu/~kauffman/Laws.pdf

Kauffman, S. 1986. "Autocatalytics sets of proteins". *Journal of Theoretical Biology*. No. 119: 1-24.

Kauffman, Stuart. 1993. *The Origins of Order: Self Organization and Selection in Evolution*. Oxford, UK: Oxford University Press. ISBN 0-19-507951-5

Kauffman, Stuart. 1995. *At Home in the Universe: The Search for Laws of Self-Organization and Complexity*. Oxford, UK: Oxford University Press. ISBN 0195111303.

Kauffman, Stuart. 2000. *Investigations*. Oxford, UK: Oxford University Press. ISBN 0199728941.

Kauffman, S, Logan, Robert Este, Robert, Hobill, David, Shmulevich I, 2008, Propagating organization: An Enquiry, Biology & Philosophy, 23:27–45, https:doi:10.1007/s10539-007-9066-x.

Kauffman, Stuart. 2008. *Reinventing the Sacred: A New View of Science, Reason, and Religion*. New York, NY: Basic Books. ISBN 0-465-00300-1

Kauffman, Stuart. 2016. *Humanity in a Creative Universe*. New York, NY: Oxford University Press.

Kestin, Joseph. 1976. The Second Law of Thermodynamics. Hoboken, NJ: John Wiley & Sons Inc.

Kierkegaard, S. 2017 revised. *Stanford Encyclopedia of Philosophy.* https://plato.stanford.edu/entries/kierkegaard/

Kline, Morris. 1980. *Mathematics: The Loss of Certainty*, New York, NY: Oxford University Press.

Koch, Christof. 2009. A "Complex" Theory of Consciousness: Is complexity the secret to sentience, to a panpsychic view of consciousness?" *Scientific American MIND*.

Koestler, Arthur 1963, *The Sleepwalkers: A History of Man's Changing Vision of the Universe*. New York: Grosset & Dunlap. ISBN 978-0-448-00159-3. Original edition published by Hutchinson (1959, London)

Kolb, DA. 1984. *Experiential Learning: Experience as the Source of Learning and Development*. Englewood Cliff, NJ: Prentice Hall.

Koshland, D.E. Jr. 2002. "The Seven Pillars of Life". *Science*. No. 295: 2215-2216.

Kuhn, T. S. 1962. *The Structure of Scientific Revolutions*. Chicago: University of Chicago Press. ISBN 0-226-45808-3

Kullback, S. 1968. *Information Theory and Statistics*. New York, NY: Dover Publications.

Kullback, S., Leibler, R. 1951. "On Information and Sufficiency*". Annals of Mathematical Statistics*, No. 22:79-86.

Lakoff, G, Núñez, R. 2000. *Where Mathematics Comes From: How the Embodied Mind Creates Mathematics.* New York, NY: Basic Books.

Lama, His Holiness the Dalai. 2006. *The Universe in a single atom: the convergence of science and spirituality*. New York, NY: Broadway Books. ISBN 0767920813

Landauer R. 1961. "Irreversibility and Heat Generation in the Computing Process". *IBM Journal of Research and Development,* No. 5 (3):183–191.

Landauer, R. 1988. "Dissipation and noise immunity in computation and communication." *Nature,* No. 335:779.784.

Langer, Monika M. 1989. *Merleau-Ponty's "Phenomenology of Perception": A Guide and Commentary.* New York, NY: Palgrave Macmillan.

Langton, Christopher G. 1990. "Computation at the edge of chaos". *Physica D*. 42:12.

Leff, Harvey S, Rex Andrew. 2014. *Maxwell's Demon: Entropy, Information, Computing*. Princeton, NJ: Princeton University Press.

Leine, R.L. 2010. "The historical development of classical stability concepts: Lagrange, Poisson and Lyapunov". *Nonlinear Dynamics.* No. 59:173-182.

Leonard, Mary. 2013. http://frontiersmagazine.org/luigi-fantappie-and-the-physics-of-Life/

Lewes, George Henry, 2016 reprint, *Problems of Life and Mind*. Boston, MA: Forgotten Books ISBN: 1330413261

Libet, B. 2002. "The Timing of Mental Events: Libet's Experimental Findings and Their Implications". *Consciousness and Cognition*, No. 11, Edition 2: 291-299.

Lotka, A.J. 1922a. "Contribution to the Energetics of Evolution". *Proceedings of the National Academy of Sciences of the United States of America*, No. 8: 147-151.

Lotka, A.J. 1922b. "Natural Selection as a Physical Principle". *Proceedings of the National Academy of Sciences of the United States of America*, No. 8: 151-154.

Luhmann, Niklas. 2012. *Introduction to Systems Theory*. Cambridge, UK: Polity.

Luigi Luisi, Pier, 2016, *The Emergence of Life: From Chemical Origins to Synthetic Biology* Cambridge, UK: Cambridge University Press; ISBN: 1107092396

Lyapunov, Aleksandr. 1892, 1992. "The General Problem of Stability of Motion". *International Journal of Control*, No. 55(3): 531-773. Transl. Davaux, E., Fuller, A.T.

Mathis, C., Bhattacharya, T., Walker, SI. 2017. "The emergence of Life as a first-order phase transition". *Astrobiology*, No. 17(3): 266-276.

Maturana Humberto, Varela, Francisco. 1980. *Autopoiesis and Cognition: The Realization of the Living*. Boston, MA: Reidel.

Maturana, H.R., Varela, F.J. 1991. *Autopoiesis and Cognition: The Realization of the Living*. New York, NY: Springer Science & Business Media.

Maxwell, James. 1867, 1995. "Letter to Peter Guthrie Tait". *The Scientific Letters and Papers of James Clerk Maxwell,* Vol. 2, 331-2.

Mayr, Ernst. 1965. *Cause and effect in biology.* In Lerner, D. (ed.). Cause and effect. New York, NY: Free Press.

McClare, C.W.F. 1971. “Chemical Machines, Maxwell’s Demon, and Living Organisms”. *Journal of Theoretical Biology,* No. 1(30):1-34

McKay, Donald M. 1972. *Information, Mechanism and Meaning.* Cambridge, MA: MIT Press.

McKay, Michael D. 1995. *Evaluating Prediction Uncertainty*. Washington, DC: U.S. Nuclear Regulatory Commission, Office of Nuclear Regulatory Research.

McShea, Daniel W., Brandon, Robert N. 2010. *Biology's First Law: The Tendency for Diversity and Complexity to Increase in Evolutionary Systems*. Chicago, IL: University Of Chicago Press.

Mingers, John. 1995, *Self-Producing Systems: Implications and Applications of Autopoiesis*, New York, NY: Plenum Press.

Mitchell, Melanie. 2011, *Complexity: A Guided Tour* 1st Edition. Oxford, UK: Oxford niversity Press; ISBN: 0199798109

Monod, J. 1997. *Chance and Necessity: Essay on the Natural Philosophy of Modern Biology.* London, England: Penguin Books Ltd.

Moreno, Alvaro, and Mossio, Matteo. 2015. *Biological Autonomy: A Philosophical and Theoretical Enquiry*. New York, NY: Springer.

Morin, Edgar. 1992. *“From the concept of system to the paradigm of complexity”. Journal of Social and Evolutionary Systems.* 15 (4): 371–385.

Morin E. 2008. *On Complexity*. Cresskill, NJ: Hampton Press.

Morowitz, Harold J. 1968. *Energy Flow in Biology: Biological Organization as a Problem in Thermal Physics*. Cambridge, MA: Academic Press.

Morowitz, H.J. 2002. *The Emergence of Everything: How the World Became Complex*. New York, NY: Oxford University Press.

Morris, William, ed. 1971. *The American Heritage Dictionary of the English Language*, New York, NY: Houghton Mifflin.

Nagel Ernest, 1961, *The Structure of Scientific Theories: Problems in the Logic of Scientific Explanation*. New York, NY: Harcourt, Brace & World, Inc.

Nagel, Ernest, and Newman, James. 2008. *Gödel's Proof.* New York, NY: New York University Press.

Newby, Gregory B. 2001. “Cognitive space and information space”. *Journal of the American Society for Information Science and Technology*. No. 52:12

Newton, Isaac. 2016. *Philosophiae Naturalis Principia Mathematica.* (English), Andrew Motte (Author). Amazon: ISBN:1536887056

Nicolis, G., Portnow, J. 1973. *Chemical Reviews*. Washington, D.C.: American Chemical Society. https:doi.org/10.1021/cr60284a003.

Nicolis, G., Prigogine, I. 1977. *Self-Organization in Nonequilibrium Systems*. New York, NY: Wiley; 1st edition, ISBN-10 0471024015
Noë, Alva. 2006. *Action in Perception.* Cambridge, MA: MIT Press, Bradford Books.
Norton J.D. 2011. "Waiting for Landauer. Studies in History and Philosophy". *Modern Physics*, No. 42:184-198.
Odum, H.T., Pinkerton, R.C. 1955. "Time's speed regulator, the optimum efficiency for maximum output in physical and biological systems". *American Scientist*, 43, 331-343.
Odum, Howard T., Odum, Elisabeth C. 1981. *Energy Basis for Man and Nature*. New York, NY: McGraw Hill.
Ogden, C.K., and Richards, I.A. 1989. *Meaning Of Meaning, First Edition*. New York, NY: Mariner Books
Oizumi M, Albantakis L, and Tononi G. "From the Phenomenology to the Mechanisms of Consciousness: Integrated Information Theory 3.0". *PLoS Computational Biology,* No. 10(5): e1003588.
Ornstein, Robert. 1991. *Evolution of Consciousness: The origins of the way we think.* New York, NY: Simon and Schuster.
Pascal, R, Pross, A. 2014. "The nature and mathematical basis for the material stability in the chemical and biological worlds". *Journal of Systems Chemistry,* No. 5:3:1-8.
Peacocke, A.R. 1989. *An Introduction to the Physical Chemistry of Biological Organization.* Oxford, UK: Oxford Science Publications, Oxford University Press.
Pearl, J. 1988. *Probabilistic Reasoning in Intelligent Systems*. Los Altos, California: Morgan Kaufmann.
Pearl, Judea. 2009. *Causality: Models, Reasoning and Inference, 2nd edition*. Cambridge, UK: Cambridge University Press.
Peirce, CS. 1931-1935. *Collected Papers, Volume I-6*. Cambridge, MA: Harvard University Press.
Penrose, R. 1989. *The emperor's new mind: Concerning computers, minds, and the laws of physics.* Oxford University Press.
Penrose, Roger. 2004. *The Road to Reality: A Complete Guide to the Laws of the Universe.* New York, NY: Alfred A. Knopf. ISBN 9780679454434.
Penrose, Roger. 2010. *Cycles of Time: An Extraordinary New View of the Universe*. New York, NY: Alfred A. Knopf. ISBN: 978-0-307-27846-3.
Pierce, J.R. 1980. *An Introduction to Information Theory: Symbols, Signals and Noise*. New York, NY: Dover Publications.
Pirie N.W. 1957, "Some assumption underlying discussion on the origins of Life". *Annals of the New York Academy of Sciences.* 69(2):369-376.

https:doi:10.1111/j.1749-6632.1957.tb49672.x.

Pirsig, R. 1974. *Zen and the Art of Motorcycle Maintenance: An Inquiry into Values.* New York, NY: Harper Torch. ISBN 0-06-095832-4

Pittendrigh, C. S. 1958. *Adaptation, natural selection, and behavior, in Behavior and Evolution*. ed. A. Roe and George Gaylord Simpson, New Haven, CT: Yale University Press. 390–416

Popa, Radu. 2010. *Between Necessity and Probability: Searching for the Definition and Origin of Life*. Berlin, Germany: Springer-Verlag.

Popper, K. 1959. *The Logic of Scientific Discovery*. New York, NY: Basic Books.

Popper, K. R. 1967a. "Structural Information and the Arrow of Time". *Nature*, 214, 322.

Popper, K. R. 1967b. "Time's Arrow and Feeding on Negentropy". *Nature*. 213, 320.

Popper, K.: 1999, All Life is Problem Solving. New York, NY: Routledge

Prigogine, I. 1961. *Introduction to Thermodynamics of Irreversible Processes.* New York, NY: John Wiley.

Prigogine, I. 1980. *From Being to Becoming: Time and Complexity in the Physical Sciences,* San Francisco, CA: W.H. Freeman.

Prigogine I, Stengers I. 1984, *Order Out of Chaos*, Brooklyn, NY: Bantam; ISBN: 0553343637

Pross, Addy. 2012. *What is Life? How Chemistry becomes Biology*. Oxford, UK: Oxford University Press.

Pykh, Yu A. 2003. "Energy Lyapunov function for generalized replicator equation". *Proceedings of International Conference Physic and Control*. IFEE Publ. 2003, 271-276.

Rao, C.R. 1992. "R. A. Fisher: The founder of modern statistics". *Statistical Science*. No. 7: 34-48. https:doi:10.1214/ss/1177011442.

Reese, William L. 1980. *Dictionary of Philosophy and Religion*. Atlantic Highlands, NJ: Humanities Press.

Edward Rosen, "Copernicus, Nicolaus", *Encyclopedia Americana*, International Edition, volume 7, Danbury, Connecticut, Grolier Incorporated, 1986, ISBN 0-7172-0117-1, pp. 755–56.

Rosen, R. 1985. *Anticipatory Systems*. London, England: Pergamon Press.

Rosen, R. 1991. *Life Itself: A Comprehensive Inquiry into the Nature, Origin and Fabrication of Life*. New York, NY: Columbia University Press.

Ross, Sydney. 1962. "Scientist: The story of a word". *Annals of Science*. No. 18(2): 65-85. https:doi:10.1080/00033796200202722.

Rovelli, Carlo, *Relative information at the foundation of physics* (Marseille, CPT & Toulon U.). October 31, 2013. 3 pp.; Published in "It from Bit or Bit from It? On Physics and information", A Aguirre, B Foster and Z Merali eds., 79-86 (Springer 2015).

Rovelli, C: Helgoland, Penguin 2021.

Rovell, C: Reality is not what it seems, Penguin 2016.

Rovelli, C. "Aristotle's Physics: A Physicist's look", Journal of the American Philosophical Association, 1 (2015) 23-40 arXiv:1312.4057.

Rovelli, Carlo. "Meaning and Intentionality = Information + Evolution", in "Wandering Towards a Goal", A. Aguirre, B. Foster and Z. Merali editors, Springer 2018. arXiv:1611.02420.

Ruiz-Mirazo, Kepa, and Moreno, Alvaro. 2011. "The Need for a Universal Definition of Life in Twenty-First-Century Biology". *Information and Living Systems: Philosophical and Scientific Perspectives*. No. 3-24. Edited by George Terzis and Robert Arp, Cambridge, MA: MIT Press. 2011, pp. 3–24. JSTOR, www.jstor.org/stable/j.ctt5hhvb.5

Russell, Bertrand. 2005. *Our Knowledge of the External World.* New York, New York, NY: Routledge – Taylor and Francis.

Rutkowski, Leszek. 2010. *Computational Intelligence. Methods and Techniques*. New York, NY: Springer. ISBN 978-3-540-76287-4

Sagan, Carl, Druyan, Ann. 2007. *The Varieties of Scientific Experience: A Personal View of the Search for God, Reprint Edition*. London, England: Penguin Books.

Salthe, S.N. 2005. "Energy and semiotics: the Second Law and the origin of Life". *Cosmos and History 1*, 128 - 145.

Salthe, S. N. 2007. "The Natural Philosophy of Work". *Entropy,* No. 9(2), 83-99. https:doi.org/10.3390/e9020083

Schneider T.D. 1991 "Theory of Molecular Machines. II. Energy Dissipation". *Molecular Machines*, No. 148(17):125-137. https://https:doi.org/10.1016/S0022-5193(05)80467-9

Schneider, E.D., Kay, J.J. 1994a. "Life as a Manifestation of the Second Law of Thermodynamics". *Mathematical and Computer Modeling*. No. 19(6-8): 25-48.

Schneider, Eric D., Kay, James J. 1994b. "Complexity and Thermodynamics: Towards a New Ecology." *Futures*, No. 26: 626-647.

Schneider, Eric D., Sagan, Dorion. 2006. *Into the Cool: Energy Flow, Thermodynamics and Life*. Chicago, IL: University of Chicago Press.

Schrödinger, Erwin. 2012. *What is Life?* Cambridge. MA: Cambridge University Press, reprint edition.

Scott, Alwyn C. 2007. *The Nonlinear Universe, Chaos, Emergence, Life*. New York, NY: Springer Verlag.

Sebok Thomas A., 1999, *Signs: An Introduction to Semiotics*, Toronto, Canada: University of Toronto.

Sellars, W. 1997. *Empiricism and the Philosophy of Mind*. Cambridge, MA: Harvard University Press.

Seth A.K., 2019. "Truth, Lies, and Uncertainty". *Scientific American:* No. 321:3.

Shalizi, Cosma Rohilla. 2001. *Thesis: Causal Architecture, Complexity and Self-Organization in Time Series and Cellular Automata*. Physics Department, University of Wisconsin, Madison, WI.

Shalizi, C.R. 2009. "Dynamics of Bayesian updating with dependent data and misspecified models". *Electronic Journal of Statistics*, No. 3:1039-1074.

Shannon, C. A. 1948. "Mathematical Theory of Communication". *Bell System Technical*, No. 27: 379–423.

Shannon, C., Weaver, W. 1949. *The Mathematical Theory of Communication*. Urbana, IL: University of Illinois Press.

Shore, J.E., Johnson, R.W. 1980. "Axiomatic Derivation of the Principle of Maximum Entropy and the Principle of Minimum Cross-Entropy". *IEEE Transactions on Information Theory*, No. 26: 26-37.

Shore, J.E. Johnson, R.W. 1981. "Properties of Cross-Entropy Minimization". *IEEE Transactions on Information Theory,* No. IT-27: 472-482.

Shore, J.E. 1986. "Relative entropy, probabilistic inference". *Uncertainty in Artificial Intelligence,* AI. L.N. Kanal, J.F. Lemmer (Eds.), North-Holland, Amsterdam. 211-215

Simon, Herbert A. 1962. "The Architecture of Complexity". *Proceedings of the American Philosophical Society*, *Vol. 106*, No. 6: 467-482.

Skarda, Christine, Freeman, W. 1987. "How Brains Make Chaos in Order to Make Sense of the World". *Behavioral and Brain Sciences*. No. 10. 161-195.

Skarda, C., Freeman, W. 1990. "Chaos and the New Science of the Brain". *Concepts in Neuroscience*. 1-2. 275-285.

Skarda, Christine. 1990. "The Neurophysiology of Consciousness and the Unconscious". *Behavioral and Brain Sciences*, No. 13:4. 625-626.

Skarda, C. 1991. "The Biology of Life and Learning". *Journal of Social and Biological Structures,* No. 14:2. 221-228.

Skarda, C. 1992. "Perception, Connectionism, and Cognitive Science", in F. Varela & J.-P. Dupuy (eds), *Understanding Origins*, Netherlands: Kluwer Academic Publishers, 265-271.

Skarda, C. 1999. "The Perceptual Form of Life" in Reclaiming Cognition": The Primacy of Action, Intention, and Emotion". *Journal of Consciousness Studies 6,* No. 11-12. 79-93.

Smith, Andrew P. 2008. *The Dimensions of Experience: A Natural History of Consciousness*. Bloomington, IN: Xlibris Corporation.

Smith, J. Maynard. 1982. *Evolution and the Theory of Games*. Cambridge, UK: Cambridge University Press. ISBN 978-0-521-28884-2

Smith, Maynard J., Burian, R., Kauffman, S., Alberch, P., Campbell J. 1985. "Developmental Constraints and Evolution". *Quarterly Review of Biology*, No. 60: 265-287. 10.1086/414425.

Smolin, Lee. 2019. "Beyond Weird". *New Scientist*. 35-37.

Soto-Andrade, Jorge, Jaramillo, Sebastián, Gutiérrez, Claudio, Letelier, Juan-Carlos. 2011. "Ouroboros avatars: A Mathematical Exploration of Self-reference and Metabolic Closure". *Advances in Artificial Life, ECAL 2011: Proceedings of the Eleventh European Conference on the Synthesis and Simulation of Living Systems*. Cambridge, MA: The MIT Press. No. 763-770.

Spencer-Brown, G. 1973. *Laws of Form*. New York, NY: Bantam.

Sperry, R.W. 1952. "Neurology and the mind-body problem". *American Scientist*. No. 40: 291–312.

Stapp, Henry P. 2011. *Mindful Universe: Quantum Mechanics and the Participating Observer*. New York, NY: Springer-Verlag.

Stoltzfus, A. 1999. "On the Possibility of Constructive Neutral Evolution". *Journal of Molecular Evolution,* No. 49: 169-181.

Stonier, Tom. 1997. *Information and Meaning An Evolutionary Perspective*. New York, NY: Springer Verlag.

Summers R.L. 1998, "Computer simulation studies and the scientific method". *J. of Applied Animal Welfare Sci.* 1(2): 119-131.

Summers R.L., *Experiences in the Biocontinuum: A New Foundation for Living Systems*. 2020; Cambridge Scholars Publishing. Newcastle upon Tyne, UK, ISBN (10): 1-5275-5547-X, ISBN (13): 978-1-5275-5547-1

Summers, R "The imperative of self-reference in a theoretical framework for biology". 2021. *Academia Letters*, Article 1093. https://doi.org/10.20935/AL1093.

Summers RL. "An Action Principle for Biological Systems". *J. Phys.: Conf. Series 2021*, 2090 012109.

Summers, RL. "Methodology for Quantifying Meaning in Biological Systems". *WASET. International Journal of Bioengineering and Life Sciences* 2023;17(1):1-4.

Summers, R.L. "Lyapunov Stability as a Metric for the Meaning of Information in Biological Systems". *Biosemiotics* 2022. https://doi.org/10.1007/s12304-022-09508-5.

Summers RL, Musshafen L, Marshall G, Bates J, Jain S, Johnston J, Castillo A, Ayyalasomayajula K, Sandlin R, Frame IJ, Lirette S, Chandra S,

Robinson DA, Garrett M. "Following the Science: the role of the academic medical center in the COVID-19 pandemic". *Journal of Public Health in the Deep South* Vol. 3 No. 1, April 2022.

Swenson R., Turvey, M.T. 1991. "Thermodynamic Reasons for Perception-Action Cycles, *Ecological Psychology*" 3:4, 317-348. https:doi:10.1207/s15326969eco0304_2.

Swenson, R. 1988. "Emergence and the principle of maximum entropy production: Multi-level system theory, evolution, and non-equilibrium thermodynamics". *Proceedings of the 32nd annual meeting of the International Society for General Systems Research*. 32: 32.

Swenson, R. 1991. "Order, evolution, and natural law: Fundamental relations in complex system theory". In C. Negoita (Ed.) *Cybernetics and Applied Systems.* 125-148. New York, NY: Marcel Dekker Inc.

Swenson, R. 2009. "The fourth law of thermodynamics: The law of maximum entropy production" *LMEP, Chemistry 18.* 333-339.

Szathmary, E. 2000. "The evolution of replicators". *Philosophical Transactions of the Royal Society*. No. 355:1669-1676.

Szent-Gyorgyi, Albert. 1977. "Drive in Living Matter to Perfect Itself". *Synthesis 1,* Vol. 1, No. 1: 14-26.

Taylor, P.D., Jonker, L. 1978. "Evolutionary Stable Strategies and Game Dynamics". *Mathematical Biosciences*, No. 40:145-156.

Teilhard de Chardin, Pierre. 1959. *The Phenomenon of Man.* trans. Bernard Wall. New York, NY: Harper & Row, p. 11.

Thalos M. 2009. "Systems". *The Monist,* No. 92(3):452-478.

Thalos, M. 2014. "The Grammar of Experience". *Philosophy*, 89(2): 223-250.

Thomas, S.R. 2007. "Robert Rosen in the Age of Systems Biology". *Chemistry and Biodiversity 2007*, No. 4:2407.

Thompson, Evan, Varela, Francisco J. 2001. "Radical Embodiment: Neural Dynamics and Consciousness," *Trends in Cognitive Sciences,* No. 5:418-425.

Thompson, Evan. 2004. "Life and Mind: From Autopoiesis to Neurophenomenology. A Tribute to Francisco Varela". *Phenomenology and the Cognitive Sciences,* No. 3:381-398.

Thompson, Evan. 2010. *Mind in Life: Biology, Phenomenology, and the Sciences of Mind.* Cambridge, MA: Belknap Press.

Todhunter, Issac. 2020. *A history of the mathematical theory of probability: from the time of Pascal to that of Laplace, Alpha Edition*. London, England: BiblioLife, Arkose Press.

Tononi G., Edelman G.M. 1998. "Consciousness and Complexity". *Science,* No. 282: 1846–1851.

Tononi, G. 2001. “Information Measures for Conscious Experience”. *Archives Italiennes de Biologie,* No. 139:367–71.

Treiakov, K.V., Szleifer, I., Grzybowski, B.A. 2013. “The Rate of Energy Dissipation Determines the Probability of Nonequilibrium Assemblies”. *Angew. Chem. Int. Ed.* 52:10304-10308

Tribus, Myron. 1961. *Tribus Thermostatics and Thermodynamics Hardcover* New York, NY: Van Nostrand Reinhold.

Turing, A.M. 1950. “Computing machines and intelligence”. *Mind*, No. 59:459–76.

Tuyls, K., Lenaerts, T., Verbeeck, K., Maes, S., Manderick, B. 2002. “Towards a relation between learning agents and evolutionary dynamics“. *Proceedings of BNAIC.* 21-22.

Tuyls, Karl, Nowe, Ann, Lenaerts, Tom, Manderick, Bernard. 2004. “An Evolutionary Game Theoretic Perspective on Learning in Multi-Agent Systems”. *Synthese, Section Knowledge, Rationality and Action.* No. 139, Issue 2. 397-330.

Ulanowicz, Robert E., Goerner, Sally J., Lietaer, Bernard, Gomez, Rocio, Conrad, M. 1972. “Can there be a Theory of Fitness?” *International Journal of Neuroscience.* No. 3:125-134.

Ulanowicz R.E., Goerner, Sally J., Lietaer, B., Gomez, R. 2009. “Quantifying sustainability: Resilience, efficiency and the return of information theory”. *Ecological complexity*. No. 6:27-36. http://dx.https:doi.org/10.1016/j.ecocom.2008.10.005.

Ulanowicz, Robert E. 2016. “Process Ecology: Making Room for Creation.” *Sophia.* 1-24.

Valiant, L. 2013. *Probably Approximately Correct: Natures Algorithms for Learning and Prospering in a Complex World.* New York, NY: Basic Books.

Varela, Francisco. 1975. “A Calculus for Self-Reference”. *International Journal of General Systems*, No. 2: 5-24.

Varela, Francisco, Maturana, Humberto. 1987. *The Tree of Knowledge: The Biological Roots of Human Understanding*. Boston, MA: Shambhala Press.

Varela, F., Thompson, E. Rosch, E. 1991. *The Embodied Mind: Cognitive Science and Human Experience.* Cambridge, MA: MIT Press.

Verlinde, Erik P. 2011. “On the Origin of Gravity and the Laws of Newton”. *Journal of High Energy Physics,* 1104:029.

Verlinde, Erik P. 2017. “Emergent Gravity and the Dark Universe”. *SciPost Phys.2*. No. 016, https:doi: 10.21468/SciPostPhys.2.3.016.

Volkenstein, Mikhair V. 1994. *Physical Approaches to Biological Evolution.* Berlin, Heidelberg, Germany: Springer-Verlag von Neumann, J. 1958. *The Computer and the Brain.* New Haven, CT: Yale University Press.

von Neumann, John. 2018. *Mathematical Foundations of Quantum Mechanics.* New Edition. Nicholas A. Wheeler (ed.). Translated by Robert T. Beyer. Princeton, NJ: Princeton University Press.

Vopson, M.M.; Lepadatu, S. The second law of information dynamics. AIP Adv. **2022**, 12, 7.

Vopson, M.M. A Possible Information Entropic Law of Genetic Mutations. Appl. Sci. **2022**, 12, 6912.https://doi.org/10.3390/app12146912

Walker, S.I., Cisneros, L, Davies, P.C.W. 2012. *Evolutionary Transitions and Top-Down Causation*. No. arXiv preprint, arXiv: 1207.4808.

Walker, S.I., Davies, P.C.W. 2013. "The Algorithmic Origins of Life". *Journal of the Royal Society Interface,* No. 10(79).

Walker, S.I., Aguirre, A., Foster, B., Merali, Z. 2016. *Trick of Truth: The Mysterious Connection between Physics and Mathematics*. New York, NY: Springer Verlag

Walker S. I. 2016. "The descent of math". In A. Aguirre, B. Foster, and Z. Merali, editors. *Trick of Truth: The Mysterious Connection between Physics and Mathematics*. New York, NY: Springer.

Watanabe, Sumio. 2009. *Algebraic Geometry and Statistical Learning Theory*. Cambridge, MA: Cambridge University Press.

Watt, Ward. 1986. "Power and Efficiency as Indexes of Fitness". *Metabolic Organization American Naturalist.* No. 127.

Weaver, Warren, 1948. "Science and Complexity". *American Scientist*. 36 (4): 536–44. PMID 18882675.

Weber, B.H. 2010. "What is Live? Defining Life in the Context of Emergent Complexity". *Origins of Life and Evolution of Biospheres*. No. 40:221-229.

Weiner, N. 1948. *Cybernetics: Or Control and Communication in the Animal and the Machine.* Cambridge, MA: MIT Press.

Wheeler John A., Patton C. M. 1978. "Is Physics Legislated by Cosmogony?" *The Encyclopedia of Ignorance: Everything you ever wanted to know about the unknown,* Ronald Duncan and Miranda Weston-Smith (Eds), New York, NY: Pergamon Press, Pocket Books (a Wallaby book)

Wheeler, J.A. 1983. "On recognizing 'law without law', oersted medal response at the joint American Association of Physics Teachers meeting New York*". American Journal of Physics, No. 51*: (5):398{404.

Wheeler, John A. and Zurek, Wojciech H. 1983. *Quantum Theory and Measurement*. Princeton, NJ: Princeton University Press.

Wheeler, J. A. 1989. "Information, physics, quantum: the search for links". *Proceedings III International Symposium on Foundations of Quantum Mechanics*. Tokyo, p.354-368.

Whitehead, Alfred North, Russell, Bertrand. 1977. 1962. 1910. *Principia Mathematica, Cambridge Mathematical Library*. Cambridge, MA: Cambridge University Press. https:doi:CB09780511623585

Whitesides, G. 2012. *The Improbability of Life, Volume 1*. Cambridge, UK: Cambridge University Press.

Wicken J.S. 1987. *Evolution, Information and Thermodynamics: Extending the Darwinian Program*, New York, NY: Oxford University Press.

Williams, Garnett P. 1997. *Chaos Theory Tamed*. Washington, DC: Joseph Henry Press.

Wolchover N. 2014. "A New Physics Theory of Life: Jeremy England". *Quanta Magazine*.

Wolf, Fred Alan. 1989. *Taking the Quantum Leap: The New Physics for Nonscientists*, Revised Edition. New York, NY: Harper Perennial. ISBN 0-06-096310-7, ISBN 978-0-06-096310-1.

Wolfram, S. 2002. *A New Kind of Science*. Wolfram Media.

Wolfram S. The Second Law: Resolving the Mystery of the Second Law of Thermodynamics. 2023, Wolfram Media, Inc.

Xu L. and Li Y. 2010. "Emerging themes on information theory and Bayesian approach". *Frontiers in Electrical and Electronic Engineering China* 5(3): 237-240.

Yates, F.E. 1994. "Order and Complexity in Dynamical Systems: Homeodynamics as Generalize as Generalized Mechanics for Biology" *Mathematical and Computer Modelling*. No. 19:49-74.

Yockey, H.Y. 1992. *Information Theory and Molecular Biology*. Cambridge, UK: Cambridge University Press.

Ziegler, H. 1983. *An Introduction to Thermomechanics*. Amsterdam, The Netherlands: North-Holland.

Zurek, W.H. 1989. "Algorithmic randomness and physical entropy". *Physical Review*. No. A40:4731-4751.

International Criminal Justice Series

Volume 9

Series Information

The *International Criminal Justice Series* aims to create a platform for publications in the whole field of international criminal justice. It, therefore, deals with issues relating, among others, to:

- the work of international criminal courts and tribunals;
- transitional justice approaches in different countries;
- international anti-corruption and anti-money laundering initiatives;
- the history of international criminal law.

The series concentrates on themes pertinent to developing countries. It is peer-reviewed and seeks to publish high-quality works emanating from excellent scholars, in particular from African countries.

Editorial Office

Prof. Dr. Gerhard Werle
Humboldt-Universität zu Berlin
Faculty of Law
Unter den Linden 6,
10099 Berlin, Germany
gerhard.werle@rewi.hu-berlin.de
moritz.vormbaum@rewi.hu-berlin.de

More information about this series at http://www.springer.com/series/13470

Cassandra Steer

Translating Guilt

Identifying Leadership Liability for Mass Atrocity Crimes

ASSER PRESS

Springer

Cassandra Steer
Faculty of Law
McGill University
Montreal, QC
Canada

ISSN 2352-6718 ISSN 2352-6726 (electronic)
International Criminal Justice Series
ISBN 978-94-6265-170-8 ISBN 978-94-6265-171-5 (eBook)
DOI 10.1007/978-94-6265-171-5

Library of Congress Control Number: 2016958989

Published by T.M.C. ASSER PRESS, The Hague, The Netherlands www.asserpress.nl
Produced and distributed for T.M.C. ASSER PRESS by Springer-Verlag Berlin Heidelberg

Printed on acid-free paper

This Springer imprint is published by Springer Nature
The registered company is Springer Science+Business Media B.V.
The registered company address is: Van Godewijckstraat 30, 3311 GX Dordrecht, The Netherlands

For my parents (all of them). This labour of love is thanks to your endless encouragement to seek out the world and ask the hardest questions.

Acknowledgements

Above my desk I have a quote from Carl Jung, to which I refer often: 'The goal is important only as an idea. The essential thing is the opus which leads to the goal: *that* is the goal of a lifetime'.[1] For their contribution to this opus I am grateful to a number of people.

My mentor, Prof. Harmen van der Wilt, whose guidance was like a lighthouse, regardless of whether the skies were clear or dark and stormy. Steady, even, always truthful, ever inspiring, and truly focused. Thank you for your commitment, your insights, and for your generosity with your time.

I was supported a great deal in the comparative research I undertook in various countries by a number of generous people. In Germany at the Institut für Kriminalwissenschaften of the Georg August Universität in Göttingen I was welcomed and supported by Alexander Heinze, and am grateful also to Prof. Kai Ambos for his guidance. In the US at Cornell Law School Jens Ohlin provided hours of investigative debate and inspiring exchange of thoughts. I am indebted to him for his philosophical contributions and his all-round support. In Argentina at the Universidad Torcuato Di Tella in Buenos Aires I was welcomed by a warm community of intellectuals, and was supported in particular by Alejandro Chehtman, Hernán Gullco and Marcello Ferrante. I appreciate all their patience with my stumbling Spanish, and their extremely useful guidance on Argentine law. In Canada I was made to feel especially welcome in what I can only describe as an academic family at McGill Faculty of Law in Montreal. There Frédéric Mégret and Nandini Ramanujam took wonderful care of me as a guest of the Centre of Human Rights and Legal Pluralism. I am also grateful to the Hon. Patrick Healy, for his time, advice and candid views. And finally at the University of British Columbia in Vancouver I was treated to rich and conceptually essential debates with James Stewart, to whom I am grateful for his time, and his willingness to help me improve on my arguments in opposition to his.

[1] From *The Psychology of Transference*, para 400.

I am also grateful to the Fulbright Association of the Netherlands and to the University of British Columbia for the Mobility Research Award.

In this opus there has been a veritable team of wise and supportive people:

To the men of Cabero Stediki, Rory Kilmartin and Benjamin Dives, I am honoured to be the recipient of such a sophisticated listening and unbending support, and just as honoured to contribute to each of you as we ensure each other's success in our respective life goals.

To Alain Youell, I am blessed to have your wisdom and guidance in raising the level of growth and awareness in each of life's challenges to one of beauty and importance.

To my touchstone Bex Hronek, I am so blessed to have you by my side, and to know there is no question we cannot ask each other. To Jakob Hronek, I am grateful for your final touch of originality to this book.

To Kees Drabbe, I am indebted to you for the commitment with which you translated summaries, and for yours and Netty's omnipresent support.

To Kathryn Greenman, thank you for your ever insightful and incredibly constructive critique.

To the women of Stepping Out, the women's network at the Universiteit van Amsterdam Faculty of Law, thank you for being such an amazing network of inspiring, elegant, intelligent and powerful lawyers, academics, colleagues and friends.

To my parents and brothers, thank you for your endless belief in me, for your encouragement for me to follow my path on the other side of the world, even when this meant missing important moments in your lives. This book is dedicated to you, my family. My success is your success.

And finally to my fiancé, Sebastien, thank you for your patience, for always standing beside me, always encouraging me, and for tolerating the love affair I was having in the creation of this book, which took me away from you so many times.

Contents

Abbreviations

BGH	*Bundesgerichtshof* (German Federal Court of Justice)
BiH	State Court of Bosnia Herzegovina
CJM	*Codigo de Justicia Militar* (Argentine Military Code of Justice)
DRC	Democratic Republic of Congo
ECCC	Extraordinary Chambers in the Courts of Cambodia
ECHR	European Convention on the Protection of Human Rights and Fundamental Freedoms
IACHR	Inter-American Court of Human Rights
ICC	International Criminal Court
ICCPR	International Covenant on Civil and Political Rights
ICJ	International Court of Justice
ICL	International Criminal Law
ICRC	International Committee for the Red Cross
ICTR	International Criminal Tribunal for Rwanda
ICTY	International Criminal Tribunal for Former Yugoslavia
IHL	International Humanitarian Law
ILC	International Law Commission
IMT	International Military Tribunal
ISIS	Islamic State of Iraq and al-Sham
JCE	Joint Criminal Enterprise
LRA	Lord's Resistance Army (Uganda)
MCP	Model Penal Code (United States of America)
OTP	Office of the Prosecutor
PTC	Pre Trial Chamber
RUF	Revolutionary United Front (Sierra Leone)
SCSL	Special Court for Sierra Leone
StGB	*Strafgezetsbuch* (German criminal code)
STL	Special Tribunal for Lebanon
UN	United Nations
USA	United States of America
WCA	War Crimes Act 2000 (Canada)

Chapter 1
The Problem of Liability in International Criminal Law

> This trial represents mankind's desperate effort to apply the discipline of the law to statesmen who have used their powers of state to attack the foundations of the world's peace and to commit aggressions against the rights of their neighbors.
>
> — Chief Prosecutor Robert Jackson's opening statement
> Nuremberg, November 21, 1945

Although mass atrocity does not always take the same form as it did during the Second World War, this book is concerned with the continuation of humankind's 'desperate effort' to apply the discipline of the law to leaders of mass atrocity crimes. In recent years this desperate effort has produced what can only be described as confusion at the various international tribunals as to how to deal with individual leadership liability for collective crimes of atrocity. How to categorise the individual liability of these leaders, and to what extent they are truly responsible, is a matter of continued debate. As an illustration, in June 2016, former vice-president of the Democratic Republic of Congo Jean-Pierre Bemba was sentenced by the International Criminal Court (ICC) to 18 years imprisonment, due to his command responsibility for failing to prevent or punish the forces under his command and control for war crimes and crimes against humanity.[1] In March 2014, Germain Katanga was convicted by the ICC for 'contributing' to war crimes by a majority of the judges, and yet three of them wrote lengthy separate, partially concurring judgments, in which they disagreed with each other on how the final judgment dealt with Katanga's exact level of responsibility.[2] In April 2012, Charles Taylor, former president of Liberia, was convicted for 'aiding and abetting' in various international crimes, and sentenced to 40 years by the Special Court for Sierra Leone.[3] In March of the same year, Thomas Lubanga was sentenced by the ICC to a mere 14 years for his role as a 'principal' in conscripting child soldiers

[1] Bemba Sentencing Judgment 2016.

[2] Katanga Trial Judgment 2014.

[3] Taylor Judgment 2012.

C. Steer, *Translating Guilt*, International Criminal Justice Series 9,
DOI 10.1007/978-94-6265-171-5_1

in the Democratic Republic of Congo.[4] And back in 1998, the seminal case at the International Criminal Tribunal for Former Yugoslavia (ICTY) saw Dusko Tadić sentenced to 25 years imprisonment for his role as a 'principal' in a joint criminal enterprise.[5] While the facts of each case are different, still the disparate conclusions as to the grounds and form of individual liability to be applied, and the appropriate sentence, leave one questioning to what extent we can speak of a coherent body of international criminal law (ICL). This book will elucidate that the problem is one of translation.

Ever since the famous axiom spoken by Justice Jackson at the closing of the trial of the major war criminals at Nuremberg that 'crimes against international law are committed by men, not by abstract entities',[6] the starting point of ICL has been that individuals should be held liable for the acts committed through the collective. Leaders could no longer hide behind the collective notion of 'acts of State'. This major paradigmatic shift in international law meant that the individual was recognised as being a subject of international law for the first time, however the reasoning was based very much in concepts born out of public international law concepts, and not out of criminal law; this is one level of translation which has since proved problematic.

In order to capture as many participants as possible in the collective crimes committed within the Nazi regime, one form of liability was intended to apply to the leaders as well as all those involved further down the ranks, but in the final analysis there was no consensus as to what form of liability this should be. Since the 1990s and the explosion in international prosecutions for mass atrocity, international tribunals have had even greater problems: not all crimes of mass atrocity are committed within such an efficient and thoroughly documented system of bureaucracy as the Nazi regime, and the mixed legal background of judges, prosecutors and defence lawyers has led to seemingly unsolvable debates as to the proper approach.

These debates on modes of liability in ICL have become more and more complex over the last 20 years. The problem has become concentrated on the question whether to widen the net of liability and treat all those involved in mass atrocities equally liable—a unitary approach—or whether to differentiate between principals and assistants, and apply different modes of liability—a differentiated approach. There appears to be a policy trend to focus on 'those most responsible', however the question remains *who* is most responsible, and what form of liability should attach. In many situations currently under investigation at the ICC, for instance, suspects with authority over others have maintained a deliberate remoteness between themselves and the actual crimes committed, and even if they have exerted explicit authority, there are often disparate and complex hierarchies and groupings among physical perpetrators, leading to further uncertainty as to liability. Those at the top do not engage in the 'dirty work'; the question is whether they are liable for the fact that they have others do the dirty work for them.

---

[4] Lubanga Trial Judgment 2012.

[5] Tadić Appeals Judgment 1999.

[6] Nuremberg Judgment Closing Statement of the Prosecution 1945.

From pragmatic, moral and legal points of view, we must answer the question whether the intellectual authors of these crimes should be held responsible in the same way as those who drove the train full of victims being taken to concentration camps, or those who kidnapped, tortured or killed with their own hands. There is another level of translation evident here, namely from collective acts to individual liability. There is much disagreement as to whether there should be a normative distinction drawn between parties to crimes of mass atrocity, to what extent causation is relevant, and whether there is a danger of guilt by association. These disagreements can be drawn back to the differences between the predominant civil law and common law traditions.

Doctrinal questions of a criminal law nature require more attention than they have received thus far, such as whether liability is based on transferring *blame* from the collective to the individual, or rather based upon attribution of the *acts* of the many within a collective to the individual. Each of these approaches leads to conflicting answers to the question of responsibility: when attributing the acts of all to each individual in the collective, everyone within the collective could be seen to carry equal responsibility. On the other hand, when focusing on the question of individual blame for the collective result, the distribution of blame means some actors within the collective could be singled out as being more responsible than others, due to their particular role. The debates and lack of clarity as to which approach should prevail in ICL are a result of the differences in approach at the domestic level and, it will be argued throughout this book, the way in which these domestic notions are being translated to the international plane.

Due to the relative infancy of ICL and the fact that these crimes are committed in vastly different ways and by different means in each unique conflict situation, the makers of law within ICL are forced to be creative in their solutions. It is the complexity of this problem and the nature of the process of patchworking a new system together that form the focus of this book. There has not been a great deal of attention paid to theorizing the most appropriate system of liability in ICL,[7] instead the gap has been filled in the most efficient way possible, namely by drawing on domestic law from various jurisdictions, and translating this to the international context.

It is therefore necessary to consider the specific context of mass atrocity crimes, and to ask a functionalist comparative question[8]: not 'which domestic jurisdiction or legal tradition has the better modes of liability?', but rather 'which approach will reflect the collective nature of mass atrocity crimes and fulfill the aims of ICL in the best way possible?'

In this book, this enquiry is undertaken in three parts. Part I lays the foundations for being able to answer this functionalist question. It begins with the pragmatic, moral and legal considerations mentioned above, arguing a rationale for holding leaders criminally liable as "those most responsible" for systemic crimes of mass atrocity. It then looks at the way in which domestic criminal law notions of liability

[7] Fletcher 2011; Ohlin 2014; Vogel 2002; Robinson 2008; van der Wilt 2012.

[8] More will be said about the meaning of functionalism in Sect. 2.5.

act as sources of ICL. Given that the legal instruments such as treaties, statutes, resolutions and customary law have little to offer on these technical criminal law questions, decision makers have had little choice but to turn to domestic criminal law for inspiration, instruction and doctrinal development. The role that domestic law plays is more than merely interpretative guidance, or the ascertainment of general principles. Because the normative content of domestic criminal law is particular to the legal tradition, historical context and policy desires of the State in which it operates, it is difficult to speak of general principles or even customary law. Rather, the way in which domestic law notions are being borrowed from and drawn up into the international context can be seen as comparative law in action.

Part II of this book undertakes to demonstrate how a comparative law perspective on this very process of law development can shed light on why this takes place, how it takes place, and why there appears to be a clash of legal cultures emerging in case law and doctrine on the question of modes of liability. Part II enters a comparative study of modes of leadership liability in various domestic jurisdictions, and -more importantly- a consideration of the reasons that have played a part thus far in choices that have been made to translate certain notions from the domestic to the international context. Understanding this process can also aid in considering how we could be making this somewhat haphazard process more consistent, in order to come to more informed and deliberate choices. The assertions here are that decision-makers in the processes of ICL look to the jurisdictions with which they are most familiar, that there is insufficient attention paid to the policy factors which have led to the development of each domestic system of liability, and further that there is insufficient attention given to the question whether these policy factors are akin to those at the international level.

A comparative law perspective can offer a method to solve the clash of legal cultures that appears on the international plane, by offering the missing analysis needed to discover which system of liability is most appropriate to the specific context of ICL and why. By posing the question in this way, the assumption that one legal tradition should prevail as a matter of superiority can be avoided. This study offers arguments as to why there should in fact be a normative distinction drawn between participants in mass atrocity crimes, and on what grounds leaders should be held responsible for the crimes committed by their subordinates. A comparative analysis allows a more neutral formulation of a rationale of responsibility of leaders of system crimes, upon which the most appropriate modes of liability for mass atrocity can be formulated, regardless of the legal tradition from which they may stem.

The intention here is not to provide a detailed outline of all the modes of liability applicable to collective criminality in ICL. There are already many excellent extensive studies on this.[9] The intention is rather to look at the policy and doctrinal reasons for whether or not there should be a normative distinction drawn between parties to a crime, and how best to design a system of liability in ICL, in order to resolve the question of leadership liability for mass atrocity crimes. This focus means that in comparing the domestic and existing international modes of liability, the historical

[9] See for example Boas et al. 2007; Olásolo 2009; van Sliedregt 2012.

and policy context is given primary attention, and the specific context of mass atrocity is the axis around which such comparison takes place.

In Part III it will become apparent that in fact this book is not only about modes of liability. It is also about understanding ICL as a process of comparative law, and understanding comparative law both as a method and a methodology applicable to ICL. As a method, comparative law is the tool with which we can analyse the factual processes that have led to the development of various modes of liability on the domestic and international planes, and to untangle the debates on the terminology which hail from differing systems of liability. As a methodology, it provides a lens through which to understand the way in which ICL has developed to date, and a context within which law-makers and law-appliers can confidently make a selection from among the existing possible systems of liability, by identifying which one is most appropriate for the context of mass atrocity. If applied properly, such a methodology can justify the very act of making a selection as an inherent part of this process of ICL law development.

The argument to be made is not that there is a particular domestic approach that deserves to win out against the other, but rather that the process by which ICL develops has always been one of comparative law. In order to resolve the debates on modes of liability, use can be made of this process to better understand the factors at play and the goals which ICL aims to serve. The methodological requirements of comparative law can strengthen the justification of selecting which system fits the context of ICL best, and the subsequent choices of modes of liability. The conclusion drawn in the end is that it would appear that an objective approach, with a normatively differentiated system of liability, and more culpability for leaders of mass atrocity crimes, does just that.

The question has been posed by one scholar in a particularly poignant way: if you had to share the dinner table with one of them, who would you prefer to dine with; the mastermind or the executioner?[10] I will return to this question in the final chapter, however it gives nothing away to state here that the underlying assertion of this entire study is that there is something more understandable and more telling of the fallibility of the human spirit in the actions of an executioner who takes part in a collective where heinous acts are not only expected of him, but they have become the norm. And there is something more distasteful, more horrendous, more blameworthy about the 'armchair killer', the leader who is in a position to influence others such that they become her instruments, such that they become capable of atrocities which serve an ideology that the leader wishes served by her subordinates. When she is fully capable of intervening, of using her influence to minimise rather than maximise the atrocities acted out by others at her behest, there is something more disturbing about the evil expressed through her when she does the opposite, than the evil expressed through her subordinates. We are all capable of the kind of evil expressed by those subordinates under the right (or wrong) circumstances. One would hope not all of

---

[10] Gideon Yaffe, Professor of Law, Professor of Philosophy, and Professor of Psychology at Yale University, posed this question during a presentation at Yale Law School. I am grateful to James Stewart for presenting the question to me during discussions at the University of British Columbia.

us are capable of the other kind of evil. This author would therefore choose to dine with the executioner, with the human being who may have done the unforgivable, but not the unthinkable.

References

Boas G, Bischoff JL, Reid NL (2007) Forms of responsibility in international criminal law, vol 1, International Criminal Law Practitioner Series. Hart Publishing, Oxford

Fletcher GP (2011) New court, old dogmatik. J Int Crim Justice 9:179–190

Ohlin JD, (2014) Co-perpetration: German dogmatik or German invasion? In: Stahn C(ed) The law and practice of the International Criminal Court: a critical account of challenges and achievements. Oxford University Press, Oxford

Olásolo H (2009) The criminal responsibility of senior political and military leaders as principals to international crimes. Hart, Portland

Robinson D (2008) The identity crisis of international criminal law. J Int Crim Justice 21:925–963

van der Wilt H (2010) Why international criminal lawyers should read Mirjan Damaška. In: Stahn C, van den Herik L (eds) Future perspectives in international criminal justice. TMC Asser Press, The Hague

van Sliedregt E (2012) Individual criminal responsibility in international law. Oxford University Press, Oxford

Vogel J (2002) How to determine individual criminal responsibility systemic contexts: twelve models. Cahiers de Défense Social, pp 151–169

Cases

Bemba Sentencing Judgment (2016) The Prosecutor v. Jean-Pierre Bemba Gombo, Judgment, ICC-01/05-01/08

Katanga Trial Judgment (2014) The Prosecutor v. Germain Katanga, Trial judgment ICC-01/04-01/07

Lubanga Trial Judgment (2012) The Prosecutor v. Thomas Lubanga Dyilo ICC-01/01-01/06

Nuremberg Judgment (1945) Judgment of the International Military Tribunal [1945] International Military Tribunal at Nuremberg

Part I
Laying the Foundations

Chapter 2
Leadership Liability for Collective Crimes

Contents

The truism that crimes of mass atrocity are by definition collective may be one of the greater banes of criminal law lawyers attempting to solve the problem of liability. The paradigmatic commitment to individual guilt that forms the very basis of ICL brings with it problems that are unique to the context of mass atrocity. In conflict situations the moral universe has shifted, and ordinary people, who would not normally commit acts of violence, become capable of heinous acts on a grand scale. Such crimes are characterised by the use of State (or State-like) apparatuses by government or military officials and superiors to mobilise masses towards grave and large scale violence, but the distance between these superiors and the bloody acts committed by the hands of others makes it difficult to untangle questions of responsibility. The greatest challenge is how to accurately reflect the collective nature of these crimes, and at the same time identify who is truly to blame for the emergence of such a situation, and for the individual crimes which take place as a result of it: the problem of collective guilt means that 'where all are guilty, nobody is.'[1]

With the Nuremberg trials, the move from state to individual responsibility for transgressing the laws of armed conflict and human rights was based on the notion that only by holding individuals responsible could these laws be upheld. This notion was in order to pierce the veil of the State entity behind which leaders could otherwise hide. It was also based on the notion of personal autonomy of the individual; only a person is capable of a moral wrong, not an abstract entity such as the State. But this also means that an individual can only be held liable to the extent that she is

[1] Arendt 1987, p. 43.

C. Steer, *Translating Guilt*, International Criminal Justice Series 9,
DOI 10.1007/978-94-6265-171-5_2

responsible for the moral wrong.[2] And as a matter of logic, we can only be held responsible for that over which we have control, in the sense of moral agency.[3]

Many scholars have pointed out the difficulty of delineating this moral and legal responsibility in the context of mass atrocity, where the extraordinary has become the norm: there is an orthodoxy of hate, and violent, systemised crimes become acceptable among members of the collective.[4] Criminologists have contributed to understanding what one scholar terms *Makrokriminalität*:

> The individual crime is conditioned by a conflict in which the whole society is involved. Hence, it is embedded into certain developments and events at the macro level. In this respect, it is not deviant, but conform behaviour.[5]

In these situations, the moral climate has shifted, and it is often the case that if an individual were to refuse to take part in specific crimes, she would risk becoming the next victim, being seen as a threat to the ideology being enforced. This is known to have occurred under regimes such as Pol Pot in Cambodia, and the Junta in Argentina, where anyone considered to be a subversive was declared an enemy to be rooted out.[6] Thus, the moral autonomy of individual 'foot soldiers' is reduced. Perhaps they are not entirely exculpated for their evil deeds, but they cannot be said to be acting truly autonomously. Their identity becomes so caught up in the masses and in the violence perpetrated on others, that individuals have been known not only to commit horrendous acts against strangers, but even to denunciate friends, neighbours and family members as enemies in order to ensure their own survival.[7]

Sociological studies of mass atrocity demonstrate that violence only becomes systematic and widespread in this way if a central authority at the very least encourages it, and more often than not those at the height of power do more than this, explicitly mobilising subordinates to support their authority and ideology by extreme means.[8] Situational aspects to collective crimes include the imposition and utilization of ideologies such as nationalism, scapegoating, and utopianism;[9] obedience to authority under situations of authorization and routinisation of violence, and dehumanization of victims;[10] and de-individuation in large groups leading to conformity with group norms.[11]

---

[2] Nino 1996, p. 136; Mégret 2013, p. 93.

[3] Duff 2007, p. 58.

[4] See for example Drumbl 2005, p. 541; Tallgren 2002, p. 573; Robinson 2012, p. 134.

[5] H. Jäger, *Makrokriminalität: Studien zur Kriminologie kollektiver Gewalt*, (Surhkamp 1989) at 12, cited in Harrendorf 2014, p. 233.

[6] Sancinetti and Ferrante 1999, p. 23; Nino 1996, p. 58; Hinton 1998, pp. 95, 113; Tallgren 2002, p. 573.

[7] Arendt 1973, p. 397.

[8] Arendt 1973, p. 311; Semelin 2007, pp. 166–168.

[9] These are the attributes of ideologies as defined by A. Alvarez, 'Destructive Beliefs: Genocide and the Role of Ideology' in: Haverman and Smeulers 2008, 216; see also Harrendorf 2014, p. 235.

[10] H.C. Kelman and V.L. Hamilton, *Crimes of Obedience: Towards a Social Psychology of Authority and Responsibility*, (Yale University Press 1989) at 16–20, cited in: Harrendorf 2014, p. 240.

[11] Harrendorf 2014, p. 243.

The question of moral responsibility therefore shifts up the hierarchy to the leaders, and it is argued here that the legal responsibility should shift in weight accordingly. At the same time, the challenge remains how to reflect the collective nature of the crimes, so that in prosecuting an individual for specific acts, the bigger picture doesn't disappear into notions of individual actions and individual blame. Each specific crime takes place as part of an organised, ideological context, as a means to a greater end. It is therefore necessary to identify modes of liability that can reflect the responsibility of leaders not only for specific crimes committed by subordinates, but for their functional role in the creation and abuse of a system that condones and even encourages atrocities to take place, while at the same time respecting the limits of individual liability according to the principle of culpability.

As a first step towards explaining this, the question of moral agency within the collective will be discussed here. This represents the first translation of guilt to be dealt with in this book: the translation from the collective to the individual and to the role of leaders in particular. There is an intuition that is expressed in the policies of international tribunals that the leaders are the 'most responsible', however the rationale behind this intuition does not receive much attention. It is therefore necessary to consider on what basis individuals are responsible within a collective, and who among the collective may be more responsible for the actions ensuing, and therefore be held to account.

2.1 Translating from the Collective to the Individual

There is an assumption made at the heart of ICL that requires more attention: that the collective guilt belonging to a nation or society for crimes of mass atrocity can in some way be translated to individual responsibility. There is an intuition that certain individuals are most responsible for the collective, and at the first international tribunal in Nuremberg the blame was presumptively placed on the leaders of the regime. While there may be a logic to this, it is necessary to clarify the rationale behind it, since we cannot assume that just because someone is in a leadership position, they automatically carry greater responsibility for the crimes committed.

Some criminology scholarship points to personality types that are predisposed to committing crimes of violence which, if combined with situational aspects described above such as a dominant ideology of nationalism or utopianism, routinisation of violence and dehumanization of victims, can lead to an escalation of the kinds of crimes committed.[12] This typology can help to understand the driving motive that certain individuals may have in taking part in collective crimes of atrocity, and particularly in understanding the role of the opportunist, the fanatic and the criminal mastermind who make use of the predisposition of certain other individuals to encourage mass

---

[12] Alette Smeulers, 'Perpetrators of International Crimes: Towards a Typology', in: Haverman and Smeulers 2008; see also Harrendorf 2014, p. 244.

group behaviour.[13] In this way, even those whose predisposition is minimal, or only triggered by external threat, can become parties to extraordinarily violent behaviour which has become normal due to the fact that so many others already take part. However this understanding does not help to isolate who is more or less responsible, and why.

The assumption that collective guilt can be translated to individual responsibility involves two conceptual shifts. Firstly, from the collective to the individual in terms of guilt, and secondly from the collective to the individual in terms of responsibility. In domestic criminal law, these two shifts occur in respect of collective crimes in different ways, depending on the legal tradition within which the system plays out. The first shift from the collective to the individual is common to most (western) domestic criminal systems because we deal with individual guilt, and not with collective guilt.[14] There must therefore be some way to link the individual suspects' actions and intentions to the crime committed by a group. How this is done is reflected in the second shift, from collective to individual responsibility, which differs depending on the emphasis placed upon either the subjective intention of the individual, or upon the objectively measured contribution to the commission of the crime.

2.2 Individual Versus Collective Guilt

With respect to the first shift from collective to individual guilt, what is happening in these interactions can perhaps better be understood in the light of what American criminal law theorist George Fletcher has described as the war between Liberals and Romantics.[15] Romanticism is associated with a strong identity with the collective, where war and militarism become a source of inspiration for taking part in an ideology worth dying for, and for accepting a role within a hierarchy and part of the fighting collective.[16] Liberalism, on the other hand, is associated with principles of voluntary choice and individual responsibility, which dominate ICL due to their roots in western criminal law systems.

When it comes to the notion of collective or individual guilt, romantics are expansionist, arguing for collective guilt, and liberals are reductionist, arguing for individual guilt.[17] The conflict between focusing on the collective or the individual is

---

[13] Alette Smeulers, 'Perpetrators of International Crimes: Towards a Typology', in: Haverman and Smeulers 2008, p. 242.

[14] Mégret 2013, p. 86. An important exception is the notion of *qasāma* in many schools of Islamic law, whereby if a victim is found dead and there is no identifiable suspect, either the owner of the property or land on which the victim was found, or all the inhabitants of the quarter in which the victim is found can be liable to pay 'blood money' as a compensation to the victim's family members. This is predicated on the notion that the landowner or the community would be more likely to ensure security in their quarter of living if they know there is a risk they will be held liable in the event of a violent crime. See Peters 2005.

[15] Fletcher 2002.

[16] Fletcher 2002, p. 1501.

[17] Fletcher 2002, p. 1508.

a foundational feature of two different views of reality. For example, the collective notion of State responsibility is central in public international law, and we consider ICL to be a part of public international law, yet we struggle with the notion that entire bodies of people can be guilty for the crimes carried out by a few in the name of the collective. As former ICTY chief prosecutor Carla Del Ponte stated:

> all Serbs, all Muslims, and all Croats are not responsible for the crimes committed by a relatively small number of offenders … I do not intend to put the whole Serbian people on trial. On the contrary, I want to help Serbia turn the page and bring to justice those who, as individuals, are responsible.[18]

The key question is therefore whether the individual is the ultimate unit of action, or whether we, as individuals, are invariably implicated by the actions of the groups of which we are a part.[19] As already pointed out, in the context of mass atrocity the extraordinary becomes the norm, and violent, systemised crimes are accepted, condoned and perhaps even expected among members of the collective.[20] This deprives people of their second-order capacity to rein in their criminal impulses: the rational choice that an individual agent can make according to either moral impulses or impulses given by their physical surroundings.[21] If an individual chooses to follow the senses which would instruct violent crime over the moral principles which would counsel against it, under normal conditions in domestic criminal law, this would lead to full criminal liability. Guilt in this sense is personal. Yet when the surrounding norm has become one of violence, the ability to make this choice may be reduced. The romantic group identity takes over from the liberal individual identity. It has been suggested that a deindividuated state can actually be induced in people where the group becomes so large that there is increased anonymity and a diffusion of responsibility; the ability of an individual to evaluate the group norm decreases as the sense of self decreases.[22] As Hannah Arendt pointed out, the normality of atrocities in these circumstances is what is so terrifying, since the crimes are committed under circumstances that make it nigh impossible for the perpetrator to know that what is being done is wrong.[23]

In ICL, the liberalist construction of the individual as the central unit of action means that a number of selected individuals are to be blamed for systemic levels of violence.[24] However at the same time the basis upon which these individuals are selected is not always clear. The agents responsible for creating a climate of hate are not easy to identify; teachers, religious leaders, politicians, policies of the state, and bureaucrats enforcing a system of supportive laws are involved. Should ICL

---

[18] Press Statement by Prosecutor Carla Del Ponte on the Occasion of her Visit to Belgrade, The Hague, 30 January 2001, P.I.S./558-E.

[19] Fletcher 2002, p. 1504.

[20] See Arendt 1973, p. 314; Drumbl 2005, p. 541; Tallgren 2002.

[21] Fletcher 2002, p. 1543.

[22] For an insightful and succinct discussion of the scholarship on this phenomenon, see Harrendorf 2014, p. 243.

[23] Arendt 1964, p. 253.

[24] Drumbl 2005, p. 568.

select all of these individuals as culpable for the crimes that ensue? Or only some of them? If the latter, which ones, and based upon what criteria? Fletcher argues that in these situations, the collective guilt could (and should) be used to mitigate individual guilt, rather than placing the full weight of the collective guilt on the shoulders of one individual.[25] In the example of Adolf Eichmann, put on trial for his role towards executing 'The Final Solution' in Nazi Germany, Fletcher would argue that the collective guilt of the nation of Germany should have mitigated Eichmann's individual guilt. This was Eichmann's own argument, that he felt he was 'being made to pay for the glass that others have broken.'[26]

Nevertheless, the preference for the reductionist, liberalist approach over the expansionist, romantic approach in applying the regime of ICL is evident, and while collective guilt in a moral sense may be seen to exist, collective responsibility in a legal sense is rejected outright. As stated by Antonio Cassese in his capacity as former President of the ICTY:

> If responsibility for the appalling crimes perpetrated in the former Yugoslavia is not attributed to individuals, then whole ethnic and religious groups will be held accountable for these crimes and branded as criminal. In other words, collective responsibility — a primitive and archaic concept — will gain the upper hand; eventually whole groups will be held guilty of massacres, torture, rape, ethnic cleansing, the wanton destruction of cities and villages.[27]

The problem remains, however, how to identify those individuals responsible within the collective for crimes committed by the collective. There is a danger of collapsing criminal liability for a single crime amounting to *an act* of genocide or a war crime, with responsibility *for* the genocide or grander scale of war crimes.[28] The other side of the coin is the tendency to use modes of liability which absorb all individuals into the collective, holding them liable for all crimes committed by the collective while avoiding the difficulty of proving who actually tortured or killed in specific cases.[29] The doctrine of Joint Criminal Enterprise (JCE) has, for example, been used to include convictions where an individual had acted in the sphere of politics, leading to a criminalisation of political behaviour and a blurring of the line between collective action and individual liability.[30]

The terms 'guilt' and 'liability' are therefore to be used with caution, since the translation of moral guilt and collective guilt to individual criminal liability is not a direct one.[31] This also points to the important relationship between guilt and agency, which will be discussed next.

---

[25] Fletcher 2002, p. 1543.

[26] Osiel 2009, p. 20, citing Enrique Gimbernat Ordeig, Autor y Complíce en Derecho Penal (1966) at p. 187.

[27] Report of the President of the ICTY 1994, para 16.

[28] Mégret 2013, p. 109.

[29] Simpson 2007, p. 71.

[30] Haan 2005, p. 173.

[31] More attention will be paid to this in Sect. 5.3.6.3 where the definitions of terminology are clarified.

2.3 Individual Versus Collective Agency

Given the preference for individual guilt in the place of collective guilt in western criminal law and in ICL, there must be clear principles based upon which it can be determined when and to what extent an individual is responsible for crimes committed through a collective. Who is most to blame for the collective crimes? The key concept is that of agency. In domestic criminal law we consider group action to be of greater danger than individual action. We have specific crimes and often higher sentences for organised group actions and organised crimes. And we have modes of liability to deal with the problem of switching back and forth between expansionist and reductionist realities. In ICL, despite individualism at its core, we still believe that crimes of mass atrocity express the actions and the implicit guilt of entire groups of people, most typically of nations that are in conflict.[32] We therefore require a theory of agency that justifies shifting the responsibility for this group action to the individual as an agent within the collective.

Complicity itself, understood as participation in collective crime, and translated into different modes of liability, deals explicitly with this shift from individual to collective and back again. Criminal law theorist Christopher Kutz explains action in terms of generality, whereby the collective exists as an agent with its own intention, in the way we see corporations, a basketball team, or an orchestra.[33] At the same time Kutz opts for a reductionist approach, in that each individual's actions within the collective agency can be seen to be caused by the collective will. Individual members of a group intentionally do their part in promoting a joint outcome, or a joint activity.[34] A board member signs a paper on behalf of the corporation, a team member shoots a basket, a violinist plays her part of a symphony. Individuals act in this context with the intention that the group perform an act, and with the expectation that other members of the group will do their part.[35] But individual intentions and beliefs can still be ascribed to the individual, based on a functionalist approach, so that even if an individual might say 'the group made me do it', there is still some individual agency possible, since it is possible to interpret our actions as our own. This is especially relevant when we are not talking about an orchestra playing a symphony, but, for example, members of an air force collectively bombing a city, an action which involves weighing up the moral choices.

Kutz argues that participatory intention entails implication, in the sense that if an individual intentionally participates in a wrongful act, this would automatically entail individual responsibility for the collectively produced result.[36] He draws a descriptive distinction between participatory, inclusive accountability, based upon the relation between an individual's will and the resulting wrong or harm, which is a subjective approach, and direct, exclusive accountability, which is based more upon

[32] Fletcher 2002, p. 1512; Drumbl 2005, p. 567.

[33] Kutz 2000, pp. 68, 75.

[34] Kutz 2000, p. 69.

[35] Kutz 2000, p. 96.

[36] Kutz 2000, p. 146.

a causal relation to the individual's actions, and is therefore predicated on an objective approach. However Kutz maintains that normatively there can be no difference between these two forms of accountability and that both complicit participants and direct actors must be seen to be equally and jointly culpable for collective crimes.[37] The individual air force members must all be held culpable for the war crime of bombing a city. In this same way, any unintended consequences of the collective action that are foreseeable, including further or different criminal acts committed by other members of the collective, should be ascribed to the group and back again to all the rest of its individual members:

> [R]uined flowers are a foreseeable part of a project of picnicking, as a product of any group member's actions. Neither of us needed to expect that we would ruin flowers, but each ought ex post to acknowledge that it was a possible consequence of what we did together. And so it is reasonable to ascribe the mess to us, and to me inclusively.[38]

This approach would lead to the conclusion that there need be no normative distinction between participants in a crime. It is a subjective approach, focusing upon the intention of the individual within the collective, regardless of any difference between the role of, for example, the violinist and conductor of an orchestra. However there are arguments for making such a distinction when it comes to mass atrocity crimes in particular, and for opting for a more objective approach to the question of participation in collective actions.

2.4 Deliberative Structures and Those Most Responsible

Legal philosopher Jens Ohlin offers a further step in this analysis when he speaks of overlapping agents,[39] a notion which Kutz also discusses, but with different conclusions as to the distribution of responsibility. The problem with ascribing the collective will to a group, and speaking of group agency in the romantic sense, is that it interferes with the liberal notion of individual liability, which is central to the criminal law paradigm. Just as Kutz's theory shows, we are left with a continuous shift between group and individual, since the group intent is said to cause the individual action, which then gets attributed to the group, and finally in terms of criminal liability back to the individual again. Ohlin agrees with Fletcher that a full reduction of the group to the individual, in the liberalist ideal which Kutz follows, is an unsatisfactory conclusion.[40] However, where Fletcher would argue for mitigation of guilt and therefore of legal responsibility of the individual, Ohlin instead says that while a group can act with a certain collective rationality, the individual still retains individual agency, even though there is a submission of some individual reason to the group.[41] The

[37] Kutz 2000, pp. 147–154.

[38] Kutz 2000, p. 155.

[39] Ohlin 2007.

[40] Ohlin 2007, p. 173.

[41] Ohlin 2007, p. 181.

basketball team member submits some reason to the team, in that it is not simply her goal to score baskets individually, but also to do what is best for the team in order to win, even if she never shoots a single basket herself. The violinist plays his part of a symphony as an individual, for if he were to fail to play, or to play a different piece of music, it would be clear that he was not subsumed into the group. At the same time he has submitted some of his reason to the collective will of the orchestra to perform a certain symphony.

In this way the group and the individual are overlapping agents.[42] This means that while the group can still be said to be acting collectively, with a collective will, the individual can still be held responsible for his or her own actions, and possibly—though not always—for the actions of the group. The determining factor is the extent to which there was a possibility for the individual to be involved in the group deliberation, which led to the collective will. A person lower down in a hierarchy within a group would be less culpable, because they had surrendered their individual reason to the group deliberative process, whereas those on equal footing with each other, or higher up within a group, can be held responsible for the collective will and action.[43] The conductor of an orchestra is more responsible for the final performance of the symphony than each of the individual musicians, even though their collective participation is necessary. Those individuals who exercise more individual agency in the deliberative process of the collective are able to exercise more control over the actions taken by the collective, since they have more control over what is done, how it is done and why it is done. With greater agency and control comes more responsibility.

This notion of overlapping agency helps to explain the problem of individual responsibility for complex collective crimes within the bureaucracy of the State or a State-like organisation of insurgents. The romantic, expansionist ideal would ascribe full liability to the collective, meaning that individuals acting through the collective would be immune to scrutiny of their actions. Avoiding such immunity was precisely the reason why the Nuremberg trials shifted the paradigm of responsibility to the individuals who were blamed for committing crimes in the name of the State. On the other hand, the liberalist ideal would place full blame on the shoulders of individual members of the collective, which would risk scapegoating and, in the case of their removal from the organisation, would leave the same bureaucratic climate in place that allowed the crimes to take place, and would do nothing to ensure prevention of further criminal activity. A point must therefore be sought in between these two ideals, where the collective nature of the crimes is accurately reflected, while at the same time individual responsibility within the collective can be identified.

The problem remains of how to identify which individual members should shoulder this blame. The notion of overlapping agency allows for clarification of the collective rationality, and at the same time not all of the individual members are subsumed into the abstract of the collective. Those lower down in the organisation can be said to play a less determinative role, since they exercise less agency over the

[42] Ohlin 2007, p. 185.

[43] Ohlin 2007, pp. 196–7.

decision-making, and since their refusal to take part would do little to change the collective actions. If one person refuses, another willing individual could take their place. Those higher up have a different role in the deliberative structure of the collective, since they exercise more control, and they are less easily replaced by others. Should a leader interfere with the collective will, this could change the course of the atrocities taking place, thus when their interference leads to the commission of crimes, they have a higher degree of moral responsibility as individuals.[44]

The notion of overlapping agency helps to identify both the collective reason and the individual reasons that lead to the commission of collective crimes. This gives justification for punishment of certain, but not all, individuals within the collective. As Ohlin points out, the more horizontal the participation of individuals in the deliberative process of the collective, the more justification there is to consider all of the members equally culpable,[45] but where there is more of a discernible hierarchy, overlapping agency describes the dynamic of collective criminality in such a way that the focus is upon those who have more influence over the crimes.

2.5 Why the Leaders?

The complexities of crimes of mass atrocity are difficult to capture in prosecutions, and yet there is a commitment to continue to build a system of ICL that can accurately reflect not only the collective nature of these crimes, but also the individual liability of certain individuals within that collective. The question posed in this chapter has been how to identify those within the collective who carry the most blame for the atrocities that have taken place; who are those 'most responsible', to be held to account? The extraordinary nature of mass atrocity comes about because a collective has begun to accept these actions as normal behaviour—but it takes some individuals with specific political aims and sufficient influence over others to convince them to give up their individual reason to the group and use violent means to fulfil these ends.

Given that criminal liability is based upon moral agency, in that we can only be held responsible for that over which we have control, the answer must be that those who have the most freedom of agency and exercise the most control over the collective must be most responsible. This is especially so since responsibility is intrinsically linked with moral agency; we can be held responsible only for that in relation to which we have some effective agency, some form of control.[46] Although thoughts and intentions are an aspect of this agency, and must be taken into account in determining criminal liability, they are not in themselves sufficient. Criminal liability requires first and foremost responsibility in this sense of agency and control.[47] As criminal law theorist Antony Duff has put it:

---

[44] Ohlin 2007, p. 197.

[45] Ohlin 2007.

[46] Duff 2007, p. 58.

[47] Duff 2007, p. 37.

> We are criminally responsible as moral agents, since we are responsible for our failures to respond appropriately to the moral reasons with which criminal law is concerned: culpable responsibility for moral wrongdoing is responsibility as a moral agent.[48]

In selecting a model for modes of liability most appropriate to ICL, it is still important to maintain some way of reflecting the collective nature of these crimes. The translation to individual criminal liability may threaten to subsume the collective context if blame is placed on the shoulders of those who took part in the masses, without identifying those who are most responsible for affecting the moral universe in which the collective will is enacted. The challenge is identifying the difference between influential leaders and those individuals lower down the hierarchy who may have reduced agency when it comes to the decisions that lead to crimes being committed, and who sometimes find their own survival dependent on their participation. In seeking a system of liability most appropriate for the context of ICL, these deliberative structures should be taken into account.

Given the specific character of mass atrocity crimes as system crimes, the notion of overlapping agency offers a way to distinguish between individuals involved in these collectives, by identifying a higher level of blameworthiness for those who exercise control or influence over the deliberative process in a collective. 'Those most responsible' are therefore those with the most influence and control. The notion of overlapping agency provides a tool for disentangling the collectivity of mass atrocity while helping to delineate criminal liability.

Overlapping agency also reflects the sociological analysis of totalitarianism according to Hannah Arendt, whereby leaders make use of propaganda to fuel ideologies beyond limited political belief, such that something even more powerful than a mob mentality arises; a mass mentality begins to take over within which moral standards have shifted drastically, and any act that ensures the survival of the perpetrator at the cost of victims' livelihood or lives becomes both acceptable and expected.[49]

It would therefore seem there are moral grounds for identifying the leaders within a hierarchy as 'those most responsible' in the translation from the collective to the individual. These are general, justifying grounds for singling out the leaders of mass atrocity, providing the rationale for the intuition that has played consistently throughout attempts to prosecute crimes of mass atrocity. But it is also necessary to consider how to best reflect these grounds in terms of criminal liability. What system of liability will best reflect the reality of collectively committed atrocities and the role of leaders therein? One of the questions faced by the International Military Tribunal (IMT) at Nuremberg is still unsettled today, namely whether there should be a selection made, either pragmatic or normative, as to who specifically to prosecute and under what mode of liability. This has to do with a policy choice, and a prosecutorial strategy to ensure a system as effective as possible, to which the next chapter will turn attention.

---

48 Duff 2007, p. 47.

49 Arendt 1973, p. 307.

References

Arendt H (1964) Eichmann in Jerusalem: a report on the banality of evil, Penguin Press, New York
Arendt H (1973) The origins of totalitarianism. Houghton Mifflin Harcourt, Orlando, Florida
Arendt H (1987) Collective responsibility. In: Bernauer SJJW (ed) Amor Mundi: Explorations in the faith and thought of Hannah Arendt. Boston College Studies in Philosophy, Boston, pp 43–50
Drumbl MA (2005) Collective violence and individual punishment: the criminality of mass atrocity. Northwestern University Law Review, Winter p 539
Duff RA (2007) Answering for crime: responsibility and liability in the criminal law. Hart Publishing
Fletcher GP (2002) The Storrs lectures: liberals and romantics at war; the problem of collective guilt. Yale Law J 111:1499–1573
Haan V (2005) The development of the concept of joint criminal enterprise at the International Tribunal for the Former Yugoslavia. Int Crim Law Rev 5:167–201
Harrendorf S (2014) How can criminology contribute to an explanation of international crimes? J Int Crim Justice 12:231–252
Haverman R, Smeulers A (eds) (2008) Supranational criminology: towards a criminology of international crimes. Intersentia, Morstel, Belgium
Hinton AL (1998) Why did you kill? The Cambodian genocide and the dark face of honour. J Asian Stud 57:93–122
Kutz C (2000) Complicity: ethics and law for a collective age. Cambridge University Press, New York
Mégret F (2013) Les angles morts de la responsabilité pénale individuelle en droit international. Revue Interdisciplinaires d'Etudes Juridiques 71:83–136
Nino C (1996) Radical evil on trial. Yale University Press, New Haven
Ohlin JD (2007) Group think: the law of conspiracy and collective reason. J Crim Law Criminol 98(1):147–206
Osiel M (2009) Making sense of mass atrocity. Cambridge University Press, Cambridge, UK
Peters R (2005) Crime and punishment in Islamic law: theory and practice from the sixteenth to the twenty-first century. Cambridge University Press, Cambridge, UK
Report of the President of the ICTY (1994) Annual report of the International Tribunal for the Prosecution of Persons Responsible for Serious Violations of International Humanitarian Law Committed in the Territory of the Former Yugoslavia since 1991. Tech. Rep. UN Doc A/49/342, President of the ICTY
Robinson D (2012) A cosmopolitan liberal account of international criminal law. J Int Crim Justice 26(1):127
Sancinetti MA, Ferrante M (1999) El Derecho Penal en la Protección de los Derechos Humanos. Hammurabi, Buenos Aires
Semelin J (ed) (2007) Purify and destroy: the political uses of massacre and genocide. Hurst and Company, London
Simpson G (2007) Law, War and Crime. Oxford University Press, Oxford
Tallgren I (2002) The sensibility and sense of international criminal law. Eur J Int Law 13(3):561–595

Chapter 3
Putting the Leaders of Mass Atrocity on Trial

Contents

Even with a rationale for why the leaders of collectives are more responsible, as identified in the previous chapter, the question arises how to design a system of liability to reflect this. The demands made of the international criminal justice project are complex, in that it must be at once effective, just in terms of outcome and fair towards defendants. This creates a series of tensions which will be discussed in this chapter. The first is that the desire for an efficient system of prosecutions requires a clarity of goals, yet there are numerous and competing goals asserted by those with different interests, including victims, defendants, and transitional justice advocates. An attempt will therefore be made in Sect. 3.1 to identify an overarching goal against which efficacy can be measured, and within which the question can be answered whether, indeed, focusing on leadership liability is prudent.

A second series of tensions has to do with how to identify specific modes of liability that can best reflect the role of leaders within the collective, while also satisfying the requirements of a criminal law system that is both fair and just. The principles of legality and culpability are imperative limiting principles in any system of criminal liability, and will therefore be discussed in Sect. 3.2.

Finally, in Sect. 3.3, the reality must be faced that all of these tensions lead to the necessity to make certain choices in designing a system of liability and identifying

C. Steer, *Translating Guilt*, International Criminal Justice Series 9,
DOI 10.1007/978-94-6265-171-5_3

specific modes of liability within that system. In the end all of this comes down to being cognisant of the policy choices to be made, and of the impact of each of the choices on the system as a whole.

3.1 Efficacy and Symbolism: The Aspirations of International Criminal Law

Given the moral rationale for holding certain leaders within a collective accountable due to their deliberative capacity and ability to exercise control, a set of principles must be identified in order to guide in the selection of a system of liability as a whole. What we seek is a system that can effectively bring those most responsible to justice, but efficacy can only be measured against a set of goals, about which there is much debate. Due to the multitude of interests in the international criminal justice project and the resulting different legal paradigms, from human rights law, to International Humanitarian Law (IHL), criminal law, transitional justice and victims interests, there is also a multitude of asserted goals and aspirations. It would be impossible to meet all the demands put forward of ICL, especially as they are sometimes contradictory. In the 'emotional minefield' that follows mass atrocity it is important to move carefully when making choices as to how a system of liability should operate, and to 'resist the clamour of raw vengeance, especially if this implies over-reaction towards lesser culprits as scapegoats for the sake of quenching public feelings of discomfort.'[1]

While it may not be possible to reconcile all these tensions, an attempt will be made here to find alignment where possible and to see what these alignments suggest in terms of a system of liability. The competing legal paradigms and various interests will be discussed with the aim of reducing these to an over-arching goal. Against this, the question of efficacy can be measured. The symbolic, narrative aspiration that is particular to prosecutions of mass atrocity will be given due attention. It would seem that not only is some alignment possible among these tensions, but that such alignment can be instructive on how a system of liability can best reflect the role of deliberative leadership in atrocity crimes.

3.1.1 Competing Paradigms: Accountability as a Central Goal

There are so many stated aims and goals of international prosecutions, some of which are inherently conflicting, that it has perhaps become impossible for an essentially horizontal international system to fulfil them all. One way of envisioning these tensions is from the perspective of the different actors involved; the victims, the defendant, and the ever elusive 'international community'.

From the victims' perspective, there are goals such as truth-telling, which can be related to the need to confront the perpetrators, and the need for some kind of

---

[1] van der Wilt 2010b, p. 56.

closure and healing. Truth-telling is particularly important where part of the atrocities included disappearances and refusal to give information as to the whereabouts or fate of the victims, such as occurred in Cambodia under the Pol Pot regime, and in many Latin American countries under military regimes in the 1970s and 1980s. It can also be a necessary part of reconciliation and collective healing for victims to be able to tell their stories as part of the historical record. However because criminal trials are selective of factual evidence based on what is *legally* relevant, the victims' experiences are likely to disappear somewhat in the legal terminology.[2] History-writing might be easier in a system where investigation is in the hands of the prosecutor on behalf of the State, and the emphasis is on the investigative phase, such as is typical in an inquisitorial tradition,[3] but in a system where the emphasis is on the trial phase, and the judges are dependent on two competing versions of the truth presented by competing parties, such as is typical in an adversarial tradition, a broader mandate of history-writing will become more difficult.[4]

From the perspective of the defence, goals include the setting of international standards of fair trial, such as expediency, the right to counsel, equality of arms and the necessity of due notice when it comes to which modes of liability will be indicted.[5] These goals relate to the requirement of fairness, discussed below in Sect. 3.2. Such goals stand in opposition to the desire for truth-telling, if the number of witness statements reaches anything representative of the horrors experienced by an entire community or society in these kinds of trials, a factor which has already led to very lengthy trials. As well, such extensive evidence could even lead to prejudice on the part of judges against the defendants, and thus the history-writing mode of international trials clashes with the probative test of applying evidence only to the question of criminal liability.[6]

From the perspective of the courts, reprobation and stigmatisation feature highly in the discourse of international tribunals, as well as putting an end to impunity, the desire to create an accurate historical record of events, and to contribute to reconciliation and the restoration of peace.[7] The preferred targets of prosecution therefore tend to be those at the top, the architects and leaders: because of their high visibility and power, their convictions produce a dramatic effect and hold the greatest promise of contributing to the strengthening of accountability for gross human rights violations.[8] However this may create a tension with the perspective of victims who may harbour stronger retributive feelings towards immediate and recognisable physical perpetrators, than towards the more distant leaders. As well, in some cases the prosecution of these leaders may in fact go against the aim of contributing to the restoration of peace or rule of law. One example is the way in

---

[2] Damaška 2008, p. 336; Simpson 2007, p. 84.

[3] Damaška 1984, p. 160.

[4] Damaška 2008, p. 337.

[5] Robinson 2013, p. 155.

[6] Damaška 2008, p. 334; Simpson 2007, p. 82.

[7] S/Res 1315 2000, para 7; Swart 2008, pp. 100, 102.

[8] Damaška 2008, p. 335; de Vlaming 2010, p. 155.

which ICC investigations were seen to be blocking peace negotiations in Uganda in 2008, when Lord's Resistance Army (LRA) leader Joseph Kony refused to take part unless the ICC agreed to drop its investigations of crimes committed by him and other members of the LRA.[9]

From the perspective of the international community, there are many goals which may themselves be internally conflicting. For example there are goals which can be said to fall under a transitional justice rhetoric, such as reconciliation, maintenance of peace and security, delegitimizing the violent regime, establishing a rule of law and prevention of further crimes.[10] Further there are those goals which fall under a criminal justice ideal, such as strengthening accountability, ending impunity, widening the net of liability to include all those involved in mass atrocity while at the same time avoiding guilt by association. The pursuance of these latter goals may conflict with the former goals of reconciliation or the maintenance of peace and security, and yet it is difficult to identify which of these should prevail.

Of particular importance is the challenge of capturing the collective nature of mass atrocity crimes, while still correctly reflecting individual liability. There is a volatile tension between the symbolism of international trials and the risk of scapegoating individuals as exemplary for the collective. It could even be said in some circumstances that suppressing collective moral responsibility could lead to sympathy for the few individuals singled out for prosecution, galvanizing large segments of society against externally imposed justice, and defeating the goals of reconciliation, collective healing, truth-telling, and the requirements of fairness and justice.[11]

Managing the tensions between all of these over-abundant goals would be easier if there were some set of priorities, perhaps based on an understanding of the relative weights of competing goals.

Since punishment must be based upon individual blameworthiness, and not upon the expression of grief by an entire community or society, the victims-oriented thinking of the drafters of the Rome Statute has received some criticism.[12] A criminal trial is first and foremost the determination of individual liability and the justification of punishment. Whatever broader social trends the crime may grow out of, and whatever ripple effects the criminal trial might have, the legal focus of the trial is narrow: determining the past acts and the ultimate fate of the individual defendant.[13] The problem with a victim-focused teleology is that it conflates the general justifying aim of criminal law, that of retribution, with the question of whether it is justified to punish a particular individual for a particular crime, and how to qualify such a

[9] Siebert 2010, pp. 94–96.

[10] See S/Res 827/1993 1993, para 7: 'Believing that the establishment of an international tribunal and the prosecution of persons responsible for ...violations of international humanitarian law will contribute to ensuring that such violations are halted and effectively redressed.'; Danner and Martinez 2005, p. 90.

[11] Damaška 2008, p. 333 (footnote 6).

[12] See for example Fletcher 2007, p. 325; Zahar and Sluiter 2008, p. 78; Robinson 2013, p. 142.

[13] Danner and Martinez 2005, p. 87; Vasiliev 2014, p. 215.

conviction.[14] This is exactly why there is so much debate on the modes of liability and the question whether leadership liability should be in any way differentiated. The pursuit of the general justifying aim must still be qualified by restraints such as the principle of culpability.[15]

A victim-oriented approach may be more suitable for a strictly human rights venue, where the identity of the perpetrator is less important as long as the victim receives some redress. Victims may be disappointed by the capacity of a criminal trial to provide a full opportunity for having their stories heard, for public mourning, for retribution, reconciliation, or for reparation. Not only is the selectivity of situations and defendants a problem in this respect, since not all atrocities will be brought to trial, but the increased role of victims at the ICC has brought many rights of the defendant under even greater strain.[16] Given that there are other avenues possible for victims to receive restorative justice, such as compensation schemes, civil litigation, and various other mechanisms, many of which have a lower burden of proof than criminal trials and are therefore more accessible to victims, it may be that these concerns should not rank above others in trying to identify a central goal of international prosecutions.[17]

The greatest tension is between the defendant's interests on the one hand, and on the other hand the aspirations from the perspective of the courts' didactic aims and from the perspective of the international community to create a historical record for posterity's sake. One way to resolve this tension is to see the narrative and didactic functions as institutional rather than procedural goals.[18] Procedural goals focus on the determination of the guilt or innocence of the defendant and should not be overridden with wider, idealistic aspirations which are better placed as institutional goals. At the same time, institutional goals cannot be swept aside when so much of the public legitimacy and perceived efficacy of the international criminal justice project depends upon them. Thus while efficacy surely must be partly measured against the extent to which trials result in procedurally fair outcomes, the overall externally perceived efficacy of the institution of international criminal justice can best be measured according to a centralising goal which takes into account as many of these tensions as possible.

The attempt to articulate a central mission for ICL that aligns many of these tensions may be ambitious, but it is submitted here that it is possible to take into account the didactic role of international prosecutions, the need to respect defendant's rights, and the ideology behind the goals asserted by those concerned for transitional justice, without giving priority to victims at the cost of other concerns, and without asking a criminal trial to be more than it is intended to be. The distinction drawn between institutional and procedural goals is important in this respect: the articulation of a central goal is not intended to give an answer to the design or reform of the trial in a procedural sense,[19] but rather to reduce the confusion and tensions caused by

---

[14] Robinson 2013, p. 142; Hart 1970, p. 8.

[15] Robinson 2013, p. 142 as will be discussed below in Sect. 3.2.

[16] Damaška 2008, p. 342; Vasiliev 2014, p. 182.

[17] Damaška 2008, p. 343; Vasiliev 2014, p. 182.

[18] Vasiliev 2014, p. 216.

[19] Vasiliev 2014, p. 202.

the competing interests discussed thus far. For this purpose, comparativist Mirjan Damaška has asserted that a central goal could be formulated as: strengthening the public sense of accountability for massive human rights breaches.[20] This is compelling as an overarching goal, since it would satisfy the requirement of fairness and justice by ensuring *accountability* in the form of criminal liability and not just collective blame, as well as being measurable in terms of effectiveness by helping to focus on *who* should be held accountable. This could be said to give priority to the expressive function of international trials as a central institutional goal, without denying the procedural goals and requirements discussed above.[21] This serves as an important guiding principle in the design of liability as a whole.

Some objections to such an expressive goal should not be overlooked, such as the problem of selectivity. The question of who is the recipient of a didactic outcome of international trials is of importance. It may be that a decision fits with the expectations of an international society, but is disappointing to the community that suffered the atrocities. It is therefore the task of participants in the processes of ICL to acknowledge the problem of selectivity, and especially for judges to make clear in their decisions when ICL deviates from local laws, why it does so. While this may raise discomfort in respect of addressing the position of those who suffered the very atrocities being prosecuted, it must be recognised that criminal prosecutions make up only one component of the transitional justice toolbox, which also includes truth commissions, amnesties, reparations, and civil proceedings seeking monetary damages.[22]

Since ICL tribunals cannot satisfy all those with an interest in the proceedings, they must first and foremost ensure a verdict that is well founded on the evidence as a procedural goal,[23] but they must do so based on a system of liability that is consistent with this central institutional goal of strengthening the public sense of accountability for mass atrocity crimes—a goal which finds some agreement across all of these tensions. The question is how to strengthen this sense of accountability across cultures, legal systems and conflicts that differ greatly. Reducing the number of different approaches to liability will fulfill this goal better than a disparate and fragmented approach.

3.1.2 Symbolism: History Writing and Truth-Telling

The central goal of increasing the public sense of accountability for mass atrocity has a particularly expressive value, as well as a normative one when it comes to a system of liability. This has to do with the particular symbolic value placed on prosecutions of mass atrocity, echoed in some of the asserted goals discussed above. There is a shared notion that ICL is an expression of global justice: that it is both multilateral, in the sense that it brings States together, and cosmopolitan, in the

20 Damaška 2008, p. 347.

21 Ohlin 2013, p. 63.

22 Danner and Martinez 2005, p. 91; Ohlin 2013, p. 76.

23 Vasiliev 2014, p. 207.

sense that it brings different legal systems together and strives to bring humanity together in a common response to atrocities.[24] The symbolic nature of international prosecutions may also lie in the very rite of the trial, a public event that serves a purpose, perhaps even a collective funeral of sorrow for victims.[25] As the ICTY Trial Chamber said in *Erdemović*: 'thwarting impunity even to a limited extent would contribute to appeasement and give the chance to the people who were sorely afflicted to mourn those among them that had been unjustly killed.'[26]

The tradition created at the IMT at Nuremberg was that the international prosecution of mass atrocity crimes was partly to lay down the facts for posterity, a historical record so that the victims' stories are heard and so that international society may never forget the atrocities that took place.[27] In this way, trials act as a kind of remembering, which may counter the forgetting that occurs with the grant of amnesties to leaders; such amnesties ensure impunity and a 'collective amnesia' as to the extent and facts of the atrocities, and the names of the victims.[28] The narrative function plays a particularly important role where the atrocities included mass disappearances, such as occurred in Argentina and many other Latin American countries under military regimes. The Inter-American Court of Human Rights (IACHR) has recognised a 'right to truth' in response to these disappearances and to the refusal to release information about the fate of the victims—a right that is considered to belong to victims, family members and society as a whole.[29]

Similarly, the failure to complete trials could be said to amount to a failure to complete history, with the risk of forgetting it. One former president of the ICTY Theodore Meron, stated that the tribunal's 'work in bringing justice and contributing to reconciliation in the region will go unfinished, and the record will remain incomplete, so long as the most prominent indictees remain at large.'[30] The untimely death of Slobodan Milosović before the conclusion of his trial left a great sense of dissatisfaction in this sense; no outcome was apparently as bad as, or worse than, an undesirable outcome.[31]

This history-writing has a didactic aspect to it as well. It has been asserted by many authors following Hannah Arendt, that the weight of evidence presented in the Eichmann trial about the existence of the Holocaust itself was more important than his individual conviction.[32] The District Court in Jerusalem itself spoke about the 'great educational values implicit in the very holding of a trial.'[33] Similarly, the

---

[24] Simpson 2007, p. 36.

[25] Tallgren 2002, p. 593.

[26] Erdemović, Sentencing Judgment 1996, para 65.

[27] Arendt 1964, p. 252.

[28] Simpson 2007, p. 81.

[29] Based on Articles 1(1), 8, 13 and 25 ACHR 1969. For examples, see Velasquez Rodriguez 1988, para 166; Barrios Altos 1985, para 45.

[30] ICTY Press Release, 11 July 2005, CVO/MOW/988e.

[31] Beaumont 2006.

[32] Arendt 1964, p. 16; Simpson 2007, p. 80; Osiel 2009, p. 148.

[33] Arendt 1964, p. 19.

trial of former Gestapo leader Klaus Barbie in France in 1987 has been described as 'pedagogical'[34] and a reminder specifically to the people of France of the need to know more about history.[35] And the trial of Saddam Hussein in Iraq in 2005, for all its flaws, was intended partly to be didactic in terms of the rule of law, including the training of Iraqi judges and lawyers in international criminal law standards.[36]

There are risks associated with the history-writing aspiration of ICL, such as the fact that history-writing can never be objective or neutral, and the historical record created by an international court may be designed to support one particular ideology, and discredit another—the familiar critique of victor's justice.[37] Such prosecutions may also allow a process of 'forgetting' certain parts of history, because these trials are by their nature selective.[38] And there is the problem of incongruity between technical legal language, into which the facts and the victims' experiences may disappear, as may the factual nature of the 'radical evil' that is on trial.[39] Furthermore, many scholars question whether a criminal court is the appropriate venue for establishing an accurate historical record, and whether judges can act as historians, given the selection of legally relevant information for the determination of a defendant's guilt or innocence.[40] The length and expanse of the trial against Milosević is a case in point, with charges covering a period of 8 years and generating a transcript of approximately forty-nine thousand pages.[41]

And there is of course the risk that such symbolic emphasis leads to exemplary prosecutions, and the scapegoating of individuals who are deemed to carry the collective guilt of all those involved on the side of the perpetrators. Given that mass atrocity crimes are highly politicised, the choices as to which atrocities to bring to court, who to prosecute, and how, are all themselves highly politicised choices.[42] The greatest tension apparent is therefore between the very notion of individual criminal lability, and the collective nature of these crimes. It is for this reason that the debate on modes of liability has recently become even more prevalent, as the ICC searches for the preferred interpretation under its statute and the need to capture the complex collective nature of mass atrocity, while protecting the principle of individual culpability, which will be discussed below in the following section.

While a criminal trial cannot satisfy all the aspirations expressed as goals, the central expressive goal identified above, that of increasing the public sense of accountability for mass atrocity, goes a long way to encompassing many of them, including the desire for a truthful record of events. It is asserted here that each conviction has

---

34 Simpson 2007, p. 80, citing Alice Kaplan in: Alain Finkielkraut, 'Remembering in Vain', p xvi.

35 Barbie 1988; The Guardian 1987.

36 Simpson 2007, p. 81; Heller 2007, p. 262.

37 Simpson 2007, p. 88.

38 Simpson 2007, p. 102; Cryer 2001, p. 31.

39 Arendt 1964, p. 271; Simpson 2007, pp. 83–84.

40 Nikolić 2003, para 122, stating that 'this Tribunal is not a final arbiter of historical facts. That is for historians.' See also Damaška 2008, p. 338; Tallgren 2002, p. 593.

41 Transcript of Record at p. 49,191, 14 March 2006, Prosecutor v. Milosević, ICTY-IT-02–54.

42 Drumbl 2005a, p. 581; Simpson 2007, p. 13; Cryer 2001, p. 31.

a story to tell, not only for the victims, but also with respect to the defendants. The conviction of leaders such as Milosević or Karadzić should express the roles they played in the full picture of events, and should not amount to the same conviction as that attached to Tadić and others lower down the hierarchy. To delineate between them and express the responsibility of the leaders as something more than the participation of the 'smaller fish' is not only an expression of historical fact that satisfies the victims' interests and the interests of transitional justice, it is also fairer to a defendant such as Tadić in terms of what goes down in history. Identifying modes of leadership liability is important in this respect. The symbolic nature of identifying those who are responsible for atrocities, aims to express the norms and values of an international society and somehow offer bereavement to the victims and survivors of such atrocities; many would agree that law can still be part of the process of closure for victims, and the creation of 'responsible memory'.[43]

The symbolic effect of prosecuting individuals, and especially the leaders who plan and instigate mass atrocity, may reach even further. Under the traditional justifications for punishment in domestic criminal law theory, three main categories stand out; retribution, consequentialism (including general deterrence, special deterrence and rehabilitation, all of which focus on social protection),[44] and norm expression. All three categories appear in various discussions as to the *raison d etre* of international prosecutions, and there is often an overlap between them.[45] As mentioned above, the ad-hoc tribunals were established by the Security Council under its Chapter VII powers to maintain peace and security, and had the intention to 'contribute to the process of national reconciliation and to the restoration and maintenance of peace.'[46] Some authors believe that the punishment of individuals in international fora has a considerable impact on the possibility of reconciliation and even on deterrence of future crimes.[47] The link between general prevention and the stabilisation or expression of norms is also very strong, and in fact it may be the expressionist function which features more heavily even where the language of deterrence is used.[48] Others see international prosecutions as a tool in negotiations, or a message to precede or accompany military intervention in a situation of grave human rights abuses.[49]

In any case there is some consensus that the intention is to disseminate information on common values, and to contribute in some way to the process of transitional justice, by holding those responsible for injustices up to public and judicial scrutiny.

Criminal trials for mass atrocity crimes may not be the perfect response, but one thing they certainly do is promote public deliberation, which can promote a coun-

---

43 Simpson 2007, p. 88.

44 Hart 1970, p. 12; Fletcher 2000, pp. 414–415.

45 Erdemović, Sentencing Judgment 1996, para 58; Delalić Trial Judgment 1998, para 1234; Krajišnik Trial Judgment 2006, para 1134; Duch, Appeal Judgment 2012, para 357; Heller 2011, p. 601; Werle and Jessberger 2014, p. 37.

46 S/Res 955/1994 1994, para 1.

47 Akhavan 2001; Cassese 2004; Meron 1998, p. 463.

48 Werle and Jessberger 2014, p. 37.

49 Tallgren 2002, p. 592.

teraction of the authoritarian tendencies which led to the weakening of democratic systems and mass atrocities. At the same time the disclosure of truth through the trials can encourage a collective consciousness and self-reflection regarding the norm of violence that arose.[50] This means that while the physical perpetrators— with whom victims will have had more personal contact— also deserve punishment, the distinction between these individuals and those who were instrumental in the authoritarian creation of the collective norm of violence is also of great importance to the narrative that arises, and to the sense of accountability for what occurred. Whether this can lead to reconciliation at a collective level is uncertain, but it can at least contribute to the expression of the norm that impunity for such crimes is no longer acceptable, and that the leaders of such system criminality play a vitally different role than the multiple actors involved either as physical perpetrators or on the periphery.

The success of fulfilling the expressionist function is predicated on accurate history writing,[51] and requires sufficient clarity in terms of what exactly is being expressed.[52] For this reason, the modes of liability applied in convictions for all those involved in mass atrocity crimes must be consistent and clear with respect to the actual role played by each individual accused. Any mode of liability that is opaque, or for which the doctrine is too fluid, cannot serve this function satisfactorily. Moreover, the differentiation between participants in a moral and legal sense would aid in expressing the facts in the convictions, and expressing the moral opprobrium of the actions of those who have more control over the circumstances.

3.1.3 Leadership Liability as a Prosecutorial Policy

If we take this central goal of strengthening the public sense of accountability for massive human rights breaches as the yardstick against which to measure the efficacy of prosecutions, the question arises how to increase this sense of accountability most effectively.

The policy of the Nuremberg trials had been to capture as many of those involved in the mass atrocity crimes as possible. At first the idea was to prosecute those at the top, and the derivative liability for those lower down in the echelons was expected to 'trickle down', making the evidentiary burden easier in terms of individual involvement, as they moved down the hierarchy of defendants.[53] However it soon became clear what an impossible task this was. Instead the IMT at Nuremberg focused only on the leaders of the Nazi regime, and some domestic military tribunals carried the work further, under Control Council Law No. 10.[54]

[50] Nino 1996, p. 147.

[51] van der Wilt 2010b, p. 55.

[52] Cryer 2009, p. 395.

[53] Ryan 2007, p. 62.

[54] Allied Control Council 1945.

However the selection of defendants at both Nuremberg and Tokyo was also influenced a great deal by the post-war policies of the allies. In particular the US had a specific de-nazification policy in mind, which meant that the top and mid-level leaders were the most important targets for prosecution. The negotiations between the Allies eventually led to a list of defendants based most importantly upon their functions within the Nazi regime.[55] At the same time there was a desire to build Japan back up as an economic ally, and the hope was that the Japanese emperor would play a key role in this, and so despite the desire of many of the Allies and the chief prosecutor at the Tokyo Tribunal to prosecute the emperor as the most responsible, it was decided to protect him from the dock.[56] In the end this selectivity led to some of the stronger criticisms against the outcome of the Tokyo tribunal, as summarised by Judge Röling: '[the] Tokyo judgment has suffered in Japan precisely because many people were not convinced that some of the statesmen found guilty were actually among those most responsible …'[57]

There was a similar political sensitivity with respect to the early prosecutions at the ICTY, though this time not under the influence of governments, but rather of the UN and the attempted peace process that was underway.[58] The tensions in respect of which goal to serve, and therefore which individuals to prosecute, were evident in the mixed aspirations that the tribunal should restore peace and the rule of law, establish the truth, provide victims with a sense of retribution, and not threaten peace negotiations with leaders such as Milosević.[59] The starting point under the mandate of the UN was to hold '*all* persons who participate in the planning, preparation or execution of serious violations of international humanitarian law in the former Yugoslavia' individually responsible.[60] However the impossibility of this task soon became apparent and over the years there was a shift towards using investigations against the lower level perpetrators instrumentally as a way of building evidence for the cases against the mid and higher level leaders.[61] Over time the prosecutorial strategy shifted to target 'those individuals who are most culpable for the atrocities committed'[62] and the goal was made explicit to move up the chain of command to find those most responsible.[63]

Despite this prosecutorial policy to focus on 'those most responsible', the system of liability favoured by prosecutors and judges at the ICTY has made no such distinction.[64] At the same time, there were difficulties faced in securing the individuals who

---

[55] de Vlaming 2010, pp. 101, 103.

[56] Boister and Cryer 2008, p. 66; de Vlaming 2010, p. 105.

[57] Judge Röling, cited in Hosoya et al 1986, p. 18.

[58] de Vlaming 2010, p. 113.

[59] Report of the Secretary General on the ICTY 1993, para 54; de Vlaming 2010, p. 117.

[60] Report of the Secretary General on the ICTY 1993, para 54. Emphasis added.

[61] Meron et al. 2002, p. 445.

[62] Office of the Prosecutor Charging and Indictment Guidelines, undated, cited in de Vlaming 2010, p. 136.

[63] Macdonald 1998, para 130.

[64] For a full discussion of the system of liability applied at the ICTY, see Sect. 8.11.4.

had held positions of political or military importance, as Serbia refused to co-operate with the ICTY in tracking them down and surrendering them to the court. The former Prosecutor turned to a political tactic herself, asking the EU to refuse membership to these new independent States unless they co-operated fully and ensured capture of the indicted individuals.[65] In 2000 the ICTY released a report stating its focus would shift to 'charges against high-level accused or against lower accused for particularly heinous crimes'.[66] And in 2004 the Security Council issued a resolution regarding its concern for the difficulty of the ad-hoc tribunals to fulfil their Completion Strategies, and urged them both to 'to ensure that any such [new] indictments concentrate on the most senior leaders suspected of being most responsible for crimes within the jurisdiction' of each tribunal.[67]

The policy of tribunals such as the Special Court for Sierra Leone (SCSL), Extraordinary Criminal Chambers of Cambodia (ECCC) and the ICC has from the outset been to focus on 'those most responsible' for the systematic crimes at hand. Article 1(1) of the Statute of the SCSL states: 'The Special Court shall ... have the power to prosecute persons who bear the greatest responsibility for serious violations for international humanitarian law and Sierraleonian law'.[68] The prosecutions at the ECCC have focused upon the leaders within the Pol Pot regime, with a clear mandate to focus on those 'most responsible'.[69] Similarly, although the ICC can investigate anyone involved in mass atrocity crimes and has various modes of liability available in the Rome Statute, the attention has been upon ending impunity for those most responsible for such crimes.

It should be emphasised that it is not the intention here to argue that *only* the leaders should be prosecuted, leaving others involved in mass atrocity crimes unpunished—especially since this would often mean impunity for the physical perpetrators. Rather the argument being made is that there are policy reasons for focusing on the leaders as the most responsible actors in orchestrating entire groups of individuals towards committing heinous crimes, since they are the actors who have the most influence over creating societal norms of violence within which individuals become perpetrators, and holding these leaders to account can satisfy the central goal identified earlier in this chapter. These policy reasons are central to the design of a system of liability, and to being able to ensure such a system is effective in achieving its goals. This was exactly the policy choice made in Argentina, when the first democratic government following the military Junta regime struggled with how to design a prosecution strategy during a delicate shift towards re-establishing a rule of law; their solution was to base the first stages of prosecution on the deliberative capacity that leaders

---

65 de Vlaming 2010, p. 145.

66 ICTY Report on NATO Bombing 2000, Chap. 90; Tallgren 2002, p. 572.

67 S/Res 1534 2004.

68 See Article 1(1) Statute of the SCSL 2000.

69 Article 1, UN ECCC Agreement 2003. See also UN Doc. A/RES/57/228B (13 May 2003)(approving draft ECCC Agreement).

in the top and mid levels of the regime had over the crimes, since these were the individuals most responsible, and whose prosecution would achieve the most.[70]

The effectiveness of convicting the 'foot soldiers' in these scenarios as equally liable is limited, precisely because these individuals at the bottom of the pyramid have less influence over the culture of hatred and violence, and less ability to act outside of it. If we are to have an effective system of international criminal justice, there must be a system of liability which allows the law to link the facts to the leaders in a meaningful way, even if the leaders did not themselves physically commit the crimes. As former chief prosecutor at the ICTY, Carla Del Ponte, has put it:

> [F]unctional responsibility may often be a good indicator of possible criminal responsibility for the crimes committed. Without attributing any systematic or group responsibility, and with respect for the presumption of innocence, it remains important to evaluate the possible participation of the persons who were occupying the highest political or military positions … Far too often, the crimes falling within the jurisdiction of the ICTY simply cannot be committed without the active complicity, if not the enthusiastic participation, of these structures, or at least without the participation of key state officials.[71]

Granted, a prosecutorial policy to focus on the leaders as a matter of efficacy does not necessarily mean that there need to be specific modes of leadership liability to differentiate between the leaders and others who are in some way complicit in atrocity crimes. It would be possible to consider leaders as accomplices in the same way as anyone lower down the hierarchy, regardless of their actual contribution to a specific crime. The question, therefore, is how to strive towards a system of liability that is not only effective in this policy sense, but at the same time also best reflects the collective nature of these crimes, as well as the moral responsibility of those who exercise more influence and control over the collective due to their functional role and deliberative capacity. The challenge is to link these leaders to the crimes in such a way that the other extreme is avoided, whereby leaders may be held strictly or vicariously liable purely due to their role, without sufficient causation and requirements of fault (mental elements such as knowledge or constructed knowledge). These requirements temper a system that aims to be effective, but which also must be fair with respect to defendant's rights and just with respect to the outcome of convictions.

Therefore, from a system of liability that identifies leaders as those most responsible, and a prosecutorial strategy that reflects this in terms of efficacy, the next question is how to translate this into specific modes of liability.

---

70 Nino 1996, p. 67.

71 Del Ponte 2004, p. 519.

3.2 Fairness and Justice: The Requirements of International Criminal Law

One of the challenges in identifying a coherent system of liability is that ICL is being built relatively rapidly, when compared with mature domestic systems which are embedded in entire legal traditions. While efficacy is an important aspiration of ICL, the incorporation of fairness and justice as core limiting principles into the design or selection of a system of liability is of utmost importance, particularly given the volatility of public prosecutions of crimes which are of an inherently political nature.

This gives rise to another set of tensions, since even with such key notions as fairness and justice we encounter a problem of translation. In lay terms, these two words may be considered to be more or less synonymous, however as legal terms of art, they are not self-evident. With regards to any violent crimes, and particularly crimes of mass atrocity, it would be reasonable to say that victims desire justice to be done, whereas defendants and their lawyers are concerned with the fairness of the trial.

The two may lead to contradictory outcomes. A fair trial does not necessarily reach a just result, in terms of acquitting the innocent and convicting the guilty, since taking care of the interests of the defendant may sometimes lead to exclusion of evidence and even acquittal for procedural or technical reasons.[72] One example is the acquittal of John Demjanjuk in Israel in 1993; although the evidence pointed to his involvement in the indicted crimes at Treblenka concentration camp during the Second World War, his identity as the indicted 'Ivan the Terrible' was not proven.[73] And yet the need to protect defendants' interests must not be underestimated, when one considers the political nature of international prosecutions, and the need to ensure that the procedure is legitimate and not merely a 'show trial'.[74] Concerns surrounding the scapegoating of a defendant for an entire society's grievances were raised in response to the trial of Adolf Eichmann in Jerusalem in 1960,[75] and similar criticisms of the nature of the trial of Saddam Hussein were raised more recently in 2007.[76]

The principles of legality and culpability play an imperative role in protecting the defendant's interests and placing limits on criminal liability. They will each be given due attention here, as well as the place of truth-finding in a criminal trial, in order to help answer what the requirements are for a system of leadership liability that is both just and fair.

---

[72] Fletcher 2000, p. 135; [para 43–44] Delalic Evidence.

[73] Demjanjuk 1993. Demjanjuk was, however, later convicted of aiding and abetting in crimes committed at another concentration camp in Sobribor, for his role as a camp guard. Demjanjuk Judgment 2011. (This judgment has not been published.) However Demjanjuk died before his appeal was decided, see Weigend 2014, p. 261.

[74] Vasiliev 2014, p. 334.

[75] Fletcher 2002, p. 1543; Nino 1996, p. 15.

[76] Peterson 2007; Tallgren 2002; Simpson 2007, p. 15.

3.2.1 Fairness and the Principle of Legality

The requirement of a fair trial when it comes to prosecutions for mass atrocity is particularly important given the political nature both of the atrocities and of the trials. The selectivity behind the prosecution of war crimes and crimes against humanity are hard to deny. For example, until recently gender was not recognised to play a specific role when it comes to vindicating victims,[77] and there are still criticisms that race plays a certain role in the selection of situations and defendants against whom to investigate. In particular the fact that the ICC has focused almost exclusively on prosecutions in Africa to the exclusion of other situations of concern in Europe, or of allegations of responsibility of leaders in the US with respect to crimes committed in Iraq and Afghanistan, has led to the critique that the ICC operates as a kind of neo-colonial instrument of the dominant powers, while ensuring impunity for their own leaders.[78] At various trials, prominent defendants such as former Japanese Prime Minister Hideki Tojo, Saddam Hussein and Slobodan Milosević have rejected the legitimacy of the tribunals and of the trial due to this political selectivity.[79]

At the same time, the fact that there is an inherent political nature to mass atrocity prosecutions is not necessarily a bad thing. Such trials aim partly to delegitimize the violent regimes under which mass atrocity took place, and to establish a rule of law that brings an end to immunity for political leaders and to impunity for any kind of leader within these regimes. It is the pursuit of a political end through jurisprudential means.[80] In this sense, fairness is rightly considered one of the imperatives in measuring the legitimacy of international prosecutions.[81]

The inherently collective nature of these crimes, and the complex ways in which individuals are manipulated or driven towards violence on a grand scale, mean that it may be tempting to blame the leaders for everything the collective enacted. This is why the clarification of modes of liability applicable to leaders is of utmost importance. One of the recurring criticisms of the lack of clarity is from the defence perspective, that it is impossible to prepare a defence to all the possible modes of liability without knowing which ones will be indicted by the prosecution, and how the judges will interpret them.[82]

In this respect, the principle of legality plays an important role. The rights of the defendants must be respected if there is to be any sense of a fair trial, and the

---

[77] See for example the ground-breaking decision to include acts of sexual violence and rape as acts of genocide, Akayesu 1998, para 731. See further Cryer et al. 2016, pp. 251, 287.

[78] See e.g. Cryer 2005, p. 202; Simpson 2007, p. 18; Heller 2010, p. 227.

[79] Nino 1996, p. 13; Peterson 2007, p. 258; Rockwood 2007, p. 126; Simpson 2007, p. 15.

[80] Simpson 2007, p. 24.

[81] See for instance former ICTY Chief Prosecutor R. Goldstone, 'Address Before the Supreme Court of the United States', 1996 CEELI Leadership Award Dinner, 2 October 1996, where he stated: 'Whether there are convictions or whether there are acquittals will not be the yardstick. The measure is going to be the fairness of the proceedings.' See also Simpson 1997, p. 30; Mégret 2010, p. 50; Robinson 2012, p. 145; Vasiliev 2015, pp. 77–78.

[82] Boas 2010, p. 502.

concern for consistency and predicability must be respected. The principle of legality, expressed as *nullum crimen, nulla poena sine lege*, is aimed at preventing arbitrary and discriminatory enforcement of criminal law, and encompasses the notions of fair notice, *lex certa*, and the prohibition on retroactivity.[83] In all the international instruments in which legality is named, including the Rome Statute, the focus is on these protections and therefore on the clear and proscriptive codification of the crimes, or at the very least, recognition of the crimes under international law, including the Nuremberg Principles.[84]

When it comes to modes of liability, the application of the principle of legality is perhaps more controversial than with respect to the definition of the crimes themselves, particularly as the definitions of the latter have crystallised and been codified fairly consistently over time since the Nuremberg trials. The question is whether the definition of modes of liability are subject to the same strict requirements. While some authors assert that they are,[85] arguably this is not the case. Modes of liability are necessarily open norms, which cannot be defined in terms as concrete as the definitions of crimes.

As H.L.A. Hart has pointed out in his ‘Concept of Law’, not only is there an inherent limit to the guidance that general language can provide, but when it comes to legal rules, as opposed to other sorts of communications, this limit and the generality of language is unavoidable.[86] Because of our inability to anticipate all the possible combinations of circumstance, there is a relative indeterminacy about legal rules, which he calls the ‘open texture’ of norms. Uncertainty is ‘the price to be paid for the use of general classifying terms in any form of communication concerning matters of fact.’[87] Because it is not only impossible but also undesirable to try and formulate legal rules to such detail that their application would always be settled in advance, this open texture means that legal judgments are conducive to casuistry.[88]

Casuistry can be described as the comparison of concrete situations and cases in order to decide whether they are governed by the same norm or legal principle.[89] Casuistry brings with it the contingency of legal concepts to particular circumstances; legal decisions will be modified by new circumstances, and the norms and principles that have yet to be tested are contextually bound to social, political and historical changes.[90] Modes of liability in domestic criminal law jurisdictions have always

---

[83] Boot 2001, p. 125.

[84] Article 11 of the Universal Declaration of Human Rights; ICCPR 1966, Article 15(2); ECHR 1950, Article 7; Articles 8 and 9 deal with the requirements of fair trial and proscription of ex post facto laws respectively. ACHR 1969; African Charter on Human Rights 1981, Article 7; Statute of the ICC 1998, Article 22; Boot 2001, pp. 160, 175.

[85] Vogel 2002, p. 160.

[86] Hart 1972, pp. 123–124.

[87] Hart 1972, p. 125.

[88] Hart 1972, p. 125; van der Wilt 2008, p. 264.

[89] van der Wilt 2008, p. 265.

[90] van der Wilt 2008, p. 266.

developed in response to social, policy and historical needs.[91] The translation of these notions from the domestic to the international does not escape such casuistry; international crimes are most certainly not context neutral, and the various responses in terms of modes of liability at the international level have been anything but uniform and universal. Indeed, standards in ICL have shifted over time, reflecting political and social changes and shifting moral perceptions, as well as shifts that follow whichever particular domestic approach is most influential in an international tribunal at any given moment. The fact that ICL has no centralised system of legislation, and that so much is left up to case law to decide, renders it particularly 'open textured', and renders casuistry the most important method by which the normative details are developed.

With respect to the principle of legality, the requirement for the elements of the crimes is that they are defined in certain terms, however because modes of responsibility are necessarily couched in more general terms, they cannot be strictly bound by a *lex certa* requirement.[92] As Hart explained, the more open a norm is, the more difficult it is to attain certainty, and the more futile it is to require it.[93]

During the drafting of the Rome Statute, there were concerns that too much judicial discretion in determining the applicable law would be a breach of the principle of *nullum crimen*.[94] However this is difficult to substantiate, given that these sorts of norms are by definition open, and *must* be determined by way of casuistry.

That is not to say that it is a free-for-all for a prosecutor or even for judges to determine which modes of liability are applicable in a given case without due notice to the defendant. The defence must be able to prepare arguments specifically against each charge, including against each mode of liability. This is part of the recent criticism against the broad interpretation given by the ICC of Regulation 55, that the mode of liability in the charge can be changed any time after the confirmation hearings, to such an extent that even during the trial the defence may essentially be expected to be prepared to argue against any and all of the modes of liability in the Statute.[95]

---

[91] The way in which this has taken place in the common law and civil law traditions is the focus of Part II of this book.

[92] van der Wilt 2010a, p. 217; van der Wilt 2008, p. 260, citing also George Fletcher (Basic Concepts of Criminal Law, Oxford, 1998) p. 107.

[93] Hart 1972, p. 125.

[94] These concerns were expressed during the drafting of the Rome Statute, Preparatory Commission Report 1996, p. 105; de Guzman 2008, pp. 702–703. However this attempt to restrict judicial discretion was not so much out of concern for the fairness for defendants, as for the desire to maintain a certain control over the formation of the law for States and State interests. Pellet 2002, p. 1052; Cryer 2009, p. 392; Heller 2011, p. 598; Sliedregt 2012, p. 13.

[95] Separate Opinon of Judge Christine van den Wyngaert, Katanga Trial Judgment 2014. For a critical discussion see also Jakobs 2014.

3.2.2 Fairness and the Principle of Culpability

Despite the unique context of mass atrocity crimes that ICL is faced with, in which the moral universe has shifted and the moral agency of individuals may be limited, the principle of culpability is still a fundamental tenet which cannot be ignored or reduced if criminal prosecutions are to have any value at all. Thus, while Mark Drumbl has argued that in the light of collective and organic violence, the recourse some international tribunals have given to 'generous—and at times somewhat vicarious—liability theories becomes eminently understandable',[96] the concern is that such developments lead to liability without (sufficient) culpability. In seeking the most appropriate system of liability for the context of ICL, it is therefore important that the collectivity of the crimes do not lead to the eradication of individual agency altogether. Thus, although the phenomenon of system criminality and mass atrocity present challenges to our understanding of conformity, 'deviant' behaviour, agency and moral choice, at the same time ICL must maintain the principle of culpability precisely because it focuses on the individual: 'once criminal law is employed, the individual is the *unit of punishment*. Once we decide to punish and stigmatize individuals for crimes, we are obliged to determine what we are punishing them *for*.'[97]

The principle of culpability acts as a limit on the reach of criminal liability, and is based on the notion of individual autonomy and rational agency, discussed above. Each individual should be held responsible for her own behaviour, but only to the extent that she could have been expected to do otherwise under the circumstances.[98] It goes beyond the scope of this study to consider the debates on determinism and free will that this statement leads to, and the defences of excuse that may arise.[99] Of importance here, with respect to modes of liability, is that we are responsible in so far as we are capable of responding to relevant reasons to act or not to act in a certain way, in other words, in so far as we are rational agents.[100] This means that strict liability is unacceptable, and that when it comes to collective criminality, the question must be asked to what extent an individual is capable of making a rational choice, and to what extent she is (morally) responsible for the outcome of her actions.

In ICL the principle of culpability has been recognized in the case law,[101] and can be read into the Rome Statute as being applicable as a general principle.[102] Deter-

---

[96] Drumbl 2005b, p. 1309.

[97] Robinson 2012, p. 140, emphasis in original.

[98] Hart 1970, p. 39; Ashworth 2009, pp. 23–24.

[99] See for example Hart 1970, p. 33 and further; Duff 2007, p. 106; Ashworth 2009, p. 196 and further.

[100] Duff 2007, p. 39.

[101] Nuremberg Judgment 1945, p. 251; Tadić Appeals Judgment 1999, para 185; Nikolić 2003, para 140; Brđanin Trial Judgment 2004, para 1090; Kordić Trial Judgment 2001, para 1075; See further Manacorda 2009, p 288.

[102] Statute of the ICC 1998, Article 21(1)(c).

mination of a sentence at the ICC must also reflect the culpability of the convicted individual.[103] Furthermore, as the ICTY Trial Chamber said in *Tadić*:

> The basic assumption must be that in international law as much as in national systems, the foundation of criminal responsibility is the principle of personal culpability: nobody may be held criminally responsible for acts or transactions in which he has not personally engaged or in some other way participated (nulla poena sine culpa).[104]

It is therefore important, in order to maintain a system that fulfils the requirement of fairness, to consider carefully how far liability extends when it comes to acts committed by others. It is true both in the domestic context and in the context of mass atrocity that collective action could be said to expand agency, since aims can be attained which cannot be attained alone. But, as discussed in Sect. 2.5, in the context of mass atrocity it is also (paradoxically) true that it reduces agency, since moral choice in situations of social pressure, propaganda and demands of authority is reduced such that an individual cannot be held responsible for everything attained by the collective.[105] An example of this reduced agency is the case of Erdemović before the ICTY, where the defendant had been threatened with his own life when he expressed resistance to taking part in the execution of hundreds of boys and men in Srebrenica.[106] To hold him liable for all the killings committed by his fellow soldiers is to go beyond the reach of his moral culpability.

Over-expansive modes of liability which threaten to convict individuals for guilt by association should therefore be avoided if the principle of culpability is to be respected.[107] This is why the notions of overlapping agency and deliberative structures have been suggested above as a way of identifying which individuals have retained their individual agency, and can be held more responsible for their role in the collective, and which individuals have reduced agency and can therefore not be held liable for the acts of others.

When it comes to collective criminality, the principle of culpability also requires that there is some correspondence between the required mental states for the crime, and the mental state of the accused; there must be sufficient knowledge and intent in relation to the conduct, such that we may find the accused 'personally reproachable'.[108]

As the Security Council report on the ICTY has put it:

> Because of the wide discretion granted to international prosecutors …because of the novelty of the international forum, because of the political nature of many of the prosecutions, and because of the lack of consensus about the meaning of some of the crimes in [ICL], it is

[103] Rules of Procedure and Evidence of the ICC 2002, Rule 145(1)(b).

[104] Tadić Appeals Judgment 1999, para 185.

[105] Robinson 2012, p. 141.

[106] Erdemović, Sentencing Judgment 1996.

[107] Danner and Martinez 2005, p. 137; Osiel 2005, p. 1772.

[108] Delalić Trial Judgment 1998, para 424; Robinson 2013, p. 118. The principle of correspondence plays an important role in domestic criminal law, and the Canadian Supreme Court has paid particular attention to how this limits complicity liability when it comes to serious crimes such as murder and war crimes. For a full discussion, see further Sect. 6.4. See also Ashworth 2009, p. 76.

especially important that international criminal judges protect defendants through careful attention to the culpability principle and similar doctrines that seek to ensure that defendants are convicted for their own conduct and not for the violent trauma experienced by entire nations.[109]

3.2.3 *Justice for Victims and for Defendants*

The desire for fairness for the defendant and the desire for justice on the part of victims or society may come into direct conflict with each other in terms of the procedural outcome of a criminal trial.[110] It may be the case that protection of the defendant's rights and of the standards of international trials lead to an acquittal or otherwise unsatisfying outcome in the eyes of victims. On the other hand there is the risk, as noted above, that the collective nature of these crimes of mass atrocity could lead to the desire to hold certain individuals accountable as exemplary, as a means to an end.

This is the common charge that the prosecution of international crimes is an exercise in 'victor's justice', or of justice politicised. The international trials following the Second World War are a clear case in point: indeed, Telford Taylor made clear in his stoical justification of the Nuremberg trials that this was a case of the victors determining the law, in order to accomplish 'not a legal thing, but a political thing'.[111] Even at the conclusion of the trials, there was uneasiness as to the extent of this victor's justice. Similarly, Judge Röling said of the Tokyo trials that the Japanese were being tried for political crimes.[112] It was in fact painfully ironic that on the very day the Allies signed the London Charter to establish the IMT, the US dropped its second atomic bomb, on Nagasaki, immediately killing at least 70,000 civilians.[113] As one minister in Churchill's war cabinet said, it if were a crime, then each of the Allies was equally as guilty as the German Nazi party.[114]

Legal historian Gerry Simpson has described the difficulty faced by ICL in this sense as being a 'negotiation between international law's universalist impulse (sometimes represented by the prosecution of those who commit "crimes against humanity") and its imperialist instinct (mostly found in the repression of "enemies of mankind").'[115] While the aspiration is to provide justice to humanity by bringing certain individuals to justice with criminal sanction, the depiction of certain indi-

[109] Report of the Secretary General on the ICTY 1993, para 34.

[110] Fletcher 2000, p. 135.

[111] Telford Taylor, An Approach to the Preparation of the Prosecution of Axis Criminality, (June, 1945). Cited in Ryan 2007, p. 70.

[112] Simpson 2007, p. 15.

[113] Simpson 2007, p. 16. In response to this, George Bernard Shaw wrote that with the dropping of the atomic bomb on Japan, none of the allies was in a position to hang Göring. George Bernard Shaw, Letter, Daily Express, 10 September 1946, cited in Cesarani 2006, p. 37.

[114] Cesarani 2006, p. 37.

[115] Simpson 2012, p. 118.

viduals as enemies of humankind is highly selective, and the trials that have taken place historically are also highly selective. No high powered western European or American has been brought before any of the modern international tribunals. Where they have been prosecuted domestically, there is a history of impunity, such as the acquittal of General Medina for the Mai-Lai massacre in Vietnam, and the failure to prosecute officers directly above and below him in the chain of command, following the elaborate cover-up that unravelled in the course of the investigations.[116]

In a similar vein, the hegemony of the most powerful western States has been noted by several scholars with respect to jurisdiction, definition of crimes and of modes of liability in ICL in general, and especially when it comes to the ICC.[117] In this sense Robert Cryer has drawn a distinction between 'safe' and 'unsafe' tribunals: the creators of these tribunals have taken a wider view of liability (and therefore a narrower one of defences) when they do not perceive a risk of their own nationals falling under the jurisdiction of a given tribunal, and conversely require more definitional precision when there is a possibility that their own nationals may fall under the jurisdiction of a tribunal.[118] During the drafting of the Nuremberg Charter, there was a great deal of controversy about the inclusion of conspiracy as a form of liability, but there was little attention paid to other forms of complicity.[119] The detail was left up to the judges, because the drafters knew they themselves were not subject to prosecution; a greater degree of judicial discretion was therefore 'safe'. There was more attention paid to the notion of complicity with the drafting of the Statute of the ICTY, but most of the work was still left to the judges to determine—arguably because, once again, those involved in the drafting of the Statute were not themselves subject to the jurisdiction of the tribunal.[120] With the drafting of the Rome Statute, principles of liability were set out in detail for the first time.[121] Given that this was the first time that all the States involved in the drafting knew that their nationals could fall under the jurisdiction of the court, and thus the ICC could be considered to be 'unsafe', the attention paid to defining and codifying the principles of liability was far greater than before.[122]

Thus while it may not be in the self-interest of powerful States to ensure a system of ICL that is as impartial and just as possible with respect to the selection of situations and defendants and the extent of liability, it is in the interest of defendants, victims, international society, and the success of the international criminal justice project as a whole. Any conviction that is considered to be exemplary rather than factually accurate, because it is overtly selective or goes beyond the princi-

---

[116] Rockwood 2007, pp. 120–126.

[117] Cryer 2001, p. 4; McCormack 1997, pp. 123, 683; Simpson 2012, p. 123; Tallgren 2002, pp. 594–595.

[118] Cryer 2005, p. 233; Cryer 2001, p. 4.

[119] Cryer 2001, pp. 20–21.

[120] Cryer 2001, p. 30.

[121] Cassese 2008, p. 153; Cryer 2001, p. 21.

[122] Cryer 2001, p. 20.

ple of culpability, undermines the project of prosecuting mass atrocity because it is institutionally unjust.[123]

The factual accuracy of a trial outcome, however, may depend in part on the style of prosecution that is informed by different legal traditions, and the role that 'truth-finding' plays in each tradition.

3.2.4 Justice and Procedural Versus Substantive Truth

The differences in approach to collective and leadership liability in the civil and common law traditions are at the root of the debates in ICL, and will be given full attention in Part II of this book. However there are some nuances in the emphasis placed on the very role of modes of liability in the outcome of a trial which also differ in these legal traditions, and which deserve some enquiry here. While it may be an over-simplification to say that the civil law tradition is characterised by an inquisitorial or investigative criminal trial, and that the common law tradition is characterised by an adversarial trial, some aspects of these divisions can explain the different degrees of attention paid to forms of complicity in each of the traditions.

A predominantly inquisitorial trial has a slightly different task underpinning it than an adversarial trial, namely the ascertainment of the truth through inquiry on the part of the judge.[124] This has been described as the search for an 'objective', 'substantive' or 'ontological truth', whereby fact-finding is assigned to court officials.[125] In an adversarial system, fact-finding is up to the parties, and two competing versions of the truth are presented to the court, from which the decision-makers (the lay jury as well as judges) must choose. Thus the task of an adversarial trial is to come to a conclusion that resolves the competition between parties in the fairest way possible, but not necessarily to ascertain the objective truth.

Due to the competition of different versions of the truth in an adversarial system, there are many procedural rules designed to protect certain 'collateral values', such as exclusionary rules which do not allow the admittance of evidence procured without a warrant, or procured by pressuring the defendant, since these things would breach the defendant's rights.[126] Such exclusionary rules are part of the design of the system, and serve different interests, such as equality of arms, which may be considered more important than the establishment of an objective truth in an adversarial trial. The exclusion of evidence could be said to result in the finding of a 'procedural truth' which in fact diverges from the substantive truth rather than establishing it.[127] This may amount to an acquittal due to procedural rules, even though the defendant was guilty of having committed the crime, or to a change in the qualification of the crime.

[123] Vasiliev 2014, p. 340.

[124] Safferling 2003, p. 217.

[125] Summers 1999, p. 498; Grande 2008, p. 146.

[126] Summers 1999, p. 500; Weigend 2003, p. 168.

[127] Summers 1999, p. 498; Weigend 2003, p. 160.

In an inquisitorial system there is less perceived need to have such extensive evidentiary rules, since the investigation into the facts is in the hands of an impartial judge rather than in the hands of the parties.[128] As well, because in an inquisitorial system the pre-trial investigation allows more evidence to be considered than only that which is presented in court on the day of trial, the risk of a party-informed 'procedural truth' is perhaps less.[129] Of course, no model is perfect, and an inquisitorial system cannot claim to establish the truth definitively.

Over time the two models have converged somewhat, and in particular those jurisdictions with a typically inquisitorial trial have had to allow for protection of defendant and witness rights, such as the right to remain silent, exclusion of illegally procured evidence, and greater participation of the defence and the victims, such that a divergent 'procedural truth' may still emerge.[130] The point, however, is that the central focus of each tradition differs; although the establishment of the guilt or innocence of the defendant is the core outcome, the different roles assigned to judge, prosecutor and defendant in each tradition leads to a different emphasis on the kind of truth that is sought in the end.

From this comparative view on the place of 'truth' in the two differing traditions, it could be said that justice is considered to be served in an inquisitorial tradition when something approximating the objective truth has been established, and when protection of witnesses and victims has been taken into account. On the other hand, in the adversarial tradition justice instead is considered to equate with fairness in the sense of equality of arms, and focuses on protection of the defendant, which may result in a procedural outcome that is fair but not necessarily objectively true.[131]

This may be one explanation for the tendency in the civil law tradition, which is more heavily characterised by the inquisitorial trial, to differ so precisely between parties to a crime according to different modes of liability; the final conviction must reflect the facts, and not just group all parties to a crime into one generic finding of guilt for the crime *in toto*, if not all the parties were factually equally responsible. The normative differentiation that occurs in systems such as those compared in Chap. 7 rests upon very specific technical definitions; the extent to which an actor has contributed factually to the completion of the crime matters a great deal for the specific qualification of the conviction. When an actor did not exercise substantial control over the commission of the crime, he or she is convicted for assisting the ones who did, but is not convicted for having committed the crime, and degrees of liability are defined carefully. By comparison, in the common law tradition where the trial is more typically adversarial in nature, the outcome of the trial is more concerned with a procedural truth and a protection of the parties' interests. It therefore matters less whether the conviction of parties describes in accurate factual terms exactly what the role of each individual was; participation in the collective crime amounts to equal

---

[128] Weigend 2003, p. 158.

[129] Weigend 2003, p. 160.

[130] Grande 2008, p. 154; Weigend 2003, p. 162.

[131] Grande 2008, p. 148.

liability for all parties, and forms of complicity are described in far less technical detail.

There is a difference in emphasis on the concern for a substantive or a procedural 'truth' depending on the style of investigation in the trial, leading to different concerns for the definitions of complicity. 'Justice' and 'truth' serve different purposes in each ideal trial model, which affects the way in which modes of liability are defined in the conviction of each individual.

Given that ICL is an amalgamation of both the inquisitorial and adversarial models, it may not be self evident whether there should be a preference for either a substantive, factually accurate outcome in terms of describing the roles played by each individual, or a procedurally just outcome which is concerned only with whether an accused was involved, but not the degree to which he was. Some would argue that the trial process at the ICTY has had more tendencies towards procedural truth, and indeed in the recent (somewhat controversial) acquittal of former President of the Serbian Radical Party Vojislav Šešelj, the Trial Chamber pointed out explicitly that its findings:

> do not claim to establish the entire truth about the events that occurred, let alone to recount the complex history of a conflict. The Chamber's role is limited to providing a legal response to the allegations made in support of the Prosecution's theory. Therefore, this judgement is dependent on the limited truth that may or may not have been established by the facts presented before the Chamber.[132]

But there is an important factor at play in ICL, namely the desire for truth-telling and history-writing in the outcome of the trials, as discussed above. The fact that victims are now an active part of the trial process at the ICC reflects the importance given to these factors at the permanent court. This desire for truth-telling and history-writing would suggest the need for more factual clarity in each conviction of the actual role played by each individual, and a differentiation of modes of liability. Particularly given the greater moral agency and influence exerted by leaders within a collective, and the role such leaders have in the very existence of atrocities, it would seem that modes of liability which can reflect this truth more accurately should be preferred.

3.3 Designing a System of Liability in International Criminal Law

The problem of how to put the leaders of mass atrocity on trial, in terms of designing a system of criminal liability based on general principles and then selecting specific modes of liability to be applied, is manifold. The discussion here has aimed to bring to the forefront the various tensions at play, and to seek alignment among them, even though a perfect agreement may not be possible. In the end, the entire process of identifying an appropriate system of liability is one of policy-making. There are many different systems of criminal justice possible, and many different modes of

[132] Šešelj Trial Judgment Summary 2016, p. 1.

liability from which to choose. Nothing is predetermined, which means that if we are to make an informed choice, we must dig deeper into the questions behind what drives these different models in terms of efficacy, symbolism, fairness, justice and truth-finding.

In the previous chapter, deliberative capacity was distinguished as a key tool for identifying leadership liability for system crimes of mass atrocity. However this can only be a useful tool if there is a clear context in which to apply it, namely a system of liability as a whole. The goals of international prosecutions, identified here as a central goal of increasing the public sense of accountability, as well as the symbolic function of ICL as norm expressionand history-writing, all express the moral and legal principles which can guide in the design of a system of liability as a whole. Identifying leaders as independent actors in relation to mass atrocity crimes is more effective both with respect to the central goal and with respect to the expressive function of trials. When it comes to the question how to give form to specific modes of liability within this system, the principles of fairness and justice are to be taken into account. The didactic, symbolic nature of international prosecutions and the expressionist function should not lead to trials that underplay the importance of the defendant's rights, nor of the 'meta principle' of fairness that acts as a tempering requirement on other aspirations and functions.[133] In this sense the selectivity and potential for exemplary prosecutions that amount to scapegoating should be avoided at all costs. By upholding the requirements of justice and fairness, while maintaining some integrity towards the aspirations of efficacy and symbolism, it would appear from the discussion here that focusing upon the leaders of mass atrocity as especially responsible is both warranted and necessary.

But these principles are only part of the puzzle, laying the foundation for more specific questions to be posed. In order to design an appropriate system of liability, and to define the modes of liability within such a system, questions as to the subjective or objective theory underpinning liability must be posed, as well as to which tests should be applied, whether in a system of collective responsibility for all individuals, or one of differentiated liability. Because domestic criminal law is the place to look for answers, and because the answers differ in different legal traditions, what is needed in order to come to a fully informed and grounded system of liability in ICL is a comparative study, which is the purpose of Part II.

However there is first a series of problems to deal with in terms of the role domestic laws and policies can play as sources of international law, to which the following chapter will now turn.

---

133 Swart 2008, p. 96.

References

ACHR (1969) American Convention on Human Rights
African Charter on Human Rights (1981) African (Banjul) Charter on Human Rights
Akhavan P (2001) Beyond impunity: can international criminal justice prevent future atrocities? Am J Int Law 95:7–31
Allied Control Council (1945) Nuremberg Trials Final Report Annex D: Control Council Law No. 10, Punishment of Persons Guilty of War Crimes, Crimes Against Peace and Against Humanity. http://avalon.law.yale.edu/imt/imt10.asp
Arendt H (1964) Eichmann in Jerusalem: a report on the banality of evil. Penguin Press, New York
Ashworth AJ (2009) Principles of criminal law, 6th edn. Oxford University Press, Oxford
Beaumont P (2006) Slobodan Milosevic dies alone with history still demanding justice. http://www.theguardian.com/world/2006/mar/12/warcrimes.milosevictrial
Boas G (2010) The difficulty with individual criminal liability in international criminal law. In: Stahn C, van den Herik L (eds) Future Perspectives on International Criminal Law. TMC Asser Press, The Hague, pp 501–519
Boister N, Cryer R (2008) The Tokyo International Military Tribunal: a reappraisal. Oxford University Press, Oxford
Boot M (2001) Nullum crimen sine lege and the subject matter jurisdiction of the International Criminal Court. Intersentia, Leiden
Cassese A (2004) The ICTY: a living and vital reality. J Int Crim Justice 2:585–597
Cassese A (2008) International criminal law. Oxford University Press, Oxford
Cesarani D (2006) The International Military Tribunal at Nuremberg: British perspectives. In: Reginbogin HR, Safferling CJ (eds) The Nuremberg Trials: International Criminal law Since 1945. K.G, Saur, Munchen, pp 31–37
Cryer R (2001) The boundaries of liability in international criminal law, or 'selectivity by stealth'. J Conflict Secur Law 6(1):3–31
Cryer R (2005) Prosecuting international crimes: Selectivity and the International Criminal Law Regime. Cambridge University Press, Cambridge
Cryer R (2009) Royalism and the King: Article 21 of the Rome Statute and the politics of sources. New Crim Law Rev 12:390–405
Cryer R, Friman H, Robinson D, Wilmshurst E (2016) An introduction to international criminal law and procedure, 3rd edn. Cambridge University Press, Cambridge
Damaška M (1984) The faces of justice and state authority: a comparative approach to the legal process. Yale University Press, New Haven
Damaška M (2008) What is the point of international criminal justice? Chicago-Kent Law Rev 83:329–364
Danner AM, Martinez JS (2005) Guilty associations: joint criminal enterprise, command responsibility, and the development of international criminal law. California Law Review pp 75–169
de Guzman MM (2008) Article 21. In: Trifterer O (ed) Commentary on the Rome Statute of the International Criminal Court: Observers' notes, article by article, 2nd edn. Beck-Hart-Nomos, Munchen
de Vlaming F (2010) De Aanklager: Het Joegoslavië Tribunaal en de Selectie van Verdachten. Boom Juridische Uitgevers, The Hague
Del Ponte C (2004) Prosecuting the individuals bearing the highest level of responsibility. J Int Crim Justice 2:516
Drumbl MA (2005a) Collective violence and individual punishment: the criminality of mass atrocity. Northwestern University Law Review, Winter p, p 539
Drumbl MA (2005b) Pluralizing international criminal justice. Mich Law Rev 103:1295
Duff RA (2007) Answering for crime: responsibility and liability in the criminal Law. Hart Publishing
ECHR (1950) European Convention for the Protection of Human Rights and Fundamental Freedoms
Fletcher GP (2000) Rethinking criminal law. Oxford University Press, USA

Fletcher GP (2002) The Storrs lectures: liberals and romantics at war; the problem of collective guilt. Yale Law J 111:1499–1573

Fletcher GP (2007) The grammar of criminal law: American, comparative and international, Vol One: Foundations. Oxford University Press, USA, New York

Grande E (2008) Dances of criminal justice: thoughts on systemic differences and the search for truth. In: Jackson J, Langer M, Tillers P (eds) Crime, procedure and evidence in a comparative and international context, Hart, Portland, Oregon, Chapter 8, pp 145–164

Hart HLA (1970) Punishment and responsibility: essays in the philosophy of law, 2nd edn. Oxford University Press

Hart HLA (1972) The concept of law, 2nd edn. Oxford University Press

Heller KJ (2007) A poisoned chalice: the substantive and procedural defects of the Iraqi High Tribunal. Case West Reserve J Int Law 39:261

Heller KJ (2010) Situational gravity under the Rome Statute. In: Stahn C, van den Herik L (eds) Future perspectives on international criminal law. TMC Asser Press, The Hague, pp 227–253

Heller KJ (2011) The Rome Statute of the International Criminal Court. In: Heller KJ, Dubber MD (eds) The handbook of comparative criminal law. Stanford University Press, Stanford, pp 593–634

Hosoya C et al (eds) (1986) The Tokyo War Crimes Trial. Kodansha, Tokyo

ICCPR (1966) International Covenant on Civil and Political Rights

ICTY Report on NATO Bombing (2000) Final Report to the Prosecutor by the Committee Established to Review the NATO Bombing Campaign Against the Federal Republic of Yugoslavia

Jakobs D (2014) The ICC Katanga Judgment: A commentary (Part 2): Regulation 55 and the modes of liability. http://dovjacobs.com/2014/03/11/the-icc-katanga-judgment-a-commentary-part-2-regulation-55-and-the-modes-of-liability/

Macdonald GK (1998) Fifth Annual Report of the International Tribunal for the Prosecution of Persons Responsible for Serious Violations of International Humanitarian Law Committed in the Territory of the Former Yugoslavia Since 1991

Manacorda S (2009) Culpability. In: Cassese A (ed) The Oxford companion to international criminal justice. Oxford Scholarly Authorities on International Law, Oxford University Press, Oxford

McCormack T (1997) Selective reaction to atrocity: war crimes and the development of International criminal law. Albany Law Rev 60:681

Mégret F (2010) Beyond fairness: Understanding the determinants of international criminal procedure. UCLA J Int Foreign Aff 14:37

Meron T (1998) War crimes law comes of age. Am J Int Law 92:462

Meron T, Goldstone R, Neier A, Anderson K, Wald P (2002) Where do we go from here? New and emerging issues in the prosecution of war crimes and acts of terrorism: a panel discussion. Soc Res 69(4):1177

Nino C (1996) Radical evil on trial. Yale University Press, New Haven

Nuremberg Judgment (1945) Judgment of the International Military Tribunal

Ohlin JD (2013) Goals of international criminal justice and international criminal procedure. In: GS et al (eds) International criminal procedure: principles and rules, Oxford University Press

Osiel M (2005) The banality of good: aligning incentives against mass atrocity. Columbia Law Rev 105:1751–1862

Osiel M (2009) Making sense of mass atrocity. Cambridge University Press, Cambridge, UK

Pellet A (2002) Applicable law. In: Cassese A, Gaeta P, Jones MJR (eds) The Rome Statute of the International Criminal Court: a commentary, Vol II. Oxford University Press, Oxford

Peterson J (2007) Unpacking show trials: situating the trial of Saddam Hussein. Harv Int Law J 48:257–292

Preparatory Commission Report (1996) Report of the UN Preparatory Commission of the International Criminal Court, Volume I

Report of the Secretary General on the ICTY (1993) Report of the Secretary-General Pursuant to Paragraph 2 of Security Council Resolution 808 (1993)

Report of the Secretary General on the ICTY (1993) Report of the Secretary General Pursuant to Paragraph 2 of Security Council Resolution 808 (1993) on the ICTY
Robinson D (2012) A cosmopolitan liberal account of international criminal law. J Int Crim Justice 26(1):127
Robinson D (2013) The two liberalisms of international criminal law. In: Stahn C, van den Herik L (eds) Future perspectives on international criminal law. Ashgate, Surrey, pp 307–322
Rockwood LP (2007) Walking away from Nuremberg: Just war and the doctrine of command responsibility. University of Massachusetts Press, Amherst
Rules of Procedure and Evidence of the ICC (2002) Rules of Procedure and Evidence of the International Criminal Court
Ryan AA (2007) Nuremberg's contribution to international law. Boston College International and Comparative Law Review 30 (Winter 2007):55
Safferling C (2003) Towards an international criminal procedure. Oxford University Press, Oxford
Siebert J (2010) The ICC's pursuit of the Lord's Resistance Army and the limits of criminal proceedings. Conrad Grebel Rev 28(3):89–107
Simpson G (1997) War crimes: a critical introduction. In: McCormack T, Simpson G (eds) The law of war crimes: national and international approaches. Kluwer, The Hague
Simpson G (2007) Law, War and Crime. Polity Press, Cambridge
Simpson G (2012) Atrocity, law, humanity: punishing human rights violators. In: Gearte C, Douzinas C (eds) The Cambridge companion to human rights law. Cambridge University Press, Cambridge, Chapter 6, pp 114–133
S/Res 1315 (2000) On the situation in Sierra Leone
S/Res 1534 (2004) On necessity of trial of persons indicted by the International Tribunal for the Former Yugoslavia and the International Criminal Tribunal for Rwanda
S/Res 827/1993 (1993) On the establishment of an International Tribunal and adoption of the Statute of the Tribunal
S/Res 955/1994 (1994) On the establishment of an International Tribunal and adoption of the Statute of the Tribunal
Statute of the ICC (1998) Statute of the International Criminal Court
Statute of the SCSL (2000) Statute of the Special Court for Sierra Leone
Summers RS (1999) Formal legal truth and substantive truth in judicial fact-finding - their justified divergence in some particular cases. Law Philos 18:497–511
Swart B (2008) Damaška and the faces of international criminal justice. J Int Crim Justice 6:87–114
Tallgren I (2002) The sensibility and sense of international criminal law. Eur J Int Law 13(3):561–595
The Guardian (1987) Nazi war criminal Klaus Barbie gets life. http://news.bbc.co.uk/onthisday/hi/dates/stories/july/3/newsid_2492000/2492285.stm
UN ECCC Agreement (2003) Agreement between the United Nations and the Royal Government of Cambodia Concerning the prosecution under Cambodian law of crimes committed during the period of Democratic Kampuchea. http://www.eccc.gov.kh/sites/default/files/legal-documents/Agreement_between_UN_and_RGC.pdf
van der Wilt H (2008) Equal standards? On the dialectics between national jurisdictions and the International Criminal Court. Int Crim Law Rev 1(2):229–272
van der Wilt H (2010a) National law: A small but neat utensil in the toolbox of international criminal tribunals. Int Crim Law Rev 10(2):209–241
van der Wilt H (2010b) Why international criminal lawyers should read Mirjan Damaška. In: Stahn C, van den Herik L (eds) Future perspectives in international criminal justice. TMC Asser Press, The Hague
van Sliedregt E (2012) Individual criminal responsibility in international law. Oxford University Press, Oxford
Vasiliev S (2014) International criminal trials: a normative theory, Volume 1: Nature. Doctoral thesis, University of Amsterdam

Vasiliev S (2015) The plural pluralisms of international criminal procedure In search of a normative identity. In: van Sliedregt E, Vasiliev S (eds) Pluralism in international criminal law, Oxford University Press, Oxford, Chapter 4
Vogel J (2002) How to determine individual criminal responsibility systemic contexts: twelve models. Cahiers de Défense Social pp 151–169
Weigend T (2003) Is the criminal process about truth? A German perspective. Harv J Law Public Policy 26:157–173
Weigend T (2014) Problems of attribution in international criminal law: a German perspective. J Int Crim Justice 12:253–266
Werle G, Jessberger F (2014) Principles of international criminal law, 3rd edn. Cambridge University Press
Zahar A, Sluiter G (2008) International criminal law: a critical introduction. Oxford University Press, Oxford

Cases

Akayesu (1998) The Prosecutor v. Akayesu, Trial Judgment ICTR-96-4-T
Barbie (1988) Klaus Barbie, Cour de Cassation, Chambre criminelle, 3 June 1988, 87-84.240
Barrios Altos (1985) Barrios Altos v. Peru No. 75, [2001] Inter-American Court of Human Rights Ser C
Brđanin Trial Judgment (2004) The Prosecutor v. Brđanin and Talić, Trial Judgment, IT-99-36
Delalić Trial Judgment (1998) The Prosecutor v. Delalić et. al, Trial Judgment IT-96-21-T
Demjanjuk (1993) Israel v. Demjanjuk criminal appeal No. 347/88, [1993] Supreme Court of Israel
Demjanjuk Judgment (2011) Demjanjuk Judgment [2011] 1 Ks 115 Js 12496/08 (Landgericht Munchen)
Duch Appeal Judgment (2012) Kaing Guek Eav. alias Duch, Appeal Judgment 001/18-07-2007/ECCC/SC
Erdemović Sentencing Judgment (1996) The Prosecutor v. Erdemović, Sentencing Judgment IT-96-22-T
Katanga Trial Judgment (2014) The Prosecutor v. Germain Katanga, Trial Judgment ICC-01/04-01/07
Kordić Trial Judgment (2001) The Prosecutor v. Dario Kordić and Mario Čerkez, Trial Judgment IT-95-14/T-2
Krajisňik Trial Judgment (2006) The Prosecutor v. Momčilo Krajisňik, Trial Judgment IT-00-39
Nikolić (2003) The Prosecutor v. Nikolić, sentencing decision IT-94-2-S
Šešelj Trial Judgment Summary (2016) The Prosecutor v. Vojislav Šešelj, Trial Judgment summary, IT-03-67-T
Tadić Appeals Judgment (1999) The Prosecutor v. Dusko Tadić, appeals judgment IT-94-1-A
Velasquez Rodriguez (1988) Velasquez Rodriguez Case, No. 4 Inter-American Court of Human Rights Ser C

Chapter 4
A Comparative Theory of International Criminal Law

Contents

So far in ICL, attempts to define a system of liability under each international tribunal's statute have been wrought with fierce debates and uncertainty. The question must be posed whether it is possible to simply pick and choose from different interpretations of liability as presented in domestic systems and in the various international tribunals. Whether or not we *should*, it will be argued in this chapter that in fact this is the way that ICL has developed to date. It is an amalgamation of notions translated from domestic criminal law; therefore a comparative look at different modes of liability is exactly what is needed to provide the foundations for a solid system of leadership liability on the international plane. If we are actively looking for legal solutions to the problem of linking group leaders to collective criminality, whether we are doing so as judges, as prosecutors, as defence lawyers or as scholars, we are taking part in a process of identifying, applying and even forming the law.

But this raises further questions as to the nature of sources of law in ICL, and the way in which the law is formed. A traditional understanding of international law denotes States as the predominant, if not sole law-makers, and considers domestic

C. Steer, *Translating Guilt*, International Criminal Justice Series 9,
DOI 10.1007/978-94-6265-171-5_4

case law to be merely a subsidiary source of law, according to Article 38 of the Statute of the International Court of Justice (ICJ). This is a static view of the law, since it does not allow for shifts in the relevance of certain players in the development of law, and does not take into account the actual processes by which certain branches of international law are formed. A more dynamic view is therefore required in order to understand and analyse the processes observable in ICL.

If we look at the practice of law, it is possible to move beyond the assumption that any notion is original or given, and instead to show how these notions are formed.[1] If the processes by which the law is formed can be understood at a deeper level, then there is the possibility of extracting ourselves from the debates on 'correct' sources and whether asserted modes of liability are based on 'correct' law according to a static perspective of the law, and of thereby coming to a considered solution as to which modes of liability are most *appropriate* for leadership liability with respect to mass atrocity. Such an analysis requires a dynamic perspective that will allow a broader consideration of which players are involved in these processes, and how law is formed. For this reason a comparative law perspective is taken here, based upon a dynamic descriptive model of law as process.

Much has been left up to practice—to judicial and academic debate—which is why observation of this practice is so essential. The assertion here, which will be explored in Part II of this book, is that in seeking instructive sources and interpretative tools, judges in both international and domestic courts turn to the notions and institutions familiar to them from their domestic training, and apply them in the international context, as do other participants involved in the development of ICL. This can be described as a process of law-creation by way of borrowing from domestic systems, which is unavoidable when we are confronted with lacunae in this nascent body of international law. As will be shown in Sect. 4.1, this is akin to the process of legal transplantation in any domestic law system, with one important difference: rather than importing notions into an existing system that follows a certain legal tradition, instead a new system is being patchworked together.

However it is not just the judges who are part of this process. It is argued in this chapter that prosecutors, defence counsel, academic commentators writing new texts on the subject matter, and even authors of *amicus* briefs to courts on the normative content of ICL, are all participants in this process of identifying, applying and forming the law. The term 'participants in the process of law' is a key one to this approach on how the law is formed and who can form it, which will be explained further in Sect. 4.2.

This process is inherently bound up in policy. On the domestic level, liability has been developed as part of a criminal law system which has specific political and social aims. In Sect. 4.3 it will be argued that the constantly shifting choice and development of modes of liability on the international plane is also always in response to a specific historical context and policy desire.

To understand ICL as comparative law is to understand the processes that have always underpinned its development. It is nothing radical to say that many concepts

[1] Orford 2012, p. 616, citing Michel Foucault.

in ICL draw on domestic criminal law notions, nor is it news to learn that there are differences in interpretation on the international level due to differences in domestic jurisdictions. However the comparative law theory of ICL developed here goes a step further. By observing the processes by which ICL develops, and the players involved in these processes, this comparative theory understands ICL to be a process of translation, and the players who are doing the translating to be forming the law. This challenges the traditional conceptualisation of international law as discoverable by application of reason, and embedded in distinct, determined sources. It is therefore perhaps the most controversial aspect of this book. However such an approach does not seek to undermine the doctrine of sources, but rather to bring a realistic and dynamic analysis to the law, to broaden the perspective and allow a wider scope of tools with which to unravel the complex arguments surrounding the legitimacy, form and content of various models of liability.

Given this dynamic perspective on law and on the ways in which the normative content of the law develops, the question returns as to whether we can pick and choose from various domestic modes of liability when applying ICL. In short, the answer is that we can indeed, but that there need to be explicit criteria by which we do this, otherwise what occurs is an unsophisticated clash of legal cultures.

4.1 International Criminal Law as Translation: From the Domestic to the International

The fact that ICL is born out of different bodies of law—International Humanitarian Law (IHL), human rights and domestic criminal law—brings it into difficulties when there are attempts to adhere to the underlying principles of each of these different paradigms. Whereas IHL and human rights both focus on improving systems to protect individuals and provide remedies for breaches, criminal law is based on assumptions about individual autonomy, and has certain conditions for punishment and stigmatisation of an individual.[2] While these are both 'liberalist' paradigms, they sometimes do not agree. Human rights is particularly concerned with protection and remedies for victims against the State, and IHL is formed mostly out of more general norms and principles regulating (in the first place) State use of force in a conflict situation. On the other hand criminal law must be at least partially concerned with the rights of the defendant. As a body of law it is particular to domestic law, contextually determined in each domestic jurisdiction, and is derived from State authority and the claim to a monopoly over force within State boundaries. It differs systemically from jurisdiction to jurisdiction, and from legal tradition to legal tradition.

In combining the international and the domestic, ICL has been built on a structure of international instruments in which the general is fairly clear, but the specific has often been left opaque due to differences in domestic criminal law. Especially during the explosion of practice of ICL in the 1990s, practitioners from human rights and

[2] Robinson 2013, pp. 116–118.

IHL took up positions of expertise in ICL, and the reasoning that emerged is often pervaded by the assumptions of human rights and IHL, and not necessarily from criminal law.[3] Whereas the elements of the crimes have typically been defined at the international level, through case law and codification, and based mainly on IHL, the normative content of specific aspects such as defences and liability have been left fairly open.

In a normative vacuum, where the content of ICL is left up to practice, it is therefore inevitable that law-makers and law-appliers will look to domestic criminal law notions to aid them. Domestic norms and institutions are drawn up to the international plane where the context may differ from the specific socio-legal and political contexts of their domestic origins. While this process may be inevitable, it is not always accepted as such. According to the traditional doctrine of sources in international law, domestic law is merely a subsidiary source, only to be applied when treaties, custom or general principles do not provide a clear answer.[4] And although general principles themselves are extrapolated from those principles which are recognised by 'civilised nations', or according to the Rome Statute, where they are also considered to be subsidiary sources, 'derived by the Court from national laws of legal systems of the world',[5] it is very difficult to ascertain a general principle when it comes to liability, since this is where domestic criminal systems differ according to their traditions and histories.

As will be described in more detail in Sect. 8.11.4, from the inception of the trials at Nuremberg to the ad-hoc tribunals of the 1990s, the statutes were drafted primarily by common law trained lawyers, leading to a more adversarial style procedure, and a more unitary approach to liability.[6] The first drafts of the International Law Commission (ILC) Draft Code of Crimes Against Peace and Mankind followed this model in the document that was to form the basis of the Rome Statute, and it wasn't until the French delegation to the ICC Preparatory Committee submitted a proposal in 1996 that there was true acceptance of a mixed system for this new permanent court.[7] This was followed in 1997 by a joint proposal between Canada, Germany, the Netherlands and the United Kingdom to re-model the very general article on liability, creating the extensive and detailed text that has now become Article 25 of the Rome Statute.[8]

This deliberate attempt to combine the civil law and common law approaches was an acknowledgment of the necessity of drawing from domestic law, but during drafting of Article 21 of the Rome Statute on the applicable sources, there was further contention. On the one hand there were concerns that too much judicial freedom would lead to a breach of *nullum crimen*, and that domestic law of the territorial State, the State of nationality of the accused, or the custodial State should prevail.[9] On

[3] Robinson 2013, p. 117; Fletcher and Ohlin 2005.

[4] According to Statute of the ICJ 1945, Article 38(1)(d).

[5] Statute of the ICC 1998, Article 21(c).

[6] Ambos 2007, p. 431.

[7] Ambos 2007, p. 433.

[8] Schabas 2010, p. 423.

[9] Preparatory Commission Report 1996, p. 105; van der Wilt 2010, p. 215.

the other hand there were those who considered that direct application of specified domestic laws would affect the coherence of ICL as a system and would lead to inconsistent outcomes, since suspects would be measured by different standards for the same conduct.[10]

The final text of Article 21 lists the applicable sources according to a hierarchy:

> (a) In the first place, [the] Statute, Elements of Crimes and its Rules of Procedure and Evidence;
> (b) In the second place, where appropriate, applicable treaties and the principles and rules of international law …
> (c) Failing that, the general principles of law derived by the Court from national laws of legal systems of the world …provided that those principles are not inconsistent with this Statute and with international law …

Section 2 of that article states that the Court 'may apply principles and rules of law as interpreted in its previous decisions.' In the end, domestic law is only acknowledged as providing support for a general principle. Thus although the translation from the domestic to the international is a constant process, it is only sometimes so in a deliberate sense; mostly it is implicitly.

Yet domestic law has become the main source from which solutions have been drawn at the international criminal tribunals, albeit re-packaged as customary law or general principles. At the ICTY, Judge Cassese went to great efforts in his dissenting opinion to the *Erdmeović* Appeals Judgment to point out that the differences between international tribunals and domestic jurisdictions in structure, aim, scope and institutional background mean that international tribunals are not disposed to the direct importation of national law into international criminal proceedings. On the other hand he considered it acceptable to 'assimilate or transform the national law notion so as to adjust it to the exigencies and basic principles of international law.'[11] And yet when he famously authored the *Tadić* decision a year later, in which the notion of JCE was asserted as a norm of customary international law, this assertion was based entirely upon domestic case law.[12] In the *Furundžija* judgment, the Trial Chamber of the ICTY, featuring Judge Cassese himself, quoted Cassese's dissenting opinion in *Erdmeović* and stated that, before determining the existence of a general principle, an explicit comparative consideration of domestic criminal law norms must take into account:

> the specificity of international criminal proceedings when utilizing national law notions. In this way a mechanical importation or transposition from national law into international criminal proceedings is avoided, as well as the attendant distortions of the unique traits of such proceedings.[13]

---

[10] van der Wilt 2010, p. 216.

[11] Separate and Dissenting Opinon of Judge Antonio Cassese, Erdemović, Sentencing Judgment 1996, §3.

[12] Tadić Appeals Judgment 1999, paras 197, 200, 246, 247. See further the discussion in Sect. 8.11.4.

[13] Furundžija 1998, para 178.

While it would appear there is awareness of the difficulties of translating directly from the domestic to the international, at the same time the inevitability of this translation is evident.

To complicate matters, the proliferation of international tribunals in the last decade of the twentieth century and the first decade of the twenty-first has meant that there are many more actors involved in this process of interpretation, translation and formation.[14] This process is sometimes referred to as hybridization, but could perhaps better be described as patchworking, as a new body of law is being created through this process of comparative law in action.[15] The question is how to ensure the patchwork pieces fit together harmoniously, and how to minimise the clash of legal cultures that has appeared with respect to some pertinent questions such as liability. Unless there can be some more grounded reasoning given for a particular solution, it is merely a case of which legal culture wins out. However, this is unsatisfactory to defendants and to international criminal lawyers alike, due to the unpredictability, arbitrariness and inconsistency of such a process. To alleviate this clash of cultures, a comparative perspective can provide an overview of the translation of domestic notions to the international plane, by ensuring a transparent process and a move toward a functioning pluralism of sources, which in turn can aid in making a selection from among all the possible translations to find which one may be most suitable for the specific context of ICL.

4.1.1 Comparative Law as a Tool of Analysis

The explicit application of comparative law as a tool for finding and interpreting legal principles and rules is a phenomenon gaining recognition in recent international law theory. As Martti Koskenniemi has pointed out, it is a mistake to consider international law to be entirely universal in its language, rules, interpretation and application, since each domestic system has its own perspective and context for approaching and applying international law.[16] As the number of regimes and judicial forums increase, so does the concern for fragmentation. But, as Koskenniemi optimistically points out, rather than operate from a conflict of laws perspective, which seeks to resolve fragmentation by asserting a procedural rule that will determine which competing law would apply, instead it is possible for international lawyers to deal with these multiple regimes by using one's own familiar domestic legal context as a point of reference.[17] Domestic courts are no longer simply the guardians of international law enforcement, rather domestic law has become a prominent source of ICL beyond its 'subsidiary' status according to the classical doctrine of sources.[18] This is all the

[14] Kingsbury 1999, p. 690; Zahar and Sluiter 2008, pp. 11–12.

[15] Steer 2015.

[16] Koskenniemi 2009, p. 6.

[17] Koskenniemi 2009, p. 7.

[18] Roberts 2011, p. 57.

more so because a growing volume of relevant case law and practice is emerging in domestic courts, to which international tribunals turn for instruction on certain issues.

The plea for more attention to be paid to comparative methodology in various legal and academic fora of international law has also been made by Mireille Delmas-Marty, not only as a cognitive tool, but also an instrument of normative integration and of critical analysis.[19] She argues that only by a comparative study can we reach common norms, a common vocabulary, and a truly pluralist or hybrid system of ICL. She is absolutely correct that without such an approach the process becomes one of unilateral transplantation of one dominant system.[20]

Different notions, institutions and rules exist in accusatory and inquisitory systems for very different reasons.[21] Therefore to transpose certain rules or substantive notions directly from domestic criminal law to the hybrid international one is to accept judicial bias and to create problems in the development of this new international body of law. Rather, what is needed is a deeper understanding of the different systems, in order to facilitate the didactic and expressionist functions of ICL asserted in Sect. 3.3, as well as a more successful process of hybridisation as ICL develops.

It is for exactly these reasons that a comparative approach is taken throughout this book to understand the shifts and debates over time with respect to a system of liability in ICL. Lawyers inevitably look to interpret international instruments through the lens of their own domestic training. When this is done without a comparative understanding there is not only a constant struggle between different interpretations, there is also very often the assumption that we all mean the same thing when we talk of complicity, or of principal and secondary liability, when in fact there are many legal languages being spoken, especially (but not only) through the translation to English from other languages. Due attention is therefore paid to the terminology used throughout this book, which will be laid out in Sect. 5.3. The point here is that a comparative perspective on ICL can help to understand the processes, and analyse the concepts being debated at a deeper level. International criminal lawyers and scholars must be comparativists in their work, and look beyond that which is familiar to them. 'A truly mixed procedure requires prosecutors, defence counsel and judges who have knowledge of both common and civil law, and who are able and willing to look beyond their own legal systems.'[22]

To this end, there is much comparative law scholarship from which we can learn that will benefit our understanding, analysis and eventually the ability to agree on a system of liability in ICL. One aspect that may be particularly useful is a more sophisticated understanding of the traditional civil law–common law divide. To this end, the notion of legal traditions will be discussed next, followed by a look at the scholarship on legal transplants that can be useful in understanding the process of borrowing that takes place in ICL. However there is one distinct difference in the

[19] Delmas-Marty 2008, p. 251; Delmas-Marty 2002; Delmas-Marty 2003.

[20] Delmas-Marty 2008, p. 251.

[21] Damaška 2004, p. 1019.

[22] Ambos 2007, p. 429.

case of ICL, which is that it is not an existing, mature system of criminal law that is consciously selecting from other, similar systems. Rather there is a process of patchworking together a new system from different elements drawn from domestic systems, which will be discussed last in this section.

4.1.2 Legal Traditions

Comparative law is sometimes seen as a project of categorisation; grouping domestic systems into legal families, with a 'mother' system at the historical root and a 'daughter' system inheriting an entire system or a set of rules.[23] However the simplified notion of 'mother' systems leads to a selective study of the main donor systems such as England or the US when one wants to know about a common law rule, and Germany, Italy, or France when one wants to know about a civil law rule, to the exclusion of other domestic systems which may have developed in interesting and new directions. More recent comparative law has moved away from this simplification and from the project of classification and taxonomy. Instead, scholars talk of legal traditions and legal cultures, which can span across societies and are not necessarily tied to historical inheritance.[24]

One useful analysis is provided by the influential comparativist, the late H.P. Glenn, who argued that the approach of classifying systems into legal families is based upon a mistaken assumption of incommensurability,[25] much like the conflict of laws. The mistaken assumption is that no two systems can co-exist, so to solve a problem of substantive law where 'foreign' laws conflict with local ones, collision rules must be created to determine which system applies to any given legal problem. Similarly, any comparative law approach that is predicated on classifying systems into legal families is asserting an 'epistemology of separation'.[26] Yet this is an inherently conflictive and static approach, which ignores processes of reciprocation, changes in influence and internal changes in systems over time, and which disallows reconciliation between systems or between rules.

Glenn also argued that this limiting methodology is 'Eurocentric', due to the inherent bias of the concept of legal families.[27] Many colonised states took on aspects of the legal systems of the colonising powers, and not always by choice. To group these systems into a legal family with their historical 'mother' systems leads to the highly selective study of systems as mentioned above, and also to a misunderstanding of the role of legal institutions in these 'daughter' or receiving systems.

---

[23] Glenn 2006, p. 424.

[24] Glenn 2006; Glenn 2010; Nelken 2006, p. 372; Hoecke and Warrington 1998. Some traditional schools of comparative law research still use the 'legal families' taxonomic approach, see e.g. Zweigert and Kötz 1998, p. 57.

[25] Glenn 2001, p. 133.

[26] Glenn 2001, p. 430.

[27] Glenn 2001, p. 435.

This selection of the most dominant systems is a phenomenon that can be witnessed in ICL, and has been criticised as a questionably selective 'survey style' approach. Often when a question arises such as whether duress is a complete defence or whether an asserted mode of liability pre-existed prosecution of a crime under customary international law, there has been heavy reliance on selected post-Second World War case law, which was predominantly decided by military tribunals following English and US law.[28] This approach has been repeated in international tribunals continuously, where a mere handful of jurisdictions are surveyed to make a general conclusion.[29] International lawyers of the 21st century—and particularly international criminal lawyers who must be sensitive to the inherent divergences in domestic criminal laws—must make a broader, more diverse and non-Euro-centric selection of jurisdictions in order to assert that a rule does or does not exist.

For these exact reasons, Glenn proposed the concept of 'legal traditions' to replace that of 'legal families' when undertaking any comparative study of rules or notions across different legal systems. Where 'legal families' lead to a static, taxonomic approach and an assumption of incompatibility, the concept of 'legal traditions provides no inherent demarcations, and can serve as a technique for reconciliation of different laws.'[30] The concept of legal traditions allows a more useful issue-by-issue basis of comparison. Glenn still used the terms 'civil law' and 'common law', which are also used throughout this book, but he spoke of these as traditions rather than families, allowing for more domestic systems to be included, as well as the possibility that one domestic system may fit into more than one tradition.[31] This is particularly helpful if we consider that ICL has characteristics of both traditions but could never be truly classed as belonging to one legal 'family' or the other.

In this sense, Glenn considered 'tradition' to be at the centre of an exchange of voices, a dialogue across cultures. Since traditions are not static, but are a product of their time and of social mores and values,[32] this dialogue itself is dynamic and can lead to a dialectic between those engaging in it. Just as domestic systems can influence each other horizontally, so too is there a dialectic between domestic systems and the international plane vertically. For example, domestic systems have taken on the obligation to implement the Rome Statute into their criminal law legislation, and the case law of international tribunals has influenced some domestic courts directly. At the same time, domestic law notions inform international criminal lawyers (in all their capacities) as they interpret and appraise international instruments and assert normative rules such as those on liability.

In Part II of this book, explicit use will be made of a comparative approach involving the notion of legal traditions rather than legal families. This will better aid to understand the processes that have led to the development of modes of liability

[28] See for example Zahar and Sluiter 2008, pp. 96, 99, 239; Heller 2011a; Steer 2015.

[29] This will be discussed in detail in Sect. 8.11.4.

[30] Glenn 2006, p. 433.

[31] Ferrante 2011, p. 12. The impact of this will be discussed in Sect. 7.4 in the comparative study of Argentina's criminal law system. See also Glenn 2010, pp. 45, 374.

[32] Glenn 2010, pp. 22–25; Glenn 2006, p. 432.

in the various domestic jurisdictions, the ways in which domestic jurisdictions have influenced each other in a horizontal relationship to each other, as well as the way in which these domestic jurisdictions have been translated to the development of ICL on the international plane in a vertical relationship. In seeking an answer to the question of the most appropriate modes of liability for the architects of mass atrocity crimes, adding the notion of legal traditions allows a more sophisticated and less biased analysis and outcome.

4.1.3 Legal Transplants

Against this background of legal traditions influencing each other horizontally and vertically, the focus here is on the resulting phenomenon of legal transplants. Academic debate in comparative law was focused for many years on the question whether it is practicably possible to pluck a legal institution from one systemic context and transplant it successfully into another. In recent years there seems to be a general agreement that transplants are possible—since they occur all the time—and the discussion now delves deeper into questions as to how this can best be done.[33]

The term 'legal transplants' was coined by legal historian and comparativist Alan Watson in the 1970s.[34] His description of customary laws in thirteenth century Germany and France, prior to the reception of Roman law, demonstrated that wherever there was a desire for some certainty and clarity, legal transplanting took place on a surprisingly large scale.[35] Watson drew some important conclusions from this:

> First, in the absence of written law and of a central authority that imposes its will, custom is likely to be very local and fragmentary. The popularity of transplants here in part results from the fact that custom does not provide an answer or one easy to find for many issues. ...Secondly ...the choice of borrowing depends in large measure on the accessibility of the donor source. ...And thirdly, the adoption of town law and use of decisions ...of other towns are indications of the importance of authority in legal borrowing and in legal growth generally.[36]

The importance of accessibility and authority as the two key factors for a system to become a donor system are recurring threads throughout Watson's work. He observed that legal transplanting throughout Europe in the thirteenth to fifteenth centuries was very common, but that it would rarely happen when the law to be borrowed was not set out in a systematic and easily accessible fashion, and that in order for a system to become a donor system, it was almost always 'in a position of political and military authority over the borrower'.[37]

---

[33] Örücü 2002; Berkowitz 2003; Kingsley 2004.

[34] Watson 1974; Watson 1983; Watson 1984.

[35] Watson 1984, pp. 26–27.

[36] Watson 1984, p. 27.

[37] Watson 1984, p. 36.

The process of creating and refining ICL as a system of law could be seen as a continuing process of legal transplantation, in the same way as the reception of Roman law throughout most of continental Europe was.[38] When we consider the three conclusions Watson drew on the way in which legal transplants have occurred historically, there is a clear parallel with the way in which ICL is developing. First, there is an absence of central authority in the ICL system. While there is certainly written law in the sense of treaties and statutes, it does not exist as one coherent body of law, but rather a fragmentary conglomeration of sources which are various in nature and have both international and domestic birthplaces. The core crimes are themselves based on the older body of IHL.[39] The institutional framework has been created, as already mentioned, by way of treaty and special agreements with the UN. But there are areas of this new body of law which are not fully regulated by these codifications, and for which custom can provide no clear answer because the law on an international level is simply too new. As the jurisprudence and case law develops from numerous domestic and international courts applying and 'finding' this law, there is a repeated dialectic between the domestic and international level.[40] Just as Watson concludes, where custom does not provide an easy answer, transplants become popular.

Secondly, as Watson points out, the principle of accessibility determines the choice of systems from which to borrow. Wherever there are gaps in the normative detail of ICL, there is evidence that participants in this process reach to the law they know from their own background training, and that reference is increasingly made to domestic law in order to extrapolate custom or general principles of international law.[41] This most recent movement has not been a process over centuries but merely decades. Following Watson's analysis, it makes sense that in a system which is already a patchwork of various sources, and which is developing as rapidly as ICL is, solutions to address any gaps or uncertainties will be sought by looking to familiar domestic systems that could act as donor systems. Which system is most accessible depends upon the linguistic capabilities of the participants seeking a solution, as well as cultural barriers or digital accessibility of domestic laws and judgments, leading to a highly selective process of surveying state practice.[42]

Put simply, the system which is most accessible is the system which is best known to the participants asserting a rule at any given time. This phenomenon will become clear as central to the development of modes of liability in ICL in Part II under Chap. 8.

Thirdly, Watson points to the authority of a donor system as a key factor in selection, and he asserts that the quality of law is less important. What matters is the

---

[38] Watson 1984, p. 25.

[39] Such as the Lieber Code 1863; as well as multilateral treaties making up the 'Hague Law' (methods and means of warfare) and 'Geneva Law' (protection of civilians and combatants).

[40] van der Wilt 2008, p. 272.

[41] See in particular *infra* Sect. 8.11.4 for a full observation and analysis of this phenomenon. See further Delmas-Marty 2003, p. 13; Steer 2011, p. 302; Roberts 2011, p. 58.

[42] As noted above, Zahar and Sluiter 2008, p. 94.

desire for clarity of substantive law, and for borrowing from the most authoritative donor systems rather than a search for greatest quality.[43] ICL is facing the same problem that domestic legal systems have faced over the centuries in terms of lack of sources, and the same approach of looking to other systems for solutions to specific normative problems prevails. Wherever there are areas of law that are unclear, the legitimacy of sources themselves is not at issue, but rather the ready availability of rules to solve the problem.[44] It is asserted here that the choice of a system to borrow from is not driven by the search for a source, be it customary law or otherwise, but rather by the search for a normative rule which can be immediately applied.

Although historically the question of which system was most authoritative was a matter of military power, the principle of authority with respect to donor systems in ICL still applies. The tradition that is most pervasive in the sharing or spreading of its information will attract the most adherence.[45] It is more a question of the ability of a given system to assert itself as influential in this process of patchworking in one of the two following ways.

One way is that the individual participants who are influential within certain tribunals manage to assert notions from their domestic systems. For example the German notion of 'perpetration by means of an organization' which, as will be discussed in detail in Sect. 8.8, was included in the ICC Pre-Trial Chamber's decision confirming charges against Katanga and Ngudjolo Chui.[46] Similarly there was explicit reliance on German theorist Roxin's writings in interpreting indirect perpetration under Article 25(3)(a) of the Rome Statute in the judgment against Lubanga.[47] Arguably, this has a lot to do with the influential role not only of civil-law trained judges at the ICC, but also of the team of lawyers whose own domestic training influences the way in which such judgments are formed—a phenomenon which has been described as the 'age of the law clerk'.[48]

The second way is that an entire legal system can assert authority over the receiving system by way of old colonial ties or through involvement in the development of new ICL institutions. This is exemplified by the influence of French law on the Cambodian legal system, which, subject to the ECCC Agreement, forms part of the applicable law of the ECCC.[49]

Once a domestic notion is transplanted through judgments into the emerging body of ICL, it is often legitimized as a part of ICL by referring to customary law, when in fact there is scant custom in this nascent field of law.[50] What is actually occurring is exactly the process of legal transplantation that Watson describes, and the most

---

43 Watson 1984, pp. 96–97.

44 See for example Miller 2003, p. 846.

45 Glenn 2010, p. 42.

46 Katanga Confirmation of Charges 2008, paras 477–518.

47 Lubanga Trial Judgment 2012, paras 997–999.

48 See further Sect. 8.11. See also Posner 2010, p. 61; van der Wilt 2009, p. 309.

49 Etcheson 2004, p. 199.

50 Zahar and Sluiter 2008, pp. 82, 85, 99.

accessible or most authoritative system wins out regardless of whether it is the most appropriate system or even whether it offers the most appropriate rule.

However this is not always done with a full understanding of how the normative rule functions in its domestic context, which is part of the reason why there has been much continued misunderstanding about modes of liability taken out of their original context. For this reason, in Part II full attention will be paid the terminology and functions of different systems of liability in each of the domestic jurisdictions compared, before analysing their adoption into the international tribunals. This will aid in understanding the full context of each of the models of liability, and will thereby facilitate the informed selection of the most appropriate system of liability and modes of leadership liability in the context of mass atrocity.

4.1.4 *Legal Patchworking*

By looking at this process of legal transplants within ICL, an implicit analogy is being made between ICL as a system of law and any domestic system of law which borrows from other domestic systems. But in fact ICL is unlike any domestic system of criminal law, since it is less developed as a system and, more importantly, it is not an instrument of a central political authority which has a recognizable legal tradition. Domestic criminal justice systems have developed over time inside the context of the nation-State, and where these systems have borrowed from other systems, legal transplants have merged into the receiving system in the way a transplanted organ merges into the recipient body, becoming a part of the original whole.[51] But ICL does not have the same relationship to public international law that domestic criminal law has to the rest of the nation-State law of which it is a part. If we see ICL as a part of a meta-system of public international law, then it has a peculiar place, if only for the fact that it deals with the punishment of individuals and the notion of individual responsibility. ICL is such a new addition to international law that the process of legal transplantation could in fact be described as one of legal patchworking. Rather than transplanting organs into an existing body, an entirely new creation is being put together piece by piece, like a patchwork quilt.

This patchworking results in a manifold system. There is much discussion on the concern of fragmentation in international law in general, and in ICL in particular, due to the increasing number of international tribunals which may decide similar questions in entirely different ways. This is amplified by the fact that international tribunals have different statutes, which contain different interpretations of, for instance, defences or modes of responsibility, and perhaps by the fact that they each deal with different types of conflict and different ways in which international crimes are being committed. On top of this, domestic courts are asked to deal with the same questions, and may look to these international tribunals for guidance, sometimes taking the entire body of case law from a tribunal to be binding law, and sometimes taking

[51] Watson 1974, p. 27.

it to be an exercise in comparative law, but in the end often applying domestic implementation laws to solve the problem.[52] A diversity of law-makers and law-appliers means an inevitable diversity in the norms being applied, with the risk that this leads to a lack of legal certainty and undermines ICL as a system.

Many scholars have attempted to alleviate the concern for fragmentation by describing this phenomenon as 'pluralism'—a move which is not merely semantic but in fact normative.[53] If fragmentation means a splintering and a loss of coherence, threatening the very legitimacy of ICL as a body of law, then pluralism is an acceptance that the diversity of tribunals and participants will lead to diverse possible interpretations of the law, but that this does not have to lead to lack of legitimacy.[54] Rather it is an inevitable attribution of a field of international law which is developing so rapidly.

Rather than engaging in the debate as to whether this diversity is destructive to the coherence of ICL as a system or whether it is to be embraced, the intention here is to remain descriptive. This makes it possible to observe the phenomenon by which ICL is constantly being developed, and to understand the players who are influencing this process. There is no single interpretation of norms precisely because there are multiple fora and multiple actors. The vertical dialectic between international tribunals and domestic legal systems, and the horizontal one between international tribunals among themselves, is taking place between the actors who are part of these institutions, and this is where the law is formed, debated and developed. A descriptive observation of the patchworking process is exactly what will be undertaken in Part II of this book, with an emphasis on who is doing the patchworking. In order to map out this patchworking process, a dynamic understanding of the processes of law and the actors involved is required, which will be established in the next section.

4.2 International Criminal Law as Process

While States have come together as law-makers to form much of the institutional framework of the ICL system, the nuances of the normative content of ICL is difficult for States to agree upon. This is because ICL is first and foremost a criminal law system, and criminal law differs greatly from State to State and from jurisdiction to jurisdiction. The functioning of international criminal justice has therefore left much to practice, and what has emerged is essentially a struggle between common law and civil law traditions, since these are the traditions represented by the majority of those involved in the creation and functioning of international criminal institutions.

The lack of precedence for international tribunals could have been considered an opportunity for a carefully considered and well founded corpus of procedural rules and a system of liability, however the patchworking that has emerged can rather

[52] For example Trapani 2011.

[53] Steer 2015.

[54] See generally Berman 2012; Delmas-Marty 2003; van der Wilt 2008.

be characterised as 'arbitrary and opportunistic'.[55] Not only has this led to shifts back and forth between the different ideologies of the civil law and common law traditions, but these shifts themselves have been dependent on shifts in interests and in the influence of certain actors.

This struggle between traditions was already apparent during the design phase of the Nuremberg Charter, when there were disagreements between US and European delegates about how to design modes of liability to apply to the leaders of the vast and complex Nazi regime.[56] A similar tension was felt throughout the drafting process of the ICC. Because so much is left up to practice, and because there is so little to draw on in terms of custom or previous practice, an opportunity arises to observe the process by which the law is developing and who is influencing this process, in a relatively contained and concentrated way. But this requires a dynamic approach to the definition of law and who the law-makers are.

A traditional approach to the questions posed here would be to describe ICL as a branch of Public International Law, and to describe Public International Law as a body of law made up of treaties, customary law and general principles, formed primarily by the expression of States' consent in their interactions with each other. According to a traditional doctrine of sources, answers to the legal question regarding appropriate modes of liability should be sought in these sources, and only if these offer no clarity can we turn to the subsidiary sources of case law and the 'teachings of the most highly qualified publicists', following the wording of Article 38 of the Statute of the ICJ. However this static approach is insufficient to describe contemporary international law, and is certainly not a realistic picture of the nature of ICL.

By understanding law as a process of decision making rather than a (pre-existing) body of rules, the content of the law can be identified by reference to what the actors on the international plane believe to be normative in their relations with each other.[57] Who is included among these actors will be discussed shortly. The point here is that there are many more moments of law-making and many more fora in which it takes place, than only the formal moments and places such as the UN Security Council promulgating a resolution which brings a tribunal into being, or the signing of a treaty by States establishing the Rome Statute.

Key to this approach is the perspective from which the law is analysed. Positivism can be said to be viewed from the perspective of the receiver of commands, the 'political inferior'.[58] From this perspective law is a body of commands, and its essential technical problem is identifying the content and meaning of the command. Positivism is concerned with sources, and with identifying that which must be obeyed as a normative rule.[59] A view of law as process, on the other hand, is taken from the perspective of the person charged with making decisions, and the essential problem is to determine which choices are appropriate for the issue at hand. The body of rules

[55] Zahar and Sluiter 2008, p. 37.

[56] Heller 2011a, p. 251.

[57] Higgins 1994, p. 18.

[58] Reisman 1992, p. 119.

[59] Reisman 1992, p. 120.

still exist, but they are complex communications, conveying authoritative information about community policy that must be assessed on a case by case basis, in order to be shaped into a decision, according to the best possible approximation of a community's social goals.[60]

The challenge, of course, is to approximate what these social goals are, particularly in the field of ICL which has many competing goals and functions. The arguments made in Chap. 3 are that these can at least be reduced to the goal of increasing accountability for mass atrocity, combined with the endeavour towards efficacy and norm expression. But in order to select and assert which norms—which modes of liability—match these goals and functions, it is necessary to understand the processes of lawmaking within which such selection is made.

Given the fact that there is a wide cast of participants involved in these processes, it is necessary to define what is meant by 'participants'; that is, the actors who have sufficient authority to assert norms as law.

4.2.1 Participants as Law-Makers

Under a traditional, static view of international law, there is a limited range of actors who are authorised to make law, according to the formal institutions and formal sources. The core assumption is that States are the only fully fledged subjects of international law, bearing rights and responsibilities without the need for domestic intervention. All other actors are, in this paradigm, necessarily 'objects' of international law, regulated by it but limited in their ability to take active part in its formation, and only having limited rights or obligations as acknowledged by States.[61] But because this does not realistically depict the wide range of actors involved in lawmaking and law application, instead a different lexicon will be used here, one that moves beyond the subject-object dichotomy.

According to a more dynamic understanding of law as process, all those actors involved in the decision making processes to varying degrees can be called 'participants'. Included among these participants are States, international organisations, NGOs, multinational corporations, and individuals.[62] Areas of international law such as conduct during hostilities and human rights have been developed largely due to the participation of non-State actors in the decision-making processes, and apply directly to individuals. As Dame Rosalyn Higgins has stated, these are not;

> exceptions conceded by historical chance within a system of rules that operate as between states. Rather, they are simply part and parcel of the fabric of international law, representing the claims that are naturally made by individual participants in contradistinction to state-participants.[63]

---

[60] Reisman 1992, pp. 119, 120.

[61] Higgins 1994, p. 49.

[62] McDougal et al. 1988, p. 807; Higgins 1994, p. 50; Reisman 2010, p. 104.

[63] Higgins 1994, p. 50.

Judges and arbiters are just one species of decision makers, and they may be restricted in the sources they can consider by the constitutive instruments of their respective fora.[64] However, a dynamic conceptualisation of law as translation and as process includes a wide spectrum of law-making modalities and necessarily includes more participants as law-makers, from the delegated powers of international organisations, to the informal yet authoritative and controlling arrangements between other non-State participants on various levels. While these dynamic and fluid processes may not lead to *lex scripta*, still they form expectations and demands, and are at the core of the debates on the content of ICL.

Given that the institutional framework of ICL is created by States, its origins fit a more traditional understanding of international law-makers and sources. However this picture is insufficient for understanding the ways in which the normative content of ICL has developed and continues to develop. When it comes to modes of liability, the current disparity and uncertainty would appear from a traditional perspective utterly chaotic, unsatisfactory and would pose a threat of undermining the predictability necessary for a functioning criminal law system. Concerns about fragmentation in ICL arise when taking this perspective.[65] A more dynamic approach offers tools to take into account many more participants, processes and factors leading to the choices being made, and provides an access to both understanding how the various modes of liability have come to be, and how to select from them in a systematic way. This may offer something more akin to the clarity and determinacy demanded of a criminal justice system than does a traditional, static view of the law.

For these reasons a wide cast of law-making participants is taken into account in the comparative study that will take place in Part II of this book, including not only States as they have expressed their will in treaties and the constitutive instruments of the various tribunals, as well as legislators and judges, but also the lawyers engaged in prosecutions of mass atrocity crimes, and the scholars engaged in appraisal, analysis and commentary on these processes.

With respect to the role of lawyers, who are not generally seen to be law-makers, their role as active participants in the processes of ICL cannot be denied. The community of lawyers appearing before domestic and international tribunals dealing with questions of ICL is important in framing the legal issues and possible choices to be made, by the way arguments are brought forward and indictments are written. One pertinent example is the definition of rape as an act of genocide, which has only become accepted law because of the *Akayesu* decision before the ICTR.[66] Originally there were no charges of rape in the indictment, until spontaneous and recurring evidence was offered by witnesses during the trial. The indictment was amended by the prosecutor and finally sufficient attention was paid by judges such as then presiding Navi Pillay, which led to the development of the law according to expectations and demands, namely the definition of rape as an act of genocide.[67]

64 Reisman 2010, p. 135.

65 Steer 2015.

66 Akayesu 1998.

67 Boyle and Chinkin 2007, p. 291.

With respect to the role of scholars, while it is wise to caution against the notion that scholars are making law when they write academic commentary,[68] at the same time there are certain areas of international law which are heavily reliant on the doctrinal development to be found in scholarly work. The pro-active role taken by scholars in their analysis and appraisal of many ICL notions, both procedural and substantive, cannot be denied, nor can the impact and acceptance of the work of many of these scholars as authoritative. For these assertions and appraisals to be considered effective in terms of impacting the law, perhaps requires the further step of being appropriated by judges in their formal decision-making. In this sense, scholars may not be directly producing law, but they are certainly pertinent participants in the processes leading to the formation of law. At the same time, the authority with which scholars are cited in increasing numbers of judgments when it comes to the interpretation of modes of liability would suggest that their role as participants in the processes leading to lawmaking may be more direct than it was in the past.[69]

4.2.1.1 Dynamic Law-Making

Both the shifting status of participants and the content of the law are temporally contextual. The processes described here, and the participation of a wide range of actors therein, are a continual response to the policies asserted at a given time by various members of the international community. While there may not be consensus on what these policies are, the continuous process of decision-making which lead to law-making in all its forms are themselves dynamic, as the asserted prescriptions or norms are debated towards acceptance or rejection. This can be observed in many other areas of international law besides ICL in which non-State participants have been the predominant lawmakers, such as climate change regulations, human rights, and even international trade regulations.[70] Because of the predominant role of non-State participants, this could be termed 'bottom-up' lawmaking, in the sense that the law is not formed in a centralised way from States 'down' to subjects, but rather from a broad base of participants 'up' to the international plane.[71]

Article 38 of the Statute if the ICJ is therefore not 'wrong', or obsolete, it is rather incomplete as a description of all the forms international law can take, and the ways it can be formed.[72] Instead an empirical approach is required, case by case, in order to ascertain whether an assertion has become law. This will give rise to an assessment of the normative content of ICL at *any given moment*, since the processes are continuous and changing over time.[73] This is exactly why international lawyers

68 See e.g. d'Aspremont 2010.

69 See further the analysis of the shifting trends at the international tribunals in Sect. 8.11.

70 Levit 2007, pp. 399–407; Nagan and Hammer 2007, p. 779.

71 Levit 2007, p. 408.

72 Reisman 1981, p. 119.

73 Reisman 1981, p. 120.

need to be able to track or 'map' the decision making processes in order to be able to identify law.

This assertion of policies by participants, and the attempt to best approximate the what the goals and policies are of a given community, brings us to the role of policy as one of the central factors of law-making in international law, and thus in ICL.

4.3 International Criminal Law as Policy

In the opening line of her book 'Problems and Process', Rosalyn Higgins, former chief justice of the ICJ, makes the bold statement that 'international law is not rules. It is a normative system.'[74] This normative system has the object of ordering conduct and avoiding chaos, and is therefore never neutral or devoid of policy. Rather it is an expression of policies transformed into rules and principles. Higgins describes law as 'decision making by authorised decision makers when authority and power coincide.'[75] To many international lawyers, conceding the inherent role of policy in law may be an anathema, however it cannot be denied that law is formed as an expression of policy. The question is to what extent 'policy' can be included as part of the very definition of law.

4.3.1 International Law as Policy

As Higgins points out, a traditional understanding of international law is that it deals with authority alone, and not with power, since the latter is considered to belong to the realm of international relations, leaving international law to be objective and neutral. However since authority cannot exist without supporting control, she rather defines law as the intersection between power and authority.[76] Of course, where power overrides authority, the final outcome is not lawful,[77] however to deny that power is an integral part of international law is to deny the latter any force as law.

The discomfort many lawyers have with conceding the inherent relationship between law and policy has to do with the assumption that policy means 'anything goes' when a particular end is desirable, leaving the assertion of norms and rules to 'naked power'.[78] However there is a distinction between the intersection of power with authority, and 'naked power', which has no regard for the expectations of

---

74 Higgins 1994, p. 1.

75 Higgins 1994, p. 15.

76 Higgins 1994, p. 15.

77 Higgins 1994, p. 4.

78 Indeed, many critics of a policy-oriented view of the law have pointed out that such a theory is at risk of being appropriated for certain State interests: 'the selective application of (this) jurisprudential theory in emphasizing certain policy objectives as fundamental goals and minimizing others has thrown a sharp light on the dangers of policy-oriented jurisprudence. If applied with a nationalist

authority and rightness.[79] Higgins states that a policy understanding of law does not lead to utter indeterminacy nor susceptibility to arbitrariness. Decisions must still be tested for validity, according to the authority (whether formal or informal) of the participant asserting the decision, and the effectiveness of the assertion.[80]

Furthermore, even if one is to take a more positivist approach, and to maintain that law is defined as a body of rules promulgated by formally authorised entities, then there is still no guarantee that these rules will be applied in an unbiased or neutral manner—in fact this assumption can be empirically debunked. All legal arrangements have been consciously created by people operating in political processes.[81] There can be no neutral 'correct legal view' on any question, discerned merely by applying rules as defined by the accumulation of past decisions; these decisions will always have been taken with regard to specific contexts and circumstances, and always with an element of (policy) choice. In making such a choice, says Higgins;

> one must inevitably have consideration for the humanitarian, moral and social purposes of the law …A refusal to acknowledge political and social factors cannot keep the law 'neutral', for even such a refusal is not without political and social consequence.[82]

Thus, even with a more formalist definition, law is still inherently caught up in policy when it comes to appraisal, interpretation and application. But there is a difference between on the one hand a formalist acceptance that law may be applied according to policies, which still insists that normativity remains immune to policy, and on the other hand the dynamic perspective of law *as* policy. The dynamic understanding applied throughout this book strives to make explicit and visible the policy-oriented aspects of the design and making of law, in order to allow both a thorough scrutiny of the assertions made, and a systematic analysis on a more abstract level as to the content of the law. By connecting laws and policy, international law is seen as reflecting the needs of international policy arguments and vice versa.[83] By bringing the wide array of participants to the forefront of a description of the processes, and analysing the policy aspects of the processes, a realistic picture can emerge, and a deeper understanding of the way in which norms operate can be nurtured.

(Footnote 78 continued)
bias, it becomes an ideological instrument to override specific restraints of law.' Comments by Richard Falk, cited in Wolff 1985, pp. 267, 273. This author would argue, however, that where this happens the descriptive model of law as process is taken a step further from being a theory 'about' law to become a prescriptive, normative theory 'of' law. Just as some other authors have asserted, this author asserts that it is possible to maintain the descriptive model as a useful tool of analysis without taking on a normative project based on a universalist assumption that there are policy values applicable in every situation which can prescribe what the outcome should be. See for example Schachter 1968; Falk 1995; d'Aspremont 2014, p. 1119.

[79] McDougal et al. 1967, p. 257; McDougal et al. 1968, p. 283.

[80] McDougal et al. 1967, p. 257; McDougal et al. 1968, p. 283; Higgins 1994, p. 5.

[81] Higgins 1994, p. 6; Reisman 2010, p. 46.

[82] Higgins 1994, p. 5.

[83] Koh 2007, p. 563.

Higgins therefore defines international law as 'an authoritative system of decision making available in a decentralised system to all authorised decision makers.'[84] In line with this definition, this book includes more than only the formally denoted entities among these authorised decision makers, and considers all those participants who are in *fact* forming the law.

4.3.2 Participants as Policy-Makers and Law-Appliers

International courts are traditionally presented as agents of the creating States; a claim that maintains the myth of State hegemony in international law. However States can regulate their creations only to a limited degree.[85] The independence and discretion of both judges and prosecutors which is necessary for these international criminal tribunals to operate, leads to a wider margin for policy-making and lawmaking without direct involvement of States. One example is the shift that was made by the ICTY with respect to the distinction between international and non-international armed conflicts for the purpose of the application of IHL. States had debated this when drafting the Additional Protocols in the 1970s, but had maintained a strict dichotomy between the two, and limited the rules of IHL with respect to non-international conflict.[86] However in the 1990s the ad-hoc tribunals 'updated' IHL to fit the circumstances of contemporary conflicts. As the ICTY Appeals Chamber put it:

> This dichotomy [between international and non-international armed conflict] was clearly sovereignty-oriented and reflected the traditional configuration of the international community, based on the coexistence of sovereign States more inclined to look after their own interests than community concerns or humanitarian demands.[87]

The move to expand the application of IHL was an express policy move made by non-State participants against traditional State interests, and outside of State regulation. At the same time, some aspects of the conflict in former Yugoslavia were internal rather than international, and thus expanding the application of some IHL rules was in fact in line with the original policy intention of the UN Security Council, and those States involved in the UN decision-making, to punish those involved in crimes in former Yugoslavia.[88] This illustrates the dynamic nature of law as process, and the inherent nature of policy within this process. The fact that this shift was later codified in the Rome Statute demonstrates that such policy was accepted by many States as having become the rule.[89]

There is some suggestion, and it would seem logical, that judicial policy making and law-making has increased dramatically with the increase in the number of

---

[84] Higgins 1994, p. 10.

[85] Danner 2006, p. 53.

[86] Henckaerts and Doswald-Beck 2005, p. xxxiv.

[87] Tadić Interlocutory Decision 1995, para 96.

[88] Danner 2006, p. 48.

[89] Statute of the ICC 1998, Article 8(2)(c)–(f); See also Danner 2006, p. 50.

international tribunals, and the increased importance of case law developments in ICL in particular.[90] Although States themselves, and those who adhere to a formalist view of the law, would not be quick to admit the role of tribunals and the judges sitting on them as policy makers and lawmakers, since this would undermine the myth surrounding States as the sole full subjects of international law, at the same time it would appear that these international judicial functions are both widespread and accepted. The most prominent example is the choice to leave the clarification of many issues to the Court when drafting the statutes of the ad-hoc tribunals and the ICC; some delegates to the ICC's Preparatory Commission argued that problems arising from ambiguity in the Rome Statute would be naturally addressed by the bench.[91]

In this respect, there may be a difference between 'judicial activism' in a domestic context, and the role of judges and tribunals as policy makers in the international context. In the domestic context, the separation between the judiciary and the legislature requires that judges maintain a certain restraint, even in the common law tradition where case law creates precedent. But in the international context, where there is no central legislator, and where the relationship between international tribunals and their creators is a very different one, the role of judges as policy makers may be less difficult to concede even from a more traditional perspective, at least when it comes to filling gaps in the law.[92] When it comes to more progressive lawmaking, a static perspective that operates from the subject-object definitions in international law would make it difficult to accept policy making by judges, since this would lead to law creation without direct participation of the 'subjects' of international law, i.e. States, thus raising questions as to legitimacy.[93] However from a dynamic perspective which moves away from the subject-object dichotomy, and instead takes into account a greater number and range of participants as law-makers, even the progressive assertions made by judges as to the content and interpretation of the law are accepted as legitimate law-making. The test, as previously noted, is whether there is sufficient authority and effectiveness, which in the case of international judges is unquestionable.

When it comes to other participants as policy makers and law-appliers, the question will turn on the extent of authority. Who has authority will depend on the nature of the claim; some actors will have more authority in ICL, and less in other fields of international law, both individually and as a group.[94] For example, States sometimes compete for influence with civil society in the form of NGOs, which claim to be representing the international community.[95] The role of NGOs in the drafting of

---

[90] One author has estimated that 63% of all international judicial activity to date has occured since the 1990s; Danner 2006, p. 4, citing Karen J. Alter, 'Do International Courts Enhance Compliance with International Law?', 25 Review Asian and Pacific Studies 51 (2003).

[91] Wessel 2006, p. 386.

[92] Wessel 2006, p. 388.

[93] Wessel 2006, p. 387.

[94] Higgins 1994, p. 50.

[95] Wessel 2006, p. 383.

the Rome Statute was very important when it came to some issues such as victim participation, and some of the crime definitions, and in this sense these participants were policymakers with sufficient authority, even though they were not the ones signing the treaty. Similarly the ILC, when drafting the Draft Code of Crimes Against Peace and Mankind that formed the basis of the Rome Statute, was not tasked with executing specific State policies, but with drafting a statute that could represent an international policy.[96] The ILC is made up of members chosen for their expertise, and while some of them are diplomats, in the end these individual members—as well as the ILC as a body of the UN—can also be seen as policy makers outside of direct State control. Furthermore scholars, the 'most highly qualified' of whom are recognised as producing subsidiary sources even according to a traditional view of law-makers, may have more authority in some contexts than others. Though scholars themselves would not often claim to be policy makers, it would seem that the direct citation of certain scholars by the ICC to support a specific policy choice in interpreting Article 25 of the Rome Statute has raised these scholars to the same level of authority internationally that they enjoy in their domestic fields of influence.[97]

4.3.3 The Policy Choice of Modes of Liability

If law is understood as a process of decision-making by authorised decision-makers, rather than a body of readily discoverable rules, then identifying the modes of liability most appropriate in the case of mass atrocity is not simply an exercise of applying legal knowledge to find a correct answer. Rather it is a matter of identifying the processes of interpretation and choice that already exist in ICL, and then identifying which of these match the values or goals which ICL seeks to promote, and the objectives it seeks to achieve.[98] The values central to ICL have been discussed in Chap. 3 and an attempt was made to reduce the diverse and sometimes competing goals to a single one: increasing the sense of accountability for mass atrocity crimes. Coupled with the importance of norm expression and the endeavour towards efficacy, these goals and objectives become the selection criteria for modes of liability, and the final choice is itself a part of the policy process that is the law.

In any given domestic criminal law system the development of a system of liability is a result of the the political-legal context of that State, the exact context of the crimes to which they are applied, and the demands of a specific society. As discussed above, the open texture of certain norms in law such as notions of liability, mean that their determination and application are contingent on socio-legal and historical

[96] Heller 2011b, p. 594.

[97] This is discussed in full in Chap. 8. For examples of the authority of some scholars, see the citations in Katanga Confirmation of Charges 2008, paras 498, 501; Lubanga Trial Judgment 2012, para 1004.

[98] With respect to this process in international law in general, see Higgins 1994, p. 10.

context.[99] Within domestic systems of liability, the specific modes of liability have been developed as policy responses to particular types of collective criminality, as will be investigated in Chaps. 6 and 7. These policy responses differ in each jurisdiction according to the types of crimes to which they have responded, and according to the socio-legal context of each jurisdiction. This is the amalgamation of law as process and law as policy, and these factors will therefore be examined in the comparison between jurisdictions in this study. In the special context of international criminal law and mass atrocity crimes, the type of collective criminality taking place may or may not match that which domestic criminal law is typically responding to. When we understand law as policy, and modes of liability as policy responses to criminal behaviour, then the translation from the domestic to the international requires a consideration of these factors.

Judicial policy-making and law-making are present in all kinds international tribunals, but what makes this particularly important in ICL is the fact that States may not be in a position to participate effectively in the development of the normative content of ICL once the constitutive instruments of the tribunals have been completed.[100] International criminal courts have the ability to take the principles negotiated by States, and develop their precision and applicability to the needs and context of ICL. This casuistic adjudication may also produce rules more attuned to the reality of large scale conflicts than the abstract formulae worked out in diplomatic conferences.[101]

4.4 The Problem of Ambiguity

There is a concern raised by some critics that an understanding of law as process and as the intersection between policy and authority leads to ambiguity as to what is law and what is not law.[102] The argument is that although a dynamic process-based approach to law allows for a wider analysis of participants involved in the formation of a norm, it still remains too open and vague when it comes to the ascertainment of that norm as law.[103] A formalist or static point of view only considers a norm to be a rule of international law if it meets a predefined formal standard, or pedigree.[104] This view strives to offer more certainty, however at the same time it limits the definition of law to the formal sources as understood according to Article 38 of the Statute of the ICJ, which is no longer a realistic reflection of the many law-applying authorities. This 'source thesis' therefore only allows for a limited formal ascertainment of law.[105]

---

[99] Hart 1972, p. 125; van der Wilt 2008, p. 263.

[100] Danner 2006, p. 52.

[101] Danner 2006, p. 51.

[102] Hathaway 2000, p. 129; D'Amato 2011, p. 66.

[103] d'Aspremont 2011b, p. 3.

[104] d'Aspremont 2014, p. 1134.

[105] d'Aspremont 2011a, p. 124; d'Aspremont 2014, p. 1145.

Hart described this pedigree approach in terms of a rule of recognition, which would provide the authoritative criteria for identifying primary rules of obligation.[106] To state that a primary rule of obligation is valid is to say that it satisfies all the criteria of a rule of recognition. In a system with a variety of sources, this rule of recognition must be more complex, involving legislation, a constitution—or in the case of ICL, various statutes—as well as case law, custom and perhaps other sources.[107] What defines a rule of recognition, however, is that it is the 'supreme and ultimate' secondary rule which is at the foundation of any legal system.[108]

The problem with the pedigree approach is that clear rules of recognition in international law are already elusive, and even more so in ICL in particular. Aside from the coming into force of treaties, even the formation of custom in ICL is debated due to the fact that there is very little historical basis which could be called State practice, and the differences between domestic criminal law systems means it is difficult to argue that any asserted rule of criminal law has been applied due to the belief that there is an international obligation to do so (*opinio juris*). Moreover, given the inherently open texture of norms in criminal law, and therefore in ICL, certainty is fairly unattainable. The open norm of liability can only be applied in concrete cases under consideration of all the facts, coupled with the desired policy response to the crime in question—which is why it is so important to identify the desired policy with more specificity than has been done to date.

The underlying ambiguity in most rules of international law is evident even when we consider the text of Article 21 of the Rome Statute, which was drafted to try and give specific and clear directions as to the applicable law before the ICC. The concern for too much judicial creativity led to the assertion of a clear hierarchy of sources in Article 21(1).[109] But this hierarchy is problematic because it does not reflect the interrelationship between sources in international law, and does not allow for the complexity of treaty law.[110] In this sense it departs from the approach taken to sources thus far in other international criminal tribunals, none of the statutes of which contain an article pertaining to applicable law, and many of which have referred to Article 38 of the ICJ Statute as their starting point.[111] Nor does it allow for the fact that open textured norms such as liability will require further interpretation by judges, who must by definition look beyond established 'principles of international law'.

The fact that there is no reference to the decisions of other international tribunals in Article 21 is noteworthy. Such decisions will not always fall under 'principles and rules of international law', since there is already much discrepancy between international tribunals on matters such as defences and modes of liability. With

---

106 Hart 1972, p. 97.

107 Hart 1972, pp. 98, 100.

108 By supreme he meant that a rule of recognition is the highest authority above all other rules, and by ultimate he meant that it is the final rule, as there are no other tests of validity. See Hart 1972, p. 103.

109 Schabas 2010, p. 384; Cryer 2009, p. 392.

110 Cryer 2009, p. 393.

111 Schabas 2010, p. 382.

respect to customary law, although it may fall under the wording of 'the principles and rules of international law', it would seem there have been some intentional derogations from custom at the ICC, such as the inclusion of indirect perpetration which had not previously found a place in the statutes of international tribunals, nor general acceptance in international case law.[112]

In its decision on the Arrest Warrant for Al Bashir in 2009, the ICC Pre-Trial Chamber stated that all other sources of law listed in Article 21 are meant only as supplementary to the Statute when there are lacunae in the Statute, the Elements of the Crimes and the Rules.[113] However this begs the question, since the question of whether there is a lacuna in these documents is itself a matter of interpretation.[114]

The problem of ambiguity is thus not unique to a policy-oriented view of the law. The problem it poses is not one that can be solved by taking a different theoretical stance, since even according to a formalist view, there is still no clarity currently on how to interpret and apply modes of liability, even with an extensive article denoting different forms of liability in the Rome Statute.

What a policy-oriented view of the law offers is a rule of recognition that is dynamic; namely, when a rule or norm is asserted by a participant with sufficient authority, then that norm can be considered law to the extent that it is accepted and applied as such. That this rule of recognition does not satisfy a positivist or formalist viewpoint is not to say that it fails as a rule of recognition.

Jeremy Waldron has pointed out that legal positivists sometimes exaggerate the need for certainty in the law, and for a rule of recognition, however when we consider that legal arguments proceed 'quite well without any great certainty about what follows or what does not follow from a given line of cases,' then it becomes apparent that the need for a rule of recognition is overplayed.[115] For instance, judges do not decide cases according to a sense of obligation to a rule of recognition in the same way that States might with respect to the drafting and signing of a treaty.[116] Even when they refer to previous case law as precedent or as instructive, judges pay attention to a vast array of normative considerations. In some cases, a commitment to integrity or reliance may be the drive behind keeping faith with previous decisions, in other cases there may be other normative considerations which carry overwhelming force, leading to a decision that departs from previous case law.[117] There may be in fact several rules of recognition, which may even come into conflict with one another, and in this sense they are not ultimate.[118]

But perhaps a middle ground is possible. Legal theorist Jean d'Aspremont suggests that a combination of the dynamic analysis of participants can be combined with a

---

112 Cryer 2009, p. 399.

113 Al Bashir Decision on Arrest Warrant 2009, para 126.

114 Cryer 2009, p. 397.

115 Waldron 2009, p. 13.

116 Waldron 2009, p. 14.

117 Waldron 2009, p. 15.

118 Waldron 2009, p. 22, citing Jospeh Raz, (Practical Reason and Norms, 2nd edition, Princeton University Press, 1999).

formalist requirement of the pedigree of a norm; that is, a combined 'source thesis' and 'social thesis', that would take into account the social practices that give rise to a norm being apprehended as a rule, while still requiring the elevation of this norm into a recognised pedigree as law.[119] An assertion by the media, for instance, would not pass this test,[120] but assertions by non-State actors such as the ICRC, or by practicing lawyers, judges, and scholars whose work is cited as authoritative interpretations of law, may well be elevated to the status of law.

Thus, instead of reading from the static, classical doctrine of sources, this dynamic model demonstrates a more realistic picture of the processes leading to the formation of different modes of liability as law.

4.5 A Comparative Law Approach to Resolving the Search for Modes of Leadership Liability

What is asserted in this chapter is both a descriptive theory about the formation of international criminal law, and a methodology for understanding and mapping out this formation. Since the continuing debates about 'correct' interpretations and applications of modes of liability on the international plane are to do with the role that domestic law plays, either directly imported or as a source of influence, it is necessary to consider what role domestic law has as a source of international law, and of ICL in particular.

In this sense there is much to be learned from the comparative law scholarship on legal traditions and legal transplants. By paying full attention to the characteristics and context of the two legal traditions which are informing the development of ICL namely the common law and civil law traditions, the analysis that ensues in Part II of this book paints a full background to understanding the debates at the international level. It may be common knowledge that there is a difference between a more unitary approach and a differentiated approach, but the historical, policy and socio-legal reasons for these differences may not be obvious without consideration of the legal traditions from which they hail. Similarly, this investigation can then aid in comparing the context of ICL, which does not comfortably fit in one tradition or the other, and in considering which factors may overlap. This process can guide the final selection of which system of liability is most appropriate to the specific context of ICL.

The express inclusion of policy as part of the definition of law, and not just as a factor in understanding how the law operates, is not unproblematic. Policy-making by judges and prosecutors is not necessarily what States had in mind when they partook in the creation of international criminal tribunals, and some criticism has been aimed at the alleged unfettered discretion of the Prosecutor to instigate politically motivated investigations. In fact, prosecutorial independence may have been one of the biggest

---

119 d'Aspremont 2011a, p. 24; d'Aspremont 2014, p. 1145.

120 Compare Michael Reisman, who argues that the media could be considered a participant with sufficient controlling intent: Reisman 2000, p. 13; Reisman 2010, p. 104.

factors in the US 'un-signing' the Rome Statute.[121] However, given the inherently political nature of prosecutions of mass atrocity, including the deligitimization of a repressive regime, the expressionist or didactic role of such prosecutions and the development of IHL and ICL norms, to deny the role of policy in lawmaking and law-applying is to deny the reality in which international law plays out.

Once again, this stance may be considered controversial to many who read it. It is one thing to say the law is informed by policy, it is another to say that this is what judges should be doing when coming to a decision in a particular case. Though it may be unsatisfactory to some, the intention of this book is not to critique the existence of this process as to whether or not it is correct. Rather it is to observe the process, and from this observation seek to understand it, and find a way to improve it. Merely criticising the fact that it exits or questioning whether it should be the way in which decisions are made and the law is formed, adds no guidance to the players involved. Placing it in a broader theoretical context provides a methodology for analysis, and in the end some tools for improvement.

The question remains how to go about selecting a system of liability that is most appropriate for mass atrocity prosecutions. The most obvious way, given the understanding of the development of ICL as comparative law in action, is to continue to undertake a comparative search and analysis. What this requires, however, is to move beyond arbitrary patchworking and towards a process that is self-conscious, thorough, and that applies a methodology of comparative law which will ensure a well grounded and non-arbitrary result. The following chapter will set out what is required by this methodology, before such a study is undertaken in Part II.

References

Ambos K (2007) The structure of international criminal procedure: 'adversarial', 'inquisitorial', or mixed? In: Bohlander M (ed) International criminal justice: a critical analysis of institutions and procedures. Cameron May, London

Berkowitz D (2003) The transplant effect. Am J Comp Law 51:163

Berman PS (2012) Global legal pluralism: a jurisprudence of law beyond borders (introduction). Cambridge University Press, New York

Boyle A, Chinkin C (2007) The making of international law. Oxford University Press, Oxford

Cryer R (2009) Royalism and the King: Article 21 of the Rome Statute and the politics of sources. New Crim Law Rev 12:390–405

Damaška M (2004) Negotiated justice in international criminal courts. J Int Crim Justice 2:1018

D'Amato A (2011) Non-state actors from the perspective of the policy-oriented school: power, law, actors and the view from New Haven. In: d'Aspremont J (ed) Participants in the international legal system, Chapter 4. Routledge, London, pp 64–75

Danner AM (2006) When courts make law: how the international criminal tribunals recast the laws of war. Vanderbilt Law Rev 59:1–64

d'Aspremont J (2010) The doctrinal illusion of the heterogeneity of international lawmaking processes. Sel Proc European Soc Int Law 2:297–312

[121] Wessel 2006, p. 380.

d'Aspremont J (2011a) Non-state actors from the perspective of positivism: the communitarian semantics for the secondary rules of international law. In: d'Aspremont J (ed) Participants in the international legal system: multiple perspectives on non-state actors in international law. Routledge, London, pp 23–40

d'Aspremont J (2011b) Non-state actors in international law: oscillating between concepts and dynamics. In: d'Aspremont J (ed) Participants in the international legal system: multiple perspectives on non-state actors in international law. Routledge, London, pp 1–21

d'Aspremont J (2014) Cognitive conflicts and the making of international law: from empirical concord to conceptual discord in legal scholarship. Vanderbilt J Trans Law 46:1119–1147

Delmas-Marty M (2002) Towards a truly common law: Europe as a laboratory for legal pluralism. Cambridge University Press, Cambridge

Delmas-Marty M (2003) The contribution of comparative law to a pluralist conception of international criminal law. J Int Crim Justice 1(1):13–25

Delmas-Marty M (2008) Reflections on the 'hybridisation' of criminal procedure. In: Jackson J, Langer M, Tillers P (eds) Crime, procedure and evidence in a comparative and international context: Essays in honour of Professor Mirjan Damaška. Hart Publishers, Oxford, pp 251–259

Etcheson C (2004) The politics of genocide justice in Cambodia. In: Romano C, Nollkaemper A, Kleffner J (eds) Internationalized criminal courts and tribunals: Sierra Leone, East Timor, Kosovo and Cambodia. Oxford Scholarship, p 181

Falk RA (1995) Casting the spell: the New Haven school of international law. Yale Law J 104:1991–2008

Ferrante M (2011) Argentina. In: Heller KJ (ed) Dubber MD. The handbook of comparative criminal law. Stanford University Press, pp 12–49

Fletcher GP, Ohlin JD (2005) Reclaiming fundamental principles of criminal law in the Darfur case. JICJ 3:539–561

Glenn HP (2001) Are legal traditions incommensurable? AmJCompL 49:133

Glenn HP (2006) Comparative legal families and comparative legal traditions. In: Zimmermann R, Reimann M (eds) Oxford handbook of comparative law. Oxford University Press, pp 421–440

Glenn HP (2010) Legal traditions of the world: sustainable diversity in law, 4th edn. Oxford University Press, USA

Hart HLA (1972) The concept of law, 2nd edn. Oxford University Press

Hathaway JC (2000) America, defender of democratic legitimacy? Eur J Int Law 11(1):121–134

Heller KJ (2011a) The Nuremberg Tribunals and the origins of international criminal law. Oxford University Press, Oxford

Heller KJ (2011b) The Rome Statute of the International Criminal Court. In: Heller KJ, Dubber MD (eds) The handbook of comparative criminal law. Stanford University Press, Stanford, pp 593–634

Henckaerts JM, Doswald-Beck L (2005) International committee of the Red Cross's study of customary international humanitarian law. Cambridge University Press, Cambridge

Higgins R (1994) Problems and process: international law and how we use it. Oxford University Press, Oxford

Hoecke MV, Warrington M (1998) Legal cultures, legal paradigms and legal doctrine: towards a new model for comparative law. Int Comp Law Q 47:495–536

Kingsbury B (1999) Foreword: Is the proliferation of international courts and tribunals a problem? NYU J Int Law Polit 31:679

Kingsley JJ (2004) Legal transplantation: is this what the doctor ordered and are the blood types compatible? Ariz J Int Comp Law 21:493

Koh HH (2007) Is there a 'new' New Haven school of international law? Yale J Int Law 32:559

Koskenniemi M (2009) The case for comparative international law. Finn Yearb Int Law 20:1–8

Levit JK (2007) Bottom-up international lawmaking: reflections on the New Haven school of international law. Yale J Int Law 32:393

Lieber Code (1863) Instructions for the Government of Armies of the United States in the Field

McDougal MS, Lasswell HD, Reisman WM (1967) The world constitutive process of international decision. J Leg Educ 19:253
McDougal MS, Lasswell HD, Reisman WM (1968) Theories about international law: prologue to a configurative jurisprudence. Va J Int Law 8(2):188
McDougal MS, Lasswell HD, Reisman WM (1988) The world community: a planetary social process. Univ Calif Davis Law Rev 21:807
Miller JM (2003) Typology of legal transplants: using sociology, legal history and Argentine examples to explain the transplant process. Am J Comp Law 51:839–885
Nagan W, Hammer C (2007) Communications theory and world public order: the anthropomorphic, jurisprudential foundations of international human rights. Va J Int Law 47:725
Nelken D (2006) Legal culture. In: Smits JM (ed) Elgar encyclopedia of comparative law. Edward Elgar Publishing, Cheltenham, UK/Northampton, MA, USA, pp 372–381
Orford A (2012) In praise of description. Leiden J Int Law 25(3):
Örücü E (2002) Law as transposition. Int Comp Law Q 51(2):205–223
Posner RA (2010) How judges think. Harvard University Press
Preparatory Commission Report (1996) Report of the UN Preparatory Commission of the International Criminal Court, Vol. I
Reisman WM (1981) International lawmaking: a process of communication. Am Soc Int Law Proc, p 101
Reisman WM (1992) The view from the New Haven school of international law. Am Soc Int Law Proc 86:118–124
Reisman WM (2000) Unilateral action and the transformations of the world constitutive process: the special problem of humanitarian intervention. Eur J Int Law 11(1):3–18
Reisman WM (2010) The quest for world order and human dignity in the twenty-first century: constitutive process and individual commitment, collected courses of the Hague Academy of International Law, vol 351. Martinus Nijhoff, Leiden
Roberts A (2011) Comparative international law? The role of national courts in creating and enforcing international law. International and Comparative Law Quarterly. ICLQ 60:57–92
Robinson D (2013) The two liberalisms of international criminal law. In: Stahn C, van den Herik L (eds) Future perspectives on international criminal law. Ashgate, Surrey, pp 307–322
Schabas W (2010) The International Criminal Court: a commentary on the Rome Statute. Oxford University Press, Oxford
Schachter O (1968) Towards a theory of international obligation. Va J Int Law 8(2):300
Statute of the ICC (1998) Statute of the International Criminal Court
Statute of the ICJ (1945) Statute of the International Court of Justice
Steer C (2011) Non-state actors in international criminal law. In: d'Aspremont J (ed) Participants in the international legal order: multiple perspectives on non-state actors in international law. Routledge, London, p 96
Steer C (2015) Legal transplants or legal patchworking? The creation of international criminal law as a pluralistic body of law. In: van Sliedregt E (ed) Harmonization and pluralism in international criminal law. Oxford University Press
Trapani A (2011) Assessing the impact of the international ad-hoc tribunals on the domestic courts of the Former Yugoslavia. http://www.domac.is/media/domac/Domac-11-BAlkan-AT-Final.pdf
van der Wilt H (2008) Equal standards? On the dialectics between national jurisdictions and the International Criminal Court. Int Crim Law Rev 1(2):229–272
van der Wilt H (2009) The continuous quest for proper modes of criminal responsibility. J Int Crim Justice 7(2):307–314
van der Wilt H (2010) National law: A small but neat utensil in the toolbox of international criminal tribunals. Int Crim Law Rev 10(2):209–241
Waldron J (2009) Who needs rules of recognition? In: Public Law and Legal Theory Workshop Paper Series (Paper No. 09–21)
Watson A (1974) Legal transplants: an approach to comparative law. Scottish Academic Press, Edinburgh, London

Watson A (1983) Legal change: sources of law and legal culture. University of Pennsylvania Law Review. UPALR 131:1121–1157
Watson A (1984) Sources of law, legal change, and ambiguity. University of Pennsylvania Press
Wessel J (2006) Judicial policy-making at the International Criminal Court: an institutional guide to analyzing international adjudication. Columbia J Trans Law 44:377–452
Wolff P (1985) McDougal's influence: utility, influence, controversy. Am Soc Int Law Proc 79:266
Zahar A, Sluiter G (2008) International criminal law: a critical introduction. Oxford University Press, Oxford
Zweigert K, Kötz H (1998) An introduction to comparative law. Oxford University Press, Oxford

Cases

Akayesu (1998) The Prosecutor v. Akayesu, Trial Judgment, ICTR-96-4-T
Al Bashir Decision on Arrest Warrant (2009) The Prosecutor v. Omar Hassan Ahmad al Bashir, second decision on the prosecution's application for a warrant of arrest against Omar Hassan Ahmad al Bashir, ICC-02/05-01/09
Erdemović, Sentencing Judgment (1996) The Prosecutor v. Erdemović, Sentencing Judgment, IT-96-22-T
Furundžija (1998) The Prosecutor v. Furundžija, Trial Judgment, IT-95-17/1-T
Katanga Confirmation of Charges (2008) The Prosecutor v. Germain Katanga and Mathieu Ngudjolo Chui, decision on the confirmation of charges, ICC-01/04-01/07
Tadić Appeals Judgment (1999) The Prosecutor v. Dusko Tadić, Appeals Judgment, IT-94-1-A
Tadić Interlocutory Decision (1995) The Prosecutor v. Dusko Tadić, decision on the defence motion for interlocutory appeal on jurisdiction, IT-94-1-A

Chapter 5
Applying the Tools

Contents

The process of legal patchworking described in the previous chapter, akin to legal transplantation, is a dynamic process involving judges, practitioner lawyers and scholars. As these participants strive to clarify which modes of liability do and should apply, comparative law is the obvious methodology. However this patchworking process has rarely taken place with attention to methodology, rather it has been arbitrary, depending upon which participants are most influential in a given forum, and what their preference is for specific domestic systems of liability. The result is a lack of uniformity among the tribunals and even within tribunals, due to a bias in selectivity and a lack of justification—or of satisfactory justification—for the importation of certain domestic notions over others. What has emerged is a lack of clarity and a clash of legal cultures, undermining both the predictability required for defendants as well as the consistency required for a stable and functioning body of criminal law.

To clarify the confusion of terminology, and to ascertain a system of liability most appropriate for leaders of mass atrocity, it would therefore be prudent to apply a comparative law method in a more deliberate and focused manner. Not only will this clear the path towards a well founded selection of a system of liability, but furthermore the application of a thorough and transparent method of comparison may also contribute

C. Steer, *Translating Guilt*, International Criminal Justice Series 9,
DOI 10.1007/978-94-6265-171-5_5

some consistent methodological restraints on those participants involved in an otherwise unfettered process of law formation and application. Although this does not amount to a determinate rule of recognition, such constraints may nonetheless aid in calming the arbitrariness that can be associated with a dynamic understanding of the law.

Precisely because of the diversity and specificity of domestic criminal law systems, and because these traditionally subsidiary sources have such an important role to play in the development of ICL, a common method can also better protect the due process rights so essential to the legitimacy of any criminal law system.

This chapter will therefore lay down the comparative methods to serve a comparative methodology.[1] First, in Sect. 5.1, a functionalist approach will be described and a series of criteria will be set out against which different modes of liability can be tested when searching for the most appropriate way in which to consider leadership liability. In Sect. 5.2, the selection of jurisdictions to be compared in this study will be explained. And in Sect. 5.3 attention will be paid to defining the most important terms to be used throughout this study, including complicity, guilt, liability and responsibility. The methods laid out here will be applied throughout Part II of this study, and will provide the platform for the conclusions drawn in Part III. Although participants such as practitioner lawyers and judges may not have the luxury of undertaking such an expansive study, in principle the methods laid out here can also provide guidance to those directly involved in the formation and application of norms in ICL.

5.1 Functionalism

Comparativists will agree there is no single correct method of comparative law; the nature of a comparative study depends upon the aims and objectives of the study, as does, to a certain extent, the selection of jurisdictions to be compared.[2] The present comparative study has a twofold aim. Firstly, to understand the processes by which modes of liability have been developed on the international plane to date, including a deeper understanding of the domestic models which have been drawn upon. This is where comparative law as a methodology, an understanding of the law as a whole, is most useful. And secondly, to offer a solution to the continued debates about which system of liability is most appropriate for leadership of mass atrocity on the international plane, based upon the analysis of these processes and the domestic models. This is where the tools of comparative law as a method will be applied.

---

[1] To be clear, what is meant by methodology here is the analysis of the principles or processes of the enquiry as a whole, and what is meant by method is the procedures or techniques employed to answer the specific question posed. So here the methodology is the framework for understanding ICL as comparative law in action, and the methods are the specific comparative tools engaged, to be described further in this chapter. For further discussion of this distinction, see Maxwell 2008, pp. 222, 233.

[2] Örücü 2006, p. 451; Oderkerk 2001, p. 310.

As has been described in the previous chapter, the process of patchworking in ICL has similarities to the processes of legal transplants that take place between domestic systems which are seeking an internal law reform. Although ICL is not a project of law reform, but rather of patchworking a new system together from IHL, international law and domestic criminal law, the choice as to which comparative methods to apply can be informed by this likeness with legal transplants as law reform. A functional or institutional approach is therefore best suited to this enquiry: rather than looking for modes of responsibility with a certain description or title, such as 'complicity', instead the search is for modes of responsibility which fulfil a certain function,[3] namely to what extent a person who has a leadership role within an organised group is culpable for the criminal acts committed by other (often lower ranking) members of that group. This is a functional rather than a technical question.[4]

Functionalism in comparative law is understood to mean many things. In 1971 the German comparativists Zweigert and Kötz claimed that the basic methodological principle of all comparative law is functionalism,[5] and since then the term has been used ubiquitously but without consistency as to its meaning. For some it is a return to a form of natural law, or *jus gentium*, in the sense that all legal problems and all their solutions are expected to be similar across all jurisdictions: solutions can be found from 'the nature of things'.[6] This interpretation of functionalism can be described as teleological, because it seeks to explain social norms and structures by the ends they serve.[7] This is not the definition of functionalism applied here; the underlying assertion of this book is that there are many possible models of liability that could be applied which may be in some ways similar, but there is not one single 'correct' model. No domestic model is inherently superior to any other, the question is which one is best suited to deal with the specificity of mass atrocities.

For others, functionalism can be used as a kind of social engineering; because the law responds to society's needs, those institutions or rules that adapt best will survive, and these are the ones that can best be transplanted.[8] Another version of this interpretation is instrumentalist functionalism; since the law fulfils functions and meets societal needs, then the lawyer's job is to develop laws to perform these tasks, and comparative law can help find the 'best' rules or institutions to do this, and thus ensure progress.[9] But this is also not the interpretation applied here, since these adaptationist or instrumentalist approaches apply functionalism as a theory rather than a method.

In this book functionalism is applied very much as a method, and in this sense could best be described as equivalence functionalism, or the idea that similar functional

[3] Zweigert and Kötz 1998, p. 34; Örücü 2006, p. 443.

[4] Örücü 2006, p. 34.

[5] Zweigert and Kötz 1998, p. 34.

[6] Michaels 2006, p. 346.

[7] Lempert 2008, p. 398.

[8] Michaels 2006, p. 348.

[9] Michaels 2006, p. 351.

needs in different societies can be fulfilled by different institutions.[10] Rather than searching for universals, or a 'common core', this functional method instead seeks to explain each institution in the jurisdictions compared as one contingent solution among several possibilities. Thus, the various systems of liability and different modes of liability within those systems are contingent on the society in which they were developed. The systems of law and the elements within them are not independent from each other.

Because the normative content of criminal law is very specific to the procedural, political and historical choices made in each domestic system, which in turn are influenced by similar factors within each legal tradition, these background factors must not go ignored. On the contrary, they must be brought to the foreground if a suitable translation is to be made of a domestic notion to the international system of ICL. Thus, while the 'data' being studied and compared consists of the positive law in selected legal systems, it is also necessary to understand a legal institution (such as a mode of criminal responsibility) in the context of the system in which it appears, including the policy choices which led to its inception and the way in which participants in the processes of law apply it. This 'mapping' process is in fact required by comparative law; such contextual analysis is needed in order to know whether translation of this institution to other jurisdictions is appropriate.[11]

Furthermore, in order to test the hypothesis that policy drives the choices made by participants when it comes to the legal institutions of modes of responsibility, it is necessary to identify the context in which these modes are rooted. Unlike elements of crimes, concepts of criminal responsibility have to 'fit into the legal texture of domestic systems where they face the competition of tried and tested equals'.[12] The legal cultural background to each of the modes of liability and their application within each jurisdiction will therefore be investigated, as well as the ways in which these have been translated to the international context.

Factors of relevance are the types of crimes for which these modes of responsibility are applied (economic crimes, crimes of violence), the historical context (for example the oppression of the social phenomena of collective crimes, the need to deal with gang violence or, on a grander scale, conflict situations), the context of legal tradition in which these modes exist (the common law or civil law traditions) and other aspects of the domestic system (for example procedural rules) which might influence the inception and application of each mode.

Even among the different interpretations of functionalism there are some core agreements among comparativists: first, functional comparison focuses not on rules, but on their effects; second, it combines a factual analysis with the theory that these facts must be understood in the light of their functional relation to society; and third, the function of the institutions being compared serves as the yardstick for comparison, or the *tertium comparationis*.[13] This is why the functional questions

[10] Michaels 2006, p. 357; Örücü 2006, p. 443.

[11] Zweigert and Kötz 1998, p. 38.

[12] van der Wilt 2008, p. 254.

[13] Michaels 2006, p. 342; Oderkerk 2001, p. 310.

posed above are so important: in each of the jurisdictions compared, to what extent are leaders responsible for the acts of others within an organisation, and based on what *functional* understandings of liability does this responsibility get translated to criminal guilt? And in the context of mass atrocities, which of these understandings and models is most appropriate?

The evaluations and conclusions drawn from the comparative aspect of this study give the full context for being able to understand the different approaches in domestic criminal law models, however a normative measure is still needed in order to make a specific selection of which modes of liability are most appropriate for leadership responsibility in the context of mass atrocity crimes. Because the functional equivalence method of comparison takes into account the social and political context of the development of modes of liability in each jurisdiction, the normative measure for selection must also include these contextual questions. A set of evaluation criteria based on this method is therefore proposed here:

i. Does the mode of responsibility reflect the collective nature of the crime?
ii. Does the mode of responsibility allow for identification of the person responsible within the group even when there is no clear hierarchical line?
iii. Could it operate in a different legal tradition (for instance where procedural rules play a more or less determinative role in the application of substantive law)?
iv. How has it been applied judicially, and what kinds of interpretation have been given to it in various social contexts?
v. Are these contexts akin to that of mass atrocity?

Some counter-indications for applying a mode of responsibility in the specific context of ICL would include the following: if the mode of liability does not serve the goals of ICL, as enumerated in Chap. 3; if it relies upon specific procedural rules particular to the legal system from which it hails; if it does not sufficiently protect the principle of culpability when applied to collective crimes of a large scale.

These criteria and counter-indications will be tested in Part III of the book, where conclusions are drawn based on the comparative study undertaken in Part II.

5.2 Selection of Jurisdictions

Any comparative study is by definition selective in that only a restricted number of jurisdictions can be compared. In order to avoid bias (as far as this is possible) or arbitrariness, the selection of jurisdictions must be made explicit and justified. One of the criticisms often levelled at the interpretation of JCE at the ICTY, or of indirect perpetration at the ICC, is that the domestic jurisdictions which were cited as authoritative were overly selective and insufficiently representative of different legal traditions.[14] The reasons for selecting the jurisdictions compared in this book are therefore made explicit here.

[14] See the discussions in Chap. 8.

Certain domestic jurisdictions have already been explicitly drawn upon in the international tribunals in an attempt to clarify and justify a particular interpretation. These jurisdictions can therefore be a starting point for comparison in this study, but given that the selection made by the tribunals is not always sufficiently broad, this must not be the end point. The next criterion is to select jurisdictions which can be instructive due to the higher level of development on the topic in question.[15] A third criterion can be to consider whether there are systems which are representative.[16] This stems from the idea that for each legal problem only a few clearly different solutions have been conceived of in law.[17] Those systems which are representative of a solution that appears in various other jurisdictions may be included. A final negative selection can be made based on language limitations and other limiting reasons.[18]

With these criteria in mind, the following jurisdictions have been selected:

Germany has been selected based on all three positive criteria. Firstly, German doctrine and case law have been cited explicitly by the ICC, and are at the centre of recent developments on the debate on modes of liability. The investigation made in this book aims to make this doctrine as clear and accessible as possible, since the historical and policy context in which it was developed has been left out of the judicial discussions at the ICC thus far. Secondly, the doctrine in this system is highly developed on the problem of holding a person in a leadership role within an organisation culpable for the criminal acts of his or her inferiors within that organisation, including for crimes which are of a similar nature to mass atrocity crimes. Thirdly, the solution offered in Germany is representative since this model has been borrowed and applied by several other domestic jurisdictions.

The second step was to ensure different legal traditions are represented in this study, and to look beyond European and North America systems, which gain the most attention in many comparative studies and in the judicial debates on modes of liability. For this reason, Argentina has been selected as a second jurisdiction representing the civil law tradition. Although the Argentine criminal law system draws heavily on the German model, it is a particularly interesting case study in the transition from conflict to criminal justice, due to the way in which the judiciary and scholars responded to the atrocities in the so-called 'Dirty War' of 1976–1983.[19] The approach in Argentina may also be considered representative for similar systems in Latin America, since it was one of the first to take this approach, and has been cited in

---

15 Oderkerk 2001, p. 313.

16 Oderkerk 2001, p. 317.

17 Oderkerk 2001, p. 317, citing U. Drobnig, in E. von Caemmerer et. al. (Jus Privatum Gentium, 1969), p. 222.

18 The scope of this book is restricted to a comparison between common law and civil law traditions, because these two traditions inform the development of international criminal law. A comparison with notions of liability in Islamic law, or traditions such as Chinese law, or indigenous traditions in countries where such mass atrocities have taken place would offer an enriching understanding of leadership responsibility, criminal liability, and guilt. However this goes beyond the scope of this study.

19 See Sect. 7.4.

subsequent case law and doctrine in many Latin American countries. Furthermore, Argentina has modelled some of the elements of its criminal law system on a common law tradition, making it a good candidate for comparing the impact of procedural rules on modes of liability, and the impact of the patchworking process taking place in ICL.

Two systems which are representative of the solutions which often appear in the common law tradition have also been selected. Firstly, the USA has been selected, in part because it may be considered representative, because of the strong influence jurists from the US have had as participants in the processes of ICL. In particular, the notion of conspiracy played a part in the drafting of the Nuremberg Charter, in such a way that suggests it stemmed more from the American application of this notion than the English application.[20] Although typically an adversarial system, which has very different rules of procedure and evidence from systems with a more inquisitorial tradition, the USA has developed its own distinct system of liability.

Canada has been selected as the second system representing the common law tradition. The intention was to choose a jurisdiction that would offer a counter-balancing perspective to the interpretation of certain modes of liability found in the USA. Due to the fact that Canada applies the notion of conspiracy only as an inchoate crime and not, in contrast to the USA, as a mode of liability, this made it a good candidate for comparison. Since both Canada and the USA have inherited their criminal law systems from English common law, this provides a wider lens on the possibilities of interpretation, and the historical, social and policy reasons for such choices at a domestic level, and how these choices may be instructive for the choices to be made at an international level. Canada may also be considered representative in that it has been pro-active at a domestic level when it comes to suspects of international crimes.[21] Some attention will therefore be paid to the role played by domestic modes of liability in application to the context of mass atrocity crimes, and how this domestic jurisdiction interprets developments at an international level.

Finally, with reference to the final criterion described above, limits of space, time and language were part of the final selection.[22] Furthermore a comparative study benefits from a reduced number of jurisdictions, since to spread the selection too wide risks lack of depth, and difficulties in the comparative analysis.[23]

---

20 See Sect. 6.5.3.

21 Lafontaine 2012, pp. 23, 31, 32.

22 With respect to language as a limitation, I used my background in Dutch and undertook a course in reading German in order to access the necessary materials, since English translations of the core materials were limited or unavailable. Many German doctrinal materials have been translated into Spanish for the benefit of Argentine and other Latin American jurisdictions heavily influenced by the German tradition. For this reason, studies were also undertaken to improve my level of Spanish in order to access these materials as well as original Argentine case law and materials. Wherever translations of original materials are my own this has been noted in the footnotes. Native speakers have been consulted on the final text, however any errors remain my responsibility.

23 Örücü 2006, p. 41.

It should be noted that an explicit effort was made not to select jurisdictions based on their supposed role as 'mother' systems within what can be termed legal families.[24] This is one reason why England, for example, was not selected, despite it being the jurisdiction from which many others have inherited their common law tradition of criminal law. The avoidance of such a selection is motivated by several reasons. Firstly, the notion of legal families tends to be more useful for a project of private law comparison than for criminal law.[25] Public law comparative projects may be less numerous due to the fact that domestic jurisdictions do not seek to model or re-model their constitutional, administrative and criminal law by borrowing from other systems in the same way as they do with respect to their private law rules, since the public law aspects of a State's legal system are inherently caught up in the identity of the State.[26]

Secondly, such taxonomies can only be relative, since they are contingent on time, and also on the purposes of each particular comparative study.[27] Since this study is interested in the social, policy and legal cultural reasons for the development and interpretation of modes of liability, choosing from a taxonomy of legal families would prove an insufficient basis for selection.

And thirdly, the preference has been to follow the notion of legal tradition over legal family, since this facilitates a comparison that is focused on the bringing together of different solutions, rather than the competition of differences.[28]

5.3 Terminology: The Problems of Translation

Because the starting point of this comparative study is a functional-institutional approach, the *tertia comparationiae*, or points of comparison must be described with a lexicon and vocabulary that represents the function, and which are therefore as neutral as possible. The greatest pitfall in any comparative study is to take the most familiar jurisdiction as the 'norm', and to use terms and terminology from that system as the point of comparison. This can result not only in a skewed analysis, but also in confusion when translating terms from different languages, as well as in cross-cultural translation between systems which are English-language, but which make different use of the terminology.[29] The problem of *faux amis* is familiar to any translator, and should be one that international criminal lawyers are more aware of, especially when using terms such as complicity or principal and secondary liability.[30] Indeed, a criminal law comparativist may need to be better at understanding

[24] Watson 1974; Zweigert and Kötz 1998, p. 64.

[25] Zweigert and Kötz 1998, p. 65.

[26] Ziller 2006, p. 607.

[27] Zweigert and Kötz 1998, p. 65.

[28] See the full discussion in Sect. 4.1. See further Glenn 2006, p. 439; Glenn 2010.

[29] Örücü 2006, p. 447.

[30] Pradel 2002, p. 41.

the nuances of the legal systems being compared, in a contextual sense, than at being a good linguist or translator.[31] As Fletcher has put it: 'Language is the paradigmatic expression of local culture. Any concept that derives from language must, by definition, be tied to the world embodied in that language.'[32]

The translations throughout this book include linguistic translations from Spanish and German to English, translation of technical terminology reflecting different theories (derivative, accessory), modes of liability (principal, secondary, accomplice), and translations of concepts of responsibility, culpability and guilt, which are not always understood nor measured in the same way from system to system. As well there are translations from the domestic context to the international, from 'normal' crimes to crimes of mass atrocity, in which the political and social context leads to different notions of what kind of behaviour has become the norm, and to complex power paradigms. This may lead to different assessments of responsibility, culpability and guilt.

For all these reasons, a lexicon will be maintained throughout this book that is as neutral as possible, in that it strives to be broad and flexible enough to embrace the linguistic and systemic differences, while still describing the functions that are being compared. This lexicon can be described as follows.

5.3.1 Unitary and Differentiated Systems

The debate as to whether or not a distinction should exist between participating members in a collective crime centres on reasoning drawn from different domestic legal traditions. Those who hail from criminal law jurisdictions in which a distinction is made, may reject a system which considers all those who are involved in a collective crime to be equally culpable and therefore equally responsible. On the other hand, those who hail from jurisdictions in which no normative distinction is made may see no point in drawing complex dogmatic lines, particularly if the sentencing does not reflect a greater degree of guilt. In the pluralistic field of ICL where this question is left to the judges in various tribunals, the search for an answer appears to give rise to a clash of legal cultures.

The question of differentiation goes to the core of the notion of complicity, and how this differs between criminal law systems.[33] The term 'complicity' will be used cautiously throughout this study, as defined below in Sect. 5.3.2, but it deserves some attention here in order to understand the differences between unitary and differentiated systems. In its simplest terms, complicity can be defined as helping another (the perpetrator) in the commission of a crime.[34] In the Nuremberg Principles formulated by the ILC in 1950, 'complicity in the commission of a crime against peace,

---

[31] Pradel 2002, p. 42.

[32] Fletcher 2007, p. 291.

[33] Dubber 2007; Stewart 2014.

[34] Fletcher 2000, p. 637; Smith 1991, p. 2; Stewart 2014, p. 1.

a war crime or a crime against humanity' was defined separately from the notion of committing one of these crimes.[35] Similarly, complicity is explicitly mentioned in the Torture Convention, where it is stated 'Each State Party shall ensure that all acts of torture are offences under its criminal law … [as well as] any act by any person which constitutes complicity or participation in torture.'[36] While complicity is clearly different from commission in some sense, the question is to what extent this difference is legally relevant: is there differentiation or not? Once again we are faced with the complexities of translation, since these international documents were written in multi-lateral contexts, and with an attempt to find English terms to cover multiple meanings of assisting in the commission of a crime.

5.3.1.1 Unitary Theories

Some systems of criminal liability start from the point that there is no need to consider degrees of personal blameworthiness, or define what is required to *complete* an act, as opposed to helping another to do so, and instead prefer a single concept: 'in or out', you were involved or you were not involved, and thus you are liable or you are not. In German theory this is known as *Einheitstäterschaft*, translated into English as a unitary theory.[37]

Under the German terminology, *Einheitstäterschaft* means that the liability of every person involved in a crime is entirely independent of any other persons; a plurality of persons amounts to a plurality of offences.[38] This could be termed a 'pure unitary' system, in which all those individuals who have causally contributed to a crime, and who have the requisite fault standard, are equally and independently liable: minimal causal contribution and blameworthy moral choice are considered to be both necessary and sufficient conditions for liability.[39] There is no difference between perpetration and complicity, and no regard for the form that participation takes. If there is any individualisation of the parties to a crime, this is left to the sentencing phase and to the discretion of the judge, dependent on the system in question.[40] In this respect, unitary systems could be called sentencing-oriented.[41]

---

[35] Principle I refers to 'any person who commits (an international crime) is responsible and therefore liable to punishment', and Principle VII refers to complicity; ILC 1950, p. 1144.

[36] Convention Against Torture 1984, Article 4.

[37] Vogel 2002b, p. 410; Vogel 2002a, p. 152, where he uses the term 'monistic'; Stewart 2014, p. 8.

[38] Vogel 2002a, p. 152; Pradel 2002, p. 274; Olásolo 2009, p. 18, where he discusses the roots of this theory in the work of D. Kienapfel, 'Der Einheitstäter im Strafrecht', 1971.

[39] Stewart 2014, p. 8.

[40] Some jurisdictions maintain this unitary model with respect to specific crimes, such as Austria, Brazil, Denmark and Germany. See e.g. van Sliedregt 2012, p. 66.

[41] Vogel 2002a, p. 159.

5.3.1.2 Differentiated Theories

A differentiated system, on the other hand, distinguishes between a principal and a person who is some way assists the principal, variously described as a participant, a secondary participant, an accessory or an accomplice, although the meaning of each of these terms may also differ, as will be discussed shortly. It could be said that differentiated models are closer to the social reality of collective criminality, especially in the context of system crimes and mass atrocity.[42] Among differentiated systems there are different forms; a dualistic system distinguishes only between principals and accomplices,[43] a pluralistic system distinguishes further between different kinds of accomplices, such as instigation versus aiding and abetting.[44]

Descriptively Differentiated

Furthermore, among differentiated systems, there are those which distinguish between all participants based on a naturalistic approach and those which are based on a normative approach. A naturalistic approach looks for the causal contribution, an act which in some way contributed to the completion of the offence, in order to determine whether a person is a principal or an accomplice (and in a pluralistic system, what kind of accomplice).[45] A principal is the physical perpetrator, who most immediately caused the criminal act, and an accomplice is the one who contributes to causing the act.[46]

A naturalistic approach could be said to be crime-oriented, or, similarly to unitary systems, sentencing-oriented, in the sense that although there is a linguistic difference between the principal and the accomplice, the accomplice is convicted for the crime proper, and is often punished 'as a principal', or at least with the same maximum sentence as the principal.[47] In this sense, under a naturalistic approach complicity is conceived of both as binary, in the sense that there is a distinction between perpetrator and helper,[48] and at the same time all-inclusive: because the accomplice is charged with the crime proper, it's as if she pulled the trigger herself.[49]

---

[42] Vogel 2002a, p. 153.

[43] Vogel 2002a, p. 152.

[44] van Sliedregt 2012, p. 66; Vogel 2002a, p. 152.

[45] van Sliedregt 2012, p. 71; Vogel 2002a, p. 155.

[46] van Sliedregt 2012, p. 72.

[47] van Sliedregt 2012, p. 70; Stewart describes this as a third, weak or soft type of unitary system, Stewart 2014, p. 9.

[48] Stewart 2014, p. 5.

[49] For a full discussion of the rationale of this type of liability, and the impact of being 'charged with the crime proper', see the comparisons made in Chap. 6. See further Smith 1991, p. 5; van Sliedregt 2012, p. 71, citing John Gardner, 'Aid, Abet, Counsel, Procure: An English View of Complicity', in: A. Eser et. al. 'Einzelverantwortung und Mitverantwortung im Strafrecht: Individual, Participatory and Collective Responsibility in Criminal Law', Freiburg, 1998 at p. 227.

Because the accomplice is still treated as equally culpable as the principal, a naturalistic approach can be seen as being limited to a descriptive or linguistic differentiation. Some examples of such systems are those that have inherited the terminology of the old English common law, such as modern English law, the US and Canada. Under the old terminology there was a mandatory differentiation with respect to sentencing between principles and accessories, but in many modern common law jurisdictions there is no difference when it comes to the charges nor the sentencing, and yet there are still efforts to define the role of a party to a crime as either principal or accomplice.[50] For example, the federal complicity statute in the US states that '[w]hoever commits an offense against the United States or aids, abets, counsels, commands, induces or procures its commission, is punishable as a principal.'[51] An accomplice is 'tried, indicted and punished as a principal offender'.[52] It could therefore be said that these systems are in fact functionally unitary, since there is no legal or normative difference between accomplices, and they are all punished equally alongside the principal or physical perpetrator, as equally responsible for the crime. The terms 'functionally unitary' and 'descriptively differentiated' will be used interchangeably throughout this book, since they are two sides of the same coin. They will be preferred over the term 'naturalistic', since they are better descriptors of what role the modes of liability play in such a system as a whole.

In complex systemic crimes a descriptively differentiated approach leads to a bottom-up system of liability. The starting point is the immediate actor, the person directly causing the criminal act, and from there liability extends upwards, meaning that a leader at the top would have a derived, accomplice-type liability.[53] This is the approach favoured at the ad-hoc tribunals, but the problem is that the longer the causal chain, the more difficult it is to prove liability of a leader.[54]

Normatively Differentiated

By comparison, a normatively differentiated system takes *responsibility* as the leading concept instead of causation, and looks to the role played by each individual within the collective.[55] This means that whereas under a naturalistic approach the principal is the physical perpetrator, and the accomplice is anyone who contributed to the completion of the crime by the physical perpetrator, under a normative approach, the principal is the person who is most responsible for the completion of the crime, the one who has a decisive influence. This extends to include those who are the mas-

---

[50] This will be described in full in Chap. 6.

[51] 18 U.S.C., § 2.

[52] See for example the English Accessories and Abettors Act 1861 (24 and 25 Vict. c. 98) (UK) s. 8; Stewart 2012, p. 9.

[53] van Sliedregt 2012, p. 72; Vogel 2002a, p. 154.

[54] Vogel 2002a, p. 155.

[55] Vogel 2002a, p. 159.

terminds or intellectual authors, even if they did not 'pull the trigger'.[56] A secondary party is anyone who has less influence over the completion of the crime, and is therefore considered to be less blameworthy. The 'mitigation principle' applies and a lesser sentence is usually codified according to this lower degree of blameworthiness.[57]

A normative approach leads by definition to a normative differentiation between principals and secondary parties. An example of this approach is evident in the German code, where there are doctrinal differences between different types of principals, such as a perpetrator, a co-perpetrator, and an indirect perpetrator, each of whom receives the same full penalty, whereas an aider or abettor is considered to have less influence, less responsibility and therefore receives a lesser punishment.[58] This is the true meaning of a secondary participant; a person who's contribution to and/or influence over the crime committed is of a lesser degree. Because a secondary party did not fulfil the elements of the crime, she is not held liable for having committed the crime, but rather for having assisted the perpetrator(s) in their commission of the crime.[59] This is an important distinction in the qualification that appears in the conviction.

This normative, responsibility-based approach means that in complex or systemic criminality, a top-down approach is applied. Now the starting point is the person with the most influence, the mastermind, and from this liability it is possible to work down the levels of authority to subordinates, even if these were the physical perpetrators.[60] This approach can be recognised in the growing case law of the ICC, though there are still judicial and academic debates as to the exact nature of leadership liability under the Rome Statute. If we are to apply the definitions here, it could be that this lack of clarity is due to the fact there appears to be a mix of a normative, responsibility-based approach combined with a sentencing approach, and it is unclear whether such combinations are systemically possible.

5.3.2 *Complicity*

This lack of clarity and uncertainty as to the whether the approaches described here can be combined is due to the fact that the very notion of complicity receives different attention and different meanings cross-jurisdictionally. In normatively differentiated systems, typical to the civil law tradition, there is much more attention paid to complicity, since it matters in what way and to what extent an actor was complicit. In the

---

[56] van Sliedregt 2012, p. 72.

[57] van Sliedregt 2012, p. 72.

[58] § 25(1), 25(2) and 26 Strafgesetzbuch apply to principals, § 49(2) determines that the sentence for an accessory shall be three quarters of the maximum sentence applicable for a principal. The actual sentence handed down must be in reference to the actual sentence given to a principal. See Sect. 7.3 for a full discussion of the German approach to liability.

[59] Smith 1991, p. 5.

[60] van Sliedregt 2012, p. 72; Vogel 2002a, p. 155.

common law tradition, and thus in much English-language literature, there is less attention paid to the nuances of complicity, since the final conviction makes only a linguistic distinction, but no normative differentiation.[61]

Even the term 'accomplice liability' can be loaded with different meanings. Some authors use it to mean *any* person involved in a collective crime; whether they be a physical perpetrator, or a 'helper' in any other form that is considered to be principal or secondary in nature in the normative sense.[62] Others refer to principals as distinct from accomplices, since to be complicit is to assist the physical actor in some way, without fulfilling all the elements of the crime.[63]

This confusion is exacerbated because 'accomplice' is often used interchangeably with the term 'accessory' and with the notion of 'secondary participation' even in systems which are not normatively differentiated. Some of this confusion may stem from the inheritance in many English-language jurisdictions of the old common law modes of liability, where a distinction was made between principal accomplices (those who assisted the physical perpetrator at the scene of the crime), and 'accessory accomplices', defined as accessories before the fact (those who assisted in some way before the fact, but were not present), and accessories after the fact.[64] For instance, Black's Law Dictionary, an authoritative text in the UK and the US, defines an accomplice according to the traditional common law terminology as 'a person who is in any way involved with another in the commission of a crime, whether as a principal in the first or second degree or as an accessory.'[65] These distinctions no longer exist in common law jurisdictions, as will be discussed in Chap. 6, however this is likely also the reason why the incorrect term 'secondary liability' is still used in many Anglo-American jurisdictions, even though no normative distinction is made.[66]

5.3.3 Accessory Liability

It is important to understand that the term 'accessory' is not synonymous with 'accomplice', despite the interchangeable use of this lexicon in much case law and literature. If complicity is defined as assisting, then it usually refers to anyone who did not physically commit the crime. Each jurisdiction may define different forms of complicity slightly differently, and the question whether some of these forms amount to secondary participation depends on whether a functionally unitary or normatively differentiated system applies. Only in the latter case is the term 'secondary participation' correct. However the term 'accessory' correctly applied refers to a theory of

---

61 Fletcher 2000, p. 637; Stewart 2014, p. 5.

62 Smith 1991, p. 1; Fletcher 2000, p. 637.

63 Dubber 2007, p. 979; Stewart 2014, p. 1.

64 Blackstone Commentaries, cited in Kadish et al. 2007, p. 590; van Sliedregt 2012, p. 113.

65 Garner 2009, p. 18, under 'accomplice'.

66 See Chap. 6.

derivative liability, and not to the question whether an accomplice is a normatively secondary participant. In its simplest terms, accessory liability means that if no crime is committed then there is no criminality from which an accomplice's liability can be derived.[67] This limited derivative nature of accomplice liability would seem to apply in most systems, but there are more complex, extended meanings in some jurisdictions.

What happens if a crime is committed, but the principal was able to argue that the act was justified, and therefore not wrongful? If the principal is no longer liable due to this justification, then 'accessorial' liability would say the accessory is also no longer liable. This would not apply in a unitary system, since each person involved in the crime is considered to be independently liable; for this reason there are no 'accessories' under a unitary system. And again, what happens if the act is wrongful, but the principal is able to give a reason why he is not punishable, due to duress, or his psychological state of being? The question refers to whether the liability of an accomplice is derivative or accessory only in a limited sense, or in an extended sense.

In the first place, derivative modes of liability are dependent on the actual commission of a crime by the principal; the physical perpetrator must have reached at least the stage of attempt for an accessory to be liable. This has been termed an external, or quantitative accessorial nature.[68] In the second place, 'accessorial' can also be read to mean that the modalities applicable to the principal actor, such as the required intent, justifications and excuses, have immediate effect for the accessory, regardless of whether they are related to the wrongfulness of the act or the culpability of a specific actor.[69] So if the principal is acquitted, then the secondary participant must also be acquitted by definition. This has been termed internal, or qualitative accessorial nature.[70]

There are examples of functionally unitary systems where a theory of vicarious liability applies, such that all actors are deemed to be liable for all acts committed by any other person in the collective;[71] in these systems, because liability of accomplices is not derivative of, or accessorial to, that of a principal, they should not be referred to as 'accessories'.[72]

This confusion has also occurred in ICL, especially when drawing on post-Second World War case law. For example the description of the 'distinction made between the perpetrator and the accomplice (accessory)',[73] where the term in brackets is added as if to clarify, when it fact it convolutes the legal meaning of accomplice as a

---

[67] Smith 1991, p. 94.

[68] Sancinetti 1997, p 59.

[69] Welzel 1970, p. 164; Bohlander 2009, p. 168; Sancinetti 1997, p. 60, citing Jakobs.

[70] Sancinetti 1997, p. 59.

[71] See for instance conspiratorial liability in the US, discussed in Sect. 6.3.

[72] In this the present author does not agree with the definition given by Marcus Dubber that the physical perpetrator's act is by definition imputed to an accomplice as his own this is one systemic approach that is not universal. See Dubber 2007, p. 978.

[73] van Sliedregt 2012, p. 35.

participant in a collective crime and the meaning of accessory as a theory of liability. The collapsing of these terms has appeared in many judicial decisions, especially where the debate has been at the forefront as to whether a normatively differentiated system applies in ICL.[74]

The definitions here will be stricter. The terms 'complicity' and 'accomplice' will be used to describe *any* person involved in a collective crime who is not the physical perpetrator. This applies both in functionally unitary systems and in normatively differentiated systems. Thus in the latter type of system there may be other persons who are not physical perpetrators, but who are considered to be principals; these too will be referred to as accomplices. The term 'accessory' will be avoided as much as possible, except where it used with regards to the theory of derivative (or accessorial) liability, as opposed to a theory of vicarious liability.[75] These differences have consequences for the different standards of the requisite mental state applicable to different individuals, and for the availability of certain defences such as justifications or excuses.[76] Note that an accessory is not always a secondary party in a normative sense; for example in Germany, the theory of derivative liability applies, and thus an instigator is derivatively liable, however she is also considered a principal, normatively distinct from secondary parties.[77]

5.3.4 Parties to a Crime

In order to avoid possible confusions on the above terminology, a neutral term must also be chosen to describe the involvement of multiple individuals in a collective crime, regardless of whether a system is differentiated, unitary, or functionally unitary (where it applies the semantics of a differentiated system, but is not normatively differentiated). International criminal lawyers sometimes talk cross-purposes when describing the notion of complicity, because there is failure to clarify the consequences of this terminology within a given system, as identified above. Consequences such as the true qualification in a conviction, fair labelling and sentencing are not

---

[74] See for example Trial Chamber II at the ICTY, which argued that the distinction between principal and accessory—meaning secondary—liability was unnecessary to make, since no such distinction is made when it comes to sentencing. It also criticised judges in fellow Trial Chamber I in *Kvočka* for distinguishing between principal and accomplice—meaning secondary—in a normative sense. It further adopted the language of co-perpetrator but defined this is a non-principal participant, Krnojelac Appeal Judgment 2003, paras 75, 77.

[75] This echoes the definition given by Fletcher that an accessory is someone who is held 'derivatively liable for another's committing the offence.', Fletcher 2000, p. 637. Although it should be noted that he goes on to specify that this includes instigators as well as aiders and abettors. That specificity is avoided here, due to the fact that there are some systems in which accessorial liability does not apply in the theoretical sense, but rather vicarious liability in its place.

[76] See for example Olásolo 2009, p. 17.

[77] See Sect. 7.3.

consistent across the systems compared here, and have led to debate and confusion at the international level.

In order to keep the comparison clean, and to avoid unwanted assumptions about consequences, the terms 'principal' and 'secondary party' will be avoided as much as possible, except where a system uses this terminology to describe a normative differentiation. For those systems, it will be clarified what the conceptual, substantive and procedural consequences are within that system. However for the purposes of comparison this will not be the general lexicon used.

One choice of terminology to resolve this would be to describe all those involved in collective crime as 'participants'. However once again in some systems this term refers to secondary participation in a normatively differentiated sense, whereas this does not apply in other systems. Moreover this term could lead to a further lack of clarity, given the policy-oriented understanding of law-making processed used in this book, which describes actors involved in the processes of the formation, appraisal and application of law as 'participants', as discussed in the previous chapter.[78]

For all these reasons, the term 'parties to a crime' will be used to refer to all individuals involved in collective crime, regardless of their role. This term is borrowed from both the English and the Canadian lexicon,[79] both systems which do not normatively differentiate between individuals. There is therefore a certain risk that this may lead to assumptions about the consequences attached to such a term. However it has been chosen as a more neutral term precisely because it has not yet been used with so many differing meanings at the international level. The aim is to introduce a more neutral term for the purposes of comparison, before arriving at a conclusion as to whether a descriptively or normatively differentiated system would be more appropriate in ICL.

5.3.5 Guilt

The title of this book is 'Translating Guilt', which refers to many layers of translation; between languages, between domestic systems, between the domestic and the international, between criminal law concepts, and between parties to a crime. However when it comes to translating the very notion of guilt as a criminal law concept, this requires particular clarity, since in some systems 'guilty' is the equivalent of criminal liability and punishability, whereas in others it denotes only a part of the determination of liability.

In Sect. 2.1 attention was paid to the question of collective guilt versus individual guilt in the context of the collective nature of mass atrocity crimes. Here it is necessary to distinguish between guilt, responsibility, liability and culpability.

---

78 See Sect. 4.2.

79 Smith 1991, p. 6; Criminal Code 1985, s. 21(1).

5.3.5.1 Guilt as Responsibility or Liability

The first question to ask is whether there is a difference between guilt as responsibility and guilt as liability. In public international law terms, there is a difference between responsibility in the sense that States and international organisations may be responsible for wrongful acts, and liability, in the sense that some compensation or reparation must be made. Liability is a second order obligation that flows from responsibility. But it would be incorrect to say a State is guilty, since this term is reserved for the criminal law. Similarly, in criminal law, responsibility is a second degree obligation that is triggered by a first degree obligation; that is, a duty to take or omit an act according to law. In this sense, responsibility presupposes a duty, but should not be confused with it: 'responsibility determines who must be held accountable for carrying out or failing to carry out the ... obligation.'[80] Criminal responsibility can be seen as the ground for criminal liability; a person is liable only if he is responsible for his (participation in) wrongful acts, but if he was unable to act otherwise—that is if there are justificatory or exculpatory reasons for his having acted in such a way—then he may still be responsible, but he is not liable.[81]

Anthony Duff defines responsibility as answerability.[82] Responsibility is relational; we are always responsible *to* someone, for instance a victim, or to society. Responsibility is also *for* something, namely for a moral and legal wrong. And finally we are responsible as moral agents; we are criminally responsible for moral wrongdoing where we fail to respond appropriately to the moral reasons with which criminal law is concerned.[83] In this sense moral responsibility a precondition for criminal liability. Anyone who is a party to a crime is therefore answerable for the wrong that occurs as a result of collective action, where she failed to do what the criminal law expects of her, and only to the extent that it expects something of her. Yet even if she is answerable, that is, responsible, she is only *liable* to the extent that she was in fact able to do otherwise, to the extent that she is not able to argue defences that would create at least a reasonable doubt as to her liability.[84]

For individual criminal responsibility, it is necessary to explicitly establish the nature and scope of a particular individual's participation in the crime.[85] This is the first step towards establishing criminal liability. For collective criminality, this becomes more complex, since 'within networks of collective action [...] the degree of criminal responsibility does not diminish as distance from the actual act increases; in fact, it often grows.'[86] In normatively differentiated systems, responsibility can

[80] Ronaldo Tamayo Salmorán, *Responsabilidad* (in: Diccionario Jurídico Mexicano, vol. 8, Mexico City: Instituto de Investigaciones Jurídicas, UNAM, 1983), p. 44. Cited in Salazar and Urquiaga 2010, p. 73.

[81] Duff 2007, pp. 20–21.

[82] Duff 2007, p. 23.

[83] Duff 2007, p. 47.

[84] Duff 2007, p. 230.

[85] Ambos 2005, p. 74; Salazar and Urquiaga 2010, p. 74.

[86] Werle and Jessberger 2014, p. 954.

therefore be measured in degrees, and is not merely a positive or negative assessment. Criminal liability, on the other hand, is either present or negated. It is a purely legal determination, which forms the grounds for punishment.

Neither criminal responsibility nor criminal liability should be equated with 'guilt' without further qualification. In those common law jurisdictions where a jury has a choice between finding a defendant 'guilty' or 'not guilty', based on whether they are convinced beyond reasonable doubt by the evidence presented to them, the finding of 'guilty' amounts to the same as criminal liability.[87] The grounds for punishment are fulfilled. In fact it's possible to conceive of a finding of 'guilty' as being both the grounds for the conclusion, and the conclusion itself.[88] For the same reason, if a defendant pleads guilty, there is no need to continue with the process.[89] However a jury does not make a legal determination, but rather a factual one. No reasons are given for a finding of 'not guilty'; there is no distinction in such a decision between whether there were justifications making the act no longer wrongful, or whether the act was wrongful but the accused was excused for some reason such as duress—both of which would leave responsibility intact but would deny liability—or whether the evidence simply didn't add up to satisfy the indictment—which would mean no responsibility was established and *therefore* no liability. In the end, a finding of 'not guilty' simply means the charges were not successful, and does not necessarily amount to innocence of the accused.[90] In fact, historically speaking, the criminal law in the common law tradition was built upon a dialogue between lawyers and judges about what it meant to be guilty or not guilty, and instruction was then passed on to juries so that they could make these decisions based on the facts.[91]

In the civil law tradition, 'guilt' is considered a subjective condition of liability, alongside the two objective criteria of an act that fits the definition of a crime and the wrongfulness of the act.[92] Thus, if the elements of the crime are fulfilled, and there are no justifications for the act (i.e. it is wrongful), but the accused can show there is a reason she is not 'guilty' for having committed it, in the sense she could not have been expected to act otherwise under the circumstances, then she cannot be held liable.[93] Although the act remains wrongful, there is something morally forgivable about the fact that *this* actor committed it, in these circumstances.[94] Excuses such as duress would fall under this category, leaving the actor responsible for her acts, but not liable.

Since in ICL a clear doctrine is lacking, neither the understanding of guilt as factual nor the understanding of guilt as a pre-condition for liability can be said to prevail. It would seem, however, that the distinction between responsibility and

---

87 Damaška 1984, p. 95.

88 Fletcher 2007, p. 302.

89 Damaška 1984, p. 95.

90 Fletcher 2007, p. 302.

91 Milsom 2003, p. 9.

92 Welzel 1970, p. 95; Ambos 2006, p. 2652. For a more detailed discussion of this, see Sect. 7.4.

93 Fletcher 2007, p. 307.

94 Duff 2007, p. 265.

liability is a useful one to make, and although 'guilt' may be translated as different aspects of these two concepts, in the end the legal finding of 'guilty' handed down by judges of a tribunal is a finding of criminal liability.

One way to conceive of this is to ask for *what* punishment is in fact imposed.[95] We can only be punished for that over which we have control.[96] Punishment could be focused on the person of the accused, or on the act. If the focus is on the *person*, either on the expected feelings of guilt or remorse, or on the dangerousness of the character of the accused, then it would seem that the requirements of the commission of a crime, and the objective elements in particular, fall away altogether. Rather the *wrongful act* must be seen as the grounds for punishment, and criminal guilt should be seen first in terms of degrees of responsibility, and second as liability.[97] This understanding is predicated on respect for the autonomy of an individual.[98] If the focus is on the person rather than the act, then guilt becomes something that is related to the character of an individual accused, leaving no room for circumstantial variables or choice on the part of the actor. Furthermore, if guilt is understood only as a factual determination of liability, in the sense that there are no degrees of responsibility, then there is also no room for considerations such as excuses negating the finding of guilt. Thus, an act-based approach puts the act of wrongdoing at the intersection between punishment and 'guilt' understood as liability: punishment is imposed for an act of wrongdoing, but is justified only if, and only to the extent that, the actor is responsible.[99] Because responsibility is measured in degrees, the level of deserved punishment increases as the gravity of wrongdoing increases, (genocide is more grave than a theft) but the level of deserved punishment also decreases where the degree of responsibility is less, due to diminished capacity, or excuses such as duress.[100]

Thus, even if the act is the focus of punishment, wrongdoing alone is not sufficient for liability; the degree of responsibility should also be considered. As Fletcher has put it, 'findings of guilt are about blaming criminals for their wrongful actions', and not about eliciting feelings of remorse, or only judging the character of an accused.[101] This brings us to the definition of guilt as culpability or blame.

5.3.5.2 Guilt as Culpability or Blame

If the legal finding of guilt, in the sense of criminal liability, is about blaming a person for wrongful actions, then it is necessary to determine when and to what extent a person is blameworthy. This is often where the term 'guilt' is used to mean culpability.

---

[95] Duff 2007, p. 79.

[96] Duff 2007, p. 57.

[97] Duff 2007, p. 81; Fletcher 2007, p. 29.

[98] Fletcher 2007, p. 37.

[99] Fletcher 2007, p. 307.

[100] Fletcher 2007, p. 28.

[101] Fletcher 2007, p. 319.

Blameworthiness is a key term in all the jurisdictions compared in this study. We punish because we can justifiably blame someone for their wrongdoing.[102] This raises the question, what is 'blameworthy' conduct? H.L.A. Hart expressed blameworthiness as the 'fundamental requirement that for criminal responsibility there must be "moral culpability".' By this he meant that there are no excusing conditions present, and that the 'morally evil mind' referred to in the maxim *actus non est reus nisi mens sit rea* is therefore present.[103] The voluntary doing of a 'morally wrongful act' is the minimum requirement, in order to ensure the morally innocent are not punished.[104] Furthermore, 'everything which is regarded as enhancing the moral guilt of a particular offence is recognised as a reason for increasing the severity of the punishment'.[105]

The translation of blame into the principle of culpability is interpreted in slightly different ways in different jurisdictions. In the US the principle of culpability is explained in the descriptive sense: it is used to denote the different fault or mental states.[106] There is always some normative aspect to this, since a person is more culpable if they act with intention, and less so if they act recklessly, with various degrees of 'culpability' (mental fault states) in between. However according to this descriptive interpretation of culpability, in order for punishment to follow, the focus is on proof of the required mental state of a given offence and not on any variable circumstances.[107] This relates to the binary understanding of liability, described above, which makes no distinction between degrees of responsibility as a preceding requirement of liability. Either a person is liable or he is not. Therefore all parties to a collective will be held liable for crimes committed by anyone in the collective, as long as they have the requisite mental state.

By comparison, typically under the civil law tradition the principle of culpability is understood in a normative sense: that is, there must first be an enquiry into the blameworthiness of the acts of the accused, measured in degrees, before the question of liability is determined.[108] Responsibility measured by degrees of blame is therefore not a binary but an incremental determination. Thus the statement that there should be no punishment without blame has the corollary that the punishment should match the degree of blame: the measure of punishment must be based on a court's conviction of the extent to which the defendant is personally reproachable for the crime (or for the assistance provided to another who committed it).[109] It should be noted that

---

102 Fletcher 2007, p. 132.

103 Hart 1970, p. 35.

104 Hart 1970, pp. 36, 39.

105 Hart 1970, p. 169, citing Justice James Fitzjames Stephens.

106 Kadish 1980, p. 10. See further Sect. 6.3.

107 Jescheck 2004, p. 44.

108 Fletcher 2000, p. 398; Bohlander 2009, p. 13; Krey 2002, p. 97; Vormbaum and Bohlander 2014, p. 132; Nino 1980, p. 418 and further; Bacigalupo 1994, p. 184. See further Sect. 7.2.

109 Jescheck 2004, p. 44.

some jurisdictions which otherwise follow the common law tradition have made a doctrinal move towards this normative understanding.[110]

The principle of culpability is recognised as being part of ICL.[111] The question is whether *nulla poena sina culpa* should be interpreted to mean that punishment should also match the degree of blame, according to a normative theory of culpability, in which case an enquiry into blameworthiness must be undertaken as a separate aspect of responsibility, or whether a descriptive theory should apply, whereby proof of the mental element suffices to determine liability, and the question of degrees of responsibility is left out.[112] This leads to the very question at the core of this book; should there be distinct modes of liability, which reflect different degrees of responsibility, or should there only be a descriptive approach to complicity, with no legal, normative distinction between parties to a crime? The argument will be made in Sect. 10.2 why a normative approach should be preferred in ICL. However this question is also necessarily tied to whether an objective or a subjective approach to liability should be preferred.

5.3.6 Subjectivity and Objectivity as Approaches to Liability

In order to identify the approach to criminal liability that will best serve the functions and central goals of ICL, it is necessary to identify the doctrinal underpinning to criminal liability that differs between the traditions of common law and civil law. This understanding is somewhat absent in the debate on appropriate modes of liability at the international level. Since neither approach is inherently superior, it is necessary to understand the ways in which they operate in order to consider which approach may be better suited to the specific context of mass atrocity crimes.

Fletcher distinguishes between two patterns of criminality, or two main theories of liability under criminal law. On the one hand he describes manifest criminality, that is, a crime is objectively discernable even if there is no explicit evidence of the perpetrator's intent; intent is subsidiary in analysing liability.[113] The arm of the law, and the exercise of a court's jurisdiction, may only take place when a manifest act has taken place. On the other hand he describes the pattern of subjective criminality, or subjectivity, the core notion of which is to protect certain interests in society, and thus intent is the primary test of liability. In order to deter people from embarking on dangerous behaviour, those who intend to do so should be punished. Clearly this means the arm of the law can intervene before any manifest act has been undertaken. A manifest act may merely be proof of the intent, needed in order to discern between

---

[110] See the discussion in Sect. 6.4 on Canada's move towards this understanding, based on the constitutional concern for protection of the morally innocent.

[111] See for example the statement by the IMT that 'criminal guilt is personal', Nuremberg Judgment 1945, p. 251; Tadić Appeals Judgment 1999, para 186; Jescheck 2004, p. 44.

[112] Manacorda 2009, p. 288.

[113] Fletcher 2000, p. 117.

those actors who have a firm intention to commit a crime, and those who merely have the fantasy to do so.[114]

Typically in the common law tradition, subjectivity has been favoured in order to allow early intervention and to resolve the law of attempts; although it is important that the actor take certain steps towards completion of a criminal intent, no specifically defined act is required for liability.[115] This means that even if the criminal plan is legally or factually impossible, the actor may still be punished. This preventive approach is also important with respect to modes of liability, especially with respect to the notion of conspiracy as one of the key policy responses to collective criminality in the US.[116] Furthermore, this approach does not give general standards, rather judges are left to extrapolate from the specific case whether criminal liability is present.

Typically in the civil law tradition, a more legalist philosophy prevails.[117] Under this philosophy, the criminal law pronounces standards of behaviour that are interpreted and applied to specific cases; the overall goals of the system guide decisions in particular cases.[118] Liability for attempts, and for participation in collective criminality, is therefore based upon the manifest act, or the extent to which specific acts have been undertaken by an actor, which fulfil the definition of the crime in the legislation. This means, for example, that (factually) impossible attempts are not punishable, since the actor has not fulfilled the preannounced definition. Such an approach leads to a different conceptualisation of liability as a whole, which is not preventative in nature, but rather serves to enforce the norms that are preannounced in the criminal legislation.

Since each of these approaches is embedded in traditions with political and legal histories, it is no easy to task to transpose an entire conceptualisation of liability to the international plane without further ado. As will become very apparent in Part II of this book, neither approach is inherently more effective or superior; the test of which is most appropriate depends upon the goals and intentions of (international) criminal law. Given the discussion in the previous chapter that the narrative or expressive function should prevail in ICL, with an eye to increasing the public sense of accountability for mass atrocity, these should be taken into account as factors in designing or selecting a system of liability. Do we seek a preventive system with a focus on the actor, or do we seek to pronounce standards and focus upon the act? This question will be discussed further in Part III of this book.

5.3.6.1 Pure Theories of Liability Versus Approaches to Liability

It is important to clarify in this respect that a subjective approach is not the same as a pure subjective theory of liability, nor is an objective approach the same as

[114] Fletcher 2000, p. 119.

[115] Fletcher 2000, p. 166; State v Wilson 1981, p. 119.

[116] For a full discussion see Chap. 6.

[117] See further Sect. 7.5.

[118] Damaška 1984, p. 168; Fletcher 2000, p. 171.

a formal objective theory of liability. These two theories are explored further in Sect. 7.5, but it is worth a brief explanation here to define the terminology that will be used throughout this study. A subjective theory of liability determines whether an accomplice in a collective crime is liable based on whether he intended the crime of the physical perpetrator to be his own.[119] According to this theory;

> if a person has the mind of a perpetrator *(animus auctoris)*, he will be treated as a principal regardless of the importance of his factual contribution; if he only wishes to help another person commit the crime *(animus socii)*, he will be convicted as an aider and abettor [as a secondary party].[120]

Subjectivity, or a subjective approach to liability, does not operate as strictly as this, but does place a focus on the intention or agreement made between parties as the starting point of liability. The main difference is that subjectivity tends to lead to a functionally unitary system of liability, whereas a pure subjective theory is applied only in a system which needs to distinguish between parties in a normative sense.

On the other hand, a formal objective theory determines liability by the actual contribution to the crime; any person who fulfils the objective requirements of the crime is a perpetrator, and any person who does less, is a secondary party.[121] This theory has some limits, as will be explained in the Chapter on the German model of liability, and as such has been rejected in recent doctrine and case law. By comparison the term 'objectivity', or an objective approach to liability, will be used to denote a broader approach, whereby the contribution is the starting point, but not the only test of liability.

In this respect, as Fletcher has put it, the terms

> "objective" and "subjective" are cast about as though they were dice with only one face of meaning. In fact they have several faces, and any discussion that fails to acknowledge this diversity of meaning proves to be loaded in favor of unarticulated and undefended theories of liability.[122]

5.3.6.2 Subjective and Objective Elements

One possible definition of these terms is that the 'the question of wrongdoing is objective, since it focuses on the act rather than the actor, whereas the question of [responsibility] is subjective in that it focuses on the actor's personal accountability for the wrongdoing'.[123] The objective standards in this sense determine what behaviours and actions are to be punished, based on external measures, or the primary legal norms which prohibit or require particular acts. These prohibited actions may have certain exceptions of justification.[124] In the civil law tradition where systematic

[119] As decided and discussed in Staschynski 1962, para 134.

[120] Weigend 2011, p. 95.

[121] Stratenwerth 2005, p. 369.

[122] Fletcher 2000, p. 504.

[123] Fletcher 2000, p. 506.

[124] Fletcher 2000, pp. 459, 491.

criminal codes are typical, the term 'objective elements of the crime' refers to the wrongfulness of the act. This is often translated to the common law lexicon of *actus reus*. On the other hand, the subjective elements determine under what circumstances the actor is or is not responsible for this legal wrongdoing, depending on whether certain excuses may be available to the individual actor. In the common law lexicon this is translated to *mens rea*, or fault elements.

However the distinction between wrongdoing and responsibility as objective and subjective respectively are not reflected in the notion of *mens rea*. This has to do with a second possible definition of the terms, that ' "objective standards" are external and apply regardless of whether an actor thinks they are doing the right thing, whereas "subjective standards" focus on the actor's state of mind.'[125] In the civil law tradition, there is little difficulty in admitting that even the subjective elements, in the sense of the actor's personal responsibility, can be tested by means of either 'subjective' standards, focusing on the actor's state of mind, or 'objective' standards that are external and apply regardless of whether the actor thinks they are doing the right thing, such as standards of due care and attention. Personal responsibility sometimes rests upon such objective standards; the failure to act according to these objective standards may lead to negligence liability. On the other hand, in the common law tradition, there has been more cause for debate as to whether objective fault standards can really be called *mens rea*, since this Latin term is supposed to represent 'the guilty mind'. Standards of fault are generally described as subjective when they are measuring an individual's intent, knowledge or recklessness with respect to the results of their acts; but standards of negligence or strict liability are not a (subjective) state of mind, and as such, although they are sometimes admitted to be fault elements, they do not fit under the category of *mens rea*.[126]

Some of the resistance to accepting such objective standards applicable to individual responsibility is due to the implication that the fault is 'social' rather than individual, and that punishment is therefore a matter of social protection rather than a distribution of responsibility.[127] At the same time, such objective standards are common in most criminal law systems: these are expected social standards of due care and attention, the departure from which renders an individual culpable. However their place in the distinction between objective *actus reus* and subjective *mens rea* remains elusive.

5.3.6.3 Subjectivity and Objectivity as a matter of tradition

In the end, criminal liability can be seen as the sum of the assessment of individual culpability (blameworthiness) for wrongdoing. This means there is both a subjective and an objective dimension; the objective dimension focuses on the act and the harm

---

[125] Fletcher 2000, p. 506.

[126] See for instance the judicial debates in Canada on this point in Sect. 6.4. See also the debates between the international tribunals in Sect. 3.11.4.

[127] Fletcher 2000, p. 505.

it may cause, the subjective dimension focuses on the actor and the question whether the particular actor is accountable for the act of wrongdoing.[128] Therefore throughout this book, when a particular approach is described as reflecting 'subjectivity' (typically in the common law tradition), it is not to say that the objective elements are disregarded, nor vice-versa if an approach is described as reflecting 'objectivity' (typically in the civil law tradition), does it suggest that the subjective elements are not an important part of determining the guilt of an individual. These terms are used here to describe the determination of liability in various jurisdictions and in the international tribunals, based on a difference in emphasis on either responsibility for the objective 'wrongdoing', or on the subjective willing participation in a collective.

As will become clear in Part II, to which we now turn, the difference in emphasis leads to different results in terms of collective criminality, and especially in terms of the responsibility of leaders therein.

References

Ambos K (2005) La Parte Peneral del Derecho Penal Internacional: Bases para una elaboración dogmática. Fundación Konrad Adenauer, Montevideo

Ambos K (2006) Toward a universal system of crime: comments on George Fletcher's grammar of criminal law. Cardozo Law Rev 28:2647–2673

Bacigalupo E (1994) Manual de Derecho Penal: parte general, segundo edn. Editorial Temis SA, Santa Fé

Bohlander M (2009) Principles of German criminal law. Hart Publishing, Oxford

Convention against Torture (1984) Convention against Torture and Other Cruel, Inhumane or Degrading Treatment or Punishment

Damaška M (1984) The faces of justice and state authority: a comparative approach to the legal process. Yale University Press, New Haven

Dubber MD (2007) Criminalizing complicity: a comparative analysis. J Int Crim justice 5:977–1001

Duff RA (2007) Answering for crime: responsibility and liability in the criminal law. Hart Publishing

Fletcher GP (2000) Rethinking criminal law. Oxford University Press, USA

Fletcher GP (2007) The grammar of criminal law: American, comparative and international, Vol. One: Foundations. Oxford University Press, New York

Garner BA (ed) (2009) Black's law dictionary, 9th edn. Thomas and Reuters, Minnesota

Glenn HP (2006) Comparative legal families and comparative legal traditions. In: Zimmermann R, Reimann M (eds) Oxford handbook of comparative law, Oxford University Press, pp 421–440

Glenn HP (2010) Legal traditions of the world: sustainable diversity in law, 4th edn. Oxford University Press, USA

Hart HLA (1970) Punishment and responsibility: essays in the philosophy of law, 2nd edn. Oxford University Press

ILC (1950) Principles of international law recognized in the Charter of the Nuremberg Tribunal and in the judgment of the tribunal

Jescheck HH (2004) The general principles of international criminal law set out in Nuremberg, as mirrored in the ICC Statute. J Int Crim Justice 2:38–55

Kadish SH (1980) Why substantive criminal law? A dialogue. Clevel St Law Rev 29:1

Kadish SH, Schulhofer SJ, Paulsen MG (2007) Criminal law and its processes: cases and materials, 8th edn. Aspen Publishers, New York

[128] Fletcher 2000, p. 510; Fletcher 2007, p. 319.

Krey V (2002) Deutsches Strafrecht Allgemeiner Teil: Lehrbuch in Deutsch und Englisch. W Kohlhammer, Stuttgart
Lafontaine F (2012) Prosecuting genocide. Crimes against humanity and war crimes in Canadian courts. Carswell, Thomson Reuters, Toronto
Lempert R (2008) Anglo-American and continental systems: marsupials and mammals of the law. In: Jackson J, Langer M, Tillers P (eds) Crime, procedure and evidence in a comparative and international context. Hart Pubishers, Portland, Oregon
Manacorda S (2009) Culpability. In: Cassese A (ed) The Oxford companion to international criminal justice. Oxford University Press, Oxford
Maxwell JA (2008) Designing a qualitative study. The Sage handbook of applied social research methods, pp 214–253
Michaels R (2006) The functional method of comparative law. In: Reimann M, Zimmerman R (eds) The Oxford handbook of comparative law. Oxford University Press, Oxford
Milsom SFC (2003) A natural history of the common law. Columbia University Press, New York
Nino C (1980) Los Limites de la Responsibilidad Penal. Editorial Astrea
Nuremberg Judgment (1945) Judgment of the International Military Tribunal
Oderkerk M (2001) The importance of context: selecting legal systems in comparative legal research. Netherlands Int Law Rev XLVIII:293–219
Olásolo H (2009) The criminal responsibility of senior political and military leaders as principals to international crimes. Hart, Portland
Örücü E (2006) Methodology of comparative law. In: Smit J (ed) Elgar encyclopedia of comparative law. Edward Elgar Publishing, Cheltenham, UK/Northampton, MA, USA pp 422–454
Pradel J (2002) Droit penal comparé. Dalloz, Paris
Salazar K, Urquiaga XM (2010) Digest of Latin American jurisprudence on international crimes Due Process of Law Foundation, Washington
Sancinetti MA (1997) Ilícito personal y participación. Ad Hoc, Buenos Aires
Smith (1991) Complicity. Oxford University Press
Stewart J (2012) The end of 'modes of liability' for international crimes. Leiden J Int Law 25(1):165
Stewart J (2014) Complicty. In: Dubber M, Hörnle T (eds) Oxford criminal law handbook. Oxford University Press, Oxford
Stratenwerth G (2005) Strafrecht, Allgemeiner Teil I: Der Straftat (Translated as: Derecho Penal, Parte General I: El hecho punible). Hammurabi, Buenos Aires
van der Wilt H (2008) Equal standards? On the dialectics between national jurisdictions and the International Criminal Court. Int Crim Law Rev 1(2):229–272
van Sliedregt E (2012) Individual criminal responsibility in international law. Oxford University Press, Oxford
Vogel J (2002a) How to determine individual criminal responsibility systemic contexts: twelve models. Cahiers de Défense Social, pp 151–169
Vogel J (2002b) Individuelle verantwortlichkeit im völkerstrafrecht. Zeitschrift für die gesamte Strafrechtswissenschaft 114(2):403–435
Vormbaum T, Bohlander M (eds) (2014) A modern history of German criminal law. Springer, Heidelberg
Watson A (1974) Legal transplants: an approach to comparative law. Scottish Academic Press, Edinburgh/London
Weigend T (2011) Perpetration through an organization: the unexpected career of a German legal doctrine. J Int Crim Justice 9:91–111
Welzel H (1970) Das Deutsche Strafrecht (Translated as Derecho Penal Aleman: Parte General). Editorial Juridica de Chile, Santiago
Werle G, Jessberger F (2014) Principles of international criminal law, 3rd edn. Cambridge University Press
Ziller J (2006) Public law. In: Smits J (ed) Elgar encyclopedia of comparative law. Edward Elgar Publishing, Cheltenham, UK/Northampton, MA, USA
Zweigert K, Kötz H (1998) An introduction to comparative law. Oxford University Press, Oxford

Cases

Krnojelac Appeal Judgment (2003) The Prosecutor v. Milorad Krnojelac, Appeal Judgment, IT-97-25-A
Staschynski (1962) Staschynski, 9 Bundesgerichtshof Strafkamer, StE 4/62
State v. Wilson (1981) 95 Wash 2d 828 (Supreme Court of Washington)
Tadić Appeals Judgment (1999) The Prosecutor v. Dusko Tadić, Appeals Judgment, IT-94-1-A

Part II
Leadership Liability Through a Comparative Lens

Chapter 6
Subjectivity Reflected in the Common Law Tradition

Contents

C. Steer, *Translating Guilt*, International Criminal Justice Series 9,
DOI 10.1007/978-94-6265-171-5_6

In order to understand the debate and loss of clarity on which modes of liability should apply in ICL for crimes of mass atrocity, and how they should operate, it is important to understand the way in which they have been developed in the domestic jurisdictions from which they hail. In turn, in order to understand this, it is necessary to consider the traditions within which each of the selected domestic jurisdictions operate. The policy role played by modes of liability in the common law tradition in determining guilt and punishment, and the actors who determine these policies, differ from those in the civil law tradition. These differences are at the core of the debate on the international level, and understanding them at a deeper level on the domestic level will aid in making the selection most appropriate for the context of ICL.

In the following analysis, it will become apparent that when it comes to determining guilt for parties in collective crimes, the common law tradition has as a characteristic subjectivity at its core. Criminal liability has both a subjective and an objective dimension; the objective dimension focuses on the act and the harm it may cause, the subjective dimension focuses on the actor and the question whether the particular actor is accountable for the act of wrongdoing. The subjectivity which is identified here as a characteristic of the common law tradition is not to say that the objective elements are disregarded when determining the existence of a punishable crime. Rather, in order to determine the attribution of collective crimes to, and the liability of, an individual accused, there is more emphasis on the intention of the accused and little on the extent to which that individual contributed to the objective 'wrongdoing' elements of the crime. The difference in emphasis leads to a particular result in terms of collective criminality, and especially in terms of the responsibility of leaders therein. Because under a subjective approach the intention to contribute is determinative, and not the extent of the actual contribution, all parties to a crime are equally liable, including for the acts of all the other parties. Thus, even where there is some terminology suggesting differentiation between principal and accessory, this is descriptive only, leading to a functionally unitary system, as will be demonstrated in this chapter.

First the common law tradition of law-making will be elaborated upon in Sect. 6.1 and 6.2, before taking a closer comparative look at the way this tradition has impacted upon modes of liability in the two selected jurisdictions, the USA in Sect. 6.3, and Canada in Sect. 6.4.

6.1 The Common Law Tradition: A Context

Historically, from the time of the Norman conquest to the nineteenth century, the common law developed essentially as a law of procedure, based on the system of writs that were handed down by the royal court to judges as instructions on how to proceed with a case,[1] and neither the judge nor the lawyers had any responsibility

[1] Glenn 2010, p. 242.

to find 'objective' fact.[2] Instead this was the role of the jury, made up of locals who could decide on a dispute according to local customs and expectations. Since the members of the jury were usually illiterate, arguments and proof had to be presented orally in the trial, in which the judge plays a commanding but distant role; quite literally a source of law.[3] Originally there were only first-instance judges, no courts of appeal, so the judges themselves were left with the task of working out what was to be allowed under the writs, and it was better not to suggest to each other that they might have had erred.[4] Hence over time a series of judgments built up to create a path of decisions that would guide future decision. This path would always represent a community law rather than a law imposed from above, and one which was distinguished from local rules by the notion of commonality.[5]

As a product of history, and since the common law was not imposed from above with a higher-order set of values, the limits of common law were societal rather than moral; the law had to express itself in terms of its surrounding society.[6] At the same time, the emergence of English common law as a system was more than the spontaneous evolution of customs; it was in fact dependent on the emergence of the English state. The promulgation of some new sense of uniformity across England was one of the vehicles for the expansion of state power from the twelfth century on.[7] The myriad customs that governed different regions in England were replaced by a single common law, and precedents formed in one part of the kingdom were deemed applicable throughout the rest of the kingdom.[8]

In the nineteenth century this led to the emergence of the concept of *stare decisis*, and of these precedents becoming formally binding.[9] There was no premium on uniformity or cohesion, and over time as judges were expected to decide cases on the merits by the application of substantive law, there was no clear source for this, other than previous decisions and the influence of the historical writs. The adversarial system was thus able to flourish, with a lot of freedom for litigators and prosecutors to conduct the content of the proceedings.[10] This predominant role of lawyers led to their influence, both direct and indirect, on the development of the normative content of the law, something which is also recognisable in ICL, and which demonstrates the multiplicity of participants in the processes of the law, beyond the formal law-makers.

The independence of judges is another key characteristic of the common law tradition that affects the process of law-making. Glenn describes the common law tradition as an ethic of adjudication, which says much about the role of judges and the courts as policy makers, in comparison to the role of the legislator. Independent

---

[2] Leflar 1971, p. 320; Glenn 2010, p. 243.

[3] Milsom 2003, p. 8.

[4] Glenn 2010, p. 244.

[5] Glenn 2010, p. 247.

[6] Lempert 2008, p. 403; Milsom 2003, p. 9; Glenn 2010, p. 257.

[7] Fukuyama 2011, p. 256.

[8] Fukuyama 2011, p. 257.

[9] Leflar 1971, p. 323; Glenn 2010, p. 252.

[10] Damaška 1984, p. 104; Glenn 2010, p. 257.

judges are not institutionally bound to apply the law in an administrative sense, in accordance with the legislator's policies, but are rather free to be law-seekers as much as law-appliers, and to develop rules according to the immediate perceived policy needs of society. Doctrinal initiatives were therefore few and far between before the nineteenth century, and even then with the development of *stare decisis*, 'doctrines' have become a reflection of case law trends, which are explained in treatises as the law of judges.[11]

This is thus a very different notion of how doctrines develop and what their roles are compared to the civil law tradition of systematic, theoretical doctrines which are reflected in a systematic code.[12] As Fletcher points out, in the civil law tradition, *Dogmatik*, or doctrine in the sense of scholarly authority, is explicitly regarded as a source of law, and the norms and principles of a legal system can be extrapolated from these teachings.[13] He suggests that the common law tradition also looks to treatises to clarify the law, but that these are not recognised as a formal source of law;[14] this is possibly because in the common law tradition the treatises are regarded as following the case law rather than guiding it. There are 'doctrines', or legal institutions, which can be said to be typical of common law systems, such as conspiracy, which will be discussed in detail in this chapter. However these doctrines are not part of a complex and coherent analytical system, but rather developed through the case law in response to immediate societal needs. Given that ICL is itself a system that is highly dependent on case law and the scholarly opinions of influential experts, and that does not have a tradition of coherent and matured doctrines in the sense of *Dogmatik*, there is a similarity with the common law tradition which may be instructive in understanding the processes of law-making and law-applying in ICL.

The comparative criminal theory of Mirjan Damaška explains the differences between the role of participants and the role of doctrine in common law and civil law traditions by considering (i) the organisation of governmental authority on one hand, which he describes as either hierarchical or co-ordinate models, and (ii) the purpose of the criminal legal process on the other hand, which he describes as oriented towards either policy implementation or individual conflict solving.[15] A hierarchical organisation of authority will lean towards a policy-implementation process in criminal law, which is reflected in the civil law tradition, whereas a horizontal or co-ordinate organisation of authority will lean towards a conflict-solving trial in criminal law, which is reflected in the common law tradition.

With respect to (i), the organisation of governmental authority, the trend towards a less hierarchical and more horizontally, coordinately organised authority is historically the trend found in the UK, where there was a preference for checks and balances on executive and legislative power, rather than an all-powerful sovereign. The US, Canada and other countries that have inherited this tradition are all built

---

[11] Glenn 2010, p. 260.

[12] See Sect. 7.5.4 on objectivity as reflected in the civil law tradition.

[13] Fletcher 2011, p. 181.

[14] Fletcher 2011, p. 180.

[15] Damaška 1984.

on an inherent distrust in government, and thus there is an important role played by lay decision makers, including, but not only, a lay jury in the courtroom.[16] One result of lay participation in the courtroom is that notions such as modes of liability must become accessible to the jury, and therefore less technical and legalistic. For example, in the US, the right to trial by jury is considered to be fundamental to the notion of justice; in the words of the Supreme Court, 'in order to prevent oppression by the Government [...] if the defendant prefers the common-sense judgment of a jury to the more tutored but perhaps less sympathetic reaction of a single judge, he is to have it.'[17] In Canada, although less than 10 % of all criminal charges end in a jury trial, this is due to the fact that many charges are for less serious 'summary' crimes for which there is no trial by jury.[18] For indictable crimes, on the other hand, there is a right to trial by jury,[19] and among these, many involve collective criminality.[20] A lay jury is unlikely to be able to fully understand and correctly apply the technical requirements of each mode of liability, and the fact that each case is decided by a different jury, a new collection of individuals, it is near impossible to maintain consistency from decision to decision.

Juries do not record or report their reasons for coming to a decision, so there is no way of ascertaining the details. The only result is binary: guilty or not guilty. There is no room for qualifications as part of the verdict. The fact that the trial is the focus of the criminal process in the common law tradition, with evidence presented orally, and multiple legal issues being dealt with in the courtroom, means that the legal question of liability has been reduced to simple terms, in order to facilitate the factual question.

This horizontal organisation of authority also means there is a place for multiple and sometimes dissenting voices in a judgment handed down from a court:[21] individual judges identify themselves and their views, and thus the law is debated and discussed over the course of several judgments and commentaries. Independent and individual judicial positions become part of the fabric of the common law when they are congruent with society's systemic tendencies and mores.[22]

When it comes to (ii) the aims of the criminal process, this horizontal distribution of authority also leads to a preference for conflict solving rather than policy implementation. Because there is less of a hierarchical approach to the creation of the system of law, and therefore internal coherence is less important than the perceived requirements of each particular case, procedural justice receives primacy over substantive justice. The trial becomes the focal point of the whole criminal process; its preparation is not the responsibility of professionalised, hierarchical authorities, and

---

16 Damaška 1984, pp. 24–25.

17 Duncan v Louisiana 1968, p. 149.

18 Criminal Code 1985, s. 495(2)(c); Statistics Canada 2010.

19 Criminal Code 1985, s. 471.

20 Statistics Canada 2010, providing data on adult criminal court statistics most recently for 2010–2011.

21 Damaška 1984, p. 24.

22 Cunningham 2006, p. 775.

its unfolding is in the hands of the parties, the prosecutor and the defence.[23] Instead of a search for objective fact, the trial is concerned with the resolution of a particular conflict between these parties, and the fact-finding becomes a competitive or adversarial one.[24] The search is rather for a procedurally just outcome in each instance. The parties have a great deal of autonomy and therefore influence over the outcome of the case,[25] and at the same time these participants have a great deal of influence over the development of the normative content of the law, as doctrine emerges from case law.

One way of describing this tradition is that it establishes path dependency, which can be understood in the following way. The adjudicative ethic described by Glenn and Damaška, and the ensuing paths of case law, can be seen as an iterative process: each case creates a particular legal result which is available for use in succeeding cases, where it can be used to reach a new legal result, which is in turn available for use in later cases. This iteration means that all cases are systematically linked to previous cases by the conditions, or facts and circumstances, which are similar among them all.[26] These conditions generate a path dependency; later courts are constrained by what earlier courts have said about those conditions, and it may be hard to stray from this.[27] An important reason why laws may vary across jurisdictions with a common law system is the dependence on initial conditions, which themselves will vary from jurisdiction to jurisdiction depending on the existing social facts and societal needs in each one.[28] Thus, even though two situations may be analytically and functionally the same, the difference in initial factual conditions will lead to the emergence of different legal rules in different jurisdictions.[29] Therefore the evolution of modes of liability in common law jurisdictions share a common systemic tradition, but may differ according to the kinds of collective crimes that are prevalent in each jurisdiction, and the policy choices in responding to these crimes. The same, of course, can be said for the different responses in ICL, given that there are some similarities in the importance of case law and the trend towards path dependency, however the problem in ICL is the lack of longer paths of case law from which to draw. Instead the paths of case drawn upon hail from domestic jurisdictions; but without awareness of the initial conditions, there is little grounding for selecting the most appropriate solution. The process of reification of new modes of liability in the international tribunals will be described in Chap. 8, and this is a perfect mimicry of this notion of path dependency, with the added factor that the paths to draw from are less defined.

Since this iterative process is driven by facts and is a reflection on and articulation of social norms, the common law process leads to a doctrinal malleability; an ability

[23] Damaška 1984, p. 65.

[24] Damaška 1984, p. 119.

[25] Damaška 1984, p. 104.

[26] Cunningham 2006, p. 748.

[27] Cunningham 2006, p. 749; Glenn 2010, p. 17.

[28] Cunningham 2006, p. 750.

[29] Cunningham 2006, p. 758.

to respond to the needs of justice in an individual case.[30] What this flexibility also means is that there is little, if any, doctrinal coherence, since individual justice is given primacy over systemic consistency. A judge may depart from a previous rule because the case requires it; the initial conditions may be similar to previous cases, but the rule articulated differs from the previous one, leading to a bifurcation from the previous path, in response to the perceived need for a different outcome to ensure a substantively just outcome. In this way, the development of common law rules can be seen as a continuously bifurcating series of lines throughout time.[31] While it may seem that this lack of doctrinal coherency would lead to instability and uncertainty, over time this iterative process and the notion of bifurcation actually leads to a continual re-stabilisation, as more new cases will follow the new paths of dependency. What emerges is a social congruence rather than a doctrinal one.[32]

The development of the normative content of laws within the common law tradition has therefore been a continuous process of reciprocity, through the process of path dependency and bifurcation, and through the influence that the common law tradition has in terms of its horizontal organisation of authority and its conflict solving trials. In the two jurisdictions compared here, this has either occurred as a matter of a historical formation of identity, as in the case of the USA,[33] or as a matter of being part of the common law of the Commonwealth, as in the case of Canada.[34] This continuing process of reciprocity has an underlying harmony, leading to stability of the law despite there being no priority given to fixed abstract rules or uniformity of result.[35]

6.2 The Old Common Law Modes of Liability

Despite the inheritance of the old English common law in modern day common law jurisdictions, over time the bifurcation process described above has led to a significant departure from the modes of liability that were the product of the English tradition. This is apparent in many jurisdictions, including the two compared in this book. However even though the old common law modes of liability have disappeared, the language used today of 'principal' and 'accessory' is a remnant of these distinctions. To the extent that they are applied, it is not always clear that they are useful, since the definitional consequences no longer apply. What used to be a normatively differentiated system has become merely semantically differentiated in many modern jurisdictions, with all parties to a crime now being held equally liable regardless of their level of participation. This is one reason why there is significant

[30] Cunningham 2006, p. 759.

[31] Cunningham 2006, p. 770.

[32] Cunningham 2006, p. 773.

[33] Watson 1983, p. 1125; Glenn 2010, p. 263.

[34] Roach 2012, p. 21; Glenn 2010, p. 267.

[35] Glenn 2010, p. 269.

debate and confusion as to the notions of leadership responsibility and complicity at the international level; when notions of liability are translated from other languages into English, the working language of ICL, the terminology chosen appears to be consistent across English language jurisdictions. However cross-jurisdictionally this terminology is no longer used consistently to mean the same thing.

For many centuries in England there was a clear normative line drawn between different parties to a crime. In the first place, an actor could be a principal in two degrees: in the first degree was the actor who was the 'absolute perpetrator of the crime',[36] in other words the person who fulfils all the elements of the crime alone. In the second degree was the actor who was 'present, aiding the fact to be done'.[37] The physical presence of this actor was essential, although constructive presence was also deemed sufficient, in the sense that someone could be keeping watch at a considerable distance from the *locus delicti*. Principal in the second degree was sometimes referred to as an accessory at the fact; according to Black's Law Dictionary this is '[some]one by whom the actual perpetrator of the felony is aided and abetted at the very time at which it is committed'.[38] The language of accessory here must be understood in the theoretical sense that this actor's liability is derived from that of the physical perpetrator (principal in the first degree). Although aiding and abetting is a lesser form of principal liability under this construction, the actor in question is still considered to be a principal. Presence of the actor at the scene of the crime still renders that person a chief actor.

The final category under the common law was the normatively distinct 'accessory', that is, someone who is 'not the chief actor, nor present at its performance, but is some way concerned therein, either before or after the fact ...'[39] This is distinct from the accessory at the fact, due essentially to fact that this normatively distinct accessory is not present at the *locus delicti*. The term 'accessory' used in this way correctly refers to the theory of derivative liability, and does not denote a distinction between a principal and a secondary party, since the accessory at the fact was considered a principal, whereas this accessory was not.[40] The distinction between this accessory and the two degrees of principal was drawn specifically to differentiate under felony law, for which the death penalty applied, so that an accessory would not receive this most severe punishment.[41]

Based on this categorisation, the principals in both the first and second degrees are the most culpable, whereas the accessory is less culpable. However not all modern, collective crimes fit this pattern. Particularly in cases of collective criminality where the conduct requires multiple participants, this can be more complex. Consider the example of an organised crime where a mob boss orders an underling to shoot a victim. Under this common law categorisation, the hit-man would be the principal

[36] Blackstone Commentaries, cited in Kadish et al. 2007, p. 590.

[37] Kadish et al. 2007, p. 590; Manning and Sankoff 2009, pp. 246–247.

[38] Garner 2009, p. 16.

[39] Blackstone Commentaries, cited in Kadish et al. 2007, p. 590.

[40] For the definition of 'accessory' liability as derivative liability, see Sect. 5.3.6.3.

[41] van Sliedregt 2012, p. 112.

perpetrator in the first degree, whereas the mob boss would be merely an accessory before the fact, since he is absent from the scene of crime. What is essentially solicitation (which is an inchoate crime in many of the state jurisdictions within the US,[42] and also in Canada,[43] for which the solicitor is 'principally' liable) amounts to mere 'assisting or encouraging' under these old categories.[44] This goes against a moral intuition that the mob boss is more culpable since the crime took place at his instigation.

Partially due to this forced dichotomy, and partly due to the lack of systemic approach to such definitions, over time the normative distinctions have been eradicated. In the UK, the Accessories and Abettors Act of 1861 stated that 'whomsoever shall aid, abet, counsel or procure the commission of any indictable offence [...] shall be liable to be tried, convicted and indicted as a principal offender.'[45] The title of the Act refers to accessories and abettors, collapsing what were originally two categories, namely principals in the first degree and the general category of accessories. While the choice to do away with the normative qualifications may have been one of policy, the conflation of all technical terms into one generic group means that continued use of these terms can be confusing. Similarly, the Criminal Code in Canada and most federal and state statutes in the USA have eradicated these normative distinctions altogether, but the terms 'principal' and 'accessory' are still used. However all parties to a crime are considered to be equally liable for the crimes committed by the collective, even if they did not physically contribute to those crimes, and even if other members of the collective commit further, unintended crimes; as long as they share some agreement. In the USA, this emerges as blanket liability for the acts of others, in Canada the test has to do with the moral blameworthiness of an individual for having had a criminal intent in the first place, as will be described in the following sections.

This departure from the original system of modes of liability over time is a result of the process of bifurcation and the role of case law in guiding the codification in the common law tradition. Judges have been picking and choosing over time in response to the societal norms and the specific factual conditions they have been faced with, demonstrating that dealing with the question of liability is a matter of policy choice. Different domestic models of liability are a result of these policy choices, and while there are systemic similarities within the common law tradition, the policy impact on the differences between specific modes will be explored next in the two jurisdictions compared here, the US and Canada.

42 This is due in large part to the formulation of solicitation as an inchoate crime in the Model Penal Code § 5.02. Many states have otherwise criminalised solicitation of specific offences such as prostitution or murder, Dressler et al. 2001, p. 421.

43 Criminal Code 1985, s. 21.

44 Garner 2009, p. 16.

45 Cited in Ashworth 2011, p. 539.

6.3 United States of America—Liability for the Acts of Others

The USA is a strong case in point as to the way in which the common law has been reinvented in many domestic jurisdictions; the development of a national identity meant that the law has become a local product.[46] This is apparent with respect to the development of a simplified system of modes of liability, moving away from the old common law differentiated system.

One important factor in this shift is the federal model that is particular to the USA. Individual states within the US have maintained a great deal of legislative autonomy which was a condition for many to join the confederation. There are therefore only a few crimes which belong to federal law, whereas most crimes, substantive norms and procedural rules fall under state jurisdiction, leading to differences from state to state.[47] There are therefore multiple bifurcations, and over time while the core concepts and terminology from the old common law concepts remain, criminal liability has developed in a very particular direction in the US. This is important to keep in mind when considering the influence of these US notions on the debates in ICL, since their particularity may make them ill-suited to the context of mass atrocity crimes.

From the mid-nineteenth century, the emerging lack of congruence between individual state criminal laws raised concerns due to the positivist trend and the ensuing push for codification across the US.[48] In fact it has been posited that there is no inherent structure to the criminal law in the US at all.[49] In response to this, in the 1950s the American Law Institute took on the task of developing a Model Penal Code (MPC), which was adopted in 1962, the aim of which was to systematize the criminal law, and thereby promote more harmony in individual state legislation.[50] To a certain extent it was a successful law reform project, as about two thirds of states endeavoured to bring their legislation in line with the MPC,[51] with the notable exception of the rule on liability. In the decades following, many individual states responded to increased waves of criminal activity with a general policy of crime control, by increasing the number and types of crimes and changing some of the general principles and definitions of liability, without regard for the systematisation for which the MPC strove.[52] The MPC remains instructive in helping to interpret some legislation, and is a commonly used teaching tool to help understand certain principles, but with respect to modes of liability its impact is limited.

If we consider the role of the different participants in the processes of law-making, as well as their perspectives and the various fora of interaction, in the US these

---

46 Glenn 2010, p. 263.

47 LaFave 1986, p. 2.

48 Kadish et al. 2007, p. 534.

49 Fletcher 2000, p. 406.

50 Dressler et al. 2001, p. 31.

51 Lynch 2003, p. 223.

52 Lynch 2003, p. 224.

processes have led to the emergence of a criminal law system which does not have any guaranteed consistency across the jurisdictions of individual states, and where the presence of a lay jury has an influence on the development of the normative content of the law in general, and modes of liability in particular. Alongside this is the strong institution of judicial review, which has been read into the United States Constitution. The judiciary thus has a limiting role on the legislature, and can respond to societal and policy needs in a more immediate way than the legislature might be able to. The case law that emerges guides future legislation, and is thus very influential on policy developments as they are codified. Precisely because knowledge of the historical development of the case law is necessary in distinguishing and clarifying the normative content of the law, judges play a particularly authoritative role in these law-making processes.

6.3.1 *Subjectivity*

The emphasis on the subjective mental element, or *mens rea*, as the determinative factor is reflected in the fact that the principle of culpability is understood in the US literature in the descriptive sense; referring to the required mental state for liability.[53] Thus, while the moral basis for punishment is the blameworthiness of the actor, this blameworthiness is measured only by the applicable mental state, rather than by the acts and contributions to the fulfilment of the crime.[54] The evidentiary test of the mental state of the accused becomes at once the moral test for blameworthiness.

The consequences of this approach are twofold. In the first place, it is difficult to prove the subjective mental state of an accused beyond reasonable doubt—which is the evidentiary standard for determining guilt. For this reason, much is inferred as a question of fact from the words and actions of the accused, through so-called circumstantial evidence.[55] This does not mean there is a shift to an objective approach, but rather that the actions taken by the accused are considered to be both an expression of the mental state and evidence of the *actus reus*. In the second place, and somewhat at odds with the first consequence, in terms of collective criminality, this subjective approach means that not only are all parties to a crime considered to be equally liable, due to their individual intention to partake in the crime, but they also become liable for each other's acts, when one or more of the parties commit crimes that go beyond the expected result. The test is not what each individual contributed to the crime, but whether each individual is culpable for the sum of all the outcomes of collective criminal activity, due to the blameworthy mental state that was present in the very

---

[53] For the distinction between a normative and descriptive understanding of the principle of culpability, see Sect. 5.3.5. See further Fletcher 2007, p. 308.

[54] Kadish 1980, p. 10; Fletcher 2000, p. 400.

[55] It should be noted that there are some constitutional limitations on the allowance of presumptions or permissive inferences as proof of the required mental state beyond reasonable doubt; Kadish et al. 2007, p. 220.

intention to take part. This, despite the difficulty of proving that mental state based mostly on inference.

By way of clarification, the requisite mental states in the US generally fall under four categories of diminishing culpability. The MPC serves as authoritative and instructive in this, since most states have followed the model in their own legislation.[56] The most culpable mental state is 'purposely', defined as being the accused's 'conscious object to engage in [the prohibited] conduct ...or to cause such a result', or with respect to attendant circumstances, being 'aware' or believing or hoping that these circumstances existed.[57] The second most culpable state is 'knowingly', defined as being aware of the nature of the conduct or the existence of attendant circumstances, or being aware that 'it is practically certain' that the conduct will bring about the prohibited result.[58] These first two levels of intent focus on the actor's goal,[59] and the distinction between them rests on the willingness versus the mere awareness of the outcome. Many state jurisdictions still prefer to use the term 'intent' to refer to both of these, but much of the statutory language has been updated to reflect the nuances in the MPC distinctions.[60]

The lower two levels of culpable mental state are 'recklessly' and 'negligently'. Recklessness is defined as conscious disregard of a 'substantial and unjustifiable risk'. The disregard must amount to a 'gross deviation from the standard of conduct that a law-abiding person would observe in the actor's circumstances.'[61] Negligence is defined objectively; the actor *should* be aware of a substantial unjustifiable risk', and the failure to perceive such a risk is the very thing that leads to gross deviation from the expected standard of conduct.[62] Rather than focusing on the actor's goal, these two lower levels of culpability focus on the risk created by the actor.[63] The distinction between the two rests on the level of awareness the actor has of the risks involved in the conduct.

These definitions are necessary in order to understand the extensive liability for the acts of other individuals involved in collective criminality in the US. The central role given to these mental states in measuring the blameworthiness of an accused, and the subjectivity this reflects, will become apparent in the following discussion on liability for collective crimes.

---

[56] The old common law crime definitions tended to distinguish between 'intentionally' on the one hand, and 'recklessly' or 'negligently' on the other hand, however this terminology was inconsistent across time and across different state jurisdictions, and led to confusion. See Dressler et al. 2001, p. 121.

[57] MPC 1962, § 2.02(2)(a).

[58] MPC 1962, § 2.02(2)(b).

[59] Fletcher 2000, p. 443.

[60] Dressler et al. 2001, p. 121.

[61] MPC 1962, § 2.02(2)(c).

[62] MPC 1962, § 2.02(2)(d), empahsis added.

[63] Fletcher 2000, p. 442.

6.3.2 Vicarious or Derivative Liability for Parties to a Crime

As mentioned, most federal and state statutes have eradicated the old common law normative distinctions between parties to a crime altogether. The only one that remains distinct is the accessory after the fact, defined as a person who 'knows that a crime has been committed and tries to help the offender escape arrest or punishment',[64] who is generally subject to a lesser punishment in most state jurisdictions. For all other parties to the crime, the conviction is for commission of the crime proper, and the punishment is the same regardless of the actual role played by each individual.[65] All parties are considered to be principals, and unlike the old English common law, there is no distinction between degrees. The fact that all parties are qualified as having committed the crime, regardless of their actual contribution to its fulfilment, is reflected in the description in some literature that party liability is an 'alternative way of committing a crime'.[66]

Textbook definitions of party or accomplice liability in the US include soliciting, requesting or commanding another actor to commit the crime, or aiding the other person in planning or committing,[67] all forms of assisting,[68] and 'all persons concerned in the commission of a crime'.[69] Thus all parties to a crime are considered to fall under one, generic group, leading to a functionally unitary system of liability. As far-reaching as these definitions are in terms of the actual contribution of a party to a crime, the one essential element in all of these definitions is the necessary intention to facilitate or promote the commission of the crime. The subjective approach is entirely determinative: because assistance (of any kind) is given intentionally, an accomplice is considered to be equally as blameworthy as the physical perpetrator, regardless of the actual contribution to completion of the crime. In fact it may not be clear from the definitions who is the physical perpetrator and who is an accomplice, and the final qualification renders such factual distinction legally irrelevant.

In practice there does appear to be some differentiation when it comes to sentencing, however this is purely discretional and not based on any doctrinal or systematic distinction.[70] It could be argued that it is precisely because all participants are collapsed into one notion of 'principal' that reflecting any difference in moral culpability is left to the discretion of the judge deciding on the sentence.[71]

The drafters of the MPC attempted to systematise this collapsing of different parties and the inconsistent use of terminology throughout individual state

---

64 According to the federal law statute 18 USC § 3. See also Garner 2009, under 'accessory after the fact', p. 16.

65 Kadish et al. 2007, p. 590.

66 Wiesberg 2000, p. 222.

67 Wiesberg 2000, p. 220.

68 Dressler et al. 2001, pp. 473–475.

69 Kadish et al. 2007, p. 590.

70 Kadish et al. 2007, p. 591.

71 This is the sentencing-oriented approach associated with a functionally unitary system, as described in Sect. 5.3.1. See further Vogel 2002, p. 159.

legislation. Section 2.06 of the MPC attempts to draw a series of complex definitional distinctions, to differentiate descriptively between parties to a crime, even though this would not amount to a normative differentiation.[72] As the later commentaries to the MPC point out, the distinctions used were 'for the purpose of developing their content . . . [but do not] have any procedural significance.'[73] However these articles on complicity are considered by many to be extremely complex and were rejected by most legislatures during the wave of law reform that followed the adoption of the MPC.[74]

The most important thing of note about section 2.06, however, is the title of the section, which is indeed reflected in the case law and statutes, namely that complicity is 'liability for the conduct of another'. Subsection (1) states that 'a person is guilty of an offence if it is committed by his own conduct or by the conduct of another person for which he is legally accountable, or both.'[75] A person is considered to be legally accountable for the conduct of another if she causes an innocent agent to commit a crime,[76] or if she is an accomplice,[77] which is further defined as soliciting, aiding or abetting.[78]

This liability for the acts of another can be defined as vicarious liability, although it differs from the kind of vicarious liability found in tort or contract law. In private law, a parent or an employer can be held vicariously liable for the conduct of a dependent or an employee, *in the place of* the person who was causally responsible for the harm, due to the fact that this dependent or employee cannot be held liable, and the burden of reparation must rest elsewhere. However in criminal law, where liability is not about restoring damages, but about who can be held accountable and punished effectively for the harm caused, it is unacceptable to shift liability to a third person in the place of an offender. An explicit theory of vicarious liability has been rejected in the case law as inappropriate for criminal law. In one case vicarious liability of an employer for his employee's criminal acts was found to be a breach of the substantive due process clauses of the constitution of the state,[79] and in another the vicarious liability of parents for their children's criminal behaviour purely on the basis of their status as parents was rejected as offensive to due process if the parents had no fault in the situation.[80] However it should be noted that in both these cases,

---

[72] For the full text of the MPC articles on liability, see Appendix A.2.

[73] American Law Institute 1985, p. 299.

[74] Lynch 2003, p. 222.

[75] MPC 1962, § 2.06(1).

[76] MPC 1962, § 2.06(2)(a).

[77] MPC 1962, § 2.06(2)(c).

[78] MPC 1962 § 2.06(3)(a)(i) and (ii) respectively. Under (iii), an ommission where there is a legal duty to prevent the commission of an offense also falls under complicity.

[79] The Supreme Court of Minnesota found that attaching vicarious liability to the owner of a restaurant for the illegal sale of alcohol to a minor by one of his employees would have far-reaching unjust consequences for the restaurant owner. It should be noted that he neither had any awareness of the actions of his employee, nor did he in any way implicitly ratify them. State v Guminga 1986.

[80] State v Akers 1979.

the absence of any fault at all on the part of the accused was a reason to reject what would otherwise amount to strict liability.

Nonetheless, the understanding of party liability as liability for the acts of others can still be best described as vicarious liability. While proxy punishment is not acceptable, liability may be distributed among more than one actor, and in the case of party liability in the US, it is distributed vicariously; every party to a crime is responsible for the acts of all other parties, but jointly, rather than in their place. This is based on the subjective approach: anyone who 'knowingly, voluntarily and intentionally unites with the principal offender'[81] is legally accountable for the acts of that principal offender.[82] A party to a crime will therefore not be convicted for assisting in the crime, but rather for the commission of the crime

One way of explaining this indirect or vicarious liability is that the physical perpetrator's acts *become* the acts of the accomplice, and the accomplice's identity as a person subject to criminal punishment is subsumed in that of the physical perpetrator.[83] In the case law this has been described as 'forfeited identity'[84] whereby the accomplice loses her right to be treated as an individual when she chooses to engage in group criminality. The acts of the physical perpetrator are attributed to the accomplice, because the accomplice is deemed to have said 'your acts are my acts', or as one judge put it in his instruction to the jury, 'the hand of one is the hand of all'.[85] However at the same time, a party to a crime can be prosecuted regardless of whether the physical perpetrator has been prosecuted,[86] and it is irrelevant for the prosecution of the accomplice whether such a physical perpetrator was convicted or acquitted.[87] The theory of forfeited identity is not entirely consistent with the procedural relationship between a physical perpetrator and an accomplice. In any case it is clear that the subjective approach in the US leads to an attribution of the acts of the physical perpetrator to all those parties who had some form of shared intent. Indeed it has been stated in the case law that the liability of a person tried as an aider and abetter 'is vicarious'.[88]

In some literature, the accomplice's liability is described in terms of derivative liability, however it is done so to clarify that an accomplice is not guilty of an independent *offence* of aiding and abetting, but rather that the acts of the physical

---

81 Garner 2009, p. 18, under 'accomplice'.

82 MPC 1962, § 2.06 (1).

83 Dressler et al. 2001, p. 466.

84 Prettyman 1996, p. 1018.

85 Dressler et al. 2001, p. 467.

86 Kadish et al. 2007, p. 590.

87 MPC 1962, § 2.06(7). For a long time, differences between justifications and excuses in this respect were disregarded in the common law. Thus even if the physical perpetrator could make a claim to a justification defence, this did not necessarily flow down as a defence for an accomplice. Once again, the presence of a lay jury, and the absence of any requirement for giving reasons for an acquittal, leads to the conflation of these two different types of defences and their consequences for an accomplice. However there does seem to appear to be a distinction made in judge's instructions in more recent case law. Fletcher 2000, pp. 652, 634.

88 Croy 1985.

perpetrator become the acts of the accomplice.[89] This is neither a true theory of derivative accessory liability, nor is it the dominant theory in the US in case law, nor in the leading literature.

Understanding the theory underpinning party liability in the US as vicarious is important for understanding the way in which liability for parties to a collective crime operates under US law, and also for understanding the way in which notions such as conspiracy and complicity have been translated to the international plane by US participants in the international processes of law.

6.3.3 Mens Rea *and* Actus reus *Requirements for Accomplice Liability*

Because accomplice liability is explained as liability for the acts of others, it must be determined whether there was a criminal mental state present on the part of the accomplice, to join with the physical perpetrator in the commission of a crime. In general the case law requires at least a 'shared common purpose', 'community of intent',[90] or criminal motive between the accomplice and other actors involved in the group criminality. However there is discrepancy in the terminology across state jurisdictions, and it is not necessary that the accomplice and the physical perpetrator share the same mental state. Because there is no uniformity in the definition of intent or purpose, an accomplice could include anyone who has knowledge of the intended crime. As it was put by one judge in a 1940 case, even the person who sells a gun to another, knowing it will be used for murder, cannot escape liability for murder by saying the gun was sold merely for profit and not for the purpose of the crime:

> Guilt as an [accomplice] depends, not on "having a stake" in the outcome of the crime …but on aiding and assisting the perpetrators; and those who make a profit by furnishing to criminals, whether by sale or otherwise, the means to carry on their nefarious undertakings, aid them just as truly as if they were actual partners with them …[91]

This somewhat loose standard was refined by Judge Learned Hand in the oft-cited *Peoni* judgment. His definition of accomplice liability requires that the accused must 'in some way associate himself with the venture, that he participates in it as something he wishes to bring about, that he seeks by his action to make it a success. All the words used […] carry an implication of purposive attitude towards it.'[92]

The original MPC draft had included 'knowingly' as a loose standard for accomplice liability, however the final draft follows Judge Learned Hand's formulation, and requires a higher standard of a 'purpose to facilitate the crime'.[93] Most federal and

[89] Dressler et al. 2001, p. 466; Kadish et al. 2007, p. 589.

[90] State v Hayes 1891.

[91] Backun v United States 1940, p. 637.

[92] United States v Peoni 1938, p. 402; For an analysis of this formulation and its subsequent application in case law, see Wiesberg 2000, p. 236.

[93] American Law Institute 1985, p. 310.

state courts follow this standard, but interpretation of the standard can sometimes differ.

The subjective approach that accomplice liability is essentially the joining of intention with the physical perpetrator, thereby creating liability for the acts of others, renders the mental state extremely important. However even Judge Learned Hand's formulation requires some form of participation in the crime, and the question arises what an accomplice must do in order to be liable for the acts of others. Due to the vicarious nature of accomplice liability, each accomplice is charged with the complete crime, regardless of his actual participation. The *actus reus* requirement is therefore relatively simple: as long as there is some actual assistance to the crime, the degree of assistance is irrelevant.[94] There is, for instance, no causal requirement. If the physical perpetrator would have committed the crime even without the assistance of the accomplice, this accomplice will still be considered fully culpable as a 'principal'. This can sometimes lead to troubling consequences in terms of attenuated liability and disproportionate punishment.[95]

In an old case that is sometimes referred to as a standard, an accomplice was convicted for murder, even where the evidence failed to show whether the defendant did anything to facilitate the murder, other than to say some vaguely encouraging words to the physical perpetrator, a fact which the defendant denied.[96] This follows the old common law doctrine that an accomplice can be guilty of aiding and abetting even if he is 'present at the scene for purposes of aiding and abetting, but refrains from aiding abetting because it is no longer necessary to assist.'[97]

In some cases involving the sale of drugs, courts have found that an accomplice can be liable even if there is no prior agreement between the actors; aiding and abetting can be spontaneous, and there is no planning necessary over time.[98] Once more it is clear that it is the intention of the accused which is determinative of his or her role as accomplice, and the only *actus reus* required is a *de minimus* standard of non-causal assistance.[99]

---

94 Dressler et al. 2001, p. 475.

95 Dressler et al. 2001, p. 477.

96 Hicks 1893.

97 Hicks 1893.

98 State v Gladstone 1970; State v McKeown 1979; State v Wilson 1981.

99 There is also a narrow possibility for complicity by omission. Some courts have been willing to define a failure to act or intervene in a crime as accomplice liability, see in which a father not only stood by while his son raped a young woman in his house, but he also told her he could not help her when she screamed for him to stop his son, State v Davis 1989; see also Stanciel 1992, in which a mother was convicted as an accomplice to her boyfriend who beat her 3 year old daughter to death, for failure to protect her child. However the MPC requires a legal duty to act before such accomplice liability by omission is punishable, MPC 1962, § 2.06(3)(a)(iii).

6.3.4 *Extended Liability: Further Crimes Committed by Others*

The subjective approach to accomplice liability raises the question what limits there are on the kinds of criminal acts of others which can be imputed to the accomplice. If there is no objective *actus reus* requirement other than some form of assistance to a crime, and assuming the test for the required mental state is a 'purpose to facilitate the crime', can an accomplice be held liable for additional crimes committed by others in furtherance of the original crime? If an accomplice has a purpose to facilitate a bank robbery, can he or she also be held liable as a principal for a murder committed by one of the other actors which was not explicitly agreed upon? It would seem that there are few limits on accomplice liability, and that as long as additional crimes are a 'natural and probable' consequence of the intended crime,[100] then an accomplice can and will be convicted for these other crimes, regardless of whether she or he intended them.

Illustrative of this extended liability is the *Luparello* case, in which the defendant enlisted the help of several violent friends to help him get information from a man who would know the whereabouts of his ex-wife. A first confrontation failed to result in any information, and when the defendant's friends visited the man a second time they did so armed with a gun and shot and killed the man. The defendant Luparello was not present and claimed he did not know about the unplanned and unintended murder, yet he was still convicted for the murder. The court reasoned that even if an accomplice does not share the intent necessary for the specific crime:

> [l]iability is extended to reach the actual crime committed, rather than the planned or "intended" crime, on the policy [that] aiders and abettors should be responsible for the criminal harms they have naturally, probably and foreseeably put in motion ...[The accomplice] is guilty not only of the offence he intended to facilitate or encourage, but also of any reasonably foreseeable offence committed by the person he aids and abets ...[101]

Noteworthy is the concurring opinion by Judge Weiner, who agreed with the conclusion of the court, yet pointed out the logical and theoretical flaws in this 'reasonably foreseeable' test. For one thing, the liability of an accomplice is no longer predicated on his or her own mental state, but on that of the physical perpetrator, a result which is founded on the 'outmoded and logically indefensible proposition that if a person exhibits some intent to violate the law, we need not be terribly concerned that the contemplated crime was far less serious than the crime which actually took place ...'[102] This test has sometimes been formulated as the 'natural and probable consequences' test;[103] defined where an accomplice to crime A is also an accomplice to crime B, if 'in the ordinary course of things' crime B 'might reasonably

---

[100] Prettyman 1996, p. 1020.

[101] People v Luparello 1987; citing Croy 1985.

[102] Concurring opinion Weiner J., People v Luparello 1987.

[103] Prettyman 1996.

ensue from the planned events'.[104] It is not enough that crime B 'might conceivably ensue' from crime A; crime B must be 'within the normal range of outcomes that may be expected to occur if nothing unusual has intervened.'[105] Where the physical perpetrator in *Luparello* had a highly culpable mental state of purposely killing the victim, Luparallo himself was, with regard to the killing, at best reckless or negligent.

Because no distinction is made between an accomplice and a principal, this inflates the participation of the accomplice, even in the absence of the requisite mental state.[106] This goes against the principles of culpability and proportionate punishment, that a person should only be punished to the degree that he or she is culpable, even where the principle is understood as a description of the mental states, as it is in the US. It has therefore received strong criticism by scholars and by the American Law Institute, in line with the criticism voiced by Judge Weiner, for being incongruous and unjust.[107] Indeed, the American Law Institute in its commentary on § 2.06 of the MPC stated that an accomplice should not be convicted of an offence unless she has the same or higher degree of culpability (meaning mental state) required to convict the physical perpetrator.[108]

Thus legally it doesn't matter who the physical perpetrator is, and all those involved in collective criminality can be convicted as 'principals'. Although this may serve to solve the evidentiary problem of over-determination if it is unclear whose bullet killed a victim, or whose blow delivered a serious injury, it creates a new problem in terms of the theory of liability. In a case where there is no individual physical perpetrator identifiable, in effect all the actors involved in a collective crime are each other's accomplice. This means the net of liability is cast very wide, and many actors on the periphery of a crime can be caught in this net as accomplices.

This far-reaching net is cast even wider when combined with the minimal *actus reus* requirement and the 'naturally foreseeable' rule, meaning that in the end anyone rendering any kind of non-causal assistance to others in committing a specific crime can be held fully liable as a principal for most any crime committed by those others. Only a crime which is completely unrelated in any way to the original criminal behaviour, and therefore not reasonably foreseeable, might be excluded from this extended liability.

There is a specific rule which extends the liability of an accomplice even further, which does away with the foreseeability requirement: the felony-murder rule. If someone dies as a result of a felony, the felony-murder rule treats the defendants as if they have committed murder rather than unintentional or reckless manslaughter. This not only steps up the culpability from an act which is committed with mere recklessness to punishment for a crime normally requiring purpose or at least knowledge, but it also amounts to a kind of strict liability, in that foreseeability of the death of the victim is not required, as long as the death was factually caused by the felonious

[104] Prettyman 1996; Roy 1995.

[105] Roy 1995, p. 1105; Dressler et al. 2001, p. 483.

[106] Dressler et al. 2001, p. 484.

[107] Dressler et al. 2001, p. 484; Comment to § 2.06, American Law Institute 1985, p. 312, note 42.

[108] Comment to § 2.06, American Law Institute 1985, p. 312, note 42.

act. This even includes the rule that a criminal actor must take the victim as she is. In one case during a bank robbery the defendant told the victim to lie still, and during the next 10 min the victim suffered a heart attack and died. The felony-murder rule applied and the defendant was convicted for murder.[109] Such an outcome is unfamiliar to most civil law type jurisdictions, and uncommon in other jurisdictions which follow the common law tradition.

This rule has been rationalised by the notion that someone engaging in a felony is engaging in inherently risky activity and must be held liable when things turn out worse than expected.[110] The question is whether such a rationale can also be applied to an accomplice. Because an accomplice is liable for the acts of a physical perpetrator, she is also liable under this strict liability rule for the death caused by the underlying felony, as if it were a murder. One could question whether it is reasonable to treat a person as a principal in a 'murder' which was not, in fact, intended nor even necessarily foreseeable, when that person aided the physical perpetrator in a distant or merely constructive manner. Although the felony-murder rule has become a minority rule and has been removed from many state statutes, its existence in the first place demonstrates the strength of the subjective approach in the US in general.

Even leaving aside the felony-murder rule, the outer limits of accomplice liability appear to be quite elastic. The policy reasons behind this have to do with the desire to prosecute risky behaviour associated with participation in group criminality, and the ensuing theory that anyone associated with a crime with the purpose of facilitating that crime is vicariously liable for the acts of others in the collective criminality. Liability is not predicated on participation but on wilful association with the physical perpetrator, and to draw a line limiting liability in any way different from the liability of the physical perpetrator would be to re-introduce the common law distinction between different participants, making evidentiary standards near impossible for the prosecution to fulfil. Instead the accomplice is treated in every way as a principal.

The subjective approach and its attendant notions of extended liability are also evident in the institution of conspiracy as it has taken specific form in the US, to which we will now turn.

6.3.5 Conspiracy: The Common Law Solution to Group Crimes

Conspiracy is a notion particular to the common law tradition. When the Nuremberg Charter was being drafted there were debates with respect to the inclusion of conspiracy due in large part to the fact that the Russian and European lawyers were entirely unfamiliar with the way this institution would work, and in some small part to the fact that even between the English and US lawyers there was a difference of

[109] People v Stamp 1969.

[110] Fletcher 1981, p. 427.

opinion.[111] With its roots in old English common law as an inchoate crime, many common law jurisdictions continue to apply it as such, as a way of responding to the difficulty of attributing acts to individuals in group crimes; however in the US alone it has developed as a mode of liability. This particularity will receive full attention in the section below, but first the history and construction of conspiracy as an inchoate crime will be discussed, since this forms the same core for conspiracy in both the jurisdictions compared here, Canada and the US.

6.3.5.1 Conspiracy as an Inchoate Crime

The doctrine of conspiracy is said to have been formulated as an inchoate crime by the Star Chamber for the first time in 1611.[112] In its simplest terms, the common law definition of conspiracy is an agreement between two or more persons to commit a criminal act, or to commit a lawful act by unlawful means.[113] Although it was originally a private law concept in seventeenth century England, namely conspiring to falsely accuse and convict an innocent person,[114] in the nineteenth century it developed into an inchoate offence, where conspiring to commit any felony was considered to be as serious as the felony itself, and punished just as heavily.[115] At this time in England it was used as an extension of criminal law liability to punish an agreement to commit any unlawful act, including civil wrongs such as torts and contractual wrongs.[116] This made it an effective social and political tool, as it was used particularly against trade unions, to convict people for agreements to strike while it was still illegal to do so, thus suppressing organised industrial action.[117] Because the acts, though unlawful, were not themselves criminal, and there was therefore no substantive offence to be prosecuted, conspiracy was brought as a charge where the agreement had actually been carried out, rather than as a preventative measure for acts not carried out.

The justification for the criminalisation of a mere agreement has generally been that it is preventative, since it enables the justice system to intervene before any actual harm has been inflicted. However it is noteworthy that criminal conspiracy goes much further than the similar notion of attempt, since all that is needed is an agreement, and unlike attempt, there is no requirement of an act that is more than merely preparatory.[118] Thus although the legal definition of conspiracy focused on

---

[111] For a full discussion of these debates, see Sect. 8.2.

[112] Harvard Note 1959, p. 923. This academic 'note' is considered an authoritative treatise on the law of conspiracy in the US which, even if somewhat dated, still serves as a background text for understanding conspiracy in American jurisprudence. See further Tigar 1986, p. 127.

[113] Harvard Note 1959, p. 922, citing an American case from 1842.

[114] LaFave 1986, p. 454.

[115] Fletcher 2000, p. 222.

[116] Ashworth 2009, p. 448.

[117] Ashworth 2009, p. 449.

[118] Ashworth 2009, p. 448; Harvard Note 1959, p. 923.

the agreement, in fact historically it was not used in its true sense as an inchoate crime, but rather as a substitute for a completed substantive crime, with a social focus on the (non-criminal) unlawful acts.

The original common law definition of a conspiracy requires 'two or more persons' to enter an agreement, which is known as the plurality rule. One consequence of this rule is that it would automatically lead to the acquittal of a co-conspirator if one of two defendants is acquitted, since nobody can conspire alone.[119] Under the plurality rule the focus is very much upon the group and not upon the blameworthiness of each individual within that group. In the US, the MPC departs from this rule and focuses instead on the culpability (i.e. the required mental state) of the individual accused, 'rather than on that of the group of which he is alleged to be a part.'[120] The approach of the MPC, which is arguably more consistent with the (descriptive) principle of culpability, has been followed by the majority of states.[121] The question whether to focus on the role of the individual with respect to the agreement, or the role of the agreement at the centre of group activity, plays out in the conceptualisation of conspiracies involving disparate groups of individuals, described below in Sect. 6.3.5.2.

In the US, conspiracy has been dubbed 'the darling of the modern prosecutor's nursery',[122] a criticism which reflects the elasticity of this doctrine. Conspiracy gained increased favour by US policy makers in the early twentieth century, mostly as a response to mafia based crimes during the era of prohibition.[123] Because it was difficult to prove who did what in these complex and secretive webs of organised crime, the best policy solution was to widen the net as far as possible by focusing on the agreement between participants. This was particularly with respect to the desire to be able to capture the leaders, or family heads, who were almost never present at the scene of a murder, nor did they do anything physically with respect to the movement of illegal alcohol, and for whom it was difficult to prove in a court that they had given explicit orders for a specific crime to be committed. Instead, all that needed to be proved was that there was an agreement between two or more people to commit a crime; thus the agreement satisfies the *actus reus* requirement, and can also be used to infer the required mental state.[124] This allowed authorities to prosecute the leaders of mob crimes despite lack of evidence of their participation, to intervene in the planning of mobster crimes before the crime took place, and also in the prevention of economic crimes such as violations of the anti-trust laws.

The underlying assumption is that collective action poses a greater risk to society than individuals ever could. A conspirator is assumed to be less likely to withdraw from a commitment made to others in the group than if it were purely an individual

[119] Morrison v. California 1934, p. 92.

[120] Comment on § 5.03 American Law Institute 1985, p. 393.

[121] Dressler et al. 2001, p. 449.

[122] Justice Jackson's expression in his critical analysis of the ever-expanding doctrine, in Harrison v the United States 1925.

[123] Harvard Note 1959.

[124] State v. Carbone 1952.

plan to commit a crime, and group criminality also allows for a division of labour, enabling grander schemes and wider-scale criminality to take place.[125] These risks to society are said to justify the early intervention in conspiracies even before any substantial act has been undertaken to begin the planned crime, let alone completion of the crime. In one case the Supreme Court stated that:

> although the law generally makes criminal only antisocial conduct, at some point in the continuum between preparation and consummation, the likelihood of a commission of an act is sufficiently great and the criminal intent sufficiently well formed to justify the intervention of the criminal law.[126]

One reason that conspiracy holds such favour among prosecutors is that although conspiracy is an inchoate crime, the so-called 'merger doctrine' which applies to other inchoate crimes such as attempt, does not apply. The merger doctrine logically says that an attempt cannot be indicted as a separate charge where a crime is in fact completed; the attempt is subsumed in the charge of the completed crime. On the other hand, conspiracy to commit a crime can be indicted cumulatively with the completed crime itself.[127] The Supreme Court has more than once stated that a conspiratorial agreement 'is a distinct evil which may exist and be punished whether or not the substantive crime ensues.'[128] The increased risk to society of collective criminality, and the state's policy interest in stamping out such planned criminal conduct, is at the root of this shift away from the double-jeopardy or *ne bis in idem*, rule.[129]

But there is some criticism of this practice, since conspiracy to commit a crime and the charge of the completed crime itself can also be said to amount to 'alternative forms of the same offense',[130] and cumulative charges lead to cumulative sentences that are more severe than the intended maximum as prescribed by the legislator for a specific substantive crime.[131] For these very reasons, the MPC departs from this rule and states that a person may not be convicted for both a conspiracy and the completed crime, unless there is evidence that the conspiracy included an agreement to commit crimes that were never committed or attempted.[132] The majority rule in most states, however, does not follow the MPC.

Another reason that conspiracy is the 'darling of the prosecutor's nursery' is that there are several evidentiary and procedural advantages. Because conspiracy is an inchoate crime the evidentiary burden is very low, since all that needs to be proven is the existence of an agreement, and there is no need to prove any causation with respect to a particular defendant. Moreover, because conspiracy is aimed at

125 Harvard Note 1959, p. 923.

126 The United States v Feola 1975, p. 694.

127 Dressler et al. 2001, p. 433.

128 Jimenez Recio 2003, p. 274, citing Salinas v United States, 522 U.S. 52, 65 (1997).

129 Harvard Note 1959, p. 928; Dressler et al. 2001, p. 433.

130 Krulewitch 1949.

131 Harrison v the United States 1925.

132 MPC 1962, § 1.07(1)(b); Comment to § 1.07, American Law Institute 1985, p. 109.

organised crime, the rationale is that the evidentiary burden should be low due to the greater dangers to society of such endeavours.[133] But the evidentiary advantages go further. For example there is an exception to the hearsay rule; contrary to this exclusionary rule, any statement made by a co-conspirator can be used against other defendants accused of participating in the conspiracy.[134] Furthermore, acts done by co-conspirators *before* another individual accused joined the conspiracy can also be used as evidence of the existence of the agreement into which that defendant later joined, and for which he becomes liable.[135] Also the determination of the *locus delecti* for purposes of jurisdiction can be much broader with a charge of conspiracy, since the venue of the actual agreement may shift over time as more people enter the agreement.[136] And finally, because conspiracy is considered to be a continuing crime, individuals can be prosecuted for having entered into the agreement at any time, even if the predicate crime has already been fulfilled.[137]

Because of these procedural advantages, and the low evidentiary burden for the prosecution, the charge of conspiracy has been continually expanded by prosecutors, demonstrating their influence as participants in the development of the law as an expression of policy. There is some resistance to the elasticity of the doctrine, as articulated for example by Justice Jackson in the Supreme Court case *Krulewitch*, where he displayed his concern for the 'tendency of the principle to expand itself to the limits of logic,' and criticised the ominous application of the doctrine as being 'so vague that it almost defies definition'.[138]

The *Actus Reus* Requirement

Conspiracy is defined as an agreement between two or more persons to commit an unlawful act. Since this may amount to the punishment of 'evil thoughts' without further action, a notion abhorrent to criminal law, the *actus reus* is construed to consist of the making of the agreement.[139] Subjectivity is especially apparent here: conspiracy is always 'predominantly mental in composition' because it consists primarily of a meeting of minds and an intent.[140] The agreement both declares and corroborates the criminal purpose of the conspirators.[141] While this could be said to be a stretch of the legal notion of an act, in terms of policy it certainly solved a great problem of linking the individual to collective criminality, and especially with respect to the more invisible leaders behind the scenes.

---

[133] Dressler et al. 2001, p. 432; Kadish et al. 2007, p. 664.

[134] Krulewitch 1949; Katyal 2003.

[135] Dressler et al. 2001, p. 435.

[136] Pilsbury 2009, p. 339.

[137] Kadish et al. 2007, p. 668.

[138] Krulewitch 1949, p. 445.

[139] State v. Carbone 1952, p. 574; See Harvard Note 1959, p. 926.

[140] Krulewitch 1949, p. 445.

[141] MPC 1962, § 5.03; American Law Institute 1985, p. 97; Fletcher 2000, p. 221.

Although the agreement is at the core of the inchoate crime, it need not be explicit, and the prosecution need not show that an accused knew the full planned scope of the conspiracy, nor even who else was participating in it. It need only be shown that the defendant knowingly agreed to be part of the core crime, a low threshold that has allowed the prosecution of conspiracy to become extremely elastic. The US Supreme Court has found that a tacit agreement is sufficient to establish a conspiracy.[142] Furthermore, knowledge of all the details or phases of a conspiracy is not required in order for an individual to be guilty; knowledge of the 'essential nature of it' is considered sufficient.[143]

One recent example is the trial of Zacarias Moussaoui who was prosecuted for multiple counts of conspiracy to commit terrorist acts, for his alleged involvement in the September 11 attacks in 2001.[144] Moussaoui's connection to the hijackers was alleged to have consisted of attending a flight school, joining a particular gym, and purchasing knives shortly before passing through airport security, all of which were parallel to actions of the known hijackers. These actions amount at most to preparation, which is not normally criminalized. He pleaded guilty to the conspiracy charge, so there was no need for proof beyond reasonable doubt of his connection to the Al Qaeda conspiracy, but the jury gave him a lesser sentence than the death penalty, apparently on the basis that there was no evidence he had any knowledge of the actual workings of Al Qaeda.

Courts have been willing to delineate conspiracy from spontaneous group responses which do not amount to co-ordination or planning, however implicit.[145] In *United States v Garcia* a gun fight between two gangs had ensued following abusive verbal attacks, and the question was whether the defendants could all be held liable for entering a conspiracy to shoot and assault members of another gang.[146] The court found that membership of a gang, and the assumed general agreement to back each other up in situations of violence or conflict, is not enough to amount to the kind of co-ordination and planning that conspiracy makes punishable. If there is no direct evidence of co-ordination and planning, then it can be inferred only if the result could not be achieved without some form of tacit agreement or shared illegal

---

142 Interstate Circuit, Inc. 1939.

143 United States v. James 1976.

144 In fact he had been arrested on unrelated immigration charges prior to the September 11 attacks and the conspiracy charges came later. United States v. Moussaoui 2006; Kadish et al. 2007, p. 692

145 A notable exception is the questionable result in Griffin v. State 1970, in which a motorist was loudly protesting the traffic violation for which a police officer had stopped him. A crowd gathered and when the officer attempted to arrest Griffin, Griffin pushed him to the ground and members of the crowd began to assault the officer. The defendant was convicted for conspiring to commit assault, but it remains unclear whether there was any tacit agreement between the individuals rather than a spontaneous 'mob mentality' response amounting to mere parallel action. See Ohlin 2011, p. 44.

146 United States v. Garcia 1998.

objective.[147] But the shootings were not an act the nature of which 'would logically require coordination and planning' and hence could not amount to a conspiracy.[148]

This minimum threshold serves to distinguish a punishable conspiracy from a coincidence of actions: on the one hand, all that is needed is sufficient evidence that the accused parties had a 'common design' and 'pursued by their acts the same object, often by the same means.'[149] On the other hand, this common design must denote some kind of coordination, a shared intention, such that coincidentally parallel actions do not become punishable, where there was no agreement at all between the actors.

The question remains whether an overt act is also required in order to fulfil the *actus reus* for conspiracy. In traditional common law definitions of conspiracy there was no requirement of an overt act, it was sufficient to infer the punishable agreement from circumstances.[150] Now in US federal law,[151] in the MPC[152] and in most states,[153] an overt act in pursuance of the conspiracy is required, however the standard is a *de minimus*: the act need not itself be a criminal act and could be as simple as picking up the phone to make a phone call regarding the agreement.[154] As well, the overt act can be performed by any member of the conspiracy in order to corroborate the agreement, rendering all participants in the conspiracy criminally liable from that moment forth, including individuals who join the conspiracy at a later moment in time.[155] A defendant need not have performed an overt act, as long as some other member of the conspiracy had done so. This last point further reflects a subjectivist reading of the overt act requirement, and of the doctrine of conspiracy in general.[156] This subjectivity allows for the extensive application of conspiracy as a substantive crime; an individual may be liable for commission of *this* inchoate crime while having done nothing but displayed some behaviour from which could be inferred that he or she was party to an agreement to commit *a* crime.

The Required Mental State

Given that conspiracy is predominantly mental in composition, with the entire focus on the agreement between co-conspirators, it is no surprise that the required mental state is often described as double-intent; the defendant must have the intention to join in the agreement, and must also share the required mental state for the intended

---

147 Such as that found in Interstate Circuit, Inc. 1939.
148 United States v. Garcia 1998, pp. 1245–1246.
149 Rex v. Murphy 1837, p. 508.
150 Kadish et al. 2007, p. 693.
151 18 USC § 371.
152 § 5.03(5) MPC 1962.
153 Pilsbury 2009, p. 344.
154 Harvard Note 1959, p. 946.
155 Harvard Note 1959, p. 946.
156 Fletcher 2000, p. 224.

crime itself.[157] As long as these can be proven, it is irrelevant whether that defendant knows who else is involved in a conspiracy and what the precise details of the planned commission of the crime are.

With respect to the first intention, to join the conspiracy, the question is whether mere knowledge of the intended crime is sufficient to construe intent. The general standard in the US can be said to follow the formulation that:

> the intent of a supplier who knows of the criminal use to which his supplies are put to participation in the criminal activity connected with the use of his supplies may be established by (1) direct evidence that he intends to participate, or (2) through any inference that he intends to participate based on (a) his special interest in the activity or (b) the aggravated nature of the crime itself.[158]

The court in this case cited two earlier cases which decided in opposite directions on this point. In *United States v. Falcone*[159] the sellers of large quantities of sugar, yeast and cans were absolved from participation in a moonshining conspiracy amongst distillers, while in *Direct Sales Co. v United States*[160] a wholesaler of morphine sulphate was convicted of conspiracy to violate federal drug laws by selling drugs in large quantities to a co-defendant physician who was supplying them to addicts. In both cases the defendant had knowledge of the crimes committed by others, but the distinction was drawn between the former case involving supply of every day goods or services which were used for unlawful acts, and the latter case of supplying dangerous goods which can only have been used for unlawful purposes in the quantities in which they were being ordered. This formulation represents some willingness to put limits on the otherwise extremely broad application of conspiracy to anyone who may enact an innocent action while having knowledge of the intention held by others to commit a crime.

6.3.5.2 Parties to a Conspiracy and Scope of Multiple Conspiracies

When circumstances show that a conspiracy to commit a crime is taking place, it may not be clear whether there is a single conspiracy involving many individuals, or several smaller conspiracies with fewer participants. This is of importance in terms of linking substantive crimes to various individuals, and in terms of the accumulation of counts against an individual.[161] Given that the common law doctrine of conspiracy has played a role in the development of modes of liability in ICL such as JCE, the

---

[157] Pilsbury 2009, p. 349.

[158] People v. Lauria 1967, p. 471.

[159] United States v. Falcone 1940.

[160] Direct Sales Co. v. United States 1943.

[161] It may also be of importance for some procedural issues such as to which hearsay evidence may or may not be accepted against a co-conspirator, the jurisdictional question of *locus delicti*, and, as discussed above, the question whether a conspiracy with respect to an individual defendant has been corroborated by an overt act of another person who is part of that same conspiracy. Dressler et al. 2001, p. 454.

conceptualisation of single or multiple conspiracies in US law may be useful in helping to untangle some of the confusion that has emerged over a few years of case law from the ad-hoc tribunals. These conceptualisations may also shine a different light on the uncertainty surrounding the notion of indirect co-perpetration which has been raised in recent case law at the ICC.[162]

One possible conceptualisation is that of the 'wheel conspiracy', whereby individuals may have no relationship with each other, but may still be participating in a single conspiracy by operating 'as separate spokes meeting at a common center';[163] for instance if there is an individual or group of individuals at the centre who co-ordinate drug or fraud crimes occurring in many different locations with people who have no contact with each other.

Another conceptualisation is the 'chain conspiracy', in which separate agreements are linked by way of passing on information or goods, such as the selling of counterfeit money to someone who then sells it on to third parties. The issue with such a conceptualisation is how many people or groups of people can be linked together for the purpose of one charge.[164] The longer the chain and the greater the number of 'links', the more tenuous the relationship between each of the individuals, and the more questionable the charge of conspiracy.

One way of solving this problem from the perspective of the prosecution is the conceptualisation of a chain-wheel conspiracy, such as in the distribution of drugs from smugglers to middlemen to sellers on the street in various states and cities. In *Bruno*,[165] the defendant was charged with eighty-seven others for conspiracy to import, sell and possess narcotics. By linking all of the defendants into a chain-wheel conspiracy, the retailers on the street who did not know each other and had no link to each other at all, were still connected by virtue of their individual links to the middlemen, who also had a shared link to the smugglers. The court found that although the disparate geographical groups of retailers did not know each other, they must have known there were others engaged in the same activity selling drugs from the same suppliers, hence they must have known they were a link in a greater conspiracy.

There is however a problem with this kind of conceptualisation, in that it can lead to an undermining of the requirement of an agreement, which is the essence of the crime of conspiracy. Effectively the conspiracy in the above scenarios is seen as a group of people rather than a criminal act, and individuals can be held liable for crimes committed by the group without there being a tacit agreement between them and other actors.[166] If a defendant does not know of the existence of others in other cities, how can that individual be said to be in agreement with them to commit a

162 See Chap. 8 on modes of liability in the international tribunals.

163 In the leading case it was found that the facts did not support the charge of a single conspiracy, however the notion of a 'wheel conspiracy' has since become a conceptual reference. Kotteakos v. United States 1946; Dressler et al. 2001, p. 454.

164 Roach 2012, p. 106.

165 United States v. Bruno 1939.

166 Harvard Note 1959, p. 934.

crime? It is not the crime of selling narcotics at issue here, but the inchoate crime of conspiring or agreeing with others to do so, regardless of whether the sales are successfully completed.

For this reason it is also difficult to say that an individual has aided and abetted a conspiracy, since this would require aiding and abetting the agreement, rather than the intended crime.[167] An individual either aids and abets the commission of a crime, or, once an individual is party to the agreement, he is liable as a principle in the inchoate crime of conspiracy. This same problem has arisen in the case law of the ICTY, where there have been assertions that an accused could have aided and abetted a JCE.[168] However this clearly collapses aiding and abetting the crimes committed by others who share a common enterprise, with aiding and abetting the agreement.

These difficult details are related to the ambiguity in the literature with respect to whether conspiracy is meant to describe the inchoate crime of agreeing to commit a further crime, or to describe a mode of responsibility. An attempt will be made in the next section to clarify the differences.

6.3.5.3 Expanding Conspiracy as a Mode of Liability

Conspiracy as a substantive inchoate crime has been described in detail here because the roots and doctrinal questions are identical for conspiracy as a mode of responsibility. It is not always clear in the literature nor in case law which of the two forms is at issue, and it would appear that not all of the law-appliers in the American criminal law system are conscious of the differences. The development of conspiracy liability is particular to the USA; conspiracy exists only as an inchoate crime in the UK, Canada and many other jurisdictions following the common law tradition. It has thus developed as a matter of policy, as a way of extending liability for co-conspirators further, without there necessarily being any deeper doctrinal grounding, which may explain why distinguishing between the two forms is unclear in the literature, in the courts, and by lawyers. In fact, conspiracy as a mode of liability is described in much of the literature as a conceptual continuation of the inchoate crime. There is also little distinction between conspiracy liability and other forms of accomplice liability (aiding and abetting), however the distinction is simple: aiding and abetting require proof that the actor at least indirectly assisted in the crime, whereas conspiratorial liability requires only an agreement.[169]

In the MPC, conspiracy as an inchoate crime is described as being 'a means of striking against the special danger incident to group activity, facilitating prosecution of the group, and yielding a basis for imposing added penalties when combination is involved.'[170] This same policy intention has been the rationale behind creating conspiratorial liability as a form of vicarious liability. The acts of all co-conspirators

---

[167] Harvard Note 1959, p. 934.

[168] See Sect. 8.6.

[169] Dressler et al. 2001, p. 493.

[170] American Law Institute 1985, Comment to § 5.03, p. 387.

are attributed to the defendant as long as the defendant intended to join in the agreement and intended to further the commission of the crime. In its most extensive form, according to the so-called *Pinkerton* rule, *all* crimes committed by a co-conspirator can be attributed to the defendant, even if these acts were not a part of the original agreement, as long as those acts are 'reasonably foreseeable'.[171]

The facts of the case are remarkable. Two brothers, Walter and Daniel Pinkerton were charged and convicted for tax evasion crimes, even though there was no evidence that Daniel had participated directly in the commission of the substantive offenses. In fact, Daniel was incarcerated for an unrelated felony at the time his brother committed the tax crimes. The question was submitted to the jury whether both brothers could be found guilty of the tax offenses, 'if it was found at the time those offenses were committed, [the two brothers] were parties to an unlawful conspiracy and the substantive offenses charged were in fact committed in furtherance of it'.[172]

There was evidence that Daniel and Walter had, over the years, worked together to unlawfully posses, transport and deal in whisky, in violation of the Internal Revenue Act. While Daniel was incarcerated, his brother continued other acts of a similar general nature, but there was no evidence that Daniel had 'counseled, advised, or had knowledge of those particular acts or offenses'.[173] He had therefore clearly not aided or abetted in these acts as an accomplice. However the court found that because there was an agreement between the brothers to commit crimes of a certain nature, and because there was an overt act taken by Walter, it was therefore justifiable to attribute the 'same or other acts in furtherance of the conspiracy [to Daniel] for the purpose of holding [him] responsible for the substantive offense'.[174] The court drew a limit to this extended liability only for offenses which would be

> committed by one of the conspirators [which] was not in fact done in furtherance of the conspiracy, did not fall within the scope of the unlawful project, or was merely a part of the ramifications of the plan which could not be reasonably foreseen as a necessary or natural consequence of the unlawful agreement.[175]

This rule is constructed as a kind of continuing conspiracy, with far-reaching consequences. No evidence is needed of further collaboration. In fact in further case law it has been determined that even acts outside the scope of the original agreement can be attributed to a co-conspirator, 'if they are reasonably foreseeable as the necessary or natural consequences of the conspiracy'.[176] This stretches the liability of a co-conspirator even further.

The only restraint on this extensive vicarious liability that appears to have developed in the case law is the one in *United States v. Alvarez*,[177] that a co-conspirator

[171] Pinkerton 1946.

[172] Pinkerton 1946, p. 642.

[173] Pinkerton 1946, p. 651.

[174] Pinkerton 1946, p. 647.

[175] Pinkerton 1946, p. 648.

[176] State v. Bridges 1993.

[177] United States v. Alvarez 1985.

who plays a minor role may be excluded from the extensive reach of *Pinkerton* liability. Two members of a drug conspiracy were charged with second degree murder, due to the fact that a co-conspirator shot and killed a federal agent during a shoot out. The murder was considered to be a reasonably foreseeable consequence of a drug conspiracy, since the court found that the co-conspirators must have been aware that it was likely at least some of their number would have been armed and would have used deadly force to protect their interests with respect to the large quantity of drugs. But it was accepted that if the appellants had played a sufficiently minor role in the original conspiracy, that they would not be liable for the murder as a crime committed in furtherance of that conspiracy. The rationale was that minor participants must not be protected only by the whim of the prosecutor who may or may not choose to pursue them for the further crimes, but also by a legal exclusion.[178]

The *Pinkerton* doctrine has received much criticism both from scholars and judges. Justice Routledge raised a serious concern in his dissent to the *Pinkerton* majority judgment, which has been echoed by many since, that even 'if it does not violate the letter of constitutional right, it fractures the spirit.'[179] As another court put it, the law 'lose[s] all sense of just proportion if simply because of the conspiracy itself each [conspirator is] held accountable for thousands of additional offences of which he was completely unaware and which he did not influence at all.'[180] Yet another court stated that:

> it is repugnant to our system of jurisprudence, where guilt is necessarily personal to the defendant [...] to impose punishment, not for the socially harmful agreement to which the defendant is a party, but for the substantive offences in which he did not participate.[181]

In at least one state there is a distinction between crimes of specific intent, for which it has been found that the *Pinkerton* rule does not apply, and crimes of general intent, for which it may apply: the court has found that allowing *Pinkerton* liability for specific intent crimes would allow the prosecution to sidestep the statutory *mens rea* requirements for those crimes.[182] Once again, this distinction could be useful when considering the limits of JCE liability in ICL, particularly with respect to the specific intent crime of genocide, given the similarities between JCE and conspiracy.[183]

Perhaps due to the concerns raised in the literature and in the state courts, the MPC has rejected the notion of *Pinkerton* liability altogether. The MPC focuses on individual culpability rather than on extended liability for crimes committed by oth-

---

[178] The facts did not support such an exclusion for the appellants in *Alvarez*, but the rule is now accepted, based also on cases cited by the court: *Government of Virgin Islands v. Dowling*, 633 F.2d 660, (3rd. Cir. 1980) and *Park v. Huff*, 506 F.2d 849, (5th Cir. 1975). See also State v Walton 1993.

[179] Justice Routledege, partial dissent Pinkerton 1946, p. 650.

[180] State v. Nevarez 2005, p. 1158; see also American Law Institute 1985, p. 307, comment to § 2.06.

[181] People v. McGee 1979, pp. 1181–1182.

[182] Bolden v. State 2005; Kadish et al. 2007, p. 683.

[183] For discussion on the similarities, and more importantly on the ways in which conspiracy as a domestic mode of liability has influenced the development of JCE on the international plane, see Sect. 8.6.

ers in the group, and § 5.03, which deals with conspiracy, limits the liability of an individual to those crimes for which a defendant had the purpose of promoting or facilitating. The MPC does allow for extended liability in that it defines the existence of a single conspiracy with 'a multiplicity of criminal objectives, as long as such crimes are the object of the same agreement of continuous conspiratorial relationship.'[184] This would allow, so the argument goes, for the finding that some of the parties in a chain conspired to commit the entire series of crimes that were actually committed, whereas others only conspired to commit some of these crimes. This is a worthy attempt to break down conspiratorial liability into a more basic form, distinguished from the extended *Pinkerton* form, however it would seem that neither courts nor legislatures have taken this opportunity to re-organise conspiratorial liability into these classifications. The most extensive form under the *Pinkerton* rule is accepted in a majority of states in the US.[185]

The development in the US is thus a unique combination of conspiracy and complicity. This is significant not only because of its uniqueness among common law jurisdictions, but also because there were attempts to introduce this combined form into the Charter of the IMT at Nuremberg. As will be discussed in Chap. 8, the conspiracy-complicity form was rejected, but may have re-emerged in the form of JCE. The debates surrounding the interpretation of this mode of liability may have something to do with the discomfort most non-US trained lawyers have with this combined form, which has been criticized for simplifying the notion of complicity by using conspiratorial agreements as a standard.[186] It goes without saying that this is an incredibly powerful doctrine; any acts committed by one of the partners to a conspiracy which are deemed by a court to be 'reasonably foreseeable' can be attributed to every individual who first entered a completely different unlawful agreement. This stretches the principle of individual culpability to an uncomfortable extreme.[187]

6.3.5.4 Conspiracy Under International Law as Assessed in the US

As Justice Jackson noted in *Krulewitch*, conspiracy is a notion particular to Anglo-American law. The way in which the US has assessed its application to war crimes prosecuted under US jurisdiction is noteworthy, since this is where the Anglo-American participants intersect with international participants and law-making processes.

In *Hamdam v Rumsfeld*, the accused had been captured in Afghanistan shortly after the attacks of September 11, 2001.[188] He had been the bodyguard and driver of Osama bin Laden, but his involvement in the attacks allegedly consisted only of joining the conspiracy. Hamdan was tried by military commission, whose jurisdiction

[184] American Law Institute 1985, p. 426.

[185] Kadish et al. 2007, p. 683.

[186] Fletcher 2000, p. 219.

[187] See Ohlin 2007, p. 160.

[188] Hamdan 2006.

is limited to the laws of armed conflict, and thus Hamdan argued in his appeal to the Supreme Court that conspiracy is not an offence under international law, and therefore not triable. The Supreme Court majority held that conspiracy is not a violation of the laws of armed conflict, since it is not recognised under the Geneva Conventions or the Hague Conventions.[189] The court cited post-World War II cases *Yamashita*[190] and *Quirin*,[191] both of which were decided under US jurisdiction, and both of which, argued the court, show that conspiracy was never considered to fall under international law. Furthermore, the refusal of the IMT to apply conspiracy to commit war crimes was considered to be important in this determination.[192]

However the court was not unanimous on this, with four judges voting for, and three in dissent. In his dissenting opinion, Justice Thomas argued that because conspiracy falls under US common law, it is also triable by a military commission: 'whether an offense is a violation of the law of war cognizable before a military commission must be determined pursuant to the system of common law applied by military tribunals.'[193]

As a matter of international law, and since the charges related to the international law of armed conflict, it would seem the majority opinion was correct.

6.3.6 Summary

The problem of linking an individual to collective criminality is solved in US criminal law with a subjective approach under all of the modes of liability described here. This subjective approach appears to be inherent to American criminal law, where the intention of an actor determines liability rather than their actual participation or contribution to the completion of a crime. Although it is difficult to classify different forms of participation in a crime due to the eradication of any normative distinction between individual actors, some observations can still be made as to the different modes of liability.

First, the subjective approach, with its focus on the intention of the accused and less concern for the actual contribution to the crime, has led to an eradication of the normative distinction between parties to a crime. Essentially the US has a functionally unitary system of liability. Especially since conspiratorial liability is now considered to be a form of complicity, the line between 'principal' and 'accomplice' has been blurred so much over time that even though these terms are still used, they have no normative meaning anymore. The role of the jury in US criminal law is an important factor in this development. A complex, legalistic system of modes of liability according to the old common law distinctions has little chance of survival

[189] Hamdan 2006, p. 40.

[190] Yamashita 1946-02-04.

[191] Ex Parte Quirin 1942.

[192] Hamdan 2006, 47.

[193] Justice Thomas, dissenting opinion Hamdan 2006, p. 12.

in a system where lay juries are dependent on instructions from the judge, and are asked to make a simple decision of guilty or not guilty based on the facts. While there is room for the judge to reflect the level of culpability in a normative sense in the sentencing, this is entirely up to discretion, and has no systematic grounding.

Second, party liability and conspiracy are both envisioned as liability for the acts of others. It is not the act of the accomplice defendant but their intention to aid and assist, or to join in the intention of others, that renders them punishable. It would seem this is most fittingly described as a form of vicarious liability (but jointly rather than in place of the liability of those others). Accomplice and conspiratorial liability rest upon the relationship between the defendant and other actors, and allow attribution to the defendant of a criminal act committed by another by coupling this relationship with the knowledge and purpose to further the crime.

Third, according to this vicarious liability for the acts of others, the theoretical underpinning of party liability is that the acts of others involved in a collective crime are attributed to an individual accused. Thus, a subjective approach re-writes the facts into a legal fiction, and a person who assists a physical perpetrator, or who merely shares an agreement but does nothing to actually assist in the crime, can be found guilty of the crime itself as if he had committed it. The hand of one becomes the hand of all.[194] A co-conspirator can be convicted as a murderer without having done anything to assist in the killing of the victim. An abettor, who encourages the physical perpetrator to commit violent crime, but does nothing violent herself, can be convicted for assault. The discussion on fair labelling applies here in the context of domestic law when it comes to serious felonies, and is one reason why the subjective approach should not be favoured in the specific context of mass atrocity crimes.[195]

Finally, with respect to conspiratorial liability in particular, this is a combination of the doctrine of conspiracy with the notion of complicity. Although such distinctions are not made in the literature, it would not be incorrect to see this as consisting of two forms. In the first place, basic conspiratorial liability would be the notion that a party to an agreement to commit a crime can be held liable for all the crimes committed within the scope of that agreement, even when the defendant did nothing to contribute to some of the crimes. This is predicated on the intention to join the agreement. In the second place, extended conspiratorial liability is characterised by the *Pinkerton* rule, whereby a co-conspirator can be held liable for crimes committed by co-conspirators even if these crimes are outside the scope of the original agreement, as long as they are objectively reasonably foreseeable. Because this goes beyond the intention of the individual, and goes against the basic tenet of criminal law that an individual should only be punishable where there is personal culpability, there is much criticism of the almost limitless reach of *Pinkerton* liability. Yet courts continue to apply it and so it remains a strong, if unique, doctrine in US criminal law. Both the extended and basic form are mirrored in the different forms of JCE, and the role that conspiracy has played in influencing the development of modes of liability on the international plane will be discussed in Sect. 8.2.

---

194 Curry 2006, p. 653.

195 For a discussion of the importance of fair labelling, see further Sect. 10.4.

6.4 Canada—Moral Blameworthiness and the Importance of Stigma

Whereas the USA was selected for comparison because of the significant contribution it has made to the patchworking process in ICL, Canada has been selected in order to deepen the understanding of the common law tradition, most especially with respect to conspiracy and complicity.

As a matter of comparative law methodology, it is most valuable to select a jurisdiction that has a high level of development on the topic in question, and is representative.[196] While an obvious choice would be the UK, since this is where the common law tradition heralds from, at the same time it is useful to look beyond this 'mother' system, to see how the traditional notions have developed in response to policy needs in different legal climates. Canada is instructive in terms of the development of the topic in question, because of the deliberate scholarly thinking that has taken place, coupled with extensive judicial consideration, due to the power of judicial review and recently the strength of the Canadian Charter of Rights and Freedoms.[197] This is of particular importance with respect to the requisite mental state for accomplice liability, and for the limits on extended liability, as will be discussed below. Canada has therefore become one of the leading jurisdictions in applying such rights-based limits to the general principles of liability.[198] The English tradition, on the other hand, is not one based on rights, but rather on 'right of reason', dependent upon the judges as law-seekers and law-finders, but historically with no consistent parameters, and no power of judicial review. Hence, Canada provides a modern representation of the common law tradition that places rights of the defendant in the spotlight, which is consistent with the goals of ICL as enumerated in Chap. 3.

Furthermore, it is useful to consider a jurisdiction that shares many cultural similarities with the US, but which offers a counter-balance to the specific doctrine of conspiracy that has emerged in the US. When compared to the US, Canada bears more similarity to many other common law jurisdictions in its more limited approach.[199] In this respect Canada may be more representative than the US of the common law tradition. At the same time, other notions of extended liability in Canada bear resemblance to those in the US, and therefore these differences and similarities can be instructive in clarifying the debates surrounding a common law reading of, and influence on, modes of liability in ICL.

Finally, in recent decades Canada has been pro-active in the prosecution of war crimes and crimes against humanity under its domestic law.[200] This is a reflection of the feedback dialectic between domestic and international law understandings of modes of liability applied to mass atrocity crimes. This may prove useful to

[196] Oderkerk 2001, p. 317.

[197] Canadian Charter 1985.

[198] Fletcher 2007, p. 101.

[199] Ashworth 2011, p. 450.

[200] Lafontaine 2012, pp. 23, 28, 32.

international tribunals in terms of considering how mass atrocity crimes are dealt with in the more mature domestic context.

6.4.1 'Principles of Fundamental Justice' and the Formation of Criminal Liability

Canada does not share the same model of federalism as the US, and instead gives the federal Parliament exclusive legislative authority when it comes to criminal law,[201] and therefore has a single Criminal Code,[202] which has allowed a fuller approach to liability to emerge than in the US. Prior to Canadian confederation in 1867, the provinces and territories had developed in many different directions and there was no coherency nor consistency between them. In 1892 the Criminal Code was enacted, the intent of which was to simplify and unify the law across the country.[203] Yet despite this central legislation, the Criminal Code had no general part, and there is no internal structural requirement for consistency. The most recent full enactment of the Criminal Code dates from 1985, although the Code has been amended every year since 1989, leading to a less coherent codification and a system with far less overview than was originally intended.[204] With regards to the law on parties to a crime, the provisions maintain much of their simplicity, however in its application there have been many nuances introduced through case law.

The role of judges as leading participants in the processes of criminal law in Canada is clear when it comes to the very strong protections offered by the Canadian Charter of Rights and Freedoms, enacted in 1982 as an addition to the Constitution.[205] The 1982 Constitution Act also strengthened the power of judicial review,[206] moving away from the British model of parliamentary sovereignty, and allowing judges of the Supreme Court to strike down any law that is deemed unconstitutional, and to quash any judgment handed down by a lower court for the same reason. This judicial review is particularly important with respect to interpretation of the Charter in criminal law matters, since both procedural and substantive rights are enumerated in the Charter rather than in the Criminal Code.[207] Many challenges have been brought to the courts

---

201 s. 91(27) of the Constitution Act of 1867 establishes sole jurisdiction of federal Parliament over criminal law.

202 Criminal Code 1985.

203 Manning and Sankoff 2009, p. 6.

204 Manning and Sankoff 2009, p. 22.

205 Enacted as part of the Constitution Act 1982, c. 11 (UK), by the United Kingdom Parliament, which entrenched the Charter in the constitution.

206 s. 52(1) Constitution Act provides that the Constitution, including the Charter, is the supreme law of the land, and that any law inconsistent with it is of no force or effect.

207 Sections 8 through 14 of the Charter list procedural rights such as the right to be free of arbitrary arrest or search and seizure, the right to a fair trial, the rules of *nemo tenitur, ne bis in idem* and others.

under constitutional review, forcing federal legislators to reform many aspects of the Code, or to enact supplementary legislation.[208]

Of particular importance for this study are the rights to substantive fairness found in s. 7, which states: 'Everyone has the right to life, liberty and security of the person and the right not to be deprived thereof except in accordance with the principles of fundamental justice.' In 1985 the important case *Reference re Motor Vehicle Act* was handed down by the Supreme Court of Canada, in which it was determined that s. 7 pertains to substantive as well as procedural fairness.[209] The Court emphasised that to draw a distinction between substantive and procedural justice is to import a terminology and jurisprudence from the US where procedural due process is of great importance, whereas it is unnecessary under Canadian criminal law to do so.[210] The words 'principles of fundamental justice' have proven to be pivotal in the interpretation of many substantive notions, including mental fault requirements and party liability in collective crimes.

Because there is no general part to the Criminal Code establishing general principles, and because many of the crimes in the Code have no required mental element as part of their definition, much has been left to the judges in terms of clarifying the law.[211] In *Reference re Motor Vehicle Act*, Justice Lamer highlighted the tension between the role of the judiciary and the legislator in determining the legal limits of justice and the shape of criminal law. Lamer wrote that the principles of fundamental justice are 'to be found in the basic tenets of our legal system', and are not part of public policy determined by a legislator, but rather are the 'inherent domain of the judiciary as the guardian of the justice system'.[212] The system of ICL, which lacks a central legislator, may well have many similarities to the Canadian system in this respect.

Aside from the judges, the existence of jury trials in this horizontal system of authority means that lay people are also participants in the way in which Canadian criminal law develops. Section 471 of the Criminal Code specifies that all indictable offences are to be tried by judge and jury.[213] Although jury trials are not the most common form of bringing a charge to a close,[214] many of those that do reach jury trial involve collective criminality. The impact that such a system has on modes of liability is great, since a jury must be able to understand and apply them when deciding on the factual question of guilt. At the same time a jury never delivers a written judgment, so there is no argumentation or justification made for why an accused is convicted or acquitted, other than that the jury was or was not convinced beyond

[208] Roach 2004, p. 54.

[209] Reference re Motor Vehicle Act 1985, p. 498.

[210] Reference re Motor Vehicle Act 1985, p. 498.

[211] Roach 2012, p. 163; Manning and Sankoff 2009, p. 148; Stuart et al 2012, p. 6.

[212] Reference re Motor Vehicle Act 1985, p. 503.

[213] Criminal Code 1985, s. 473 For various offences the accused may choose to waive this right and be tried by a judge only.

[214] Less than 10 % of all criminal charges end in a jury trial. See Statistics Canada 2010, providing data on adult criminal court statistics most recently for 2010–2011.

reasonable doubt of the facts presented to them. There is therefore no guarantee of consistency when it comes to understanding liability. Just as in the US, a system has therefore been developed over time that may be descriptively, but not normatively differentiated, since when it comes to considering the different roles of parties to a crime, a simpler system will be more easily applicable than a complex technical taxonomy with different requirements for each party.

6.4.2 Moral Innocence

Moral blameworthiness has developed as the central underpinning notion in determining guilt. This is partly a result of the fact that the historical development of crimes under the common law is itself a reflection of what judges deem to be morally reprehensible in a given society at a given time.[215] Moreover, recently it has been deemed a constitutional requirement under s. 7 of the Charter. In the above mentioned case *Motor Vehicle Act*, the issue at hand was whether a statute in the province of British Columbia was unconstitutional, because it had created an absolute liability offence (the accused need not have any knowledge of the wrongfulness of the act) with an automatic minimum prison sentence of 7 days.[216] The Supreme Court found that it violates fundamental principles of justice under s. 7 of the Charter to deprive someone of liberty in the absence of knowledge of the wrongfulness of the act.[217]

This particular 'principle of fundamental of justice' has been formulated as: 'the morally innocent shall not be subject to criminal penalty.'[218] The Supreme Court had decided that absolute liability was unacceptable under criminal law, since there must be some inclusion of *mens rea* in the determination of guilt.

6.4.3 Standards of Fault for Party Liability

Because the Criminal Code has no general part, and there is no internal structural requirement for consistency, the legislation of the specific offences has been very piecemeal over time, and many do not specify a *mens rea* requirement at all, leaving it to the judges to determine.[219] For a while the Law Reform Commission carried

---

215 Roach 2012, p. 22; Stuart 2007, p. 66.

216 The Motor Vehicles Act created a regulatory offence for driving a vehicle while the person is prohibited from driving or has had their licence suspended, whether or not the driver knew of the prohibition or suspension. Reference re Motor Vehicle Act 1985.

217 The Court thereby created the possibility to strike down legislation that it finds is not in accordance with unstated 'principles of fundamental justice', even if it does not violate any specific provision of the Charter: Manning and Sankoff 2009, p. 64.

218 This follows the wording of the Supreme Court in a case decided before the enactment of the Charter, which nonetheless concerned the same issue: Sault Ste Marie 1978.

219 Roach 2012, p. 163; Manning and Sankoff 2009, p. 148.

the task of re-drafting the Criminal Code, including designing a general part and simplifying the modes of liability, however this project was disbanded in 1993.[220]

Despite the lack of a general part, there is agreement on the definition of the possible mental states. The Supreme Court has affirmed the principle that moral culpability varies according to the level of fault, and that therefore acts undertaken intentionally are to be punished most severely.[221] The mental states can be defined in decreasing order of culpability and therefore punishment.[222]

First, the highest level of awareness and the most commonly involved in the commission of a crime is intention.[223] Second is knowledge of the circumstances, with respect to conduct crimes.[224] Third, wilful blindness, which is rarely made explicit in the Criminal Code. It sits somewhere between knowledge and recklessness, but wherever knowledge is a required standard wilful blindness is also considered sufficient, and can be described as constructed knowledge: where the accused denies actual knowledge but the circumstances required an enquiry into the facts and the accused deliberately refrained from ascertaining the truth.[225] This can be seen as equivalent to the civil law notion of *dolus eventualis*. And finally there is recklessness, which is defined as the subjective awareness that the conduct of the accused could bring about the results prohibited in law, and the decision to act nonetheless.[226] The risks involved must be subjectively foreseeable and unjustifiable. It is the choice to act which renders a person morally culpable.[227]

Since the introduction of the Charter, the tension between the judiciary and the legislator as it was expressed in *Reference re Motor Vehicle Act* has been apparent in the debates surrounding constitutional limits and protections for an accused when it comes to required fault standards. For most of the twentieth century, Canadian courts interpreted all criminal offences as having one of these subjective fault elements, even if there were no specific fault element given in the legislation.[228] There was a general resistance against objective standards of care, because of the underlying principle of autonomy as the foundation of criminal responsibility.[229] However since

---

220 The Law Reform Commission itself existed temporarily from 1971 to 1993 and from 1997 to 2006, with many successful law reform projects. However the project to review the Criminal Code was never completed. See Bailey 2009. See also Roach 2012, p. 7.

221 Creighton 1993.

222 Roach 2012, p. 165; Stuart 2007, p. 219.

223 Manning and Sankoff 2009, p. 162.

224 In the Criminal Code there are some offences which stipulate knowledge as the required mental state, such as defamatory libel, which requires that the accused knows the information is false, see s. 300 Criminal Code 1985. However many offences which do not include any required mental state have been interpreted as requiring knowledge of the circumstances, such as possession of narcotics. For example s. 4 Controlled Drugs and Substances Act, (S.C. 1996) c. 19, see Manning and Sankoff 2009, p. 178.

225 Manning and Sankoff 2009, p. 180.

226 Sansregret 1985.

227 Manning and Sankoff 2009, p. 174.

228 Manning and Sankoff 2009, pp. 152, 186.

229 Ashworth 2009, p. 186; Healy 1993, p. 266.

the codification of criminal offences has never been consistent, the legislator started to introduce some crimes with an element of criminal negligence, using terms such as 'reasonable', 'careless' and 'negligent'. The question arose whether negligence—an objective test of fault—was constitutionally acceptable. The legislator's focus had shifted in criminal law from intention to reasonableness, and from the actor to the act,[230] so courts began to shift their interpretations accordingly; whereas the tradition had been one of restraint in applying criminal sanctions, it moved towards an objective test, sanctioning an actor for departing from 'due care' expectations.[231]

The judicial debates that ensued over the course of several decades between subjectivists and so-called constructivists are noteworthy because they are much more nuanced than any debates that have taken place in the US or perhaps even the UK.[232] They are relevant here because they provide both the grounds and the limits for extended party liability, which is important when considering the translation of the common law approach to ICL, especially with respect to JCE and its extended form.

The shift back and forth between a subjectivist approach and a constructivist one, and the policy shifts these approaches represent, can be tracked in a series of cases decided by the Supreme Court. In 1978 in the *Sault Ste. Marie* case, the most renowned subjectivist statement was handed down,[233] requiring a clear subjective fault standard of intention, wilful blindness or recklessness for all criminal offences, since such standards standards prevent punishment of morally innocent individuals.[234] In the aforementioned *Motor Vehicle Act* decision, the Supreme Court affirmed in 1985 that the morally innocent must be prevented from punishment, and ruled out absolute liability.[235] The Supreme Court went a step further when it determined in 1987 that s. 7 of the Charter would require, as a principle of fundamental justice, 'a minimum mental state which is an essential element of the offence'.[236] For a small number of offences, in any case including murder and war crimes, a minimum mental state is constitutionally required that 'reflects their particular nature', in order to justify the heavier sentences and heightened stigma attached to them.[237] Recklessness is therefore ruled out for murder charges, including parties to a murder.

It is noteworthy, with respect to the influence of individual participants in the processes of the law, that Justice Lamer wrote all the majority decisions for the

230 Healy 1993, p. 266.

231 Stuart et al 2012, p. 361.

232 Fletcher 2007, p. 101.

233 Paciocco 2014, p. 11.

234 Sault Ste Marie 1978; Confirmed in Théroux 1993. This presumptive principle was the basis of a controversial decision in 1980, where a defendant accused of rape was acquitted because he argued that he honestly believed the victim was consenting, despite the fact that there was serious doubt as to whether this was a 'reasonable belief' in the circumstances. The Criminal Code was later modified to reject this defence of mistaken belief, unless the defendant can prove having taken reasonable steps to ascertain whether the complainant was in fact consenting. See Pappajohn 1980; see also Criminal Code 1985, s. 273.2.

235 Reference re Motor Vehicle Act 1985.

236 Vaillancourt 1987, para 652.

237 Vaillancourt 1987, para 653.

cases which followed a subjectivist approach, in two of them in his capacity as Chief Justice,[238] and his influence on this line of reasoning has been recognised by scholars.[239] The expectation was that this subjectivist approach would be constitutionally entrenched as a required principle of fundamental justice under s. 7 of the Charter for all indictable crimes.[240] However the Supreme Court retreated from this stance in a series of cases which broadened the fault requirement beyond subjective *mens rea* standards; on the one hand by including objective tests as constitutionally acceptable, such as constructive liability for parties to a crime, and on the other hand by condoning criminal negligence.[241] In 1992 in a majority judgment the Supreme Court stated explicitly that an accused need not actually intend the consequences of an otherwise blameworthy act, as long as it was a crime with a lesser stigma than murder.[242] Because criminalisation of the offence in this case, namely causing serious bodily harm, aims to prevent objectively dangerous acts, it was reasoned that this did not contravene the principle that the morally innocent must not be punished.[243]

In a series of cases known as the 'Creighton quartet',[244] this so-called modified objective test was established as the new norm. Although blameworthiness is primarily a function of the accused's mental state, an objective standard became acceptable for crimes other than murder, attempted murder and war crimes.[245] In the landmark case, Creighton and his friend were charged under s. 222(5)(a) and (b) of the Criminal Code for manslaughter by negligence, after having injected the victim at her request with a large dose of cocaine, and then leaving her in a coma, rather than calling for help, as a result of which the victim died. The Supreme Court upheld the conviction of the two parties to the crime under an objective test as being constitutionally acceptable, since manslaughter apparently did not carry the same stigma as murder. A very low standard of objective *mens rea* was thereby accepted, since the foreseeability needed only to be proven with respect to serious bodily harm, and not with respect to death.

It is no surprise that the subjectivist Justice Lamer dissented from this judgment and the decision has been heavily criticised for breaching the outer limit of foreseeability and changing a fault-based crime into an absolute liability crime.[246] The line of reasoning in the majority opinion, however, is helpful in understanding the relatively low fault standards currently applicable—which are often objective—when it comes to extended party liability.

---

238 Martineau 1990; Logan 1990.

239 Manning and Sankoff 2009, p. 161; Paciocco 2014, p. 12; Roach 2012, p. 71.

240 Paciocco 2014, p. 12.

241 Wholesale Travel 1991; Nova Scotia Pharmaceutical 1992.

242 De Sousa 1992.

243 A similar conclusion was drawn in 1993, with respect to the offence of dangerous driving, for which the standard was determined to be 'a marked departure from the standard of care of a reasonable person'; Hundal 1993, p. 884.

244 Healy 1993; Paciocco 2014, p. 13; Thomson 1994, p. 400.

245 Creighton 1993; Naglik 1993.

246 Healy 1993, pp. 3, 4; Paciocco 2014.

Justice Healy, then professor of criminal law, suggested that the Supreme Court majority opinion shifted towards a narrow understanding of the s. 7 Charter requirements as a reflection of the belief that the purposes of the Charter have to be interpreted flexibly according to specific legal contexts and policy.[247] In other words, even constitutional interpretations of criminal liability are contextual and contingent on policy shifts.

6.4.4 Basic Party Liability

With this debate on standards of fault and stigma in the background, the construction of party liability under Canadian criminal law can be better understood. The Criminal Code specifically did away with the old common law divisions between principal and accessory already in its first incarnation in 1892, when Canadian law-makers chose to depart from the English law and instead include the liability of all those participating in a collective crime under a single mode of liability with the title 'parties to a crime'.[248] Under the current Criminal Code, the wording of the main provision dealing with party liability remains relatively unchanged from the original drafting, in what is now s. 21(1):

> Every one is a party to an offence who
> (a) actually commits it;
> (b) does or omits to do anything for the purpose of aiding any person to commit it; or
> (c) abets any person in committing it.

The notion of common intention and liability for further crimes is covered in s. 21(2), and counselling is covered in s. 22. Both of these will be explained in more detail below.[249]

Important to note is that neither the alternative forms of contributing to a crime under subsections (b) and (c) of s. 21, nor s. 22 are considered to be derivative of the liability of a physical perpetrator;[250] they are all considered to be original forms of liability, which is in-keeping with a system that makes no differentiation between parties to a crime.[251] This is affirmed in s. 23(1) which was inserted into the Code in 1985, and which reads: 'For greater certainty, sections 21 to 23 apply in respect of an accused notwithstanding the fact that the person whom the accused aids or abets, counsels or procures or receives, comforts or assists cannot be convicted of the offence.'

---

247 Healy 1993, p. 2.

248 Berryman 1990, Part IV.

249 The old common law distinction of accessory after the fact has maintained a place in the Criminal Code in s. 23, but will not receive attention here. For the full text of the Criminal Code articles on party liability, see the Appendix A.3.

250 Roach 2013, p. 310.

251 As affirmed in much case law, see for instance Briscoe 2010, para 13.

This broad theory of criminal liability has been identified by the Supreme Court as a deliberate policy move on the part of the legislator in the *Thatcher* case;

> [this provision] is designed to make the difference between aiding and abetting and personally committing an offence legally irrelevant. It provides that either mode of committing an offence is equally culpable, and, indeed, that whether a person personally commits, or only aids or abets, he is guilty of that offence.[252]

In practice this means that the prosecutor does not need to make a choice between modes of liability in the charge or evidence presented, since all parties are equally culpable.[253] It also means that a lay jury does not need to distinguish between different legal requirements for different modes of liability. It has been asserted by judges and scholars alike that s. 21(1) has a specific policy design, to prevent acquittal where a specific mode of liability cannot be proven, and to put aiders and abettors on the same footing as principals.[254]

Thus anyone convicted as a party to a crime is convicted of that crime as if they had perpetrated it themselves. Just as in the US, and in-keeping with the subjective approach, a fictitious attribution of the acts of others to the accused takes place. Not only is a derivative theory explicitly ruled out by s. 23(1), but it would seem that vicarious liability is the best way to describe party liability in Canada, just as it is in the USA. Once again, this vicarious attribution of acts is not the same as private law vicarious liability, since the parties are not considered liable *in the place of* each other, but rather jointly with each other for each other's acts.[255] In Canadian case law, vicarious liability in the sense of *respondeat superior*, or the liability of one person for the acts of another purely on the basis of their relationship (master/servant or parent/child) has been explicitly rejected under s. 7 of the Charter, for offences which threaten to punish by deprivation of life, liberty or security.[256] However as a theory to explain the nature of party liability, vicarious is simply a more correct description than derivative.

Some would argue that even though the old common law distinctions have disappeared, the principles are still applied,[257] and certainly some scholars and judges still talk of principal versus secondary modes of liability. Indeed, in some cases judges still cite the work of the Law Reform Commission on this point:

---

[252] Thatcher 1987, p. 690.

[253] However if the prosecution does include a specific mode of liability in the indictment, the specific elements of that mode must be proven beyond reasonable doubt. See Roach 2013, p. 307.

[254] Chow Bew 1956; Thatcher 1987, p. 655, citing *Chow* and English common law dating from 1846; see also Roach 2013, p. 309.

[255] Vicarous liability does however apply in the sense of delegating the liability of one actor to another when it comes to corporate liability for the acts or omissions of an organisation's members. See Bill C-45 2003; Manning and Sankoff 2009, p. 239.

[256] Recall that this was the rule according to Reference re Motor Vehicle Act 1985, which means that smaller administrative offences such as traffic infractions can be based on absolute vicarious liability for the owner of a vehicle, even if the owner was not the offender; see Manning and Sankoff 2009, p. 277.

[257] Manning and Sankoff 2009, p. 247.

> Primary liability attaches to the commission of acts which are outlawed as being harmful, as infringing important human interests and as violating basic social values [...] Secondary liability attaches on the same ground to their attempted commission, to counselling their commission and to assisting their commission.[258]

However as *Thatcher* demonstrates, this distinction is purely semantic and no longer legal nor normative in nature. The term 'accessory' is particularly misleading, given the fact that party liability under the Criminal Code is very clearly *not* derivative or accessorial in nature, other than in the most limited sense that a crime must have been committed, which an accomplice counselled or assisted.[259]

Nonetheless there are efforts in the case law and in the literature to distinguish the various modes of liability from each other. This is what distinguishes a functionally unitary system, which differentiates descriptively, from a pure unitary system.[260] This might be one reason why those who advocate a functionally unitary interpretation of the Rome Statue are still willing to pay attention to the definitional differences between the modes of liability, even though they are seen only as 'alternative forms' from which the prosecutor can choose.[261] Some brief attention will therefore be paid here to the various definitions of party liability.

6.4.4.1 Commission and Joint Commission

Under s. 21(1)(a) of the Criminal Code, anyone who 'actually commits' the crime in question is liable as a party to the offence.[262] Joint commission is considered to fall under this title, and has been referred to variously in the case law as 'co-principals', 'joint principals', 'co-perpetrators' and 'joint perpetrators'.[263] Under this provision, no proof of a common intention is needed, nor even knowledge of the actions of the other parties, as long as some active participation and the requisite *mens rea* for each individual party can be proven.[264] According to the leading case on this provision, if it cannot be proven that an accused actually committed the crime, it is sufficient if there is proof she was 'otherwise an active participant'.[265]

In at least one case co-perpetration has been the grounds for a guilty verdict for causing death by negligence; in other words co-perpetration with no *mens rea* standard required at all.[266] It has been remarked that although this case law represents

---

[258] Law Reform Commission of Canada 1985, p. 5; cited in Hamilton 2005, para 65.

[259] Recall s. 23.1 Criminal Code 1985.

[260] For these definitions, see Sect. 5.3.6.3.

[261] See for example the arguments made by Judge Fulford in his dissenting opinion, Lubanga Trial Judgment 2012.

[262] In some literature and in some judgments this is still referred to as principal liability, although all this means is that it is the physical perpetrator; Manning and Sankoff 2009, p. 248.

[263] Ball 2011, para 23.

[264] Ball 2011, paras 23–25.

[265] Pickton 2010, para 12.

[266] Hughes 2011.

a broad causal expansion of liability as a 'principal' offender, in the end it makes no difference since any of these accused could have been convicted anyway under other forms of party liability under s. 21(1) of the Criminal Code.[267] For this reason not much attention is paid to the notion of commission and joint commission in the literature.

6.4.4.2 Aiding and Abetting

While the terms 'aiding and abetting' are often used together, they are in fact defined under s. 21(1)(b) and (c) respectively as two separate modes of party liability, and they are generally understood as having two different *actus reus* requirements. Aiding has been defined in the case law as 'assisting or helping the actor', whereas to abet has been defined as 'encouraging, instigating, promoting or procuring the crime to be committed',[268] whether by words or actions.[269]

Generally there are high levels of *mens rea* required for parties to a crime who do not actually commit the crime, such as knowledge or purpose, regardless of the *mens rea* requirement for the crime committed, as a way of placing some limits on the otherwise deliberate policy attempt to broaden the net of liability.[270] There must be a double intent, that is an intent to assist the physical perpetrator, as well as knowledge of at least the type of crime that is to be committed, though not necessarily the exact crime.[271] The Supreme Court has cited with approval Professor Roach's assertion that 'the more peripheral the accused's involvement to the completed crime, the more sense it makes to require a higher form of subjective *mens rea*.'[272]

For aiding there is a a particularly high *mens rea* threshold, due to the wording of s 21(1)(b), that everyone who does or omits to do anything '*for the purpose*' of aiding another to commit the crime is liable. In *Briscoe*,[273] the accused had been acquitted at trial for both aiding and abetting in the gang rape and murder of a woman. He had driven the other members of the group to and from the scene, had provided weapons and at one point had held the victim down and told her to shut up, and also threatened the victim's friend, but he did not physically rape the victim, nor did he enact any physical violence that led to her death, and he testified that he did not know what the other members of the group actually wanted to do with the victim. However in appeal the Supreme Court found that the requirement of his having acted 'for the purpose of' aiding those who physically committed the crime would be fulfilled if he had knowledge of both the crime, and of the intent of the perpetrators to commit

[267] Roach 2013, p. 312.

[268] Greyeyes 1993, para 26; Briscoe 2010, para 14, citing Greyeyes.

[269] Roach 2012, p. 148.

[270] Roach 2012, p. 124.

[271] Hibbert 1995, para 17; Roach 2012, p. 150; Manning and Sankoff 2009, p. 257.

[272] Roach 2010, para 36.

[273] Briscoe 2010.

the crime.[274] At the same time, wilful blindness would also suffice in the place of actual 'knowledge'; for instance, the fact that Briscoe had reason to believe that the other members of the group wanted to commit rape and inflict serious violence, and had refused to enquire further as to their intent, was deemed sufficient to evidence aiding.[275]

Liability for aiding therefore rests less on the question whether the accused's conduct actually assists the physical perpetrator, and more on the intention of the accused to bring about the consequences of the physical perpetrator's actions,[276] once again demonstrating the prevalence of the subjective approach.

An abettor, however, must have actually encouraged the physical perpetrator with their acts or words. Since abetting is generally understood to entail instigating or promoting, it implies some intention to see the crime committed.[277] However this standard is somewhat relaxed in the case of extended liability. If the person assisting the physical perpetrator knowingly assists in an act that is *objectively* likely to cause harm, and in fact death ensues, then the aider or abettor is also liable for manslaughter.[278] This is not at odds with the otherwise high subjective *mens rea* threshold for aiding and abetting, since the double intent is still required with respect to the predicate crime that the person aiding or abetting actually intends to assist.[279] Given the broad theory of liability under Canadian criminal law it is in fact to be expected that where this objective test is applicable to the physical perpetrator, it would also be applicable to aiders and abettors.

6.4.5 Constructive Extended Liability

Extended party liability and constructive liability are the notions that have been most controversial over time, and around which the debates on the subjectivist versus the constructivist debate have centred. Moral blameworthiness, combined with the stigma of specific serious crimes such as murder and war crimes, are the two tests for when constructed or extended liability are constitutionally unacceptable.

Historically, under the so-called malice principle, a person who chose to engage in an illegal act was liable for whatever consequences ensued, regardless of whether these consequences were foreseen by the accused on not.[280] In the case of predicate and aggravated offences, this objective approach meant that a person who assaulted

[274] Briscoe 2010, para 16.

[275] Because the trial judge had failed to instruct the jury as to the possibility of wilful blindness being sufficient grounds for proving constructed knowledge, the Supreme Court ordered a re-trial: Briscoe 2010, para 25.

[276] Morgan 1993, para 17.

[277] Manning and Sankoff 2009, p. 261.

[278] Jackson 1993; Manning and Sankoff 2009, p. 259; Roach 2013, p. 315.

[279] Manning and Sankoff 2009, p. 260.

[280] See Manning and Sankoff 2009, p. 156, citing Blackstone.

a victim would be liable for the ensuing death of the victim, even if the death was not (subjectively) foreseeable. This also had consequences for parties to a crime, where a party may have had one intention, to assault, whereas collectively the level of violence was raised by the intention of another party to cause serious injury or death. Later, however, the principle of correspondence, or symmetry, developed as a result of a liberal, subjective interpretation of *mens rea*; criminal liability should be based on what a defendant knows or believes to be doing or risking, and not on actual facts which may not have been known at the time.[281] This means there must be correspondence between the fault element for a crime and the conduct element; if the conduct element of a crime is 'causing serious injury', then the defendant must have intended serious injury and should not be held liable if unintended death ensues.[282] Under a less strict or moderate application of the principle of correspondence, at the very least some kind of objective foreseeability of the likelihood of death resulting from the specific assault would be required.[283]

The notion of constructive liability has intruded on this subjectivist approach, even in its moderate form. Until recently the Canadian criminal law followed the English common law that murder and manslaughter could be committed by way of grievous bodily harm, even if the accused only intended this harm and did not intend the death, and a person could therefore be convicted for a worse crime than contemplated.[284] This constructive liability is based on a moderate version of the notion *versari in re illicita*, which resembles the malice principle, in that anyone who transgresses the law shifts their normative position and is considered liable for all the consequences, even if these were unintended.[285] Keeping in mind the policy desire to expand party liability for crimes of mass atrocity, it is possible that this principle underlies a functionally unitary approach to liability in ICL recognisable in the way JCE is applied. However it should again be noted that Canadian law has departed from this constructive liability with respect to murder and war crimes, due to the high stigma associated with this crime, which is a limit we don't see in ICL. It does still apply to manslaughter and other lesser crimes in the Canadian context.[286]

There are two forms of extended party liability in Canadian criminal law, namely common intention, sometimes called joint criminal enterprise, and counselling. The former is of particular interest for the way in which it represents the traditional common law approach that has been drawn upon in the development of JCE at the international level. The latter is of interest by way of comparison as to the grounds and limits of extended liability.

---

281 Ashworth 2009, p. 155.

282 Ashworth 2009, p. 155.

283 Manning and Sankoff 2009, p. 157.

284 Ashworth 2009, p. 76.

285 However there is still the requirement of a subjective *mens rea*, and constructed liability is limited to crimes which are in the same family as the intended crime. See Ashworth 2009, p. 77.

286 Mills 2013.

6.4.5.1 Common Intention

Whereas liability under s. 21(1) deals with intended crimes and their consequences (as discussed above with respect to aiding and abetting), here s. 21(2) deals with liability for further unintended crimes, according to the notion of common intention. This differs from unintended consequences of an intended crime.[287]

If two or more people form a shared intention to commit a crime, and that crime is all that in fact happens, then the rules of party liability under s. 21(1) would apply, and all the parties would be considered either physical co-perpetrators or aiders or abetters. Common intention liability is specifically for cases where, in the course of an agreed collective crime being committed, one or more of the parties goes further and commits other crimes. It requires almost no action on the part of an accused in order to be liable for further crimes committed by other people; the very act of coming to agreement with one another fulfils the *actus reus*, there need not be any actual assistance carried out.[288] In this sense it shares some similarities with the conspiratorial mode of liability in the US, except that it is restricted to cases where a party goes further than the original intention, and does not go as far as the *Pinkerton* doctrine in the US.

There are two distinct fault elements for proving a common intention, which can be broken down into further aspects. Firstly there must be a common intent; this consists of (a) a shared intent to (b) carry out an unlawful purpose and (c) to assist each other in that purpose. It is enough that this intent is to commit a type of crime, such as 'a robbery' or 'an assault', without there being specific agreement as to where, how or against whom.[289]

Secondly, there must be either subjective knowledge or objective foreseeability of the likelihood of the crimes actually carried out by any or all members of the group. This broad definition amounts to a constructive vicarious liability,[290] and shares similarities with the extended form of JCE liability, which has been described as being based upon such domestic forms of liability.[291] The limits of this were tested in the case against *Logan*, in which the accused and two others had a common intention to carry out an armed robbery.[292] In the course of the robbery the accused's brother shot the clerk, but she survived. The shooter was convicted of attempted murder, since the intention to kill had been proven. The question was whether Logan could also be convicted of attempted murder based on s. 21(2), given that this was a crime that was a probable consequence of carrying out an armed robbery. Constructed

---

[287] S. 21(1) reads: 'Where two or more persons form an intention in common to carry out an unlawful purpose and to assist each other therein and any one of them, in carrying out the common purpose, commits an offence, each of them who knew or ought to have known that the commission of the offence would be a probable consequence of carrying out the common purpose is a party to that offence'.

[288] Moore 1973; Roach 2012, p. 154; Manning and Sankoff 2009, p. 265.

[289] Manning and Sankoff 2009, p. 263.

[290] See above, p. 46

[291] See Sect. 8.6.

[292] Logan 1990.

liability is based on either subjective knowledge or an objective foreseeability of the further crime, however, given the case law regarding the minimum level of subjective *mens rea* required for the high stigma crimes of (attempted) murder and war crimes according to the Charter,[293] the Supreme Court found that only subjective knowledge on the part of the accused would suffice. The words 'or ought to have known' are simply considered to be inoperative when it comes to the high stigma crimes of murder or war crimes.[294] It is worth noting that this judgment was written by Chief Justice Lamer, who was instrumental in securing this subjectivist limit to constructive liability for crimes of severe stigma.[295]

The Canadian parliament's stated policy decision with respect to common intention is to 'impose liability on persons who engage in criminal conduct with others for additional criminal acts that could be foreseen as likely to be committed in the furtherance of the underlying offence'.[296] For this reason, it has been determined by the Supreme Court that the requirement of shared intent need not include the requirement of shared motive or desire: in other words, an accused could not argue that she sold weapons to others for the purpose of profit, rather than with the shared intention to facilitate the crime.[297] The Supreme Court has continued this policy argument in other cases, stating that the intention of s. 21(2) is to widen the net of liability beyond aiding and abetting, when the physical perpetrator has committed crimes beyond which the parties had intended or agreed to aid and abet.[298]

The legislative aim of deterring and minimising joint criminal enterprises was recognised as justifying the far-reaching consequences of this form of liability in most cases.[299] In fact it was stated that there is no fundamental principle of justice requiring that all parties to a crime share the same *mens rea* as the physical perpetrator: for crimes other than murder, the mere test of 'objectively foreseeable' is acceptable.[300]

On the other hand, Justice Lamer noted that there *is* a principle of fundamental justice that there is proportionality between the sentence and the degree of a person's responsibility. So although a person may be convicted of a crime they did not intend, they should receive a lesser sentence. The sentencing scheme in Canada is apparently flexible enough to accommodate the differing degrees of responsibility which are not otherwise recognised by application of common intention liability.[301]

---

293 Martineau 1990; Vaillancourt 1987.

294 Logan 1990, p. 745.

295 Lamer highlighted the fact that this finding is only applicable in limited cases, since for most crimes there is no constitutional minimum requirement for a subjective *mens rea*; Logan 1990, p. 741.

296 Hibbert 1995, p. 1007.

297 Pacquette 1976; Hibbert 1995, p. 1004.

298 Simpson 1988.

299 Logan 1990, p. 747.

300 Logan 1990, p. 741.

301 Logan 1990, p. 741.

Alternatively, it is also possible to convict a party of a lesser crime than that which the physical perpetrator has committed, as long as that lesser crime is encompassed by the definition of the crime that the physical perpetrator actually committed. For example, because manslaughter is encompassed in the definition of murder as culpable homicide, minus the intent to kill, it would be possible to convict the party of manslaughter under s. 21(2) if the physical perpetrator were convicted of murder.[302] The Supreme Court has upheld the finding that this maintains an equilibrium between moral culpability and the offence for which the party is finally convicted.[303] This alternative appears not to have found a place in ICL, possibly because it is very difficult to determine a hierarchy of gravity between different crimes.

While the evidentiary standards for proving liability under s. 21(2) may be more difficult than under s. 21(1), because the former requires proof of a common intention having been formed, whereas general party liability under s. 21(1) does not, there is still some procedural advantage in asserting common intention liability. Once a common intention has been proven, anything said or done by any individual in furtherance of the unlawful purpose can be used as evidence against all other parties. Under aiding and abetting, on the other hand, if there is no proof of a shared intent, then the words and deeds of one person are only admissible as evidence against that individual and cannot count against other parties.[304] It is unclear whether such procedural advantages can be assumed to attach to the notion of JCE in ICL, but it would seem that the procedural aspects have hardly played a role in the policy arguments behind the assertion of a common intention mode of liability in ICL.

6.4.5.2 Counselling

The grounds and limits of counselling as a form of extended liability under Canadian law may be helpful in clarifying the parameters of a common law based interpretation of the Rome Statute when it comes to the 'masterminds' of collective crimes, especially since a clear policy rationale has been provided in Canadian jurisprudence.

Counselling appears as a form of party liability under s. 22 of the Criminal Code, but it also appears as an inchoate crime in s. 464 of the Criminal Code, punishable only when the crime that was counselled was not committed; a kind of 'attempted counselling'. As a mode of liability it is defined as procuring, inciting or soliciting,[305] and has a far-reaching liability in two senses. In the first sense, the person who counselled is party to the offence that is committed 'notwithstanding that the offence was committed in a way different from that which was counselled.'[306] So if a person counsels another to kill a third person by using a knife, and the perpetrator in fact

---

[302] Jackson 1993, citing Creighton 1993 and De Sousa 1992.

[303] Jackson 1993, p. 585.

[304] Manning and Sankoff 2009, p. 262.

[305] Criminal Code 1985, s. 22(3).

[306] Criminal Code 1985, s. 22(1).

uses a gun, the person who counselled is still liable for the murder committed.[307] In the second sense, a kind of constructive liability is codified in s. 22(2), whereby:

> Every one who counsels another person to be a party to an offence is a party to every offence that the other commits in consequence of the counselling that the person who counselled knew or *ought to have known* was likely to be committed in consequence of the counselling. (Emphasis added.)

The crime committed need not even be the crime that was solicited; as long as the actual crime was (objectively) foreseeable, the person who counselled or incited is liable. This extensive liability goes back to the policy of wanting to deter risk-inducing behaviour, and the notion that there is moral blame in initiating a crime, even when the initiator did not herself physically commit it, or if the crime committed was not exactly what she intended. In fact the blameworthiness extends even further, since it has been found that even if the counsellor was not the initiator, and the physical perpetrator already had the will or intent to commit the crime, then the counsellor is still liable. The British Columbia Court of Appeal stated that the policy intention of s. 22 is 'to discourage anyone from bringing about the commission of an offence'.[308]

The practical role for counselling as a mode of liability is, however, limited due to the fact that the prosecutor can also rely on the charge of an inchoate crime such as conspiracy, which does not require distinguishing any *actus reus* other than an agreement.

6.4.6 *Conspiracy: The Limited Common Law Solution to Group Crimes*

In Canada, just as in the UK and most other common law jurisdictions, conspiracy is only known in its form as an inchoate crime, and not as a mode of liability for completed crimes.[309] The rationale often given in Canadian jurisprudence for making a mere agreement punishable is prevention. Just as with attempts, creating an inchoate crime of conspiracy allows the police to intervene at an early stage, since no actual preparatory act nor even an attempt is required; the very act of agreeing is sufficient.[310] However, since most agreements to commit a crime will be made behind closed doors, the preventive effects of criminalising them are questionable. Instead the charge of conspiracy has some other procedural advantages, similar to those in the US, such as the fact that statements made by one co-conspirator are admissible as evidence against all co-conspirators, whereas generally any admission made by one co-defendant in a case of collective crime cannot be used against any other accused. The only other

---

307 Manning and Sankoff 2009, p. 325.

308 Glubisz 1979. Note that although this is a provincial Court of Appeal and not a federal one, such case law is considered to be instructive throughout Canada, because the Criminal Code is federally applicable. Manning and Sankoff 2009, p. 16.

309 Ashworth 2009, p. 448.

310 O'Brien 1954, p. 669.

exception to this exclusionary rule of evidence is common intention.[311] As well, it is possible to charge accused persons both with conspiracy and with the substantive crime itself to which was conspired, if steps are actually taken to complete that crime.[312] Just as in the US, the Supreme Court has found that the 'merger doctrine' which applies to attempt, by which completion of the crime subsumes attempt, does not apply to conspiracy, and there is no breach of the double jeopardy rule, or *ne bis in idem*.[313]

As discussed at the beginning of this Chapter, historically conspiracy was used as an extension of criminal law liability to punish an agreement to commit any unlawful act, including civil wrongs such as torts and contractual wrongs.[314] This made it an effective social and political tool, which was used in the UK and in Canada particularly against trade unions, to convict people for agreements to strike while it was still illegal to do so, thus suppressing organised industrial action.[315] In 1976, the law in the UK was changed such that conspiracies should only be criminal if the conduct agreed upon itself was also a criminal offences.[316] It took another decade for Canada to follow suit.[317] The main provision on conspiracy in the Canadian Criminal Code includes a general definition as well as some specific forms of conspiracy, with their own applicable maximum sentences.[318] The general provision states that the maximum sentence is the same as that applicable to the predicate crime. Conspiracy is therefore treated as seriously as the commission of the predicate crimes themselves.

The rationale behind using the conspiracy charge as a tool of social control is based on a subjectivist rationale that a shared intent creates a greater risk that should be criminalised, rather than being based on any actual social harm. As the Canadian Supreme Court has put it:

> The very fact that several persons in combination agree to do something has for many years been considered to constitute "a menace to society"[...] In fact, the scale of injury that might be caused to the fabric of society can be far greater when two or more persons conspire to commit a crime than when an individual sets out alone to do an unlawful act.[319]

---

311 Ashworth 2009, p. 450.

312 Manning and Sankoff 2009, p. 310.

313 Sheppe 1980; Grewall 2001, para 20. There is authority in England to support the same conclusion there, see Stuart 2007, p. 707.

314 Ashworth 2009, p. 448; Stuart 2007, p. 683.

315 Ashworth 2009, p. 449.

316 Law Commission of the UK 1976 Report on Conspiracy, cited in Ashworth 2009, p. 449.

317 Criminal Code 1985, s. 645(1). The general provision in the Criminal Code had made it an indictable offence to conspire 'to effect an unlawful purpose', but this was repealed in 1985, see Stuart 2007, p. 686.

318 For instance, conspiracy to commit murder is punishable by a maximum imprisonment term for life, which, according to s. 235(1), is the mandatory sentence for a person convicted of murder: Criminal Code 1985, s. 465(1)(a). As well there are some specific offences listed separately, which include conspiracy as part of their definition, or as an extension of the jurisdictional limits, such as hijacking an aircraft, s. 7(2)(e); offences involving nuclear materials, s. 7(3.3)(a); terrorism offences, s. 7(3.73); and sedition, s. 59(3).

319 Dynar 1997, para 89.

According to the Supreme Court, conspiracy is an even more preliminary offence than attempt, since 'the offence is considered to be complete before any acts are taken that go beyond mere preparation to put the common design into effect.'[320] Because conspiracy is merely the agreement to commit a crime, the Supreme Court also found that even in a case where it is factually impossible to fulfil the crime to which was conspired, this is no barrier to punishment, once again reflecting the subjectivist approach underlying liability.[321] The same rule applies to impossible attempts.[322]

This subjective approach is further reflected in the fact that, although a conspiracy is an agreement between two or more people, there is no accessorial relationship between the co-conspirators, and if one co-conspirator is not convicted for any reason, or even not prosecuted, then this does not preclude conviction for any other co-conspirator.[323] As well, if it is not possible to identify the co-conspirators, but an agreement is possible to prove, then it is possible to convict one person of conspiring with one or more 'unknown parties'.[324]

As to the requirements to prove an agreement, this need not be concrete; an implicit or tacit agreement is sufficient.[325] The plan or agreement itself may even change over time; in Canadian case law the rule is that 'so long as there is a continuing, overall, dominant plan there may be changes in methods of operation, personnel or victims, without bringing the conspiracy to an end.'[326] There must be evidence of an intent to enter the agreement, and a shared intent to carry out the agreed common design, even though there is no requirement of any overt action in furtherance of this common design.[327]

The question as to whether it is possible to be party to a conspiracy, in the sense of aiding or abetting a conspiracy, is not entirely clear. Some case law suggests that a person may be charged for abetting or encouraging the conspirators to pursue the object of the conspiracy, however it would seem this is an incorrect approach to the notion of conspiracy. Either a person has joined the conspiracy, by forming an agreement, in which case they become a co-conspirator, or the person aids and abets the fulfilment of the substantive crime, in which case they have not joined the conspiracy but they have become party to the commission of the crime.[328] This lack of clarity may explain the confusion that has arisen with respect to the question of parties to a JCE at the international level, since JCE is grounded in the notion of conspiracy. It would appear that JCE follows the US model of being a form of

---

320 Dynar 1997, para 87.

321 Dynar 1997, paras 96, 105.

322 Ashworth 2009, p. 450; Stuart 2007, p. 707.

323 Cotroni 1979; Roach 2012, p. 137; Manning and Sankoff 2009, p. 309.

324 Manning and Sankoff 2009, p. 315, citing a case from the British Columbia Court of Appeals, R. v MacDonald (1963) B.C.J. No. 79, 10 C.C.C. (2d) 488 (B.C.C.A.).

325 Roach 2012, p. 137.

326 Cotroni 1979, pp. 17–18.

327 O'Brien 1954;Dynar 1997.

328 Manning and Sankoff 2009, p. 316.

complicity, rather than the general common law model of being an inchoate crime, thus aiding and abetting a JCE would not appear congruent.

6.4.7 Application of Canadian Modes of Liability to the War Crimes Act

The final aspect of party liability under Canadian criminal law that is of interest is its direct application to war crimes and crimes against humanity. Following a period of reticence to be proactive in prosecuting parties to war crimes after the Second World War,[329] the Canadian government made an explicit turn-around in 1985 due to allegations that many Second World War criminals had immigrated to Canada.[330] The nature of the crimes was deemed to justify retro-active legislative reform, including criminalisation of war crimes, crimes against humanity and war crimes under the Criminal Code. Article 11 (g) of the Canadian Charter of Rights and Freedoms grants Canadian courts jurisdiction over international crimes, regardless of where they were committed, as long as the defendant is under Canadian jurisdiction. The existing Criminal Code remains the basis for general notions such as liability.[331]

The combination of international customary standards and domestic criminal law definitions proved problematic in the first war crimes case following these reforms. Former Hungarian national Imre Finta was charged with crimes against humanity and war crimes committed during his time in the Royal Hungarian Gendarmerie during the Second World War.[332] Finta had been acquitted at trial and the prosecution's appeals were denied at every level, including the Supreme Court, due to the controversial findings that a crime against humanity required additional elements over and above the general definitions that had developed under international law. The Supreme Court majority decision has been criticised often for this result.[333]

One of these unnecessarily high thresholds was requirement that in order to achieve the level of crimes against humanity, the acts should be considered inhumane and particularly shocking to society.[334] This goes beyond general understandings of crimes against humanity, for which the focus tends to be on the 'widespread or systematic' nature of the crimes, rather than any heightened requirement as to the gravity.[335]

---

[329] Lafontaine 2012, p. 20.

[330] In 1985 the Deschênes Commission of Enquiry was established, and its extensive report 2 years later formed the basis of legislative reform and the prosecutorial basis for pursuing a series of cases. Trapani 2009, p. 11; Lafontaine 2012, p. 25.

[331] Lafontaine 2012, p. 25.

[332] Finta 1994.

[333] Cotler 1997, p. 265; Lafontaine 2012, p. 27; Trapani 2009, p. 14.

[334] At trial the judge had instructed the jury that this could be defined as 'barbarous cruelty', 'brutal', and 'not having qualities proper or natural to a human being'; Finta 1994, paras 90–91.

[335] ILC Draft Code of Crimes 1996; Trapani 2009, p. 16; Indeed, the *Finta* judgment has since been cited by the ICTY was being exceptional to the general understanding of the *actus reus* requirements

With respect to the *mens rea* requirement the threshold was also set unnecessarily high. The special stigma of war crimes and crimes against humanity was again at the centre of this controversial finding, since according to the majority judgment the accused must have 'an awareness of [or alternatively wilful blindness with respect to] the facts or circumstances which would bring the acts within the definition of a crime against humanity'.[336] No doubt this heightened subjective *mens rea* requirement was intended to be in line with the *Creighton* rule that any crimes with a sufficiently high stigma, including murder and war crimes, require a subjective *mens rea*.[337] However this severe burden of proof for the prosecution meant not only that convictions became near impossible, but it also implied a possible defence of ignorance of the law for war crimes and crimes against humanity, which is unanimously rejected among international criminal lawyers. As pointed out by Justice La Forest in his dissenting opinion, 'to forcibly confine or kidnap 8,617 people is equally blameworthy whether he knew or did not know they were Jews.'[338] The controversial findings as to the requirements of party liability seemed to take the lines of reasoning under Canadian criminal law and inflate them with respect to an international crime of mass atrocity.

Following the disappointing outcome of *Finta*, Canada introduced the War Crimes Act (WCA) in 2000, after ratification of the Rome Statute, and a more pro-active trend began with respect to prosecuting war crimes and crimes against humanity. Section 4(4) of the WCA stipulates that the definitions of the crimes are to follow customary international law, which it deems to include the definitions in the Rome Statute.[339] However when it comes to modes of liability, the WCA ignores the Rome Statute and instead uses terminology from the Criminal Code, yet with a different structure than the Code. Section 4(1) states that 'every person is guilty of an indictable offence who *commits* (a) genocide; b) a crime against humanity; or (c) a war crime.' (Emphasis added) section 4(2) determines that: 'Every person who conspires or attempts to commit, is an accessory after the fact in relation to, or counsels in relation to, an offence referred to in subsection (1) is guilty of an indictable offence.'[340] There is no mention of the party liability forms of aiding, abetting and common intention, which are part of both the Rome Statute and the Canadian Criminal Code. As well, the inchoate crimes of conspiracy, attempt and counselling are listed separately the way they are under the Criminal Code, whereas conspiracy is not part of the Rome Statute, and in the WCA these are listed under 4(2) as a form of commission.

The lack of clarity in this drafting was argued in a case on appeal before the Federal Court, by the accused, Nasrullah Zazai, who had applied for refugee status in Canada. Zazai's application had been refused because there were reasons to believe

(Footnote 335 continued)
for crimes against humanity. See Tadić Appeals Judgment 1999, paras 265–268; Blaškić Trial 2000, para 248; Cotler 1997, p. 262.

336 Finta 1994, paras 95–97.

337 See the discussion on the 'Creigton Quartet' above; Creighton 1993.

338 Dissenting Opinion of J La Forest, Finta 1994, para 296; Trapani 2011, p 17.

339 s. 4(4) War Crimes Act 2000.

340 s. 4(2) War Crimes Act 2000.

he had been complicit in crimes against humanity in Afghanistan, which would make him inadmissable for refugee status due to a breach of the WCA.[341] Zazai argued that although 'complicity' in the sense of aiding and abetting was criminalised under international law, including the Rome Statute, it was not criminalised under the WCA. The Federal Court responded by referring to s. 34(2) of the Interpretation Act, which provides that where Canadian legislation creates an offence, all the provisions of the Criminal Code apply to this offence and that therefore the Code 'including the aiding and abetting provisions in s. 21 [applies] to the War Crimes Act.'[342]

In the follow-up case, the Federal Court posed the question slightly differently: whether the definition of crimes against humanity in the WCA included complicity therein.[343] This time, rather than looking to the Interpretation Act, the Federal Court read the notion of complicity into the word 'commits' in the WCA. The Court found that:

> at common law and under Canadian criminal law, [complicity] was and still is a mode of commission of a crime. It refers to the act or omission of a person that helps, or is done for the purpose of helping, the furtherance of a crime. An accomplice is then *charged with, and tried for, the crime that was actually committed* and that he assisted or furthered. In other words, whether one looks at it from the perspective of our domestic law or of international law, complicity contemplates a contribution to the commission of a crime.[344]

This analysis can be considered correct to the extent that under Canadian law it does not matter which mode of liability is proven in the end, since following *Thatcher* they all amount to a conviction for the crime in full.[345] However it has also been pointed out that if the prosecutor alleges a specific mode of liability, then the elements of this mode must still be proven.[346]

It is therefore noteworthy that the Federal Court chose to characterise complicity as a 'mode of commission' rather than a form of party liability. As one author has pointed out, this makes it problematic given that under the Canadian Criminal Code, commission is listed alongside aiding and abetting as a form of party liability in s. 21, but this does not make aiding and abetting a *form* of commission.[347] The fact that the term 'party' was excluded from the War Crimes Act makes this interpretation misleading, since it would seem that aiding and abetting were thereby also excluded, which would make the Court's reasoning a circumnavigation of legislation. On the other hand, the drafting history of the War Crimes Act, including the parliamentary debates, suggests that the legislator's intention was in fact to include these forms of party liability.[348]

---

[341] Zazai No. 1 2005.
[342] Zazai No. 1 2005, para 47.
[343] Zazai No. 2 2005.
[344] Zazai No. 2 2005, para 13, emphasis added.
[345] Thatcher 1987.
[346] Roach 2013, p. 307.
[347] Bayley 2011, p. 57.
[348] Bayley 2011, p. 56.

More problematic is the fact that the Federal Court turned to the case law of the ICTY to give authority to the assertion that complicity is an inherent part of the definition of commission of crimes against humanity. Once again this approach conflates a mode of liability with a definition of the crime, which would lead to problems of ascertaining which *mens rea* and *actus reus* requirements apply. Moreover, the Federal Court cited *Kvočka*, in which the ICTY had discussed the possibility of being an aider or abettor to a JCE. This conflates even further the place of party liability, since the ICTY was discussing being a party *to a party* to a crime.[349] Furthermore, since the Canadian mode of liability most similar to JCE, that of common intention, was not even at issue for Zazai, the various forms of party liability under the umbrella of 'complicity' seems to have become utterly confused.

It could be argued that the ICTY has a broader mandate for prosecutorial innovation in order to solve the problem of complex crimes of mass atrocity, while the Canadian Federal Court of Appeal is bound to the Canadian system and to Parliament's exclusive power to legislate criminal laws.[350] On the other hand, when it comes to interpreting liability for international crimes, perhaps it is inevitable that Canadian courts will turn to the case law of international tribunals as instructive. The Supreme Court has continued the trend of using and (re-)interpreting decisions of international tribunals, even while it asserted that it is bound by Canadian criminal law.

One such example is the case against Mugasera, whose refugee status was overturned due to reasonable belief that the applicant had been involved in crimes against humanity and genocide in his home country of Rwanda.[351] At issue were the *mens rea* requirements for incitement to genocide, and counselling a crime against humanity, crimes that carry an extremely high stigma. Mugasera had fled Rwanda after a warrant for his arrest had been issued, pursuant to his speech to about 1,000 people in the extremist Hutu political party of which he was a member, and had applied for refugee status in Canada.

The Supreme Court held that with respect to a crime committed abroad, a conclusion that the *actus reus* elements of the crime according to Canadian criminal law have been made out will be determinative.[352] For these purposes the crime of incitement to (or counselling) murder was used, under s. 464(a) of the Criminal Code.[353] Yet at the same time, with respect to the *mens rea* requirements, because genocide is a crime specific to international law, the Supreme Court turned to international law 'to play a crucial role as an aid in interpreting domestic law.'[354] The Court relied exclusively on case law from the ICTR, which may be understandable with respect to

---

[349] Bayley 2011, p. 57.

[350] Bayley 2011, p. 58.

[351] Mugasera 2005-06-28; Under Article 1F(a) of the UN Refugee Convention, a person will be denied refugee status if there are serious reasons for considering that they committed a crime against peace, a war crime or a crime against humanity, UN Refugee Convention 1951, Article 1F(a).

[352] Mugasera 2005-06-28, para 58.

[353] Mugasera 2005-06-28, para 64.

[354] Mugasera 2005-06-28, para 82.

incitement. However with respect to 'counselling', which is a mode of liability under the Criminal Code but not under the statutes of the ad-hoc tribunals, the Court decided to interpret this by analogy, and applied the notion of 'instigation'. According to the case law of the ICTR instigation requires proof of the materialisation of the predicate crime—that is, proof of genocide as a result of the instigation—contrary to the inchoate crime of incitement. The Supreme Court found that there was insufficient evidence to show that murders had occurred as a result of Mugasera's speech.[355]

Given that counselling under the Canadian Criminal Code has some more flexible aspects to it, in that the crime committed need not exactly match the crime counselled, and that counselling a crime which is not committed is also available as an inchoate charge, it is unclear if the result would have been the same if the Supreme Court had applied domestic modes of liability. What is clear is that international case law is considered instructive when it comes to the determination of complicity in international crimes, both with respect to prosecutions under the WCA and with respect to decisions under Article 1F(a) of the Refugee Convention in Canada, leading to inconsistencies with domestic criminal law concepts.

In another recent decision hinging on a refugee application, the Supreme Court was drawn once again to interpret complicity under ICL and the question whether the applicant had been complicit in war crimes in his home country of the DRC.[356] Ezekola had worked for the DRC government for many years, and was leading the DRC Permanent Mission to the UN in New York when he resigned and fled to Canada, claiming refugee status for himself and his family. He claimed that his resignation from a government he considered corrupt, anti-democratic and violent would be considered an act of treason in the DRC.[357] The question before the Supreme Court was to what extent senior public officials could be held liable for international crimes committed by the governments on whose behalf they perform official duties, or as the Court put it: 'when does mere association become culpable complicity?'[358]

The Court rightly pointed out that 'complicity is a defining characteristic of crimes in the international context, where some of the world's worst crimes are committed often at a distance, by a multitude of actors.'[359] The Court stated that determination of whether an individual is complicit in an international crime cannot be decided with reference to only one legal system. Not only does the Refugee Convention require the determination to be made according to international law, but the nature of these crimes 'transcends domestic norms.'[360] It subsequently found that to be excluded from the definition of refugee protection, there must be evidence that the individual knowingly made a 'significant contribution' to the group's crime or criminal purpose, citing the UNHCR, the agreement of States party to the UN Refugee Convention, the

355 Mugasera 2005-06-28, para 136; van der Wilt 2008, p. 244.
356 Ezekola 2013.
357 Ezekola 2013, para 14.
358 Ezekola 2013, para 4.
359 Ezekola 2013, para 1.
360 Ezekola 2013, para 44.

Rome Statute, and case law of the ad-hoc tribunals.[361] An individual can be complicit without being present at the crime and without physically contributing to the crime, but whereas

> senior officials may be held criminally responsible for crimes committed by their government if they are aware of the crimes being committed yet remain in their position without protest and continue to defend the interests of the government …this does not mean that high-ranking government officials are exposed to a form of complicity by association. Complicity arises by contribution.[362]

The Court added that 'neither mere membership in a government that had committed international crimes nor knowledge of those crimes is enough to establish complicity' and that 'where the individual has committed no guilty act and has no criminal knowledge or intent, beyond a mere awareness that other members of the government have committed illegal acts', there can be no complicity.[363]

Although some human rights groups may not be happy with such a conclusion, since it potentially allows a door to impunity to be opened to those in senior government positions, at the same time it would seem a sensible limitation on the extent of complicity in international crimes, removing the threat of guilt by association. This finding was not based on domestic modes of liability, but rather on an awareness of the specific context of mass atrocity crimes, and of developments in many different jurisdictions, in the search for the most appropriate definition of liability for these crimes. It is therefore a carefully constructed and advanced judgement which may be instructive to the study at hand.

6.4.8 Summary

Typical of the common law tradition, the Canadian criminal law system places no primacy on legalistic coherence, but rather on the ability of the judiciary to respond to immediate policy concerns, and thus to create law in a flexible way. There is therefore no internal consistency in the Criminal Code, which has no general part, and contains many crimes without any defined *mens rea* standards. Despite repeated attempts of a principled, coherent law reform that would focus on creating a general part to the Criminal Code, it would seem that legislators in the Anglo-American tradition are more drawn to the expediency of ad-hoc reforms, and the 'good politics of a quick fix' solution.[364]

Liability in Canadian criminal law is extended by a broad theory which deliberately places all parties to a crime on equal footing, regardless of the level of their involvement in the crime. It is therefore descriptively differentiated, but not normatively. It has been emphasised repeatedly in the case law and the literature that all

[361] Ezekola 2013, paras 22, 35, 48, 51.

[362] Ezekola 2013, paras 7–8.

[363] Ezekola 2013, paras 78, 80.

[364] Stuart 2007, p. vii.

parties are considered equally liable. Although there is still some use of the terminology of 'principal' and 'secondary participant', these terms have no legal meaning in Canadian criminal law, and have not had any since the first Code of 1892. This broad theory is further underpinned by the choice to ensure there is no derivative relationship between parties, nor between parties and a physical perpetrator.

Essentially this means that a kind of vicarious liability is at the core of most party liability, whereby the acts of each party are fictitiously attributed to all other parties. This is particularly evident when it comes to extended liability such as counselling and common intention. When the physical perpetrator commits further crimes that were at least objectively foreseeable, then parties are still liable and convicted as if they had committed the further crime, even if they did not share the same *mens rea*, or did not have the requisite *mens rea* for the crime actually committed. This is in line with the finding by the Supreme Court that it is not a fundamental principle of justice that parties are required to have the same *mens rea* as the physical perpetrator. The exception to this objective test for extended party liability is when another party has committed murder or attempted murder, due to the high stigma attached to this crime, and the constitutional breach of a 'fundamental principle of justice' when punishing someone for murder when they did not, in fact, commit it. This same line of reasoning followed in the *Ezekola* decision with respect to complicity in war crimes. It could be questioned, and will later be discussed, whether such limits should be placed on extended liability in ICL in general. The answer depends partly on the policy aims behind the system of modes of liability in ICL.

With respect to aiding and abetting the objective test does not apply at all, and there is a particularly high standard of subjective *mens rea*, namely that the party has the *purpose* of assisting the physical perpetrator, regardless of what the *mens rea* requirement is for the particular crime. The scholarly view on this has been affirmed in case law; due to the fact that party liability enlarges the scope of criminal liability, it is a necessary counter-balance to require a higher fault standard for these peripheral actors.

On the other hand, it has also been affirmed that there *is* a fundamental principle of justice that sentences are proportionate to the level of responsibility, according to s. 718.1 of the Criminal Code. As such, even though the degree of involvement by each individual in a crime is not reflected in a conviction, since every party is equally liable, this degree of involvement should be reflected in the sentencing, however this is entirely up to the discretion of the judge. This conclusion rests on the premise that the moral blameworthiness of an accused is the *ground* for punishment, which is equal for all parties, but the measure of the punishment may be individualised.

For those who 'mastermind' a crime and gather, direct or encourage others to commit it, there is no specific category under Canadian criminal law which recognises such a role. However the term 'mastermind' has been used occasionally in judgments, and although this is purely in a descriptive sense, such a description may be an aggravating factor in determining the sentence, due to the aforementioned principle that punishment must be proportionate to the level of responsibility.[365]

The preference for a subjective approach in the Criminal Code is reflected in several ways. First with respect to the inchoate crimes of attempt, conspiracy, and counselling a crime that was not committed, punishment is based on concern for the intent and not on actual social harm.[366] Second, this subjectivity is further reflected in the fact that where a person has the intention to carry out an act which is factually impossible, the attempt or conspiracy is still punishable, due to the criminal intent.[367] The predominant theory of punishment is related to moral blameworthiness. Third, the preference for a subjective approach is reflected in the judicial and scholarly debates on whether subjectivity is constitutionally required when it comes to fault standards. Fourth, the subjective approach to party liability is also prevalent, whereby all parties are considered to be equally liable due to their intention to join in a collective crime. And finally the approach to extended liability is clearly built on subjectivity.

When it comes to the required *mens rea* standards, while absolute liability has been outright rejected because it has no requirement of fault at all, constructive liability may be equally objectionable because it 'rests on the fiction that liability for the whole of an offence can be predicated upon proof of guilt for part of it.'[368] For Healy there are few concepts more noxious to the supreme law of Canada; the fictional construction of liability departs from the very core tenets of criminal law. Although constructive liability is accepted, it has been further criticised because it breaches the principle of correspondence; if an accused can be held liable for unintended consequences based on objective foreseeability standards, then there is less protection of the morally innocent when it comes to the more serious crimes.

Given concerns for the protection of the morally innocent, it is clear that conspiracy plays a very specific role in the common law tradition outside of the unique form it has gained in the US as a mode of liability. Generally, it has many procedural advantages including evidentiary; the possibility of early, preventive intervention; and the ability to cumulate charges of conspiracy with the actual attempted or completed crimes. However outside of the context of these domestic criminal law procedural rules, it may be less desirable to punish individuals for merely entering an agreement, without there being any requirement of an act. The decisions made by the Supreme Court recently to limit 'guilt by association' are a reflection of the constitutional limits placed on party liability in this way.

---

365 See for example Amara 2010, where a life sentence was justified by the fact that the accused had been the 'mastermind' of a terrorist plot.

366 Manning and Sankoff 2009, p. 324; Roach 2012, p. 161.

367 Dynar 1997, para 67.

368 Healy 1993, p. 4.

6.5 Legal Culture and Policy Choices in the Common Law Tradition

Four general conclusions can be drawn about the jurisdictions compared here. First, there is an inherently subjective approach in determining an individual's culpability: there is less emphasis on the acts and contributions of an individual to a crime (the objective aspect, or *actus reus*), and far more focus on their mental state (the subjective aspect, or fault standards referred to by *mens rea*) as the test of culpability. Second, this subjective approach leads to a theoretical underpinning of vicarious liability for participation in collective crimes; all parties to a crime are considered responsible for the acts of all other parties. Third, as a consequence of this theoretical underpinning of the relationship between the group and the individual, there is no normative distinction between parties to the collective. All parties are punished as principals, regardless of whether the language of 'accessory' is still used in some jurisdictions. And fourth, there is an inherent relationship between the character of criminal trials in the common law tradition, and the system of liability that ensues, due at least in part to the role of participants in developing the law. Each of these conclusions will be discussed further.

6.5.1 Subjectivity

Subjectivity is reflected in many aspects of the approach to liability in the common law tradition. In the jurisdictions compared here, the emphasis on the risk posed by criminal intentionality, rather than the actual harm brought about by any one individual's contribution to a crime, can be seen in the fact that impossible attempts are also criminalized, as well as in the fact that all parties to a crime are considered equally liable precisely because there was a shared intent.

Indeed, in the US the principle of culpability is understood as referring to the required mental state as an element of a specific crime,[369] which is described by Fletcher as a descriptive understanding of culpability.[370] In Canada the principle of culpability is understood in a more normative sense, placing moral blameworthiness at the centre of culpability. This blameworthiness is tested by the intention to commit or join in the commission of criminal activity. In an effort to protect the morally innocent, this blameworthiness test has led to certain limits on extended liability. Especially where there is a high stigma attached to a crime, such as murder or war crimes, it has been determined that only when an accused had the requisite intent, or could at least subjectively foresee that a party with whom she had engaged in criminal activity would commit murder (or a war crime), can liability attach for her.

[369] Kadish 1980, p. 10.

[370] Fletcher 2007, p. 288.

Any lower threshold, such as objective foresight, is deemed unacceptable according to 'principles of fundamental justice'.

Subjectivity can be seen as an expansion of liability, since even actors on the periphery who may have contributed little or nothing to the actual crime are considered to be on the same footing as the physical perpetrator. This is expressed clearly in certain modes of liability which are particular to these jurisdictions.

6.5.1.1 Conspiracy and Common Intention

The most important examples of this deliberate expansion of liability are common intention in Canada and conspiracy on the US, both of which provide the basis for the common intention or joint criminal enterprise doctrine in ICL. Common intention was designed by legislators in Canada to 'impose liability on persons who engage in criminal conduct with others for additional criminal acts that could be foreseen as likely to be committed in the furtherance of the underlying offence'.[371] The Supreme Court has confirmed that the policy intention of this mode of liability is to widen the net of liability beyond aiding and abetting, when the physical perpetrator has committed crimes beyond which the parties had intended or agreed to aid and abet.[372]

Similarly, explicit policy intentions can be traced to explain the emergence of conspiracy as a mode of liability in the US. The word 'conspire' derives from two Latin words, 'con' and 'spirare', meaning to breathe together; to conspire is to agree.[373] In Canada, as in the UK from where Canada inherited the notion of conspiracy, it is only known in its form as an inchoate crime. In the US, it has gained its own form as a combination of conspiracy and complicity, creating a far-reaching doctrine of conspiratorial liability which gained popularity during the era of prohibition, as a policy response to the difficulty of prosecuting the leaders, or family heads of organised crime.[374] These individuals were almost never present at the scene of a murder, nor did they do anything physically with respect to the movement of illegal alcohol, and for whom it was difficult to prove that they had given explicit orders for a specific crime to be committed. Instead, with the doctrine of conspiracy liability, all that needed to be proven was that there was an agreement to commit a crime.

The *Pinkerton* doctrine extended this liability even further, to include liability for all co-conspirators when one co-conspirator commits crimes that were not part of the agreement, but were objectively foreseeable.[375] The extent of the rule may seem far-reaching to anyone not used to operating with a subjective approach to liability, and is perhaps even far-reaching when compared to most other common law jurisdictions. But it has become the norm in the US, and is the preferred tool for dealing with complex organised crime and collective criminality. In this sense

---

[371] Criminal Code 1985, s. 21(2); Hibbert 1995, p. 1007.

[372] Simpson 1988.

[373] Dynar 1997, para 87; Cotroni 1979, p. 276.

[374] Harvard Note 1959, p. 923.

[375] Pinkerton 1946.

it is at least more successful than the attempts under Canadian law to prosecute for participation in a criminal organisation, which is based on the same policy, but has suffered severe procedural problems leading to disastrous and lengthy mega-trials, such as the Hell's Angels and Manitoba Warriors cases.[376]

However even judges of the US Supreme Court have agreed in the majority that conspiratorial liability is not accepted as a mode of liability in international law. The fact that it is particular to the US does not preclude it *per se* from application at the international level, however this factor should be taken into account when translating it directly or in new forms. Most importantly, it should be considered whether this domestic solution to mafia-type crimes is appropriate for mass atrocities.

The rationale often given for making a mere agreement punishable is prevention. Just as with attempts, the inchoate crime of conspiracy allows the police to intervene at an early stage, since no actual preparatory act nor even an attempt is required; the very act of agreeing is sufficient. However, since most agreements to commit a crime will be made behind closed doors, the preventive effects of criminalizing them are questionable. In the context of ICL, where international tribunals only react *ex post facto* to events of atrocity, this preventive, early intervention is not a factor at all, and so the translation of this approach to ICL does not carry with it the advantages which were a large policy factor in its creation at the domestic level.

6.5.1.2 Limits on Extended Liability

Despite the success of conspiratorial liability as the 'prosecutor's darling' in the US,[377] the policy choice *not* to create a form of conspiratorial liability in Canada may be due to the extremely valuable attention that has been paid in Canadian judicial debates and scholarly commentary to the notion of constructive liability, and the need to restrict it. Since conspiratorial liability constructs liability for every party to an agreement for all of the acts committed by any other party, this means that in effect guilt for the whole of an offence is predicated upon proof of guilt for just part of it.[378] In Canada the constitutional limits provided by the notion of 'principles of fundamental justice' have allowed judges to limit the reach of constructive liability, at least in the case of high stigma crimes such as murder and war crimes, by requiring proof that the accused had the highest requisite mental state of intent.[379] Moral blameworthiness is at the core of this test, since it is considered unfair to label a person as a murderer who did not intend to commit murder.[380] The principle of correspondence between the crime and the fault or mental state of the accused prevails

---

376 Lesage and Code 2008, p. 16; Stuart 2007, p. 735.

377 Justice Jackson's expression in his critical analysis of the ever-expanding doctrine, in Harrison v the United States 1925.

378 Healy 1993, p. 4.

379 Creighton 1993; Naglik 1993; Healy 1993.

380 Ashworth 2009, p. 78; Paciocco 2014, p. 13; Chalmers and Leverick 2008, p. 222.

for these high stigma crimes, due to the strong relationship to moral blame and fair labelling.

These limits according to 'principles of fundamental justice' can be instructive when it comes to modes of liability under ICL. However it is not to say that there must always be absolute correspondence between the mental state of an accused party and the requisite fault standard (*mens rea*) for a crime. In the US, all forms of accomplice liability require a high standard of 'knowingly' or for the purpose of facilitating' the physical perpetrator.[381] In Canada this only seems to apply to aiding and abetting, due to the premise that widening the net of liability requires a counter-balancing test, a higher fault standard, in order to protect the morally innocent.[382] At the other extreme, for crimes that have a lower stigma, mere objective foreseeability is deemed to be sufficient for party liability.[383] This brings the minimum standard of fault very close to recklessness for party liability for lesser crimes. One could argue that for the purposes of this enquiry the concern is only with high stigma crimes such as murder committed as part of a culture of mass atrocity, and war crimes, and that the domestic common law tradition would therefore instruct either a requirement of correspondence, or that the accomplice must have acted 'for the purpose of facilitating' the physical perpetrator. However this is not the pattern we see in the subjective element tests for party liability in ICL, which appear to allow objective foreseeability as a sufficient test. Nonetheless it should be kept in mind if a subjective approach is asserted in ICL, since this would be consistent with the limits to liability that have been created under the subjective approach at the core of the common law tradition.

With respect to the notion of fair labelling, despite the fact that the subjectivity prevalent in the common law tradition leads to a functionally unitary system of liability, there is still some judicial discussion in Canada as to whether it is fair to convict a party for a crime she did not intend, based only on the shared intention she had to commit a lesser collective crime. Although there is no fundamental principle of justice requiring that all parties to a crime share the same *mens rea* as the physical perpetrator,[384] judges on the Canadian Supreme Court have stated that there *is* a principle of fundamental justice requiring proportionality between the sentence and the degree of a person's responsibility; so although a person may be convicted of a crime they did not intend, they should receive a lesser sentence than the physical perpetrator.[385]

---

[381] American Law Institute 1985, p. 310.

[382] Hibbert 1995, para 17; Roach 2012, p. 150; Manning and Sankoff 2009, p. 257; Roach 2010, para 36.

[383] Mills 2013; Manning and Sankoff 2009, p. 157.

[384] Logan 1990, p. 741.

[385] Alternatively, it is also possible to convict a party of a lesser crime than that which the physical perpetrator has committed, as long as that lesser crime is encompassed by the definition of the crime that was actually committed. For example, it would be possible to convict a party of manslaughter if he was party to a collective violent crime where the physical perpetrator was convicted of murder, but the party himself did not intend murder. It has been stated that this maintains an equilibrium

The discussion as to fair labelling for parties to a crime appears absent in the US. The policy choice to emphasise the subjectivity of liability, most especially with respect to the extended conspiratorial liability under the *Pinkteron* rule may explain this, since all parties and co-conspirators are always considered to be equally liable. There are no such constitutional 'principles of fundamental justice' in the US to restrict this in any way.

6.5.2 Vicarious Liability

The fact that subjectivity leads to equal liability of all parties to a crime, regardless of their actual contribution, is linked to the third general conclusion that can be drawn, namely that the theory underpinning liability for collective crimes in this tradition is vicarious liability. In criminal law a notion of vicarious liability as recognized in private law is deemed unacceptable, since we should not be punished for acts over which we have no control, and for which we are not culpable.[386] However the way in which the subjective approach bases liability on the joining of intention, and attributes the acts of all parties to a crime to each of the parties, can best be described as vicarious. The physical perpetrator's acts *become* the acts of the accomplice, or as it has been described in the case law in the US, the accomplice 'forfeits [his] identity',[387] and loses his right to be treated as an individual when he chooses to engage in group criminality. The acts of the physical perpetrator are attributed to the accomplice, because the accomplice is deemed to have said 'your acts are my acts', or as one judge has put it, 'the hand of one is the hand of all'.[388]

This is prevalent most especially in the US with respect to conspiratorial liability, but is also apparent in the various modes of extended liability in Canada and in the common law tradition in general, such as common intention (sometimes known as joint criminal enterprise) and counselling, and in the fact that even aiders and abettors are convicted as if they had committed the crime themselves. This is reflected in the policy response to gang-type crimes, however it is questionable whether this is an appropriate way to approach the acts of many individuals involved in potentially very diffuse groups which take part in different aspects of mass atrocity crimes.

As a consequence of this theoretical underpinning of the relationship between the group and the individual, there is no normative distinction between parties to a collective. All parties are convicted as principles, regardless of whether the historical language of 'accessory' is still used in some jurisdictions. This can be seen as a functionally unitary (descriptively differentiated) system, where the language of the old common law distinctions still remains in the legislation and in some of the case

between moral culpability and the offence for which the party is finally convicted. See Jackson 1993, p. 585; Creighton 1993; De Sousa 1992, which affirms *Creighton*.

386 State v Akers 1979; State v Guminga 1986; Duff 2007, p. 20; Ashworth 2009, p. 149.

387 Prettyman 1996, p. 1018; Dressler et al. 2001, p. 466.

388 Curry 2006, p. 653; Dressler et al. 2001, p. 467.

law, but where this language has no purpose when it comes to determining culpability. When there is no normative distinction between parties to a crime, the test for liability is not in degrees, but is rather binary; either an individual is liable, or she is not. A semantic distinction may still have a role to play, even if it is not a legal normative one. It may aid a judge in determining a sentence, and in Canada the Supreme Court has found that there is a principle of fundamental justice that requires sentencing to be proportionate to the level of responsibility: moral blameworthiness is the ground for punishment, which is equal for all parties, but the measure of the punishment may be individualised. However this is entirely up to an unchecked judicial discretion, which means there is no consistency in sentencing.

6.5.3 The Relationship Between the Character of Criminal Trials and a System of Liability

As highlighted at the beginning of this chapter, one explanation for the election of a system that does not differentiate between complex categories of liability is the presence of juries as fact triers, which is typical to the common law tradition. Since it is a jury of lay people who decide the question of guilt, and judges have to instruct juries as to the correct law, it has been emphasised more than once that there are policy reasons for keeping the system singular in its effects. Even though the prosecution may choose to charge an individual based on a specific mode of liability it is not necessary, since any mode of liability will lead to a verdict of guilt, and a jury will not require the specific elements of each mode to be made out. In Canada the rule is that if the prosecution charges a specific mode of liability, it must prove the elements of that mode of liability, but if it charges multiple or alternative modes of liability, it need not prove all the different elements for each mode, since any of them will suffice. As has been pointed out, it is mostly for sentencing purposes that any attention is given at all to modes of liability. In terms of proving guilt beyond reasonable doubt, a simpler task for juries is created by maintaining a system of liability that does not differentiate normatively.

The existence of lay juries as important participants in common law trials has been explained by Damaška by the horizontally organised governmental authority, where judges are conflict solvers, and where the trial is thus not so concerned with consistent application of a legislator's policies, but rather with reaching a result that satisfies the immediate societal concerns. This horizontal approach to authority could furthermore be said to be reflected in the horizontal relationship constructed by the law between all parties to a crime. It should be of little surprise that in a system where authority is conceived of horizontally, the accountability for decisions made in groups is also conceived of horizontally.

The conflict-solving type of proceeding described by Damaška allows a process of law-making that has been described above as iterative. Over time, although domestic jurisdictions following the common law tradition will retain certain characteristics such as the subjective approach to culpability, and its correlated functionally unitary system, judicial decisions will lead to new solutions to the problem of collective criminality based on societal needs. By way of casuistry and bifurcation, then, the judiciary in the US has been able to develop conspiratorial liability, based on a policy need to widen the net of liability as far as possible. This may explain why it is solely in the US that conspiracy exists as mode of liability, whereas exists only as an inchoate crime in other subjectivist, common law jurisdictions.

Finally, the background notions of justice, fairness and truth are interpreted in a particular way that impacts on the criminal trial outcome. One useful way to consider this is through Damaška's analysis of the common law tradition expressing a horizontal organisation of authority, whereby lay participants are entrusted with factual decisions, and judges have the task of adjudicating between two equal parties—the defence and the prosecution.[389] Because the defence and the prosecution have a great deal of autonomy in in a typically adversarial court, the common law tradition judge acts as an 'umpire' between two equal competing versions of the truth, and must ensure a fair outcome.[390] In this sense, justice is equated with procedural fairness: an outcome that is procedurally fair, protecting both parties equally, is considered to be just. Thus the final legal decision is less concerned with substantive truth, and more concerned with equality of arms.[391] The outcome in terms of liability is therefore a symbolic one, or a 'procedural truth' that may or may not fully align with the substantive truth; all parties to a crime are considered equally liable, even if their factual roles differed. As long as the outcome is fair in a procedural sense, a detailed factual and moral story-telling is less important in the conviction, and a delineation between parties is therefore unnecessary.

These conclusions can help shed light on the 'clash of legal cultures' that occurs in ICL in the search for modes of liability, especially when compared with the characteristics of a more objective approach, as will be discussed in the following chapter.

References

American Law Institute (1985) Model Penal Code and commentaries: official draft and revised commentaries

Ashworth AJ (2009) Principles of criminal law, 6th edn. Oxford University Press, Oxford

Ashworth AJ (2011) United Kingdom. In: Heller KJ, Dubber MD (eds) The handbook of comparative criminal law. Stanford University Press, Stanford, California, p 531

Bailey PG (2009) Law Reform Commission. http://www.thecanadianencyclopedia.com/en/article/law-reform-commission-of-canada/

Bayley D (2011) Six degrees of separation: Canadian accessory liability in Afghan war crimes. Dalhousie J Legal Stud 20:43–70

Bill C-45 (2003) Bill C-45: An Act to amend the Criminal Code (criminal liability of organisations)

389 Damaška 1984, p. 104.

390 Damaška 1984, pp. 65, 119.

391 Grande 2008, p. 154; Weigend 2003, p. 162.

Canadian Charter (1985) Canadian Charter of Rights and Freedoms
Chalmers J, Leverick F (2008) Fair labelling in criminal law. Mod Law Rev 71(2):217–246
Cotler I (1997) Bringing nazi war criminals in Canada to justice: a case study. Am Soc Int Law Proc 91:262
Criminal Code (1985) Criminal Code of Canada
Cunningham LA (2006) The common law as an iterative process: a preliminary enquiry. Notre Dame Law Rev 81:747–781
Damaška M (1984) The faces of justice and state authority: a comparative approach to the legal process. Yale University Press, New Haven
Dressler J, Strong FR, Moritz ME (2001) Understanding criminal law. Lexis Publishers
Duff RA (2007) Answering for crime: responsibility and liability in the criminal law. Hart Publishing
Fletcher GP (1981) Reflections on felony-murder. Southwestern Univ Law Rev 12:413
Fletcher GP (2000) Rethinking criminal law. Oxford University Press, USA
Fletcher GP (2007) The grammar of criminal law: American, comparative and international, Vol One: Foundations. Oxford University Press, New York
Fletcher GP (2011) New court, old dogmatik. J Int Crim Justice 9:179–190
Fukuyama F (2011) The origins of political order: from prehuman times to the French revolution. Farrar, Straus and Giroux, New York
Garner BA (ed) (2009) Black's law dictionary, 9th edn. Thomas and Reuters, Minnesota
Glenn HP (2010) Legal traditions of the world: sustainable diversity in law, 4th edn. Oxford University Press, USA
Grande E (2008) Dances of criminal justice: thoughts on systemic differences and the search for truth. In: Jackson J, Langer M, Tillers P (eds) Crime, procedure and evidence in a comparative and international context. Hart, Portland, Oregon, Chapter 8, pp 145–164
Harvard Note (1959) Developments in the law: criminal conspiracy. Harvard Law Rev 72:920
Healy P (1993) The Creighton Quartet: Enigma Variations in a lower key. Crim Rep 23:265
ILC Draft Code of Crimes (1996) Draft Code of Crimes against the Peace and Security of Mankind. http://untreaty.un.org/ilc/publications/yearbooks/Ybkvolumes(e)/ILC_1996_v2_p2_e.pdf
Kadish SH (1980) Why substantive criminal law? A dialogue. Clevel St Law Rev 29:1
Kadish SH, Schulhofer SJ, Paulsen MG (2007) Criminal law and its processes: cases and materials, 8th edn. Aspen Publishers, New York
Katyal NK (2003) Conspiracy theory. Yale Law J 112:1307
LaFave WR (1986) A handbook on criminal law. University of California Press, California
Lafontaine F (2012) Prosecuting genocide. Crimes against humanity and war crimes in Canadian courts. Carswell, Thomson Reuters, Toronto
Law Reform Commission of Canada (1985) Secondary liability: participation in crime and inchoate offences
Leflar RA (1971) Sources of judge-made law. Oklahoma Law Rev 24:319–336
Lempert R (2008) Anglo-American and continental systems: marsupials and mammals of the law. In: Jackson J, Langer M, Tillers P (eds) Crime, procedure and evidence in a comparative and international context. Hart Publishers, Portland, Oregon
Lesage HPJ, Code M (2008) Report of the review of large and complex criminal case procedures. www.attorneygeneral.jus.gov.on.ca
Lynch G (2003) Revising the Model Penal Code: keeping it real. Ohio State Crim Law Rev 1:219
Manning M, Sankoff P (2009) Criminal law, 4th edn. Lexis Nexis, Ontario, Canada
Milsom SFC (2003) A natural history of the common law. Columbia University Press, New York
Moore SF (1973) Law and social change: the semi-autonomous social field as an appropriate subject of study. Law Soc Rev 7(4):719–746
MPC (1962) Model Penal Code
Oderkerk M (2001) The importance of context: selecting legal systems in comparative legal research. Netherlands Int Law Rev XLVIII:293–219
Ohlin JD (2007) Group think: the law of conspiracy and collective reason. J Crim Law Criminol 98(1):147–206

Ohlin JD (2011) Joint intentions to commit international crimes. Chicago J Int Law 11:2
Paciocco PJ (2014) Objective mens rea and attenuated subjectivism: Guidance from Justice Louise Charron in R v Beatty. In: Oliver P, Mayeda G (eds) Principles and pragmatism: essays in honour of Louise Charron. Lexis Nexis
Pilsbury SH (2009) How criminal law works: a conceptual and practical guide. Carolina Academic Press, North Carolina
Roach K (2004) Dialogic judicial review and its critics. Supreme Court Law Rev 23:49
Roach K (2012) Criminal Law, 5th edn. Irwin Law, Toronto
Roach K (2013) Canada. In: Reed A, Bohlander M (eds) Participation in crime: domestic and comparative perspectives. Substantive issues in criminal law. Ashgate, Surrey, Chapter 17, pp 307–322
Statistics Canada (2010) Adult Criminal Court statistics. http://www.statcan.gc.ca/pub/85-002-x/2010002/article/11293-eng.htm#a24
Stuart D (2007) Canadian criminal law: a treatise, 5th edn. Carswell, Toronto
Stuart D, Coughlan S, Delisle RJ (2012) Learning Canadian criminal law, 12th edn. Carswell, Toronto
Thomson N (1994) Fundamental justice, stigma and fault. Toronto Fac Law Rev 52:379–404
Tigar ME (1986) Crime talk, rights talk and double-talk: thoughts on reading the Encyclopedia of crime and justice. Texas Law Rev 65:101
Trapani A (2009) Comparative analysis of prosecutions for mass atrocity crimes in Canada, the Netherlands and Australia. http://www.domac.is/reports/
Trapani A (2011) Assessing the impact of the international ad-hoc tribunals on the domestic courts of the Former Yugoslavia. http://www.domac.is/media/domac/Domac-11-BAlkan-AT-Final.pdf
UN Refugee Convention (1951) United Nations Convention Relating to the Status of Refugees
van der Wilt H (2008) Equal standards? On the dialectics between national jurisdictions and the International Criminal Court. Int Crim Law Rev 1(2):229–272
van Sliedregt E (2012) Individual criminal responsibility in international law. Oxford University Press, Oxford
Vogel J (2002) How to determine individual criminal responsibility systemic contexts: twelve models. Cahiers de Défense Social, pp 151–169
War Crimes Act (2000) Crimes against Humanity and War Crimes Act
Watson A (1983) Legal change: sources of law and legal culture. University of Pennsylvania Law Review. UPALR 131:1121–1157
Weigend T (2003) Is the criminal process about truth? A German perspective. Harvard J Law Public Policy 26:157–173
Weisberg R (2000) Reappraising complicity. Buffalo Criminal Law Review 4:217–282

Cases

Amara (2010) R. v. Amara ONCA 858 (Ontario Court of Appeal)
Backun v. United States (1940) Backun v. United States 112 F2d 635 (Fourth Circuit Appeals Court)
Ball (2011) R. v. Ball 82 CR (6th) 72 (British Columbia Court of Appeals)
Berryman (1990) R. v. Berryman 78 CR (3d) 376 (British Columbia Court of Appeal)
Blaškić Trial (2000) The Prosecutor v. Tihomir Blaškić, Trial Judgment IT-95-14/1-T
Bolden v. State (2005) Bolden v. State 124 P3d 191 (Nevada Supreme Court)
Briscoe (2010) R. v. Briscoe 1 SCR 411 (Supreme Court of Canada)
Chow Bew (1956) Chow Bew v. the Queen SCR 124 (Supreme Court of Canada)
Cotroni (1979) Papalia v. the Queen, the Queen v. Cotroni 2 SCR 256 (Supreme Court of Canada)
Creighton (1993) R. v. Creighton 91, 3 SCR 3
Croy (1985) The People v. Croy 41 Cal 3d 1 (California Supreme Court)

Curry (2006) The State v. Curry 636 SE2d 649 (South Carolina Court of Appeals)
De Sousa (1992) R. v. de Sousa 2 SCR 944 (Supreme Court of Canada)
Direct Sales Co. v. United States (1943) 219 US 703 (Supreme Court of the United States)
Grewall (2001) R. v. Grewall, Grewall and Toor 48 BCSC (Supreme Court of British Columbia)
Greyeyes (1993) R. v. Greyeyes 3 SCR 76 (Supreme Court of Canada)
Griffin v. State (1970) Griffin v. the State 455 S.W. 2d 822 Supreme Court of Arkansas
Hamdan (2006) Hamdan v. Rumsfeld 126 S.Ct. 2749 (2006), (2006) 548 US 557
Hamilton (2005) R. v. Hamilton S.C.J. No. 48, 2 SCR 432
Harrison v. the United States (1925) 7 F.2d 259
Hibbert (1995) R. v. Hibbert 99 CCC (3d) 193 (Supreme Court of Canada)
Hicks (1893) Hicks v. the United States 150 US 442 (Supreme Court of the United States)
Hughes (2011) R. v. Hughes BCCA 220 (British Columbia Court of Appeal)
Hundal (1993) R. v. Hundal 1 SCR 867 (Supreme Court of Canada)
Interstate Circuit, Inc. (1939) Interstate Circuit, inc. v. United States 306 US 208 (Supreme Court of the United States)
Jackson (1993) R. v. Jackson S.C.J. No 134, 4 SCR 573
Jimenez Recio (2003) United States v. Jimenez Recio 537 US 270 (Supreme Court of the United States)
Kotteakos v. United States (1946) Kotteakos v. United States 328 US 750 (Supreme Court of the United States)
Krulewitch (1949) Krulewitch v. United States 366 US 440
Logan (1990) R. v. Logan 2 SCR 731 (Supreme Court of Canada)
Lubanga Trial Judgment (2012) Prosecutor v. Thomas Lubanga Dyilo ICC-01/01-01/06
Martineau (1990) R. v. Martineau 2 SCR 633 (Supreme Court of Canada)
Mills (2013) Regina v. Mills, Hylton, Dennis [2013] ONSC 5051 (Ontario Superior Court of Justice)
Morgan (1993) R v. Morgan 80 CCC (3d) 16 (Ontario Court of Appeals)
Morrison v. California (1934) Morrison v. California 291 US 82 (United States Supreme Court)
Mugasera (2005-06-28) Mugasera v. Canada (Minister of Citizenship and Immigration) (Supreme Court of Canada, 28 June 2005)
Naglik (1993) R. v. Naglik 3 SCR 122 (Supreme Court of Canada)
Nova Scotia Pharmaceutical (1992) R v. Nova Scotia Pharmaceutical Society (1992) 2 SCR 606 (Supreme Court of Canada)
O'Brien (1954) R v. O'Brien (1954) 110 CCC 1 (Supreme Court of Canada)
Pacquette (1976) R v. Pacquette (1976) 30 CCC (2d) 417 (Supreme Court of Canada)
Pappajohn (1980) R v. Pappajohn (1980) 2 SCR 120 (Supreme Court of Canada)
People v. Lauria (1967) The People v. 251 Cal App 2d 471 (California District Court of Appeals)
People v. Luparello (1987) The People v. Luparello 187 Cal App 3d 410 (California Court of Appeal, 4th District)
People v. McGee (1979) The People v. McGee 49 NY 2d 48 (New York Court of Appeals)
People v. Stamp (1969) The People v. Stamp 2 Cal App 3d 203 (California Court of Appeals)
Pickton (2010) R. v. Pickton SCC 32, [2010] 2 S.C.R. 198
Pinkerton (1946) Pinkerton v. United States 328 US 640 (Supreme Court of the United States)
Prettyman (1996) The People v. Prettyman 926 P2d 1013 (California District Court)
Reference re Motor Vehicle Act (1985) Reference re Motor Vehicle Act (Britsh Colombia) s 94(2) S.C.J. No. 73, 2 SCR 486
Rex v. Murphy (1837) Rex v. Murphy 173 English Reports 502
Roach (2010) R. v. Roach 2 SCR 98 (Supreme Court of Canada)
Roy (1995) Roy v. United States 652 A2d 1098 (Washington DC Court of Appeals)
Sansregret (1985) R. v. Sansregret SCR 570 (Supreme Court of Canada)
Sault Ste Marie (1978) R. v. Sault Ste Marie (City) S.C.J. No. 59, 2 SCR 1299
Sheppe (1980) R. v. Sheppe S.C.J. No. 39, 2 SCR 22
Simpson (1988) 38 CCC (3d) 481 (Supreme Court of Canada)
Stanciel (1992) 606 NE2d 1201 (Supreme Court of Illinois)

State v. Akers (1979) 400 A2d 38 (Supreme Court of New Hampshire)
State v. Bridges (1993) (1993) 133 NJ 447 (New Jersey Court of Appeals)
State v. Carbone (1952) (1952) 10 NJ 329 (Appeals Court of New Jersey)
State v. Davis (1989) 182 W Va 482 (West Virginia Supreme Court of Appeals)
State v. Gladstone (1970) 78 Wash 2d 306 (Supreme Court of Washington)
State v. Guminga (1986) 395 NW2d 344 (Supreme Court of Minnesota)
State v. Hayes (1891) 105 Mo 76 (Supreme Court of Missouri)
State v. McKeown (1979) 23 Wash App 582 (Washington Court of Appeals)
State v. Nevarez (2005) 130 P3d 1154 (Idaho Court of Appeals)
State v. Walton (1993) 227 Conn 32 (Connecticut Supreme Court)
State v. Wilson (1981) 95 Wash 2d 828 (Supreme Court of Washington)
Tadić Appeals Judgment (1999) The Prosecutor v. Dusko Tadić, Appeals Judgment IT-94-1-A
Thatcher (1987) R. v. Thatcher, S.C.J. No. 22, 1 SCR 652
The United States v. Feola (1975) 420 US 671 (Supreme Court of the United States)
Théroux (1993) R. v. Théroux 2 SCR 5 (Supreme Court of Canada)
United States v. Alvarez (1985) 755 F2d 830 (Federal Court of Appeals, 11th Circuit)
United States v. Bruno (1939) 100 F. 2d 921 (Federal Court of Appeals, 2nd Circuit)
United States v. Falcone (1940) 311 U.S. 205 (Supreme Court of the United States)
United States v. Garcia (1998) 151 F.3d 1243 (Federal Court of Appeals, 9th Circuit)
United States v. James (1976) 528 F.2d 999 (Federal Court of Appeals, Fifth Circuit)
United States v. Moussaoui (2006) 1:01CR00455-001, (2006) AO 245 S (Rev. 2/99) EDVA
United States v. Peoni (1938) 100 F.2d 401 (Second Circuit Court of Appeals)
Vaillancourt (1987) R. v. Vaillancourt, 2 S.C.R. 636 (Supreme Court of Canada)
Wholesale Travel (1991) R. v. Wholesale Travel, S.C.J. No. 79, 3 SCR 154
Yamashita (1946-02-04) Trial of General Tomoyuki Yamashita, Law Reports of Trials of War Criminals IV, Vol 1; Confirmed by the Supreme Court of United States, In re: Yamashita, 4 February 1946, 327 U.S
Zazai No. 1 (2005) Zazai v. Canada (Minister of Citizenship and Immigration) (2005) 2 FCR 78 (Federal Court of Canada)
Zazai No. 2 (2005) Zazai v. Canada (Minister of Citizenship and Immigration) 259 DLR (4th) 281, 50 ImmLR (3d) 107 [2005] FCA 303 (Federal Court of Canada)

Chapter 7
Objectivity Reflected in the Civil Law Tradition

Contents

C. Steer, *Translating Guilt*, International Criminal Justice Series 9,
DOI 10.1007/978-94-6265-171-5_7

This chapter aims to demonstrate the systemic and conceptual reasons for, and consequences of, the existence of objectivity underlying concepts of liability typical to the civil law tradition, by way of comparison the subjectivity that was evident throughout the previous chapter. In particular the influence of different participants within the civil law tradition on the formation and development of the law will be considered, since in this tradition the highly influential role of doctrine (and therefore of scholars) differs from the common law tradition. As a result the policy choices made in responding to collective criminality differ in some important ways. While it may be stating the obvious that this is where the dividing opinions on the international level stem from, it is worth taking a closer look at these differences and their historical and socio-legal reasons, since this may help in understanding the disparities between the traditions, and between the positions taken in debates surrounding which approach is most appropriate to the context of mass atrocity crimes.

First, Sect. 7.1 and 7.2 will explore the civil law tradition in terms of law-making and law-application, and the role of different participants. Following from this, the two jurisdictions which have been selected for comparison will be discussed, namely Germany in Sect. 7.3 and Argentina in Sect. 7.4.

It will become apparent in this analysis that an objective approach—where there is emphasis placed upon the degree of contribution by an actor to the fulfilment of the crime—leads to a normatively differentiated system. 'Guilt' in the sense of blameworthiness is the key concept in this differentiation. Because blameworthiness rests on the degree of contribution, blameworthiness is itself also measured in degrees. The final qualification of each party to a crime specifies to what degree he or she contributed, leading to a conviction that reflects this measure expressed as differentiated classes of conviction: principal or secondary.

7.1 The Civil Law Tradition: A Context

In order to understand the reasons for the normatively differentiated approach to liability that is typical in the civil law tradition,[1] it is necessary to consider the historical and systemic factors that have led to the very formation of that tradition. Although the civil law is usually associated with a greater commitment to consistency and systemic stability than the common law tradition, in many ways it could be said that there is less of a consistent tradition of the civil law. Rather than emerging over time as a conglomeration of customs, it has had several stops and starts throughout history, and has been described as a constructed tradition rather than an organic

---

[1] To say it is typical is not to assert that all domestic jurisdictions that follow the civil law tradition have the same approach to liability. One exception is the French criminal law which, although it resembles other continental European systems in many ways, has developed a notion of liability for collective crimes which is descriptively differentiated, but not normatively. All forms of contributing to a principal perpetrator are considered to fall under the term *complicité*. See Article 121 of the *Code Pénal*; see further Pradel 2002, p. 278; van Sliedregt 2012, p. 67.

one.[2] The reception of Roman Law throughout western continental Europe was not seamless, and in fact some jurisdictions such as France never truly accepted it, even though the Napoleonic codes are often considered to be modern representatives of the codex tradition.[3] Instead Roman law was transplanted into various European systems in 'blocks'; rather than the ancient system as a whole influencing modern jurisdictions, or even the selection of specific rules, self-contained sections of the books of the Justinian *Corpus juris civilis* were imported.[4] In this way Roman law shaped the modern civil law tradition by way of shaping the systematisation and legal attitudes, without it necessarily leading to uniformity throughout Europe.[5]

The reception of Roman law in these blocks had important consequences for certain elements of the civil law tradition. For instance, the systematisation of such blocks meant that Roman law was the perfect vehicle for education in legal rules, regardless of the quality of the law, and regardless of the original historical context of any specific institution or rule.[6] Hence Roman law dominated legal education through Europe, based on the *Institutes*, which were a textbook introduction to the *Corpus juris* hailing from the 6th century A.D.[7] The interpretation and teaching of these books throughout Europe, in different jurisdictions, were kept relatively uniform and international in the hands of academic experts. Thus the legal elite—those participants with the most influence on the teaching and development of the law—were the academic jurists rather than the practitioners.[8]

In the 5th century A.D., in attempt to systematize the multiple sources of law, there were a select few authors whose writings were deemed to be authoritative.[9] Similarly, later with the 're-discovery' of Roman law in the late middle ages, and the establishment of universities around Europe, there emerged strong schools of interpretative authority: the glossarists in Italy stemming from the university of Bologna, whose main focus was to glossarize and annotate the ancient texts in order to define terminology and individual rules; the scholastics in France and some parts of Germany who combined canonic law with the commentators on Roman law; and the German Pandectists whose main focus was to interpret the relationship between the self-contained blocks within the *Corpus juris* as a whole, and to apply the ancient texts to contemporary needs.[10] Anyone wishing to study the law would follow one of these traditions depending on which university he attended, since these were the only interpretations deemed authoritative.

All in all, the establishment of codified systems of law required learned, academic interpretations, rather than training in inns of courts, which was the place of learning

---

[2] Glenn 2010, p. 134.

[3] Zweigert and Kötz 1998, p. 75; Koschaker 2000, p. 76; Glenn 2010, p. 142.

[4] Watson 1981, p. 15.

[5] Watson 1981, p. 15.

[6] Watson 1981, p. 19.

[7] Watson 1981, p. 12; Koschaker 2000, p. 64.

[8] Watson 1981, pp. 62, 188; Watson 1995, p. 472; Glenn 2010.

[9] Watson 1981, p. 36; Glenn 2010, p. 139.

[10] Koschaker 2000, pp. 58, 90, 266; Schermaier 2006, p. 284.

English common law.[11] Development of the law in the civil law tradition therefore relies on doctrinal understanding, and is an interpretative process, rather than the iterative process described in the previous chapter. This interpretative process brings with it consistency and a tradition of deductive thought and logic, which translates into the legalism of national criminal codes. The *Institutes* and the rest of the *Corpus juris* lent themselves to philosophical and scientific approaches, and doctrinal advances were, and still are, in the hands of legal scholars whose concern is for systematic consistency and a 'law of reason', rather than casuistic problem solving.[12]

Roman law has left behind its strongest mark in Germany. The modern doctrinal interpretations in German legal thought could be said to be reminiscent of the authoritative Pandectist interpretations of the twelfth century, whose writings were an analytic approach to understanding the *Corpus juris* as a systemic whole. In the nineteenth century a dogmatic, historical approach became prominent in Germany, based largely on the historical school of Friedrich von Savigny and his scholarship on the Pandectist analysis of Roman law.[13] This approach was extremely influential on the German criminal code in particular, which has a deliberate, systematized and logical construction and is interpreted according to principles of logic.[14]

The commitment to an internally coherent system in criminal law can also be historically traced in terms of procedural developments. Procedures became open to everyone, and were not guarded by a system of writs such as in the common law. Because the substantive law had to be interpreted, the focus in legal education was on the substantive, and the procedure was considered to be a matter for practitioners.[15] Procedure was generally investigative rather than adversarial, and was therefore not subject to a vast array of technical rules. Furthermore, courts of appeal took on the task of testing the entirety of the case decided by the lower instance, and not just the contested issues, since an erroneous decision was illegal and had to be quashed; this also contributed to a certain systemic consistency and harmony.[16]

This systemic consistency has meant that legislation and doctrine in general have primacy over case law in the civil law tradition, though the case law is certainly important in establishing the correct interpretation of the codified law. When judges seek to interpret the law they turn primarily to recognised scholastic opinion.

As discussed in the previous chapter, Damaška explains the essential differences between the common law and civil law traditions by considering (i) the organisation of governmental authority on one hand, which he describes as either hierarchical or co-ordinate models, and (ii) the purpose of the criminal legal process on the other hand, which he describes as oriented towards either policy implementation or individual conflict solving.[17] Whereas a hierarchical organisation of authority will

---

[11] Watson 1981, p. 25.

[12] Watson 1981, p. 85; Damaška 1984, p. 168.

[13] Watson 1981, p. 127.

[14] Koschaker 2000, p. 263; Glenn 2010, p. 153.

[15] Watson 1981, p. 32; Glenn 2010, p. 144.

[16] Glenn 2010, p. 141.

[17] Damaška 1984.

lean towards a policy-implementation process in criminal law, which is reflected in the civil law tradition, a horizontal organisation of authority will lean towards a conflict-solving trial in criminal law, which is reflected in the common law tradition. It should be noted that Damaška intends these models to be ideal types, rather than descriptive depictions of any particular system. However as comparative analytical tools, they are useful to help understand the socio-legal reasons for the underlying characteristics observable in the common law and civil law traditions.

The historical development of the civil law tradition described here demonstrates a preference for a hierarchical organisation of authority, which requires the professionalisation of officials, trained in the technical language of the law and in the scholastic interpretation of the codes.[18] Considering the preference for centralised power in ancient Rome, and later in modern Europe under the Roman Catholic Church, this hierarchical organisation of power may come as no surprise. By comparison, in England the feudal system had developed as a way of protecting the individual against threats to village boundaries, in return for service to the richer and more powerful protecting lords.[19] But this led to bias on the part of lords when it came to legal conflicts, and the only way to receive a less biased adjudication was to go to the king's court (or the king's representatives at the royal courts). This piecemeal distribution of power led to a more horizontal organisation of authority, but at the same time to a 'common law' that became applicable throughout the land.[20]

In Germany, on the other hand, it was the very weakness of imperial power in the early fifteenth century that led to territorial power and local laws separating from each other. For a time there was no influential class of imperial jurists to ensure a uniformity throughout the empire, and conflicts were resolved before a judge based on discretion and 'practical wisdom' rather than any consistent or rational basis. The rediscovery of Roman law in the late middle ages was therefore welcomed and Roman law was received wholesale in many parts of the country.[21] Later, in the fifteenth and sixteenth centuries, with the re-centralisation of power, Roman law became part of the training of judges and jurists.[22] This became the guiding path towards codification in the eighteenth and nineteenth centuries, and the law of reason became a system of principles to be taught and applied.[23]

One characteristic of a centralised and hierarchical organisation of authority is the univocal court, speaking as the 'mouthpiece of the law' in the words of Montesqueue; dissenting voices are silenced and the court hands down a judgment as a single institution, rather than as a conglomerate of individual judges. In fact Damaška

---

[18] Damaška 1984, pp. 18, 32.

[19] Fukuyama 2011, p. 236.

[20] Zweigert and Kötz 1998, p. 136.

[21] Zweigert and Kötz 1998, pp. 134–135.

[22] Watson 1981, p. 8.

[23] Zweigert and Kötz 1998, p. 136.

depicts hierarchically ordered authority as leaning towards a particular kind of logical legalism, by which the subtle differences between concrete cases become less important than the greater consistency of decision making, and thus justice is not individualised, but is rather an outcome of applying general standards consistently.[24]

The hierarchical ideal places trust in the ability of higher authority to order society. A lay jury is therefore not typical in the civil law tradition. (Again, there are exceptions, such as in Belgium and France, but these are atypical.) At the same time, the integrity of centralized authority requires adherence to rules, and thus when it comes to the criminal law procedure, there is a strong tendency towards a process of policy-implementation.[25]

The resulting criminal trial has a great emphasis on the gathering of information and examination of evidence by an official prior to the actual trial in court, and the intention is to establish the truth as objectively as possible, rather than based on a competition of evidence between two parties.[26] An adversarial style of trying an accused does not sit well with this model, and instead parties to a case are considered to be suppliers of information, and the judge is expected to be activist on the part of the State, in the sense that she should go beyond the arguments presented by the parties in order to attain the 'right' and reasoned result.[27] In this way, the 'right' result will be deemed to be that which ascertains a conviction or acquittal based on the application of the system of liability in a consistent and coherent manner. The judge is not expected to make new law, but rather to apply the law in the way in which he has been trained as a professional authority. This is why Damaška suggests that a hierarchical approach to authority will tend towards policy-implementation as a characteristic of the criminal trial.

Thus, when it comes to the development of modes of liability in the civil law tradition, the most influential participants in the interpretative processes of the law are those who are regarded as having a trained and learned understanding of the doctrines and the system as a whole, and who have the authority to interpret and develop, namely the scholars. The lack of lay participation also means that more complex doctrines of liability have been able to develop, since the deciders of both fact and law in the trial are educated in these doctrines and can ensure they are implemented consistently. There are therefore different conceptions of justice and fact-finding in the civil law tradition than we saw in the common law tradition, coupled with an emphasis on recording a substantive rather than a procedural truth.[28]

---

[24] Damaška 1984, p. 22.

[25] Damaška 1984, p. 33.

[26] Damaška 1984, p. 162; Safferling 2003, p. 267.

[27] Damaška 1984, p. 168.

[28] For a fuller discussion of different conceptions of procedural and substantive truth in the two traditions, see Sect. 3.2.4.

7.2 A Normative Theory of Culpability

In the previous chapter it became apparent that although the principle of culpability plays a significant role in the common law tradition, the notion of culpability is itself descriptive in nature. It refers to the different mental states; acting purposefully or knowingly are 'kinds of culpability' according to the Model Penal Code of the US,[29] in a purely descriptive sense, regardless of whether there is also a notion of blame. In the civil law tradition, culpability is understood in a normative sense because there is a distinction between wrongdoing and culpability in the very definition of criminal liability, as will become apparent in the comparative study below. The degree to which a person is to blame for their behaviour determines the degree to which are they considered to be culpable, which translates to degrees of criminal liability.

This distinction matters for a number of reasons. Firstly it matters in terms of whether negligence is properly understood as a form of criminal fault; a descriptive theory cannot find a place for negligence under fault or *mens rea*, since it is not a mental state, but an absence of a standard of care. A normative theory, on the other hand, views negligence as a proper ground for *blaming* an actor for making a mistake, or acting with undue care with regards to a foreseeable outcome.[30] This is important when we consider the negligent liability of a party to a collective crime when other parties to the crime have committed further, unplanned but foreseeable crimes.

Secondly the distinction matters when it comes to contextualising liability, such that the surrounding circumstances of a criminal act can be taken into account. Under a descriptive understanding of culpability, which is found in systems where there is no definitional distinction between wrongdoing and culpability, as soon as the required elements of the crime are fulfilled, including the required mental states, an actor is liable unless a complete defence is available. But in systems where liability is determined by not only the presence of wrongdoing, but also culpability in the sense of blame (and not just the presence of the requisite mental state), a normative theory of culpability makes it possible to ask the question 'what would be reasonably expected of the actor under all of the surrounding circumstances?' Even if they acted *intentionally*, thereby fulfilling the requisite mental state, perhaps they did not act in a blameworthy way. Perhaps nothing else could have been expected of them in the circumstances.[31] This matters when we consider the context of mass atrocity, in which violence has become the norm, and perpetrators may be faced with the dilemma of enacting atrocities on others or becoming the next victim. While it is wrongful to do so, perhaps it is also a lot to expect them to become sacrificial heroes, and the degree of blame may be therefore less, though not null.

This separation between wrongdoing and culpability in the sense of blame has historical roots as well. Under Roman law liability was by definition strict: there was no general principle of liability according to fault. It was sufficient to establish that the causal conditions of liability were met, according to the objective descriptions

[29] MPC 1962, § 2.02(2).

[30] Fletcher 2000, p. 400.

[31] Fletcher 2007, pp. 320, 322.

of how damage was inflicted (theft, fire, breakage, physical harm etc.).[32] While modern criminal codes based on the Roman tradition are no longer as simplistically objective, in that they do also require some mental 'fault' element to be fulfilled, the separation between wrongdoing and culpability (understood as blame) stems from this objective approach. Such a normative theory was developed explicitly in the doctrine in Germany throughout the twentieth century, and is apparent in many other systems which follow this tradition. This structure will be examined in more detail in the two systems compared in this chapter.

7.3 Germany: Responsibility for Control over the Act and Control over an Organisation

The selection of Germany as one of the jurisdictions to be compared in this study may be self evident, given the direct reference to German theories of liability in recent case law of the ICC. Two other reasons for selecting a jurisdiction for comparison are whether it can be considered representative, and the extent to which it is developed in the area of comparison.[33] The German system can be considered to be representative due to the extent of its influence on other domestic jurisdictions,[34] including Argentina, where the same theories of liability were applied to cases of mass atrocity within national borders. Furthermore, doctrines built upon the German Pandectist schools of the nineteenth century have been influential throughout Europe, including in the very notion of unifying Europe under certain universal legal values.[35]

With respect to the extent of development in the area of comparison, it cannot be denied that when it comes to the complex questions surrounding liability for collective crimes, Germany has a highly developed and internally coherent system. The systematic analysis of criminal law issues and the theories and doctrines underpinning them, which stem from the Roman law tradition, have become particularly characteristic of German law.[36] This is also characteristic of the methods taught in German law schools in general, with a technical approach to legal problem solving. In particular the method of subsumption dominates such teachings: a set of facts must be subsumed under, or allocated to, a legal rule, and the interpretation of the rule must be consistent with the system as a whole.[37] The modes of liability have been developed starting with a thorough and systematic codified taxonomy of different forms of perpetration and assisting, followed by a careful doctrinal expansion

[32] Glenn 2010, p. 138.

[33] See further Sect. 5.2. See also Oderkerk 2001.

[34] Particularly with respect to codification of private law, in countries as diverse as Austria, Brazil, China, Italy, Greece, Japan, Switzerland and Turkey, although criminal law notions have also been influential in Argentina, the Netherlands, Peru and others. See Schermaier 2006, p. 284.

[35] Koschaker 2000, p. 277.

[36] Weigend 2006, p. 221.

[37] Schermaier 2006, p. 281.

built upon the codified system. The solution offered by German doctrine to the specific problem of leadership liability within an organisation has been cited by other domestic jurisdictions and by the ICC precisely because it is so highly developed and doctrinally solid.

Furthermore, the German doctrine on leadership liability developed in the second half of the twentieth century specifically as a policy response to the involvement of those in senior political positions in atrocities such as the Berlin Wall shootings. The positivism of the early twentieth century had suffered some loss of legitimacy after the Nazi party promulgated several laws according to correct procedure, but with offensive moral content, yet one of the reasons the German legal system survived this period so well is because of the strong doctrinal traditions and systemic commitment to internal coherence. By continuing to build upon these foundations, German criminal law was restored in the second half of the twentieth century to a system that served new policies which aimed to punish those who abused their organisational power, rather than those who were society's persecuted.[38] This historical fact alone is a reason to consider German criminal law modes of liability as instructive for the prosecution of international mass atrocity.

The structure and core principles of the system are important to consider as a basis. Germany is a federation, but one which resembles Canada more than the US in that it has striven for national unity in its laws, in particular due to the founding of the German Empire in 1871. The *Strafgesetzbuch*, or criminal code (hereinafter StGB), was created in 1870, soon followed by the criminal procedure code in 1877.[39] Though amended and expanded over time, these have remained for the most part unchanged. The main codes implemented in this time were heavily influenced by Roman law, which had formed an important part of German law from the twelfth century, when the German emperors implemented it as a way of lending authority to their claim that they were the successors to the Roman empire.[40] The hierarchical ordering of authority described by Damaška was thus historically embedded in German legal institutions.

The StGB is divided into two parts: the general part (*Allgemeiner Teil*) which enumerates the basic principals and elements which are applicable to all offences, such as modes of liability; and the special part (*Besonderer Teil*), in which the individual offences are set out each with their own specific required elements.

As one of the core principles of the StGB, the *nullum crimen* principal has a strict application, in the sense that laws must be codified in order to be a valid basis for a crime. However there is an exception to this prohibition of retroactivity when it comes to the 'most serious international crimes'.[41] From the fact that modes of liability have also been developed by way of doctrine, beyond the strict wording of the StGB, it would also seem that the *nullum crimen* principle is not considered to

---

[38] Weigend 2011, p. 94.

[39] Schermaier 2006, p. 275.

[40] Schermaier 2006, p. 273.

[41] Bohlander 2009, p. 26; This echoes the intention of art 7(2) ECHR and art 15(2) ICCPR, which are taught in German text books as the 'Nuremberg clauses', Bohlander 2009, p. 16.

be of strict application to these notions in the same way as they are to the definition of crimes.

7.3.1 *The Importance of Doctrine*

Because it follows the Roman codex tradition historically, the German legal system has maintained a fairly strict positivism, and the principal of legality is enumerated as *nullum crimen sine lege*,[42] but is generally understood to be augmented by an emphasis on *lex scripta*.[43] This falls under what Damaška describes as logical legalism, whereby individualised justice is readily exchanged for a greater degree of consistency across a wider class of cases.[44] The approach to law as a science, with a correct answer to each legal problem (rather than an individualised response to each legal problem) is typical of the civil law tradition in general, and German law in particular.[45] Much of the concern of German criminal law science is the correct categorisation of a particular problem in order to resolve it, according to the method of subsumption. This is also the case with the development of modes of liability, which have had to find a place within the existing system in order to be accepted and applied.[46]

The notion of justice in the German tradition is therefore less concerned with procedural rules, and instead could be said to be embedded in a kind of natural law, which works as a safety valve in a legal system which is otherwise strictly positivistic. 'Natural justice' is considered to be the guiding principle in applying any method of interpretation that may lead to departure from a strict textual reading.[47] It has been claimed that Germany is not only ruled by law and order, but that Germans themselves identify very strongly with the importance of the principals at the core of their system, and with the commitment to (natural) justice underlying it.[48] If the fundamental principles of justice and fairness are violated in a given case, a defendant can bring an appeal to the constitutional court, or make prejudicial requests for preliminary rulings.[49] The Constitutional Court (*Bundesverfassingsgericht*) has the

---

[42] *'Eine tat kann nur bestraft werden, wenn die strafbarkeit gesetzlich bistimt war, bevor die Tat begangen wurde*, 'An act can only be punished if it has been lawfully criminalised prior to the commission of the act' (Author's own translation.) Grundgesetz 1949, § 103.

[43] For discussion of the reading of *lex scripta* as part of this requirement, see Schumann 2005, p. 387; Bohlander 2009, p. 11.

[44] Damaška 1984, p. 22.

[45] Dubber 2005, p. 1053.

[46] Dubber 2005, p. 1067.

[47] The notion of 'natural justice' refers to a background reliance on morality and universal principles to guide the content of the law, and is not intended to refer to 'natural law' in the Lockean sense, whereby the ultimate source of the normative content of the law and of primary rules can be traced to divinity. See Bohlander 2009, p. 13.

[48] Schermaier 2006, p. 273.

[49] Bohlander 2009, p. 24.

power to strike down acts of parliament which are deemed unconstitutional, which may be the case if one of these fundamental principals is violated.[50]

In order to fill in this guiding principle of justice, judges look to the works of scholars, and the primacy of codified law is thus nuanced by the importance of doctrine in developing the law. Typically a system of 'professor's law', German academics produce extensive commentaries on the laws, analyses and critiques of judgments, which are then cited in later judgments. Through a combination of natural law and the historical school developed by Savigny in the nineteenth century, the development of a particular German *Rechtswissenschaft* has emerged over two centuries as a scientific and dogmatic approach to the law as an inherently logically structured system.[51] Although professors may have less influence today than in the past, the logic of analysis offered by scholars and their development of doctrine can convince judges to depart from previous lines of judgment, and informs the judiciary both formally and informally. Judgments in appellate courts often contain long passages debating theories asserted by various law professors to justify their decisions, and many important notions have been developed by scholars in the literature and subsequently applied in the courts,[52] not the least of which is the notion of *Organisationsherrschaft*, which will be discussed in detail below. This demonstrates the important role played by academic participants in the process of the law: academia is seen as the 'guiding light' for the judicial process.[53]

The primacy of legislation is therefore nuanced by the role of doctrine. Indeed, while there were suggestions made during one wave of legislative reform to include definitions in the StGB regarding different mental states, these were rejected explicitly in order to allow continued development through the legal science of doctrine.[54]

Furthermore the judicial case law can be seen as a supplementary source of law: case law can never form new criminal offences because of the emphasis on the *scripta* requirement of *nullum crimen*. A judge can therefore only apply new doctrine or existing interpretation of a legal concept, by citing recognised authorities. In other words, case law can be instructive in the affirmation of doctrine and the interpretation of modes of responsibility, but it is not the case law itself which is considered to develop new law. In terms of mapping who makes the law and where it is made, this provides a very different picture from the iterative process observable in the common law tradition. The authority given to scholars and the processes by which these participants develop the law should be kept in mind when considering the multitude of participants in the processes of ICL.

[50] Grundgesetz 1949, Articles 20(3), and 101–104; Reiterated in Strafgesetzbuch 1870, § 1 and § 2.

[51] Dubber 2005, p. 1051; Koschaker 2000, p. 276.

[52] Dubber 2005, p. 1057; Weigend 2006, p. 211.

[53] Bohlander 2009, p. 9.

[54] Dubber 2005, p. 1054.

7.3.2 *Culpability as Blameworthiness*

A scientific, dogmatic approach to the system of criminal law inevitably leads to a carefully considered system of liability, however it has also been subject to shifts in the doctrine due to historical shifts in the 'science' of law in Germany. In the nineteenth century, the doctrine favoured a descriptive, or 'psychological' understanding of blameworthiness, which was understood to denote intention or negligence.[55] This had some similarities to the descriptive understanding of the principle of culpability in the US as referring to the requisite mental state, but it went further doctrinally. The tripartite definition of liability was understood as distinguishing between unlawfulness and blameworthiness, (the objective and subjective elements of the crime, respectively), and unlawfulness was further divided into the definition of the offence (*Tatbestand*) and the wrongfulness of the specific act (*Rechtswidrigkeit*). Responsibility was seen as a precondition for blameworthiness, but not as a component of it.[56]

This changed in the twentieth century, when scholars began to criticise the psychological, descriptive understanding of culpability. This was partly because it was problematic for the notion of negligence, which is not a mental state at all, and partly because it does not take into account actors who may commit a crime with the requisite intent, but under duress.[57] For the same reason, understanding responsibility as a precondition for blameworthiness, but not as a component of it, is problematic in cases where a person may act with intent but be mentally ill, and therefore not 'responsible' in any moral sense.[58] These criticisms led to the development of a normative understanding of culpability; blame became synonymous with culpability, and lead to the understanding of responsibility as a reflection of the moral position of the actor. Heavily based on the scholarship of Hans Welzel, who stated that 'the will is the backbone of the act', this became known as the finalist approach.[59] According to this understanding, the act is still seen as the most important starting point of criminal responsibility, but it has both a subjective and an objective side, and blameworthiness is a separate test: *Vorwerfbarkeit*, or whether a person can fairly be be blamed for their actions.[60] Thus blameworthiness in the sense of moral culpability *leads to* responsibility, rather than responsibility being a precondition for blame.

The tripartite definition of liability is therefore now understood in a slightly different construction from when culpability was interpreted to be descriptive. A charge against a defendant is now subject to a three stage examination consisting of (1) the *Tatbestand*, or the elements of the crime (the subjective mental element and the

---

[55] Vormbaum and Bohlander 2014, p. 131.

[56] Vormbaum and Bohlander 2014, p. 131.

[57] Vormbaum and Bohlander 2014, p. 132; Ambos 2006, p. 2652.

[58] Fletcher 2000, p. 455; Vormbaum and Bohlander 2014, p. 132.

[59] Welzel 1970, p. 33; Ambos 2006, p. 2650. The ontological assumptions of 'finalist' thinking have been since rejected, a discussion which goes beyond the scope of this study, however the systematic consequences are still considered to be correct. See Ambos 2006, pp. 2653, 2655.

[60] Vormbaum and Bohlander 2014, p. 132.

objective act requirements); (2) the *Rechtswidrigkeit*, or (objective) wrongfulness of the act; and (3) *Schuld*, or the (subjective) blameworthiness of the actor, including the absence of any exculpatory defence such as insanity, duress, mistake of law etc.[61]

This shift in thinking from a descriptive understanding of blameworthiness as reflecting mental states, to a normative understanding of blameworthiness as moral culpability, led to a refinement in the criminal offences and in the definition of liability, since more individual problems could now be resolved within the system as a whole.[62] It also represented an acceptance of the moral content of the law, both in terms of the definitions and in terms of the decisions to be made by judges in determining liability. The system became more clearly differentiated and also more flexible.[63] Liability, which justifies punishment, is determined according to degrees of blameworthiness. As a reflection of this, the principle of culpability is understood to mean 'no punishment without guilt' in the sense of blameworthiness.[64] It is argued by various authors that the punishment must never exceed the measure of the actor's guilt, (*Schuld*) or blameworthiness,[65] but that it must also not fall short of the degree of blameworthiness, because the punishment that does not live up to the extent of culpability is not appropriate in terms of retribution and compensation towards society.[66]

One characteristic of a tripartite system is that the three elements must be addressed in the specific order given; first the definition of the crime must be fulfilled by the acts and the mental state of the accused, second the wrongfulness of the act (the absence of any justification), and only finally the question of culpability, or *Schuld*, (and the absence of any excuses). Thus, moral culpability as blameworthiness is conceived of as separate from wrongdoing, but dependent upon it. A person can only be culpable for a wrongful act, and if the act is justified and therefore not wrongful, there can be no culpability, i.e. an acquittal. This means culpability is a question of attribution;[67] not of the act as we saw in the common law tradition, but rather of responsibility *for* the act.

As the *Bundesgerichtshof* (BGH) has put it:

> Punishment is premised on guilt. Guilt means blameworthiness. By finding a defendant guilty, we blame him for not having acted lawfully, for having chosen to break the law although he could have acted lawfully, could have chosen to abide by the law. The inner reason for the judgment of guilt lies in the fact that human nature is grounded in the freedom and responsibility of moral self-determination, and that we are therefore capable of deciding for the law and against injustice, of modelling our actions on the norms of the legal commands and avoiding that which is forbidden by law.[68]

---

[61] Bohlander 2009, p. 16.

[62] Vormbaum and Bohlander 2014, p. 132.

[63] Vormbaum and Bohlander 2014, p. 148.

[64] Krey 2002, p. 97.

[65] Roxin 2005, § 3.

[66] Krey 2002, p. 113; Jescheck and Weigend 1996, p. 82.

[67] Fletcher 2000, p. 459; Fletcher 2007, p. 218.

[68] Schuldprinzip 1952, p. 201 (Author's own translation).

7.3.3 *The German System of Modes of Liability*

With this normative notion of personal blameworthiness at the core of German criminal law, the existence of a categorical normative division between principal (*Täter*) and secondary participant (*Teilnahmer*) is perhaps easy to understand. Actors can be complicit in the commission of a crime in many ways, but the degree of their blameworthiness, measured by the extent of their contribution to the crime, will determine whether they are principals or 'mere' participants in a normatively secondary sense. The culpability of an accomplice is independent of that of the principal or any other participant in the crime; each participant is punished according to his or her own guilt.[69] This means that only a limited derivative or accessorial theory applies, that is to say, the principal must at least have committed an act that amounts to an attempt of the crime in question, but the liability of a participant is not dependent on the liability of the principal.[70] The sentencing is, however, interdependent. The maximum sentence for a secondary party is three quarters of the statutory maximum for the crime in question, and the actual sentence for a person convicted as a secondary party is determined with reference to the actual sentence given to the principal.[71]

As noted above, the concern for categorisation in the German legal science tradition leads to a complex taxonomy of modes of liability. Only professional judges schooled in the law can make determinations on both fact and law, and their decisions are informed by scholarly doctrine authored by professors specialised in criminal law science. Compared with lay juries, judges can be expected to provide the correct reasoning in their judgments, and to apply the system in a consistent manner.

7.3.3.1 Principals

There are various ways in which a party can fulfil the definition of principal actor.[72] §25(1) StGB describes the *Täter* as the person who, alone or through another, commits the crime.[73] This short sentence, which was modified in 1975 to include the words 'or through another',[74] includes two forms of principal liability; direct and indirect perpetration. The person who fulfils all the elements of the crime alone is obviously a direct (sole) perpetrator. The indirect perpetrator will be given due attention below, since it is a mode of participation highly relevant to the central problem of this enquiry.

---

[69] *Jeder Beteiligte wird onhe Rücksicht auf die Schuld des anderen nach seiner Schuld bestraft*; Strafgesetzbuch 1870, § 29.

[70] See below Sect. 7.3.3.3 for a more detailed discussion of the limited accessorial theory.

[71] Strafgesetzbuch 1870, § 49(1).2.

[72] For the full text of the German StGB and a diagram of the relationship between the modes of liability, see the appendix A.4.

[73] *Als Täter wird bestraft, wer die Straftat selbst oder durch einem anderen begeht;* Strafgesetzbuch 1870, § 25(1).

[74] Nack 2006, pp. 342–345.

§25(2) describes the co-perpetrator, or *Mittäter*. The relationship between co-perpetrators is a horizontal one, where all the participants have a more or less equal and hypothetically interchangeable role.[75] If a crime is committed by several actors jointly, they are each fully culpable as principal perpetrators.[76] This form of co-perpetration is recognisable in many criminal law systems which share the civil law tradition, and is a way to lift the burden of evidence from the prosecutor who can show that the accused were involved in the crime, but is unable to prove exactly who did what. Under German law, all the actors are jointly culpable if (a) there is a shared plan to commit the crime in question, (b) the contribution by each of the actors has a causal impact on the commission of the crime, and (c) the intention of each actor is to influence the commission of the crime.[77]

7.3.3.2 *Teilnahme*: Secondary Participation

There are two forms of secondary party liability under German criminal law. The first is found in Section 27 StGB: the 'aider' is defined as the person who 'intentionally renders assistance to enable another to intentionally commit a crime.'[78] The second is the *Anstifter*, or instigator, sometimes translated as abettor, defined as someone who 'intentionally incites another person to intentionally commit a crime'.[79] Both of these forms of liability require that the actor intentionally contributed to the crime, though the contribution is less than the causal impact of the principal's acts.[80]

According to the wording of §26, the *Anstifter* or instigator can receive the same maximum sentence 'as an equal to principal actors'. However in terms of liability, in fact this form of participation is considered to be 'lesser' than that of principal, because it is less blameworthy behaviour, and therefore it is classified a form of

[75] There is also a form of co-perpetration known as *Nebentäterschaft*, or multiple independent perpetrators who, independently from each other and without a common plan, embark upon the same criminal act and it is not clear whose actions led to the fulfilment of the crime. This is a specific form of perpetration which does not occur very frequently and which is not distinguished in the StGB, but only in the literature. It will receive no further attention here. See Bohlander 2009, p. 160.

[76] *Begehen mehrere die Straftat gemeinschaftlich, so wird jeder als Täter bestraft (Mittäter).* Strafgesetzbuch 1870, § 25(2).

[77] Bohlander 2009, p. 162.

[78] Author's own translation. The original reads: *Als Gehilfe wird bestraft, wer vorsätzlich einem anderen zu dessen vorsätzlich begangener rechtswidriger Tat, Hilfe geleistet hat*; Strafgesetzbuch 1870, § 27.

[79] Author's own translation. The original reads: *Als Anstifter wird gleich einem Täter bestraft wer vorsätzlich einem anderen zu dessen vorsätzlich begangener rechtswidriger Tat bestimmt hat*. Strafgesetzbuch 1870, § 26. Note that the terms instigator, abettor and the verb 'incites' are used interchangeably in English translations of this secondary mode of liability, though there seems to be preference for the term 'instigator', and for the Spanish term '*instigación*'; see Welzel 1970, p. 166; Jescheck and Weigend 1996, p. 677; Bohlander 2009, p. 161; Stratenwerth 2005, p. 413; Hamdorf 2007, p. 210.

[80] Bohlander 2009, p. 168.

Teilnahme, or secondary participation. The difference is one of qualification rather than sentencing,[81] which matters in terms of the moral re-telling of the facts in a conviction, and in terms of the elements required to prove a principal versus a secondary mode of liability. Although generally the sentencing of a participant in a crime is also meant to reflect the personal blameworthiness of the individual in the commission of the crime, in the end the qualification of a convicted person as principal or secondary party is where this reflection is strongest: a conviction as *Anstifter* does not carry the same moral weight as a conviction as principal perpetrator would, and the basis differs in important ways. The person who instigates or abets does not have the same causal impact on the crime as a principal perpetrator.[82]

This is demonstrated clearly in the *Mauerschützen* decision (the Berlin Wall shootings), discussed in detail below. The court went to great lengths to justify convictions of the members of the National Defence Committee as (indirect) perpetrators rather than mere instigators, even though this could have led to the same maximum sentence.[83] The difference may only have been 'in the name' as Professor Jakobs argued in his commentary on the case,[84] but there was clearly an intention to reach a judgement that would reflect the seriousness of the abuse of power on the part of the accused individuals which would not have been satisfied by qualifying the actors according to the mere secondary participation of *Anstifter*.

7.3.3.3 Derivative Nature of *Teilnahme* Liability

In terms of the nature of the liability of a secondary party for the acts of others involved in a collective crime, and especially when it comes to further, unplanned crimes committed by members of the collective, the question is raised to what extent the secondary party's liability is derivative of the principal's liability. As we saw under the jurisdictions compared in the common law tradition, there liability of parties to a crime amounts to a kind of vicarious liability, whereby any act of another member of the collective can be attributed to all other members of the collective. The derivative nature of party liability is very limited in the sense that all is required is a crime (or an attempted crime) on the part of any of the actors.

In a normatively differentiated system, however, the question must be answered differently. In the German StGB, all secondary participants are defined as either inducing (abetting) the principal to commit a crime, or assisting the principal in some way.[85] Thus, their liability is dependent on the commission (or attempt) of a crime by another, the principal.[86] In general, a 'limited' external understanding of

---

[81] Bohlander 2009, p. 161; Hamdorf 2007, p. 210.

[82] Jescheck and Weigend 1996, p. 677; Hamdorf 2007, p. 210.

[83] Mauerschützen case 1994, para (bb).

[84] Jakobs 1995, p. 26.

[85] Stratenwerth 2005, p. 413; Strafgesetzbuch 1870, §26 and §27.

[86] As discussed in Sect. 5.3.3, this is termed an external or quantitative form of accessory liability; as opposed to an internal or qualitative derivation of accessory liability, where all modalities

the derivative nature of party liability is accepted as being applicable in Germany: that is, the wrongfulness of the act of the principal is necessary in order for a secondary party to be liable, and this wrongfulness, or counter-factually, a ground for justification, would also apply to the accessorial secondary party automatically.[87] This does not reach as far as an 'internal' understanding of accessory liability:[88] the guilt of the principal need not be established in order for the guilt of an accomplice to be established; if the principal has not been charged, or if there is an excuse such as insanity or duress applicable, this does not affect the charge against a secondary party for assisting the crime. This is reflected also in §29 of the StGB which states that the guilt of each party to a crime is independent of the guilt of others.[89] Guilt is understood here to mean personal blameworthiness, and not the totality of liability.

7.3.4 *Objectivity and Control over the Crime*

As with any differentiated system, the line between co-perpetratorship and merely assisting in a secondary sense is difficult to draw; in some circumstances the driver of the getaway car is merely an assistant to others who commit a crime such as armed robbery, and in other circumstances the same driver may have had a key role in designing the criminal plan, which would raise the status from mere participation to co-perpetration.

The doctrine and the debates have shifted over time with respect to the doctrinal approach underlying a theory of liability. In the nineteenth century and early twentieth century a more subjective theory prevailed, whereby the question was asked whether the accused *intended* the crime as his or her own act, or whether the intention was to support someone else's crime.[90] If the latter was the case, then the actor would be qualified as a secondary party, and not as a perpetrator. It should be noted that this subjective theory is not the same as the subjectivity reflected in the common law tradition, which does not ask whether an actor intended the crime as her own, but merely asks if there was any shared intention to take part in the crime in any way. This subjectivity is a horizontal expansion of liability to include all actors as equally liable for all acts. The pure subjective theory that was applied in German criminal law, on the other hand, was a very specific theory applied to a normatively differentiated system of liability; a subjective test for discerning which actors should be punished

(Footnote 86 continued)
applicable to the principal such as the required intent, justifications and excuses, have immediate effect for the accessory. See further Welzel 1970, p. 165; Stratenwerth 2005, p. 416.

[87] Welzel 1970, p. 164; Bohlander 2009, p. 168; Sancinetti 1997, p. 60, citing Jakobs.

[88] See Sect. 5.3.3 for the full definitions.

[89] Strafgesetzbuch 1870, § 29.

[90] As decided and discussed in Staschynski 1962, para 134; Weigend 2011, p. 95.

as principals and which as secondary parties, and thus a restriction, rather than an expansion of liability.[91]

One extreme and somewhat controversial example of this is the *Staschynski* case.[92] In 1962 a Russian KGB agent was prosecuted before the BGH for the murder of two exiled Russian politicians in West Germany in 1957. Staschynski admitted to having murdered the two men, but claimed as a defence that he did so under direct order of his superiors in the KGB. The BGH applied a fairly extensive interpretation of the pure subjective test, which by then had already come into some disfavour,[93] and it determined that because Staschynski did not intend the crime as his own, described as '*animus auctoris*', but was carrying it out on behalf of another, described as '*animus socii*', he was merely a secondary party. He was therefore convicted as a secondary party for aiding rather than as a principal, and was given a sentence of 8 years, rather than the mandatory life sentence that would have otherwise applied for murder.[94] It is highly likely that there was a political impetus for this conclusion, since Staschynski had defected to West Germany in 1961 and the authorities may have wanted to demonstrate to other possible defectors that they would not receive the full punishment of the law, in the hope that they would divulge certain information about their actions and Russian intentions.[95] For this the court has been criticised, since it in fact shifted the defining line between principal and secondary party by applying an extensive pure subjective test for policy reasons that did not align with the system of liability as it was designed.[96]

Around the time of the *Staschynski* case, and certainly after this decision, there was an emerging preference in the doctrine for a more objective approach to liability. The so-called formal-objective theory dominated the thinking for some time, according to which liability is determined by the actual contribution to the crime; any person who fulfils the objective requirements of the crime is a perpetrator, and any person who does less, is a secondary party.[97] However this theory is limited because it does not explain how a person who uses another party as an innocent instrument to achieve the execution of a crime could be held liable, since such a person does not contribute to the crime through his or her own actions directly. Also, a formal-objective theory would hold an innocent agent liable as a principle, since her own actions are the physical causation fulfilling the crime, despite the fact that she is not fully to blame. This is the problem of the *Hintermann*, or the actor behind the physical actor. Yet most followers of the formal-objective theory will admit that such a phenomenon occurs and that the physical actor being used as an instrument is less blameworthy, whereas the actor behind the innocent agent is blameworthy for the crime and should

---

[91] Sancinetti 1991, p. 634, citing German theorist Edmund Mezger.

[92] Staschynski 1962.

[93] Bohlander 2009, p. 162; Hamdorf 2007, p. 210.

[94] Staschynski 1962, para 152. See also Roxin 2005, p. 563.

[95] Bohlander 2009, p. 162.

[96] Roxin 2006, p. 400.

[97] Stratenwerth 2005, p. 369.

be fully punished as a principal.[98] In a system which preferences internal consistency of the system as a whole, this contradiction is problematic.

7.3.4.1 Defining Indirect Perpetration

Indeed in 1975, the legislators amended §25 of the StGB to include the person who commits the crime indirectly 'through another' in the definition of a (principal) perpetrator.[99] During the drafting process there was disagreement as to whether indirect perpetration should be defined extensively, to include cases in which the physical actor is fully culpable, or restricted to cases where the physical actor is an innocent agent. The open-ended formulation eventually decided upon was a compromise, partly due to the doctrinal debates that were taking place at the time, and which had not yet been resolved one way or the other.[100] Interpretation of this amended legislation therefore needed a doctrine more sophisticated than the formal-objective theory, which required actual contribution to the crime.

Some German theorists had attempted to solve this problem by expanding upon the notion of what constitutes committing a crime, by saying that the author of a crime is the person who executes the crime himself *or* through another; the indirect perpetrator therefore executes the crime in their own name but through another person.[101] However this interpretation resembled the pure subjective theory, since it relied upon the intention of the *Hintermann* to have the crime committed in his own name. This could be seen as a subjective-objective approach, but it didn't do much to solve the objections to the pure subjective theory.[102]

Others considered the answer to be in a re-classification of the *Hintermann* as a co-perpetrator, since the contribution of this intellectual author could be considered to be sufficiently 'significant' to amount to joint control and co-perpetration together with the physical perpetrator.[103] These authors do not consider indirect perpetration to be applicable in scenarios where the physical perpetrator is fully culpable for her actions, since this would be extending the principle of responsibility too far. Rather this more classical understanding of perpetration 'through another' required the innocence of the physical actor and would render the *Hintermann* a secondary party in cases where the physical actor is fully responsible.[104] However it could be argued that there is not the requisite horizontal relationship between the *Hintermann* and the physical

[98] Sancinetti 1991, p. 641; Stratenwerth 2005, p. 370.

[99] Nack 2006, p. 342.

[100] Mauerschützen case 1994, para aa.

[101] Nack 2006, p. 342. Nineteenth-century authors such as Frans von Liszt and Wilhelm Sauer were followed in this, and they had also found a following among some Spanish authors, and some acceptance in Argentinean doctrine in the early twentieth century, see Sancinetti 1991, pp. 642–643

[102] Roxin 2005, p. 58.

[103] Jakobs 1995, p. 26; Jescheck and Weigend 1996, p. 670.

[104] Jescheck and Weigend 1996, p. 670.

perpetrator for there to be a relationship of co-perpetration.[105] There is no common decision to jointly commit the crime, which is an important requirement for co-perpetration.[106] In a situation where the physical perpetrator is acting under control of the *Hintermann*, the latter is the one with the most influence, and the physical actor can no longer be said to be equally blameworthy, acting as a co-perpetrator.[107]

In the end, the interpretation that won out was informed by authors such as Hans Welzel,[108] Günter Stratenwerth[109] and Claus Roxin,[110] all of whom had written extensively about a more expansive material-objective theory, which became known as *Tatherrschaft*, or 'control over the crime'. Under this theory the test of liability is a more functional one; who was in a position to ensure the fulfilment of the crime, either by being the physical perpetrator, or indirectly, by exercising sufficient control and influence over the will of the physical actor?[111] In order to determine the liability of an indirect perpetrator as distinct from an instigator or abettor, the normative theory of culpability is applied. The physical actor is the one who carries out the objective elements of the crime. However if she was acting without blame (remembering that *Schuld* is the key notion in establishing culpability) due to an excuse such as duress or mistake of law, the physical actor is not punishable and is therefore an 'innocent agent'.[112] Indirect perpetration can also occur if the physical actor is fully responsible for her acts, and therefore not an innocent agent, but she is excused by way of a mistake of law. If A is the manager or owner of a small business and she uses her employees to carry out fraudulent deliveries or fraudulent payments to clients, and the employees are unaware of the unlawful nature of the instructions they are carrying out, then the employees are the innocent agents despite being the physical actors, and A is the indirect perpetrator due to her intention and control over the act, according to the test of (*Tatherrschaft*).

Because in a normatively differentiated system the qualification is of importance in terms of the moral weight of the conviction, in the German literature, and in the case law that applies the literature, the doctrine of *Tatherrschaft*, or control over the act, has been developed to draw a line of distinction. When an actor has sufficient influence or control over the execution of the crime, this is considered to be a form of perpetration as a principal, even if the physical contribution may have been nihil.[113] Thus, where the *Hintermann* has such influence or control over the will of the physical actor that the the blame rightfully falls on the former's shoulders, a new explanation for the liability that should arise for this culpability was needed.

---

105 Roxin 2005, § 25, No. 120.

106 Roxin 2005, § 25 No. 107.

107 Ambos 2008, p. 28; Olásolo 2009, p. 286.

108 Welzel 1970.

109 Stratenwerth 2005.

110 Roxin 2005.

111 Stratenwerth 2005, p. 374, citing Welzel.

112 Welzel 1970, p. 171; Roxin 2005, p. 80; Hamdorf 2007, p. 216.

113 Hamdorf 2007, p. 210; Weigend 2011, p. 95.

7.3.5 Roxin's Theory of Organisationsherrschaft

The academic doctrinal debate on this question had already been posed in the 1960s, and it was in fact due to the debate that the legislators in 1975 were faced with the challenge how to define the indirect perpetrator. The final wording inserted into §25(1) that a principal could commit an act alone 'or through another' was a compromise, leaving the refinement to doctrine, and in the end it was Roxin's scholarly contributions which were the most influential.[114]

Roxin had first articulated his theory of *Organisationsherrschaft*, or 'perpetration by means of an organisation', in a lecture titled 'Crime in the Framework of Organized Apparatuses of Power,' given in Hamburg in 1963. Forty years later, he recalled this when lecturing at Lusíada University in Lisbon, Portugal, when he said:

> the historical example that came to mind as I was developing this form of perpetration-by-means was the violent power of National Socialism. Whenever Hitler or Himmler or Eichmann gave orders to kill, they could be sure they would be carried out .. In my conception, here, a perpetrator-by-means is anyone who is situated at the helm of a power apparatus—regardless of the hierarchical level—and is able, by giving orders, to cause crimes to be committed, notwithstanding the individual identity of the direct perpetrator.[115]

The trial against Eichmann in Israel was a particular inspiration for Roxin to develop his theory more substantially.[116] Eichmann was *Obersturmbannführer* (equivalent to Lieutenant Colonel) in the SS and was prosecuted before the Israeli courts for coordination of the registration and deportation of Jews from all over Europe to concentration camps in Russia and Eastern Europe. His involvement in the 'Final Solution' was not that of primary policy-maker of the highest rank, but rather an administrator and coordinator of a middle rank.[117] The charges against him were nonetheless aimed at holding him responsible as decision-maker and overseer, as he was convicted as a principal (co-)perpetrator for his acts 'together with others, caus[ing] the killing of millions of Jews, executing the plan known as "the final solution of the Jewish problem" with intent to exterminate the Jewish people'.[118]

While the District Court in Israel admitted that it would not be proper to put responsibility for the entire holocaust on the shoulders of one individual, it also asserted that 'the guilt of the others does not lessen by one jot the guilt of the accused'.[119] The conviction of Eichmann as principal co-perpetrator was based on the effective control he had over those under him. Both at trial and in appeal, the courts explicitly excluded the accessory modes of responsibility of instigator, aider or abettor as inappropriate.[120] Instead the emphasis was placed on the fact that these

---

[114] Nack 2006, p. 343.

[115] Cited in Etchecolatz 2006, para 1d.

[116] Rotsch 2009, p. 551.

[117] Urban 2004, p. 29.

[118] Eichmann 1961, p. 271.

[119] Eichmann 1961, p. 272.

[120] Eichmann 1961, p. 272; Eichmann Appeal 1962, p. 332.

were crimes of mass atrocity, and that Eichmann's role in the systematic organisation and execution thereof was of such importance that he had full responsibility for these mass crimes.[121] It was from this reasoning that Roxin took inspiration.

In appeal Eichmann contended that he was only a small cog in the Nazi machine and that 'he functioned as a minor official without any independent initiative, and that nothing but the compulsion of orders and blind obedience to commands from above guided him in carrying out his work in all stages.'[122] While it is possible that the District and Supreme Courts of Israel exalted him to a higher and more powerful level than his position in fact allowed him in terms of decision-making and influence,[123] his position as head of the Department for Jewish Affairs and Evacuation Affairs clearly meant Eichmann was no mere accessory.[124] The physical actors effectively carried out Eichmann's orders and wishes as if they were agents of the *Hintermann*, and it was considered irrelevant what the intent or responsibility was of these agents.[125] It is a case illustrative of the dynamic whereby the physical distance between the accused and the execution of the crimes in fact increased his responsibility due to his role as coordinator.[126]

When Roxin developed this theory extensively in his writings in 1965, he made a tripartite distinction between the direct control a perpetrator can have over the other actors, the functional control he can have over the act itself, and the control he can have over the will of the physical actors. In the first form, when a person controls the other actors, when there is duress or other means of manipulating the physical actor or the circumstances, then this could either amount to *Anstifter* (a mode of secondary participation) or to indirect perpetration based on §25 StGB.[127] In the second form, when a person has functional control over the act, this refers to a planning or co-ordinating role in the fulfilment of a crime committed jointly with others, even if this person is not present at the scene of the crime.[128] Such a person is considered to be co-perpetrator. The third form of effective control distinguished by Roxin, that of effective control over the *will* of the physical actors, and not necessarily over their actions or the act itself, is the key to his notion of indirect perpetration by virtue of a hierarchical organisation.[129]

Under all three forms of effective control, if the physical actor acts under duress or due to mistake of law, the *Hintermann* is criminally liable specifically because the physical actor is not.[130] This was the undisputed interpretation of §25 StGB, since the physical actor is then an innocent agent. However this was too restrictive an

121 Roxin 2005, p. 247.

122 Eichmann Appeal 1962, p. 278.

123 Arendt 1964, p. 40.

124 Eichmann Appeal 1962, p. 330.

125 Roxin 2005, p. 272.

126 Urban 2004, p. 29.

127 Roxin 2005, p. 132.

128 Roxin 2005, p. 301.

129 Roxin 2005, p. 242.

130 Roxin 2005, p. 243.

interpretation according to many academic commentators and judges, since it leaves out the problematic cases discussed in the *Mauerschützen* decision (see below). Attempts had already been made to extend this to cases where the physical actor is fully culpable for her own acts and therefore not an 'innocent agent', by way of the subjective theory of *animus auctoris*. But Roxin pointed out that this is a false solution to the problem. The fact that the person who manipulates the situation has an interest in the fulfilment of the crime is not a sufficient criterion to distinguish the indirect perpetrator from the instigator or abettor, who also has the same interest, yet is a 'mere' accessory. And at the same time the intent of the physical actor should not determine the responsibility of the *Hintermann*, especially given the fact that under the German criminal code the guilt of each participant is independent of the guilt of others.[131] The predisposition of the physical actors to carry out the crime is also an insufficient criterion according to Roxin, for the very same reason.[132]

Instead Roxin argued that the differentiation can only be made based on the notion of control over the crime by looking at the actual influence and effective control that the *Hintermann* has over the way in which subordinates within the organisation operate. This is not a purely objective theory, since it goes beyond the direct, physical causation that a formal-objective theory requires, yet looking to 'control over the crime' as the determinative factor still reflects objectivity as a central approach. The way in which a leader such as Hitler controls the structural organisation in order to commit certain crimes, is very different from the way in which an instigator influences individuals. Of central importance is the fact that an organisation has a virtual life of its own, independent of the individuals of which it consists.

The very interchangeability, or fungibility, of these individuals means that the leader is not instigating a criminal act in agreement with certain individuals, dependent upon their willingness or predisposition to carry out the crimes, but rather the leader is using her influence over the organisational apparatus as a whole. Should one of the subordinates refuse to participate, or withdraw from the organisation as a whole, this would not influence the commission of the crime since any other individual can step into the place of he who refuses.[133] In such a way the indirect perpetrator exercises effective control over the will of the organisation as a faceless apparatus.

The fungibility of the physical actors is in fact the deciding factor in determining whether the *Hintermann* has effective control over the will of the physical actors. At first sight this may seem like a very abstract link to make, but Roxin argued there are only three ways in which an indirect perpetrator can ensure the crimes she intends to be committed will actually be fulfilled without using her own hand; by way of duress, by way of putting into place the required circumstances for the fulfillment of the crime in such a way that the physical actors are unaware of this (manipulating the situation, such that the physical actors would act under mistake of fact, becoming innocent agents), or the physical actors must be replaceable and acting

[131] Strafgesetzbuch 1870, §29.

[132] Roxin 2005, p. 244.

[133] Roxin 2005, p. 245.

as mere cogs in the machine.[134] In this third case, where there is no defence of duress or mistake, the culpability of the physical actor who carries out the crime in full knowledge and with full responsibility is irrelevant for the culpability of the indirect perpetrator. The *Hintermann* is the central player in the occurrence of the crimes, and uses the structural apparatus of the organisation, which operates independently from its individual members, to commit the crime.[135]

Uncoupling the criminal liability of the *Hintermann* from the liability of the physical actors is the most important aspect of Roxin's theory. The grounds upon which he based this theory have been most at issue. Friedrich-Christian Schroeder, an influential professor of criminal law, gave a slightly different definition of the same notion of *Organisationsherrschaft*.[136] Essentially the indirect perpetrator's use of the systematic hierarchy to commit a crime remained central to Schroeder's analysis, but rather than the fungibility of the subordinates as a determining factor, Schroeder focused on the predisposition of the subordinates within an organisation to carry out the crime. Without this willingness on the part of the physical actors, according to Schroeder the *Hintermann* can only be considered to be an instigator rather than an indirect perpetrator punishable as a principal.[137]

In his later writings, Roxin conceded that the requirement of the replaceability of the subordinates within an organisation was perhaps too strict when it came to commercial and corporate organisations. While he maintained this as a sufficient condition, in the end he in fact agreed with Schroeder that it is not a necessary one and could be supplemented by showing the predisposition of the subordinates to carry out the crimes in question.[138] In his first conception of *Organisationsherrschaft*, Roxin had specific political and military organisations in mind, following the atrocities of the Second World War. It is likely that the concessions he made in his later writings were a result of the fact that his theory was being applied in a more flexible manner by the German courts to corporate criminality.

7.3.5.1 Roxin's Theory Applied in the *Mauerschützen* case

In 1994 the BGH was forced to take an explicit stance in resolving this dilemma, when the *Mauerschützen* case was brought before it.[139] It was clear to the judges that their decision would be of great doctrinal and political importance. The members of the National Defence Committee had been prosecuted for the multiple murders of East German citizens who were either shot, or killed by mine explosions upon attempting to cross the internal East-West German border. The members of the Committee had handed down an administrative decision in which the policy of 'shoot to kill' was

---

134 Roxin 2005, p. 245.

135 Rotsch 2009, p. 549.

136 Friederich-Christian Schroeder, '*Der Täter hinter dem Täter*', cited in Rotsch 2009, p. 550.

137 Rotsch 2009, p. 550.

138 Claus Roxin, 'Strafrecht, Allgemeiner Teil, Band 2' (2003), cited in Rotsch 2009, p. 551.

139 Mauerschützen case 1994.

determined, and the instruction to lay mines along some areas of the border was given. The members knew of the deaths that ensued, since they were presented with reports confirming that their decision was being carried out, and furthermore the soldiers were informed that any successful escape across the border would have consequences for them.[140] When it came to the question of qualification, the problem in applying the definition of the modes of criminal liability as they were understood at the time, was that there were political and moral pressures to convict those most responsible according to the full extent of their role.[141] *Anstifter*, being a mere secondary form of being party to the crime, was an unsatisfactory classification.

One key problem was that the soldiers who shot or who laid the mines under instruction were also fully aware of the consequences of their actions, and were fully culpable, with no defences available to them which would lead to their status shifting to that of 'innocent agent'. This would have rendered the members of the Committee merely secondary parties. Moreover, in their search for a solution that would be doctrinally sound and still ensure that those most responsible were convicted as principals, the judges were aware that the reasons they gave would have farther reaching consequences than this one judgment. They were in search of a formulation that could also be applied to other cases, such as the rising number of fraud cases where the manager or owner of a business 'used' the employees to carry out fraudulent deliveries, but where the employees were also fully aware of these activities.[142] These actors were not being used as innocent agents, yet they were clearly not the persons most responsible for these white collar crimes. Thus the judges turned to a development in the doctrine in order to expand upon the application of indirect perpetration, and used the theory of *Organisationsherrschaft* to solve their doctrinal and practical problem. In doing so they recognised this form of perpetration formally in German law.[143]

The judges began by pointing out that the question whether the *Hintermann* can be punishable as indirect perpetrator when the physical actor is fully responsible and culpable, was left deliberately open by the legislator when §25 StGB was modified in 1975 to include 'perpetration by means of another'.[144] In the academic debates that followed there was some consensus that an accused who uses the 'conditions of organisational power structures' in order to ensure a crime is committed, should fall under the definition of indirect perpetrator, regardless of whether the subordinates in the organisation who were physical actors were culpable and responsible for their acts.[145] In considering the possible interpretations of §25 StGB and the notion of perpetration by means of another, the BGH referred to the writings of many well-known and well-respected scholars,[146] however it was Roxin's theory which

---

140 Hamdorf 2007, p. 211.

141 Hamdorf 2007, p. 211; Weigend 2011, p. 101.

142 Nack 2006, p. 343.

143 Rotsch 2009, p. 550.

144 Mauerschützen case 1994, p. 2704.

145 Mauerschützen case 1994, p. 2706.

146 Including Jescheck, Wessels and Stratenwerth: Mauerschützen case 1994, p. 2706.

offered the most comprehensive basis for the notion of perpetration by means of an organisation.

The distinction made by the BGH was the following: in general when the physical actor fully controls and intends to control the events which have to do with fulfilling the crime in question, the *Hintermann* cannot be considered to be the one who controls the act. In this case the *Hintermann* is a mere *Anstifter*. There are, however, cases in which the contribution on the part of the *Hintermann* more or less automatically leads to the fulfilment of the crime he or she intends. This is the case when the *Hintermann* takes advantage of the hierarchical structures and conditions of the organisation, 'within which the contribution [of the *Hintermann* towards the crime] triggers a certain chain of systematic processes'.[147] In particular the court pointed to State, business and corporate organisational structures, as well as chain of command hierarchies. According to the court, if the *Hintermann* acts in full knowledge of these circumstances, exploits the willingness of the physical actor to carry out the illegal act, and intends the fulfilment of these acts to be a result of his or her own actions, then the *Hintermann* must be considered to be an indirect perpetrator, because he or she has control over the act (*Tatherrschaft*).[148]

In fact, the court pointed out, in such cases the *Hintermann* has more control over such acts than in other cases where indirect perpetration is assumed without further question, for instance for tax fraud or economic crimes where the only reason the physical actor is considered to be an 'innocent agent' is because one of the requisite personal characteristics for the crime is lacking (the physical actor is not the addressee of the statutory provision and therefore an element of the crime is missing).

> Not to treat the *Hintermann* as principal actor in such cases is to do injustice to the objective weight of his contribution, especially considering the fact that responsibility is not reduced, but rather increased, when there is a greater distance [between the *Hintermann* and] the scene of the crime.[149]

With this extensive interpretation of the indirect perpetrator, the judges were able to cover various other hypothetical crimes for which it had been difficult to hold the most responsible actors accountable. Examples they suggested were crimes committed by mafia syndicates, and white-collar economic crimes.[150] Applying the mode of responsibility 'co-perpetrator' would mean considering the division of labour, which would leave the leaders of the organisation—those who give orders or determine policy—free of principal culpabilitydue to the spatial, temporal and hierarchical

---

[147] Author's own translation. Original text reads: '*Solches kann vorliegen, wenn der Hintermann durch Organisationsstrukturen bestimmte Rahmenbedingungen ausnutzt, innerhalb derer sein Teilbeitrag regelhafte Abläufe auslöst.* Mauerschützen case 1994, p. 2706, para bb.

[148] Mauerschützen case 1994, p. 2706, para bb.

[149] Author's own translation. Original text reads: '*Den Hintermann in solchen Fällen nicht als Täter zu behandeln, würde dem objecktiven Gewicht seines Tatbeitrags nicht gerecht, zumal häufig die Verantwortlichkeit mit gröβen Abstand zum Tatort nicht ab-, sondern zunimmt.* Mauerschützen case 1994, p. 2706.

[150] Mauerschützen case 1994, p. 2707.

distance between them and the physical actors. A conviction as indirect perpetrator would, according to the judges of the BGH, better reflect the role of such leaders and their use of the organisational hierarchies and power structures to commit a crime.

The key to this form of indirect perpetration is the extent to which the *Hintermann* has the capacity to exercise effective control over the organisation. What is being attributed to the indirect perpetrator is the responsibility for the crime,[151] not the act itself nor the personal (subjective) guilt of the physical actor. Since personal blameworthiness is the pivotal notion upon which the criminal liability of an actor turns in German law, it appears that the guilt of the indirect perpetrator is constructed from the attribution of responsibility, due to the measure of effective control this indirect perpetrator has over the organisation which is used to commit the crime. In other words, the *Hintermann* does not have to issue specific instructions, nor have direct control over the physical actors themselves in order to be considered criminally responsible as a principal. Where there is a lack of control over the individual physical actors, there may be sufficient control over the power apparatus such that the *Hintermann* can orchestrate the commission of crimes by subordinates.[152] It is in fact control over the act 'by virtue of' or 'by proxy through' a hierarchical organisation.

7.3.5.2 The Impact of *Organisationsherrschaft*

The application of Roxin's theory by the BGH made a significant impact on German law in the way the court had predicted, in terms of crystallising the doctrine and affirming its place in the law. However, despite it being the first time it was applied in a German court, it was not the first time it had been applied in a judicial context. The influence of German law and theory on other countries borrowing and transplanting into their own domestic systems was mentioned at the beginning of this chapter, and the application of the doctrine of *Organisationsherrschaft* is a striking example of this, since the theory was applied by the Argentinean judiciary already in 1983.[153] In fact it could be said that a kind of dialectic between these domestic jurisdictions took place, since after the successful application of this doctrine in Argentina, the German courts probably had more confidence in accepting the doctrine.

This doctrine has further been received and applied in various Latin American jurisdictions to deal with the specific problem of holding accountable those political leaders who have a great deal of influence and control over the creation of a system of mass violence, but whose direct involvement in the crimes would be more difficult to prove.[154] Because it has been applied specifically to mass atrocity crimes, it has also now been applied in the ICC, with direct citation of Roxin's writings.[155]

---

[151] Ambos 2007, p. 183.

[152] Ambos 2007, p. 183.

[153] See the following section on Argentinean law.

[154] For the impact on other Latin American States, see the next section on Argentina.

[155] See Chap. 8, below, on modes of liability applied in International Tribunals.

Within Germany, not only has the 'control over the crime' test become the accepted test for principal liability, but the doctrine of *Organisationsherrschaft* has now become accepted by the majority of scholars as part of the system of modes of liability,[156] and has been applied in various cases, many of which involve further convictions for killings of civilians attempting to cross the Berlin Wall.[157]

7.3.6 Summary

Germany provides the quintessential example of a criminal law system expressing the civil law tradition of a highly systemised codification of liability, which is built upon and developed by way of sophisticated doctrine, operating as a legal science. Although there are shifts in the main lines of thought over time, depending on which schools of thought dominate at any given time, the majority opinion at any point in history will always be representative of the *zeitgeist* in philosophical, psychological and social understandings of the law and of liability. Thus, while a subjective theory of liability was prevalent in the nineteenth century, over time it was considered that this approach did not sufficiently deal with individual problems of negligence, the phenomenon of the use of an innocent agent, and the problem of the *Hintermann*. In order to respond to these problems in the correct way conceptually, and in such a way that the system of liability would be kept intact, objectivity is now reflected in the notion of *Tatherrschaft*, or 'control over the crime'. A more functional understanding of liability is now considered to be appropriate.

As a result of this shift in thinking, responsibility is now conceived of as a component of blameworthiness, rather than an external precondition of blame. Culpability, or *Schuld* is understood normatively; that is, culpability is the measure of the degree of blame that can be attributed to an actor for a wrongful act, whether the act is committed by herself, by another person who she assists, or by another person over whom she exercises control. Culpability is attribution of blame based on the contribution made by an individual, including the control he exercises over the act, rather than a fictitious attribution of the acts of others to him.

That there are degrees of blame is reflected in the fact that there is a normatively differentiated system of liability. In order to determine the guilt and appropriate punishment of each individual involved in a collective crime, it is important to determine whether the contribution (either causal or functional, in terms of control over the crime) amounts to a significant degree, and therefore to conviction as a principal, or to a lesser degree, and therefore to conviction as a secondary party. This qualification matters for the moral re-telling of the facts in the conviction: was the accused a murderer, or did he 'merely' assist a murderer?

---

[156] Jescheck and Weigend 1996, p. 670; Ambos 2008, p. 26.

[157] Berlin Wall II 1999; Berlin Wall III 2002; and further the doctrine has also been applied for 'white collar crimes', such as fraud committed through a company: Verfahrenseinstellung 2003; Apothek 2003; Betrug 2007.

For the most part this qualification will also have an impact upon the sentencing, in that there is mandatory mitigation for those who are convicted of having aided a crime, however in the case of instigating, the sentencing does not differ from a principal. Yet the moral re-telling of the facts remains important. Apparently the instigator is considered to be more blameworthy than the aider, but less blameworthy than a principal. This is therefore a system which regards the expressive function of modes of liability to be more important than the expressive function of sentencing, since the conviction is what stands over time, and since the sentence may take into account other individual factors which are not related directly to the degree of responsibility for crime.

In terms of identifying the role of participants in the law-making processes in Germany, the fact that there is no lay jury as a decision-maker in this criminal process is one factor in the development of such a taxonomy of modes of liability. Since none of these notions need to be translated to lay-people in the process, there is room for the doctrinal thinking to go further into technical detail, and the coherence and consistency of the system can be guaranteed by the fact that the decision-makers are professionally trained lawyers who are informed by the scholarly writings of the professors who develop the law. These scholars are the most influential participants in the development of the law, responding to societal needs as well as the internal systematic structures of the law, and in doing so, forming policy.

It was in this way that the doctrine of *Organisationsherrschaft* was developed to deal specifically with the problem of system crimes committed by agents who are fully culpable for their actions, but who act under the influence of those who exercise more control than they do as individuals within the system. The *Hintermann* the actor behind the physical perpetrator, who is able to make use of the systemic conditions within an organisation to have crimes be committed on behalf of the system, cannot be appropriately punished as a mere abettor or instigator under a system of liability which differentiates morally between actors. Such an architect of system crimes should be held culpable as a principal actor, due to the control and influence exercised over those lower down in the machine. But it took an extensive theory, which was built carefully upon the principles of liability within a complex and internally coherent system, to reach such a conclusion.

7.4 Argentina: The German Theory Applied to Leaders of Organised Mass Atrocity

Argentina has been selected as one of the jurisdictions for comparison in this study for several reasons. Firstly it is important to look beyond the European civil law jurisdictions for different legal-political contexts. From a comparative law perspective, and particularly given that the notion of legal traditions is followed here rather than the traditional notion of legal families,[158] it is necessary to consider a greater

[158] Glenn 2010.

variation of jurisdictions beyond the traditional 'donating' jurisdictions in Europe. In order to gain a better picture of some of the domestic criminal law responses to mass atrocity, it is useful to look to Latin America, where many States have suffered mass atrocity committed under military regimes against their own populations. Different policy reactions and specific modes of liability have been applied, which, in the case of Argentina, has been influential for other domestic jurisdictions, and which may therefore be instructive in the deliberate patchworking process of forming and developing ICL.

Secondly, since the extent to which the law is developed is a criterion for selection of jurisdictions in a comparative study,[159] Argentina has been selected from among various Latin American jurisdictions which could have been compared because it has been a leading jurisdiction in prosecuting leaders for mass atrocity. The German doctrine was applied directly and further developed in the case law. There has also been much written about the Argentine application of the doctrine in socio-legal literature.[160]

Finally, Argentina is highly relevant to this enquiry, as there have been several waves of criminal prosecutions over several years, with great controversy as to the selection of suspects according to their rank within the hierarchy of the military *de facto* government. These debates demonstrate the legal, political and moral factors which are at play in such situations, and why the applied modes of liability are so important to the results in the eyes of the legal community as well as in the eyes of victims.

7.4.1 Legal Transplants in Argentine Law

Argentina is an interesting case study in legal transplants, since it has borrowed heavily from the US model of constitutional law[161] while at the same time modeling its private and criminal law on the civil law tradition and drawing predominantly from French, German, Italian and Spanish doctrine and legislation.[162] For example, although the *Constitución Nacional de Argentina* contains a provision replicating the US constitutional guarantee of a jury trial for all crimes,[163] the procedural law is left entirely to the independent legislation of the provinces, none of which have actually implemented a trial by jury.[164]

---

[159] See further Sect. 5.2. See Oderkerk 2001.

[160] See for example O'Donnell 2009; Osiel 2001, 2009.

[161] Ferrante 2011a, p. 14; Miller 2003, p. 846; Kleinheisterkamp 2006, p. 286.

[162] Kleinheisterkamp 2006, pp. 267, 271, 293.

[163] Constitución Nacional de Argentina 1994, § 118.

[164] Ferrante 2011a, p. 15. Some individual defendants have filed a petition to the Supreme Court demanding a trial by jury based on the constitutional right, however the Supreme Court has consistently referred to the practice of the provinces and their freedom to give effect to this right or not.

The absence of lay juries, despite the legislative intention to import this institution from the common law tradition, is one factor in the development of criminal law along lines that resemble the civil law tradition more strongly. Since there is no need to translate different requirements of liability into terms understandable for a lay jury, there is more room for a technical classification of modes of liability (and other legal institutions), and as a result there is more expectation of coherent and consistent reasoning and application of the criminal law system by the judiciary.

The predominance of the civil law tradition in the Argentine criminal law can be attributed to the legislative history of the *Codigo Penal*, which was given a deliberately simple construction. Before the *Constitución Nacional* was passed in 1853, there was no single system of criminal law, but rather a multiplicity of sources and a great deal of judicial discretion.[165] After the constitution was passed, attempts were made to draft a single, coherent criminal code, which led to the 1887 code, heavily influenced by the Bavarian Criminal Code,[166] and a later draft in 1891 which drew significantly from the contemporary Italian Criminal Code.[167]

The core of the current *Codigo Penal* emerged with the 1921 enactment, the intention of which was to reach compromises which would satisfy practitioners and academics of differing schools of thought, while still capturing the core principles of the Italian and Bavarian codes which had formed the base of the two previous drafts.[168] The *Codigo Penal* is therefore by design a relatively simplified code, not drafted under the direction of a single jurist, but rather the result of a collection of different approaches. This minimalist drafting and hybridity is reflected especially in the articles relating to forms of participation in collective crimes, as will be discussed below.

Doctrine plays an important part in the development, application and interpretation of the criminal law, drawn as much from Argentine scholars as from influential foreign commentators in the civil law tradition. In this sense, comparative law is an important interpretative tool.[169] Today the Argentine criminal law system is most heavily influenced by twentieth century German doctrines; most Argentine criminal law theory is taught from translations of works by German theorists.[170] In order to interpret the minimalist and sometimes cryptic articles of the *Codigo Penal* in practice, much use is made of the German literature which, although it interprets the German criminal code, is taken as guidance for interpretation of the Argentine criminal code.

(Footnote 164 continued)
The province of Cordóba has in fact legislated to provide for a trial by jury, however the practice has remained limited. See e.g. D'Alessio AJ 2008, p. 7.

165 Ferrante 2011a, p. 43.

166 Ferrante 2011b, p. 13.

167 Ferrante 2011a, p. 13.

168 Ferrante 2011a, p. 13.

169 Kleinheisterkamp 2006, pp. 295–299.

170 See for examples of influential texts Welzel 1970; Jakobs 1997; Roxin 2005; Stratenwerth 2005.

The amendment of the constitution in 1994 made a fairly radical step by integrating ten international human rights treaties to become directly part of Argentine law.[171] The influence of these treaties on criminal law was in effect to codify the norm of *nullum crimen*, including *lex certa*, *lex scripta*, and *lex stricta*, without having to amend the text of the *Codigo Penal*.[172] Similarly, other core procedural and substantive rights are embedded in Article 18 of the *Constitución*, rather than in the *Codigo Penal* itself.

The operation of the criminal justice system in practice, however, does not always express these familiar general principles, and it appears to be part of the popular understanding that there is a certain level of corruption linking judges, police agents and members of organized crime groups.[173] Nonetheless the turn to criminal prosecutions of members of the military regime and other associated actors implicated in the 'Dirty War', about which more detail below, was considered to be a strong move towards delegitimising the actions and actors involved.

Argentina, like all the jurisdictions compared here, is a federation, and although the constitution grants broad powers to the autonomous provinces,[174] they cannot pass new criminal laws[175] and they all follow the national *Codigo Penal* when it comes to substantive law. However because of the independence of the provinces, there is still some divergence in application by the courts, while criminal procedure is entirely a matter for the provinces and differs greatly from one province to another.

7.4.2 Objectivity and Control over the Crime

The *Codigo Penal* is divided into two books, the first one dealing with *Disposiciones General*, or 'general laws', and the second one being the special part, enumerating the crimes and their elements, under the title *De los delitos*. There is no explicit design of liability such as is evident in the German criminal code. There are no requirements that the objective and subjective elements, the wrongfulness of the act or the blameworthiness of the suspect be satisfied in evidence. Under Title V, titled *Imputibilidad* (imputability or attribution), Article 34 lists defences that are both objective (for example self defence or defence of another) and subjective (for example insanity). Other articles under this title deal with personal circumstances to be taken into account when determining the sentence, including the level of participation in

---

[171] Although constitutional law generally trumps international law, it has been agreed that in the case of these treaties, any conflict with the constitution would result in the treaties being directly applicable. Constitución Nacional de Argentina 1994, Article 75.22.

[172] Article 18 of the *Constitución Nacional* reads 'No inhabitant of the Nation shall be punished without previous trial based upon a law existing prior to the act which gave rise to the proceedings [...]' (author's own translation). The original text reads: '*Ningún habitante de la Nación puede ser penado sin juicio previo fundado en ley anterior al hecho del proceso*'.

[173] See for example Informe sobre la Justicia 2012.

[174] Constitución Nacional de Argentina 1994, Article 121.

[175] Constitución Nacional de Argentina 1994, Article 126.

the crime, and aggravating and mitigating circumstances. This broad categorisation of attribution does give systematic guidance to the judge as to how to determine liability, however both scholars and judges tend to seek interpretative tools in the literature surrounding the German *Strafgesetzbuch*, due to the continuing influence of this criminal law regime on the Argentine one, despite the fact that the German code is much more explicit in its systematic design.

Leading Argentine theoreticians such as Bacigalupo,[176] Sancinetti[177] and Zaffaroni,[178] base much of their writings on German authors such as Jescheck, Liszt, Sauer, Stratenwerth, Roxin and Welzel. Core concepts relating to criminal responsibility such as the principle of culpability and wrongfulness, or *antijuridicidad*, are therefore considered central when it comes to the theoretical underpinning of criminal responsibility, despite not being anchored in the Argentine legislation.[179] One important difference is that culpability is understood in the descriptive sense, referring to the distinction between two levels of subjective state of mind; either *dolus* (intent) or *imprudencia* (negligence). In the special part of the *Codigo Penal* the definitions of the crimes include one or other of these subjective elements. This differs from the normative understanding of culpability as blameworthiness in the German doctrine.

Despite the fact that a descriptive understanding of culpability is usually associated with a subjective approach to liability in the common law tradition, generally the subjective theory has not been received in Hispanic legal doctrine; Argentina, Colombia, Mexico, Spain and Venezuela have all followed the German doctrine.[180] For example twentieth century jurists Sebastian Soler,[181] Carlos Nino[182] and Euginio Zaffaroni[183] all developed extensive theories of responsibility which assert that an objective approach to attribution is to be favoured, with emphasis upon the contribution made by an individual towards a collective crime rather than on the agreement or motive. These authors all agree that the subjective element must still be present, however the doctrinal leaning toward the emphasis on the objective element is uncontroversial.

Although a subjective theory has been considered by some tribunals, in general it has also been rejected in the case law.[184] As Sancinetti points out, a subjective theory coincides with a (functionally) unitary theory of participation and with an emphasis on the intent (or knowledge) of the actors involved.[185] In Argentine jurisprudence there is a preference for emphasising causation and distinguishing this from other

---

[176] Bacigalupo 1994.
[177] Sancinetti 1991.
[178] Zaffaroni 2005.
[179] Nino 1980, p. 418 and further.
[180] Bacigalupo 1994, p. 184.
[181] Soler 1992.
[182] Nino 1980.
[183] Zaffaroni 2005.
[184] Sancinetti 1991, p. 641.
[185] Sancinetti 1991, p. 638.

conditions, and for distinguishing between participants in collective crimes following the German model, and thus an objective theory of liability has prevailed.[186]

In discussing the objective approach, much of the literature draws the same distinction between a formal-objective theory and a material-objective theory as is drawn in the German literature.[187] In Argentina, just as in Germany, the formal-objective theory dominated the thinking in the first half of the twentieth century, defining liability based upon the actual contribution to the crime, or in the terminology of the Argentine code, the execution of the crime.[188] However, it was also rejected for the same reasons over time: this theory is limited because it does not explain how a person who uses another party as an instrument to achieve the execution of a crime could be held liable. The formal-objective theory was superseded by the more flexible material-objective approach developed in German doctrine by Hans Welzel and later by Claus Roxin, and introduced into Argentine doctrine by way of translation in 1960 by Jiménez de Asúa. In the 1970s and 1980s the material-objective approach was further developed by Enrique Bacigalupo,[189] who later became a judge on the Spanish Supreme Court, and by Eugenio Zaffaroni.[190]

Following the German tradition, this material-objective theory has become better known as the 'control over the crime' theory, or *dominio del hecho* in Spanish, although Zaffaroni asserted that this theory had roots antecedent in Argentine doctrine in the nineteenth century as well.[191] It is the test used to delineate between principals and secondary parties, but also to identify the various forms of principal liability. Following the tripartite distinction made by Roxin, a distinction has been made in Argentine literature between control over the action, which is achieved by a direct perpetrator; functional control over the crime, which identifies who was a co-perpetrator; and control over the will, which describes the indirect perpetrator.[192] The benefit of a theory of objective attribution over a 'personal' or subjective theory is that it answers the question whether and how to expand the notion of 'author' or 'perpetrator', when it comes to collective crimes. Specifically, a material-objective theory of 'control over the crime' allows a better understanding of how collective crimes are often committed, especially when it comes to those such as fraud, handling stolen goods, many commercial and drug crimes, all of which involve an individual

186 Though the subjective theory received renewed interest the late twentieth century, it has not been accepted as the dominant theory for some time. See Zaffaroni 2005, p. 607.

187 Sancinetti 1991, p. 640; Bacigalupo 1994, pp. 180–182.

188 The definition of a principal in the Argentine criminal code is '*Los que tomasen parte al ejecución del hecho*', or 'those who take part in the execution of the crime', see Codigo Penal 1984, Article 45.

189 See for an updated edition, Bacigalupo 1994.

190 Originally in his works in 1968. See for a more recent application of his theory Zaffaroni 2005.

191 Zaffaroni 2005, p. 606.

192 Bacigalupo 1994, p. 188; Sancinetti 1991, p. 645; Zaffaroni 2005, p. 606.

who has the 'reins' over the completion of the crime through collaboration with or use of other individuals.[193]

The application of this theory to the simplified text of the *Codigo Penal* has led to a doctrinal distinction where there was no legislative one, as will be discussed next.

7.4.3 Modes of Individual Criminal Liability

The relevant section of the Argentine criminal code was designed as a system of penalties rather than an extensive classification of modes of liability, which was another consequence of the simplified drafting of the *Codigo Penal*.[194] There is no explicit choice for either a theory differentiating between participants, or for a functionally unitary theory.[195] In fact the term '*partícipes*' is used both in the broad, neutral sense, referring to all parties who take part in a collective crime, as well as in the narrow, normative sense, referring to secondary participants.[196] On the other hand, due to the aforementioned influence of the German doctrine in interpreting the legislation, a more complex, normative classification has emerged in the literature. The dogmatic choice to interpret the text of the code as distinguishing between principal and secondary parties follows from the preference for the material-objective, or control over the crime, theory.

7.4.3.1 Principals

According to Article 45 of the *Codigo Penal*, receiving the maximum penalty are those who 'execute' the crime, those whose co-operation was so necessary that the crime could not have been committed without them, and those who instructed another person to commit the crime.[197] In the doctrine this single article has been extrapolated to a more complex division, under the heading *autoría*, or principal. The *autor* acts with autonomy and is capable of executing the crime independently, and is divided into three subcategories: (a) the *autoría directa*, or direct perpetrator, (b) *autoría mediata*, or indirect perpetrator, about which will be said more below in Sect. 7.4.4,

---

[193] Sancinetti recognises that a formal objective theory is better at delimiting when the execution of a crime actually commences, however since this is more a problem of attempt than of participation, he still prefers the material objective approach; Sancinetti 1991, p. 646.

[194] Ferrante 2011a, p. 62.

[195] Bacigalupo 1994, p. 179.

[196] Zaffaroni 2005, p. 601.

[197] The original text reads *Los que tomasen parte en la ejecución del hecho o prestasen al autor o autores un auxilio o cooperación sin los cuales no habría podido cometerse, tendrán la pena establecida para el delito. En la misma pena incurrirán los que hubiesen determinado directamente a otro a cometerlo.* For the full text of the *Codigo Penal* articles on party liability, and a diagram of the relationship between the modes of liability, see the appendix A.5.

and (c) the *coautor*, defined as all those whose co-operation is considered to be necessary.

With respect to the latter, despite the necessary nature of their contribution and the corresponding maximum penalty, the *coautor*, or co-perpetrator, is sometimes classified as accomplice in the first degree rather than as a principal.[198] This is because the accomplice is considered to be someone who cooperates in the crime of another (the physical perpetrator, or *autor*) and while their contribution may be indispensable for the realisation of a concrete plan, they lack the required characteristic to be considered author of the crime.[199]

7.4.3.2 *Cómplices* in the First and Second Degree

The last person mentioned in Article 45, the person who 'determines directly' or 'directly brings about' that another commit a crime ('*los que hubiesen determinado directamente a otro a cometer [el hecho]*'), is described in the literature as *instigador*, or instigator. This actor receives the maximum penalty applicable to principals, but is considered to be an accomplice in the first degree, and therefore to fall under the term *participación* rather than *autoría*.[200] All of the actors who fall under Article 45 are considered to have a sufficient degree of autonomy to warrant the full penalty applicable to principals, even where they are considered accomplices 'in the first degree'.

Article 46 of the *Codigo Penal* enumerates the *cómplices secundarios*, or accomplices in the second degree, whose role is derivative of those named under Article 45; those who aided the execution of the crime 'in any other way' and those who provided assistance after the fact. The only real definition of these secondary participants is a negative one: all those whose collaboration was not indispensable, or not essential.[201] This distinction between *cómplices* in the first and the second degree is only seen in the literature, and has echoes of the old English common law distinctions, although the grounds for distinguishing are slightly more complex in the Argentine doctrine.

The distinction between such a co-acting principle under Article 45 and a *cómplice* in the first degree under Article 46 is extremely difficult to draw, since their contribution must both be of an essential character.[202] It would appear that the degree to which the actors have control over the crime is determinative. One way to test this is that co-perpetrators are not subordinate to another party, since this would mean they are no longer in control of the crime.[203]

The 'non-essential' nature of the contribution of secondary parties (*cómplices* in the second degree), justifies the reduced sentence of either one third or half of

---

[198] Ferrante 2011a, p. 31; Bacigalupo 1994, p. 199. But see Sancinetti 1991, p. 648.

[199] Zaffaroni 2005, pp. 602–603.

[200] Sancinetti 1991, p. 649; Ferrante 2011a, p. 31.

[201] Bacigalupo 1994, p. 211; Sancinetti 1991, p. 651.

[202] Sancinetti 1991, p. 651.

[203] Bacigalupo 1994, p. 197.

the sentence applicable to the principal, depending which of the two forms an actor falls under according to Article 46 of the *Codigo Penal*. It would seem this is a reflection of the lower level of blameworthiness, although such terminology is not very prevalent in the literature. Following the wording of the *Codigo Penal*, the emphasis appears to be on the objective test of the actual contribution made by a participant to the 'execution' of the crime, and how essential this contribution was: the principal perpetrator 'executes' the crime, those who are secondary parties *contribute* to its execution. As Zaffaroni has put it, the full penalty applies to executors of the crime, those who determine the crime, and those whose cooperation is necessary, whereas a lesser penalty applies to those whose cooperation is not necessary for the fulfilment of the crime.[204] However the complexity and lack of clarity surrounding the classification of participants makes a consistent distribution problematic.[205]

7.4.4 Autoría Mediata: *Indirect Perpetration*

Although Article 45 of the *Codigo Penal* does not explicitly name indirect perpetration, this notion is covered by the wording of the final sentence of the article, namely 'those who directly cause another to commit the act'.[206] There is consensus that both the notion of instigation and *autoría mediata* fall under this definition, since what it describes is the ability of one person to determine the will of another, and this receives the same full punishment as the *autor*, or direct perpetrator.[207]

There is a certain inconsistency in interpreting this sentence to include both indirect perpetration and instigation, since the latter is considered to be a form of complicity in the first degree, whereas *autoría mediata* is considered to be a form of *autoría*, or principal perpetration. Some would argue that given there is no difference in the sentencing, it does not matter that a distinction is difficult to draw, or that they are both read as falling under the same article.[208] However the difference lies in the normative qualification and the moral story it tells about the role played by an individual in a collective crime. The reason an indirect perpetrator is considered to be *autor* is because according to the theory of liability of control over the crime (*dominio del hecho*), there is final control over the 'whether' and 'how' of the

[204] Zaffaroni 2005, p. 603.

[205] Nino in fact argues that the complex classification of participants taken from the German doctrine is not a necessary reading of the *Codigo Penal*, and that gradations of punishment should rather be based upon an individual participant's actual contribution to the damage or danger caused by a criminal act. His reading of the *Codigo Penal* is based upon his consensual theory of punishment. While this is a possible reading of the code, it is not the leading nor generally accepted one. Nino 1980, pp. 453–456.

[206] Author's own translation. The last sentence of this article reads *En la misma pena incurrirán los que hubiesen determinado directamente a otro a cometerlo.*.

[207] Zaffaroni 2005, p. 616; Bacigalupo 1994, p. 189; Sancinetti 1991, p. 691; Ferrante 2011a, pp. 32–33.

[208] Sancinetti 1991, p. 650.

commission of the crime.[209] An instigator, on the other hand, is someone who sets the wheels in motion, but lacks such control and cannot affect the way in which the wheels turn.

However since the wording of the article was kept deliberately broad enough to encompass various actions, and was drafted without a specific design of modes of liability, it has been left to doctrine—a recognised authoritative source of law, in the dynamic sense—to develop.

Thus, due to a gap or lack of clarity in the legislation, the Argentine scholars and judges have looked to foreign doctrine for solutions, in a typical process of legal transplantation. The theory of *autoría mediata* was forged along the same lines as the theory of 'control over the crime' in Germany, in order to avoid wrong labelling and a conviction that failed to represent the roles of those with more influence over an act of collective criminality.[210]

As Sancinetti put it, there is no need for a legal definition of indirect perpetration in the criminal code, since the interpretation of the 'execution' of a crime allows for the conclusion that a person can realise a crime *through* another. It is simply a matter of hermeneutics.[211] When the physical perpetrator lacks the qualities necessary to be considered a principal, then the *hombre de atrás*, or actor behind the physical perpetrator, must be seen as a principal in order to fill a lacuna in the legal construction of liability.[212]

Since *autoría mediata* is a form of principal perpetration, it is clearly not dependent on the commission of the crime by other physical (principal) perpetrators. In this sense it is an autonomous mode of liability.[213] A consequence of this is that the criminal liability of an indirect perpetrator's actions begin at the moment an order or direction is given to the physical perpetrator, whereas an instigator's criminal liability only commences at the moment that the physical perpetrator commences the act itself.[214]

The moment at which an indirect perpetrator's liability is triggered has another consequence, namely that this liability extends to all those acts committed by the subordinate physical perpetrator which are a result of the indirect perpetrator's exercise of control. This means that if an indirect perpetrator directs subordinate actors to kidnap or murder insurgents, which was the case under the military regime during the 'Dirty War', it is not necessary that the indirect perpetrator know exactly which individuals are to be the victims, nor how many kidnappings or murders would eventually take place, nor even how they would take place.[215] It is enough that there is an intent on the part of the *autor mediato* that these acts should take place, and that there is exercise of control over the will of the subordinates either directly or through the organisational hierarchy.

---

209 Zaffaroni 2005, pp. 185, 189.

210 Bacigalupo 1994, p. 190.

211 Sancinetti 1991, p. 691.

212 Sancinetti 1991, p. 693.

213 Sancinetti 1991, p. 718.

214 Zaffaroni 2005, p. 611.

215 Sancinetti and Ferrante 1999, p. 29; Malarino 2008, p. 61.

In the literature there are several categories of *autoría mediata*, based on the conditions of intervention, which can be simplified into three basic categories. The first category covers the cases where the crime cannot be imputed to the physical perpetrator due to the absence of the required mental state, for instance where the physical perpetrator was operating under a mistake of fact, whereas the *hombre de atrás* was fully aware of the actual situation.[216]

The second category of *autoría mediata* is where the physical perpetrator lacked full control over the crime, because she had subordinated her will to another who *did* exercise such control,[217] or because the *hombre de atrás* manipulated the physical perpetrator in order to realise the crime. Examples are duress, deliberate creation of a situation of necessity, and use of an agent who is innocent due to mental incapacity or age.[218]

The third category is the use or manipulation of the existing control over a hierarchical organisation by the *hombre de atrás* to realise a crime. There are two main ways this can occur, the first of which is the use of an instrument who acts legitimately, for instance a soldier who is given the order to trespass on private land. A soldier who does so knowing that it is illegal, would act because the damage caused by such an act would be minimal, whereas the obligation to follow orders should weigh more heavily. To disobey orders would be a greater risk, and as such the soldier acts with 'personal justification', whereas the superior is *autor mediato*.[219]

The second way this category can occur is by use of a fungible agent within the organisation. In many cases the ability to exercise psychological force over subordinates within an apparatus of power would be sufficient, but where it is not, if a subordinate were to refuse to carry out the crime, the very fact of the replaceability of such an individual with another subordinate in order to realise the crime, allows the 'author behind the author' to maintain control.[220]

In the examples given, the role played by a superior is also considered by some authors to be that of *co-autor* (co-perpetrator), since there is control over the crime at both the top and to a certain extent at the bottom with the physical perpetrator who decides whether or not to cooperate.[221] This expands on Roxin's analysis, and draws on alternative reasoning offered by German scholars Jescheck and Jakobs. Because the *autor mediato* is at the top of an organisation, leaving the execution of the crime in the hands of others but maintaining the power over the decision-making, she could be seen as a co-author in a vertical rather than horizontal relationship to the physical perpetrators. The physical perpetrators are co-authors in a horizontal relationship to each other.[222] This combination of all the actors as co-perpetrators was somewhat revolutionary, since many authors considered that if the indirect perpetrator is liable

[216] Bacigalupo 1994, p. 191.

[217] Bacigalupo 1994, p. 191.

[218] Bacigalupo 1994, p. 193, citing Roxin and Stratenwerth.

[219] Sancinetti 1991, p. 712.

[220] Sancinetti 1991, p. 713; Bacigalupo 1994, p. 196.

[221] Sancinetti 1991, p. 714, citing German scholar Jescheck; Bacigalupo 1994, p. 192.

[222] Sancinetti 1991, p. 714.

as principal, then the physical perpetrator could not also be liable as a principal, reducing the latter to secondary participation.

Argentine authors writing on *autoría mediata* cite Roxin, Stratenwerth and Welzel directly, and the German model has been applied by the courts to cases of organised crime, and most famously in the case of the *Junta* totalitarian regime under Jorge Videla during the 'Dirty War' years in Argentina. The political context of this totalitarian military regime will be given here as a background to the policy shifts back and forth on how to prosecute those involved at various levels of the hierarchy in the commission of mass atrocity.

7.4.5 The 'Dirty War' and the Criminal Responsibility of Its Leaders

In the 1970 s and 1980s, most of the governments of Latin-American countries had come to power by usurpation or force, however in the ensuing transition to democracy, Argentina was the first to prosecute the leaders of these regimes for the numerous human rights abuses that took place.[223] When Raúl Alfonsín was democratically elected as Argentine president in 1983, one of his strongest campaigns had been to put in place prosecutions in order to vilify the previous military regime's actions, and establish the importance of human rights in Argentina. Although this process became more and more complex under political pressure, due in part to the continuing influence of the military in Argentina throughout history,[224] nonetheless, the prosecutions that took place established important precedence in applying modes of liability to leadership cases.

What has become known as the 'Dirty War' took place between 1976 and 1983, beginning when the *Junta*, or joined armed forces, took over power of the country with a *coup d'etat*. In the early 1970 s there had been a rise in left-wing terrorism, and violent responses from right-wing anti-communist groups. The government under President Isabel Perón was unable to respond effectively, calling on the military to assist in every way possible, declaring a state of seige in 1974, enacting decrees giving responsibility for the campaign against leftist subversion to the military, and finally conceding power altogether in 1976.[225] At first the move to power by the military was seen as the way to resolve the civilian violence that had been growing, however the right-wing military soon proved itself to be just as extreme in its responses as the guerilla movements they were attempting to quash. The Junta leaders issued a series of statutes and decrees giving themselves executive and legislative powers,

---

[223] Garro and Dahl 1987, p. 284.

[224] Nino 1996, p. 41; Osiel 2001, p. 12.

[225] For a comprehensive overview of the situation prior to and during the Dirty War, see Nino 1996; Garro and Dahl 1987.

and replaced members of the Supreme Court with judges who were sympathetic to their political plans.[226]

The military regime consisted of three different juntas. The first, which enacted the *coup d'etat*, was under General Jorge Videla, commander of the army and self-proclaimed president; Admiral Emilio Massera, commander of the navy; and Brigadier General Orlando Agosti, commander of the air force.[227] Under these leaders a consistent pattern of 'disappearances', kidnappings, torture and killings was discernable around the country, with those targeted allegedly having some ties to leftist groups, however tenuous. At the beginning of the 1980s, by the time the third Junta took power, this '*Proceso*' had led to at least 10,000 disappearances, almost half of which had disappeared in 1976in the first year of the regime[228]—although some human rights groups estimate the numbers to have reached 300,000.[229]

Once the military began to lose its authority, international human rights groups began to successfully draw attention to the disappearances, and finally after the disastrous defeat of Argentina in the Malvinas-Falklans war in 1982, the third Junta regime had to concede power to a democratically elected president, a position which was won by Alfonsín. However immediately prior to the election, a fourth transitional Junta was put in place, and a number of decrees were enacted in order to protect the military, including the '*Documento Final*'[230] and the '*Ley de de Pacificación Nacional*'.[231] The former conceded that some human rights abuses had occurred, but insisted that these were necessary in the line of duty in the war against subversion and terrorism, and that those who had been killed had died in open combat.[232] The latter conferred a blanket self-amnesty to every member of the military forces for crimes committed between May 25, 1973 and June 17, 1982, as well as to suspected terrorists.[233] As well decrees were issued ordering the destruction of all documents relating to the military regime of repression.[234]

7.4.5.1 Political Challenges for the Prosecutions of Junta Leaders

When Alfonsín's was elected, his legal advisors argued that the laws conferred by the transitional Junta were in fact passed by a *de facto* government, and were therefore unconstitutional. Eventually a bill was passed nullifying the self-amnesty law.[235] In the first month of his presidency, Alfonsín issued decrees ordering the arrest and

226 Garro and Dahl 1987, p. 290.

227 Garro and Dahl 1987, pp. 289–290.

228 CONADEP 1984, part I(B) and part II(A).

229 Hernandez 2013.

230 Documento Final 1983.

231 Self-Amnesty Law 1983.

232 Documento Final 1983, pp. 7, 9, 12.

233 Self-Amnesty Law 1983, Article 1.

234 Garro and Dahl 1987, p. 301.

235 Garro and Dahl 1987, p. 305; Osiel 2001, p. 18.

prosecution of the nine military officers who had made up the three juntas, as well as of members of the left-wing guerilla group the Montoneros.[236] He established CONADEP, the *Comisión Nacional sobre la Desaparición de Personas*, or National Commission on the Disappearance of Persons, whose report *Nunca Más* (Never Again) documented thousands of testimonies and a great deal of detail about abductions, unlawful detainments in clandestine centres, tortures, dumping of bodies and removal of babies born in detention who were subsequently given to couples who were sympathetic to the military regime to be raised as their own children.[237] This report, presented to the government in September 1984, formed the basis of the leadership prosecutions.

The core question was how to apply this retroactive justice: should it be retributive, and applicable to all the individuals who were involved in the atrocities, at all levels? This was the stance of many human rights groups.[238] Or should it be limited to punishment of the leaders, in order to send a more instrumentalist message regarding the architects of the regime? Alfonsín preferred the latter approach, since he intended to delimit the categories of responsibility and limit the duration of the trials. He also intended to have a 'symmetrical' approach to liability, by including prosecutions of members of the leftist guerilla groups as well, and was intent on recording history through the CONADEP report and the truth-finding during the trial investigations and presentation of evidence.[239] Thus, under protest of the very human rights groups that had called for prosecutions, he embarked upon a difficult balancing act.

Added to this, due to constitutional requirements and the *Codigo de Justicia Militar* (CJM), military personnel had to be tried by a military tribunal, and not by a civilian court. However the impartiality of military judges was highly questionable, and much of the population was crying out for public trials under the criminal code.[240] When it appeared that the Military Supreme Council was unwilling to conduct a speedy trial, and the judges in fact had let it be known that they considered the repressive acts of the former regime to be justified in a war on terror, jurisdiction was handed over to the *Cámera Nacional de Apelaciones* (Federal Appeals Court) in October 1984.[241]

This move, coupled with the fact that by early 1985 there were approximately seventeen hundred prosecutions filed around the country against about five hundred members of the military and security forces, led to tensions with standing members of the military.[242] Alfonsín's planned strategy had been to distinguish three levels of responsibility: those who created and orchestrated the military regime and gave the orders; the physical perpetrators of abhorrent or excessive crimes; and those at mid- or low-level ranks who limited themselves to complying with orders. The

236 Garro and Dahl 1987, pp. 302–303.

237 CONADEP 1984.

238 Nino 1996, p. 112.

239 Nino 1996, p. 113.

240 Garro and Dahl 1987, p. 311.

241 Garro and Dahl 1987, p. 320; Nino 1996, p. 78.

242 Garro and Dahl 1987, p. 314.

idea was that the latter group would be excused from prosecution, due to the climate of compulsion and an intense propaganda campaign condoning violence,[243] in accordance with Article 34(5) of the *Codigo Penal*, which states that a person acting with due obedience is not punishable. This echoes the analysis of writers such as Arendt, Drumbl and Tallgren, who have argued that it is difficult to blame the physical perpetrators in cases of mass atrocity to the same extent as the architects of the circumstances under which the crimes take place, since the physical perpetrators are acting within a norm of violence, and to refuse to comply would be to risk their own safety or lives.[244]

In the CJM there was a corresponding article which limited responsibility of a subordinate member of the military when acting under orders; the superior giving the order was to be the only criminally responsible party, unless the subordinate had exceeded the order, in which case the soldier would be an accomplice.[245] However this Article 514 had always been read to exclude manifestly illegal orders, in which case a soldier complying with such orders would be considered liable as a direct perpetrator.[246]

In 1984 the CJM was amended to enact the policy of limiting liability of lower-ranking subordinates even when the orders were manifestly illegal, due to the norm of violence that prevailed during the Dirty War and the inability of lower-ranking individuals to exercise any deliberative authority or control over the crime. These individuals would instead receive the opportunity to cooperate in the transitional democratisation, and the higher-ranking officials who did have deliberative capacity would be punished under a policy of 'protecting the future social order'.[247] The amendment inserted a rebuttable assumption that these subordinates had acted under mistake of law, 'unless they committed atrocious or abhorrent acts.'[248]

The question that gave rise to further unrest, both on the side of human rights groups and of the military, was which acts would fall under the 'atrocious or abhorrent' exception. Human rights groups were concerned that the article could be read to excuse most of the physical perpetrators who could claim they were acting under orders, and the military were concerned that due to the nature of the allegations, all acts could be considered to be 'atrocious or abhorrent', and an endless spate of prosecutions would ensue.[249]

---

243 Nino 1996, p. 64; Osiel 2001, p. 18.

244 Arendt 1964, pp. 31–33; Drumbl 2005, p. 561; Tallgren 2002, p. 580; Osiel 2001, p. 26.

245 CJM 1951, Article 514.

246 Garro and Dahl 1987, p. 315.

247 Nino 1996, pp. 66, 68; Osiel 2001, p. 18.

248 *'A ese efecto podrá presumirse, salvo evidencia en contrario que se obró con error insalvable sobre la legitimidad de la orden recibida, excepto cuando consistiera en la comisión de hechos atroces o aberrantes*. CJM Amended 1984, Article 11.

249 Nino 1996, p. 179.

7.4.5.2 Federal Appeals Court: Autoría Mediata and Use of the State Apparatus of Power

When the Federal Appeals Court took over jurisdiction of the case against the leaders of the Junta, it embarked on a historical project involving 700 counts against the nine defendants, and the testimony of over 800 witnesses in the '*Cause 13*' trial.[250] It was a landmark decision because it was the first time the notion of perpetration through means of an organisation was applied in a court.[251] Although the court found that there had been a secret joint agreement between the commanders of each of the branches of the military, creating a criminal method for suppressing terrorism, in the end the collaboration did not amount to co-perpetration, since no powers were vested in the Junta as such, and the commanders each remained responsible for their own respective military branch.[252] The commanders themselves claimed absolute autonomy in their decision-making and control over their respective forces.[253] A theory of joint control was therefore out of the question.

Due to the decree enacted by the transitional Junta that all documents should be destroyed, there was little written evidence, however the court adduced from the testimonies and the CONADEP report that the issuance of unlawful orders was evident, in light of various established facts. Among the factors deemed to be evidence of these orders were the effective control exercised by the commanders over their forces, the fact that the installation of clandestine detention and torture centres would not have been possible without express orders, and the consistency of the acts of criminal repression around the country.[254] At the same time, ample freedom was given to the subordinates in terms of what would be done with victims who had been kidnapped; whether they would be freed, handed over to the executive, subjected to trial or killed was up to the physical perpetrators.[255] As a 'natural consequence' of the system, many subordinates went further and committed crimes that had not been ordered.[256] In this respect a causal link was established between the crimes committed and the commanders. The notion of indirect perpetration was invoked by the prosecution based on both Article 514 of the CJM and Article 45 of the *Codigo Penal.*

The defence denied that the theory of control over the crime was the dominant theory in Argentina, and that in any case the notion of control over the acts based on an organised system of power was not applicable where the physical perpetrators were themselves culpable.[257] However the court found that the theory of *dominio del hecho*, or control over the crime, was in fact the prevailing theory in Argentina,

[250] Garro and Dahl 1987, pp. 321–322.

[251] Munoz-Conde and Olásolo 2011, p. 116.

[252] Causa 13 Trial 1985, paras 11, 13, 27.

[253] Causa 13 Trial 1985, para 15.

[254] Causa 13 Trial 1985, paras 35–43.

[255] Causa 13 Trial 1985, para 269.

[256] Causa 13 Trial 1985, para 271.

[257] Causa 13 Trial 1985, paras 275, 277, 278.

rejecting the formal objective theory in favour of more contemporary legal scholarship, citing Bacigalupo and Zaffaroni among others.[258] Since criminal liability for acts committed under orders related to the military service under the CJM was placed entirely on the shoulders of the superior who gave the orders, unless the subordinate officer exceeded such orders, the court determined that the CJM was predicated on a superior-subordinate relationship and a functional notion of liability rather than the formal objective theory.[259] This was deemed to be a statutory acceptance of the control over the crime theory, and also of the distinction between principal and secondary party, based on the fact that the superior who orders a crime is in control of the events.[260] It is worth remembering that this finding was after the amendment to the CJM.

The court looked to other domestic jurisdictions for further support for the theory of control over the crime, including Italian, Spanish and German legislation and literature, all of which had been influential on the drafting of the Argentine criminal code.[261] The conclusion was that the *Codigo* makes a distinction between the 'material' or physical perpetrators, and the 'moral' perpetrators, who ordered and 'determined' or caused the crimes to be committed.[262]

The court followed the doctrinal literature on *autoría mediata*, and reiterated that the indirect perpetrator controls the criminal act by controlling the will of the physical perpetrator. Therefore, the greater the freedom of the direct perpetrator, the less control the *hombre de atrás* will have over the act.[263] But since the defendants were always in control of the system and regime as a whole, even in cases where the physical perpetrators could be said to be liable, their acts were still under control of the defendants to some extent.[264]

The court found that the defendants had given themselves absolute control by way of force and through the decrees instilling full executive power in their hands. The defendants therefore controlled the machinery of people and property through which crimes were perpetrated, and the fact that they assured impunity to the direct perpetrators compounded this control. Under these circumstances, it was deemed to be unimportant who actually executed the crimes, due also to the fungibility of the subordinates; if someone refused to obey an order, he would be immediately replaced by another who would conform, and with the likely result that the person refusing would become the next target. The plan of the defendants could thus not be frustrated by the will of the physical perpetrators.[265] As the court put it, a person

[258] Causa 13 Trial 1985, para 282.

[259] Causa 13 Trial 1985, para 285.

[260] Causa 13 Trial 1985, paras 287, 288.

[261] Causa 13 Trial 1985, paras 288, 292.

[262] Causa 13 Trial 1985, para 421.

[263] Causa 13 Trial 1985, paras 293, 295.

[264] Causa 13 Trial 1985, para 301; Malarino 2008, p. 60.

[265] Causa 13 Trial 1985, paras 301, 304–305, 307–308.

who has control over the system has control over the will of all those who make up the system.[266]

The fact that the commanders did not know the identity of the specific perpetrators, nor of all of the victims, and where not aware of the existence of each individual criminal act, was deemed unimportant in determining their liability. This was because the identity of the physical perpetrator becomes unimportant, since control of the wills and therefore of the system as a whole rested with the defendants. Their orders were generally to detain 'all subversives', but they always retained the power to shift or stop the crimes, and in fact they intervened in the crimes directly through contributing the necessary means, such as clothes, vehicles, fuel, weapons, detention centres etc.[267]

In a somewhat confusing passage the court decided that it was also possible the defendants could be convicted as secondary parties, assisting in the crime as accomplices in the second degree, declaring it was possible to be both a principal and a secondary party, since they both fall under Article 45 of the *Codigo Penal*. The court reasoned that secondary participation would be subsumed by principal liability. Following a detailed exploration and reasoned application of *autoría mediata*, this conflation of modes of liability is disappointing, and is partly what led to the re-categorisation by the Supreme Court at appeal.

One thing that is clear is that a combined notion of indirect co-perpetration was rejected, due to the exclusive command each defendant had over his own forces. This is important with respect to the ICC's combination of indirect perpetration and co-perpetration and the factual difficulty of proving both a horizontal and a vertical relationship of control.[268] In the end, of the original Junta leaders, Videla and Massera received life sentences, and Agosti received four and a half years in prison. Two of the members of the subsequent juntas were also convicted, but four of them were acquitted. The careful reasoning of the court was generally well received, as it had set forth principles essential to re-establishing the rule of law, however there were many points of criticism, including the failure to better define the defence of due obedience.[269]

7.4.5.3 Supreme Court: Co-perpetration and Necessary Cooperation

Both the prosecution and the defence appealed the decision of the Federal Appeal Court, and a year later in December 1986 the Supreme Court of Argentina reconsidered the convictions. In general the Supreme Court judges agreed with the findings

---

[266] Author's own translation. The original text reads '*Quien domina el sistema domina la anónima voluntad de todos los hombres que lo integran.*' Causa 13 Trial 1985, para 310.

[267] Causa 13 Trial 1985, paras 309–311.

[268] Katanga Confirmation of Charges 2008, para 492; Separate Opinion of Judge Christina van den Wyngaert Ngudjolo Trial Judgment 2012, para 62. See further the discussion in Sect. 8.9.

[269] Garro and Dahl 1987, p. 344; Nino 1996, pp. 89–90.

of the Federal Appeals Court, and the convictions and sentences were maintained.[270] However the mode of liability applicable was a point of contention, and the majority overturned the analysis by the Federal Appeals court of the CJM, determining that instead the normal modes of liability under Article 45 of the *Codigo Penal* should be applied.[271] In overturning the finding on indirect perpetration, the Supreme Court decided that the mental control exercised by the defendants and their abuse of power for the purpose of committing crimes through others amounted to intellectual and factual cooperation with the subordinates. Imparting orders and supplying the means necessary to perpetrate the crimes rendered them *cooperadores necesarios*, or accomplices in the first degree, and not *autores*, or principals.[272]

Although the convictions and sentences remained mostly intact with this different reasoning, the majority finding that the Junta leaders were accomplices received strong dissent from Judges Petracchi and Bacque. They were in full agreement with the Federal Appeal Court that even where the physical perpetrator is fully criminally responsible, superior officers may retain control of the acts through the machinery of the State and thus be liable as indirect perpetrators. They saw in this a reasonable interpretation of the law, in line with Roxin's theory. The interchangeability of the subordinates was considered by them to be a fundamental aspect of this liability, as well as the discrete exercise of control over the general will of all of those who make up the machinery. The judges were of the opinion that indirect perpetration was suitable to the facts of the case because of the vertical structure of power in a military institution, and the possibility this creates for those at the top to make use of part or all of the forces under their control to perpetrate crimes.[273]

It must be emphasised that the decision of the Supreme Court was based upon a specific legislative interpretation of the amended CJM. Despite this finding, the original conviction of the Junta leaders as *autores mediatos* was a turning point in Argentine case law, and courts later returned to this approach in subsequent prosecutions of those involved in the crimes committed during the Dirty War, including those who had certain control over the crimes, but were not the first line of command. One prominent example was the case against General Suarez-Mason, who had been the head of the military zone which included Buenos Aires, under the first Junta, placing him directly below the Junta leaders. However he died before his trial completed.[274] In particular, principal liability has been confirmed as being based upon *dominio del hecho*, or 'control over the crime', as asserted by the Federal Appeals Court in the *Causa 13* trial judgment.[275]

---

270 With the exception of the sentences for Viola and Agosti, which were reduced by 6 months and 9 months respectively, following acquittals on certain counts. Causa 13 Appeal 1986, para 49.

271 Causa 13 Appeal 1986, para 17.

272 Causa 13 Appeal 1986, paras 23, 28.

273 Dissenting Opinion Justice Petracchi, Judge Baque joining, Causa 13 Appeal 1986, para 15.

274 Gibney 1992, p. 169; Suárez Mason Extradition Case 1988, p. 683.

275 Simón 2005; Malarino 2008, p. 61.

7.4.5.4 The *Punto Final* and Due Obedience Laws

Following the conviction of Videla and other members of the juntas by the Federal Appeals Court, and the arrest of other standing members of the military, the delicate balance that Alfonsín strove for was under threat. In 1986, under pressure from those wanting more convictions on the one hand, and concerned military personnel on the other, instructions were given by the minister of defense to the judges of the Supreme Council to expedite the cases in which officers with decision-making capacity were on trial, and to acquit all other defendants.[276] But this only led to more criticism, and many individual officers refused to cooperate with the courts, threatening the process of transitional justice. In response, the parliament enacted a law in December 1986 known as the *Punto Final*, or 'full-stop' law.[277] This law imposed a cut-off date for all trials, effectively a new statute of limitations, in an attempt to reach a compromise toward the resistant military by ending further prosecutions while also being able to maintain the triumph of the conviction of the Junta leaders. However in the short period before the the new cut-off date, several hundred depositions were entered and a new spate of prosecutions were filed by pro-active lawyers, countering the political intention of compromise.[278] As a result, Alfonsín lost credibility both with the military and with the general population.

The *Punto Final* law also had an unintended consequence, that it placed the burden of responsibility for the crimes of mass atrocity on the shoulders of officers who had held lower and mid-level positions during the military regime, since they were the ones targeted in the flurry of new depositions.[279]

In 1987 refusal by standing military officers to cooperate became even stronger, some of them garrisoning themselves and threatening a military revolt, which led to a crisis of confidence in the civilian government.[280] Under this increased pressure Alfonsín came up with the solution which really put the breaks on the prosecutions, since it eradicated the original policy of drawing a distinction between different levels of deliberative capacity and culpability. A new law on the 'due obedience' defence was passed in June of that year, which determined that all military and police officers acted following orders and were unable to question the authority and legitimacy of the orders they received.[281] This covered officers of all ranks, except the highest echelons, such as commanders-in-chief, chiefs of defense zones and chiefs of federal or provincial police forces. It also erased the previous amendment to the CJM, by allowing the due obedience defence to apply even to acts that were 'atrocious or abhorrent'.[282]

---

276 Nino 1996, p. 91.

277 Punto Final 1986.

278 Garro and Dahl 1987, p. 334.

279 Potash 1993, p. 57; Osiel 2001, pp. 18–19.

280 This was known as the 'Easter crisis'. Nino 1996, pp. 97–99; Parenti 2010, p. 493; Potash 1993, pp. 59–61.

281 Obedencia Debida 1987.

282 Obedencia Debida 1987, Article 1.

All of this brought the intended wider policy of prosecutions to a halt, and eventually to the defeat of Alfonsín by his opponent, Carlos Menem, in 1989. Menem ceased all prosecutions and enacted a general pardon, undoing what had been achieved in the previous years, in the name of reconciliation. Many would argue that this returned Argentina to a model of impunity, and of an attempt to forget the past and appease those who were threatened with prosecution.[283]

7.4.6 *Autoría Mediata in Subsequent Case Law*

As the political climate shifted again over time, coupled with a reduced threat of rebellion from the military, there was a willingness to return to the prosecutions that had first been envisaged. The *Punto Final* and *Obedencia Debida* laws were revoked in 2003,[284] however there remained some uncertainty as to whether the court had the power to retroactively declare laws null, until in 2005 the Supreme Court confirmed that the laws meant a failure to fulfil the duty to investigate and bring to justice serious violations of human rights under the Inter-American Convention on Human Rights.[285] In 2001 the IACHR had determined that similar pardons granted in Peru were a breach of the Convention on Human Rights as well as other international obligations to investigate and prosecute serious violations of human rights,[286] as a result of which the pardons that had been issued by President Menem to the Junta leaders were declared unconstitutional,[287] and the pardons that were extended to those who had been indicted but not yet prosecuted were declared null by the successor judges of the Federal Appeals Court in Buenos Aires and the National Criminal Appeal Court respectively.[288] All of this opened the doors again to continue prosecutions after nearly two decades of standstill, and the Argentine courts have continued to apply the theory of control over the crime as articulated in the original Junta conviction.[289]

As Olásolo points out, not all the potential cases of indirect perpetration followed this trend, and there are some examples of decisions by lower courts that only the Junta leaders were the proper indirect perpetrators, whereas any mid-level leaders were co-perpetrators of the acts together with their subordinates, since they provided the necessary conditions and assistance for the crimes to be committed.[290]

---

283 Sancinetti and Ferrante 1999, p. 51; Osiel 2001, p. 20; Palermo 2010, p. 603.

284 Law Revoking Punto Final 2003; Parenti 2010, pp. 495–496.

285 Bakker 2005, p. 1107; Simón 2005.

286 Barrios altos 2001.

287 Causa 13 Revoke of Pardons 2007; Causa 13 Casación 2009.

288 Mazzeo 2007; Suárez Mason 2003; Parenti 2010, p. 495.

289 Munoz-Conde and Olásolo 2011, p. 118.

290 Olásolo cites the *Floreal Avellanda* case, where three military officers who oversaw the Campo de Mayo Infantry School were indicted as indirect perpetrators, but were convicted as co-perpetrators instead. Munoz-Conde and Olásolo 2011, p. 119.

But in cases where *autoría mediata* has been applied, the reasoning of the Federal Appeals Court has been followed almost without exception, and today the notion of indirect perpetration by way of controlling an organisational hierarchy has been fully accepted in the case law.[291] Prominent cases include General Harguindeguy, who served under Videla,[292] and General Nicolaides, leader of the fourth Junta,[293] both of whom had benefited from Menem's pardons and were later prosecuted after the pardons were declared null. Similarly in the *Etchecolatz* case, in which the defendant was convicted for numerous counts of kidnappings, torture and killings during his time as Commissioner General of Police in Buenos Aires from 1976 to 1977, his role as indirect perpetrator was found to be integral in the crimes committed at numerous clandestine centres. The court cited the original conviction of the Junta leaders extensively as authoritative in defining this mode of liability, and cited Roxin directly as well.[294] And in 2008 the former head of the Argentine Army Third Corps Luciano Menéndez, and former head of Army Intelligence Hermes Rodríguez were convicted as indirect perpetrators, together with six physical perpetrators, for kidnapping, torturing and murdering numerous civilians in the province of Córdoba.[295] Both Menéndez and Rodríguez were leaders controlling the crimes executed by others, as they gave orders and instructions and created the necessary conditions for the anti-subversive repression to take place, and supervised their subordinates' criminal activities directly.

In 2013 there were approximately 1050 prosecutions open against individuals involved in the crimes committed during the Dirty War, of which approximately 560 have gone to trial.[296] The number of convictions has grown exponentially since the revocation of the *Punto Final* law, and the most prominent central mass cases utilise the notion of *autoría mediata*, such as the operations at the large ESMA clandestine torture and detention centre in Buenos Aires, and the La Perla case in the province of Cordóba.[297]

It is noteworthy that under *Codigo Penal* and the CJM there is no notion of superior responsibility, and that even with implementation of the statute of the ICC into Argentine law in 2000 there was no introduction of a notion to resemble Article 28 of the Rome Statute.[298] This may be one factor in the judicial preference for the doctrine which provides a foundation for the liability of superiors for the crimes of their subordinates.

---

291 Malarino 2008, p. 61.

292 Harguindeguy 2004.

293 Nicolaides 2000.

294 Etchecolatz 2006.

295 Menéndez 2008.

296 Fiscal 2013, p. 2.

297 Fiscal 2013, p. 4.

298 Malarino 2008, p. 66.

7.4.7 *The Impact of Argentine Case Law in the Latin American Region*

In other Latin American countries as well the notion of *autoría mediata* has been applied by courts, with explicit reference both to Roxin's writings and to the Argentine precedent. In Peru this mode of liability has been used in application both to a non-State group which has a hierarchical organisation, in the case against members of the violent Maoist party known as 'Shining Path', and also against former president Alberto Fujimori for his control over the State apparatus.

In the case against members of the 'Shining Path', the *Sala Penal Nacional* convicted members of this party for perpetration by means of an organised apparatus of power, despite there being no explicit codification of such a mode of liability. The Court stated that:

> [b]ecause of their dimensions, capacity, wherewithal, strategies and resources, adaptability, and coverage, these organizations are in a position to commit serious crimes with devastating consequences and to fulfill their objectives more efficiently and quickly [...] The concept of domination over the organization has emerged because other criminal categories, specifically the rules of perpetration and participation, are inadequate to explain and resolve cases involving those who direct and control an organization.[299]

As such, the Court found in a procedural decision that it is not necessary to have an explicit codification of this mode of liability, since it is compatible with the criminal definition of committing a crime.[300] It cited Welzel and Roxin directly, and emphasised the requirement of a hierarchical structure and the fungibility of the subordinate physical perpetrators within that structure.[301] However the Court did not follow the Argentine interpretation which allowed both the indirect perpetrator and the physical perpetrator to be considered principals. Instead it determined that:

> one of the advantages of the control-over-the-crime theory is that it distinguishes more clearly between the perpetrator and the participant: the perpetrator shall be the one with control over the configuration of the wrongdoing and the participant the one who does not exercise such control and only collaborates in the intentional wrongdoing of another.[302]

In the renowned case against former Peruvian president Alberto Fujimori,[303] the Court again cited Roxin directly, and referred to the way in which he developed his theory of *Organisationsherrshaft* based on an analysis of the Staschynski and Eichmann cases. The conclusion to which the court drew attention was that the different actors exercised different forms of control over the crime: the direct perpetrators of the crime have control over the action, but the mid-level commanders and the 'central organ of the power structure that had ordered their commission' have control over

[299] Manuel rubén abimael guzmán reynoso 2006.

[300] Shining Path, Motion for annulment and consults 2007, para 4.5.4.

[301] Shining Path, Motion for annulment and consults 2007, para 5.5.4.

[302] Shining Path, Motion for annulment and consults 2007, para 5.5.4.

[303] Fujimori 2009.

the organisation.[304] Based on this, the Court reasoned, it was not necessary that a commander exercise control over, nor even have any personal relationship with, the physical perpetrator. What is important is that the commander exercises control over the apparatus and its structure, to which the physical perpetrator belongs.[305] In this respect, the Court also referred to the Junta trials in Argentina as exemplary.

Since Fujimori had restructured and centralized the intelligence agencies, the armed forces and the police, thereby placing them under his direct command, while also relying on a complex system of corruption, embezzlement and bribery to finance secret military operations and to secure political support for his government, he had exercised sufficient control over the apparatus as a whole to warrant conviction as perpetrator-by-means.[306]

Other countries, too, have taken the Argentine example and interpreted their criminal codes as implicitly including the notion of *autoría mediata*, and applied this to leaders within organisational hierarchies. Already the Chilean Supreme Court had done so in its decision on the extradition request made by Peru for the prosecution of Fujimori, which was necessary under the 'double criminality' requirement for a grant of extradition.[307] The Chilean court determined that there was probable cause for an indictment against Fujimori as indirect perpetrator for a number of charges, including the infamous extra-judicial killings at Barrios Altos and the forced disappearances of 10 people at La Cantuta University in Lima, all of which were committed by members of the *Colina* group, which Fujimori had created as a clandestine unit of his military intelligence.[308]

Furthermore in Uruguay and Bolivia the notion of *autoría mediata* has been applied in situations involving control over an organisational hierarchy; in Bolivia with respect to paramilitary groups, over which the accused had full control and direct links,[309] and in Uruguay with respect to military leaders of the 'Condor Plan', involving coordination between the dictatorial regimes of Argentina, Brazil, Bolivia, Chile, Paraguay, and Uruguay.[310] In both of these cases Roxin is cited directly, and in the 'Condor Plan' case the Argentinean judgments are also referred to as exemplary.

One exception to this trend is Colombia, where the courts have referred to the notion of indirect perpetration by means of an organisational hierarchy, but have explicitly chosen to reject this in favour of convictions based either on the simpler form of co-perpetration or on instigation as a mode of secondary liability, in order to follow the wording of the Colombian criminal code more faithfully.[311] For co-perpetration, as long as it can be shown that the accused made a contribution at the stage of execution of the crime, even if this contribution was not 'essential',

---

304 Fujimori 2009, p 724.

305 Fujimori 2009, p. 725.

306 Ambos 2011, p. 138.

307 Ambos 2011, p. 142.

308 Ambos 2011, pp. 138,142; Fujimori Extradition Request 2007.

309 Leaders of the Left Revolutionary Movement 1993; Salazar and Urquiaga 2010, p. 95.

310 Condor Plan 2009; Salazar and Urquiaga 2010, p. 84.

311 Munoz-Conde and Olásolo 2011, p. 123.

this suffices to fulfil the legislative definition of a common criminal plan. In this way members of the National Liberation Army Central Command were convicted as co-perpetrators for a bombing for which there were no clearly identified physical perpetrators.[312] With respect to instigation, the Colombian Supreme Court has found that any communication from the *hombre de atrás* suffices, in the form of mandate, induction, advice, coercion, order or any other feasible means that effect the commission of a crime by the physical perpetrator who is fully culpable. The difference between co-perpetration and instigation according to the Colombian courts is that the instigator has no control over the final decision of the physical perpetrator, and thus should be convicted as a secondary participant.[313]

One reason Colombia decided to reject the notion of indirect perpetration by means of an organisational hierarchy is because the sentencing for an instigator is the same as for a principal, and thus it was not necessary to depart from the strict wording of the criminal code. This is the same argument offered by courts in Spain, and scholars who have rejected Roxin's theory in general.[314] However the general trend towards applying this mode of liability to leaders of structural organisations for gross human rights violations in Latin America can also be explained precisely because conviction as a secondary party does not adequately reflect the role of these superiors and does not correspond to their actual relevance in the occurrence of such crimes.[315] It would appear that the moral re-telling of the facts in the final conviction is considered to be more important in terms of truth-telling, than in terms of retribution when it comes to sentencing.

7.4.8 Summary

Although Roxin's theory of *Organisationsherrschaft* had not yet been applied by the courts in Germany, the fact that it had already been received into the scholarly doctrine in Latin America to some extent meant that in the period immediately following the resignation of the military regime in Argentina in 1983, there were already some solutions available as to how to hold the leaders liable for the crimes of mass atrocity that had taken place. It was a policy choice on the part of newly democratically elected President Alfonsín to focus on the leadership of the regime. Although some aspects of this policy were a result of pressure from a military not willing to co-operate with the prosecutions, it was also a deliberate choice on Alfonsín's part to aim in the first place for conviction of the leaders. This was for symbolic as well as retributive purposes, and to make a move towards reconciling the past by ensuring a public truth-telling in the trials and in the convictions.

312 Machuca 2007; Díaz 2008, p. 175.

313 Díaz 2008, p. 170; Munoz-Conde and Olásolo 2011, p. 123.

314 Munoz-Conde and Olásolo 2011, p. 123; Díaz 2008, p. 169. See also the reasoning in Causa 13 Appeal 1986.

315 Munoz-Conde and Olásolo 2011, p. 115.

The German system had already influenced legislation in Argentina, so it made sense to make full use of the doctrinal scholarship as an interpretative source of law in difficult cases of participation in collective crimes. In this respect it was also an important factor in this process that legal transplantations were already an integral part of the formation of Argentine legislation and doctrine throughout the twentieth century. The fact that it was a process of discussion and dissent shows that Argentina has been able to develop its own line of jurisprudence with respect to this legal transplant, formed by judges and scholars as the most prominent participants in this dynamic process of law.

Despite the fact that the Supreme Court majority decision overturned the findings on *autoría mediata* with respect to the Junta leaders, and replaced this mode of liability with the notion of necessary cooperation, or accomplice in the first degree, still the judgment of the Federal Appeals Court is regularly cited as the leading case on *autoría mediata* in Argentina, and in cases that have followed in other Latin American countries. In fact it has been asserted by the Peruvian Supreme Court that interpreting this type of participation in system criminality as co-perpetration, incitement or any other form of complicity (in the sense of secondary participation) is a confusion of the theory and an incorrect analysis of the modes of liability applied to the facts.[316] Perhaps the resistant authority of the Argentine Federal Appeals Court in the Junta case despite the Supreme Court decision, is because of the way Roxin's theory was applied for the first time by judges in exactly the way it had been envisioned; to close the legal gap where the physical perpetrators were fully aware of the unlawfulness of their actions and therefore culpable, but where it was a larger state apparatus and the policy and orders given by the leaders that formed the architecture of mass human rights violations. This demonstrates the inherent role of policy in forming the law, in response to systemic collective crimes.

7.5 Legal Culture and Policy Choices in the Civil Law Tradition

Generally, four distinguishing characteristics can be identified in the civil law tradition represented here by the comparison between Germany and Argentina. First, just as in the common law tradition, there is an inherent relationship between the character of criminal trials in the civil law tradition, and the system of liability that ensues, due at least in part to the role of participants in developing the law. Second, a heavily leaning objective approach can be recognised. In the jurisdictions compared here the notion of 'control over the crime' has been developed as an extension of the objective approach. This objective orientation leads to the conceptualisation of individual responsibility for collective crime as a distribution or attribution of blame for the crime, rather than vicarious liability as if the co-perpetrator had committed the act herself. Third, this distribution of blameworthiness leads to a normative dis-

[316] Fujimori 2009, p. 719.

tinction between parties; the individual is blameworthy to the extent of his or her co-operation, contribution or control. Fourth, the policy desire to maintain a moral distinction between principal and secondary parties when it comes to leaders of collective crimes has led to development of the notion of an indirect perpetrator, whose physical contribution may have been nil, but whose control or dominion over the crime leads to the classification of 'principal'. Each of these characteristics is summarised below.

7.5.1 *The Relationship Between the Character of Criminal Trials and a System of Liability*

The notions of justice, fairness and truth in the criminal trial could be said to differ from the common law tradition is some general ways. If we once again apply Damaška's analysis, the civil law tradition expresses a hierarchical organisation of authority, whereby the legislator and the professional judge both may form policy, but adjudication is less casuistic, and more inductive, applying a general rule to resolve the criminal law problems raised in a specific case.[317] The judge therefore has an active role, leading an enquiry, and the parties in a criminal trial take part in a process where they are considered to be suppliers of information.

Such a process is concerned most of all with establishing the substantive or objective truth as far as possible.[318] A 'procedural truth' that departs from the substantive truth still sometimes emerges, for example if evidence were to be excluded for technical reasons, however the aim of the enquiry is to secure the substantive truth as far as possible.[319] Modes of liability have the task of representing this truth in the convictions; thus the conviction of a person who has contributed in a more influential way is for the crime proper, whereas someone who has contributed in a less substantial way is convicted for assisting the other in the crime, but not for having committed the crime proper. This normative differentiation in qualification is reflected in the record-writing through the conviction, and has a moral and factual story to tell. Modes of liability thus become a shorthand for the moral responsibility of each actor. Usually (though not always) this leads to a mandatory differentiation in sentencing as well.

There is also typically little room for lay participation in the decision-making processes of a hierarchical organisation of authority,[320] which means that the development of a more complex taxonomy of modes of liability to reflect this moral, factual and legal differentiation is possible. Since the participants in a criminal trial are all professionals trained in the criminal law system, and in some way dedicated to its consistent application, there are no procedural accommodations needed to translate

317 Watson 1981, p. 32; Damaška 1984, p. 22.

318 Damaška 1984, p. 162; Safferling 2003, p. 267; Weigend 2003, p. 158.

319 Summers 1999, p. 498; Weigend 2003, p. 168.

320 Damaška 1984, p. 22.

legal technical concepts to lay decision-makers, and it is possible to ensure that the system of liability is applied as a part of the coherent whole.[321] A reasoned judgment is given by the judge who acts as both factual and legal decision-maker, in which the applied modes of liability are explained and justified.

This goes to the question of which participants in the processes of lawmaking and law-application are the most influential. Precisely because of their knowledge and expertise, much authority is bestowed on scholars as participants whose voices are instructive to legislators and judges alike. Doctrine provides an important source and guiding line in policy implementation. It could be said that scholars have this authority bestowed upon them because they have been able to build upon centuries of thinking and analysis of the existing criminal law system, maintaining a tradition of formal rationality.[322] In Germany, Roxin built upon the theories of Welzel, who had built upon theories of the nineteenth century, and in each doctrinal development there has been a careful and thorough analysis of the place such a theory can take in the system as a whole. The inheritance of such doctrine in Latin American countries in particular, but also in many European jurisdictions which follow the same tradition, has meant that the trends in these countries towards identifying leadership liability based on a functional rather than a physical test of contribution have been fairly similar.

Finally, at the risk of simplifying the character of a criminal trial in these systems, the tendency towards a more inquisitorial and policy-implementing style has impacted on the judicial responses to particular types of collective criminality. Where mass atrocity has occurred at a domestic level, the factual and moral story told by convicting those who were the intellectual authors and decision-makers as mere accessories was considered to be not only insufficient, but incorrect. These were the discussions made explicit in the *Junta* cases in Argentina and the *Mauerschützen* case in Germany.[323] In a system where the criminal trial can express policy implementation rather than being bound by the arguments presented by two parties, there has been a deliberate policy response to this problem. Judges have applied the doctrines that were developed by scholars as resolutions to this very problem. Leadership liability has thus developed to reflect a greater degree of blameworthiness for those who controlled the organisational apparatus through which mass atrocity was orchestrated, even when—or especially when—these leaders were physically distanced from the physical perpetration of the atrocities. In turn, the doctrinal developments have ensured internal consistency within the existing system of liability, even where they expand on the interpretation of the legislation.

---

321 Damaška 1984, p. 168.

322 Watson 1981, p. 23; Bohlander 2009, p. 9.

323 Causa 13 Trial 1985; Mauerschützen case 1994.

7.5.2 *Objectivity: Derivative Liability and Control over the Crime*

The prominent influence of scholars as participants in the law has led to more explicit policy shifts in the theories underpinning criminal liability over time, as demonstrated by the shift over the course of about 80 years in Germany and Argentina from a subjective, to a formal-objective, and more recently to a material-objective test of liability.[324]

While a shared intention is always a prerequisite for holding individuals responsible for collective crimes, the emphasis of the objective approach is the actual contribution to the fulfilment of the crime. Fletcher denotes an objective approach as manifest criminality, since its starting point is a wrongful act, and proceeds from there to ask who is culpable (blameworthy) and may therefore be liable for the act. Rather than attributing *acts* of parties to one another, objectivity attributes *responsibility*, according to a normative theory of culpability. Culpability is understood as blameworthiness, measured according to the degree of contribution to the crime. In this way, liability of parties is derivative and not vicarious.[325] The term 'accessory liability' is therefore correctly applied in these systems.

The derivative nature of liability for parties to a collective crime leads to a distinction between parties. Attribution of responsibility to a leader for crimes committed as part of an organised collective is based on contribution, but since physical contribution was deemed insufficient to explain the degree of blameworthiness of a leader, the functional test of 'control over the act' has become the prevalent test of leadership liability.[326] Whomever exercises the most control, the most influence, over the fulfilment of a criminal act is considered to be the most culpable (the most blameworthy), and to be punishable as a principal.

While some would argue that 'control over the crime' is not an objective test of liability, but rather a third alternative to the subjective and objective approaches,[327] in fact it is a test of liability which is predicated on an objective approach. Since the test looks to contribution of an actor in terms of influence and control, and aims to attribute responsibility for the acts of subordinates, it reflects the objectivity typical in the civil law tradition. It is true it departs from the formal-objective theory which requires an accused to have fulfilled the objective elements of the crime, but it does not depart from objectivity as an approach, and for this reason was also termed a material-objective approach.[328]

---

324 Staschynski 1962, para 134; Roxin 2006, p. 400; Stratenwerth 2005, p. 369; Bohlander 2009, p. 162; Hamdorf 2007, p. 210; Weigend 2011, p. 95.

325 Hamdorf 2007, p. 212.

326 Jescheck and Weigend 1996, p. 670; Hamdorf 2007, p. 210; Ambos 2008, p. 95; Weigend 2011, p. 26.

327 Katanga Confirmation of Charges 2008, para 484; Shining Path, Motion for annulment and consults 2007, para 5.5.4.

328 The nuances between different uses of the term 'objectivity' are discussed in full in Sect. 5.3.6.

7.5.3 Normative Differentiation Between Parties

This functional or material-objective approach in fact limits liability because it helps to delineate between principal and secondary liability. When the acts or will of a physical actor cannot be truly seen to be under the control of the actor 'behind' the physical perpetrator, then this renders the former a secondary party, because she is less blameworthy.[329] Conversely, where there is evidence of sufficient control on the part of this actor over the physical perpetrator, or over the way in which the crime is committed, then she is a principal because she is equally as blameworthy as the physical perpetrator, and in some cases even more blameworthy. Culpability is thus a normative concept, and not a descriptive one, since it goes beyond the different levels of mental elements applicable to a crime, and instead denotes the degree of blame according to the contribution made by an individual.

Given Damaška's analysis of the tendency toward a hierarchical organisation of authority in government in the civil law tradition, which is what leads to a criminal trial that is more focused on policy-implementation than conflict resolution,[330] then it is perhaps no surprise that the analysis of power within collective criminal activity is also hierarchical. Normative differentiation between parties to a crime could be said to follow from the understanding of differentiation in authority among State apparatuses. Similarly, in the common law tradition if it is true there is a tendency towards a horizontal organisation of authority, leading to lay participation in courtroom decision making, then it is also logical that the analysis of power and responsibility within collective criminality takes a similar horizontal approach to liability, holding all participants equally liable regardless of their actual contribution to the crime.

7.5.4 Indirect Perpetration and Collective Atrocity

The policy intentions behind the development of modes of liability in the domestic context need to be taken into account when considering the possible translation to the international plane. For this reason, the development of indirect perpetration by means of an organisation is of great interest when it comes to leadership liability for mass atrocity crimes, because of the way in which it was formed as a policy response to atrocities committed at both the domestic and international level. Roxin built his theory upon the earlier work of Welzel, but he developed it specifically in response to the conviction of Eichmann in Israel, the outcome of which he and many others found unsatisfactory because it did not properly reflect the moral and factual role played by Eichmann.[331] When his theory was applied by Argentine courts in

[329] Sancinetti 1991, p. 634, citing German theorist Edmund Mezger.

[330] Damaška 1984, p. 33.

[331] Eichmann 1961; Etchecolatz 2006; Rotsch 2009, p. 551.

response to atrocities committed domestically as part of the Dirty War,[332] its utility in delineating between parties to a crime became apparent, and not only did many other Latin American courts follow this example,[333] but the German courts later did the same in response to atrocities committed at its East and West borders.[334]

Roxin's theory of perpetration by means of exercising influence over an organisational hierarchy has clearly been very influential in jurisdictions which have taken the German model of criminal law as a base, particularly in Latin America. The suitability of this theory to deal with the moral culpability of the intellectual authors of systems of mass violence is illustrated by the very fact that it has been applied specifically to cases of mass atrocity and complex system criminality in Argentina, later in Germany and in various other jurisdictions in South America. The breach of human rights on a mass scale does not take place at the instigation of a few, but through the creation of a norm of violence in which many will take part precisely because it has become a system unto which they submit their individual will. It is for this reason that the ICC has also looked to this theory to resolve the problem of complex system criminality. Although the criticism it has received for importing directly from a domestic jurisdiction may be understandable, it will be discussed in the following chapter why the final result is truly appropriate for the specific context of ICL.

References

Ambos K (2006) Toward a universal system of crime: comments on George Fletcher's grammar of criminal law. Cardozo Law Rev 28:2647–2673

Ambos K (2007) Joint criminal enterprise and command responsibility. J Int Crim Justice 5:159–183

Ambos K (ed) (2008) Imputacíon de Crímenes de los Subordinos al Dirigente: Un Estudio Comparado. Editorial Temis S.A, Bogotá

Ambos K (2011) The Fujimori Judgment: a president's responsibility for crimes against humanity as indirect perpetrator by virtue of an organized power apparatus. J Int Crim Justice 9:137–158

Arendt H (1964) Eichmann in Jerusalem: a report on the banality of evil. Penguin Press, New York

Bacigalupo E (1994) Manual de Derecho Penal: Parte General, segundo edn. Editorial Temis S.A, Santa Fé

Bakker CA (2005) A full stop to amnesty in Argentina: the Simón case. J Int Crim Justice 3:1106–1120

Bohlander M (2009) Principles of German criminal law. Hart Publishing, Oxford

CJM (1951) Codigo de justicia militar

CJM Amended (1984) Enmiendas codigo de justicia militar

Codigo Penal (1984) Codigo penal de la nacion Argentina

CONADEP (1984) Nunca más (Never again): Report of the Comisión Nacional sobre la Desaparición de Personas (CONADEP). http://www.desaparecidos.org/nuncamas/web/index2.htm

---

[332] Causa 13 Trial 1985; Simón 2005; Etchecolatz 2006; Menéndez 2008; Malarino 2008, p. 61; Munoz-Conde and Olásolo 2011, p. 116.

[333] Case Against Leaders of the Shining Path 2006; Fujimori 2009; Leaders of the Left Revolutionary Movement 1993; Condor Plan 2009.

[334] Mauerschützen case 1994; Berlin Wall II 1999, 2002.

Condor Plan (2009) José Nino Gavazzo Pereira y Otros
Constitución Nacional de Argentina (1994) Convención Nacional Constituyente
D'Alessio AJ (2008) La resistencia a un proceso penal acorde con la constitución Argentina. Revista Argentina de Teoria Juridica 9(1)
Damaška M (1984) The faces of justice and state authority: a comparative approach to the legal process. Yale University Press, New Haven
Díaz CL (2008) El case Colombiano. In: Amobs K (ed) Imputacíon de Crímenes de los Subordinos al Dirigente: Un Estudio Comparado. Editorial Temis S.A., Bogotá
Documento Final (1983) Documento final de la junta militar sobre la guerra contra el subversion y el terrorismo. http://issuu.com/memoriaabierta/docs/documento_final_de_la_junta_militar/9
Drumbl MA (2005) Collective violence and individual punishment: the criminality of mass atrocity. Northwestern University Law Review, Winter, p 539
Dubber MD (2005) The promise of German criminal law: a science of crime and punishment. German Law J 6(7):1049–1071
Ferrante M (2011a) Argentina. In: Heller KJ, Dubber MD (eds) The handbook of comparative criminal law. Stanford University Press, pp 12–49
Ferrante M (2011b) Introduccion al Derecho Penal Argentino. Ad Hoc, Argentina
Fiscal MP (2013) Informe sobre el estado de las causas por violaciones a los derechos humanos cometidos durante el terrorismo de estado. https://docs.google.com/file/d/0B8_g3dCcTjYJVEhEakVTV0xscWc/edit
Fletcher GP (2000) Rethinking criminal law. Oxford University Press, USA
Fletcher GP (2007) The grammar of criminal law: American, comparative and international, Vol One: Foundations. Oxford University Press, New York
Fujimori Extradition Request (2007) Decision on Extradition Request Fujimori. www.latinreporters.com/perouFujimoriFalloExtradicionChile21092007.pdf
Fukuyama F (2011) The origins of political order: from prehuman times to the French revolution. Farrar, Straus and Giroux, New York
Garro AM, Dahl H (1987) Legal accountability for human rights violations in Argentina: one step forward and two steps backward. Hum Rights Law J 8:283–344
Gibney M (1992) The implementation of human rights as an international concern: the case of Argentine General Suarez-Mason and lessons for the world community. Case West Reserve J Int Law 24:165–198
Glenn HP (2010) Legal traditions of the world: sustainable diversity in law, 4th edn. Oxford University Press, Oxford
Grundgesetz (1949) Grundgesetz
Hamdorf K (2007) The concept of joint criminal enterprise and domestic modes of liability for parties to a crime: a comparison of German and English law. J Int Crim Justice 5:208
Hernandez V (2013) Painful search for Argentine's disappeared. http://www.bbc.co.uk/news/world-latin-america-21884147
Informe sobre la Justicia (2012) Informe sobre la justicia en lo criminal y correccional federal y sobre la justicia administrativa en la republica Argentina. http://www.oas.org/juridico/PDFs/mesicic4_arg_sc_inf_fores.pdf
Jakobs G (1995) Mittelbare täterschaft der mitglieder des nationalen verteidigungsrat, aanmerkung. Neue Zeitschrift für Strafrecht p 26
Jakobs G (1997) La Imputacion Objetiva en Derecho Penal. Ad Hoc, Buenos Aires
Jescheck HH, Weigend T (1996) Lehrbuch des Strafrecht, Allgemeiner Teil, 5th edn. Duncker & Humbolt, Berlin
Kleinheisterkamp J (2006) Development of comparative law in Latin America. In: Reimann M, Zimmerman R (eds) Oxford handbook of comparative law, Oxford
Koschaker P (2000) Europa en het Romeinse Recht, 3rd edn. W.E.J, Tjeenk Willink, Deventer
Krey V (2002) Deutsches Strafrecht Allgemeiner Teil: Lehrbuch in Deutsch und Englisch. W Kohlhammer, Stuttgart
Law Revoking Punto Final (2003) Declárase insanablemente nulas las Leyes Nros. 23.492 y 23.521

Malarino E (2008) Argentina. In: Ambos K (ed) Imputacíon de Crímenes de los Subordinos al Dirigente: Un Estudio Comparado. Editorial Temis S.A., Bogotá
Miller JM (2003) Typology of legal transplants: using sociology, legal history and Argentine examples to explain the transplant process, a. Am J Comp L 51:839–885
MPC (1962) Model Penal Code
Munoz-Conde F, Olásolo H (2011) The application of the notion of indirect perpetration through organized structures of power in Latin America and Spain. J Int Crim Justice 9:113–135
Nack A (2006) Mittelbare täterschaft durch ausnutzung regehafter abläufe. GA pp 342–345
Nino C (1980) Los Limites de la Responsibilidad Penal. Editorial Astrea
Nino C (1996) Radical evil on trial. Yale University Press, New Haven
Obedencia Debida (1987) Ley de obediencia debida
Oderkerk M (2001) The importance of context: selecting legal systems in comparative legal research Neth Int Law Rev XLVIII:293–219
O'Donnell MK (2009) The banality of good: aligning incentives against mass atrocity. NY Univ Law Rev 83:333–374
Olásolo H (2009) The criminal responsibility of senior political and military leaders as principals to international crimes. Hart, Portland, Oregon
Osiel M (2001) Mass atrocity, ordinary evil, and Hannah Arendt: criminal consciousness in Argentina's dirty war. Yale University Press, Boston
Osiel M (2009) Making sense of mass atrocity. Cambridge University Press, Cambridge
Palermo PG (2010) The prosecution of international crimes in Uruguay. Int Crim Law Rev 10:601–618
Parenti PF (2010) The prosecution of international crimes in Argentina. Int Crim Law Rev 10:491–507
Potash RA (1993) The military under Alfonsín and Menem: the search for a new role. In: Lewis CM, Torrents N (eds) Argentina in the crisis years. From Alfonsín to Menem, Institute of Latin American Studies, London, pp 1983–1990
Pradel J (2002) Droit Penal Comparé. Dalloz, Paris
Punto Final (1986) Ley de punto final
Rotsch T (2009) Von Eichmann bis Fujimori - zur rezeption der organisationsherrschaft nach dem urteil des obersten strafgerichtshofs perus. Zeitschrift für Internationale Strafrechtsdogmatik, pp 549–551
Roxin C (2005) Täterschaft und Tatherrschaft, 8th edn. De Gruyter Recht, Berlin
Roxin C (2006) Voluntad de dominio de la acción mediate aparatos de poder organizados. In: Nez RCN (ed) Doctrina Penal: Tería y Práctica en las Ciencias Penales. Edward Elgar Publishing, Cheltenham, UK/Northampton, MA, USA, pp 273–288
Safferling C (2003) Towards an international criminal procedure. Oxford University Press, Oxford
Salazar K, Urquiaga XM (2010) Digest of Latin American jurisprudence on international crimes. Due Process of Law Foundation, Washington
Sancinetti MA (1991) Teoría del Delito y Disvalor de Acción. Hammurabi, Buenos Aires
Sancinetti MA (1997) Ilícito Personal y Participación. Ad Hoc, Buenos Aires
Sancinetti MA, Ferrante M (1999) El Derecho Penal en la Protección de los Derechos Humanos. Hammurabi, Buenos Aires
Schermaier MJ (2006) German law. In: Smits JM (ed) Elgar encyclopedia of comparative law. Edward Elgar Publishing, Cheltenham, UK/Northampton, MA, USA, pp 273–288 (Chapter 26)
Schumann H (2005) Criminal law. In: Zekoll J, Reimann M (eds) Introduction to German law. Kluwer Law International, The Hague
Self-amnesty law (1983) Ley de pacificación nacional
Soler S (1992) Derecho Penal Argentino, 10th edn. Tipografica Editoria Argentina
Strafgesetzbuch (1870) Strafgesetzbuch
Stratenwerth G (2005) Strafrecht, Allgemeiner Teil I: Der Straftat (Translated as: Derecho Penal, Parte General I: El hecho punible). Hammurabi, Buenos Aires

Summers RS (1999) Formal legal truth and substantive truth in judicial fact-finding—their justified divergence in some particular cases. Law Philos 18:497–511

Tallgren I (2002) The sensibility and sense of international criminal law. Eur J Int Law 13(3):561–595

Urban C (2004) Mittlebare Täterschaft kraft Organisationsherrschaft. V & R Unipress, Göttingen

van Sliedregt E (2012) Individual criminal responsibility in international law. Oxford University Press, Oxford

Vormbaum T, Bohlander M (eds) (2014) A modern history of German criminal law. Springer, Heidelberg

Watson A (1981) The making of the civil law. Harvard University Press, Massachusetts

Watson A (1995) From legal transplants to legal formants. Am J Comp Law, AMJCL 43:469–476

Weigend T (2003) Is the criminal process about truth? A German perspective. Harv J Law Public Policy 26:157–173

Weigend T (2006) Criminal law and criminal procedure. In: Smits JM (ed) Elgar encyclopaedia of comparative law. Edward Elgar Publishing, Cheltenham, UK/Northampton, MA, USA

Weigend T (2011) Perpetration through an organization: the unexpected career of a German legal doctrine. J Int Crim Justice 9:91–111

Welzel H (1970) Das Deutsche Strafrecht (Translated as Derecho Penal Aleman: Parte General). Editorial Juridica de Chile, Santiago

Zaffaroni ER (2005) Manual Derecho Penal: Parte General. Sociedad Anónima Editoria, Buenos Aires

Zweigert K, Kötz H (1998) An introduction to comparative law. Oxford University Press, Oxford

Cases

Apothek (2003) 1 StR 453/02

Barrios Altos (2001) Barrios Altos v. Peru No. 75, Inter-American Court of Human Rights Ser C

Berlin Wall II (1999) Berlin Wall II, 5 StR 632/98

Berlin Wall III (2002) 5 StR 282/01, BeckRS 200230291953

Betrug (2007) 2 StR 385/07

Case against Leaders of the Shining Path (2006) Manuel Rubén Abimael Guzmán Reynoso, y otros, Expediente 560-03

Causa 13 Appeal (1986) Causa originariamente instruida por el consejo supremo de las fuerzas armada en cumplimiento del decreto 158/83 del poder exejutivo naciónal, Corte Suprema de Justicia de la Nación

Causa 13 Casación (2009) Jorge Rafael Videla y Emilio Eduardo Massera S/Rec de Casación, Cámera Nacional de Casación

Causa 13 Revoke of Pardons (2007) Jorge Rafael Videla y Emilio Eduardo Massera, 13/84, Cámera Nacional de Apelaciones en lo Criminal y Correccional Federal de Buenos Aires

Causa 13 Trial (1985) Sentencía del juicia a las juntas (1985) Cámera Nacional de Apelaciones en lo Criminal y Correccional Federal de Buenos Aires

Eichmann (1961) Attorney General of Israel v. Eichmann 36 (1968) ILR (District Court of Israel)

Eichmann Appeal (1962) Attorney General of Israel v. Eichmann, appeal, 36 (1968) ILR (Supreme Court of Israel)

Etchecolatz (2006) Miguel Osvaldo Etchecolatz, 2251/06, Tribunal Oral en lo Criminal Federal (La Plata)

Fujimori (2009) Exp. No. AV 19-2001

Harguindeguy (2004) Eduardo Harguindeguy y otros, causa 35/799, Cámera Nacional de Apelaciones en lo Criminal y Correccional Federal de la Capital Federal

Katanga Confirmation of Charges (2008) The Prosecutor v. Germain Katanga and Mathieu Ngudjolo, Decision on the confirmation of charges, ICC-01/04-01/07

Leaders of the Left Revolutionary Movement (1993) Luis García Meza Tejada y otros, Corte Suprema de Justicia de la Nación, 23 April 1993

Machuca (2007) El caso Machuca, No. 23825

Mauerschützen case, (1994) 5 StR 98/94 (LG Berlin). NJW 1994:2703

Mazzeo (2007) Sr. Fiscal General solicita des archivo de causa que tramitaron por arto 10 ley 23.049 S, 387, XLIII, Corte Suprema de Justicia de la Nación

Menéndez (2008) Menéndez Luciano Benjamin, Rodríguez Hermes Oscar et. al (2008) Tribunal Oral en lo Criminal Federal de Córdoba No 1

Ngudjolo Trial Judgment (2012) Prosecutor v. Mathieu Ngudjolo Chui, ICC-01/04-02/12

Nicolaides (2000) Cristino Nicolaides y otros, Causa 31369, Cámera Nacional de Apelaciones en lo Criminal y Correccional Federal de la Capital Federal

Schuldprinzip (1952) BGSt 2 194 (Bundesgerichtshof Strafkamer)

Shining Path, Motion for annulment and consults (2007) Manuel Rubén Abimael Guzmán Reynoso y otros, motion for annulment and consults, R.N. no. 5385–2006, Segunda Sala Penal Transitoria, Suprema Corte de Justicia

Simón (2005) Simón, Julio Héctor y otros s/ privación ilegítima de la libertad, etc, Causa 17.768, Corte Suprema de Justicia de la Nación

Staschynski (1962) StE 4/62

Suárez Mason (2003) Suárez Mason, Guillermo y otros, Causa 14.216/2003, Criminal y Correccional Federal 3 de la Capital Federal

Suárez Mason Extradition Case (1988) Matter of extradition of Suarez-Mason F Supp 694 676 (District Court of California)

Verfahrenseinstellung (2003) BGH 5 Str 145/03

Chapter 8
Shifting Trends in International Tribunals

Contents

C. Steer, *Translating Guilt*, International Criminal Justice Series 9,
DOI 10.1007/978-94-6265-171-5_8

> ...(C)reativity was the mantra of the new Courts. It had to be. Any court, no matter what its origin or placement in the wider justice system, confronts daily myriad questions for which there is no ready answer in its statute or precedent ... In brand new courts they must be invented from scratch.
>
> — Judge Patricia Wald, 'Notes from the Bench'[1]

In this chapter a historical perspective will be taken on the process by which the modes of responsibility have been created and recognised in international and internationalised courts and tribunals in the last seventy years, with an eye to whether a more objective or subjective approach appears to prevail in the design and application of each one. Attention will be paid to the discussions which led to the inclusion of each mode of responsibility, and it will be considered whether there are distinguishable factors which may have determined their inclusion in a given statute, or their selection in judgments. It will therefore be of great interest to further consider the debates among participants as to the acceptance or controversy of each mode of liability, and to what extent the background legal training of these participants influences their perspective.

One thing that cannot be ignored is the central role played by these international tribunals, and the individual judges on the benches of these tribunals, in re-interpreting IHL and ICL, and in creating new law. Although at the inception of the ICTY, members of the Security Council declared that the tribunal should not and would not create new law, it seems that States have since embraced the innovations and new interpretations made by prosecutors and judges at the ad-hoc tribunals, including codifying some of these new legal notions in the treaty which brought the ICC and its statute to life.[2]

It would be difficult to draw a conclusion as to whether each of the modes of liability discussed below is 'good' or 'bad' law, in the sense that these terms are sometimes used to assess judicial findings. The intention is rather to document the process by which these modes of liability have developed, to map the interactions

---

[1] Darcy and Powderly 2010, p. xxxvi.

[2] Danner 2006, p. 4.

between those participants involved in this process, and to consider their applicability to the specific context of mass atrocity crimes, particularly the deliberative processes that lead to collective violent behaviour. These modes of liability will be discussed here in the order of historical importance, beginning with the first prosecutions for individual criminal liability for international crimes after World War II, and moving through the twentieth and twenty-first centuries to date. First, however, the context of legal tradition will be discussed, since this is a unique question for ICL which shows similarities with both traditions compared here, but is not truly representative of either.

8.1 Which Tradition? The Context of ICL

Whereas each of the domestic jurisdictions compared in the previous chapters were selected as representative of the respective legal traditions they follow, ICL cannot be located in any particular tradition. Since the creation of the Nuremberg tribunal, the attempt has been to combine elements of both the civil law and common law traditions, and in more recent years with the aim to combine the best of both worlds. However as many experts in criminal procedure would testify, the result is sometimes the worst of both worlds, or at best a clash of two worlds.[3] This clash of legal cultures takes place in substantive law as well, and it would appear that the search for the most appropriate modes of liability is where this clash and debate is highly concentrated.[4] By borrowing from familiar rhetoric in domestic criminal law systems, the international criminal justice system is assumed to function following the mechanisms of an ideal national system that cannot be localised anywhere.[5]

This assumption may lead to disappointment if what we seek is uniformity. Indeed, Fletcher has argued that a fully functioning system of ICL may be very difficult to achieve because a universal set of values and principles does not exist.[6] However, as was argued in Chap. 4, if we instead embrace the inherent pluralistic nature of this international system of law and its multiple participants, then it may be possible to seek harmony within the differences, and over time more consensus may be found. This is possible if the patchworking process is made transparent, if the bias of selectivity from among domestic jurisdictions is reduced, and if choices made to import certain domestic notions is justified in the context of mass atrocity crimes. It may indeed be possible to combine the 'best of both worlds', and come to a predictable system of liability that fulfils the goals and functions of ICL and provides some certainty for defendants.[7]

---

[3] Cassese 2008b, p. 377; Zahar and Sluiter 2008, p. 37; Mégret 2010, p. 43.

[4] Stewart 2012; van Sliedregt 2012b; Steer 2011; Steer 2015.

[5] Tallgren 2002, p. 567.

[6] Fletcher 2011, p. 179.

[7] Steer 2015, p. 35.

But such a process takes time. In any case it is possible to observe, as the historical investigation in this chapter will demonstrate, that there are elements of both the common law and civil law traditional processes in the development of modes of liability on the international plane. For instance, the process of iteration and bifurcation described in Chap. 6 as typical of the common law tradition can be observed in the reliance on both domestic and international case law. This may be partly due to the fact that judges with a common law background automatically look to previous decisions as 'precedent', without necessarily considering whether these previous cases in fact demonstrate (the formation of) custom.[8] However the tendency to refer to previous case law appears to be prevalent across the board; through repetition and reference to previous case law a reification takes place of certain interpretations of modes of liability. This is particularly evident at the ad-hoc tribunals. On the other hand, particularly at the ICC, there is an increasing reliance on scholars' interpretations of international Statutes, and also on domestic law scholarship, reflecting the tradition in civil law systems of relying on those who can find a synthesis among the conflicting formal sources of law and make a consistent, teachable claim about what the law is.

What also becomes evident in this comparison on the international plane is that modes of liability are always formed as a response to a historical reality, just as they are in the domestic jurisdictions compared in the previous chapters. At Nuremberg and Tokyo, the focus was on bureaucratic State organisation and the hierarchies within them that led to mass atrocity. Thus, the first models of individual criminal liability at the international level were an attempt to encompass as many parties within the hierarchies as possible. As will be discussed in Sect. 8.2, the Nuremberg Charter and the Tokyo Charter were both functionally unitary models of liability, and the Nuremberg Principles followed suit. When the International Law Commission (ILC) was given the task of drafting a Draft Code of Crimes Against Peace and Mankind (ILC Draft Code), it also began with a simple, functionally unitary model, and there was no article on individual criminal liability.[9] The 1994 and 1996 Draft Codes introduced a single article on liability, which encompassed intentional commission of a crime, superior responsibility for ordering or failing 'to prevent or repress the commission' of a crime; knowingly aiding, abetting 'or otherwise assist[ing], directly and substantially'; responsibility for high and mid-level superiors for 'directly participat[ing] in planning or conspiring to commit' a crime; direct public incitement; and attempt.[10] From the commentary it could be inferred that a theory of derivative liability would apply to accomplice liability, rather than vicarious liability, since 'an individual is held responsible for his own conduct which contributed to the commission of the crime, notwithstanding the fact that the criminal act was carried out by another individual',[11] however it also appears that all these modes were intended as alternate

[8] Cassese 2008b, p. 18.

[9] Schabas 2010, p. 423.

[10] ILC Draft Code of Crimes 1996, Article 2(3).

[11] ILC Draft Code of Crimes 1996, para 11.

descriptions of participating in a crime, in a functionally unitary sense, and that no normative differentiation was intended.[12]

In 1997 a joint proposal was submitted by Canada, Germany, the Netherlands and the United Kingdom, described as an informal group representing various legal systems, the text of which became the basis of Article 25 of the Rome Statute.[13] It received fairly rapid and wide agreement among the States Parties involved in the drafting process, and for the first time on the international plane, an extensive description of differentiated modes of liability appeared. The debate on the interpretation and application of this article will be discussed at length below in Sects. 8.8–8.10. The point here is that the drafters of the Rome Statute learned much from the discrepancies and lack of consensus at the ad-hoc tribunals and among scholars, and there was an attempt to pay more attention to this specifically technical criminal law aspect of an international trial. Whereas all the ad-hoc tribunals, from Nuremberg on, were spontaneous creations in response to specific conflicts, whose statutes were drafted hastily and without careful deliberation as to the system of liability, the ICC has been a creation over time, with more opportunities to consider, debate and learn from recent developments.[14]

Although the Rome Statute was still not a creation of criminal law experts, but primarily of international law experts and diplomats,[15] the joint proposal submitted in 1997 was a moment at which the coming together of both civil law and common law traditions was shown to be possible, to create a system of liability that was more carefully considered. This, too is a result of a historical reality, however the debates still continue as to how this article should be interpreted and applied. The shifts in approaches from Nuremberg to the ICC will be mapped out in the following sections.

8.2 Conspiracy-Complicity: Prevalence of the Subjective Approach

While the Nuremberg and Tokyo tribunals are often looked to as the beginnings of ICL, there is nothing intrinsic in the procedures and substantive law that was applied. Although the new crimes and modes of responsibility were justified as being based on customary law, the entire process and the content of the norms was unprecedented and therefore a process of inception rather than discovery. The content of the norms was determined along the way according to what the political and legal players of the allied countries could agree upon. The inclusion and development of the modes of responsibility of conspiracy, complicity and criminal enterprise is therefore by no means pre-determined.

12 ILC Draft Code of Crimes 1996, para 6.

13 ICC Working Paper 1997.

14 Werle and Jessberger 2014, p. 25.

15 Werle and Jessberger 2014, p. 136; Boas et al. 2007, p. 300; Fletcher 2007, p. 324; Schabas 2010, p. 291.

The World War II trials, including the prosecutions under Control Council Law No. 10, are a point of reference for the development of modes of liability in ICL ever since. Due attention to detail is therefore given in this section to these trials in particular.

8.2.1 The International Military Tribunal at Nuremberg

Perhaps one of the most famous statements to emerge from the Nuremberg process was Prosecutor Jackson's axiom that 'crimes against international law are committed by men, not by abstract entities, and only by punishing individuals who commit such crimes can the provisions of international law be enforced.'[16] It was a paradigmatic shift away from the act-of-State doctrine,[17] and towards individual criminal liability, which required some rapid and deep doctrinal thinking. This process laid bare the differences in approach of law-makers and policy makers with differing domestic law backgrounds, throughout the design and drafting process of creating the International Military Tribunal at Nuremberg (IMT), as well as during the trials themselves, and the final judgments. How should the individuals who were clearly the leaders of the entire regime be punished with respect to massive crimes committed across broad geographical expanses by thousands of other individuals?

8.2.1.1 Political Will and the Debate Surrounding Criminal Trials

It was never a given that the political and military leaders within the Nazi regime would be put on trial. The difference of opinion was also one of difference of perspective, political interest and legal analysis.[18] In the negotiations among the Allies as to whether to put Nazi leaders on trial, the French and English representatives were convinced it would be unprecedented and too complex, and argued for summary execution. The focus, according to these parties, should be on de-Nazification and re-building the countries that had been dishevelled by occupation.[19] On the other hand, many Americans advocated a military trial, in particular Secretary of War Henry Stimson.[20]

When President Truman endorsed the plan of an international military tribunal, resistance from some of the allies continued, particularly as they could not see under what laws such prosecutions could be established, and argued it would be in breach of the principle of *nullum crimen*.[21] Churchill remained so convinced of his position

---

[16] Closing statement, Nuremberg Judgment 1945.

[17] Zoller and Reshetov 1990, p. 117.

[18] See e.g. Ascensio 2006; Brown 2006a, b; Cesarani 2006.

[19] Ginsburgs 1990, p. 23.

[20] Ryan 2007, p. 59.

[21] Lafontaine 2012, p. 29.

against prosecutions that even after the UN War Crimes Commission had been established to begin work on setting up the trials, he ordered his law officers to draw up a list of offenders who were to be summarily shot. His War Office and Foreign Office moved to prevent this, and convinced Churchill to cooperate with the American-led plan.[22]

Eventually, public opinion gave sway. Indeed, Telford Taylor is quoted as having said that the allies were justified in creating this tribunal and its applicable law, in order to create 'not a legal thing, but a political thing'.[23] Questions such as whether an international military tribunal could be established according to international law, and what the legal basis was for the crimes and modes of participation, would be left to the lawyers and judges once the trials had begun. This moral justification was criticised, and the military tribunal was accused of being an instrument of victor's justice. As one minister in Churchill's war cabinet said, it if were a crime, then each of the Allies was equally as guilty in recent history.[24] However the sense of moral high ground prevailed, given the extent of the horrors committed during the Nazi regime.

This process may well be typical of ICL; following the formation of a policy of prosecution, determination of the law is left up to the institutions. The entire process is heavily under the influence of *realpolitiek*.[25] Many would argue that the prosecutions at Nuremberg were a breach of the principle of legality, because the crimes were retroactively defined.[26] As will be discussed later in this chapter, the inception of the ICTY and, moreover, of the mode of responsibility known as JCE, which is a form of conspiracy-complicity, have followed a very similar process. In the end, the law is shaped to fit the facts of the situation and arguments are formed thereafter to ground these solutions in so-called customary law.

8.2.1.2 The Inception of Individual Responsibility

There was a combined approach throughout the creation and procedure of the IMT; on the one hand there was a radical move to prosecute and hold liable *individual* members of a regime. On the other hand there was still a focus on the concerted efforts of the Nazi regime as a whole, and the policies beyond the individual crimes.[27] This combination of perspectives involving both individual and collective agency's meant that during the policy design of this new tribunal, disagreement prevailed as to the kinds of liability that should apply. The importance of agency was one of the

---

[22] Cesarani 2006, p. 34.

[23] Telford Taylor, 'An Approach to the Preparation of the Prosecution of Axis Criminality', (June, 1945), cited in Ryan 2007, p. 70.

[24] Cesarani 2006, p. 37.

[25] Ambos 2007a.

[26] See e.g. Lafontaine 2012, p. 29.

[27] Ryan 2007, p. 61.

core problems to be dealt with as individual criminal responsibility was being forged for the first time under ICL.

In January 1942, early in the proposal stages for the IMT, an Inter-Allied Declaration was signed by the representatives of nine countries that had been occupied by Germany, in which the States resolved to '…place amongst their principal war aims punishment through the channel of organized justice of those guilty and responsible for these crimes, whether they have ordered them, perpetrated them or in any way participated in them.'[28] This brief declaration threw all individuals implicated in the crimes together, regardless of whether they were part of the Nazi leadership, or were soldiers on the ground.

Following the Moscow Declaration, signed by Churchill, Roosevelt and Stalin in 1943, it was necessary to consider what form an international trial would take, based on a 'joint decision' of the governments of the Allies.[29] The Russian proposals focused upon the fact that such a large scale of organised atrocity could only come from the higher echelons of the Hitler regime; 'directly implicating the men at the summit of the official hierarchy as guilty accomplices in the misdeeds of the lower echelons'.[30] The standpoint of the Kremlin was that the punishment of *all* concerned would be necessary, regardless of rank or title, including individual employers within corporations and companies that were implicated, as well as individual economic players who invested in and profited from the Nazi regime exploits.[31]

The US proposed a two-stage system, beginning with a main trial against the Nazi party, its agencies such as the SS, as well as the individual leading officers, for 'conspiracy to commit murder, terrorism, and the destruction of peaceful populations in violation of the laws of war'.[32] Thus rather than conceiving of the leaders as accomplices in the crimes of their subordinates, this approach defined all parties as equal co-conspirators. The second stage of prosecution would then rest upon the evidence brought to bare in the main trial, and would make it possible to convict any member of the government, the Nazi party, or its agencies of conspiracy merely upon proof of membership in the organisation.[33] To some extent this second stage took place under the auspices of Control Council Law No. 10, and the Nuremberg Military Trials (NMTs), which followed the model of the IMT but were essentially prosecutions under the domestic laws of the US and the UK in the territories they occupied.[34] The strategic approach to liability appears to have been a combination of conspiracy as a mode of liability, conspiracy as an inchoate crime, membership of a criminal organisation as an inchoate crime, and complicity a notion of participation in collective crimes, following the unique US model.

[28] Inter-Allied Resolution on German War Crimes 1942.

[29] Moscow Declaration 1943.

[30] See Ginsburgs 1990, p. 11, citing the People's Commissar for Foreign Affairs.

[31] Ginsburgs 1990, p. 13.

[32] Colonel Murray C. Bernays, G-1, Subject: Trial of European War Criminals (Sept. 15, 1944), in Smith 1981, p. 33.

[33] Ryan 2007, p. 62.

[34] Heller 2011a, p. 9.

8.2.1.3 Conspiracy as a Mode of Liability and as a Crime

The US War Department's draft proposal focused on the notion of conspiracy, which was drawn from the US case law. The benefits cited were that an individual could be found guilty of conspiracy to commit the greater crimes based on evidence of a single overt act in furtherance of the conspiracy. Furthermore, evidence of one conspirator's evil acts could be used against all the other defendants, including those who did not participate in those acts or even know of them, so long as each defendant played some overt role toward the accomplishment of the conspiracy's overall objective.[35] It also provided a framework for prosecuting for the crimes committed by German citizens against other German citizens, which was conceptually difficult under the traditional understanding of war crimes, and even under the newly conceived crimes against humanity.[36]

Despite the design of US proposal, the drafters of the Charter never had the intention to condone prosecution of criminal groups, nor collective punishment of its members, and so any focus on the organisation rather than individual responsibility was rejected.[37] Yet the US design of a two-stage system prevailed in the drafting of the Charter of the IMT, which led to some confusing elements. One example is the inclusion of Article 10, according to which an organisation could be declared criminal, with the explicit intention that domestic criminal prosecutions of individuals as members of the criminal organisation could follow, based upon the final and non-justiciable finding that the organisation was criminal. Furthermore, the judgment of the IMT claimed that a conspiracy was analogous to a criminal organisation,[38] however this did not seem in line with the design of the Charter of the IMT.

At the same time, there was a lack of clarity as to whether the term 'conspiracy' was intended to be a mode of liability or an inchoate crime. The French and Russian delegations were resistant to the inclusion of conspiracy in Article 6, as they argued that it was in violation of the principle of *nullum crimen*, due mostly to the fact that 'conspiracy' was a notion unfamiliar to their domestic criminal law both as a mode of liability, and as a stand-alone crime. It was reported that during the negotiation process, the Russian and French delegations struggled to grasp the implications of the concept, and once they did, were shocked, and argued that it would lead to confusion not only for them, but also for the Germans being prosecuted.[39] Although the British lawyers were more familiar with the term, they were unwilling to accept it as a separate, fourth crime in the Charter.[40]

As stated by Professor Herbert Wechsler, who was then the Assistant Attorney General in the US, the problem was that although conspiracy as a mode of liability

35 Draft Memorandum for the President from the Secretaries of State, War and Navy, Subject: Trial and Punishment of European War Criminals (Nov. 11, 1944), cited in Smith 1981, pp. 41–44.

36 Pomorski 1990, p. 217.

37 Woetzel 1960, p. 208.

38 Nuremberg Judgment 1945, para 299. See also Woetzel 1960, p. 209.

39 Pomorski 1990, p. 219.

40 Ryan 2007, p. 73.

was extremely useful, its operation as such was unclear, and as an inchoate crime it was not only difficult to place within the ambit of crimes 'punishable as violations of the laws of war', it was, moreover, 'not universally accepted in the civilized world'.[41] In fact, the crime of conspiracy as applied in the US had even been criticised by Justice Jackson himself, in his capacity as US Supreme Court judge, for being so vague that it 'almost defies definition'.[42]

Instead, 'conspiracy' made its way into the Charter only as a mode of liability and only in relation to Crimes Against Peace; it was accepted solely in relation to holding individuals responsible for their role in waging aggressive war,[43] which conceivably can only be achieved as a matter of organisational policy. It appears to have been included in the general text of Article 6 of the Charter as a 'catch-all' form of liability.[44] Its interpretation was left up to the judges, but the wording of the final sentence of Article 6 of the Charter suggests a theory of vicarious liability, similar to that recognisable in the common law tradition:

> leaders, organizers, instigators, and accomplices participating in the formulation or execution of a common plan or conspiracy to commit any of the foregoing crimes are responsible for *all* acts performed by *any* persons in execution of such plan.[45]

Including this conception of conspiracy as complicity in the acts of others in the Charter brought the conspiracy-complicity design of liability to the international plane, even though conspiracy is only known in other common law jurisdictions as an inchoate crime. The influence of the American participants in this drafting process is clear, as is the intended policy use of such a notion based on its use in the US.

The influence of participants in the law-making process is evidenced again in the judgment of the IMT, where there was debate among the mixed bench of judges as to the meaning and applicability of conspiracy. The Russian judge General Nikitchenko, who had originally opposed the inclusion of conspiracy in the Charter, in fact embraced the charge in the indictment and argued to have it applied as broadly as possible, whereas the British and French judges wanted to discount the charges of conspiracy *in toto*.[46] In the section of the judgment dealing with the charge of a common plan conspiracy, there seems to be confusion even in the findings as to whether this notion was to operate as an inchoate crime or a mode of liability. On the one had it was stated that conspiracy was only considered to be a separate crime under the Charter with respect to aggressive war;[47] on the other hand it was reiterated that 'in the opinion of the Tribunal [the words of Article 6 of the Charter] do not add a new and separate crime to those already listed. The words are designed to establish the responsibility of persons participating in a common plan.'[48] The conflation

[41] Memorandum for the Attorney General, December 29 1944, cited in Smith 1981, pp. 86–87.

[42] Justice Jackson concurring, Krulewitch 1949, pp. 446.

[43] Kudriavtsev 1990, p. 4.

[44] Heller 2011a, p. 275.

[45] Charter of the IMT 1945, Article 6, (Emphasis added).

[46] Pomorski 1990, p. 227.

[47] Nuremberg Judgment 1945, para 40; Woetzel 1960, p. 209.

[48] Nuremberg Judgment 1945, para 44.

of the two conceptualisations reflect the resistance and misunderstanding by judges who were unfamiliar with this essentially American concept. It also reflects their influence in forming the law, in this case by rejecting a notion and preventing it from being accepted as international law.

8.2.2 The Tokyo International Military Tribunal

The Tokyo Tribunal further followed the precedent set at Nuremberg that all leaders, organizers, instigators and accomplices 'participating in the formulation or execution of a common plan or conspiracy to commit' crimes against peace are responsible for all all acts performed by any person in the execution of such plan related to it.[49] This language demonstrates a theory of vicarious liability.[50] Of course, this was limited to crimes against peace, just as at the IMT, because this was the only crime for which conspiracy was included in the Tokyo Charter.

The impact of the Tokyo Tribunal was to affirm the radical new steps taken by the IMT at Nuremberg as international law. It received similar criticisms of victor's justice as the IMT at Nuremberg, including from one of its own judges, Judge Röling, who felt that the defendants were being prosecuted for political crimes.[51]

But the IMT at Tokyo was criticised for taking a much broader approach to conspiracy than had the IMT at Nuremberg,[52] since it held that even those individuals who were not part of the original agreement, but were later adherents to a conspiracy, could be held liable for all of the crimes committed in furtherance of the common plan.[53] Notably this extensive interpretation of conspiracy was in-keeping with the reasoning behind the *Pinkerton* decision in the US, which was the influential interpretation of conspiracy liability at the time.[54]

There had been significant controversy over the very manner in which the tribunal was established, far more so than with respect to the IMT. Criticisms of 'victor's justice' were even more vehement than for its predecessor, since there was a much more explicitly political selection of defendants, notably excluding the Japanese emperor. On top of this, US General MacArthur was granted, and maintained, exclusive direction over the operation of the trials, giving it the appearance of being a US tribunal rather than an international one.[55] This demonstrates again the role of participants

[49] UN Report on Nürnberg 1949, p. 84.

[50] Just as in the common law tradition, this vicarious liability does not amount to the kind of civil liability where a defendant is liable in the place of an actor, but rather it is vicarious in the sense that all those involved in this collective crime are responsible for all the acts of others involved, without further distinction.

[51] Simpson 2007, p. 15.

[52] Cryer et al. 2016, p. 305; Cryer 2001, p. 20.

[53] UN Report on Nürnberg 1949, p. 84.

[54] For a full discussion on Pinkerton see Sect. 8.5 above; Pinkerton 1946.

[55] Bassiouni 1999, p. 211; Maogoto 2004, p. 103.

in the process of the law, and it should be of no surprise that the interpretation of conspiracy was more extensive than at the IMT, and were modelled on the unique practice of the US.

8.2.3 *Military Trials Under Control Council Law No. 10*

Unlike the Charter of the IMT, Control Council Law No. 10 did not refer explicitly to conspiracy at all. Article II(2) of the law stated in very broad terms that a person is deemed to have 'committed' a crime:

> if he was (a) a principal or (b) was an accessory to the commission of any such crime or ordered or abetted the same or (c) took a consenting part therein or (d) was connected with plans or enterprises involving its commission or (e) was a member of any organization or group connected with the commission of any such crime or (f) [...] if he held a high political, civil or military (including General Staff) position in Germany or in one of its Allies, co-belligerents or satellites or held high position in the financial, industrial or economic life of any such country.[56]

Most of the judgments handed down by the NMTs did not specify which mode of liability was applicable,[57] suggesting a functionally unitary approach whereby Article II was considered to be simply a description of all the possible ways in which an individual could take part in a crime. Any individual who knowingly participated was considered to be criminally responsible, and the principle/accessory distinction was therefore irrelevant in any normative sense.[58] At the same time it was considered that subsection (a) was rarely applicable since most of the defendants occupied high positions in the government or economic structure of Germany, and thus 'very few of them ever committed a murder or crime with their own hands.'[59] The NMTs also disregarded subsection(f) of this article entirely, because it implied strict liability for crimes against peace based purely on the function of an individual.[60] In this sense the principle of culpability acted as a limitation on liability, since, as the tribunal stated: 'guilt must be personal'.[61]

Subsections (c) and (e) have had a larger life in ICL, as they can be recognized in the modern notion of JCE, to be discussed in full below in Sect. 8.6. This leaves subsection (d), which has some similarities to conspiracy, even though it makes no explicit mention of this form of liability. In the *Medical, Justice* and *Pohl* cases, the prosecution included counts such as 'common design or conspiracy',[62] however the defence argued that because the IMT had limited conspiracy to crimes against

---

56 Control Council Law No. 10 1945, Article II(2).

57 Heller 2011a, p. 252.

58 Farben 1952, para 1141.

59 Telford Taylor's Third Trial Program, page 4. Cited in Heller 2011a, p. 251.

60 Farben 1952, para 1299; Heller 2011a, p. 186.

61 Krupp 1948, para 1448.

62 Medical case 1946, para 10; Farben 1952, para 59; Heller 2011a, p. 276.

peace, it should not be applied in these cases.[63] Although Chief Prosecutor Telford Taylor argued that conspiracy could be considered a mode of liability and not an independent, inchoate crime, the NMT dismissed the conspiracy counts altogether, arguing that subsection (d) of Control Council Law No. 10 specified liability for being connected with plans or enterprises. However once again the confusion as to the status of conspiracy as a mode of liability or as an inchoate crime was evident in the statement by the NMT that there was no jurisdiction to try 'any defendant upon a charge of conspiracy considered as a separate substantive crime.'[64]

8.2.4 After the Second World War Trials: Rejection of Conspiracy

Conspiracy as an inchoate crime has maintained a specific place in ICL with respect to genocide and aggression, but has been rejected with respect to all other crimes, and has generally been rejected as a mode of liability. The Nuremberg Principles were drawn up by the ILC in 1950, as a way of synthesising the law that had emerged from these trials, and with an eye to future codification. As such they followed the IMT Judgment by including conspiracy with respect to aggression, but excluding it from the definition of the other crimes.[65]

Notably, the ILC did include conspiracy to commit genocide as a specific crime in Article 3 of the 1948 Genocide Convention.[66] It seems clear that it was intended as an inchoate crime, just as incitement and attempt which are also enumerated in Article 3.[67] The drafting history of the Genocide Convention indicates that the notion of conspiracy was based on the common law definition of an agreement between two or more people, regardless of whether the predicate crime is committed.[68] In the 1990s the ad-hoc tribunals followed this model and criminalised conspiracy to commit genocide explicitly, but did not include conspiracy with respect to any other crimes.[69]

Despite the fact that the inchoate crime of conspiracy is not recognised in civil law jurisdictions, the criminalisation of this specific offence with regard to genocide has been recognised in the case law of the ICTR as being justified by the especially reprehensible nature of genocide,[70] and the fact that one of the main objectives of

---

63 Heller 2011a, p. 276.

64 See for example Justice Case 1947, para 235; High Command Case 1948, para 483; Heller 2011a, p. 280.

65 ILC 1950, Principle VI.

66 Genocide Convention 1948, Article 3.

67 Boas et al. 2007, p. 282; Ohlin 2009, p. 187; Werle and Jessberger 2014, p. 228.

68 Schabas 2010, p. 438.

69 Statute of the ICTY 1993, Article 4(3)(b); Statute of the ICTY 1994, Article 2(3)(b); Statute of the ECCC 2004, Article 4; Werle and Jessberger 2014, p. 228.

70 Akayesu 1998; Ohlin 2009, p. 186.

the Genocide convention was to create the legal duty to prevent it and punish even the inchoate steps towards it.[71]

The Rome Statute does not include conspiracy at all. The 1991 Draft Code of Crimes Against Peace and Mankind, which was the precursor to the Rome Statute, had one fairly general article dealing with responsibility and punishment: Article 3 enumerated commission, aiding and abetting, conspiracy, incitement and attempt.[72] Conspiracy had been included because it was in the 1954 Draft, following the Nuremberg Principles, however with later drafting revisions, a shift was made and in the 1991 commentary it was made explicit that it was intended to be included as a mode of liability, and not as an inchoate crime.[73] This shift represents a move towards the concept unique to US law, and away from the tacit acceptance until then that conspiracy was only considered to be an inchoate crime and only for the crime of aggression. The ILC noted that in the French version it had chosen to move away from the term *complot*, which had been in the French texts of the Nuremberg Principles and Article 3 of the Genocide Convention, and instead chose the word *entente*, in order to reflect that it was a mode of liability rather than a stand-alone crime.[74] Following this, the joint proposal that was submitted in 1997 by Canada, Germany the Netherlands and the United Kingdom omitted conspiracy altogether, including with respect to genocide.[75]

Schabas asserts that exclusion of conspiracy to commit genocide was almost certainly an oversight rather than an intentional omission,[76] but if it was indeed intended as a mode of liability instead of an inchoate crime, one may wonder whether one factor in its omission was the fact that none of the States which had submitted the joint proposal recognises conspiracy liability, and the US remains alone in this. The *traveaux preparatoirs* suggest that the States Party could not agree on its inclusion, and some representatives of States following the civil law tradition explicitly announced their objection to it.[77] In a more conceptual interpretation, Fletcher sees it as a deliberate move away from the model of the ad-hoc tribunals, which were established during impassioned moments in history, reflecting 'the afterglow of a dying concept'.[78] In his eyes there was nothing accidental, but rather a development at a calmer time in history, allowing deeper consideration of the (un)desirability of the doctrine of conspiracy.

Either way, it is clear that with the exception of conspiracy to commit genocide, the notion of conspiracy has been rejected in international law. Although the crime of aggression historically was the only other exception, it would seem that the fact the Rome Statute does not recognise conspiracy at all would mean this does not apply

---

[71] Musema 2000; Cassese 2008b, p. 228; Ohlin 2009, p. 192.

[72] ILC Draft Code of Crimes 1996, Article 3.

[73] ILC Draft Code of Crimes 1996, p. 99.

[74] ILC Draft Code of Crimes 1996, p. 99.

[75] Schabas 2010, p. 423.

[76] Schabas 2010, p. 438.

[77] Preparatory Commission Working Group on General Principles 1998; Ohlin 2009, p. 199.

[78] Hamdan v Rumsfeld Amicus Brief 2006-01-06, p. 12.

to the new crime of aggression, just as it does not apply to genocide under the Rome Statute.[79] While the notion of a common purpose remains punishable under the Rome Statute, which admittedly shares some similarities with the notion of conspiracy, as will be described below, reading it in relation to the other modes of liability suggests it has both a derivative and more objective nature than is associated with conspiracy.

8.3 Command and Superior Responsibility: Leadership Liability as Omission

Command responsibility and its sibling superior responsibility are unique in this process in that they were not born out of domestic criminal law notions, but more directly out of IHL.[80] They apply only to the kinds of crimes that arise in the course of large scale human conflict, and they reflect structured hierarchies and chains of command, which are particular to the armed forces and other State and State-like organisations of power.

A debate exists as to whether command and superior responsibility are modes of liability or rather stand-alone crimes of omission. The position taken here is that they are independent crimes.[81] In the *High Command* case, the American NMT stated that for command responsibility 'there must be personal dereliction. That can only occur where the act is directly traceable to [the commander], or where his failure to properly supervise his subordinates constitutes criminal negligence on his part.'[82] Indeed, command responsibility was not included in Control Council Law No. 10 as a mode of liability, and where the NMTs applied it as imposing an affirmative duty on a commander to take steps to control those under his command, they intended this as a criminal omission.[83] As the ICTY Appeals Chamber in *Krnojelac* stated:

> [i]t cannot be overemphasised that, where superior responsibility is concerned, an accused is not charged with crimes of his subordinates but with his failure to carry out his duty as a superior to exercise control.'[84]

This approach was favoured by Judges Shahabuddeen and Hunt in the *Hadžihasanović* case. As Shahabuddeen put it: 'Reading the provision reasonably it could not have been designed to make the commander a party to the particular crime com-

---

[79] See e.g. Hamdan 2006; Cassese 2008b, p. 227.

[80] Ambos 2005, p. 296; Werle and Jessberger 2014, p. 187.

[81] For a discussion on the debate surrounding command responsibility as a mode of liability or as a crime of omission, see Steer 2013a.

[82] High Command Case 1948, p. 543.

[83] See e.g. Medical case 1946, p. 207, however it should be noted that the only source cited for the status of command responsibility in international law was the Yamashita case, to be discussed below. Heller 2011a, p. 263.

[84] Krnojelac Appeal Judgment 2003, para 171.

mitted by his subordinates.'[85] Although in the minority, this view was later affirmed by both another Trial Chamber and the Appeals Chamber in *Orić*[86] and *Halilović*.[87] As it was formulated in *Orić*: 'the superior cannot be considered as if he had committed the crime himself, but merely for his neglect of duty with regard to crimes committed by subordinates.'[88]

The ICTY case law would appear to assert that command and superior responsibility are distinct from the modes of liability under Article 7 of the Statute; on the one hand because the commander bears responsibility for failure to act, rather than for the acts of the subordinates, and on the other hand because there is no causal link required.[89] Whatever counter-factual impact the commander's omission may have on the predicate crime, in that it was not prevented, causation is not the basis for the criminalisation of his or her conduct, but rather the basis lies in the personal and culpable dereliction of duty.[90]

It is possible that the wording of Article 28 of the Rome Statute suggests a move away from this understanding, in that it requires a causal link, and it states that this form of responsibility is 'in addition to other grounds', referring to the modes of individual criminal liability listed under Article 25. In its central case on command responsibility, the Trial Chamber judges have stated in the judgment convicting Bemba in June 2016 that command responsibility is a *sui generis* mode of liability, rather than an independent crime.[91] However if this was the intention of the drafters, then it appears to have been a significant move away from previous codifications of command responsibility, including in the ad-hoc tribunal statutes, and from previous doctrinal analyses of command responsibility as liability for dereliction of duty. If this is so, then it has become a mere 'corollary of subordinate liability'.[92]

Even if command and superior responsibility have been given the form of modes of liability under the Rome Statute, deontologically, there is a reason why command and superior responsibility are not at the centre of debates on leadership liability. Whether or not one considers them to be a stand-alone crime or a mode of liability, either way they are based on omission, which entails passivity. They are entirely appropriate criminal law responses when the aim is to reproach a person in a leadership situation for not *intervening* when subordinates are acting in an unruly manner, for not exercising control when that is what is required. However when it is a case of grander scale atrocity, and when a leader has taken a more active role, and is involved in the intellectual authorship and orchestration, prosecution for command or superior

[85] Partial Dissenting Opinon of Judge Shahabuddeen, Hadžihasanović Appeal Judgment 2008, para 32.

[86] Declaration of Judge Shuhabadeen, Orić Trial Judgment 2006, para 15.

[87] Halilović Appeal Judgment 2007, para 54; see also Cryer et al. 2016, p. 329.

[88] Orić Trial Judgment 2006, para 293.

[89] Halilović Appeal Judgment 2007, para 54; Hadžihasanović Appeal Judgment 2008, para 75; Olásolo 2009, p. 106; Werle and Jessberger 2014, p. 188.

[90] Mettreaux 2008, p. 40; Boas et al. 2007, p. 144.

[91] Bemba Judgment 2016, paras 171, 174.

[92] van Sliedregt 2012a, p. 197.

responsibility would be an inappropriate and insufficient response. It would send the wrong signal in terms of expressing accountability and responsibility.

Given that command and superior responsibility have developed as crimes of omission rather than more complex modes of liability, and given their limited function in dealing with active leadership liability for mass atrocity crimes, their place in this chapter may not be obvious. However the processes of judicial creativity in developing these notions is still instructive in terms of the processes by which certain gaps in the normative content of the law are filled in response to historical context and policy goals. They will therefore each receive further attention in this mapping of the process of law.

8.3.1 Command Responsibility

Throughout the history of regulation of warfare, harking back at least as far as Chinese military tactician Sun-Tzu's 'The Art of War' in the 5th century B C.,[93] the focus has been on minimising its egregious effects by clarifying ethically acceptable conduct of belligerents and their superiors. The Augustinian notion of 'just war' included the concept of competent authority; soldiers did not act as individuals but as agents of a higher authority, implying that responsibility lay with those in the higher echelons of command.[94] Grotius put it very clearly when he wrote 'we must accept the principle that he who knows of a crime, and is able and bound to prevent it but fails to do so, himself commits a crime.'[95] This notion of the crime of superior responsibility, together with the Augustinian tradition of ethical warfare and competent authority was carried into the Lieber Code of the 19th century,[96] and the Hague Convention of 1907.[97] At the turn of the twentieth century there was a growing acceptance of individual culpability for war crimes, and an increase in domestic punishment at the command level for violations of the laws of war.[98]

Although some authors suggest that the first recognition of command responsibility in an international context was in the Allied Powers' Commission on the Responsibility of the Authors of the War, following the First Word War,[99] the fact that the intended tribunal was never realised means that as a legal doctrine command responsibility really began to take form in the wake of the Second World War NMT trials,[100] in which the emphasis was very much upon the role of the defendants as military superiors. There was little attention paid to the elements of their conduct,

---

[93] Rockwood 2007, p. 20; Boas et al. 2007, p. 145.

[94] Rockwood 2007, p. 28.

[95] *De Jure Belli ac Pacis* (1615), cited in van Sliedregt 2012a, p. 184.

[96] Lieber Code 1863.

[97] Fourth Hague Convention 1907.

[98] Boas et al. 2007, p. 147.

[99] As cited by the ICTY in Čelebići Appeal Judgment 2001, para 335; Boas et al. 2007, p. 147.

[100] Pohl 1947; High Command Case 1948; Hostages Case 1948.

other than that they held positions of power within a hierarchy, which led to convictions based on something akin to strict liability. The drive to find a moral scapegoat for the atrocities which took place during the war led to questionable beginnings of this doctrine, which has been said to have been 'born in sin'.[101] In this respect, the conviction of General Yamashita for atrocities committed by his troops in Manila was heavily criticised for placing the weight of these crimes on the shoulders of one individual who could hardly have been expected to intervene in circumstances beyond his control and of which he had no demonstrable knowledge.[102]

Furthermore, as two judges in the US Supreme Court's review of the Yamashita decision pointed out, it was impossible to identify the source of such a mode of liability in international law—in Justice Murphy's dissenting opinion he asserted no such mode of liability existed at all.[103] Although the doctrine's roots as a dereliction of duty lie in IHL, the strict liability interpretation in Yamashita appears to have been a judicial development, to place legal responsibility where many felt there was a moral one.

Upon reflecting on the Yamashita conviction at Tokyo, Judge Röling said that command responsibility is too susceptible to being applied 'purely as a vengeful pretext for victorious powers to punish vanquished foes'.[104] This susceptibility is illustrated by the highly political outcome of the 1971 US Court Martial against General Medina for the My Lai massacre in Viet Nam.[105] Medina was acquitted and many officers directly above and below him in the chain of command were not even prosecuted for the elaborate cover-up that unravelled in the course of the investigations.

The definitive role played by judges of international tribunals in forming and developing the law is once again evident, since command responsibility has no doctrinal history in domestic systems. Judges have been entirely creative in refining the notion. As one author has put it;

> A great deal of the law-making activity of international courts and tribunals was necessary to transform the doctrine of command responsibility from an anarchy of sometimes contradictory precedents into a consistent body of law.[106]

Since the Second World War trials, command responsibility was included in many domestic military manuals, and was codified at the international level in the Additional Protocol of 1977.[107] It has since been recognised as customary law by the ICRC study on customary IHL.[108] More recently in the statute of the ICTY, command responsibility and superior responsibility are melded into one form under Arti-

---

101 Mettreaux 2008, p. 5.

102 van der Wilt 2010a, p. 1134; Cassese 2008b, p. 239.

103 Yamashita 1945, p. 522.

104 van der Wilt 2010a, p. 1134.

105 Rockwood 2007, pp. 120–126.

106 Mettreaux 2008, p. 9.

107 Additional Protocol I 1977, Article 6 and 87; Cryer et al. 2016, p. 321.

108 Henckaerts and Doswald-Beck 2005, p. 558.

cle 7(3), where it is stated that a superior remains responsible for crimes committed by subordinates;

> if he or she knew or had reason to know that the subordinate was about to commit such acts or had done so and the superior failed to take the necessary and reasonable measures to prevent such acts or to punish the perpetrators thereof.[109]

The first case involving command responsibility before the ICTY was the *Celebići* case,[110] in which the Trial Chamber established the three core elements of command responsibility: proof is required of (i) the existence of a superior-subordinate relationship, requiring the capacity to control or influence the subordinates and the power *de jure* or *de facto* to give orders; (ii) actual or constructive knowledge that the subordinates had committed or were about to commit the crime; and (iii) failure to take necessary and reasonable measures to prevent the criminal act, or to punish the perpetrators.[111] The very nature of command responsibility rests upon the role of the accused as a leader, yet at the same time the subjective element of knowledge, actual or constructive, balances out the risk of strict liability.[112]

This balancing out between the objective and subjective elements is an important factor in determining responsibility for a crime of omission. But it is the superior-subordinate relationship which is at the centre of command responsibility. This entails not merely influence or psychological pressure, but rather explicit command and control on the part of the commander.[113] The ICC Pre-Trial and Trial chambers agree with this definition, and have added that 'the term "command" is defined as "authority, especially over armed forces", and the expression "authority" refers to the "power or right to give orders and enforce obedience".'[114]

Jean-Pierre Bemba Gombo was the first to be indicted before the ICC for command responsibility, albeit after the Pre-Trial Chamber encouraged the Prosecution to adjust its original indictment which was based on direct perpetration under Article 25(3)(a).[115] In the confirmation of charges it was noted that the specificity of the subjective element prevents convictions based on strict liability.[116] The principle of culpability is thus given due attention.

---

[109] Statute of the ICTY. The exact wording is mirrored in the Statute of the ICTR in Article 6(3)

[110] Čelebići Appeal Judgment 2001.

[111] Čelebići Appeal Judgment 2001, para 346. These elements have been repeated in cases such as Blaškić Appeal 2004 and Delić Trial Judgment 2008, para 56.

[112] Steer 2013a.

[113] Čelebići Appeal Judgment 2001, para 265, citing some jurisprudence from the ICTR. See also van Sliedregt 2012a, p. 207.

[114] Bemba Judgment 2016, para 180.

[115] Bemba, Decision Adjourning the Hearing 2009, para 19.

[116] Bemba Decision on Charges 2009, para 434.

8.3.2 Superior Responsibility

Superior responsibility was first extended to non-military leaders in the NMT trials and the Tokyo Tribunal following the Second World War,[117] and in its earliest conception suffered similar criticisms as command responsibility for verging on strict liability. In the renowned *Medical* case at the American NMT, twenty three medical doctors and government administrators stood trial for involvement in human experimentation. The senior medical officer of the German Government during the war, Karl Brandt, was convicted of war crimes and crimes against humanity for his 'special responsibility for, and participation in' numerous experiments on prisoners of war.[118] The judgment focused upon his position, ranking highest in the medical field directly under Hitler, and his duty to investigate and intervene in the medical experiments which he knew were taking place in concentration camps. Yet the Court offered no basis for the source of duties imposed on Brandt. Since he was a civilian superior with no *de jure* command role, the lack of examination or articulation on the source, scope and content of this new international legal obligation to intervene exposes a serious weakness in the development of the doctrine.[119]

The assertion that superior responsibility verged on strict liability was repeated with respect to a number of defendants, including ministers and cabinet members, most prominently the former Foreign Minister, Kiko Hirota, who was held to have failed in his duty to take adequate steps to prevent breaches of the laws of war by Japanese troops.[120] Judge Röling dissented against this conviction, and his criticisms were echoed in many later commentaries, that Hirota's function as foreign minister was taken to be sufficient for his conviction, ignoring his lack of control over the physical perpetrators, particularly as the crimes were committed by the subordinates of another ministry over which Hirota had no control.[121]

It is noteworthy that no industrialists were prosecuted before the IMT, and civilian superior responsibility was only indicted before the domestic NMTs under Control Council No. 10. It has been asserted that this was a calculated and politically motivated move by the US and Britain, who were keen to rebuild Germany in their zones of occupation, as a way of asserting a Western political and economic order over which they could exercise sufficient influence.[122] This later took the form of the Marshall Plan, geared towards reconstruction of Europe and containment of commu-

---

117 Boas et al. 2007, p. 154.

118 Medical case 1946, p. 193.

119 Boas et al. 2007, p. 155.

120 Boas et al. 2007, p. 156.

121 Boas et al. 2007, p. 156; Röling and Rüter 1977, p. 448; van Sliedregt 2012a, p. 129. The test of actual control had been applied more precisely in some of the NMT cases where superior responsibility had been indicted, which limited the risk of strict liability. In the *Ministries* case, for example, Berger was acquitted because although he was formally superior to the perpetrators who had conducted so-called racial examinations, he did not exercise control over them: Ministries case 1949, pp. 518, 546; Heller 2011a, p. 266.

122 Maogoto 2004, p. 100.

nism. But this required building alliances with German industry, which may explain why major German industrialists retained their industries, and those tried by US courts were handed relatively light sentences and allowed to return to Germany and to their leadership roles within their businesses.[123]

That there were similar political motivations behind the convictions at the Tokyo Tribunal is nearly certain, given the controversy over the very manner in which the tribunal was established, the political criteria for selecting defendants, and the final authority which US General MacArthur maintained over the operation of the trials.[124]

In its next guise, in the Statutes of the ICTY and ICTR, no distinction was made between (military) command and (civil) superior responsibility. However the Rome Statute does separate them again under Article 28, with the only difference being that a commander's knowledge can be constructed based on what she 'should have known', whereas the requirement for a superior is slightly higher

Despite such questionable beginnings, superior responsibility has found a place in the codification and the case law of ICL, as it has filled a gap in responsibility in more diffuse kinds of atrocity. At the ICTY, the *Čelibići* Trial Chamber drew from second World War case law in order to interpret the rather vague codification of superior responsibility under its statute, and at the ICTR, where there were many more civilians facing prosecution for their involvement in the large scale genocide, including Prime Minster Kambanda, superior responsibility has proved to be an important addition.[125] The explicit role of participants in the inception, application and development of civilian superior responsibility is clear both from the criticisms against it, and from the acceptance it has received in recent years through application by the ad-hoc tribunals.

8.4 Planning, Instigating and Ordering: Forms of Encouragement

The modes of liability which are perhaps the least controversial in ICL are those of planning, instigating and ordering, most likely because they describe in relatively simple terms the ways in which a person can take direct part in a crime without physically committing it. Planning and instigating replicate domestic modes of liability which do not differ a great deal across jurisdictions of various traditions. On the other hand, ordering was born out of IHL in a similar way to command and superior responsibility.[126] These three all appeared in the IMT Charters at Nuremberg and Tokyo, but were not explicitly mentioned in Control Council Law No. 10, and have not all maintained their place in statutes of the modern tribunals. However their appli-

[123] Maogoto 2004, p. 100.

[124] Bassiouni 1999, p. 211; Maogoto 2004, p. 103.

[125] Kambanda 1998; Karemera 2012; Kayishema 1999.

[126] Cryer et al. 2016, p. 312; van Sliedregt 2012a, p. 105.

cation has remained fairly standard, and there appears to be less dynamic interaction between domestic and international interpretations, and less judicial creativity.

These modes of liability can all be considered to be more 'active' in nature than command or superior responsibility, because they are based on active participation in a crime, rather than on responsibility for dereliction in duty to intervene. However in many domestic jurisdictions, and it would seem at the ICC as well, they are considered to be secondary modes of liability. Therefore they do not satisfy the desired deontological reflection of the responsibility of leaders of mass atrocity. Furthermore, at Nuremberg there was a very clear paper trail provided by the Nazi government and military organs, who recorded every bureaucratic interaction meticulously, and these modes of liability were therefore very useful ways to widen the net of liability. However not all mass atrocity is committed in the context of such efficient bureaucracy, and therefore these modes of liability are not always relevant.[127]

From the case law at the ad-hoc tribunals it would appear there is agreement that an accused cannot be convicted of one of these modes of liability and of physically committing the same crime.[128] It would appear from the definitions in the case law that these modes of liability are not considered to be imputed liability nor vicarious liability, but rather they are derivative of the physical perpetrator's liability, and ways in which parties on the periphery can encourage physical perpetrators.

8.4.1 Planning

In the Charters for both the IMTs at Nuremberg and Tokyo, planning appeared as a mode of liability only with respect to Crimes Against Peace.[129] It appeared alongside the other modes of liability such as conspiracy and instigation, without distinction. In the Memorandum submitted by the UN Secretary General in 1949, in preparation for formulating the Nuremberg Principles, it was noted that 'common planning' was an aspect of a conspiracy to wage aggressive war, and that contributions to the plan could vary in importance, from a dictator to anyone who knowingly cooperates with the plan.[130] On the other hand, a distinction was made between contributing to a common plan as part of a conspiracy, and the 'crimes' of planning or preparation:

> A distinctive feature of these crimes as compared to the crime of conspiracy thus is that they are connected with particular wars of aggression, or aggressive wars against particular countries, whereas the common plan or conspiracy ...has as its objective aggressive war as a general policy.[131]

In this respect it is also relevant to note that some of the defendants at the IMT at Nuremberg were acquitted of the count of conspiracy, for example Funk who was

[127] Osiel 2005, p. 1798.

[128] Stakić Trial Judgment 2003, para 445; Blaškić Trial 2000, para 278; Akayesu 1998, para 468.

[129] Charter of the IMT 1945, Article 6(a); Tokyo Charter 1946, Article 5(a).

[130] UN Report on Nürnberg 1949, p. 53.

[131] UN Report on Nürnberg 1949, pp. 55–56.

not considered to be sufficiently involved with Hitler's greater plan for territorial expansion, but convicted on the count of planning and preparation of 'certain of the aggressive wars'.[132]

In the ILC's Draft Code of 1996, both planning an conspiracy had been included as a mode of liability,[133] and in the ICL's commentary it was stated that these modes of liability were intended for high level government officials or military leaders, due to the fact that the core crimes often require the formulation of a plan or policy formulated by senior officials.[134] Following this model, at the ICTR it has served this very purpose, for instance in holding Jean-Bosco Barayagwiza, one of the leaders of a radical Hutu party, liable for his 'critical role' in planning the extermination of Tutsi civilians.[135]

At the ICTY it would appear that planning is considered to be a form of accomplice liability on par with aiding and abetting.[136] However 'planning' did not make it into the Rome Statute. Although it is unclear from the *traveaux preparatoires* why it was excluded, since it was specifically enumerated in the 1997 joint proposal which formed the basis of the final wording of Article 25,[137] in the end it may have less importance for future interpretations of leadership liability for mass atrocity crimes at an international level. Moreover, if it is indeed a secondary mode of liability, then it fails to reflect the deontological role of leaders who are the architects of systemic crimes, as outlined in Chap. 2.

8.4.2 Instigating/Soliciting/Inducing

Although it did not appear in the ILC's Draft Code, the mode of liability known as instigation under the ad-hoc tribunals has appeared as soliciting or inducing under the Rome Statute,[138] and all these terms are interchangeable.[139] This variation in terminology is a reflection of the varying terminology in domestic criminal law systems, and the fact that there has been a translation into the statutes of international tribunals. There appears to be general agreement as to the definition of this mode of

132 UN Report on Nürnberg 1949, p. 56.

133 ILC Draft Code of Crimes 1996, Article 2(3)(e).

134 ILC Draft Code of Crimes 1996, p. 21.

135 Nahimana 2003, para 697.

136 Brđanin Decision on Amended Indictment 2001, para 43; Kordić Trial Judgment 2001, para 367.

137 Schabas 2010, p. 423.

138 Statute of the ICTY 1993, Article 7(1); Statute of the ICTY 1994, Article 6(1); Werle and Jessberger 2014, p. 125; Cryer et al. 2016, p. 314; van Sliedregt 2012a, p. 107.

139 Schabas 2001, p. 432; van Sliedregt 2012a, p. 108.

liability, even though it has been given different names, however it does differ from incitement, which is an inchoate crime.[140]

At the ICTY instigating has been defined as 'prompting',[141] and at the ICTR as 'urging or encouraging'[142] and it has been determined that the instigator must have some kind of influence over the physical perpetrator.[143] In terms of the objective elements, the actions of the instigator must have caused the physical perpetrator to commit the crime in question, at least to the extent that there is evidence of a 'clear contributing factor',[144] or that the instigator 'substantially contributed to the crime'.[145] Although these definitions suggest a person in a position of influence, they are forms of encouragement towards the physical perpetrator, so while there is some causal link, the instigator does not exercise control over whether and how the crime is committed.

One could wonder why this mode of liability has not been embraced with the same enthusiasm as JCE at the ad-hoc tribunals, or indirect perpetration at the ICC. One reason may be that although it has been determined that the instigator need only accept a risk of the crimes that are committed, still there must be a relatively direct relationship between the accused and the physical perpetrator(s). While this has been the case in some of those indicted before the ICTY and ICTR, in the more complex relationships of power and larger, more disparate organisations and collectives which enact crimes of mass atrocity, a direct relationship may be harder to prove.[146]

8.4.3 Ordering

Ordering has a similar background and nature to command and superior responsibility, since it is related to collective crimes committed by way of hierarchical relationships. It has been likened to instigation by abuse of power in domestic law,[147] however it has arisen from the specific context of grave breaches of IHL, and the nature of crimes of mass atrocity.[148]

---

[140] See e.g. Genocide Convention 1948, Article 3(c); Statute of the ICC 1998, Article 25(3)(e); Akayesu 1998, para 482; Boas et al. 2007, p. 346. But compare Cryer et al. 2016, p. 315.

[141] Blaškić Trial 2000, para 280; Kordić Trial Judgment 2001, para 27.

[142] Cryer et al. 2016, p. 314.

[143] Orić Trial Judgment 2006, para 271.

[144] Kvočka Trial Judgment 2001, para 390.

[145] Kordić Trial Judgment 2001, para 27; Gacumbitsi Appeal 2006, para 129; Ndindabahizi 2004, para 463.

[146] Consider the arguments made by Osiel 2009, p. 45. Even though he makes these points in respect of superior responsibility, they are equally applicable to the notion of instigation.

[147] van Sliedregt 2012a, p. 102.

[148] Cryer et al. 2016, p. 312; van Sliedregt 2012a, p. 105.

In the definition of the 1996 ILC Draft Code, a person must be 'in a position of authority and use his authority to compel another individual to commit a crime',[149] in order to be liable for crimes committed in pursuance of orders given. The position of authority, however, need not be one of effective control, which is the test applicable to command responsibility. Although the ICTY Trial Chamber has pointed out that ordering implies a superior-subordinate relationship,[150] in fact no formal relationship is necessary. An informal position of power could be sufficient, as long as the person has authority to order.[151] The test of whether a person has sufficient authority may even be satisfied based on a subjective test of the perceptions of the individual receiving the order.[152] The essence of this form of liability is that an individual in a position of power abuses this power in giving an order to commit an international crime.[153]

Just as instigation or soliciting seems only to require a low standard for the subjective element, the *mens rea* for ordering is also very elastic and appears to be that of *dolus eventualis*. The Appeal Chamber of the ICTY convicted Blaskić of crimes he had not directly ordered, because he had 'awareness of the substantial likelihood' that in ordering an attack on a village to prevent a counter-attack, a massacre would occur; 'ordering with such awareness has to be regarded as accepting the crime.'[154]

The discussion as to whether ordering is a principal or secondary mode of liability has not been as heated in the case law, but continues among commentators. Ambos argues that it properly belongs in the first subparagraph of Article 25 of the Rome Statute alongside the other modes of liability that are principal in nature, rather than where it appears in the second subparagraph alongside secondary modes of liability.[155] Sliedregt insists that it is accessorial, since it is derivative of the crime that is committed, and from this she reads that it is also secondary in nature.[156] In this respect she is correct, and it seems likely that there is agreement that ordering is indeed both derivative and a secondary mode of liability.[157] However this author does not agree with the comparison between instigation, ordering, soliciting and inducing on the one hand with aiding and abetting on the other.[158] The latter will be discussed in the next section, and are forms of (secondary) assistance to a perpetrator who may commit the crime even without the aid of these parties, whereas the former require by definition a certain level of power or influence over the physical perpetrator.

---

149 ILC Draft Code of Crimes 1996, Article 2(8).

150 Blaškić Trial 2000, para 474.

151 Kamuhanda 2004, para 75; Kordić Trial Judgment 2001, para 28; Cryer et al. 2016, p. 313; Olásolo 2009, p. 136.

152 Gacumbitsi Appeal 2006, para 182; Cryer et al. 2016, p. 313; van Sliedregt 2012a, p. 106.

153 Blaškić Trial 2000, para 601. This has been confirmed by the ICTR in Akayesu 1998, para 483.

154 Blaškić Appeal 2004, para 42.

155 Ambos 2008b, p. 480.

156 van Sliedregt 2012a, pp. 108–109.

157 Cryer et al. 2016, p. 315; Olásolo 2009, p. 142.

158 Schabas 2001, p. 431; van Sliedregt 2012a, pp. 102, 109.

Some also assert that ordering may be seen as similar to indirect perpetration,[159] however it is distinct in the statutes of the ICTY, ICTR, the Special Tribunal for Lebanon (STL) and the ICC and therefore should be interpreted as such.[160] The Pre-Trial Chamber of the ICC has made this distinction in one of its early decisions on indirect perpetration, that '[t]he highest authority does not merely order the commission of a crime, but through his control over the organisation, essentially decides whether and how the crime would be committed.'[161]

In line with this distinction, Hector Olásolo points out, citing former Chief Prosecutor of the ICTY Carla Del Ponte, that though ordering reflects the responsibility of an individual for their influential role in specific crimes that are committed by others, it does not sufficiently reflect the liability of senior political and military leaders who work together to achieve a campaign of mass violence.[162] Once again it is deontologically unsatisfying for those who are 'most responsible' to be prosecuted under this mode of liability if their authorship goes further—and especially if planning, instigating and ordering are seen as secondary modes of liability. Indeed, Werle understands ordering to be a form of encouragement, and to sit somewhere between commission as the highest form of culpability, and aiding and abetting as the lowest forms.[163] In this sense it is perhaps best suited for those who are at an intermediate level of command, who both receive and give orders.[164]

8.5 Aiding and Abetting: Hidden Controversies

Aiding and abetting are the modes of liability that differ the least both at the domestic level and in the statutory definitions at the international level, and in general there is agreement that since they are modes of assisting the physical perpetrator, they are not on par with principal liability, even in systems where there is no legal effect of making such a distinction. It would be easy to think that aiding and abetting are therefore uncontroversial. Yet the ways in which aiding and abetting have been applied to civil and military superiors by international tribunals in recent years demonstrates quite something else. In the first place, there has been some confusion over terminology included in the ad-hoc tribunal statutes, as to whether there is a difference between complicity on the one hand and aiding and abetting on the other hand. In the second place, the policy-driven nature of utilising modes of liability as catch-alls has become particularly evident with the conviction of former Liberian president Charles Taylor for aiding and abetting. In the third place, and most concerning, the potential

[159] Ambos 2008b, p. 491.

[160] Statute of the ICTY 1993, Article 7(1); Statute of the ICTY 1994, Article 6(1); Statute of the STL 2007, Article 6(1); Cryer et al. 2016, p. 314.

[161] Katanga Confirmation of Charges 2008, para 518.

[162] Olásolo 2009, p. 142, citing Del Ponte 2006.

[163] Werle and Jessberger 2014, pp. 181–182.

[164] Milutinović 2009, para 87; Werle and Jessberger 2014, p. 182.

manipulation by those participants who influence the development, application and interpretation of the law has been exposed in a most uncomfortable way at the ICTY. These three issues will receive due attention here.

8.5.1 Complicity and Aiding and Abetting

First, with respect to terminology, historically there was little attention paid at the international level to defining forms of assisting, since the Nuremberg Charter and the Tokyo Tribunal Charter referred to 'accomplices' as a residual term, but did not further define it, and the judgments did not focus on this kind of contribution to a crime.[165] The fact that these Charters did not distinguish between modes of liability reflected a functionally unitary approach, and may have had something to do with this vagueness of terminology.

When the ICTY Trial Chamber referred to the post-Second World War case law in in *Furundžija*, it was careful to state that since most of the case law was in fact based on domestic law, any conclusions drawn from this as to customary international law should be cautious.[166] There was some inconsistency in the post-war domestic case law, and it was clear that this was due to the fact that each tribunal applied concepts familiar to it depending on the domestic jurisdiction it was part of. Some of the Control Council No. 10 cases based on UK law dealt with liability of those who were 'concerned in killings',[167] for providing information, driving a car, or providing a service to the perpetrators towards the crime. The same language can be recognised in some cases based on German law which included moral encouragement as culpable assistance.[168] In the *Dachau* case decided by the US Military Tribunal, reference was made to US law on aiding and abetting.[169] Yet the ICTY Trial Chamber judges went on to rely on this domestic based case law precisely to justify their findings on customary law, by selectively 'picking and choosing' which decisions to cite,[170] and demonstrated very clearly the patchworking process by which modes of liability are developed and established in ICL.

The ICTR Trial Chamber stated that there are at least three recognised forms of complicity under the civil law tradition, to which Rwandan law belongs, namely instigation, aiding and abetting, and procuring means.[171] However as Sliedregt points out, the terms aiding and abetting are specifically of common law origin. According to the old modes of liability at the time when English law differentiated between

[165] Charter of the IMT 1945, Article 5; Tokyo Charter 1946, Article 4.

[166] Furundžija 1998, para 194.

[167] Almelo Trial 1945; Rohde 1946; Schonfeld 1946.

[168] The German *Synagogue* and *Pig Cart Parade* cases were cited by the ICTY Trial Chamber in Furundžija 1998, para 207–209.

[169] Dachau Case 1945.

[170] van Sliedregt 2012a, p. 121.

[171] Akayesu 1998, para 533.

facilitators who were present at the scene of the crime (principal accomplice) and those who were not (accessory before the fact), the distinction served to differentiate between those who would receive the death penalty, and those who would not, due to a lower degree of culpability. When these distinctions disappeared under legislative reform, the different forms of facilitation or assistance were grouped together under the term 'aider and abettor'.[172]

The problems of translating English language terms to new or different jurisdictions thus become apparent: some judges and scholars use the term complicity to mean aiding and abetting as an alternative to (but not differentiated from) committing, such as under the Nuremberg and Tokyo Charters; some use complicity as a broad term to mean assisting the physical perpetrator, under which aiding and abetting may fall alongside instigating or soliciting, as the ICTR did; and some use it to denote specifically secondary modes of liability, which aiding and abetting may or may not be, as will be discussed shortly.

This discrepancy was part of the cause of confusion at the ICTY and ICTR over whether there is a difference between complicity in genocide and aiding and abetting genocide. This question was forced upon the judges by the fact that the common article defining the crime of genocide in both of the statutes lists acts which are punishable, including conspiracy and complicity, following the wording of the Genocide Convention.[173] It would seem complicity is listed as an act of genocide *separate* from the modes of liability 'aiding and abetting', which are listed in the common article on individual criminal liability.[174] Various Trial and Appeals Chambers of both ad-hoc tribunals have attempted to reconcile this apparent repetition, alternately by insisting they are two separate charges with differing subjective and objective requirements, whereby complicity in genocide is a broader charge but requires specific intent,[175] or by seeing them as overlapping or identical.[176] Similarly, scholarly opinion has been divided on the correct interpretation, and some authors have asserted that the statutes must be interpreted strictly and exhaustively in order to distinguish the crime of complicity in genocide from the mode of liability of aiding and abetting genocide.[177]

However the most logical view is that the overlap is a result of inadvertence in the drafting of the the statutes, which were not drafted by criminal law experts, but by international law experts and diplomats who had copy-pasted the wording of the Genocide Convention into the definition of genocide, but had not seen the overlap of complicity in this definition with the modes of liability in another article.[178] Once

---

[172] van Sliedregt 2012a, p. 112. For a fuller discussion on the old common law modes of liability, and the reasons they were eradicated in the legislation of modern jurisdictions which follow the common law tradition, see Chap. 6.

[173] Statute of the ICTY 1993, Article 4(3)(b) and (e); Statute of the ICTY 1994, Article 2(3)(b) and (e).

[174] Statute of the ICTY 1993, Article 7; Statute of the ICTY 1994, Article 6.

[175] Akayesu 1998, paras 538, 548; Krstić Appeal Judgment 2004, para 139.

[176] Semanza 2003, para 394; Stakić Trial Judgment 2003, para 531.

[177] van Sliedregt 2012a, p. 123; Boas et al. 2007, p. 300, citing Chili Eboe-Osuji.

[178] Boas et al. 2007, p. 300; Schabas 2008, p. 291; Akhavan 2005, p. 994.

again the role of participants in the process of law have been influential, even where this influence is inadvertent. Other participants, namely judges and scholars, are further left with the task to clarify this terminology and untangle the haphazard patchworking process.

The clearest path would be to define complicity as all forms of participation in a collective crime other than physical perpetration. Complicity in genocide should be understood in this way too, and not as a separate definition. Conspiracy to commit genocide can therefore be considered an inchoate crime, just as incitement to genocide, which is specific to this crime.

8.5.2 *Aiding and Abetting as a Catch-All for Leadership Liability*

Apart from the terminology, another issue is the the policy-driven nature of utilising aiding and abetting as a catch-all mode of liability in order to ensure a conviction in some cases where leadership liability may have been more appropriate, but the evidence did not fit the bill.

In domestic systems which differentiate normatively, aiding and abetting are secondary in nature, and even in those systems which only differentiate semantically but which focus on sentencing to reflect degrees of blame, they are often seen as grounds for lowering the sentence. This is reflected at the ICTY, for example in *Vasiljević* where the Appeals Chamber stated that 'aiding and abetting is a form of responsibility which generally warrants a lower sentence than is appropriate to responsibility as a co-perpetrator.[179] And in Appeal, Krstić's sentence was reduced from 46 to 35 years because the conviction was 'merely' for aiding and abetting.[180]

Much criticism was raised against the conviction of former Liberian president Charles Taylor for 'merely' planning, aiding and abetting the crimes committed by members of the Revolutionary United Front (RUF) in Sierra Leone, rather than for taking part in a JCE as a principal perpetrator.[181] The fact that he was sentenced 50 years imprisonment also challenges the general understanding that 'aiding and abetting' are secondary modes of liability, or at least warrant lower sentences. Either he is deserving of a severe punishment due to the importance of his role, or his role was that of 'merely' assisting the physical perpetrators, by providing arms in exchange for so-called blood diamonds. Although the Statute and Rules of the SCSL make no reference to the mode of liability as a factor in the sentencing,[182] unlike those

---

179 Vasiljević Appeal 2004, para 182.

180 Krstić Appeal Judgment 2004, para 268.

181 Taylor Judgment 2012.

182 Article 19 and Rule 101(B) instruct the court to take into account the gravity of the offence, the individual circumstances of the convicted person, any aggravating and mitigating circumstances, and where appropriate the general practice of the ICTR and of Sierra Leonean national courts. Statute of the SCSL 2000; Taylor Sentencing Judgment 2012, para 18.

of the ICTR and ICTY, in the sentencing judgment Taylor's Trial Chamber referred to the case law of these two ad-hoc tribunals as authoritative for the principle that the sentence should take into consideration the 'form and degree of the participation of the Accused.'[183] While the Trial Chamber judgment had found that aiding and abetting did not warrant a lower sentence than other modes of liability, this was corrected by the Appeal Chamber, which otherwise upheld the conviction of Taylor.[184] Following this correction, in the Sentencing Judgment the Trial Chamber noted that the case law of the ICTR and ICTY indicate that aiding and abetting generally warrants a lesser sentence than for 'more direct forms of participation', however it determined that the unique circumstances of the case warranted the severe sentence of 50 years: in particular it considered the leadership position of Taylor as Head of State to be an 'aggravating circumstance'.[185] It is noteworthy that this was not pleaded as an aggravating circumstance in the indictment, and is one that the judges conceived of themselves to justify the sentence despite the lesser mode of liability.[186]

The construction of this reasoning is selective and arguably artificial, in order to justify the final outcome which in fact departed from the trends at the other tribunals upon which the judges of the Trial Chamber relied.

Similarly, though perhaps with less controversy, Jean-Paul Akayesu, who was the *bourgmestre* or mayor of the Commune in Rwanda in which many atrocities had taken place, was convicted at the ICTR for aiding and abetting partly *because* of his leadership position:

> [the] accused, having had reason to know that sexual violence was occurring, aided and abetted the following acts of sexual violence, by allowing them to take place on or near the premises of the bureau communal and by facilitating the commission of such sexual violence through his words of encouragement in other acts of sexual violence which, by virtue of his authority, sent a clear signal of official tolerance for sexual violence, without which these acts would not have taken place.[187]

8.5.3 Manipulation of the Law-Making Process

This brings us to the third issue identified here, namely the potentially over-selective and politically driven influence of participants in developing and applying modes of liability. Following the controversial acquittal of Momčilo Perišić at the ICTY in 2013, what emerged was an apparent manipulation of the inherently creative patchworking process. This is where the political nature of mass atrocity prosecutions becomes volatile.

---

183 Taylor Sentencing Judgment 2012, para 9.

184 Taylor Appeals Judgment 2013, para 707.

185 Taylor Sentencing Judgment 2012, para 21, 29.

186 Taylor Sentencing Judgment 2012, para 30.

187 Akayesu 1998, para 692.

Perišić had been the Chief of General Staff of the Yugoslav army, and was convicted of aiding and abetting crimes committed by the Army of Republika Srpska in Sarajevo and Srebrenica, for providing logistical and financial assistance that had 'substantial effect' on the crimes perpetrated,[188] as well as for superior responsibility, due to failing to prevent his own subordinates from committing crimes in Croatia. In appeal, the bench presided over by Theodor Meron came to a controversial decision to acquit Perišić of all charges because it found that although he had knowledge of the crimes, he did not 'specifically direct' his assistance toward the crimes.[189] The judgment referred to the mention of specific direction in the *Tadić* case, and explicitly departed from its own previous judgment in *Mrkšić and Šljivančanin*, in which it had stated that specific direction is *not* a requirement for aiding and abetting.[190] It referred to other case law in which the terms 'specifically' or 'directed' were used, even though most of these judgments did not use the combination of terms, and it disregarded any judgments in which these terms were *not* used as being incomplete definitions.[191]

This finding has been criticised for two reasons: in the first place, that the 'specific direction' requirement is a novel addition to aiding and abetting, which has no basis at all in customary law, nor in the case law of international and domestic tribunals.[192] By introducing this requirement, the line is now blurred between aiding and abetting, JCE and many forms of principal perpetration. Certainly this new definition impacted on the *Taylor* Appeal judgment. Whereas the SCSL Trial Chamber had relied in part on the *Perišić* Trial Judgment, the SCSL Appeals Chamber explicitly refused to follow the ICTY's new definition when it upheld Taylor's conviction. Instead it restated that all that is required for aiding and abetting is that 'the accused's acts and conduct have a substantial effect on the commission of the crime', since this ensures sufficient causal link between the accused and the crime.[193] If it had followed the *Perišić* Appeal definition, one wonders whether Taylor too would had to have been acquitted.

In the second place, questions have been raised as to the motivation behind the *Perišić* Appeal judgment, in particular following the controversial letter written by Danish ICTY Judge Frederik Harhoff to 56 of his colleagues and friends, which was subsequently published and made available online.[194] Judge Harhoff not only publicly criticised the Appeals Chamber's departure from set practice, but also accused the President of the ICTY Judge Meron of having pressured his fellow judges into

[188] Perišić Trial Judgment 2011, para 1627.

[189] Perišić Appeal Judgment 2013.

[190] Perišić Appeal Judgment 2013, para 26–34.

[191] Perišić Appeal Judgment 2013, para 30.

[192] James Stewart undertook an extensive empirical analysis of judgments rendered by the ICTY ICTR and other ad-hoc tribunals, and found that of the 362 judgments in which aiding and abetting was indicted, less than one third of them refer to either 'specifically' or 'directed', that all of them cite *Tadić* only in passing, and that in only 2% of the cases is it included in the conviction: Stewart 2013. See also Coco and Gal 2013; Jenks 2013.

[193] Taylor Appeals Judgment 2013, para 480.

[194] Anderson 2013.

agreeing to the decision. He further noted three other recent acquittals penned by Meron which turned back the case law on leadership liability, and alluded to political pressures by Israel and the US on Meron to do so.

While the publication of such a letter was surely unprofessional, the accusations within have never been investigated, and instead Judge Harhoff found himself disqualified from the bench of an unrelated case, ironically, for bias, and was effectively forced to leave the ICTY.[195] At the beginning of 2014 the Appeals Chamber, this time without Meron on the bench, wrote in no uncertain terms that it 'unequivocally rejects the approach adopted in [Perišić] as it is in direct and material conflict with the prevailing jurisprudence ...and with customary international law.'[196]

It would therefore appear that the introduction of the 'specific direction' requirement was a temporary sidetrack on the part of the ICTY. Aiding and abetting remains a lesser form of liability, and its application to leaders such as Charles Taylor remain questionable, especially considering the sentence that he received. What is most concerning in this turn of events is that the policy aspect of developing and applying modes of liability seems to have gotten out of hand: rather than a reasoned policy response to events, and the careful development of appropriate modes of liability, instead it would seem the law was changed to reach an outcome desired by certain participants. Even though this has been corrected, the role of individual judges as participants in this process was first made evident and then subsequently silently denied, rather than brought to the forefront. The assertion in this study is that the dynamic patchworking process by which ICL is being developed is a result of the active role of such participants, but also that this process should be made explicit and harnessed, such that the legal results are better grounded in careful doctrinal consideration, and are legitimate and trustworthy. By sweeping aside the mess that came to the surface in this process, the opposite has occurred, and the legitimacy of the international legal project has been brought into question.

8.6 Joint Criminal Enterprise: Extended and Constructive Liability

These debates between participants have been even more prevalent with respect to JCE. Because the process of developing modes of liability for mass atrocity stood fairly still for several decades following the Nuremberg trials, it is no surprise that when the first ad-hoc tribunals appeared in the 1990s, a reprise could be observed of the same processes of debate among jurists from different legal traditions on how to resolve the problem of individual liability for collective crimes. As has already been pointed out, the Charters and Statutes of all of these tribunals were drafted predominantly by international lawyers and not by criminal lawyers, and thus less attention was paid to the legal detail of criminal liability. The inception and development of

[195] Decision on Disqualification of Judge Frederik Harhoff 2013.

[196] Sainović Appeal Judgment 2014, para 1650.

the mode of liability known as JCE has therefore gained much attention over the years.

It is not the intention to give a full overview of the literature here,[197] but rather to concentrate on the process by which JCE came into being as a judicial invention, and the influence of domestic criminal law notions on its inception. First this process will be discussed critically. Second, the process by which it became such an integral part of the jurisprudence of the ICTY will be discussed. Finally, some attention will be paid to its application in some other tribunals, and the fact that despite this wider reception, it still remains somewhat controversial.

8.6.1 *The Inception of a New Mode of Liability*

In its seminal *Tadić* decision in 1999, the judges of the ICTY Appeals Chamber took great pains to resolve a serious problem of individual criminal liability. The defendant had been convicted of committing various war crimes and crimes against humanity. In appeal one of the questions raised was whether it could be proven beyond reasonable doubt that he was also responsible for the murder of five Bosnian men.[198] While Tadić's presence in the village in question was verified, as was his participation in forcefully removing the victims from their homes, his participation in, and responsibility for, the deaths of the victims was not certain. The Trial Chamber had found the 'bare possibility' that it had been him was not enough to secure a conviction on this count.[199] However the Appeals Chamber took a different line of reasoning, and considered the legal question to be 'whether the acts of one person can give rise to the culpability of another where both participate in the execution of a common criminal plan'.[200] Framing the legal question this way immediately placed the focus upon a plan or agreement, thus more on the subjective rather than the objective elements, and also suggested an approach similar to the vicarious theory of liability.

Because Tadić was proven to have been present as a member of the Serb forces and to have been involved in other related crimes, the Appeals Chamber found that he could also be held responsible for further crimes which were a foreseeable consequence of the actions of the group, and with regard to which he was reckless or had an indifferent state of mind.[201] This represents a very low threshold with respect to the principle of culpability.

---

[197] For excellent discussions of JCE, including its development and debates regarding the required subjective and objective elements, see e.g. Fletcher and Ohlin 2005; Haan 2005; Haffajee 2006; Ambos 2007b; Ohlin 2007; van der Wilt 2007; Zahar and Sluiter 2008; Olásolo 2009; van Sliedregt 2012a.

[198] Tadić Appeals Judgment 1999, para 175.

[199] Tadić Appeals Judgment 1999, paras 181–183.

[200] Tadić Appeals Judgment 1999, para 185.

[201] Tadić Appeals Judgment 1999, para 237.

It also represents an expansive reading of the ICTY Statute. Article 7(1) of the Statute lists planning, instigating, ordering, committing or otherwise aiding and abetting as alternative modes of liability. The wording mirrors the IMT Charter and the Nuremberg Principles, and no explicit normative distinction appears to be drawn between parties to a crime according to their different degrees of actual contribution. Because none of these modes of liability fit the charge against Tadić comfortably, the judges of the Appeals Chamber bench entered a process of creative interpretation and creation in order to reach the conviction that was sought, in much the same way as the judges at the Nuremberg and Tokyo IMTs had done.

This marked the birth of JCE as a wide net with which to catch as many individuals as possible who had some connection to a collective crime, even when that connection is very remote and the subjective mental element is at a very low threshold. JCE was considered to be a form of 'commission' under Article 7 of the Statute, due to the argument that the moral gravity of participation is no less than when an individual participates in the crime directly, even when the actual role played is difficult to prove.[202] Thus the Appeals Chamber asserted a system that could be described as functionally unitary; different ways in which a party can contribute to a crime are enumerated, but no distinction is made between principal or secondary liability. Yet at the same time, the Appeals Chamber had also stated that JCE is distinct from aiding and abetting, and that the latter would 'understate the degree of [Tadic's] responsibility',[203] thereby asserting a normative distinction between these modes of liability. This contradictory definition has led to a continual shift in interpretation throughout the case law of the ICTY, as will become apparent in the debates discussed here.

The doctrines of common purpose which can be recognised in various domestic criminal law systems are reflected in the language of JCE as it has developed since *Tadić*. In fact, the ICTY Appeals Chamber has remarked that in its own judgments the terms 'common purpose' and JCE have been used interchangeably, and that they refer to one and the same thing.[204] The Appeals Chamber sought authority in the case law following Second World War; the *Almelo Trial*[205] and the *Hoelzer*[206] case were cited,[207] both of which involved convictions of Germans for the killing of, respectively, a British soldier and a Canadian prisoner of war, by 'taking part in a common enterprise' with a shared common purpose. The Appeals Chamber cited these and other cases in detail, focusing on the language used of 'common design',[208]

---

[202] Tadić Appeals Judgment 1999, para 188.

[203] Tadić Appeals Judgment 1999, para 192.

[204] Ojdanić Decision on JCE 2003, para 36.

[205] Almelo Trial 1945.

[206] Hoelzer 1946.

[207] Tadić Appeals Judgment 1999, para 197.

[208] Einsatzgruppen Case 1951, cited in Tadić Appeals Judgment 1999, para 200.

'common object',[209] 'cog in the wheel',[210] and reference to the guilt shared by anyone assisting in the act.[211]

All of the cases cited in hailed from the common law jurisdictions of Canada, the UK and the USA, all of which share the tradition of the conspiracy doctrine, which is based upon a notion of common plan. They also share the subjective conceptual link between an individual and a collective crime whereby the acts of all other participants in the crime are attributed to an accused. In relying on these jurisdictions so heavily, the Appeals Chamber began to construct a notion of vicarious liability as underpinning JCE.

By comparison, a brief footnote reference to cases from German and Italian post-Second World War trials was made, in which the notion of co-perpetration was applied.[212] Here the Appeals Chamber in *Tadić* asserted that co-perpetration and the notion of common purpose are the same approach, yet at the same time acknowledged that co-perpetration distinguishes different degrees of involvement.[213] The Appeals Chamber appears to have preferred the subjective approach in order to widen the net of liability while loosening the evidentiary burden. While it was free to choose a subjective over an objective approach, the assertions that they are the same, and that this amounts to a general or customary law approach, are highly questionable.

8.6.2 *The Question of Sources for JCE*

An analysis of this process of inception reveals that a classical understanding of sources in international law has been turned upside down. Because ICL is considered to be a branch of public international law, and because the ICTY statute does not specify sources of law applicable, the tribunal must look to the sources of public international law in order to interpret its own statute, which can give rise to complexities. It has been pointed out that '[i]t is the very ad-hocness of international tribunals (the fact that they post-date the alleged crimes) that places them at a disadvantage in relation to the sources of law.'[214] Because the Appeals Chamber could not find a clear answer in treaties or in the text of the statute, and because it would go too far to suggest that JCE is a general principle of criminal law,[215] the court was left with customary law and the so-called subsidiary sources of judicial decisions and authoritative opinions. The existence of the former is debatable and the prevalence of the latter as sources cited is noteworthy.

---

209 Schonfeld 1946, cited in Tadić Appeals Judgment 1999, para 198.

210 Ponzano case 1946, cited in Tadić Appeals Judgment 1999, para 199.

211 Jepsen 1946, cited in Tadić Appeals Judgment 1999, para 198.

212 Tadić Appeals Judgment 1999, footnote 246 and 247.

213 Tadić Appeals Judgment 1999, para 201.

214 Zahar and Sluiter 2008, p. 80.

215 Steer 2012, p. 152.

In the *Furundžija* decision, the Trial Chamber at the ICTY gave three preconditions for relying on national laws and judicial decisions in order to ascertain whether a norm can be said to be customary law.[216] The first is that reliance upon national legislation is justified only when international criminal rules do not clearly define a notion of criminal law; the second is that reference should be made to more than one national legal system (unless otherwise indicated, for example due to the applicable law restrictions in a statute); the third is that 'account must be taken of the specificity of international criminal proceedings when utilising national law notions',[217] due to the unique traits of international proceedings and the procedural differences in national systems. The notion that more than one legal system should be referred to is understood generally to be a condition of ascertaining a customary law norm.[218]

While the common purpose doctrine could be seen as a common element in some jurisdictions which share a particular legal tradition, it is artificial to draw the conclusion that it is a norm of customary law, since there are so many differences in the customs and practices of States.[219] The *Tadić* Appeals Chamber concentrated heavily on post Second World War trials decided by tribunals of common law jurisdictions, and the analogies drawn with a select few civil law jurisdictions are questionable in their correctness. Even more disturbing is that recently Judge Antonetti attempted to find the Italian cases that had been footnoted in the original *Tadić* decision, in preparing for the judgment of former President of the Serbian Radical Party, Vojislav Šešelj, and could only find some of them in handwritten form, while others were not publicly available at all. As he himself pointed out;

> Given the importance of this issue, it is unfortunate that the Judges in the *Tadić* Chamber, who apparently decided to rely on the Italian Judge, did not have the handwritten appeal judgements translated into French and English, or even into B/C/S, and subsequently archived in the ICTY Library.[220]

Many critics of JCE have derided this 'survey style' of ascertaining a customary norm, since language barriers, inaccessibility of materials and lack of understanding of the legal system from whence a particular law or decision hails, all lead to a highly selective survey of state practice.[221] European systems dominate the results of such case law survey, rarely is reference made to Russian, Chinese, Indian or African state practice.[222]

The status of Second World War case law as customary law is itself questionable, since these military courts had no precedent, and were at the time hotly debated.[223] They also applied national laws, as there was no agreed international framework

216 Furundžija 1998, para 175.

217 Furundžija 1998, para 175.

218 Brownlie 1998, p. 21.

219 Steer 2012, p. 150.

220 Šešelj Judgment, Concurring Opinion of Judge Antonetti 2016, p. 162.

221 See e.g. Zahar and Sluiter 2008, p. 94; Danner and Martinez 2005, p. 112.

222 Zahar and Sluiter 2008, p. 94.

223 Cesarani 2006, pp. 34–37.

under which to operate. While this demonstrates a historical precedent for decisions taken in the ad hoc tribunals of the end of the 20th century, it demonstrates the patchworking process in action, drawing from domestic law notions to create new international ones, however this does not amount to a customary norm. The inconsistencies are too great, not only between domestic laws, but also among the case law of the post-Second World War trials, which, being decided by numerous different military tribunals, had no reason to be internally consistent.[224]

The influence of the domestic law notion of conspiracy or of common plan in this process is clear, however it is the extent and manner in which JCE has been developed which is controversial. There have been suggestions that JCE is the same as the notion of membership in a criminal organisation, under Control Council Law No. 10,[225] however it differs significantly in that it is a form of participation in a predicate crime, whereas membership of a criminal organisation was punishable in and of itself. Furthermore, this has since been rejected under international law. The three categories of JCE which have been developed at the ICTY through the case law also seem to go beyond a recognisable common purpose doctrine.

For all three categories, the objective elements (actus reus) remain the same: there must be (i) a plurality of persons, (ii) the existence of a common plan, design or purpose which amounts to or involves the commission of a crime and (iii) participation of the accused in the common design involving the perpetration of one of the crimes which fall under the jurisdiction of the ICTY.[226] What exactly amounts to 'participation' of the accused seems to differ slightly for the three categories in some later decisions of the ICTY.

The first category, later termed JCE I, is termed the basic form of JCE; all that is required is that the accused shares the intent to perpetrate a certain crime with the other perpetrators. This form resembles the US notion of conspiracy as a mode of liability, or the common law doctrine of common purpose, very closely.

The second category, later termed JCE II, is referred to as the systemic form of JCE,[227] which is really just a variant of the first category. Inspiration for this category was drawn from the concentration camp cases following Second World War, where officers working at the camps were held responsible for crimes committed within the system of the camps, regardless of their actual participation in the commission of these crimes.[228] For this category it is required that the accused has a personal knowledge of the system of ill-treatment, as well as the intent to further this common concerted system of ill-treatment.

---

224 Zahar and Sluiter 2008, p. 99.

225 See e.g. Šešelj Judgment, Concurring Opinion of Judge Antonetti 2016, p. 151.

226 Tadić Appeals Judgment 1999, para 227.

227 See e.g. Vasiljević Appeal 2004.

228 The *Tadić* bench cited the *Belsen* case, decided by a British military tribunal, and the *Dachau Concentration Camp*, decided by a US military tribunal. See Tadić Appeals Judgment 1999, para 202.

The third category, later termed JCE III or the extended form of JCE,[229] is the most controversial, whereby an accused can be held liable for a crime that was committed by other members of the group outside the ambit of the agreed common purpose, as long as (i) it was foreseeable that such a crime might be perpetrated by other members of the group and (ii) the accused willingly took the risk (described in some cases as *dolus eventualis*)[230] that such a crime would be committed. The crime must be shown to have been (subjectively) foreseeable to the accused in particular.[231] Professor Cassese, who had presided over the *Tadić* bench, has analysed his own judgment in his writings as a professor of law, and has defined this as 'incidental criminal liability based on foresight and voluntary assumption of risk'.[232]

This foreseeability and risk construction has received criticism for being too expansive and generating convictions based on guilt by association.[233] It has also recently been rejected by the Supreme Court of the United Kingdom, overturning previous law on joint enterprise, since a distinction must be drawn between foresight and intention: if the joint defendant did not act with the requisite mental element of the offence, then he cannot be held liable for the crime of another, even if it is foreseeable in the course of the original criminal agreement.[234] It is without irony that joint enterprise was previously known as 'parasitic accessorial liability' in the UK.[235]

Debates among judges at the ICTY have been continuous since the inception of JCE, and its development has been inconsistent and incongruent. Indeed, Judge Antonetti made this very statement in explaining the recent acquittal Šešelj: 'We have no choice but to conclude that the *Tadić* jurisprudence and the concept of JCE to which it gave rise have produced a degree of legal uncertainty brought about by the ambiguity of this concept.'[236]

8.6.3 The Reification of JCE

Despite the position that JCE is the child of the ICTY and should not be considered broadly applicable as a norm of ICL, it cannot be denied that it has gained solid ground as its own legal concept. This has occurred despite—or perhaps as a result of—the controversy and heated debates among judges, prosecutors, defence lawyers

---

229 See e.g. Vasiljević Appeal 2004; Stakić Trial Judgment 2003, para 246.

230 See e.g. Brđanin Appeal Judgment 2007, para 365.

231 See e.g. Brđanin Decision on Amended Indictment 2001, para 29.

232 Cassese 2007, p. 113.

233 Badar 2006; Danner and Martinez 2005; Zahar and Sluiter 2008, p. 233; Ambos 2007b, p. 169; van der Wilt 2007, p. 99; Ohlin 2007, p. 79; Werle and Jessberger 2014, p. 176.

234 R. v Jogee and Ruddock 2016, para 90.

235 Laffan 2016.

236 Šešelj Judgment, Concurring Opinion of Judge Antonetti 2016, p. 154.

and academics. And it has occurred through a process of assertion, repetition and reification.

One example is the consistent rejection by various chambers of the ICTY of any arguments that dispute the validity of JCE, simply by way of referring back to other decisions of the ICTY, in what amounts to circular reasoning. For example, in an appeal against his conviction based on participation in a JCE, Ojdanić argued that the drafters of the ICTY statute had intended to exclude a mode of liability such as JCE, since the *traveaux* of the Statute of the ICTY shows that conspiracy was rejected as a form of liability and that JCE is based on the notion of conspiracy and must therefore also be rejected.[237] The Appeals Chamber rejected this argument, relying on the Tadić judgment as its sole reason for doing so.[238]

At the ICTY, JCE has dominated the approach taken to collective criminality.[239] Some would argue that the application of JCE at other tribunals is inappropriate, since the development of and preference for JCE at the ICTY is closely connected with the specific fact situations of the cases tried before that tribunal and the limitations of the ICTY statute.[240] By way of comparison, many other ad-hoc tribunals have continued the process of reification through repetition, demonstrating the way in which a notion can become law due to its application by the participants in the process.

One exceptional application of JCE outside the ICTY has been in the State court of Bosnia and Herzegovina, (BiH), the internationlised domestic court developed to prosecute the majority of the war crimes cases in BiH.[241] The Court stated that Article 7 of the Statute is binding international law 'by virtue of its having been drafted pursuant to the powers of the United Nations.'[242] Clearly this is a misunderstanding of the kind of binding law the Security Council can affect, according to Article 103 if the UN Charter. Although the Statute of the ICTY was annexed to a Security Council resolution, it creates no obligations for States and cannot bind States in any way; it only binds the judges of the court to apply the Statute. The Court also determined that because Article 180(1) of the BiH Criminal Code was a direct derivation of Article 7(1) of the ICTY Statute, it must include all definitions and interpretations from ICTY case law, including the findings that the elements of JCE are established in customary international law.[243] This reasoning is also problematic, since there is no formal relationship between the ICTY and the State Court of BiH. The interpretations given in ICTY case law may well be instructive, but can never be binding.

---

237 Ojdanić Decision on JCE 2003, para 14.

238 Ojdanić Decision on JCE 2003, para 18.

239 According to Danner and Martinez, from June 2001 when the first indictment charging JCE explicitly was confirmed, to January 2004, 64 % of all indictments relied explicitly on JCE and 84 % of all indictments incorporated charges including phrases such as 'acting in concert', which could also be construed as a JCE. See Danner and Martinez 2005, p. 108.

240 Haan 2005, p. 167; Fletcher and Ohlin 2005, p. 550.

241 Rašević, First Instance Verdict 2008.

242 Rašević, First Instance Verdict 2008, para 103.

243 Rašević, First Instance Verdict 2008, para 104.

Article 6(1) of the ICTR Statute follows the exact wording of Article 7(1) of the Statute of the ICTY, both of them being based on the Nuremberg Charter, and therefore does not explicitly include JCE. The ICTR has often referred to ICTY case law as instructive in its own decisions, however prosecutors were slow to embrace JCE with the same enthusiasm as their colleagues in The Hague, perhaps because the context of the conflict being adjudicated was entirely different. The context in which the Rwandan genocide took place, and the multitude of individuals involved at all levels, in a less bureaucratically organised manner, led to a preference for other modes of liability such as incitement, ordering, aiding and abetting. Nevertheless, the ICTR followed the line of the ICTY in a small number of cases, making particular use of JCE III to cover the 'vast scope' of collective activity involved in the genocidal massacres that took place.[244] In all the decisions and judgments handed down by various chambers of the ICTR on JCE, extensive citation of the case law of the ICTY is given as the source of law on JCE, and it has been affirmed repeatedly that all three forms of JCE are applicable before the ICTR as customary law.[245]

In the case against Karemera, the defendant was found to have shared a common purpose with the *chef de cabinet* of the Ministry of Defense, Bagosora, to prompt non-members of the JCE to perpetrate killings,[246] and because Bagosora had distributed weapons to civilians, the rape of a woman by a citizen was imputed to him, and via the JCE also to Karemera.[247] The nexus between the accused individual and the crimes committed by the physical perpetrators becomes extremely remote with this reasoning. It is quite possible that this was a result of the policy desire to include crimes of sexual violence which had otherwise lacked sufficient attention. Such gender crimes may not have been explicitly agreed upon, but they are also not merely incidental in the course of such genocidal violence.[248] However, although it was necessary to form some kind of policy response to these gender crimes which would have otherwise remained invisible, over-extension of the principle of culpability makes this solution questionable.

Article 6 of the Statute of the SCSL mirrors the wording of the ICTY and ICTR Statutes.[249] However the prosecutor has made welcome use of JCE in the indictments, and where JCE has been applied it has been with specific reference to ICTY case law. Given that the SCSL began its work approximately one year after the *Tadić* judgment, this is of no surprise.

In its indictment of Charles Taylor, the prosecutor used the language of JCE, and embraced the doctrine of the ICTY, including the far-reaching JCE III.[250] However this was in the alternative, alongside the modes of liability planning, instigating,

---

[244] See e.g. Kayishema 1999; Nzirorera Decision 2006, para 15; Karemera 2012, para 1435.

[245] See e.g. Rwamakuba Decision on JCE 2004, para 31; Ntakirutimana Appeal 2004, paras 461–462; Simba 2005, para 385.

[246] Karemera 2012, para 1455.

[247] Karemera 2012, para 1451.

[248] Haffajee 2006, pp. 211–212.

[249] Statute of the SCSL 2000, Article 6.

[250] Taylor Indictment 2003, para 33.

ordering, committing and aiding and abetting, as well as a separate charge under superior responsibility.[251] It was perhaps a policy choice to reject JCE and opt for 'aiding and abetting' in the final conviction, even with the controversy this brought with it, as discussed above.

Of all the other cases before the SCSL in which JCE was charged in the indictment, only one led to a conviction based on this mode of liability. In the case against the leaders of the Revolutionary United Front (RUF), Augustine Gbao was convicted at trial as the RUF's 'ideological trainer'. It was found that he significantly contributed to the JCE, as the leadership of the RUF relied on this ideology to ensure and enforce the discipline and obedience of its forces.[252] As Justice Shireen Fisher stated in her dissent, this conviction 'abandons the keystone of JCE liability as it exists in customary international law', since the Trial Chamber had found that Gbao was individually criminally responsible for the crimes committed in furtherance of a common criminal purpose that he did not himself share.[253] It is noteworthy that Gbao was sentenced to 20 years under this mode of liability that is supposedly a form of principal liability, whereas Taylor received 50 years for 'merely' aiding and abetting.[254]

This reification through repetition is most evident in the processes before the Extraordinary Chambers in the Courts of Cambodia (ECCC). The trial judges had asked for instruction by way of *amicus* briefs, as to (1) the development of the theory of JCE, and (2) whether JCE was a mode of liability under customary international law during the relevant time period of 1975–1979, when the justiciable crimes were committed under the Pol Pot regime.[255] The submitted briefs reflect the disparate and opposing views on the doctrine and on the question of its customary law status.

On the one hand, Professor Antonio Cassese argued that the Tadić Appeals Chamber did not develop the doctrine itself but, rather, elucidated a mode of liability that had been a part of customary international law since the aftermath of the Second World War.[256] Cassese cited the *Tadić* decision, in the writing of which he himself had been instrumental, but he offered no further supporting material. He also asserted that the element of collective criminality inherent to international crimes invokes a

---

[251] Taylor Indictment 2003, para 34. In the amended indictment changes were made to the structure but not the essence of the charges, maintaining the language of JCE I and III, and charging these as an alternative mode of liability, Boas et al. 2007, p. 130.

[252] RUF Case, Appeal 2009, para 27. Although the notion of 'ideological trainer' was rejected in Appeal, his conviction as a member of the JCE still stood, based on the assertion that there is no requirement of a necessary relationship between the objective of a common purpose and its criminal means: RUF Case, Appeal 2009, para 296–297.

[253] Partially Dissenting and Concurring Opinion of Justice Shireen Fisher, RUF Case, Appeal 2009, para 2, 14.

[254] RUF Case, Appeal 2009, para 1320.

[255] On 5 September 2008, the Office of the Co-Prosecutors filed an Appeal brief requesting the Pre-Trial Chamber to amend the Closing Order of the Co-Investigative Judges to include Joint Criminal Enterprise as a mode of liability in the indictment against Kaing Guek Eav, aka 'Duch'. Following the Appeal Brief, the Pre-Trial Chamber invited *amici curiae* on the subject.

[256] Cassese 2008a, para 20.

specific need for JCE and that considerations of public policy aid in justifying its use.[257]

On the other hand, in his brief, Professor Kai Ambos argued that JCE is not universally accepted, pointing out that of the mixed tribunals established in the aftermath of the ICTY, only the East Timorese Special Panel for Serious Crimes had applied the doctrine,[258] and that the ICC has dissociated itself from JCE.[259] In contrast to Cassese's determination, Ambos found that JCE I and II are the only forms that can be said to have a basis in customary international law while JCE III cannot: thus, only the first two forms can be applied in cases before the ECCC.[260]

Moreover, Ambos took issue with the Co-Prosecutor's argument that JCE must be applied in order to meet the goal of prosecuting the senior leaders of the Democratic Kampuchea, since such application does not differentiate between the categories of JCE, and furthermore there are alternative modes of liability which could be applied, which have better grounding in international law.[261]

What is most interesting is the authority given to these *amicus* briefs and the individuals who authored them. It should be noted that the defence had sought a motion to disqualify Cassese from being able to provide an *amicus* brief, due to his previous role as judge at the ICTY where he was responsible for the inception of JCE, on the basis that he could never provide an impartial opinion.[262] The Pre-Trial Chamber, however, maintained that they were asking his opinion in his capacity as editor of the Journal of International Criminal Law, and not in his capacity as a judge. However it is impossible to separate these roles and pretend one individual is more than one participant in the process of the law.[263] A great deal of the academic commentary on JCE that has received attention in case law and in further academic debate is authored by Cassese. He refers to the judgments of the ICTY which confirm the stance of the judgments which he himself co-authored,[264] or refers to his further published work. This process continues even with respect to the debate as to whether JCE is the same as conspiracy, since Professor Cassese was co-author of the *amicus* brief before the US Supreme Court in *Hamdan v Rumsfeld*, where it was reiterated that conspiracy has no place in contemporary ICL, but JCE does.[265] In this self-referencing, the individual has taken an unusually influential role in the development of a normative notion in ICL.

---

257 Cassese referred to the Martens Clause and argued that it allows for social and moral needs as a basis for observance of the expression of legal views by a number of states or international entities. He argued that such expressions are sufficient for the establishment of a customary rule and its binding value, even if there is no widespread or consistent State practice: Cassese 2008a, para 35.

258 Ambos 2008a, para I.1.

259 Ambos 2008a, para II.2.

260 Ambos 2008a, para II.3.3.

261 Ambos 2008a, para II.2.

262 Iang Sary, Motion to Disqualify Cassese 2007.

263 Iang Sary, Motion to Disqualify Cassese 2007.

264 For example Cassese 2008b, 2004, p. 592.

265 Hamdan v Rumsfeld Amicus Brief 2006-01-06, para 19.

Based on these briefs, the ECCC has accepted JCE I and II as applicable international law (though not a part of Cambodian law), but held that JCE III was not customary law at the time of the crimes committed in the 1970s.[266] It has since been applied in other cases.[267] In all of these decisions, extensive and exclusive reference is made to ICTY case law, despite the fact that the crimes in Cambodia took place twenty years earlier than the first decisions of the ICTY. The reasoning was similar to that of the BiH State court mentioned above, that everything decided by the ICTY is automatically binding law due to the fact that new statutes are modelled on that of the ICTY.[268]

Under the statute of the ECCC, Cambodian criminal law must be applied in the first place, and international law in the second place.[269] Yet when the ECCC Trial Chamber compared JCE with the notion of co-perpetration under the Cambodian Criminal Code, as well as that found in the French *Code Penal* that shaped the Cambodian law, it opted for because it allows liability even when the accused is quite remote from the actual crimes and physical perpetrators.[270] A deliberate policy choice was made to allow the application of a mode of liability that was not part of Cambodian law, in order to secure a conviction.

The question of the applicability of JCE has also been raised at the Special Tribunal for Lebanon (STL). Although the criminal code of Lebanon is the primary law before the STL, still the article on modes of individual liability in the Statute of the STL follows the wording of the ad-hoc tribunals.[271] When the Appeals Chamber handed down its decision on the relevant interlocutory appeal, it made a similar, highly questionable argument to that given by the State Court of the BiH: because the wording of the Statute is the same as the ad-hoc tribunals, it represents customary law, and as such JCE, which has been embraced by these other tribunals as customary law, is also applicable.[272]

The Appeals Chamber did go to efforts to limit the application of JCE III, stating that even though Lebanese domestic law recognises a similar principle, it would not be appropriate to apply this to the crime of terrorism, since terrorism is a specific intent crime whereas JCE III allows a low subjective *mens rea* standard of *dolus eventualis*.[273] The judges recognised they were departing from the standard of the ad-hoc tribunals, which has allowed JCE III to be applied to the specific intent crime of genocide,[274] however in a clear policy move on the part of the judges, it was

[266] Duch, Trial Judgment 2010, para 511; Case 002, Pre-Trial Decision on JCE 2010, paras 69, 77, 88; Case 002, Trail Decision on JCE 2011, para 22.

[267] For example Case 002/01, Judgment 2014, at the time of writing under appeal, based largely on objections to the application of JCE.

[268] Duch, Trial Judgment 2010, para 511.

[269] Statute of the ECCC 2004, Article 1.

[270] Duch, Trial Judgment 2010, para 510.

[271] Statute of the STL 2007, Article 3(1)(b).

[272] Interlocutory Decision on Applicable Law 2011, para 206.

[273] Interlocutory Decision on Applicable Law 2011, paras 248–249.

[274] Interlocutory Decision on Applicable Law 2011, para 249.

determined that the better approach is to treat the secondary offender connected to terrorism as a 'mere' aider and abetter rather than 'pin on him the stigma of full perpetratorship.'[275] It should be pointed out that JCE's grandfather and strongest proponent, Professor Cassese, was also presiding judge at the STL when this decision was handed down.

8.6.4 Rejection of JCE at the ICC

Although Article 25(3)(d) of the Rome Statute refers to the notion of common purpose, the wording differs significantly from the various formulations of JCE.[276] Indeed, although the choice lay open to judges to introduce the doctrine of JCE into the ICC jurisprudence via this article, it would appear there has been a very distinct choice not to do so. For instance in the report made to the UN Security Council by the International Commission of Inquiry regarding the referral of the situation in Dafur to the ICC, many of the findings were termed in the language of JCE with reference to the case law of the ICTY.[277] But the preference thus far has been to favour the notions of co-perpetration and perpetrations by means, as discussed further below.

Article 25(3)(d) on common purpose has only been utilised in one judgment, against Germain Katanga, and it has not been interpreted along the same lines of JCE.[278] Originally Katanga had been charged under Article 25(3)(a) as a (co)-perpetrator by means, together with Mathieu Ngudjolo for several counts of murder, direct attacks on civilians, conscription of child soldiers and sexual slavery, based on their respective roles as leaders of two different armed groups in the Democratic Republic of Congo (DRC).[279] When the two cases were severed in November 2012, in order to continue the prosecution against Katanga in more detail, the Pre-Trial Chamber notified Katanga that the mode of liability under which he initially stood charged as a principal would be re-characterised in accordance with regulation 55 of the Regulations of the Court, so as to consider his responsibility as an 'accessory'—meaning secondary participant—under Article 25(3)(d), referring to the furtherance of a common purpose.[280]

In the end Katanga was sentenced to 12 years according to his conviction as a secondary participant under Article 25(3)(d),[281] which indicates the preference for a normative differentiated system of liability, and one that delineates between modes of liability in a more consistent manner than the application of JCE would allow.

---

275 Interlocutory Decision on Applicable Law 2011, para 249.

276 For the full text of Article 25 of the Rome Statute, see the appendix A.6.

277 International Commission of Inquiry on Dafur 2005.

278 Katanga Trial Judgment 2014.

279 Katanga Confirmation of Charges 2008, paras 573–582.

280 Katanga Severance of Charges Decision 2012, para 7.

281 Katanga Sentencing Judgment 2014.

Since the ICC has made an explicit move away from the case law on JCE, the ICTY appears also to have begun to reject, or at least limit, its own invention. In *Brđanin* the Appeals Chamber stated that JCE should not be used as 'an open-ended concept that permits convictions based on guilt by association.'[282] The acquittal of Šešelj in March 2016 is also case in point. Judge Antonetti explained the majority judgment further in his lengthy concurring opinion, in he questioned lamented the over-extensive development of JCE, criticised the fact that it has created a presumption of guilt to hang over leaders, and even questioned the very roots of this mode of liability.[283] The majority of the judges rejected the notion that the intention to build a 'new Serb-dominated State' was enough to prove a shared criminal purpose, and referred to the similar problematic arguments made in the case against Milosević—indeed it seems the judges were suggesting that had the case against Milosević come to completion, the accusations based on JCE would have failed.[284]

Whether or not there is agreement on the customary status of JCE during or prior to the Yugoslav conflict, in answer to which it was formulated in the ICTY, it seems to have gained a place in ICL through a process of repetition and reification. The question remains open whether this entails the formation of new customary law, however the policy process itself can be readily identified. Given that the ICC thus far has favoured other modes of liability, such as co-perpetration, it's possible that JCE is a solution that was formed in response to the problem of a lack of specificity in the ad-hoc tribunal statutes, and which is no longer needed in the international adjudication of mass atrocity. This would further evidence the fact that the process of developing modes of liability is always a response to historical context and policy needs.

8.7 Co-perpetration: Towards an Objective Approach

Although the debate surrounding whether co-perpetration differs from the common purpose doctrine and JCE has not been resolved among the participants in the law-making and law-applying processes, it will be argued here that in fact the language of these two doctrines can be distinguished. This has to do once again with the roots of the two doctrines in domestic modes of liability, and the deciding factor is therefore whether the approach is more subjective or objective.

The essence of co-perpetration is that the objective elements of crimes are carried out by a plurality of persons in co-performance and in furtherance of a shared criminal plan, based on a division of tasks.[285] It differs somewhat from the definition of JCE, which does not require a *conditio sine qua non* contribution of each individual involved, and which is based mainly upon the agreement to jointly carry out

[282] Brđanin Appeal Judgment 2007, para 428.

[283] Šešelj Judgment, Concurring Opinion of Judge Antonetti 2016, pp. 155, 164.

[284] Šešelj Judgment 2016, paras 227, 280.

[285] Olásolo 2009, p. 265.

a common criminal plan that *results* in the fulfilment of the objective elements of the crime, regardless of the actual contribution of each individual to that result.[286] In this respect, co-perpetration mirrors the more objective 'manifest criminality' model described by Fletcher,[287] whereby the act is determinative, and intent is subsidiary in analysing liability.[288] JCE, on the other hand, reflects the subjective model of criminality, whereby intent is the primary test of liability, and a manifest act is merely proof of the intent.[289] JCE therefore includes many more participants in a collective crime in the sense of equal liability, than does co-perpetration, which seeks to distinguish between parties to a crime based on the degree of their contribution.

In *Furundžija*, a case concerning collective acts of torture which pre-dated *Tadić*, the ICTY Trial Chamber applied the notion of co-perpetration, identifying this as a'group of people pursuing a common design to commit crimes', where 'the accused must participate in an integral part of the torture and partake of the purpose behind the torture'.[290] This reading fits the definition given here of co-perpetration.

As with the other modes of liability in this chapter, the focus here is on a chronological investigation of the policy reasons behind the application and development of co-perpetration on the international plane, and the role of individual participants in this process. For this reason the discussion begins with the ad-hoc tribunals, even though co-perpetration has generally not been favoured by the prosecutors and judges at these tribunals.

8.7.1 Co-perpetration at the Ad-Hoc Tribunals

The debates as to whether co-perpetration differs from JCE have been fuelled mainly by the differing legal traditions. It has been argued that JCE is exactly the same as co-perpetration, which hails from the civil law tradition civil law; that it can be combined with co-perpetration; and that they are alternative modes of liability. From the first camp, an oft-quoted critique of JCE has come from Judge Per Lindholm, a civil law trained judge, that JCE 'is nothing more than a new label affixed to [...] co-perpetration.'[291]

A second approach has been an attempt to combine both modes of liability, without insisting they are the same. For example in the indictment against Milosević this combined reading was evident, where alongside all the other possible modes of liability under Article 7(1), the Prosecution specified that by 'committing', it meant 'participation in a joint criminal enterprise as co-perpetrator.'[292] This could be read

286 Fletcher 2000, p. 659; van Sliedregt 2012a, p. 96.

287 See the full discussion in Sect. 5.3.6.

288 Fletcher 2000, p. 117.

289 Fletcher 2000, p. 119.

290 Furundžija 1998, para 210.

291 Separate opinion of Judge Lindholm, Simić Trial Judgment 2002, §2.

292 Milosević Indictment 2004, para 5.

as a definition of JCE as a species of co-perpetration, and some authors make a distinction between co-perpetration based on JCE and co-perpetration based on control over the crime, as two different theories underpinning a single overarching form of liability.[293] The Appeals Chamber in *Milutinović* also took this position, affirming that JCE was a form of co-perpetration.[294]

A third approach has been to insist on the difference between JCE and co-perpetration, which has been coupled with the assertion that JCE is grounded in customary law, whereas co-perpetration apparently is not. The turn-around between the Trial and Appeals Chambers at the ICTY in *Stakić* is illustrative. Trial Chamber II had asserted that JCE was inappropriate for the facts of the case because Stakić was a senior political leader who co-ordinated many crimes at a high level against civilians and also in detention camps. The Trial Chamber decided that JCE should be replaced with co-perpetration, since 'a more direct reference to "commission" in its traditional sense should be given priority before considering responsibility under the judicial term "joint criminal enterprise".'[295] The Trial Chamber cited Roxin directly,[296] and it is noteworthy that the chamber consisted exclusively of judges from civil law backgrounds.[297] The Appeals Chamber dismissed this viewpoint, arguing that co-perpetration has no support in customary international law, whereas JCE does[298] and and stating that '[t]he introduction of new modes of liability into the jurisprudence of the Tribunal may generate uncertainty, if not confusion, in the determination of the law by parties to cases before the Tribunal as well as in the application of the law by Trial Chambers'.[299] The fact that JCE could be said to be doing just that was ignored. It is important to note is that the Appeals Chamber bench consisted of judges from both civil law and common law backgrounds.[300]

Co-perpetration has been asserted in other cases before the ICTY, but has often been rejected in favour of JCE.[301] On the other hand it has received more positive discussion at the ICTR. In *Gacumbitsi* the Appeals Chamber found that, at least with respect to genocide, 'committing' is defined as 'direct and physical perpetration', but need not consist only of physical killings. Many other acts can constitute direct participation in genocide, including the kind of direction given by the accused, and

---

293 See generally Olásolo 2009; van Sliedregt 2012a, pp. 97, 101; Ambos 2005, p. 176; Cupido 2015, p. 22.

294 Ojdanić Decision on JCE 2003, paras 18–20.

295 Stakić Trial Judgment 2003, para 438.

296 Stakić Trial Judgment 2003, paras 436–438.

297 Judge Schomburg, presiding, from Germany, Judge Vassylenko from Ukraine and Judge Argibay from Argentina.

298 Stakić Appeals Judgment 2006, para 62.

299 Stakić Appeals Judgment 2006, para 59.

300 Judges Mohamed Shahabuddeen (Guyana) and Theodor Meron (USA) are from common law backgrounds; Judges Fausto Pocar (presiding) (Italy), Andrésia Vaz (Senegal) and Mehmet Güney (Turkey) are from civil law backgrounds.

301 van Sliedregt 2012a, p. 98.

his leading role in conducting and supervising the massacre at Nyarubuye Parish.[302] The intention that there should be a normative distinction between co-perpetration and other modes of liability was made clear in the Appeals Chamber's finding that:

> The Appellant did not simply 'order' or 'plan' genocide from a distance and leave it to others to ensure that his orders and plans were carried out; nor did he merely 'instigate' the killings. Rather, he was present at the crime scene to supervise and direct the massacre, and participated in it actively by separating the Tutsi refugees so that they could be killed.[303]

Thus, the policy choice towards a mode of liability that is normatively differentiated in order to reflect the role and level of culpability of the perpetrator was made explicit. Similarly in *Seromba* the ICTR Appeals Chamber, citing *Gacumbitsi*, confirmed that commission of genocide as a co-perpetrator included various acts which are 'as much an integral part of the genocide as were the killings which they enabled',[304] including the authority given by Seromba to a bulldozer driver to destroy the Nyange Parish church while approximately 1,500 Tustsis were seeking refuge inside.

With respect to the influence of individual participants in the process of the law, it is noteworthy that Judge Schomburg, a civil law trained judge, was an appellate judge in both the *Gacumbitsi* and *Seromba* cases at the ICTR, and was the presiding trial judge in *Stakić* at the ICTY where co-perpetration had been asserted in place of JCE. He has also written a plea for the ICTY to drop the notion of JCE III altogether and make better use of co-perpetration in its case law,[305] and has made clear he is an advocate of this mode of liability.[306] Aside from Judge Schomburg, the *Gacumbitsi* bench was made up of a majority of judges from civil law backgrounds, though interestingly Judge Günay from Turkey, a civil law jurisdiction, dissented with respect to the interpretation of co-perpetration.[307] The *Seromba* bench was further made up of judges from both civil law and common law jurisdictions, and Judge Liu from China was the only dissenting voice on the application and interpretation of co-perpetration.[308] While no hard conclusions can be drawn, the majority civil law make-up of these benches and the preference for co-perpetration over JCE may once again give insight into the origins of the debates on the applicability of modes of liability.

---

[302] Gacumbitsi Appeal 2006, para 60. It should be noted that co-perpetration was not applied as a replacement for JCE, since JCE was asserted in the indictment for other crimes.

[303] Gacumbitsi Appeal 2006, para 61.

[304] Seromba Appeal 2008, para 161; Gacumbitsi Appeal 2006, citing para 60.

[305] Schomburg 2010.

[306] van Sliedregt 2012a, p. 98.

[307] Partially Dissenting Judgment of Judge Günay, Gacumbitsi Appeal 2006, p. 116.

[308] Seromba Appeal 2008, para 161.

8.7.2 Co-perpetration at the ICC: Control over the Crime

The first decisions to come out of the ICC showed a clear preference for the application of co-perpetration rather than a theory of JCE, and the judges have gone to great lengths to define co-perpetration as an autonomous mode of principal liability, distinct from the other modes of liability laid out in the other subsections of Article 25.[309]

In the *Lubanga* Confirmation of Charges decision, the Pre-Trial Chamber explicitly rejected the prosecutor's invitation to consider JCE as applicable under the Rome Statute, and stated that the detail of Article 25(3) differs greatly from the non-differentiated model in the ad-hoc statutes.[310] The judgment first considered and rejected the formal objective approach, whereby only those who physically carry out an objective element of the crime could be considered to be co-perpetrators.[311] It then considered and rejected the subjective approach, which it stated had been favoured by the ICTY in its case law on JCE, whereby those who share the intent to commit the offence, regardless of the level of their actual contribution, can be considered principals to the crime.[312] Both approaches were rejected because they could not be reconciled with the wording of Article 25(3)(a) of the Rome Statute, which led to the conclusion that JCE, because of its subjective approach, would also be excluded before the ICC.[313] Instead the Chamber took the approach of control over the crime, and cited Judge Schomburg's separate opinion in *Gacumbitsi*, as well as the authors Gerhard Werle and George Fletcher, for authority on the fact this approach exists in numerous domestic jurisdictions.[314] In doing so, the Trial Chamber also justified its finding that there is a hierarchical normative distinction between the modes of liability enumerated in the Article 25 of the Rome Statute, taking a clear stance on this question for the first time.

The authority given to scholars in this respect is illustrative of the role these participants play in the civil law tradition. Of course, these scholars were all selected because of the preferences shown by the German-trained jurists Schomburg and Werle for the control over the crime theory based on German doctrine, and the sympathy Fletcher has shown in much of his comparative writings for the same.[315] This interpretation has since been confirmed since in the *Lubanga* judgment,[316] and the *Katanga* Confirmation of Charges,[317] both of which directly cite Welzel, Roxin, Stratenwerth, Jescheck and other German scholars, as well as Argentine

---

309 Ambos 2008b, p. 479; Schabas 2010, p. 428.

310 Lubanga Confirmation of Charges 2007, para 325.

311 Lubanga Confirmation of Charges 2007, para 328.

312 Lubanga Confirmation of Charges 2007, para 329.

313 Lubanga Confirmation of Charges 2007, para 335.

314 Lubanga Confirmation of Charges 2007, para 330.

315 See e.g. Fletcher 2002, 2000, p. 553, 2011; Schomburg 2010; Werle and Jessberger 2014, p. 176.

316 Lubanga Trial Judgment 2012, paras 997–999.

317 Katanga Confirmation of Charges 2008, para 484.

scholars Sancinetti and Bacigalupo, among others. It is also possible to recognise language very similar to that used by the ICTY Trial Chamber in the overturned *Stakić* judgment.[318] The patchworking process is evident in this selective borrowing from domestic doctrine in order to support interpretation of the Rome Statute. It is therefore important to note that the majority of judges on the bench come from civil-law jurisdictions.[319]

One statement made by the Pre-Trial Chamber appears to be slightly out of line with the control over the crime approach, in that it would suggest a subjective approach to attribution. In its decision confirming the arrest warrant against Sudanese president Omar Al Bashir, Pre-Trial Chamber I stated that:

> the concept of co-perpetration is originally rooted in the idea that when the sum of the coordinated individual contributions of a plurality of persons results in the realisation of all the objective elements of a crime, any person making a contribution can be held *vicariously responsible for the contributions of all the others* and, as a result, can be considered as a principal to the whole crime.[320]

As has been noted in the chapters on domestic modes of liability, a vicarious theory of liability is usually associated with a subjective approach, whereby the shared intention is the basis of culpability leading to liability, crimes committed collectively are attributed equally to all those who are party to the shared intention. However an objective approach, which emphasises the extent to which each party has contributed in a material way to the crime, rather attributes *responsibility* according to the level of contribution. This is not based on vicarious liability, but rather on derivative liability. Those who have not themselves fulfilled the objective elements of the crime are liable based on a distribution of responsibility, but not an attribution of the acts.

The theory of control over the crime applied by the ICC is a material-objective theory,[321] according to which the effective control that a person exercises over the commission of the crime determines the level of responsibility.[322] It would be incorrect to say that any person making a contribution is vicariously responsible for the contribution of others. Such an approach would suggest that any further crimes committed by other parties would also become the responsibility of an accused who exercised no control over these crimes, which goes beyond the limits of liability under this approach. This is exactly one of the reasons that Ngudjolo was acquitted by the Trial Chamber, and that Katanga was found to be liable as a secondary participant under Article 25(3)(d), and not as a principle under co-perpetration: neither was

[318] Weigend 2008, p. 479.

[319] Judge Jorda, presiding, from France, and Judge Steiner from Brazil, both of which are civil law jurisdictions. It should be noted that in France while there is a notion of co-perpetration similar to the one applied by the Pre-trial Chamber, there is no such clear distinction between principals and secondary participants; see van Sliedregt 2012a, p. 96. The third judge was Judge Kuanyehia from Ghana, a common law jurisdiction.

[320] Al Bashir Decision on Arrest Warrant 2009, para 326, emphasis added.. This was cited with approval in Katanga Confirmation of Charges 2008, para 520.

[321] For a full explanation of this theory as developed by Welzel and Roxin, see Sect. 7.3.4.

[322] Katanga Trial Judgment 2014, para 1412.

found to exercise sufficient control over the physical perpetrators who committed the crimes.[323] If a theory of vicarious liability had been applied, based on a subjective approach, this may well have led to a conviction on grounds which the Pre Trial Chamber has found objectionable.[324]

8.7.3 Balancing the Subjective and Objective Elements of 'Control over the Crime'

In interpreting the theory of control over the crime, the Pre-trial Chamber determined that there are two objective and three subjective elements required. The first objective element is that there is an agreement or common plan between two or more persons, requiring coordinated action.[325] The fact that an agreement is considered an objective rather than a subjective element could be questioned,[326] however this is in line with the way in which conspiracy is often constructed: the agreement denotes the subjective element *and* the objective element, since the action of agreeing is considered an *actus reus*. In any case there is a mix of objective and subjective components to this element.[327] The second objective element is that a person must have joint control over the crime, defined as an 'essential' contribution to the crime, in the sense that the person has the power to frustrate the execution of the crime.[328]

The subjective elements are (i) that the individual must have fulfilled the requisite mental element for the crime charged;[329] (ii) that the suspect and all other co-perpetrators must have (a) all been mutually aware of the risk that their actions would result in the fulfilment of the objective elements of the crime and (b) all mutually accepted such a result;[330] and (iii) that the suspect was aware that his or her role was essential in the commission of the crime and that he or she could frustrate the implementation of the common plan and thus of the commission of the crime.[331] This third subjective element has since been excluded by the *Katanga* Pre-Trial Chamber, when it decided that such awareness of one's own role is only necessary to prove

[323] Ngudjolo Trial Judgment 2012, para 516; Katanga Trial Judgment 2014, para 1412: *'seules les personnes qui contrôlent, effectivement et sans interférence possible, une partie au moins d'un appareil de pouvoir peuvent présider à l'exécution d'une activité criminelle'.*

[324] The Pre-Trial bench that issued the arrest warrant for Al Bashir was made up of Judge Akua Kuenyehia, presiding, from Ghana, which is a common law system, and two judges from civil law systems, Judges Anita Usacka (Latvia) and Syliva Steiner (Brazil). It is unclear whether the characterisation of co-perpetration as vicarious liability stemmed from any of these domestic systems.

[325] Lubanga Confirmation of Charges 2007, para 343.

[326] van Sliedregt 2012a, p. 99.

[327] Ambos 2009, p. 722.

[328] Lubanga Confirmation of Charges 2007, paras 342, 347.

[329] Lubanga Confirmation of Charges 2007, para 349.

[330] Lubanga Confirmation of Charges 2007, para 361.

[331] Lubanga Confirmation of Charges 2007, paras 366–377.

with respect to indirect co-perpetration, to be discussed in the next section, and not with respect to co-perpetration simpliciter.[332]

In the confirmation of charges against Jean-Pierre Bemba Gombo, the former president of the rebel *Mouvement de Libération du Congo*, and commander in chief of its military wing, the Pre-Trial Chamber replaced the prosecution's charge of co-perpetration with that of command responsibility, based on the fact that the first subjective element had not been fulfilled: it found that there was insufficient evidence that Bemba had sufficient awareness that the crimes in question would be committed.[333] There seems to be a contradiction in the finding that there was insufficient evidence of awareness, and yet sufficient evidence that Bemba 'knew' of the crimes, as required by command responsibility.[334] However this question was not further debated, and Bemba was convicted under command responsibility in June 2016.[335]

The 'control over the crime' theory was applied in the first judgment to come from the ICC, convicting Thomas Lubanga as a co-perpetrator of conscripting child soldiers. The Trial Chamber emphasised that a person can be a co-perpetrator even when she does not physically perpetrate any of the elements of the crime in question, and even if she is not present at the scene of the crime.[336] Once again the Trial Chamber cited Welzel and Roxin among other German scholars, and some Argentine and Spanish scholars.

The key test is that there must be an 'essential' contribution to the crime, and this can consist of control over the crime, though it is not necessary that any of the co-perpetrators have complete control over the crime as a whole.[337] The Trial Chamber took a very broad approach with respect to the necessary agreement between parties; it need not be specifically aimed at the crime committed, nor need the prosecution prove that it was an intrinsically criminal plan. Instead, the minimum requirement was a 'critical element of criminality': namely that implementation of the plan embodied a sufficient risk that a crime 'will be committed in the ordinary course of events'.[338] Furthermore, the agreement between co-perpetrators need not be explicit, and could be inferred from the circumstances.[339] While these latter aspects may remind one of the extended subjective approach seen in the interpretation of JCE, the emphasis remained on the 'control over the crime' theory as it had been applied by the Pre-Trial Chamber. In the end, the language used in the conviction of Lubanga is telling: those who control or 'mastermind' the crime due to the fact that they are able to decide

[332] Katanga Confirmation of Charges 2008, para 535. See also Ambos 2009, p. 21; Weigend 2008, p. 485.

[333] Bemba Decision on Charges 2009, paras 400–401.

[334] Bemba Decision on Charges 2009, para 478.

[335] Bemba Judgment 2016.

[336] Lubanga Trial Judgment 2012, para 1004.

[337] Lubanga Trial Judgment 2012, para 994.

[338] Lubanga Trial Judgment 2012, para 984.

[339] Lubanga Trial Judgment 2012, para 988.

whether and how the crime is committed, are also considered co-perpetrators and therefore principals.[340]

8.8 Indirect Perpetration/Perpetration by Means: A Normative Differentiation

The direct importation of domestic doctrine with respect to 'control over the crime' continued as ICC judges further interpreted Article 25(3)(a). From its early case law it was determined that there are three categories of principal offenders under this article: those who physically commit the crime (direct perpetration); those who control the offence because of essential tasks assigned to them (co-perpetration); and those who control the will of the physical perpetrators (commission through another person or indirect perpetration).[341] This latter form was introduced as a mode of liability new to the international plane, drawn once again from domestic doctrines.[342]

8.8.1 A New Mode of Liability on the International Plane

A mode of liability similar to indirect perpetration can be recognised in many domestic jurisdictions which follow the civil law tradition, and it also has a sibling in the common law notion of innocent agency.[343] The rationale behind the interpretation of the Rome Statute goes much further than the common law domestic variation, according to which if the physical perpetrator is fully responsible for her acts, then the actor 'behind' this physical perpetrator is considered to be a 'mere' instigator, or any such equivalent.[344] Instead, the interpretation applied by the ICC follows the notion developed by Roxin that it is possible to hold the 'perpetrator behind the perpetrator' responsible for exercising a certain control over the crime, regardless of whether the physical perpetrator is herself fully criminally responsible or not. The decision to take this broader stance in the Rome Statute reflects the specific nature of systemic, collective crimes in international law,[345] and its interpretation by the judges at the ICC has rested directly upon the domestic law developments of this notion in response to domestic atrocities.

---

[340] Lubanga Trial Judgment 2012, para 1003.

[341] Schabas 2010, p. 428; Katanga Confirmation of Charges 2008, para 488; Al Bashir Decision on Arrest Warrant 2009, para 210.

[342] For the full text of Article 25 of the Rome Statute, see the appendix A.6.

[343] See e.g. MPC 1962, §2.06, where vicarious liability results from the conduct of another person 'for whom the accused is legally accountable', such as causing an innocent person to commit a crime; Fletcher 2000, p. 639.

[344] Ambos 2008b, pp. 479–480; Werle and Burghardt 2011, p. 86; Steer 2013b, p. 177.

[345] Ambos 2008b, p. 479.

Indirect perpetration had been asserted before the ICTY, but although it was recognised to exist in many domestic criminal law systems, it was dismissed because there was insufficient evidence of its customary status.[346] Notably the ICTY bench rejecting this mode of liability was made up of a majority of judges hailing from the common law tradition, where indirect perpetration tends not to have a place.[347] Once again it would appear that it is rather a matter of which modes of liability are more familiar to the participants in the deciding processes, than which could be said to be grounded in 'customary law'.

Similarly, civil-law trained Judge Schomburg's analysis of co-perpetration *Gacumbitsi* also discussed the notion of perpetration by means of an organisation. He cited Roxin as a 'world wide accepted legal scholar' and he gave the examples of domestic courts that have applied this doctrine in cases of organised crime, terrorism, white collar crime, or state induced criminality, citing among others the *Junta* trials in Argentina and the *Maurschützen* case in Germany.[348] He stated that this doctrine is particularly applicable to ICL, since '[i]t is a means to bridge any potential physical distance from the crime scene of persons who must be regarded as main perpetrators because of their overall involvement and control over the crimes committed.'[349]

Despite the fact that Schomburg's position was very much in the minority at the ad-hoc tribunals, he cited the wording of the Rome Statute, and this is exactly the interpretation given to Article 25(3)(a) of the Rome Statute by the Pre-Trial and Trial Chambers of the ICC to date.[350] In fact the Pre-Trial Chamber added this mode of liability to the charges against Lubanga and against Katanga and Chui, since it had not been included in the indictments by the Office of the Prosecutor (OTP).[351]

The Pre-Trial and Trial chambers have followed the doctrine as applied by Pre-Trial Chamber I in *Lubanga*, and have gone on to reiterate that indirect perpetration includes perpetration by means of a state apparatus or organisation, and not just of another individual. This development is discussed next.

8.8.2 Perpetration by Means of an Organisation

In the *Katanga* Confirmation of Charges, the Pre-Trial Chamber took the control over the crime test that had been laid out in the *Lubanga* Confirmation of Charges, and applied this to the situation where a person has 'control over the will of those

---

[346] See e.g. Ojdanić Decision on Indirect Co-perpetration 2006, para 39.

[347] Judge Robinson, presiding, from Jamaica, and Judge Bonomy from the UK, both common law systems. The third judge was Judge Kwon from Korea, a system that resembles the civil law tradition strongly, but has imported many aspects of the US legal system as well.

[348] Gacumbitsi Appeal, Schomburg Separate Opinion 2006, paras 18–21. The cases cited are discussed in detail in Chap. 7.

[349] Gacumbitsi Appeal, Schomburg Separate Opinion 2006, para 21.

[350] Lubanga Confirmation of Charges 2007, para 339; Werle and Burghardt 2011, p. 86.

[351] Lubanga Decision Regarding Witness Proofing 2007, para 342; Ngudjolo Warrant of Arrest Decision 2007, para 55; Katanga Warrant of Arrest Decision 2007, para 54.

who carry out the objective elements of the offence',[352] including where that control is exerted over the physical perpetrators by means of an organisation. The PTC emphasises that these are 'the cases most relevant to international criminal law'.[353] Although Katanga was eventually convicted under the common purpose mode of liability, and not under Article 25(3)(a), the extensive and didactic reasoning in both the Confirmation of Charges and the Trial judgment is instructive in terms of the approach of the ICC judges and the patchworking process that was evident.

The fact that the wording of Article 25(3)(a) already incorporates liability for acts of another person even where that physical perpetrator is also liable, is already reminiscent of the doctrine that emerged from Roxin's theory. Thus to take the interpretation a step further to include his theory of organisations is not unthinkable.[354] The Trial Chamber recognised in the judgment against Katanga that the modalities of control exercised by leaders in these situations are more and more diverse and sophisticated, and that a sophisticated interpretation of Article 25(3) is therefore needed.[355] In the Confirmation of Charges Decision, the PTC had reasoned that since the crimes that will come before the ICC 'will almost inevitably concern collective or mass criminality', the Statute therefore specifically targets the kinds of cases which involve an indirect perpetrator's control over an organisation, citing Roxin directly and using his German term, *Organisationsherrschaft*, to describe this phenomenon.[356] It then went on to enumerate the same criteria as Roxin had developed in his theory of *Organisationsherrschaft*, namely that the organisation must be composed of sufficient subordinates that the orders of the superior will be 'automatically' carried out, and if one member should refuse to obey, then another member will take over.[357] This is Roxin's criterion of fungibility of the subordinates.

This automatic compliance with the intentions of the leader is at the heart of the leader's liability as a principal, and can be secured, according to the Pre-Trial Chamber, by way of intensive and violent training of the subordinates.[358] It is not necessary that the leader controls the will of each individual in the collective, only that she has control over the collective organisation and that there is interchangeability of those lower down the hierarchy.[359] This matches the arguments made by several authors cited in Chap. 2, that the blameworthiness of the physical perpetrators may be less in situations where their second-order ability to make moral choices has been

[352] Katanga Confirmation of Charges 2008, para 488.

[353] Katanga Confirmation of Charges 2008, para 498.

[354] See this same justification in Katanga Trial Judgment 2014, paras 1403–1404.

[355] Katanga Trial Judgment 2014, para 1410.

[356] Katanga Confirmation of Charges 2008, paras 498, 501; reiterated in Katanga Trial Judgment 2014, para 1403.

[357] Katanga Confirmation of Charges 2008, paras 515–517; confirmed in Katanga Trial Judgment 2014, para 1403.

[358] Katanga Confirmation of Charges 2008, para 518.

[359] Katanga Trial Judgment 2014, para 1408.

limited, and where their own safety or survival may rely on compliance with the societal norm of violence.[360]

The judges cited Roxin directly, as well as the *Junta* trials in Argentina, which in turn had applied Roxin's theory directly. For this reason, it has been pointed out by some authors that the doctrine may be limited when there is not an orderly, bureaucratic dictatorship or organisation, but a more diffusely organised militia or rebel group, where the authority of a leader may be contingent and temporary, and the extent of control may be ever-changing.[361] Some commentators have interpreted the statement by the PTC in the *Katanga* Confirmation of Charges to mean that attribution of responsibility through indirect perpetration can *only* be determined where there is a system of coercion of violence towards the subordinates.[362] However as has been discussed at length in the previous chapters on domestic criminal law, the opposite is also true: perpetration by means focuses on the extent to which the accused individual exerts control over the crime or crimes being committed, and there is nothing to say that this control could not also be exerted with positive incentive rather than coercion. There may in fact be many cases where the subordinates express a willingness to engage in the criminal acts—as was the case in the Argentinean *Junta* trials cited by the PTC. It was indeed one of Roxin's suggested, though not necessary, elements.[363]

The Pre-Trial Chamber asserted that there are numerous domestic jurisdictions that apply a similar notion of perpetration by means of an organisation, however there were only five cited, all of which are heavily influenced by the German doctrine in criminal law.[364] This selective survey of domestic jurisdictions has been criticised,[365] in much the same way as the same process undertaken by the chambers at the ICTY in asserting the customary status of JCE was criticised. However the important difference here is that the ICC Pre-Trial Chambers did not assert that perpetration by means of an organisation is found in customary law, but was slightly more transparent about the fact it was drawing on domestic law. It also asserted that this concept was already recognised in international law, referring to *Gacumbitsi* and *Stakić*, despite these being, respectively, an opinion in the minority and a judgment overturned at appeal. Once again the domestic law background of the judges on the Pre-Trial bench may be an indicative reason for this preference: there was a majority of civil law trained judges among the three women who handed down the decision; Judges Anita Usacka (Latvia) and Syliva Steiner (Brazil) both hailing from the civil law tradition, and Judge Akua Kuenyehia, presiding, from Ghana, which is a common law system. This was the same bench that delivered the Arrest Warrant for Al

---

360 Semelin 2007; van der Wilt 2014; Drumbl 2005; Tallgren 2002.

361 van der Wilt 2009, p. 311; Weigend 2011, p. 107; Osiel 2009, p. 99.

362 Eldar 2012, pp. 211, 214. Eldar also asserts that perpetration by means imputes the actual crime to the leader of the collective, rather than imputing responsibility.

363 Roxin 2005, p. 244.

364 The jurisdictions cited were Argentina, Chile, Germany, Peru and Spain; Katanga Confirmation of Charges 2008, para 502.

365 Weigend 2011, p. 105; Jain 2012, p. 186.

Bashir in which the control over the crime theory was asserted.[366] In any case the patchworking process is evident here, too, as an explicit choice was made to draw upon a particular legal tradition, due to its applicability to mass atrocity crimes.

After the *Katanga* and *Chui* cases were severed in 2012, Mathieu Ngudjolo Chui was acquitted of this charge at trial, because it was found on the facts that it could not be proven beyond reasonable doubt that he exercised sufficient control over the crimes committed.[367] In Judge Van den Wyngaert's separate opinion she questioned whether the wording of Article 25 (3)(a), which refers quite clearly to perpetration 'by means of another person', could in fact be interpreted to include 'by means of an organisation'.[368] She conceded that the concept of *Organisationsherrschaft* may be an important evidentiary factor in determining that an accused did dominate the will of certain individuals who were part of the organisation, and who were the physical perpetrators of the crimes in question, however she correctly pointed out that what is at issue is the 'level of control or influence of the indirect perpetrator over the physical perpetrator'.[369]

In a move towards protecting the defendant against the creation of new modes of liability, Judge van den Wyngaert criticised the fact that by loosening the link between the accused and the physical perpetrators, this in fact dehumanises the relationship between them, and dilutes the level of personal influence that the indirect perpetrator must exercise.[370] While such criticism should be kept in mind when applying this mode of liability on a case by case basis, it could be argued that attributing responsibility to a senior leader of an organisation in this way is *less* dehumanizing than the institution of JCE, which removes all requirements of actual links between an accused and the physical perpetrators, and rests liability upon a common purpose at an abstracted level, as if there is a collective will.[371] This overlap between the individual will and the collective will requires clarification, in order to ascertain the limits of individual agency and therefore individual liability.[372]

In the *Katanga* judgment rendered early in 2014, the Trial Chamber stated that the existence of a hierarchical structure of power 'authorises the judge' to ordain that the responsibility of the leaders of these power structures is that of author, or principal, due to the 'quasi-automatic' way in which the subordinates fulfil the instructions of

---

366 Al Bashir Decision on Arrest Warrant 2009.

367 Ngudjolo Trial Judgment 2012. The Prosecutor filed an appeal against this acquittal, on the basis of procedural error, in part due to untrustworthy witness statements. In the appeal the Prosecutor attempted to include 'common purpose' under Article 25(3)(d) as an alternate mode of liability. This appeal was rejected, maintaining the original acquittal, Ngudjolo Judgment on Appeal 2015.

368 Separate Opinion of Judge Christine van den Wyngaert, Ngudjolo Trial Judgment 2012, para 52.

369 Separate Opinion of Judge Christine van den Wyngaert, Ngudjolo Trial Judgment 2012, paras 54–55.

370 Separate Opinion of Judge Christine van den Wyngaert, Ngudjolo Trial Judgment 2012, para 53.

371 Danner and Martinez 2005, p. 150; Ohlin 2007, p. 86; van der Wilt 2007, p. 100.

372 See the discussion of overlapping agency in Sect. 2.4.

the leaders and of the collective.[373] The following three elements were laid out for this form of liability, which are drawn from Roxin's theory: the accused must (i) exercise control over the crime even though the material elements of the crime are fulfilled by one or more other persons, (ii) have the requisite psychological (mental) elements referred to in Article 30 of the Statute, and thereby of the crime in question and (iii) have awareness of the factual circumstances enabling him to exercise control over the crime.[374]

8.8.3 Towards a Reification of Perpetration by Means of an Organisation

Just as JCE underwent a process of reification, whereby the question as to its true roots in customary law has become more or less redundant, and it has since found its place in ICL through judicial and scholarly repetition, the beginnings of a similar process can already be witnessed with respect to the notion of perpetration by means of an organisation. Though this mode of liability is not found in the wording of the Rome Statute, the patchworking process in action is evidenced by the assertion by (predominantly civil law trained) judges that it can be interpreted under Article 25(3)(a), and the use of domestic law and domestic scholarly authorship to support this. Just as in the domestic jurisdictions in which perpetration by means of an organisation was developed, at the ICC the underlying reasoning is that there must be a normative differentiation between parties to a crime, and the actor who manipulates an organisation and manipulates those who are subordinate within it to effect atrocity, should be punished as a principal.

The ICC Chambers and OTP have continued this line of reasoning, and have begun a process of reification by way of self-referencing similar to the process by which JCE became a fixture of the jurisprudence of the ICTY. By referencing the *Lubanga* and *Katanga* decisions, and continuing to build on the German doctrine that has developed indirect perpetration the most thoroughly, the ICC has begun to carve a definite path which in turn allows it, and scholarly authors commenting on this phenomenon, to assert that there is now support for this notion in international law.

In its first indictment against a sitting head of state, Sudanese President Omar Hassan Ahmad Al Bashir, the OTP took up the *modus operandus* of the Pre-Trial Chambers, and for the first time pleaded indirect perpetration as the—exclusively—applicable mode of liability. In its request for an arrest warrant, the OTP alleged Al

---

[373] Katanga Trial Judgment 2014, para 1410.

[374] Author's own translation. The original text reads: *'pour pouvoir être reconnue pénalement responsable en tant qu'auteur indirect, une personne doit exercer un contrôle sur le crime dont les éléments matériels ont été réalisés par une ou plusieurs autres personnes; réunir les éléments psychologiques visés à l'article 30 du Statut ainsi que les éléments psychologiques propres au crime dont il est question; et avoir connaissance des circonstances de fait lui permettant d'exercer un contrôle sur ce crime.'* Katanga Trial Judgment 2014, para 1399.

Bashir's responsibility for genocide, crimes against humanity and war crimes committed against members of the Fur, Masalit and Zaghawa groups in Darfur, arguing that he had 'committed [these] crimes through members of the state apparatus, the army and the Militia/Janjaweed.'[375]

Given his ubiquitous power, it is noteworthy that indirect perpetration was the preferred mode of liability, and not co-perpetration or ordering, nor command or superior responsibility under Article 28. While co-perpetration was also a possibility, it is highly likely that the main motivation for choosing indirect perpetration is that it had already received such positive attention by the ICC Chambers, and had been established as the favoured mode of liability.[376] Subsequent indictments in the Kenya cases have followed this same model, arguably for the same reasons.[377]

It would appear that the interpretation of Article 25 as denoting a hierarchy between modes of liability, and as including perpetration by means of an organisation, is a policy choice enacted by the majority judges. The argument made by the *Katanga* PTC that the Statute must be understood to 'target the category of cases which involves a perpetrator's control over the crime' due to the fact that the crimes coming before the ICC will 'almost inevitably concern collective or mass criminality'[378] may not necessarily lead to the interpretation given that 'perpetration by means of another' includes perpetration by means of an organisation.[379] Rather this is a policy choice, in the guise of a teleological interpretation, which may in itself be justified, but is not therefore a logical outcome of interpreting the text of the treaty.

This was the very stance taken by Judges Fulford and Van den Wyngaert in their dissenting opinions of the Trial Chamber in *Lubanga* and *Katanga*. Fulford criticised the importation of German law doctrine directly and the use of Roxin and other German scholars as authority, without further justification of how these amount to sources under international law.[380] In her separate opinion to the acquittal of Mathieu Ngudjolo Chui, Judge Van den Wyngaert agreed with this position taken by Fulford in *Lubanga*. She criticised the Majority for drawing on domestic law, which she said should only be a subsidiary source of law according to Article 21(1)(b) and (c) of the Rome Statute, and that reliance upon the German theory of control over the crime could only be possible to the extent that it represents a general principle of law.[381] However a classical doctrine of sources created for public international law appears not to be sufficient for ICL, with its need for technical criminal law detail. Nor does it fit the process as a an accurate descriptive theory of law. Domestic law has become a necessary and primary source; what is needed now is a recognition of

---

[375] Al Bashir Application for Warrant of Arrest 2008, para 39.

[376] Jessberger and Geneuss 2008, p. 864.

[377] Muthaura Confirmation of Charges 2012; Ruto Confirmation of Charges 2012.

[378] Katanga Confirmation of Charges 2008, para 501.

[379] Weigend 2011, p. 106.

[380] Judge Fulford, Separate Opinion, Lubanga Trial Judgment 2012, para 10.

[381] Separate Opinion of Judge Christine van den Wyngaert, Ngudjolo Trial Judgment 2012, para 17.

this patchworking process so that it can be made transparent and be improved upon, such that the most appropriate selections are made.

Fulford stated that he considered it a 'danger' to import a domestic interpretation where the language of the Rome Statute is similar, though what that danger may be is not clear.[382] Given that this patchworking process is at the very heart of the development of the normative content of ICL, there can only be a danger if there is not sufficient doctrinal consideration imported together with such domestic notions. In this respect, although the Pre-Trial and Trial Chambers have not justified their preference for the German doctrine (and supporting Argentine and Spanish literature), there has at least been a thorough importation of the supporting doctrine. Furthermore, one could argue that the policy considerations which led to the development of the control over the crime theory in domestic law were also related to the need to respond to mass atrocity.

Many scholarly authors assert that indirect perpetration is a better solution than JCE to the problem of how to hold those leaders criminally responsible for crimes of mass atrocity who exert influence yet never harm a victim personally, because it distinguishes these leaders as principals, and attributes degrees of responsibility that fits the intuition that the superior bears greatest responsibility.[383] Not because the leader failed to prevent or punish, nor simply because there was a joining of the minds in a common plan, such as we see with command responsibility and JCE, but rather because the leader orchestrated the actions of the physical perpetrators of the crime.[384] This results in a normative story-telling about the facts, a norm expression in the final qualification. Yet at the same time, it would not be sufficient to refer solely to the position of leadership held by the accused: evidence must be procured of the superior-subordinate relationships, and of whether the superior could indeed mastermind the crimes committed by the subordinates, or replace those who were not willing with others who were. The factual control over the crime must be established, otherwise we would revert to a strict liability standard witnessed in the earliest command responsibility decisions.

As to whether this process of reification will be a repeat process of that observable with respect to JCE, it remains to be seen whether other tribunals will follow suit. The specific wording of Article 25(3)(a) is unique to the Rome Statute, however it will be interesting to observe whether future international or domestic tribunals will consider this mode of liability to have become a part of international law through the case law of the ICC.

---

382 Judge Fulford, Separate Opinion, Lubanga Trial Judgment 2012, para 10.

383 Jessberger and Geneuss 2008, 866; van der Wilt 2009, p. 314; Werle and Burghardt 2011, p. 88; Ohlin et al. 2013, p. 8.

384 Werle and Burghardt 2011, p. 88.

8.9 Indirect Co-perpetration by Means of an Organisation: Extended Liability

Even more controversial than the importation of 'perpetration by means of an organisation' from domestic law has been the combination of this mode of liability with co-perpetration. Although it had already been suggested in the *Lubanga* Decision on Confirmation of Charges, the notion of indirect co-perpetration, or co-perpetration by means of another, was not fully explored until the later *Katanga* Confirmation of Charges. This inception combines the 'control over the crime' theory as applied to joint control, with the way in which Roxin had applied it in his theory of *Organisationsherrschaft* as applied to indirect perpetration. What the judges came up with was something beyond what Roxin had envisioned, to create a kind of cross-liability, whereby it was asserted that Katanga and Ngudjolo both allegedly had control over and manipulated separate rebel organisations, but only by combining forces could they perpetrate the atrocities in the Democratic Republic of Congo. This led to the allegation that both Katanga and Ngudjolo were responsible not only for the actions of their own troops, but also for the actions of the other's troops.[385] Although Ngudjolo was acquitted, and Katanga was convicted for contributing to a common purpose rather than for indirect co-perpetration, both of these findings were on the facts, and this mode of liability has proved pivotal in many of the recent investigations and cases before the ICC.

This combination of modes of liability leads to an extension of leadership liability beyond those subordinates over whom the accused has actual control, to include those over whom a co-accused has direct control. This has been described as a diagonal axis of liability, combining vertical and horizontal relationships.[386] The connection between those at leadership level, and the connection between one leader and his or her subordinates, is deemed sufficient for the liability of all those at the top for all crimes committed at the bottom. The question arises whether it is possible to combine modes of liability in this manner, or whether a new, fourth mode of liability has been added judicially to the text of Article 25(3)(a).

The Pre-trial Chamber asserted that there is no legal rule preventing a combination of two modes of liability,[387] and indeed it would appear that this is possible in domestic jurisdictions, if we consider the possibility of co-perpetration of inducing, or even co-perpetration of indirect perpetration *simpliciter*. Two people could jointly exercise influence over a third person, who is the physical perpetrator. However by creating a combination between co-perpetration and the already extensively interpreted notion of perpetration by means of an organisation, there are many moving parts to each mode of liability which have to be considered. Judge Van den Wyngaert has pointed this out in her separate and dissenting opinions: while it is possible to combine different forms of liability under the Statute, it would be necessary to prove

[385] Katanga Confirmation of Charges 2008, para 484.

[386] Separate Opinion of Judge Christine van den Wyngaert, Ngudjolo Trial Judgment 2012, para 52; Ohlin 2012, p. 785.

[387] Katanga Confirmation of Charges 2008, para 492.

the elements of each of them on the facts. Partial fulfilment of the requirements of each should not be a reason to attempt to combined them.[388]

In two related cases against Blé Goudé and Laurent Gbagbo, both situated in Cŏte d'Ivoir, the Prosecution constructed the combination of co-perpetration and indirect perpetration by means based based on the engagement of an 'inner circle' and members of the pro-Gbagbo forces.[389] The PTC gave a five point definition of indirect co-perpetration; (i) a common plan and shared the intent together with the physical perpetrators to commit the crimes; (ii) use by the accused of others in an organisation to carry out the material elements of the crimes, while at the same time without the accused's actions, the crimes would not have been committed or would have been committed in a significantly different way; (iii) joint control over the will of the members of the forces who were the direct perpetrators of the crimes, (by virtue of control over these forces); (iv) this joint control is possible due to the organised and hierarchical nature of the organisation, which secures an almost automatic compliance with orders given; and (v) the accused must have the requisite intent and knowledge pursuant to Article 30 of the Statute with respect to the crimes, as well as awareness of the factual circumstances that enabled them to exercise joint control over the commission of the crimes through members of the forces.[390]

There are several important things to note in this definition. The first is the emphasis on the counter-factual test as to whether the crimes would have been committed without the actions of the defendant. In the Blé Goudé Confirmation of Charges the judges found that this is the 'decisive consideration'.[391] Judge van de Wyngaert refines this further, pointing out that in *Lubanga* the Appeals Chamber had emphasised there must be control over the crime, and thereby a contribution to the crime; contribution to a plan is not enough.[392] She was unconvinced that the common plan to keep Gbagbo in power was in and of itself a criminal plan.

On the facts, Van de Wyngaert was not convinced that there was sufficient evidence that Blé Goudé had enough control over the physical perpetrators to warrant this mode of liability.[393] Her cautious application of this requirement should be heeded, in order to avoid the extension of indirect co-perpetration towards a JCE III formulation.

---

[388] Separate Opinion of Judge Christine van den Wyngaert, Ngudjolo Trial Judgment 2012, paras 62–63; Dissenting Opinion of Judge Christine van den Wyngaert, Blé Goudé, Decision on the Confirmation of Charges 2014, para 7; Dissenting Opinion of Judge Christine van den Wyngaert, Gbagbo, Decision on the Confirmation of Charges 2014, paras 3–4; Ohlin et al. 2013, p. 14.

[389] Blé Goudé, Decision on the Confirmation of Charges 2014; Gbagbo, Decision on the Confirmation of Charges 2014. Due to the intertwined nature of the facts and the charges of crimes against humanity, it was decided to join the two cases together, although the defendants did not have identical charges: Blé Goudé and Gbagbo Joinder Decision 2015, para 48.

[390] Gbagbo, Decision on the Confirmation of Charges 2014, para 230; Blé Goudé, Decision on the Confirmation of Charges 2014, para 135.

[391] Blé Goudé, Decision on the Confirmation of Charges 2014, para 135.

[392] Dissenting Opinion of Judge van den Wyngaert, Blé Goudé, Decision on the Confirmation of Charges 2014, para 8.

[393] Dissenting Opinion of Judge van den Wyngaert, Blé Goudé, Decision on the Confirmation of Charges 2014, para 7.

Perhaps ironically, it would seem more likely that Simone Gbagbo could be convicted under this mode of liability than her husband and his co-accused, if the assertions in her Arrest Warrant are proven at trial, that she played key role in enlisting individuals to the pro-Gbagbo forces which committed the crimes, armed them, gave direct orders, approved suggestions of certain crimes, and requested to have troops at her disposal.[394]

As the Prosecuter has rightly pointed out, the requirement that the crime would not have taken place without the actions of the accused goes much further than the original formulation given in *Lubanga*, which only required an 'essential' contribution, and determined that the 'co-perpetrator's role is to be determined on a case by case basis', involving a 'flexible approach'.[395] The Prosecutor also referred to Roxin's foundational theory, in which he only required a 'substantial' contribution to the crime, since in practice it would often be impossible to determine whether the crime would have been committed absent a contribution of the accused.[396] In doing so, she has done what the judges at the ICC have often failed to do, carefully applying the domestic theory that has been drawn from, rather than doing so in a piecemeal fashion, and stating there is a reason to apply this theory to international crimes of mass atrocity.

The second issue of concern in the definition given by the Pre-Trial Chamber is the formulation that the organised and hierarchical nature of the organisation 'secures an almost automatic compliance with orders given'. This formulation was repeated by another bench of Pre-Trial judges in *Ruto and Sang*,[397] and in Simone Gbagbo's Arrest Warrant.[398] Again, it is the Prosecutor who has corrected this over-zealous definition. Not only is it not necessary for an indirect co-perpetrator to give direct orders, since there are other ways to mobilise members of an organisation to commit crimes, including non-coercive measures such as training and providing resources,[399] but there is nothing in Roxin's theory nor in the early *Katanga* and *Lubanga* decisions which suggest the requirement of 'almost automatic compliance'. This is either a misunderstanding of the notion of the fungibility of members of the group, or of the requirement that an indirect perpetrator has the ability to cause the organisation to commit the crimes.[400] It is possible the PTC judges failed to understand the early decisions on indirect perpetration and on co-perpetration, as well as Roxin's theory from which they both draw. It is also possible that the creativity of the judges in combining these modes of liability has expanded towards creative licence. This once again threatens to bring indirect co-perpetration into line with JCE III, something that Judge van den Wyngaert keeps implicitly cautioning against.

---

394 Simone Gbagbo Arrest Warrant 2012, paras 12–13.

395 Ruto Prosecution Submission on Co-Perpetration 2012, para 11; citing Lubanga Trial Judgment 2012, para 1001.

396 Ruto Prosecution Submission on Co-Perpetration 2012, para 12.

397 Ruto Confirmation of Charges 2012, para 285.

398 Simone Gbagbo Arrest Warrant 2012, para 14.

399 Ruto Prosecution Submission on Co-Perpetration 2012, para 15.

400 Ruto Prosecution Submission on Co-Perpetration 2012, paras 14–15.

What distinguishes the 'control over the crime' test from the subjective approach of JCE, and is at the heart of a normative differentiation between parties, is the emphasis on the actual contribution of the accused, even when the mode of liability is a combined one. As we learned in the chapters comparing the domestic modes of liability, a subjective approach attributes the acts of all to the individual accused, as if they were his own. By comparison, the objective approach looks to the degree of 'control over the crime'—in the case of indirect co-perpetration this includes control over the will of those who commit the crime—and on this basis attributes responsibility, either as principal or as secondary participant.

Unfortunately this distinction has been lost in recent decisions at the ICC. In *Blé Goudé*, the PTC stated that the accused 'jointly control[s] the action of another person to such a degree that the will of that person becomes irrelevant, and her/his action must be attributed to the co-perpetrators *as if it were their own*.[401] While this might seem a matter of mere theoretical concern, once the objective approach gets lost in the application of these modes of liability, there is no inherent doctrinal limit anymore on extended liability for crimes committed by others. If the acts of subordinates become the acts of leaders, there is nothing separating their roles, and evidence of the acts and intentions of one is likely to be used to infer the actions and intentions of the other, as we have seen with JCE. Although judicial creativity is an inherent aspect of ICL, it is argued throughout this book that these participants must become more cognisant of the processes in which they take part, and more diligent in their comparative law exercises. By applying an objective approach as it is meant, doctrinal consistency can be found, providing more predictability for defendants, and more coherence in the emerging norms of ICL.

Given this, attention should also be paid to the structural differences in the various relationships between the accused and the groups committing the crimes. The *junta* model of indirect co-perpetration which was applied in Argentina can be recognised in the charges against Al Bashir, as well as Goudé and Gbagbo, where there is one group of subordinates subject to the control of a group of leaders working together, or where there were two separate groups cooperating in one operation based on an alleged agreement between the two leaders. This *junta* model could be conceived of as a pyramid, with more than one leader at the top, sharing control over the group. A second model can be conceived of as a diagonal axis, or a reverse-L, where there is a relationship between individuals at a leadership level, and a vertical relationship between each leader and his own subordinates, and the combination of modes of liability leads to attribution responsibility of one leader for the other leader's subordinates, perhaps intended in the original charges against Katanga and Ngudjolo.[402] A third construction has been described by one author, as a horizontal group of leaders who each control a vertical branch of governmental or military authority, though only some of the vertical branches actually engage in the crimes

401 Blé Goudé, Decision on the Confirmation of Charges 2014, para 136, emphasis added. This was repeated in Ongwen Confirmation of Charges 2016, para 39.

402 Ohlin 2012, p. 785.

in question.[403] This could describe the *Stakić* and *Bemba* cases. While some authors see no doctrinal obstacle to any of these models falling under the ambit of Article 25(3)(a) of the Rome Statute,[404] others consider that a fourth mode of liability has been created, and that this creates a danger for the principle of legality.[405]

Despite these concerns, a process of reification may already be underway with respect to this judicial inception as well, which is all the more reason to demand diligence and consistency in the doctrine. Defence lawyers have submitted numerous times that 'indirect co-perpetration' was never intended to fall under the wording of Article 25(3)(a) and should therefore be dismissed, but the majority of the judges have continued to insist that it applies.[406]

8.10 The Question of Differentiation Between Parties to a Crime

It has already been observed that the Nuremberg Charter and its successor instruments, the NMTs and the Nuremberg Principles, all followed a functionally unitary approach to liability. Although the subjective approach has prevailed at the ad-hoc tribunals, it is unclear from the language used in various judicial decisions and scholarly commentary whether these tribunals have also operated under a functionally unitary approach, or whether there is indeed a difference between principals and secondary parties. The preference for such differentiation is clear at the ICC, though still debated in the minority.

8.10.1 Functional Unitarity at the Ad-Hoc Tribunals

The simple wording of Article 7 of the Statute of the ICTY suggests a functionally unitary system, mirroring that of the Nuremberg Charter. Yet JCE was conceived of under the notion of 'committing' the crime, and the moral gravity of such participation is, according to the Chambers, no less than that of carrying out the act in question. Each participant is therefore a principal and not 'merely' an aider or abettor.[407]

There are, however, some who consider JCE III to be a form of aiding and abetting, and not of principal liability, despite this being contrary to the very intention and

[403] Ohlin 2012, p. 779.

[404] Weigend 2011, p. 111; van der Wilt 2009, p. 312; Ohlin 2012, p. 781.

[405] van Sliedregt 2012a, p. 169; Separate Opinion of Judge Christine van den Wyngaert, Ngudjolo Trial Judgment 2012, para 61.

[406] Katanga Confirmation of Charges 2008, para 490; Ruto Confirmation of Charges 2012, paras 286–290; Ongwen Confirmation of Charges 2016, para 41.

[407] Krnojelac Appeal Judgment 2003, para 84.

reasoning in *Tadić*.[408] The case law itself is inconsistent in this respect. Originally in the *Tadić* Appeal judgment, the notion of common design had been described as a form of 'accomplice liability',[409] yet at the same time this doctrine was found to give rise to principal liability, as opposed to accessorial or derivative liability.[410]

The focus shifted increasingly from the low ranked members of a group,[411] to mid-level military and political leaders,[412] and later to the highest echelons of power, such as Slobodan Milosović,[413] and Radoslav Brđanin.[414] However in *Krstic*, the Trial Chamber emphasised the difference between principal and secondary liability,[415] and determined that individuals participating in a JCE who did not physically carry out the objective elements of the crime could be either principals or secondary aiders and abettors, depending on their level of contribution.[416]

Trial Chamber II determined in *Krnojelac* that the distinction between principal and accessory—meaning secondary—liability was unnecessary to make, since no such distinction is made when it comes to sentencing.[417] It criticised Trial Chamber I in *Kvočka* for distinguishing between principal and accomplices—meaning secondary—in a normative sense. Trial Chamber II then chose explicitly to adopt the language of co-perpetrator to mean the participant in a JCE who was not the 'principal'—meaning physical—perpetrator.[418] The inconsistent use of terminology causes more confusion. This is also likely a result of the mixed domestic legal backgrounds of the judges making up these benches: the bench of Trial Chamber I in *Kvočka* was made up of a majority of judges hailing from the civil law tradition, in which a distinction is made between participants in a collective crime.[419] By comparison the bench of Trial Chamber II in *Krnojelac* was made up of a majority of judges from the common law tradition, where no such normative distinction is made between participants.[420]

---

408 Ambos 2008a, para I.4, in which Professor Ambos argues that only JCE I can be properly considered to be grounded in customary law, and then because it resembles co-perpetration as recognised in domestic criminal law jurisdictions; Ohlin 2007, p. 77; van Sliedregt 2012a, p. 139.

409 Tadić Appeals Judgment 1999, para 220.

410 Tadić Appeals Judgment 1999, paras 192, 229.

411 See e.g. Tadić Appeals Judgment 1999; Vasiljević Appeal 2004; Kvočka Trial Judgment 2001; Krnojelac Appeal Judgment 2003.

412 See e.g. Krstić Trial Judgment 2001; Stakić Trial Judgment 2003; Simić Trial Judgment 2002.

413 Milosević Indictment 2004.

414 Brđanin Appeal Judgment 2007.

415 Krstić Trial Judgment 2001, para 643.

416 Krstić Trial Judgment 2001, para 601.

417 Krnojelac Appeal Judgment 2003, para 75.

418 Krnojelac Appeal Judgment 2003, para 77.

419 Presiding Judge Rodrigues from Portugal, which follows the civil law tradition, Judge Riad from Egypt, which also follows civil law tradition, and Judge Wald from the USA.

420 Presiding Judge Hunt from Australia, which is a common law jurisdiction, Judge Ndepele Mwachande Mumba from Zambia, which is based primarily on the English common law tradition, and Judge Liu Daqun from China.

In the *Ojdanić* decision on the challenge to the jurisdiction of the ICTY over JCE, the Appeal Chamber attempted to clarify this confusion, by determining that JCE is a form of co-perpetration in all its three forms, which gives rise to principal liability, as opposed to secondary liability.[421] Similarly, Cassese insists that JCE is not akin to 'merely' aiding and abetting, since JCE requires a shared intent, whereas, according to him, it is possible to aid and abet a crime without sharing the intent of the common criminal purpose.[422]

The lack of clarity results from the spontaneous way in which JCE was judicially created, and the fact that its further elaboration has been left up to the case law of ad-hoc institutions whose judges come from mixed domestic backgrounds.

8.10.2 Normative Differentiation at the ICC

Even with a statute that differs substantially from the functionally unitary Nuremberg model, there has been a voice of dissent as to whether a normative system of differentiation should apply or not at the ICC.

In the Lubanga trial judgment, the only judge on the bench who hailed from a common law jurisdiction, Judge Adrian Fulford,[423] conceded that the Pre-Trial Chamber had already determined the definition of co-perpetration, and to change that at the stage of the merits would lead to an unfair trial,[424] however his dissent was on principle. According to Fulford, there was no reason to interpret the Rome Statute with anything other than a 'plain text' reading, which would lead to the conclusion that the modes of liability under Article 25(3) are simply overlapping alternatives, without any normative hierarchy between them, and they are not meant to be mutually exclusive.[425] This dissent can be understood in light of the preference for a functionally unitary approach to modes of liability, more typical of the common law tradition from which Judge Fulford hails, which would therefore lead to a dissolving of requirements such as 'control over the crime' and 'essential contribution'. In fact Fulford makes explicit that his preferred approach would be to place 'appropriate emphasis on the accused's state of mind', or the subjective element, rather than on the objective elements of contribution to or control over the crime.

Judge Van den Wyngaert agreed with Fulford's dissent and she questioned whether distinguishing between principal and accessorial—by which she correctly means derivative—responsibility would lead to different legal treatment, in the sense of principal and secondary parties.[426] In this respect she rejected an objective approach

---

421 Ojdanić Decision on JCE 2003, para 31.

422 Cassese 2008b, p. 211.

423 The other judges were Judge Benito from Costa Rica and Judge Blatman from Germany.

424 Judge Fulford, Separate Opinion, Lubanga Trial Judgment 2012, paras 2–3.

425 Judge Fulford, Separate Opinion, Lubanga Trial Judgment 2012, para 7.

426 Separate Opinion of Judge Christine van den Wyngaert, Ngudjolo Trial Judgment 2012, para 22.

and deemed the blameworthiness of an accused individual to be dependent on the factual circumstances, rather than on 'abstract legal consequences'.[427] For her, there is no reason why someone who orders or induces a crime is any less morally reprehensible than the person who physically commits it, or even the person who aids and abets it. She agreed with Judge Fulford in this respect that the sentencing does not necessarily have to reflect a particular legal qualification, and that even if there is a policy to focus on the 'masterminds' of international crimes, this would not necessitate creating such distinct legal categories.[428] Judge van den Wyngaert departs from her own domestic legal background in this respect, since her native Belgian criminal law system does make such a distinction both in terms of the legal qualification and in terms of the sentencing.[429]

The drafting history of the Rome Statute does not give any clear answer as to whether a normative differentiation should prevail, since it was a joint proposal from various countries representing both sides of the debate which led to the final adopted text, and in fact it is suggested that there was a compromise made, leaving the court to choose which mode of liability to apply in a given case.[430] Thus far the Pre-Trial and Trial Chambers of the ICC have consistently applied the control over the crime approach, and confirmed that there is a normative differentiation between the modes of liability in the Rome Statute, and this appears to be a policy choice on the part of the judges in majority, but one that is clearly informed by the civil law tradition.[431]

When it comes to proving the elements of modes of liability, Judge Schomburg has argued that it is possible for the prosecution to plead multiple modes of liability which fall under the broad wording of the ICTY statute, without having to plead a theory for these modes of liability, and this would still give sufficient notice to individuals accused of the charges against them.[432] This approach would seem acceptable in a system where there is no normative differentiation between parties to a crime, since a conviction on one mode of liability would subsume all other modes of liability, as has been established in Canadian case law.[433] Since the ICTY appears to have a descriptively differentiated system, but not a normatively differentiated one, this would be acceptable.

However in a system of liability that differentiates normatively and has different subjective and objective requirements for each mode of liability, more care is

[427] Separate Opinion of Judge Christine van den Wyngaert, Ngudjolo Trial Judgment 2012, para 24.

[428] Separate Opinion of Judge Christine van den Wyngaert, Ngudjolo Trial Judgment 2012, paras 26, 27, 29.

[429] Articles 66, 67 and 69 of the Belgian *Strafwetboek* differentiate between *daders*, or perpetrators, and *medeplichtigen*, or secondary parties, and determine that the latter will receive a lower sentence than the perpetrator.

[430] Ohlin et al. 2013, p. 22; Schabas 2010, p. 423.

[431] Ruto Confirmation of Charges 2012, para 292; Muthaura Confirmation of Charges 2012, para 296; Banda Decision on Charges Corrigendum 2011, para 126; Bemba Decision on Charges 2009, para 348.

[432] Gacumbitsi Appeal, Schomburg Separate Opinion 2006, paras 7–8.

[433] See Sect. 6.4.

needed. The conviction under different modes of liability lead to a different factual qualification, and the elements of each specific mode must be proven in order for that mode to apply. Furthermore,during trial, a defendant must know which charges will apply, in order to be given the full possibility of preparing a defence. The ICC Pre-Trial Chamber affirmed this when it stated that if multiple or alternative modes of liability are indicted, one of them must be proven fully by the prosecution, and that if Article 25(3)(a) were to be proven, then there would be no need to revert to the secondary modes of liability in the other subparagraphs.[434] This has been one of the sticking points for Judge van den Wyngaert, in her objection to the confirmation of charges that appear too flexible with respect to modes of liability.

8.11 Legal Culture and Policy Choices in the International Tribunals

From the observations made throughout this chapter of the processes in the international tribunals, it is clear that ICL cannot be located in one tradition. Both in terms of procedure and substantive law, the creation of this international body of law has led to a hybridity which may even be considered *sui generis* as it continues to develop. While ICL is built upon, and shares similarities with, both traditions compared here, it does not represent an ideal of either, and there is therefore no singular basis for preferring one system of liability over another. There has not been a specific constitutive moment at which the choice of a foundational system of liability took place, rather it is a flow of legal patchworking between traditions, within which certain choices must be made. This patchworking process has been a continual response to specific conflicts, especially in the cases of the IMT, the post-Second World War case law and the ad hoc tribunals. The establishment of the ICC offered an opportunity to pay more attention to these doctrinal questions, but the drafting process was a series of compromises between State representatives who had certain interests, and who asserted the concepts most familiar to them.[435] Thus, although Article 25 on the modes of liability was a joint proposal made by a group of States representing both the civil law and common law traditions,[436] the debates that have continued among judges at the ICC over the correct interpretation of this article demonstrate that the opportunity to make an informed, considered and deliberate choice with respect to the applicable system of liability still remains.

---

[434] Katanga Confirmation of Charges 2008, para 471. This was one of the controversies over the application of Regulation 55 in the Katanga case, since the judges departed from the indicted charges as a principal under Article 25(3)(a) and notified the defendant that other modes of liability 'may' apply. Indeed, in the end he was convicted under Article 25(3)(d) for common purpose. Katanga Severance of Charges Decision 2012, para 7.

[435] Pellet 2002, p. 1059; de Guzman 2008, p. 705; Heller 2011b, p. 598.

[436] ICC Working Paper 1997; Schabas 2010, p. 423.

8.11.1 The Relationship Between the Character of International Criminal Trials and a System of Liability

In the two previous chapters, it was demonstrated that there may be a significant relationship between on the one hand the character of domestic criminal trials as either typically focused on conflict resolution or on policy-implementation, and on the other hand the development of a system of liability that is functionally unitary or normatively differentiated. Because international criminal trials reflect a combination of both trail characteristics, the choice for a system of liability does not necessarily flow from this in the same way as it appears to at the domestic level.

It is important in this respect to note that Damaška's conceptual framework of the character of a criminal trial and its relationship to a hierarchically or horizontally organisation of authority is a description of ideal types, for analytical purposes. No single domestic criminal justice system fits these ideal types perfectly. Applying this conceptual framework to ICL may therefore require some translation; Damaška himself questioned whether his framework can be applied directly to ICL, due to its inherent hybridity.[437] Still, some lessons can be drawn from his framework in understanding certain tendencies in ICL, and at the very least in ensuring that where choices are to be made in the development of ICL, there are no mismatches between the goals of ICL and the institutions and norms being transposed from the domestic to the international.[438]

For instance, case law is important in developing the content of the law, and judges therefore have a heightened authority as influential participants in the processes of the law, similar to the common law tradition. ICL could also be said to have an iterative process, with paths of judicial decisions reflecting societal norms and leading to clarification and development of the law. Yet at the same time judges increasingly refer to scholarly opinions as authoritative, as evidenced in the decisions of the ICC on liability thus far. Furthermore the lack of lay participation in international criminal trials means that there is room for the development of a technical taxonomy of modes of liability, and a more thorough doctrinal underpinning based on the expertise of these authoritative scholars.

Given that there is no central authority in international law, ICL it cannot be compared with a system of domestic criminal law under a single State which can determine the content of the law, and thus neither the goals of conflict-resolution nor of policy-implementation will logically prevail. It may therefore be necessary to seek solutions which are *sui generis*.

---

437 Damaška 2004a, b; van der Wilt 2010b, p. 45.

438 Swart 2008, p. 114.

8.11.2 Policy Shifts

Over time the policy trends of the international tribunals have been an influencing factor in the desire to widen the net of modes of liability. The earliest policy moves to attach criminal liability to war crimes came in the nineteenth and twentieth centuries, and focused specifically on the role of commanders and superiors.[439] Following the First World War, the report of the Allied Powers' Commission on the Responsibility of Authors of the War and on the Enforcement of Penalties had recommended the establishment of an international tribunal to prosecute individuals who 'ordered, or, with knowledge thereof and with power to intervene' failed to do so, in order to repress violations of the laws of war.[440] Although this tribunal never eventuated, it was one of the first steps in moving towards the international prosecution of mass atrocity, with an explicit focus upon those who were in positions of control and authority.

As discussed in this chapter, following the Second World War, the policy of the Nuremberg trials had been to capture as many of those involved in the mass atrocity crimes as possible. At first the idea was to prosecute those at the top, and the derivative liability for those lower down in the echelons was expected to 'trickle down', making evidence of facts and of individual involvement easier as they moved down the hierarchy of defendants. However it soon became clear what an impossible task this was.[441] Instead the IMT at Nuremberg focused only the leaders of the Nazi regime, and the domestic military tribunals under Control Council Law No. 10 carried the work further, still focusing on mid and high level leadership cases, though mostly for pragmatic reasons.[442]

This same broad intention was the starting point under the mandate and policy of the ICTY to hold '*all* persons who participate in the planning, preparation or execution of serious violations of international humanitarian law in the former Yugoslavia' individually responsible.[443] Again this has proved to be an impossible task, and the policy of the prosecutor shifted over the years to focus on the leaders. This appears to have been partly a pragmatic shift, due to the difficulties of undertaking investigations while the war was still raging in former Yugoslavia.[444] While it is difficult to ascertain the prosecutor's policy at the ICTY, due to the fact there are very few publicly available documents regarding this, some statements from the chief prosecutors over time demonstrate the shift towards focusing on the 'persons who were occupying the highest political or military positions' and their 'functional responsibility' for

[439] Werle and Jessberger 2014, p. 4.

[440] Boas et al. 2007, p. 147.

[441] Ryan 2007, p. 62; Heller 2011a, p. 15.

[442] Heller 2011a, p. 24.

[443] Report of the Secretary General on the ICTY 1993, para 54.

[444] de Vlaming 2010, p. 108.

the crimes committed by their subordinates.[445] The OTP Charging and Indictment Guidelines states:

> Our investigations are directed exclusively to establishing personal responsibility for those individuals who are most culpable for the atrocities committed in the former Yugoslavia. To that end, we should exercise internal discipline to pursue only the most serious charges and the most culpable individuals.[446]

The policy of tribunals such as the ECCC, the SCSL and the ICC has from the outset expressly been to focus on 'those most responsible' for the systemic crimes at hand.[447] Although JCE has been applied at the ECCC, the prosecutions have focused upon the leaders within the Pol Pot regime, with strong narratives supporting this choice based on the theories of retribution and norm expression.[448] This is not to say that only those in the highest positions are considered to be liable. The ECCC has rejected the arguments made by Duch that his responsibility should be either mitigated or negated by the fact that there were other senior leaders higher up the hierarchy than him who were more responsible than him.[449] Similarly, although the ICC can investigate anyone involved in mass atrocity crimes and has various modes of liability available in the Rome Statute, the attention continues to be on those who bear the greatest responsibility for such crimes, particularly in the highest echelons of power, 'including those who ordered, financed, or otherwise organized the alleged crimes.'[450]

With these shifts comes the desire to apply modes of liability that reflect the need either to include every individual involved in crimes of mass atrocity, or to focus on the role of the leaders. The patchworking process shows that judges and other authoritative voices look to domestic modes of liability. However the policy behind the domestic modes of liability has not necessarily been part of the consideration in choosing to follow one tradition or another. For example, at Nuremberg the push to include conspiracy on the part of the American participants was because they had seen the success of this in response to mafia crimes in the US, but the design and structure of the Nazi regime did not necessarily reflect the more diffuse family-based structures seen in the mafia. Similarly, JCE was based on the case law of the NMTs, however these tribunals also applied the domestic law of their own common law

---

[445] Del Ponte 2004, p. 516. See also statements made by Del Ponte's predecessors Richard Goldstone and Louise Arbour, cited in de Vlaming 2010, pp. 120, 137, 154.

[446] Cited in de Vlaming 2010, p. 138. De Vlaming received this confidential document from a former employee of the ICTY, and has the document on file, however it is not publicly available.

[447] See e.g. ICC Office of the Prosecutor Policy Report 2003, p. 7: 'The Office of the Prosecutor should focus its investigative and prosecutorial efforts and resources on those who bear the greatest responsibility, such as the leaders of the State or organisation allegedly responsible for those crimes.'; Kamara 2004, p. 13: 'The SCSL was established for the sole purpose of prosecuting persons who bear the greatest responsibility for serious violations of international Humanitarian law'. See also Duch, Trial Judgment 2010 p. 56; van der Wilt 2010b.

[448] Statute of the ECCC 2004, Article 1; Duch, Trial Judgment 2010, para 2.

[449] Duch, Trial Judgment 2010, para 20. Affirmed in Duch, Appeal Judgment 2012.

[450] OTP Prosecutorial Strategy 2010, para 19; OTP Stategic Plan 2013, p. 13.

jurisdictions, which had been developed in response to these gang and mafia type crimes.

Judge Fulford argued the same point in his separate opinion to the *Lubanga* judgment, that the way in which the 'control over the crime' theory developed in Germany and other jurisdictions as a response to systemic criminality may be a result of different policy considerations than are applicable in ICL.[451] However this theory was developed exactly in response to the need to link leaders as decision-makers to crimes of mass atrocity committed by their subordinates in domestic situations. Hence although we should heed his warning in choosing modes of liability, Judge Fulford missed the very point of the policy likeness that would in fact be positively instructive in choosing to apply the 'control over the crime' doctrine in ICL.

Unlike in a domestic system where a legislature and a judiciary can together form a more or less explicit desired policy response to collective criminality, according to which the modes of liability develop in the processes of the law, in ICL there is not only a lack of consensus, there is also a lack of a centralised forum in order to formulate such an explicit policy. As a result, the policies have shifted on a more ad-hoc basis, and the participants who are most influential in this are the less traditional law-makers; the judges, the indicting prosecutors, defence lawyers and the scholars who are cited as authoritative. An international 'system' of liability has therefore swung back and forth between subjective and objective approaches, and it is now the task of these participants to make an informed and deliberate choice.

8.11.3 The Influence of Participants on the Shifts Between Subjectivity and Objectivity

While there have been fears expressed and accusations made of judicial activism leading to unfounded or 'bad' law, the prevalent role of judges as participants in ICL is clear from the observations in this chapter. This fact can be seen in a more positive light as judicial creativity. As Judge Shahabuddeen has put it, to recognise this:

> is more satisfactory than the pretence that the many propositions produced by an international criminal tribunal for its work flow directly from customary international law. That seldom convinces anyone. All that is found is a chance statement by the Nuremberg Tribunal or by an author. If it is asked what is behind the statement, one is answered with generalities. Save where conventional law or customary law was clearly at work, it is more satisfactory to cite judicial creativity.[452]

Just as the process of patchworking demonstrates, where there is no positive law supporting the adoption of a specific doctrine or mode of liability, judicial creativity is not only permissible and valid, it is inevitable and, perhaps, necessary.[453] This is despite the concerns expressed at the inception of the ICTY that it should not create

[451] Judge Fulford, Separate Opinion, Lubanga Trial Judgment 2012, para 10.

[452] Shahabuddeen 2010, p. 186.

[453] Shahabuddeen 2010, pp. 197, 203.

any new law,[454] and should limit itself to applying notions that were unquestionably part of customary international law.[455] Yet it became clear with the inception of the ICTR that a certain amount of law-creation was inevitable, since this new tribunal was to deal with genocide as part of a non-international conflict, a new context which lacked any customary law content, and it received a broader mandate to apply humanitarian law instruments which were not beyond doubt customary in nature.[456]

The necessity to fill the gaps in ICL by way of reference to domestic law more directly is a function of the uncertainty and hybrid nature of ICL in general. For this reason the ad-hoc tribunals in particular have been described as 'hunter-gatherers in a legal wilderness.'[457] In order for the tribunals, including the ICC, to be effective, it was necessary to place as few formal constraints as possible on the judges whose task it has become to develop ICL actively and creatively. There appears to be more agreement recently that as long as the system is fair, not in flagrant violation of the principle of legality, and effective, then it is an acceptable process by which to come to new law.[458]

This judicial creativity helps to explain the influence of the more subjective approach at the ad-hoc tribunals and the more objective approach at the ICC. For example, the influence of the post-Second World War case law was prominent in the ICTY jurisprudence surrounding JCE, and this early case law was itself heavily driven by the common law tradition from which the subjective approach hails.[459] At the same time, much of the criticism levelled at the notion of JCE with respect to the fact that it fails to distinguish between individuals normatively, and that it fails to capture the collective nature of crimes of mass atrocity, tends to come from jurists who hail from the civil law tradition.[460]

Similarly, at the ICC the Germanic tradition has been driven predominantly by jurists who hail from the civil law tradition. The dissent heard against this interpretation can be traced mainly—though not exclusively—to those who hail from functionally unitary systems; systems that are semantically differentiated but not normatively differentiated.[461] In the international system, lawyers inevitably read the provisions of ICL through the familiar domestic lens with which they were trained.[462]

---

[454] Many delegations were explicit in their insistence that the new tribunal should not create any new law, and that the Statute itself should do nothing more than reflect existing norms of IHL. See Darcy and Powderly 2010, p. 24; Danner 2006, p. 22.

[455] Report of the Secretary General on the ICTY 1993, para 44.

[456] UN Report on the ICTR 1994, cited in Darcy and Powderly 2010, p. 24.

[457] Zahar and Sluiter 2008, p. 80.

[458] Darcy and Powderly 2010, p. 43.

[459] van Sliedregt 2012a, p. 96; Werle and Jessberger 2014, p. 137.

[460] Separate opinion of Judge Lindholm, Simić Trial Judgment 2002, para 2; Ambos 2007b; van der Wilt 2007; Schomburg 2010.

[461] Judge Fulford, Separate Opinion, Lubanga Trial Judgment 2012, para 9; Separate Opinion of Judge Christine van den Wyngaert (who hails from a civil law background, but whose dissent is in line with the subjective approach), Ngudjolo Trial Judgment 2012, para 27; Stewart 2012, p. 207.

[462] Werle and Jessberger 2014, p. 137; Steer 2015.

Judge Van de Wyngaert has repeatedly warned against an over-extensive interpretation of Article 25 of the Rome Statute, a provision that was created as a mix of domestic traditions and international instruments.[463] Her cautions are well founded, and point to the tension in the fact that it is the task of the judges to form a deliberate foundation for a system of liability before this permanent court. In shifting towards the objective approach, the majority judges have been criticised for importing blindly from German law. However something that should be made more explicit in these interpretations is that while it is true domestic law has become an important source, and there is risk of bias or arbitrariness, *Organisationsherrshaft* or indirect perpetration by means of an organisation was developed specifically in relation to mass atrocity crimes. It is therefore pertinent to the interpretation of this mode of liability at the international level.

8.11.4 Vicarious or Derivative Liability?

Another aspect of identifying what system of liability would be most appropriate for responding to mass atrocity in ICL is the theoretical underpinning, in terms of whether party liability is derivative or vicarious in nature. The shifting trends in international tribunals shows that no clear choice has been made thus far, since it depends on whether a subjective or an objective approach has been prevalent at any given moment, and on the preference of individual participants in each specific tribunal.

For example, JCE is predicated on a subjective approach, taking the common intention or shared agreement to be the basis of liability. In its most extended form as JCE III, liability for the unplanned but objectively foreseeable acts of other parties is shared by all according to the same vicarious attribution of all acts to all parties.[464]

On the other hand, the approach preferred by the majority of judges at the ICC to date has been predicated on an objective test of liability, which brings with it an attribution of responsibility, rather than imputation of another's acts.[465] Since an objective test of liability, including the notion of 'control over the crime', looks to the contribution of each individual party to the crime, and measures culpability by degrees through a differentiation of liability, it is impossible to attribute the act or contribution of one individual to another. Instead the liability of secondary participants is derived from the liability of the principal, depending on the degree of their contribution, thus making normatively secondary participants true 'accessories'.

---

[463] Separate Opinion of Judge Christine van den Wyngaert, Ngudjolo Trial Judgment 2012, para 13; Dissenting Opinion of Judge Christine van den Wyngaert, Blé Goudé, Decision on the Confirmation of Charges 2014, para 6.

[464] See e.g. Vasiljević Appeal 2004; Stakić Trial Judgment 2003, para 246.

[465] Lubanga Confirmation of Charges 2007, para 330; Lubanga Trial Judgment 2012, paras 997–999; Katanga Confirmation of Charges 2008, para 484; Werle and Burghardt 2011, p. 86.

It would therefore seem that the theoretical underpinning of liability follows from the choice of a subjective or objective approach, and not vice-versa. A theory of vicarious liability is associated with a subjective approach, and leads to an expansion of liability, since all parties fall under a constructed attribution of the acts of all other parties, even when one party goes further than was intended or agreed. A theory of derivative liability, on the other hand, is associated with an objective approach and leads to a normatively differentiated system of liability, which limits principal liability by defining when a party is considered to be 'merely' accessorial, receiving normatively secondary liability. Here too there is a policy choice to be made, namely whether the desired system of liability is one that extends or limits party liability.

It is clear from the analysis here that ICL develops as much through case law and academic discussion as it does through law-making treaties and statutes. It is also clear it will continue to do so. The question remains how to harness this process such that a system of liability can emerge that reflects the specific nature of mass atrocity crimes, and serves the functions and goals of ICL. It is to these final questions that Part III of this book will now turn.

References

Additional Protocol I (1977) Protocol additional to the Geneva Conventions of 12 August 1949, and relating to the protection of victims of international armed conflicts (Protocol I)

Akhavan P (2005) The crime of genocide in ICTR jurisprudence. J Int Crim Justice 3:989

Ambos K (2005) La Parte Peneral del Derecho Penal Internacional: Bases para una elaboración dogmática. Fundación Konrad Adenauer, Montevideo

Ambos K (2007a) Enjuiciamento de crimenes internacionales en el nivel nacional e internacional: entre justicia y realpolitik. Politica Criminal 4:1

Ambos K (2007b) Joint criminal enterprise and command responsibility. J Int Crim Justice 5:159–183

Ambos K (2008a) Amicus Curiae concerning Criminal Case File No. 001/18-07-2007-ECCC/OCIJ (PTC 02)

Ambos K (2008b) Article 25. In: Trifterer O (ed) Commentary on the Rome Statute of the International Criminal Court, observers' notes article by article. Beck Verlag, München

Ambos K (2009) Critical Issues in the Bemba Confirmation Decision. Leiden J Int Law 22(4):715–726

Anderson S (2013) English version: Murderers are being allowed to go free. http://www.bt.dk/udland/english-version-murderers-are-being-allowed-to-go-free

Ascensio H (2006) The French perspective. In: Reginbogin HR, Safferling CJM (eds) The Nuremberg Trials: International Criminal Law since 1945. K.B. Saur, München

Badar ME (2006) Just convict everyone!-Joint perpetration: from Tadić to Stakić and back again. Int Crim Law Rev 6:292–302

Bassiouni MC (1999) Crimes against Humanity in International Criminal Law. Martinus Nijhoff Publishers

Boas G, Bischoff JL, Reid NL (2007) Forms of Responsibility in International Criminal Law. International Criminal Law Practitioner Series, vol I. Hart Publishing, Oxford

Brown RM (2006a) The American perspective on Nuremberg: a case of cascading ironies. In: Reginbogin HR, Safferling CJM (eds) The Nuremberg Trials: International Criminal Law since 1945. K.B. Saur, Munchen

Brown RM (2006b) The role of the Soviet Union in the International Military Tribunal at Nuremberg. In: Reginbogin HR, Safferling CJM (eds) The Nuremberg Trials: International Criminal Law since 1945. K.B. Saur, München
Brownlie I (1998) The rule of law in international affairs: International Law at the fiftieth anniversary of the United Nations. Martinus Nijhoff, The Hague
Cassese A (2004) The ICTY: a living and vital reality. J Int Crim Justice 2:585–597
Cassese A (2007) The proper limits of individual responsibility under the doctrine of joint criminal enterprise. J Int Crim Justice 5:109
Cassese A (2008a) Amicus Curiae Brief of Professor Antonio Cassese and members of the Journal of International Criminal Justice on joint criminal enterprise, for Case File No. 001/18-07-2007-ECCC/OCIJ (ptc 02)
Cassese A (2008b) International Criminal Law. Oxford University Press, Oxford
Cesarani D (2006) The International Military Tribunal at Nuremberg: British perspectives. In: Reginbogin HR, Safferling CJ (eds) The Nuremberg Trials: International Criminal Law since 1945. K.G, Saur, Munchen, pp 31–37
Charter of the IMT (1945) Charter of the International Military Tribunal at Nuremberg
Coco A, Gal T (2013) Losing direction: the ICTY Appeals Chamber's controversial approach to aiding and abetting in Perišić. J Int Crim Justice 12:345–366
Control Council Law No. 10 (1945) Allied Control Council Law No. 10
Cryer R (2001) The boundaries of liability in international criminal law, or 'selectivity by stealth'. J Conflict Secur Law 6(1):3–31
Cryer R, Friman H, Robinson D, Wilmshurst E (2016) An introduction to international criminal law and procedure, 3rd edn. Cambridge University Press, Cambridge
Cupido M (2015) Pluralism in theories of liability: joint criminal enterprise versus joint perpetration (forthcoming). In: van Sliedregt E (ed) Harmonization and pluralism in international criminal law. Oxford University Press
Damaška M (2004a) Negotiated justice in international criminal courts. J Int Crim Justice 2:1018
Damaška M (2004b) The uncertain self-identity of international criminal courts. http://scholarship.kentlaw.iit.edu/morris/6/
Danner AM (2006) When courts make law: how the international criminal tribunals recast the laws of war. Vanderbilt Law Rev 59:1–64
Danner AM, Martinez JS (2005) Guilty associations: joint criminal enterprise, command responsibility, and the development of international criminal law. Calif Law Rev 75–169
Darcy S, Powderly J (eds) (2010) Judicial creativity at the international criminal tribunals. Oxford University Press
de Guzman MM (2008) Article 21. In: Trifterer O (ed) Commentary on the Rome Statute of the International Criminal Court: observers' notes, article by article, 2nd edn. Beck-Hart-Nomos, Munchen
de Vlaming F (2010) De Aanklager: Het Joegoslavië Tribunaal en de Selectie van Verdachten. Boom Juridische Uitgevers, The Hague
Del Ponte C (2004) Prosecuting the individuals bearing the highest level of responsibility. J Int Crim Justice 2:516
Del Ponte C (2006) Investigation and prosecution of large-scale crimes at the international level: the experience of the ICTY. J Int Crim Justice 4:549
Drumbl MA (2005) Collective violence and individual punishment: the criminality of mass atrocity. Northwest Univ, Law Rev, p 539
Eldar S (2012) Holding organized crime leaders accountable for the crimes of their subordinates. Crim Law Philos 1–19
Fletcher GP (2000) Rethinking criminal law. Oxford University Press, Oxford
Fletcher GP (2002) The Storrs lectures: liberals and romantics at war; the problem of collective guilt. Yale Law J 111:1499–1573
Fletcher GP (2007) The grammar of criminal law: American, comparative and international, Vol One: Foundations. Oxford University Press, New York

Fletcher GP (2011) New court, old dogmatik. J Int Crim Justice 9:179–190
Fletcher GP, Ohlin JD (2005) Reclaiming fundamental principles of criminal law in the Darfur case. JICJ 3:539–561
Fourth Hague Convention (1907) Convention Respecting the Laws and Customs of War on Land
Genocide Convention (1948) Convention on the Prevention and Punishment of the Crime of Genocide
Ginsburgs G (1990) The Nuremberg Trial: background. In: Ginsburgs G, Kudriavtsev V (eds) The Nuremberg Trial and international law. Martinus Nijhoff Publishers, Dordrecht, p 1
Haan V (2005) The development of the concept of joint criminal enterprise at the International Tribunal for the Former Yugoslavia. Int Crim Law Rev 5:167–201
Haffajee RL (2006) Prosecuting crimes of rape and sexual violence at the ICTR: the application of joint criminal enterprise theory. Harvard J Law Gender 201–221:167–201
Heller KJ (2011a) The Nuremberg Tribunals and the origins of International Criminal Law. Oxford University Press, Oxford
Heller KJ (2011b) The Rome Statute of the International Criminal Court. In: Heller KJ, Dubber MD (eds) The handbook of comparative criminal law. Stanford University Press, Stanford, pp 593–634
Henckaerts JM, Doswald-Beck L (2005) International Committee of the Red Cross's study of Customary International Humanitarian Law. Cambridge University Press, Cambridge
ICC Office of the Prosecutor Policy Report (2003) Paper on some policy issues before the Office of the Prosecutor. http://icc-cpi.int/en_menus/icc/structure/office_of_the_prosecutor/policies_and_strategies/Pages/paper_on_some_policy_issues_before_the_office_of_the_prosecutor.aspx
ICC Working Paper (1997) Working Paper submitted by Canada, Germany, The Netherlands and the United Kingdom
ILC (1950) Principles of international law recognized in the Charter of the Nuremberg Tribunal and in the judgment of the Tribunal
ILC Draft Code of Crimes (1996) Draft Code of Crimes against the Peace and Security of Mankind. http://untreaty.un.org/ilc/publications/yearbooks/Ybkvolumes(e)/ILC_1996_v2_p2_e.pdf
Inter-Allied Resolution on German War Crimes (1942) Resolution on German War Crimes signed by representatives of nine occupied countries. http://www.ibiblio.org/pha/policy/1942/420112a.html
Interlocutory Decision on Applicable Law (2011) Interlocutory decision on the applicable law: terrorism, conspiracy, homicide, perpetration, cumulative charging. http://www.stl-tsl.org/x/file/TheRegistry/Library/CaseFiles/chambers/20110216_STL-11-01_R176bis_F0010_AC_Interlocutory_Decision_Filed_EN.pdf
International Commission of Inquiry on Darfur (2005) Report of the International Commission of Inquiry on Darfur to the United Nations Secretary General, pursuant to Resolution 1564. http://www.securitycouncilreport.org/atf/cf/%7B65BFCF9B-6D27-4E9C-8CD3-CF6E4FF96FF9%7D/WPS%20%202005%2060.pdf
Jain N (2012) The control theory of perpetration in international criminal law. Chicago J Int Law 12:159
Jenks C (2013) Prosecutor versus Perišić. Am J Int Law 107:622
Jessberger F, Geneuss J (2008) On the application of a theory of indirect perpetration in Al Bashir. J Int Crim Justice 6:853–869
Kudriavtsev V (1990) The Nuremberg Trial and problems of strengthening the international legal order. In: Ginsburgs G, Kudriavtsev V (eds) The Nuremberg Trial and international law. Martinus Nijhoff, Dordrecht
Laffan D (2016) Supreme Court abolishes 'wrong turn' joint enterprise law. https://ukhumanrightsblog.com/2016/02/18/supreme-court-abolishes-wrong-turn-joint-enterprise-law-diarmaid-laffan/
Lafontaine F (2012) Prosecuting genocide, crimes against humanity and war crimes in Canadian Courts. Carswell, Thomson Reuters, Toronto
Lieber Code (1863) Instructions for the Government of Armies of the United States in the Field

Maogoto JN (2004) War crimes and realpolitik. Lynne Reiner Publishers, London
Mégret F (2010) Beyond fairness: understanding the determinants of international criminal procedure. UCLA J Int Foreign Aff 14:37
Mettreaux G (2008) The law of command responsibility. Oxford University Press, Oxford
Moscow Declaration (1943) Declaration Concerning Atrocities. http://www.ess.uwe.ac.uk/documents/moscwdecl.htm
MPC (1962) Model Penal Code
Ohlin JD (2007) Three conceptual problems with the doctrine of joint criminal enterprise. J Int Crim Justice 5:69–90
Ohlin JD (2009) Attempt, conspiracy, and incitement to commit genocide. Cornell Law Faculty Publications, pp 173–339
Ohlin JD (2012) Second order linking principles: combining vertical and horizontal modes of liability. Leiden J Int Law 25(3):771–797
Ohlin JD, van Sliedregt E, Weigend T (2013) Assessing the control theory. Leiden J Int Law 26(3):725–746
Olásolo H (2009) The criminal responsibility of senior political and military leaders as principals to international crimes. Hart, Portland, Oregon
Osiel M (2005) The banality of good: aligning incentives against mass atrocity. Columbia Law Rev 105:1751–1862
Osiel M (2009) Making sense of mass atrocity. Cambridge University Press, Cambridge
OTP Prosecutorial Strategy (2010) OTP Prosecutorial Strategy 2009–2012. http://www.icc-cpi.int/NR/rdonlyres/66A8DCDC-3650-4514-AA62-D229D1128F65/281506/OTPProsecutorialStrategy20092013.pdf
OTP Stategic Plan (2013) OTP Stategic Plan June 2012–2015. http://www.icc-cpi.int/en_menus/icc/structureofthecourt/officeoftheprosecutor/reportsandstatements/statement/Documents/OTPStrategicPlan.pdf
Pellet A (2002) Applicable law. In: Cassese A, Gaeta P, Jones MJR (eds) The Rome Statute of the International Criminal Court: a commentary, vol II. Oxford University Press, Oxford
Pomorski S (1990) Conspiracy and criminal organization. In: Ginsburgs G Kudriavtsev V (eds) The Nuremberg Trial and international law. Martinus Nijhoff, Dordrecht
Preparatory Commission Working Group on General Principles (1998) Committee of the Whole 'Working Group on General Principles of Criminal Law.'
Report of the Secretary General on the ICTY (1993) Report of the Secretary General pursuant to Paragraph 2 of Security Council Resolution 808 (1993) on the ICTY
Rockwood LP (2007) Walking away from Nuremberg: Just war and the doctrine of command responsibility. University of Massachusetts Press, Amherst
Röling BVA, Rüter CF (1977) The Tokyo Judgment: The International Military Tribunal for the Far East (IMTFE), 29 (April), 1946–12 (November 1948), vol 2. APA-University Press Amsterdam, Amsterdam
Roxin C (2005) Täterschaft und Tatherrschaft, 8th edn. De Gruyter Recht, Berlin
Ryan AA (2007) Nuremberg's contribution to international law. Boston College Int Comp Law Rev 30 (Winter 2007):55
Schabas W (2001) Commentary on Prosecutor v Akayesu. In: Klip A, Sluiter G (eds) Annotated leading cases of international criminal tribunals, vol 2. Hart, Oxford/Munich
Schabas W (2008) Article 6 Genocide. In: Trifterer O (ed) Commentary on the Rome Statute of the International Criminal Court, 2nd edn. Hart, Oxford/Munich
Schabas W (2010) The International Criminal Court: A commentary on the Rome Statute. Oxford University Press, Oxford
Schomburg W (2010) Jurisprudence on JCE–revisiting a never-ending story. http://www.cambodiatribunal.org/sites/default/files/resources/ctm_blog_6_1_2010.pdf
Semelin J (ed) (2007) Purify and destroy: the political uses of massacre and genocide. Hurst and Company, London

Shahabuddeen M (2010) Judicial creativity and joint criminal enterprise. In: Powderly J, Darcy S (eds) Judicial creativity at the international criminal tribunals. Oxford University Press, Oxford
Simpson G (2007) Law, War and Crime. Oxford University Press, Oxford
Smith BF (ed) (1981) The American road to Nuremberg: the documentary record 1944–1945. Hoover Press
Statute of the ECCC (2004) Law on the establishment of the Extraordinary Chambers in the Courts of Cambodia for the prosecution of crimes committed during the period of Democratic Kampuchea
Statute of the ICC (1998) Statute of the International Criminal Court
Statute of the ICTY (1993) Statute of the International Criminal Tribunal for the prosecution of persons responsible for serious violations of international humanitarian law committed in the territory of the Former Yugoslavia
Statute of the ICTY (1994) Statute of the International Criminal Tribunal for Rwanda
Statute of the SCSL (2000) Statute of the Special Court for Sierra Leone
Statute of the STL (2007) Statute of the Special Tribunal for Lebanon
Steer C (2011) Non-state actors in international criminal law. In: d'Aspremont J (ed) Participants in the international legal order: multiple perspectives on non-state actors in International Law. Routledge, London, p 96
Steer C (2012) What makes valid law? Shifting modes of responsibility in international criminal law. In: van Voorhout JC et al (eds) Shifting responsibilities in criminal justice: critical portrayals of the changing role and content of a fragmented globalizing law domain. Eleven International Publishing
Steer C (2013a) The Prosecutor v Delic: what are the limits of command responsibility? In: Klip A, Sluiter G (eds) Annotated leading cases of international criminal tribunals, vol XXXVII. Intersentia
Steer C (2013b) Ranking responsibility: why we should distinguish between participants in mass atrocity crimes. In: Dolmann M, Abels D (eds) Dialektiek in het Nationale en Internationale Strafrecht. Boom Juridische Uitgevers
Steer C (2015) Legal transplants or legal patchworking? The creation of international criminal law as a pluralistic body of law. In: van Sliedregt E (ed) Harmonization and Pluralism in International Criminal Law. Oxford University Press
Stewart J (2012) The end of 'modes of liability' for international crimes. Leiden J Int Law 25(1):165
Stewart J (2013) "Specific direction" is unprecedented: results from two empirical studies. http://www.ejiltalk.org/specific-direction-is-unprecedented-results-from-two-empirical-studies/
Swart B (2008) Damaška and the faces of international criminal justice. J Int Crim Justice 6:87–114
Tallgren I (2002) The sensibility and sense of international criminal law. Eur J Int Law 13(3):561–595
Tokyo Charter (1946) Charter of the International Military Tribunal for the Far East
UN Report on Nürnberg (1949) The Charter and Judgment of the Nürnberg Tribunal -history and analysis: Memorandum submitted by the Secretary-General. Topic: Formulation of the Nürnberg Principles
UN Report on the ICTR (1994) Report of the Secretary General pursuant to Paragraph 5 of Security Council Resolution 955 (1994) on the ICTR
van der Wilt H (2007) Joint criminal enterprise, possibilities and limitations. J Int Crim Justice 5:91–108
van der Wilt H (2009) The continuous quest for proper modes of criminal responsibility. J Int Crim Justice 7(2):307–314
van der Wilt H (2010a) A valiant champion of equity and humaneness: the legacy of Bert Röling for international criminal law. J Int Crim Justice 8:1127
van der Wilt H (2010b) Why international criminal lawyers should read Mirjan Damaška. In: Stahn C, van den Herik L (eds) Future Perspectives in international criminal justice. TMC Asser Press, The Hague
van der Wilt H (2014) The spider and the system: Milošević and joint criminal enterprise. In: Waters T (ed) The Milošević trial: an autopsy. Oxford University Press, Oxford

van Sliedregt E (2012a) Individual criminal responsibility in international law. Oxford University Press, Oxford
van Sliedregt E (2012b) Pluralism in international criminal law. Leiden J Int Law 25
Weigend T (2008) Intent, mistake of law, and co-perpetration in the Lubanga Decision on Confirmation of Charges. J Int Crim Justice, 471–487
Weigend T (2011) Perpetration through an organization: the unexpected career of a German legal doctrine. J Int Crim Justice 9:91–111
Werle G, Burghardt B (2011) Indirect perpetration: a perfect fit for international prosecution of armchair killers? Foreword. J Int Crim Justice 9:85–89
Werle G, Jessberger F (2014) Principles of International Criminal Law, 3rd edn. Cambridge University Press
Woetzel RK (1960) The Nuremberg Trials in international law. Stevens and Sons Ltd., London
Yamashita (1945) Trial of General Tomoyuki Yamashita
Zahar A, Sluiter G (2008) International Criminal Law: a critical introduction. Oxford University Press, Oxford
Zoller E, Reshetov I (1990) International criminal responsibility of individuals for international crimes. In: Ginsburgs G, Kudriavtsev VN (eds) The Nuremberg Trial and international law. Martinus Nijhoff, Dordrecht, pp 99–120

Cases

Akayesu (1998) The Prosecutor v. Akayesu, Trial Judgment, ICTR-96-4-T
Al Bashir Application for Warrant of Arrest (2008) Public redacted version of Prosecutor's application for warrant of arrest under art. 58, ICC-02/05-157
Al Bashir Decision on Arrest Warrant (2009) The Prosecutor v. Omar Hassan Ahmad Al Bashir, second decision on the Prosecution's application for a warrant of arrest 02/05- 01/09
Almelo Trial (1945) Trial of Otto Sandrock and three others, British Military Court for the Trial of War Criminals 1 UNWCC 35
Banda Decision on Charges Corrigendum (2011) The Prosecutor v. Banda and Jerbo, corrigendum of the decision on the confirmation of charges ICC02/05-03/09-121-Conf-Corr
Bemba, Decision Adjourning the Hearing (2009) The Prosecutor v. Jean-Pierre Bemba Gombo, decision adjourning the hearing pursuant to article 61(7)(c)(ii) of the Rome Statute ICC-01/05-01/08-388
Bemba Decision on Charges (2009) The Prosecutor v. Jean-Pierre Bemba Gombo, decision on the confirmation of charges ICC-01/05-01/08
Bemba Judgment (2016) The Prosecutor v. Jean-Pierre Bemba Gombo, judgment ICC-01/05-01/08
Blaškić Appeal (2004) The Prosecutor v. Tihomir Blaškić, Appeal Judgment IT-95-14-A
Blaškić Trial (2000) The Prosecutor v. Tihomir Blaškić, Trial Judgment IT-95-14/1-T
Blé Goudé and Gbagbo Joinder Decision (2015) Decision on prosecution requests to join the cases of the Prosecutor v. Laurent Gbagbo and the Prosecutor v. Charles Blé Goudé and related matters ICC-02/11-01/11 and ICC-02/11-02/11
Blé Goudé, Decision on the Confirmation of Charges (2014) The Prosecutor v. Charles Blé Goudé, decision on the confirmation of charges ICC-02/11-02/11
Brđanin Appeal Judgment (2007) The Prosecutor v. Brđanin and Talić, Appeal Judgment IT-99-36
Brđanin Decision on Amended Indictment (2001) Decision on form of further amended indictment and prosecution application to amend, Brđanin and Talić IT-99-36
Case 002, Pre-Trial Decision on JCE (2010) Ieng Thirith, Ieng Sary and Kheiu Samphan, decision on the appeals against the co-investigating judges' order on joint criminal enterprise 002/19-09-2007-ECCC/TC

Case 002, Trial Decision on JCE (2011) Nuon Chea, Ieng Thirith, Ieng Sary and Kheiu Samphan, decision on the application of joint criminal enterprise 002/19-09-2007-ECCC/TC
Case 002/01, Judgment (2014) Nuon Chea, Ieng Thirith, Ieng Sary and Kheiu Samphan, judgment IT-96-21-A
Čelebići Appeal Judgment (2001) The Prosecutor v. Mucić et al, Appeal Judgment
Dachau Case (1945) The Dachau Concentration Camp Trial, Trial of Martin Gottfried Weiss and thirty-nine others, XVI Trials of War Criminals Before the Nuremberg Military Tribunals Under Control Council Law No 10 5. General Military Government Court of the United States Zone, Germany
Decision on Disqualification of Judge Frederik Harhoff (2013) The Prosecutor v. Vojislav Šešelj, decision on defence motion for disqualification of Judge Frederik Harhoff and report to the vice-president, IT-03-67-T
Delić Trial Judgment (2008) The Prosecutor v. Rasim Delić, Trial Judgment IT-04-83-T
Duch, Appeal Judgment (2012) Kaing Guek Eav alias Duch, Appeal Judgment 001/18-07-2007/ECCC/SC
Duch, Trial Judgment (2010) Kaing Guek Eav alias Duch, Trial Judgment 001/18-07-2007/ECCC/TC
Einsatzgruppen Case (1951) The United States of America v. Otto Ohlendorf et al, 4 Trials of War Criminals Before the Nuremberg Military Tribunals Under Control Council Law No 10 3 (Military Tribunal of the United States at Nuremberg, Military Tribunal IV)
Farben (1952) United States v. Krauch et al, 8 Trials of War Criminals Before the Nuremberg Military Tribunals Under Control Council Law No 10 (Military Tribunal of the United States at Nuremberg, Military Tribunal IV)
Furundžija (1998) The Prosecutor v. Furundžija, Trial Judgment IT-95-17/1-T
Gacumbitsi Appeal (2006) The Prosecutor v. Sylvestre Gacumbitsi, Appeal Judgment ICTR-2001-64-A
Gacumbitsi Appeal, Schomburg Separate Opinion (2006) The Prosecutor v. Sylvestre Gacumbitsi, Appeal Judgment, separate opinion of Judge Schomburg ICTR-2001-64-A
Gbagbo, Decision on the Confirmation of Charges (2014) The Prosecutor v. Laurent Gbagbo, decision on the confirmation of charges ICC-02/11-01/11
Hadžihasanović Appeal Judgment (2008) The Prosecutor v. Hadžihasanović, Appeal Judgment IT-01-47-A
Halilović Appeal Judgment (2007) The Prosecutor v. Halilović, Appeal Judgment IT-01-48
Hamdan (2006) Hamdan v. Rumsfeld 126 S.Ct. 2749 (2006) (Supreme Court of the United States)
Hamdan v Rumsfeld Amicus Brief (2006-01-06) *Amicus Curiae* brief of specialists in conspiracy and international law in support of petitioner (conspiracy not a triable offence), Hamdan v. Rumsfeld 126 S.Ct. 2749 (2006) 548 U.S. 557
High Command Case (1948) USA v. Wilhelm von Leeb et al, Military Tribunal of the United States at Nuremberg
Hoelzer (1946) Hoelzer et al, D2474 RCAF Binder 181.009 (Canadian Military Court)
Hostages Case (1948) USA v. Wilhelm List et al, Military Tribunal of the United States at Nuremberg
Iang Sary, Motion to Disqualify Cassese (2007) Iang Sary's motion to disqualify Professor Antonio Cassese and selected members of the board of editors and editorial committee of the Journal of International Criminal Justice from submitting a written amicus curiae brief on the issue of joint criminal enterprise, D99/3118-ECCC/PTC
Jepsen (1946) Trial of Jepsen and others, Proceedings of a War Crimes Tribunal at Luneberg, Germany
Justice Case (1947) United States v. Josef Altstoetter et al Military Tribunal of the United States at Nuremberg
Kambanda (1998) The Prosecutor v. Jean Kambanda ICTR-97-23
Kamuhanda (2004) The Prosecutor v. Jean de Dieu Kamuhanda, Trial Judgment ICTR-95-54A-T
Karemera (2012) The Prosecutor v. Édouard Karemera and Matthieu Ngirumpatse, Trial Judgment ICTR-98-44-I

Katanga Confirmation of Charges (2008) The Prosecutor v Germain Katanga and Mathieu Ngudjolo Chui, decision on the confirmation of charges ICC-01/04-01/07
Katanga Sentencing Judgment (2014) The Prosecutor v. Germain Katanga, décision relatief a la peine ICC-01/04-01/07
Katanga Severance of Charges Decision (2012) The Prosecutor v. Germain Katanga and Mathieu Ngudjolo Chui, decision on the implementation of regulation 55 of the regulations of the court and severing the charges against the accused persons ICC-01/04-01/07
Katanga Trial Judgment (2014) The Prosecutor v. Germain Katanga, Trial Judgment ICC-01/04-01/07
Katanga Warrant of Arrest Decision (2007) The Prosecutor v. Germain Katanga, decision on the evidence and information provided by the prosecution for the issuance of a warrant of arrest ICC-01/04-02/07
Kayishema (1999) The Prosecutor v. Kayishema and Ruzindana, Trial Judgment ICTR-95-1
Kordić Trial Judgment (2001) The Prosecutor v. Dario Kordić and Mario Čerkez, Trial Judgment IT-95-14/T-2
Krnojelac Appeal Judgment (2003) The Prosecutor v. Milorad Krnojelac, Appeal Judgment IT-97-25-A
Krstić Appeal Judgment (2004) The Prosecutor v. Radislav Krstić, Appeal Judgment IT-98-33-A
Krstić Trial Judgment (2001) The Prosecutor v. Radislav Krstić Trial Judgment IT-98-33
Krulewitch (1949) Krulewitch v. United States 366 US 440
Krupp (1948) United States v. Alfried Krupp et al, Trials of War Criminals Before the Nuremberg Military Tribunals Under Control Council Law No 10 (Military Tribunal of the United States at Nuremberg)
Kvočka Trial Judgment (2001) The Prosecutor v. Miroslav Kvočka et al, Trial Judgment IT-98-30-1/T
Lubanga Confirmation of Charges (2007) Prosecutor v. Thomas Lubanga Dyilo, decision on the confirmation of charges ICC-01/01-01/06-803
Lubanga Decision Regarding Witness Proofing (2007) Prosecutor v. Thomas Lubanga Dyilo, decision regarding the practices used to prepare and familiarise witnesses for giving testimony at trial ICC-01/01-01/06
Lubanga Trial Judgment (2012) Prosecutor v. Thomas Lubanga Dyilo ICC-01/01-01/06
Medical case (1946) United States v. Karl Brandt et al, II Trials of War Criminals Before the Nuremberg Military Tribunals Under Control Council Law No 10 (Military Tribunal of the United States at Nuremberg)
Milosević Indictment (2004) The Prosecutor v. Slobodan Milosević, second amended indictment, 'Croatia' IT-02- 54
Milutinović (2009) The Prosecutor v. Milan Milutinović, Judgment IT-05-87
Ministries case (1949) United States v. Ernst von Weizaecker et al, II Trials of War Criminals Before the Nuremberg Military Tribunals Under Control Council Law No 10 (Military Tribunal of the United States at Nuremberg)
Musema (2000) The Prosecutor v. Alfred Musema, Trial Judgment ICTR-96-13
Muthaura Confirmation of Charges (2012) Prosecutor v. Francis Kirimi Muthaura and Uhru Muigai Kenyatta, decision on the confirmation of charges ICC-01/09-02/11
Nahimana (2003) The Prosecutor v. Nahimana, Barayagwiza and Ngeze, Trial Judgment ICTR-99-52-T
Ndindabahizi (2004) The Prosecutor v. Ndindabahizi, Trial Judgment ICTR-2001-74
Ngudjolo Judgment on Appeal (2015) Prosecutor v. Mathieu Ngudjolo Chui, judgment on the prosecutor's appeal against the decision of Trial Chamber II entitled 'Judgment Pursuant to article 74 of the Statute' ICC-01/04-02/12A
Ngudjolo Trial Judgment (2012) Prosecutor v. Mathieu Ngudjolo Chui ICC-01/04-02/12
Ngudjolo Warrant of Arrest Decision (2007) Prosecutor v. Mathieu Ngudjolo Chui, decision on the evidence and information provided by the prosecution for the issuance of a warrant of arrest ICC-01/04-02/07

Ntakirutimana Appeal (2004) The Prosecutor v. Ntakirutimana and Ntakirutimana, Appeals Judgment ICTR-96-10-A
Nuremberg Judgment (1945) Judgment of the International Military Tribunal. Am J f Int L 51:172–333
Nzirorera Decision (2006) Joseph Nzirorera et al v. the Prosecutor, decision on jurisdictional appeals: joint criminal enterprise
Ojdanić Decision on Indirect Co-perpetration (2006) The Prosecutor v. Milan Milutinović and Dragoljub Ojdanić, decision on Dragoljub Ojdanić's motion challenging jurisdiction—indirect co-perpetration IT-05-87-PT
Ojdanić Decision on JCE (2003) The Prosecutor v. Milan Milutinović and Dragoljub Ojdanić, decision on Dragoljub Ojdanić's motion challenging jurisdiction—joint criminal enterprise IT-99-37-AR72
Ongwen Confirmation of Charges (2016) Prosecutor v. Dominic Ongwen, decision on confirmation of charges, ICC-02/04-01/15
Orić Trial Judgment (2006) The Prosecutor v. Naser Orić, Trial Judgment IT-03-68-T
Perišić Appeal Judgment (2013) The Prosecutor v. Momčilo Perišić, Appeal Judgment IT-04-81-A
Perišić Trial Judgment (2011) The Prosecutor v. Momčilo Perišić, Trial Judgment IT-04-81-T
Pinkerton (1946) Pinkerton v. United States 328 US 640 (Supreme Court of the United States)
Pohl (1947) USA v. Pohl et al, Trials of War Criminals Before the Nuremberg Military Tribunals Under Control Council Law No 10 (Military Tribunal of the United States at Nuremberg)
Ponzano case (1946) The Trial of Feurstein and others, Proceedings of a War Crimes Tribunal at Hamburg
R. v Jogee and Ruddock (2016) R. v Jogee and Ruddock 2016 UKSC 8
Rašević, First Instance Verdict (2008) Prosecutor v. Mitar Rašević and Savo Todović IT-98-32-A
Rohde (1946) Trial of Werne Rohde and eight others, V Trials of War Criminals Before the Nuremberg Military Tribunals Under Control Council Law No 10 56 (British Military Court at Wuppertal)
RUF Case, Appeal (2009) The Prosecutor v. Sesay, Kallon and Gbao, Appeal Judgment SCSL-04-15-A
Ruto Confirmation of Charges (2012) Prosecutor v. William Samoei Ruto and Joshua Arap Sang, decision on the confirmation of charges pursuant to article 61(7)(a) and (b) of the Rome Statute ICC-01/09-01/11
Ruto Prosecution Submission on Co-Perpetration (2012) Prosecutor v. William Samoei Ruto and Joshua Arap Sang, Prosecution's submissions on the law of indirect co-perpetration under article 25(3)(a) of the Statute ICC-01/09-01/11
Rwamakuba Decision on JCE (2004) The Prosecutor v. Rwamakuba, decision on interlocutory appeal regarding application of joint criminal enterprise to the crime of genocide ICTR-98-44-AR72.4
Sainović Appeal Judgment (2014) Prosecutor v. Nikola Sainović, Nebojsa Pavkovic, Vladimir Lazarevic and Sreten Lukic, Appeals Judgment IT-05-87-A
Schonfeld (1946) Trial of Franz Schonfeld and others, XI UN War Crimes Commission 68 (British Military Court at Essen)
Semanza (2003) The Prosecutor v. Semanza, Trial Judgment ICTR-97-20-T
Seromba Appeal (2008) The Prosecutor v. Athanase Seromba, Appeal Judgment ICTR-2001-66-A
Šešelj Judgment (2016) The Prosecutor v. Vojislav Šešelj, Judgment IT-03-67-T
Šešelj Judgment, Concurring Opinion of Judge Antonetti (2016) The Prosecutor v. Vojislav Šešelj, concurring opinion of presiding judge Jean-Claude Antonetti attached to the Judgment IT-03-67-T
Simba (2005) The Prosecutor v. Aloys Simba, Trial Judgment ICTR-01-76-T
Simić Trial Judgment (2002) The Prosecutor v. Milan Simić, Trial Judgment IT-95-9/2

Simone Gbagbo Arrest Warrant (2012) Prosecutor v. Simone Gbagbo, warrant of arrest ICC-02/11-01/12
Stakić Appeals Judgment (2006) The Prosecutor v. Milomir Stakić, Appeals Judgment IT-97-24-A
Stakić Trial Judgment (2003) The Prosecutor v. Milomir Stakić, Trial Judgment IT-97-24-T
Tadić Appeals Judgment (1999) The Prosecutor v. Dusko Tadić, Appeals Judgment IT-94-1-A
Taylor Appeals Judgment (2013) The Prosecutor v. Charles Ghankay Taylor, Appeal Judgment SCSL-03-01-A
Taylor Indictment (2003) The Prosecutor v. Charles Ghankay Taylor, indictment SCSL-03-01-PT
Taylor Judgment (2012) The Prosecutor v. Charles Ghankay Taylor SCSL-03-01-T
Taylor Sentencing Judgment (2012) The Prosecutor v. Charles Ghankay Taylor, sentencing judgment SCSL-03-01-T
Vasiljević Appeal (2004) The Prosecutor v. Mitar Vasiljević, Appeals Judgment IT-98-32-A

Part III
Patchworking a Solution to Leadership Liability

Chapter 9
Applying a Comparative Theory: Beyond Legal Transplants, Toward Legal Patchworking

Contents

At the centre of the comparative study undertaken in the previous chapters is the translation of criminal law institutions from the domestic to the international. Although the focus has been to identify which systems and modes of liability are most appropriate to the context of mass atrocity in international prosecutions, the process of translation is important in and of itself. It has a lot to teach about the nature of ICL, and understanding it better can in fact aid in coming to an answer to that question of liability. The comparative theory laid out in Chap. 4 has a twofold aim, to be explored here; first to better understand the processes by which the normative content of ICL is being developed, and second to offer some tools to improve and strengthen this process.

C. Steer, *Translating Guilt*, International Criminal Justice Series 9,
DOI 10.1007/978-94-6265-171-5_9

9.1 Observing the Patchworking Process

The first aim is a more descriptive one, namely to understand the process by which ICL is developed. A formalist or static approach to understanding law, which assumes that there is an answer to every legal problem to be discovered with enough applied reasoning, and which assumes that the doctrine of sources satisfies the question as to where such answers can be found, is insufficient to comprehend the processes at play. Rather a dynamic approach has been applied here, according to which the notions of 'subject' and 'object' of the law, and of formally endowed law-makers, are replaced with some more realist observations. The cast of participants involved in the processes that lead to law creation and application is much larger, and includes not only States and international organisations, but also individual judges, prosecutors, defence lawyers and scholars, all of whom have a very influential role in ICL.

Two main conclusions can be drawn from a descriptive application of a comparative approach to ICL: first, the role of policy in the development, interpretation and application of modes of liability is key; and second, the role of individual participants is determinative in the selection of policies leading to such development, interpretation and application.

With respect to the role of policy, in Sect. 6.3 it was shown that in the US the policy desire was to be able to capture all those involved in mafia-type crimes, by widening the net of liability as far as possible based on the existence of an agreement to take part in a criminal plan.[1] This has led to the unique development of conspiracy as a mode of liability.[2] As shown in Sect. 6.4, although a similar policy desire can be observed with respect to gang-type crimes in Canada, there have also been limits placed on party liability with respect to high stigma crimes such as murder and war crimes. Based on the constitutional limits of 'principles of fundamental justice' it has been determined that an accomplice must have intended the further crimes and outcomes that are a result of a criminal plan, in order to be held liable for them.[3] Objective foreseeability that the further crimes may occur is not enough, contrary to the far-reaching notion of conspiratorial liability in the US.

In Sect. 7.5 it was shown that the policy intentions in Germany and Argentina align with each other, and differ importantly from those in the two countries compared under the common law tradition. In these jurisdictions the policy has been to delineate between parties to collective crimes of atrocity, particularly where these were related to State organs, or to those who had taken control of the State. The explicit policy was to punish the leaders for their control and influence over what their subordinates had done, and to punish them as principals, in order to reflect the high moral blameworthiness attached to their roles. The leaders of the *junta* military regime in Argentina were considered to be the most responsible, even though they had not

[1] Dressler et al. 2001, p. 493.

[2] The quintessential authoritative case for this is Pinkerton 1946. See also People v Luparello 1987; Croy 1985; State v. Nevarez 2005, p. 1158; American Law Institute 1985, p. 307, comment to § 2.06.

[3] Creighton 1993. See also Healy 1993; Paciocco 2014, p. 13; Thomson 1994, p. 400.

given direct orders to commit any specific crime such against any specific victim,[4] and similarly the members of the National Defence Committee of East Germany were held to be liable as the most responsible for creating and enforcing the policy to kill citizens who tried to cross the Berlin Wall.[5] One could argue that this policy desire is attached by definition to a system of liability that differentiates normatively between parties, and is therefore not necessary if there is a choice made to follow a functionally unitary system of liability. While this may be true, the important point is that there was a perceived need to identify the leaders of the atrocities as those most responsible, and to record this finding in a legal sense by classifying their guilt as that of a principal.

When it comes to international prosecutions, the lack of central authority in the international system means that there is no central agreed policy. This is one of the main contributing factors to the continued debate on systems and modes of liability. As was discussed in Chap. 4, the opportunity for a constitutive moment in international criminal justice of considering which policy should prevail has to some extent been missed. From the Nuremberg Trials right up to the ad hoc tribunals that are still being created today, the design of the system as a whole is reactive to each situation of human conflict, and while the debates have arisen, they have not been the resolved with any sense of cohesion or consensus. Instead a process of arbitrary patchworking has taken place. The Rome Statute was more carefully drafted, however this constitutive moment was also a missed opportunity to give proper attention to the policy and theory underpinning a system of liability.

This leads to the second conclusion, that the role of individual participants is determinative in the selection of policies that underly a particular development of leadership liability. In the context of ICL, the participants influence the very choice of legal tradition, and which correlative system of liability is asserted as being applicable. In Chap. 8 the translation or transplantation of these domestic approaches to the international plane was shown to result in a process of patchworking, forming a new body of ICL by piecing together different elements from the domestic systems of liability. Where there has been a majority of judges, lawyers and scholars who are influenced by the common law tradition, a system of liability has prevailed which is functionally unitary and which is predicated on a subjective approach. Debates on the content of this approach have surfaced where judges, lawyers and scholars who are influenced by the civil law tradition have made their voices heard, such as the debate on the difference between JCE and co-perpetration at the ICTY.[6] Conversely, where participants who are influenced by the civil law tradition have had the leading role, such as in some chambers at the ICC, there has been a preference for an objective approach to liability and for a normative differentiation between parties to a crime. At the ICC the dissenting minority voice has been that of participants who prefer the

---

4 Causa 13 Trial 1985.

5 Mauerschützen case 1994.

6 Separate Opinion of Judge Per Lindholm, Simić Trial Judgment 2002, para 2; Stakić Appeals Judgment 2006, para 62; Schomburg 2010.

subjective approach, inevitably influenced by the common law lens.[7] The scholarly debates have followed similar lines.

This patchworking process can be seen as a form of comparative law in action, since the participants have followed the same patterns observed by comparative scholars with respect to legal transplantations between domestic systems: where there are gaps in the law, participants borrow notions from jurisdictions which are the most accessible and the most authoritative.[8] In ICL this has led to the selection by participants asserting a rule or a mode of liability from the jurisdictions which are most familiar to them, and to the greatest influence coming from the US and, more recently, German doctrine.

These two stages of comparative investigation in Part II—of domestic jurisdictions and of the international tribunals—show the processes by which ICL develops, and as such demonstrates the characterisation of law as both process and as policy. The root of the continued debates and lack of clarity on modes of liability applicable for mass atrocity are partly a result of the fact that the patchworking processes have taken place arbitrarily, depending upon the legal identity of the participants asserting one system or another, and without sufficient regard for the requirements and aims of international prosecutions, nor for the policy factors behind the development of the domestic models from which have been drawn.

Given the inevitability of this patchworking process, the most prudent question to ask is not how to change it, but rather how to improve it. Even if there are legitimate objections raised against the volatility of this dynamic process, against the damage it does to a traditional doctrine of sources, or against its legitimacy, it would seem the existence of this process is inescapable. A normative application can offer a useful step forward from these observations.

9.2 Strengthening the Patchworking Process

The contribution that a comparative approach can make to ICL has been argued by some authors, not only as a cognitive tool, but also an instrument of normative integration and of critical analysis.[9] A comparative study can aid in the creation and identification of common norms, a common vocabulary, and a truly pluralist or hybrid system of ICL. Without such an approach the process becomes one of unilateral transplantation of dominant systems.[10] Not only can a comparative perspective on ICL help to understand the processes, and analyse the concepts being debated

---

7 Judge Fulford, Separate Opinion, Lubanga Trial Judgment 2012, para 7; Separate Opinion of Judge Christine van den Wyngaert, Ngudjolo Trial Judgment 2012, para 22. It should be noted that Judge van den Wyngaert hails from the civil law tradition in her home country of Belgium, but her dissent has been fully in line with a subjective approach.

8 Watson 1984, p. 27.

9 Delmas-Marty 2008, p. 251; Delmas-Marty 2003; Koskenniemi 2009, p. 7.

10 Delmas-Marty 2008, p. 251.

at a deeper level, it can also aid in strengthening the very same processes by contributing some methodology that could bolster the predictability and coherence of the processes and the outcomes. International criminal lawyers and scholars must be comparativists in their work, and look beyond that which is familiar to them.[11]

The concern that the process of law involves either arbitrary or biased selectivity of domestic models from which participants draw their claims, is that the foundation of any asserted model of liability becomes fairly shaky, because it is both unpredictable and difficult to prove the customary or general nature of any asserted outcome. However if we embrace the observations made throughout the comparative part of this study, and accept the patchworking processes by which ICL has always been developed as inherent to the hybrid nature of this body of law, then there are possibilities to improve the processes and the strength of the outcomes. This can be achieved by applying the 'rules of the game' of comparative law methodology; since ICL is comparative law in action, let us apply the methods that stem from this methodology, in order to ensure a well informed outcome.

The first of these methods is functionalism.[12] Where participants have failed somewhat in the transplantation or patchworking of domestic models of liability to the context of international mass atrocity prosecutions is in the formulation of the question. By beginning the enquiry into collective criminality and leadership liability with the use of terminology such as 'principal', 'accomplice' and 'accessory' liability, a series of mistakes are made which lead to utter confusion and to debates that talk cross purpose. As has been demonstrated in Sect. 4.3, and throughout the comparative study in Part II, these terms have different connotations and legal meanings in different systems. Thus to pose the question in terms of the search for the 'correct' form of complicity liability for leaders of collective crimes, is to obfuscate important factors of comparison and reduce the chances of a useful and meaningful answer. Instead the question can be posed as the search for modes of liability which fulfil a certain function, namely to what extent individuals can be culpable for another's acts. This can be applied to individuals at all levels within a collective, and the specific question can be posed to what extent a person who has a leadership role within an organised group is culpable for the criminal acts committed by other members of that group. This is a functional equivalence question rather than a technical one.[13]

Such a functional question gives rise to a series of new questions which must be asked in order to resolve this central question, and which can guide in the functional and contextual comparison of systems of liability. Functional equivalence is a method employed to compare the way different societies have answered similar legal needs in different ways.[14] Thus the political and historical factors that have influenced the development of a system of liability in a domestic jurisdiction must be investigated as part of a functional enquiry, including the policy choices which led to its inception and the way in which participants in the processes of law apply it. These questions

---

11 Ambos 2007, p. 429.

12 See Sect. 5.1 for the definition of functionalism as method and not as theory.

13 Örücü 2006, p. 34; Michaels 2006, p. 357.

14 Michaels 2006, p. 357.

have been integrated into the criteria which were set out in Sect. 5.1. Such contextual analysis is also needed in order to know whether translation of a particular legal institution to other jurisdictions is appropriate.[15]

In order to facilitate a functional comparison in these terms, the selection of jurisdictions to be compared must also be given consideration. When comparing the possible systems of liability in terms of the functional role they play, and their relationship to the goals and functions of punishment of ICL, a comparative law method would require a broad and transparent selection of domestic jurisdictions. An arbitrary or selectively biased group of jurisdictions is unsatisfactory, regardless of whether the argument made is that the result is customary law, a general principle, or even a domestic model from which to choose as a source unto itself. In all these cases it is necessary to compare a broad selection representing both the civil law and common law traditions, and, if appropriate, representing the domestic laws of the country where the crimes took place.[16]

Furthermore, when selecting jurisdictions as representative of the common law and civil law traditions, it is important to justify the selection explicitly.[17] One of the critiques of JCE is that the assertion that it has roots in customary law is based on a limited selection of NMT judgments, all of which were decided by military tribunals applying domestic law from their own— exclusively common law—jurisdictions.[18] Although it is true there were no other historical precedents to draw from, the existence of these decisions does not amount to custom, and the justification of this limited selection should have been made transparent. Similarly the dependence on German and Argentine case law and scholars at the ICC requires further justification in order to provide an acceptable basis for the importation of this model to ICL, as does the interpretation given to the limits of aiding and abetting and party liability for further, unplanned crimes committed by other members of a collective.

The role of judges and scholars has become of primary importance in the development and application of ICL norms. These participants therefore need to pay more attention to the comparative work that is taking place when they look to familiar domestic law solutions. While an extensive enquiry such as the one at hand is not a luxury that judges can afford, nor that all scholars can spend time on, at the same time if these participants were to take more effort to make explicit the choices being made, and to justify why they are being made, then this would aid in the predictability and transparency of the process and of the outcomes. It would also aid in bolstering the legitimacy of the outcome: once again, if the judges in the majority in the *Lubanga* and *Katanga* judgments had justified the reasons they were looking to Roxin's theory and to the case law in Germany and Argentina for authority, drawing parallels to the context of mass atrocity in ICL, many of the reasons for dissent and continued debate would disappear. Attention should also be paid to the policy reasons behind

[15] Zweigert and Kötz 1998, p. 38.

[16] van der Wilt 2010, p. 236.

[17] Oderkerk 2001, p. 222; Glenn 2006, p. 422.

[18] See e.g. Zahar and Sluiter 2008, p. 94; Danner and Martinez 2005, p. 112; Steer 2012, p. 150.

the domestic models being compared, and the policy desires in ICL, as some of the factors informing the final outcome.

9.3 Reversing the Traditional Doctrine of Sources: How the Subsidiary Sources Have Become Primary

Given the scope of participants taking part in the formation, application and appraisal of law that shows up in the comparisons made in Part II, and given the dynamism of the patchworking process at hand in ICL, the question arises what these processes mean for an international law doctrine of sources applied to ICL. If judges, prosecutors, defence lawyers and scholars are the most influential participants when it comes to developing the normative content of ICL, such as modes of liability, how does this impact on the understanding of different sources of law? In Chap. 4 it was asserted that a dynamic, comparative theory of law is needed in order to realistically capture the nature of ICL. The question now is whether this theory can be reconciled with the formal sources of international law and ICL.

Article 38 of the Statute of the ICJ is the starting point of the doctrine of sources, but it is in fact procedural in nature, addressing the Court in terms of applicable law, and was not necessarily intended to be a didactic or prescriptive list of sources beyond the jurisdiction of the ICJ. Still it is generally considered to be a statement of the sources of international law and the basis for a doctrine of sources, and has been referred to as such directly by the ICTY in respect of ICL.[19] Although generally there is agreement that this article does not designate a hierarchy of sources,[20] in any case it clearly states that 'judicial decisions and the teachings of the most highly qualified publicists of the various nations' may be resorted to only 'as subsidiary means for the determination of rules of law', with explicit reference to the fact that previous judgements of the ICJ are only binding *intra partes*.[21] Thus previous judgments of the ICJ are not even considered binding sources, but merely instructional.

Neither the ICTY nor the ICTR statutes make any reference to applicable law at all. Many hybrid tribunals refer to domestic law,[22] with further reference to the breach of IHL treaties when defining the crimes.[23] But none of them deal with sources or applicable law in a general sense, in terms of how to settle an uncertain question of law. However they all make direct or implicit reference to the applicability of treaties,

[19] Joint Separate Opinion of Judges McDonald and Vohra, Erdemović, Appeal Judgment 1999, para 40; Kupreškić Trial Judgment 2001, p. 540; Schabas 2010, p. 382.

[20] Cryer 2009, p. 394; Schabas 2010, p. 382.

[21] Statute of the ICJ 1945, Articles 38 and 59.

[22] Statute of the ECCC 2004, Article 1; Statute of the SCSL 2000, Article 1; Statute of the STL 2007, Article 24.

[23] Statute of the ECCC 2004, Article 33; Statute of the SCSL 2000, Article 1; Statute of the STL 2007, Article 1.

custom, general principles of international law, and domestic law as sources.[24] The conclusion may be drawn that beyond the definition of crimes, there has been a reliance on a general doctrine of sources of international law following the model of Article 38 of the Statue of the ICJ.

The Rome Statute is the first international criminal tribunal statute to dedicate attention to applicable law, in Article 21.[25] Although it differs in some important ways from Article 38 of the ICJ Statute, for instance in the respect that it determines a hierarchy between the sources in no uncertain terms, it reflects a similar approach to a formal doctrine of sources of international law, though tailored more to the criminal trial.

Each of the applicable sources will be critically treated here in the order of their ranking under Article 21 of the Rome Statute, with reference to their place and definition under Article 38 of the Statute of the ICL.

9.3.1 The Rome Statute as an Internally Referencing Source

In the first place, Article 21(1)(a) of the Rome Statute states that the ICC is to apply the Statute, the Elements of the Crimes, and the Rules of Procedure and Evidence. Despite the fact that Article 9(3) states that the Elements are not binding, but 'shall assist the court', this primacy allotted to the additional instruments was perhaps an attempt to minimise unwanted judicial freedom.[26] The very inclusion of Article 9, requiring the Elements of the Crimes to be included as an instrument, and the primary reference to them in Article 21, have both been criticised as a concession made by the drafters of the Rome Statute at the insistence of the delegation of the US, which had a particular concern about unfettered judicial discretion when it came to the extent of the jurisdiction of the court.[27] This concern has to do with the fact that the ICC was the first international tribunal that would have jurisdiction over the nationals of the very States designing it, rendering it 'unsafe' and therefore leading States to demand more restrictions on judicial discretion.[28]

But this express hierarchy of sources fails to recognise the inter-connectedness of the different sources, since in order to interpret the Statute, it will be necessary to turn to other sources. Furthermore the Elements will only aid in the definitions of the crimes, but not in any other interpretative question such as modes of liability.[29]

---

[24] Statute of the ICTY 1993, Article 2; Statute of the ICTY 1994, Articles 4, 27; Statute of the STL 2007, Article 3; Also Rules of Procedure and Evidence of the ICTY, Rule 89(B); Rules of Procedure and Evidence of the ICTR, Rule 89(B); Rules of Procedure and Evidence of the SCSL, Rule 89(B).

[25] Statute of the ICC 1998, Article 21; Schabas 2010, p. 382; Heller 2011, p. 597.

[26] Pellet 2002, p. 1052; Cryer 2009, p. 392; Heller 2011, p. 598.

[27] Pellet 2002, p. 1059; de Guzman 2008, p. 705; Heller 2011, p. 598.

[28] Cryer 2001, p. 30.

[29] Cryer 2009, p. 393.

9.3.2 Treaties as Primary Sources?

In the second place, Article 21(1)(b) of the Rome Statute states that the ICC is to apply 'where appropriate, applicable treaties'. The word 'applicable' was chosen in favour of 'relevant' during the final drafting, which raises the question whether there was an intention to limit treaties to be considered only to those that are specifically referenced in the Rome Statute, such as the 1949 Geneva Conventions, or whether other relevant public international law treaties should be included, such as the International Covenant on Civil and Political Rights.[30]

Treaty law falls under section (a) of Article 38 of the Statute of the ICJ, which may or may not place it hierarchically above other sources, but in any case this is considered to be the starting point for clarifying international law. However when it comes to the normative content of ICL this begs the question, since treaties covering IHL lack the necessary detail on notions such as modes of responsibility or defences. The tribunal judges are therefore left to come up with their own resolutions to these questions,[31] and thus they often turn to the notions familiar to them from their domestic law backgrounds, or to the expertise of other participants to aid them. However there is a risk of arbitrariness and selectivity in this, which could undermine the legitimacy of international trials.[32] This is one of the reasons why a deliberate comparative approach is important to improving the inherent process of patchworking.

9.3.3 The Limits of Custom and Principles of International Law

Beyond the applicable treaties, Article 21(1)(b) also mentions 'principles and rules of international law', but does not use the term 'customary law' explicitly. The question as to why this wording was chosen, and why 'the established principles of the international law of armed conflict' are added as if they would not fall under the 'principles and rules of international law' is unanswered by the *traveaux preparatoires*. It has been suggested that this was a result of the fact that criminal law lawyers involved in the drafting were concerned that the term 'customary law' would lead to a breach of *nullum crimen*.[33]

This is exactly where the principle of legality as an important principle of criminal law comes into tension with the dynamism of international law. The normative content of international law, including ICL, is in a constant state of flux and change, especially those norms which fall under 'customary law'. To expect the definitive-

---

30 de Guzman 2008, p. 705; Heller 2011, p. 598.

31 See e.g. Report of the Secretary General on the ICTY 1993, para 58, where it is stated that 'the tribunal will have to decide on various personal defences'.

32 Cryer 2005, p. 239; Danner and Martinez 2005, p. 138.

33 Report of ILC Working Group 1993, p. 111; Pellet 2002, p. 1070; Schabas 2010, p. 384.

ness of the principle of legality to match this is to attempt to freeze definitions which are in a process of evolution.[34] While it is indeed a requirement that the definitions of the crimes that fall under the jurisdiction of any criminal tribunal are as clear as possible, as stated in Article 11(2) of the Universal Declaration of Human Rights, and while *nullum crimen* even appears in the Rome Statute as a general principle under Article 22, at the same time there are several caveats in the Rome Statute to allow for developments in international law beyond the scope of the statute.[35] Furthermore, as has been argued in Sect. 3.2, the principle of *nullum crimen* may not strictly be applicable to modes of liability, since these differ significantly in the domestic jurisdictions from which they hail, and since they do not require the same clear notice to individuals about what kind of behaviour is criminalised, as the definitions of the crimes do.[36]

In other tribunals, where there are gaps in the treaty law, or uncertainties such as those surrounding modes of liability, there has been an apparent preference to refer to customary law as the predominant source, in order to remain faithful to the traditional doctrine of sources. The ICTY Appeals Chamber has stated that 'any time the Statute does not regulate a specific matter ...it falls to the International Tribunal to draw upon ...rules of customary international law.'[37] Indeed many judges have justified the existence and applicability of JCE as being grounded in customary law. However while there may be custom surrounding definitions of crimes and other notions in IHL, the existence of custom is highly debatable in terms of modes of liability, due to the differences between States when it comes to domestic practice, and due especially to the lack of any *opino juris*; States do not develop their criminal law systems based on the notion that there is an international obligation to do so in a certain way. Moreover, the very case law which the ad-hoc tribunals have drawn upon, in the form of post-Second World War trials, were all predicated on domestic law according to the jurisdictions under which they fell. These tribunals had nothing more to draw on at the time, but it does not mean these decisions automatically gain the status of customary law simply because they were the first decisions of their kind.

It is difficult to see how judicial decisions on the content of criminal law norms could be considered to fulfil either of the definitional requirements of customary law, that there must be a combination of (i) (sufficient) State practice demonstrating the content of a rule, and (ii) *opino juris*, or the belief held by these States that they are acting in such a way as to conform to a legal obligation.[38] A domestic judge cannot be said to be announcing State practice when she is interpreting the criminal law of her State. Nor can a collection of domestic judicial decisions amount to *opinio juris* that there is a legal obligation to apply criminal law in a particular way, given that there will be different decisions depending upon the different laws applicable in each jurisdiction.

[34] Pellet 2002, p. 1056.

[35] Statute of the ICC 1998, Articles 10, 22(3), 80.

[36] Swart 2009, p. 93; van der Wilt 2010, p. 217.

[37] Kupreškić Trial Judgment 2001, para 591.

[38] North Sea Continental Shelf Case 1969, para 77; Nicaragua 1986, para 184.

The problem of veiling new developments in terms of custom was apparent even with the establishment of the IMT at Nuremberg, where individual criminal responsibility was applied for the first time at the international level, and where 'crimes against humanity' was a newly defined crime.[39] There was no customary law whatsoever to cover many of the new developments made; what was present was rather the political will to make this move in international law. The policy response to the atrocities of the Second World War, though not consensual among all participating States, was eventually agreed upon, and the judges were left to interpret and create new law. Subsequently the new case law that developed led to the formulation of the Nuremberg Principles. Whether the formation of this political will amounts to custom is the question. When it comes to the normative detail of ICL, it is even more difficult to assert that a mode of liability or a defence is 'custom' when these notions differ from system to system and from tradition to tradition. Indeed, the *Furundzija* Trial Chamber at the ICTY itself pointed out that 'account must be taken of the specificity of international criminal proceedings when utilising national law notions', due to the unique traits of international proceedings and the differences in national systems.[40]

Some judges of the ICJ, whose case law is referred to as evidence or clarification on the existence of customary rules, consider their decisions to be declaratory rather than constitutive of a customary rule.[41] But more often we have evidence that the ICTY and ICTR have divined for themselves a power of creating customary law, simply self-referencing their own case law as well as endowing the case law of the post-Second World War tribunals with customary status. However this does not make it true that their decisions have magically become customary law. Although the creation of custom may include participants other than just States,[42] and although it is a continuous process over time, it cannot be so that judicial decisions which establish new *actus reus* or *mens rea* requirements for a crime, or for modes of liability, create immediate new custom, given that there is continuous dissent and debate among judges on these very points.[43]

Furthermore, when asserting that a rule or a mode of liability is customary law, language barriers, inaccessibility of materials, and lack of understanding of the legal system from whence a particular law or decision hails, all lead to a highly selective overview of state practice. This mirrors the explanation given by comparativist Alan Watson, that when seeking to fill a gap in law, decision-makers will select those systems that are both authoritative and accessible.[44] The result has been criticised as

---

[39] Simpson 2007, p. 19; Lafontaine 2012, p. 29.

[40] Furundžija 1998, para 175.

[41] Degan 2005, p. 75.

[42] D'Amato 2011, p. 152; Separate Opinion of Judge Shahabuddeen, Stakić Trial Judgment 2003, paras 39–40; Simma and Alston 1999, p. 99.

[43] See for example the lengthy debates and shifts on the elements of rape as an international crime, Akayesu 1998, para 688; Furundžija 1998, paras 162, 185; Kvočka Trial Judgment (2001), para 178. See also the discussions on the requirements of JCE, discussed at length in Sect. 8.6. See further Degan 2005, p. 58.

[44] Watson 1984, p. 27.

a 'survey style' of deriving a customary norm,[45] which is dubious both in terms of process and in terms of outcome, given that such a selection can never be representative of a custom upheld by a sufficiently broad number of States, when numerous other States maintain different approaches.

9.3.4 The Relationship Between General Principles and Domestic Law

Article 21(1)(c) mentions in the third place 'general principles of law derived by the Court from national laws of legal systems of the world'.[46] These general principles of law which are reflected in domestic law are not the same as general principles of *international* law, which fall under Article 21(1)(b).[47] In this sense Article 21 bears resemblance again with Article 38 of the ICJ Statute, which mentions general principles in the third place, though the Rome Statute has omitted the dated definition that these must be 'recognised by civilised nations'.[48] General principles are also an acknowledged source in some other treaties in terms of recognising international crimes.[49] Similarly, the ICTY was left to decide which defences would be applicable according to 'general principles of law recognised by all nations.'[50]

The wording of Article 21(1)(c) makes clear that these general principles are subsidiary (they are applicable 'failing' the applicability of all other sources mentioned prior in the article), but at the same time it is a clear recognition of the importance of domestic criminal law in helping to clarify and develop ICL; in fact, Schabas has suggested this part of the article can be considered an invitation to make use of comparative criminal law.[51] This would seem in line with the Working Group report made during drafting at the Rome Conference that 'the Court ought to derive its principles from a general survey of legal systems and their respective national laws.'[52] However it may also be the case that the wording of this part of the article was a compromise between those State parties who believed that interpretative questions on criminal law matters should be answered with reference to domestic laws, by including these as part of the definition of general principles, and those who wanted

[45] Zahar and Sluiter 2008, p. 94.

[46] Statute of the ICC 1998, Article 21(1)(c).

[47] Pellet 2002, p. 1073; But compare Heller 2011, p. 598, who argues they are the same.

[48] Statute of the ICJ 1945, Article 38; Pellet 2002, p. 1073.

[49] ECHR 1950, Article 7 (2), which states 'This Article shall not prejudice the trial and punishment of a person for any act or omission which at the time when it was committed, was criminal according to the general principles of law recognized by civilised nations'; and the nearly identical ICCPR 1985, Article 15(2), which uses the term 'community of nations' instead of 'civilised nations'.

[50] Report of the Secretary General on the ICTY 1993, para 58.

[51] Schabas 2010, p. 393.

[52] Preparatory Commission Working Group on General Principles 1998, p. 2.

to sever ICL from domestic law as much as possible, by giving them only subsidiary status in the article.[53]

Article 21 also invites specific application of the domestic law of the State that would normally have jurisdiction over the crime at issue, but only 'where appropriate'.[54] The question remains, of course, when it is appropriate to do so. Although this would seem to open the door to more explicit dialogue or dialectic between domestic and international criminal law, it is unclear what situations this would apply to and the *traveaux* offer no further clarity.[55] It is likely yet another compromise that was made during the drafting, which moved from the extremely broad and explicit 1994 draft which allowed 'to the extent applicable, any rule of national law' as a subsidiary source, to this more restricted attempt to include relevant domestic laws.[56]

Ironically, despite the importance of domestic law as a source of criminal law notions, and despite the direct importation of the German doctrine on indirect perpetration by judges at the ICC, the Court has been reluctant to explicitly recognise domestic law as a source, and has rather stated repeatedly that it is not bound by national law.[57] In one example, the Appeals Chamber, with two civil law trained judges and Judge Sir Adrian Fulford as the only common law trained judge, rejected arguments regarding witness proofing practices because although it had been shown these practices exited in two common law jurisdictions, it did not exist in civil law jurisdictions and could therefore not be considered a general principle.[58] The majority civil law domestic law background of the judges on this bench may once again have influenced this finding. This demonstrates the difficulty of clarifying or interpreting the criminal law in ICL based on 'general principles', for the same reason that customary law will offer very little solace; the differences between domestic law traditions differ on exactly these points.

9.3.5 *The Role of Judicial Decisions*

If domestic laws and judicial decisions are not custom, and are not 'general' enough in nature to be referred to as general principles, what are they? From the observations made of the international tribunals in Chap. 8, it is apparent that trials of the post-Second World War era were all based on domestic law, and that these were in turn the primary sources drawn upon by both the ICTY and ICTR. The ICC has chosen not to follow trends emerging from these tribunals, but it too has drawn on domestic case law to support its findings.

---

[53] Heller 2011, p. 599; Pellet 2002, p. 1075.

[54] Statute of the ICC 1998, Article 21(1)(c).

[55] ILC Draft Code of Crimes 1996, p. 51; Schabas 2010, p. 393; van der Wilt 2010, p. 216.

[56] ILC Draft Code of Crimes 1996, p. 51; Pellet 2002, p. 1074.

[57] Lubanga Confirmation of Charges 2007, para 69; Katanga Confirmation of Charges 2008, para 91.

[58] Lubanga Decision Regarding Witness Proofing 2007, para 41.

A more realistic reading of the role played by case law, whether hailing from international tribunals or from domestic courts, is that it has become an extremely important source of law unto itself. This is particularly necessary in ICL, given that there are many technical criminal law notions which have been left up to the judiciary to refine, interpret and apply as law makers. The question is what place this source can take in a traditional doctrine of sources. In Article 38 of the ICJ Statute they are designated as subsidiary sources, and the same could be read into the wording of Article 21(2) of the Rome Statute, which states *after* listing the other sources in subsection 1, that 'the Court may apply principles and rules of law as interpreted in its previous decisions'. It is possible that this was deemed a necessary addition precisely because there is no rule of *stare decisis*, and the case law cannot be assumed to be binding beyond the case at issue.[59] Because there is also no hierarchical relationship between international tribunals, this may further suggest that the case law of an international tribunal is not binding on any other tribunal, but since so may pertinent questions on the normative content of ICL is left up to the courts, the case law is an extremely important source unto itself.

Even if they are not directly binding, judicial decisions have an important role to play in the comparative law development of ICL, which is why a dynamic theory of law as process can help in understanding how they operate as sources outside a formalist definition of sources.

9.3.6 Scholars as Participants and Doctrine as a Source

With respect to the role of doctrine as a source, and of scholars as authoritative participants in these processes, this is one noteworthy divergence between Article 21 of the Rome Statute and Article 38 of the ICJ Statute, since the latter acknowledges 'the opinions of highly qualified publicists' as a subsidiary source alongside case law, but the former makes no mention of this at all. It is apparent from the description of the process of law-making in ICL thus far that the content of the law is not an *a priori*. Moreover it is subject to the coincidence of the individual participants involved in a given case, including the authority given to academic commentators. The importance of doctrine as an essential contribution to the development of criminal law in the civil law tradition, and the increasing importance of this in the common law tradition when it comes to interpreting ICL in domestic courts, renders scholars particularly authoritative participants.[60] Similarly, in ICL, doctrine formed by well-cited academics has become an important interpretative tool, but also a source unto itself in many decisions handed down by various tribunals, as is clear from observations made in Chap. 8.

Furthermore, in the text books which teach this relatively new branch of international law, these same scholars are the ones shaping the legal education and under-

---

59 Schabas 2010, p. 394.

60 Fletcher 2007, p. 93; Bohlander 2009, p. 9.

standing of the form and content of ICL. The reification and self-referencing that has taken place with respect to the development of modes of liability demonstrates this particularly vividly.[61] Given the authority bestowed upon certain individuals such as the Secretary General of the UN or the head of the International Committee for the Red Cross (ICRC),[62] statements and publications by these participants are also given considerable weight. The assumption is that these individuals are both neutral and well advised by other individual participants,[63] thus is it noteworthy that the content of the assertions made by these participants is heavily informed by scholars, whose role in developing ICL is becoming extremely influential.

9.3.7 A Different Hierarchy of Sources

It has been pointed out that it is 'the very ad-hocness of international tribunals (the fact that they post-date the alleged crimes) that places them at a disadvantage in relation to sources of law.'[64] With respect to human rights, many scholars have argued that the ability to create custom should be accorded to international organisations and 'certain non-governmental organisations [that] have a distinct, measurable impact on international affairs.'[65] It would appear that with respect to ICL there is a similar need to recognize sources other than those created solely by States.

What does this mean for the doctrine of sources as applied to ICL? We can safely say that if a hierarchy of sources exists in general public international law, ICL does not follow the same hierarchy. This is not problematic, but is of importance in protecting the due process rights of defendants in this ever shifting field. Despite the wording of Article 21 of the Rome Statute, which appears to have been drafted based on a distrust of judges on the part of some States,[66] when it comes to interpreting any unclear provisions or gaps in the law which cannot be answered by referring to treaties, the next most important sources are judicial decisions of both domestic and international tribunals, and authoritative scholarly opinions. These must be interpreted in accordance with the general principles such as *nullum crimen*, and, according to Article 21(3) of the Rome Statute, in accordance with human rights. Customary law therefore inevitably takes last place, at least until this branch of law

[61] See e.g. Cassese 2004, p. 592, 2008; Hamdan v Rumsfeld Amicus Brief 2006-01-06, para 19 Iang Sary, Motion to Disqualify Cassese 2007. For a deeper discussion on this reification process, see Steer 2011, p. 303.

[62] For example the authors of the ICRC study on customary law in IHL, Henckaerts and Doswald-Beck 2005.

[63] Zahar and Sluiter 2008, p. 83.

[64] Zahar and Sluiter 2008, p. 80.

[65] Gunning 1991, p. 213; Simma and Alston 1999, p. 99.

[66] Pellet 2002, p. 1057; van der Wilt 2010, p. 216; Cryer 2009, p. 392.

matures to a level where its content is far less disputed with respect to notions such as defences and modes of liability. But even then it is drawn from domestic criminal law notions, and is likely to keep developing and shifting in terms of normative content. Custom can only ever operate as a snapshot in time of what the law is understood to be.

The dynamic, comparative law perspective of ICL helps to understand that Articles 38 of the ICJ Statute and 21 of the Rome Statute, although they are constitutive documents, are incomplete when it comes to a realistic picture of the actual sources applied by judges. A rule or norm may fall under one of the formal sources enumerated under this article, yet it may have ceased to function as law if it is no longer applied as such.[67] Whatever status JCE has as part of the case law of the ad-hoc tribunals, it may cease to be applicable outside of these jurisdictions given that the ICC has rejected it altogether, and if other tribunals then follow the ICC's trend. What is needed is an empirical approach, where assertions are tested case by case as to whether they have become—or remain—prescriptions through acceptance and repeated application by judges and other participants.[68] This allows a more sophisticated knowledge of what actually belongs to the corpus of international law *at any given moment*.[69]

ICL is casuistic by definition, which means that case law from domestic, international and hybrid courts will become increasingly important as a source for clarifying notions in ICL. But does this mean case law prevails as a source? Anthony D'Amato argues that in public international law in general, any case decisions from the ICJ may be authoritative and objective validators of a norm in international law, since this world court has an international de facto status of recognition, but that domestic courts are only contributing to international law through custom, since they have no international acknowledgment.[70] And yet in ICL there seems to be a feed-back process at hand, whereby an international tribunal such as the ICTY looks to domestic case law to derive the source of a notion such as JCE, following which domestic judges refer to the case law of this ad hoc tribunal as a source of law. Domestic case law therefore plays an extremely important part in the process of decision-making and lawmaking by judges of international tribunals. In fact it is argued here that the so-called subsidiary sources, the case law and the scholarly commentary upon this case law, have become the most important sources of law in ICL. This is why a dynamic perspective has been applied throughout this study; in order to understand the place of those sources which matter most in terms of the development of the technical criminal law content of ICL as a process of patchworking.

---

67 Reisman 1981, p. 119.

68 Reisman 1981, p. 118; Higgins 1994, p. 24; Steer 2012, p. 161.

69 Reisman 1981, p. 120.

70 D'Amato 2004, p. 24.

9.4 Are Modes of Liability in International Criminal Law *Sui Generis*?

Given the patchworking process that is evident in the international tribunals, the question arises whether modes of liability in ICL are unique when compared to domestic criminal law solutions to collective criminality. Certainly the ICC has said that command responsibility is *sui generis*.[71] But are modes of liability in general *sui generis*, or should they be considered to be so, developing into their own body of law distinct from the domestic traditions which contribute to their content?

Within any given jurisdiction, the choice to differentiate between participants or not, and which modes of liability to apply to the leaders of mass atrocity crimes, is driven by the greater context of that jurisdiction and the legal tradition which it follows. In other words, as has been made clear here, it is not a matter of choosing to follow one domestic tradition over another, but rather of choosing the doctrinal and policy approach which matches the specificity of mass atrocity.

In this respect, Fletcher has posed the question whether there can be an effective ICL without a universal set of principles and norms, or a universal *Dogmatik*.[72] This term should not be confused with the English word 'dogma', which tends to have a somewhat negative connotation that it represents a set of principles or values which are incontrovertibly true, as the basis for an ideology, religion or belief system, and which cannot be changed without undermining the very paradigm it supports. While a *Dogmatik* does indeed strive for the expression of principles and values, it is rather a doctrinal or theoretical approach to a system of law that may be changeable over time according to shifts in societal values with respect to justice.[73] It is not static, and though it seeks to maintain coherence, a legal *Dogmatik* is receptive to policy shifts and allows the development and integration of new notions into the system as a whole. Such a doctrinal or theoretical approach is necessary to underpin an entire system of responsibility and punishment. A conceptual *Dogmatik* encompasses scholarly authority as an important source of law. Presently the debates and dissent that show up in international tribunals and in scholarly publications are evidence of the difficulty of developing a universal *Dogmatik* for ICL.

Fletcher suggests there is a difference between a pro-defendant criminal law, which is predominant in domestic criminal law, and a pro-victim 'anti' *Dogmatik* present in the case law of the ad-hoc tribunals, which secures justice for victims by increasing prosecutorial advantages at trial, sometimes at the cost of defendant-oriented concerns such as the principle of legality.[74] This approach is what has allowed the conspiracy doctrine and a return to vicarious liability to enter the case law of the ICTY via JCE, despite its rejection at Nuremberg and in the case law and general legal opinion since. The policy behind JCE is to widen the net of liability as

---

[71] Bemba Judgment 2016, paras 171, 174.

[72] Fletcher 2011, p. 181.

[73] Fletcher 2011, p. 181.

[74] Fletcher 2011, pp. 185, 187.

much as possible, and the subjective approach has aided this, but there was no deeper *Dogmatik* underpinning it, and little concern for the protection of defendant rights. The Rome Statute, on the other hand, appears to return to a more traditional *Dogmatik*, including more deeply rooted principles of culpability, with a more detailed approach to the modes of liability under Article 25.

The introduction of an explicit system of liability requires more exploration of the doctrine underpinning it. This has opened the door to an interpretation of the statute by judges, guided by authoritative scholars who instruct on the theory behind such a distinction. While this is may be considered judicial creativity, it also allows an approach different from the anti-dogmatik of the ICTY, namely the protection of individual suspects from the demands of the victims who suffer from group massacres.[75] This may allow for greater differentiation and avoidance of the desire to punish an individual for the collective wrongdoing without sufficient attention to the degree of his responsibility.[76] The difficulty is that some participants in the processes of law would argue that this is does not match a strict interpretation of the text of the Statute, precisely because these participants are more reluctant to accept the authority of scholars and do not share the desire for a deeper *Dogmatik*.[77]

It would appear that as the ICC has begun to play a more central role in developing ICL, and has taken certain turns expressly different from the case law and theoretical underpinnings that appear in the ad-hoc tribunals; there may now be a move towards a more centralised system. There is also some suggestion that the Rome Statute is more of a self-contained regime than the statutes of the ad-hoc tribunals, partly because of its internally referencing article on applicable law.[78] At the same time, because of the way in which the growing case law of the ICC is looked to for guidance on the development of ICL, the current divergence and lack of uniformity among the various international tribunals may over time become more harmonised.[79]

Many authors argue that the combination of procedural elements from the common law tradition and the civil law tradition in the Rules of Procedure and Evidence of the ad-hoc tribunals has led to more than hybridity, and to something beyond a mix of adversarial and inquisitorial procedure, to become procedure *sui generis*.[80] Judges on the ICTY and ICTR Appeals Chambers benches have themselves said as much, and have regarded the statutes as *sui generis* international law documents.[81] What makes international criminal procedure *sui generis* is the fact that the translation from domestic to international creates new meaning, new definitions, new procedures and

---

[75] Fletcher 2011, p. 190.

[76] Fletcher 2002.

[77] See e.g. Separate Opinion of Judge Christine van den Wyngaert, Ngudjolo Trial Judgment 2012, para 68.

[78] Schabas 2010, p. 382.

[79] Steer 2015.

[80] Robinson 2000, p. 580; Safferling 2003; Mégret 2005; Drumbl 2007, p. 8; Vasiliev 2009; Mégret 2010.

[81] Joint and Separate Opinion of Judges McDonald and Vohra, Kanyabashi 1999, para 15; Tadić Decision on Protective Measures 1995, p. 10; Joint and Separate Opinion of Judges McDonald and Vohra, Erdemović, Appeal Judgment 1999, para 3.

new structures specific to the tribunals. It could be said that the same is happening with respect to many substantive notions, since they are an amalgamation of definitions and institutions from the civil law tradition, the common law tradition, and IHL. Thus it is of even greater importance that the participants in this *sui generis* context are cognisant of the patchworking processes at hand, and of their duty to strengthen and improve these processes as best they can through a deliberative comparative approach.

References

Ambos K (2007) The structure of international criminal procedure: 'adversarial', 'inquisitorial', or mixed? In: Bohlander M (ed) International criminal justice: a critical analysis of institutions and procedures. Cameron May, London

American Law Institute (1985) Model Penal Code and commentaries: official draft and revised commentaries

Bohlander M (2009) Principles of German criminal law. Hart Publishing, Oxford

Cassese A (2004) The ICTY: a living and vital reality. J Int Crim Justice 2:585–597

Cassese A (2008) International criminal law. Oxford University Press, Oxford

Cryer R (2001) The boundaries of liability in international criminal law, or 'selectivity by stealth'. J Confl Secur Law 6(1):3–31

Cryer R (2005) Prosecuting international crimes. Selectivity and the international criminal law regime. Cambridge University Press, Cambridge

Cryer R (2009) Royalism and the king: Article 21 of the Rome Statute and the politics of sources. New Crim Law Rev 12:390–405

D'Amato A (2004) International law sources: collected papers, vol 3. Martinus Nijhoff, Leiden

D'Amato A (2011) Non-state actors from the perspective of the policy-oriented school: power, law, actors and the view from New Haven. In: d'Aspremont J (ed) Participants in the international legal system. Routledge, London, chapter 4, pp 64–75

Danner AM, Martinez JS (2005) Guilty associations: joint criminal enterprise, command responsibility, and the development of international criminal law. California Law Rev 75–169

Degan VD (2005) On the sources of international criminal law. Chinese J Int Law 4(1):45–83

de Guzman MM (2008) Article 21. In: Trifterer O (ed) Commentary on the Rome Statute of the International Criminal Court: observers' notes, article by article, 2nd edn. Beck-Hart-Nomos, Munchen

Delmas-Marty M (2003) The contribution of comparative law to a pluralist conception of international criminal law. J Int Crim Justice 1(1):13–25

Delmas-Marty M (2008) Reflections on the 'hybridisation' of criminal procedure. In: Jackson J, Langer M, Tillers P (eds) Crime, procedure and evidence in a comparative and international context: Essays in honour of Professor Mirjan Damaška. Hart Publishers, Oxford, pp 251–259

Dressler J, Strong FR, Moritz ME (2001) Understanding criminal law. Lexis Publishers

Drumbl MA (2007) Atrocity, Punishment, and International Law. Cambridge University Press, Cambridge

ECHR (1950) European Convention for the Protection of Human Rights and Fundamental Freedoms

Fletcher GP (2002) The Storrs lectures: liberals and romantics at war; the problem of collective guilt. Yale Law J 111:1499–1573

Fletcher GP (2007) The grammar of criminal law: American, comparative and international, Vol One: Foundations. Oxford University Press, USA, New York

Fletcher GP (2011) New court, old dogmatik. J Int Crim Justice 9:179–190

Glenn HP (2006) Comparative legal families and comparative legal traditions. In: Zimmermann R, Reimann M (eds) Oxford handbook of comparative law, Oxford University Press, pp 421–440
Gunning I (1991) Modernizing customary international law: the challenge of human rights. Virgina J Int Law 31:211
Healy P (1993) The Creighton Quartet: Enigma Variations in a lower key. Crim Rep 23:265
Heller KJ (2011) The Rome Statute of the International Criminal Court. In: Heller KJ, Dubber MD (eds) The handbook of comparative criminal law. Stanford University Press, Stanford, pp 593–634
Henckaerts JM, Doswald-Beck L (2005) International Committee of the Red Cross's study of customary international humanitarian law. Cambridge University Press, Cambridge
Higgins R (1994) Problems and process: International law and how we use it. Oxford University Press, Oxford
ICCPR (1966) International Covenant on Civil and Political Rights
ILC Draft Code of Crimes (1996) Draft Code of Crimes Against the Peace and Security of Mankind. http://untreaty.un.org/ilc/publications/yearbooks/Ybkvolumes(e)/ILC_1996_v2_p2_e.pdf
Koskenniemi M (2009) The case for comparative international law. Finnish Yearbook Int Law 20:1–8
Lafontaine F (2012) Prosecuting genocide. Crimes against humanity and war crimes in Canadian courts. Carswell, Thomson Reuters, Toronto
Mégret F (2005) In defense of hybridity: towards a representational theory of international criminal justice. Cornell Int Law J 38:725
Mégret F (2010) Beyond fairness: understanding the determinants of international criminal procedure. UCLA J Int Foreign Aff 14:37
Michaels R (2006) The functional method of comparative law. In: Reimann M, Zimmerman R (eds) The Oxford handbook of comparative law. Oxford University Press, Oxford
Oderkerk M (2001) The importance of context: selecting legal systems in comparative legal research. Neth Int Law Rev XLVIII:293–219
Örücü E (2006) Methodology of comparative law. In: Smit J (ed) Elgar encyclopedia of comparative law. Edward Elgar Publishing, Cheltenham, UK/Northampton, MA, USA, pp 422–454
Paciocco PJ (2014) Objective mens rea and attenuated subjectivism: guidance from Justice Louise Charron in R. v Beatty. In: Oliver P, Mayeda G (eds) Principles and pragmatism: essays in honour of Louise Charron. Lexis Nexis
Pellet A (2002) Applicable law. In: Cassese A, Gaeta P, Jones MJR (eds) The Rome Statute of the International Criminal Court: A commentary, vol II. Oxford University Press, Oxford
Preparatory Commission Working Group on General Principles (1998) Committee of the Whole, 'Working Group on General Principles of Criminal Law'
Reisman WM (1981) International lawmaking: a process of communication. Am Soc Int Law Proc 101
Report of ILC Working Group (1993) Report of the Working Group on a Draft Statute for an international criminal court
Report of the Secretary General on the ICTY (1993) Report of the Secretary General pursuant to Paragraph 2 of Security Council Resolution 808
Robinson PL (2000) Ensuring fair and expeditious trials at the International Criminal Tribunal for Yugoslavia. Eur J Int Law 11(3):569–589
Safferling C (2003) Towards an international criminal procedure. Oxford University Press, Oxford
Schabas W (2010) The International Criminal Court: A commentary on the Rome Statute. Oxford University Press, Oxford
Schomburg W (2010) Jurisprudence on JCE—revisiting a never-ending story. http://www.cambodiatribunal.org/sites/default/files/resources/ctm_blog_6_1_2010.pdf
Simma B, Alston P (1999) The sources of human rights law: Custom, jus cogens and general principles. In: Australian Yearbook of International Law. Australian National University
Simpson G (2007) Law, War and Crime. Oxford University Press, Oxford

Statute of the ECCC (2004) Law on the establishment of the Extraordinary Chambers in the Courts of Cambodia for the prosecution of crimes committed during the period of Democratic Kampuchea
Statute of the ICC (1998) Statute of the International Criminal Court
Statute of the ICJ (1945) Statute of the International Court of Justice
Statute of the ICTY (1993) Statute of the International Criminal Tribunal for the prosecution of persons responsible for serious violations of international humanitarian law committed in the territory of the former Yugoslavia
Statute of the ICTY (1994) Statute of the International Criminal Tribunal for Rwanda
Statute of the SCSL (2000) Statute of the Special Court for Sierra Leone
Statute of the STL (2007) Statute of the Special Tribunal for Lebanon
Steer C (2011) Non-state actors in international criminal law. In: D'Aspremont J (ed) Participants in the international legal order: Multiple perspectives on non-state actors in international law. Routledge, London, p 96
Steer C (2012) What makes valid law? Shifting modes of responsibility in international criminal law. In: van Voorhout et al JC (ed) Shifting responsibilities in criminal justice: Critical portrayals of the changing role and content of a fragmented globalizing law domain. Eleven International Publishing
Steer C (2015) Legal transplants or legal patchworking? The creation of international criminal law as a pluralistic body of law. In: van Sliedregt E (ed) Harmonization and pluralism in international criminal law. Oxford University Press
Swart B (2009) Modes of international criminal liability. In: Cassese A (ed) The Oxford companion to international criminal justice. Oxford University Press
Thomson N (1994) Fundamental justice, stigma and fault. Toronto Facul Law Rev 52:379–404
Vasiliev S (2009) General rules and principles of international criminal procedure: Definition, legal nature and identification. In: Sluiter G, Vasiliev S (eds) International criminal procedure: Towards a coherent body of law. Cameron May Publishing, London
Watson A (1984) Sources of law, legal change, and ambiguity. University of Pennsylvania Press
van der Wilt H (2010) National law: A small but neat utensil in the toolbox of international criminal tribunals. Int Crim Law Rev 10(2):209–241
Zahar A, Sluiter G (2008) International criminal law: A critical introduction. Oxford University Press, Oxford
Zweigert K, Kötz H (1998) An introduction to comparative law. Oxford University Press, Oxford

Cases

Akayesu (1998) The Prosecutor v. Akayesu, Trial Judgment ICTR-96-4-T
Bemba Judgment (2016) The Prosecutor v. Jean-Pierre Bemba Gombo, Judgment ICC-01/05-01/08
Causa 13 Trial (1985) Sentencía del juicia a las juntas, Cámera Nacional de Apelaciones en lo Criminal y Correccional Federal de Buenos Aires
Creighton (1993) R. v. Creighton 91, 3 SCR 3
Croy (1985) The People v. Croy 41 Cal 3d 1 (California Supreme Court)
Erdemović, Appeal Judgment (1999) The Prosecutor v. Erdemović, Appeal Judgment IT-96-22-A
Furundžija (1998) The Prosecutor v. Furundžija, Trial Judgment IT-95-17/1-T
Hamdan v Rumsfeld Amicus Brief (2006-01-06) Amicus Curiae brief of specialists in conspiracy and international law in support of petitioner (conspiracy not a triable offence), Hamdan v. Rumsfeld 126 S.Ct. 2749 (2006), (2006) 548 US 557
Iang Sary, Motion to Disqualify Cassese (2007) Iang Sary's motion to disqualify Professor Antonio Cassese and selected members of the board of editors and editorial committee of the Journal of International Criminal Justice from submitting a written amicus curiae brief on the issue of joint criminal enterprise D99/3118

Kanyabashi (1999) Joseph Kanyabashi v. the Prosecutor ICTR-96-15-A

Katanga Confirmation of Charges (2008) The Prosecutor v. Germain Katanga and Mathieu Ngudjolo Chui, decision on the confirmation of charges ICC-01/04-01/07

Kupreškić Trial Judgment (2001) The Prosecutor v. Zoran Kupreškić, Trial Judgment IT-95-15/T

Kvočka Trial Judgment (2001) The Prosecutor v. Miroslav Kvočka, Trial Judgment IT-98-30-1/T

Lubanga Confirmation of Charges (2007) Prosecutor v. Thomas Lubanga Dyilo, decision on the confirmation of charges ICC- 01/01-01/06-803

Lubanga Decision Regarding Witness Proofing (2007) Prosecutor v. Thomas Lubanga Dyilo, decision regarding the practices used to prepare and familiarise witnesses for giving testimony at trial ICC-01/01-01/06

Lubanga Trial Judgment (2012) Prosecutor v. Thomas Lubanga Dyilo ICC-01/01-01/06

Mauerschützen case (1994) 5 StR 98/94 (LG Berlin), [1994] NJW 1994 2703

Ngudjolo Trial Judgment (2012) Prosecutor v. Mathieu Ngudjolo Chui ICC-01/04-02/12

Nicaragua (1986) Nicaragua v. United States, ICJ Reports (1986) 98

North Sea Continental Shelf Case (1969) Federal Republic of Germany v. the Netherlands, ICJ Reports (1969) 44

People v Luparello (1987) The People v. Luparello 187 Cal App 3d 410 (California Court of Appeal, 4th District)

Pinkerton (1946) Pinkerton v. United States, 328 US 640 (Supreme Court of the United States)

Simić Trial Judgment (2002) The Prosecutor v. Milan Simić, Trial Judgment IT-95-9/2

Stakić Appeals Judgment (2006) The Prosecutor v. Milomir Stakić, Appeals Judgment IT-97-24-A

Stakić Trial Judgment (2003) The Prosecutor v. Milomir Stakić, Trial Judgment IT-97-24-T

State v. Nevarez (2005) State v. Nevarez 130 P3d 1154 (Idaho Court of Appeals)

Tadić Decision on Protective Measures (1995) The Prosecutor v. Dusko Tadić, decision on the Prosecutor's motion for protective measures for victims and witnesses IT-94-1-A

Chapter 10
Translating the Guilt of Leaders of Mass Atrocity

Contents

The title of this study 'Translating Guilt' refers to the many layers of translation that take place in the search for the most appropriate modes of liability for mass atrocity. There is a translation from the collective to the individual, in terms of guilt and agency. There is a translation from the notion of collective criminality under circumstances of 'every day' deviant criminal behaviour, to the circumstances of mass atrocity, where deviation from moral behaviour becomes the norm. There is a translation of criminal law terminology from languages including German, French and Spanish to English, and even translation between English-language jurisdictions which have different understandings of terminology such as responsibility, liability culpability, and the various forms of complicity expressed as modes of liability. And there is a translation from the domestic to the international; since the international instruments such as treaties, statutes and even customary law offer very little when it comes to more technical questions of substantive criminal law, it is necessary to look to domestic criminal law notions.

C. Steer, *Translating Guilt*, International Criminal Justice Series 9,
DOI 10.1007/978-94-6265-171-5_10

As with any process of translation, the meaning of the original term or phrase will be changed slightly in order to make sense in a new language or context. Certain choices are made by the translator, and the receiver of the translation will interpret this information. The translator must therefore make careful choices and consider the new context into which is being translated. So too with legal transplants and the process of patchworking in ICL. International criminal lawyers must become more sophisticated comparativists in this sense. They must rise above the clashes between legal cultures which insist upon identifying the 'correct' interpretation of liability, and focus rather on the question of the 'most appropriate' system and specific modes of liability. It is argued here that of the possible models to choose from, an objective approach, and a correlative normative differentiation between parties to a crime, is the most appropriate when applied to the specific context of mass atrocity crimes.

In Sect. 5.1, a set of criteria was asserted for testing which system of liability is most appropriate for ICL, based on the dynamic understanding of law as process and policy, and taking into account the tools of a comparative method: (i) Does the mode of liability reflect the collective nature of mass atrocity crimes? (ii) Does it identify an individual responsible within the group even when there is no strict chain of command between parties? (iii) To what extent can the system or mode of liability which is based on domestic systems operate in a different legal context (for instance where procedural rules play a more or less determinative role in the application of substantive law)? (iv) How has a given mode of liability been applied judicially, including what kinds of interpretation it has been given in various social contexts? And (v) are these contexts are akin to that of mass atrocity? These criteria play a role in many of the factors discussed in the conclusions that follow.

10.1 Layers of Translation

In terms of the first layer of translation, from the collective guilt to individual guilt, and from collective agency to individual agency, it has been argued throughout this enquiry that the liberal foundation of criminal law must be maintained, especially the principle of culpability. It is important that an individual should only be punished for that which she has done, or participated in, and not for the actions of others for which she has no responsibility.[1]

The principle of culpability means that modes of liability which extend to 'guilt by association' should be avoided. Collective criminality makes individual criminal liability complex, especially in the context of mass atrocity where it may be difficult to prove who did what to contribute to a specific crime. However the sole fact of intentionally taking part in group violence should not mean that every party in the group becomes responsible for every act committed by another party. Where liability is predicated on a subjective approach, which focuses on the intention or agreement to take part in a criminal enterprise, it leads to a kind of vicarious attribution of the

[1] Hart 1970, p. 39; Ashworth 2009, pp. 23–24; Robinson 2013, p. 118.

acts of all to an individual. In a domestic context this may satisfy the desire to punish for collective crimes, by capturing all parties to a crime under one form of liability regardless of the difficulty of proving who did what. However in the context of mass atrocity it can easily overreach the limits of moral culpability, and can lead to the punishment of an individual for the crimes of an entire society.

In a system predicated on a subjective theory of liability, where a vicarious theory of liability prevails, it would seem that accomplice liability is only 'accessory' in a very limited way; that is, the crime must have bee completed, but the liability of each party is independent of the liability of any other party, and every party is considered to be equally liable as a principal. The term 'accessory' is therefore misleading in these systems.

Where liability is predicated on an objective approach, culpability is translated as blameworthiness for a wrongful act, and not (only) blame for intentional participation in a plan or agreement. Blameworthiness is measured in degrees, based on the test of causal contribution. Rather than attributing the acts of everyone in a collective back to each individual under a vicarious theory of liability, instead liability is based on personal autonomy, and attribution of blame. The measure of blame is determined depending upon the extent to which an individual had sufficient control over the crime, either by physical contribution, or by control over the circumstances and over the physical perpetrators. 'Guilt by association' is thus avoided, and the translation from collective guilt to individual guilt rests upon a closer focus on individual agency. In this sense, an accomplice truly is an accessory, since the liability is accessory to, or derivative of, liability of the principle.

This is where the second layer of translation is important, from the notion of collective criminality under 'normal' circumstances, to the phenomenon of mass atrocity where deviation has become the norm. Individual criminal liability in any liberal system is predicated on rational and moral agency: an individual is punishable for acting contrary to the law where he is blameworthy for having acted against the norm that dictated he should have done otherwise (or for omitting from doing something where he was expected to do so) and where he had the choice and ability to have done otherwise.[2] Where there are exculpatory reasons, he is not punishable. While it is not argued that an individual should be excused for committing acts such as kidnapping, torture, rape, or killing simply because everyone else around him was doing the same, at the same time in the context of mass atrocity the moral universe has been turned upside down.[3] An ideology has been created under which it is expected that individuals will take part in the oppression of a designated group by extreme means, under which rational choice has been reduced, and, as discussed in Chap. 3, the moral choice to refuse to take part may become extremely difficult, lest the dissenter become the next target.

The third layer of translation must therefore take into account these circumstances, together with the notions of moral and rational agency, in order to translate guilt to responsibility and to culpability, and to express these relationships as modes of

---

[2] Fletcher 2002, p. 1543; Drumbl 2005, p. 541; Duff 2007, p. 47.

[3] Semelin 2007, p. 113.

liability. The answer is not to exculpate physical perpetrators altogether from punishment, something more is needed in response to these circumstances. As discussed in Sect. 5.3, guilt is necessarily personal and relational: a person is answerable *for* her own acts, and is answerable *to* someone for them.[4] She is answerable when she is to blame, and she is to blame when she had the ability to weigh up her acts against their consequences both for the potential victim and for herself, and to make a choice. In other words, when she is answerable she is responsible, and when she is capable of feeling remorse for her wrongful actions (whether or not she actually feels remorse), she is capable of being guilty. But this answerability, responsibility and capacity for guilt does not yet amount to liability. The relational aspect of guilt is that an individual is answerable for his criminal acts to the victim, but also to society. Criminal liability is the calling of society, represented by the State or by an international tribunal, to the individual to answer for his acts. He is criminally liable when he cannot offer an adequately exculpatory reason for the acts for which he is responsible.[5]

Within these limits, liability can be determined and expressed as different modes. A system of liability is either subjective or objective in nature, as described in Sect. 5.3 and as evidenced in the comparative chapters in Part II. The subjective approach prevails in the common law tradition, where modes of liability express a functionally unitary system, such that there is no legal difference between a 'principal' and an 'accomplice', and all parties to a crime are considered equally liable. The objective approach prevails in the civil law tradition, where modes of liability express a normative differentiation, such that a principal is considered to carry more moral blame and therefore to be more culpable than an accomplice, who is culpable for assisting in the crime, but not for having committed the crime.

With respect to liability for crimes of mass atrocity on the international plane, these two systems of liability have been influential in different ways and at different times in different tribunals. It has been argued here, especially in Chap. 8, that it is the active role of specific individual participants in the processes of law development which has led to one system or the other being more influential at any given time.

This comes back full circle to the fourth layer of translation, from the domestic to the international. In making the translation of domestic modes of liability at various international tribunals, there have often been assumptions made about the transferability of domestic notions, and confusions have arisen with respect to the terminology surrounding culpability, accomplice and accessory liability, and even the legal meaning of the term 'principal'. Since many domestic systems retain the terminology of 'principal' and 'accessory', but without this necessarily denoting a normative differentiation nor a legal effect with respect to the conviction, the translation to the international context has often lacked a functional analysis, and a clear choice for one approach to liability. In the context of mass atrocity, the limits of culpability are important to keep in mind, and, in-keeping with the third criterion

---

[4] Duff 2007, p 23.

[5] Duff 2007, p. 37.

against with a system of liability is to be tested, the effect of transplanting from one legal system to another must be given more attention.

10.2 In Defense of a Normatively Differentiated System of Liability

Two important conclusions can be drawn from observing the patchworking process by which ICL develops: the first is that there is no 'correct' answer to the question of leadership liability, but rather there is an explicit policy choice to be made by the participants in the process. Neither the civil law nor the common law tradition has a superior solution to the problem; both the notion of equal party liability and differentiated party liability serve the purposes they were created for. They developed as policy responses to certain types of collective crimes in specific socio-political contexts and legal traditions. However the debates on the international plane suggest both some misunderstandings about the context and function of each of these solutions, and a certain unwillingness on the part of many participants to consider that there might be more than one solution to the problem. The question is not which system of liability is better, but whether there is one that is most appropriate for the specific context of ICL. A clear policy choice must be made in this specific context.

The second conclusion is that whereas the patchworking process has been relatively arbitrary for the most part, a more deliberate process is in fact possible, which could lead to a better founded and more widely accepted solution. Given that domestic case law and doctrine have become the most important sources in the active patchworking that already takes place, due to the lack of international law instruments or custom that deal with the necessary criminal law details, it is not only inevitable but even advisable that participants continue to look to domestic law solutions in order to resolve the question of leadership liability in ICL. But if these participants are cognizant of what the process entails, and apply a wider lens of analysis, including the tools of comparative law methodology, it may be possible to arrive at a choice that fits the context of ICL and satisfies the critique and concerns of international criminal lawyers, regardless of their domestic backgrounds. The task of international criminal lawyers, therefore, is to take into account all of the factors that can be identified at the domestic level, and to make an explicitly justified policy choice based on which would be the most appropriate for the specific context of prosecuting mass atrocity.

It has been pointed out throughout this book that the context of mass atrocity crimes must be kept in mind when making the translation from the domestic to the international. Indeed, it is the final criterion to keep in mind when testing the appropriateness of a specific model for application in ICL. It takes a certain select group of individuals—usually an intellectual elite—to exercise influence and control over the collective, in order to lead the collective to commit atrocities in the name of whatever ideology the leaders represent. A moral universe arises in which unrestrained passions and destructive impulses are authorised: 'a universe of destruction

[that] comes into being out of an inversion of norms.'[6] This deprives people of their second-order capacity to rein in their criminal impulses, the rational choice that an individual agent can make according to either moral impulses or impulses given by their physical surroundings. 'Protest becomes the more difficult as ...each individual is summoned to show solidarity in the face of the enemy' and war can radicalise human behaviour.[7] In this way the individual rationality of subordinates within the collective is reduced, as is their agency. Not only are they expected to take part, but if they refuse to, they are replaceable and risk becoming victims themselves.[8] Those higher up in the hierarchy, however, maintain their agency and remain irreplaceable. The notion of overlapping agency was therefore asserted in Sect. 2.4 as a way of identifying which individuals within the collective can be said to be most responsible.[9] This is especially so if there is the risk that their refusal to take part may well render them the next victim, such as has been witnessed under the Pol Pot regime in Cambodia and under the Junta regime in Argentina.[10]

This specific nature of mass atrocity calls for a system of liability that can identify and express these factors by differentiating between actors in a normative sense. Such a policy choice aligns with the central goals of ICL, the role of the principle of culpability, and the way in which deliberative structures in collective crimes of atrocity impact the decision-making capacity of various actors. The objective approach which underpins a normatively differentiated system also aligns better with the kinds of crimes being committed and the ways in which they are committed. Each of these factors is discussed here.

10.2.1 Liability and the Goals of International Criminal Law

Since policy is what drives the development of a system of liability, then we must return to the central policies of ICL in order to identify which system is a better fit. In Chap. 3 the argument was made that the many competing asserted goals of international prosecutions for mass atrocity crimes could be formulated as one overarching central goal: that of increasing the sense of accountability for massive human rights breaches.[11] This is compelling as an overarching goal, since it would satisfy the desire for effectiveness by helping to identify *who* should be held accountable, as well as the requirement of fairness and justice by ensuring *accountability* in the form of criminal liability and not just collective blame. In order to increase this sense of accountability a system of liability must consistently express the relationships

---

[6] Semelin 2007, p. 238.

[7] Semelin 2007, p. 199.

[8] Hinton 1998, p. 113; Semelin 2007, p. 199.

[9] Ohlin 2007a, p. 197.

[10] Sancinetti and Ferrante 1999, p. 23; Nino 1996, p. 58; Hinton 1998, pp. 95, 113; Tallgren 2002, p. 573.

[11] Damaška 2008, p. 347.

between the accused and the crimes. It should also move beyond a determination of guilt based on tacit agreements and loose connections between all those tied to a collective, who may be far removed in a causal sense from the crimes committed. An objective test can better express these relationships since it requires at least some significant causal nexus if someone is to be blamed for committing the crime proper, rather than for merely assisting in the crime.

Expressing the relationships between actors in a diffuse collective in this can way also fulfil the functions of punishment at the international level. Although there is not a great deal of theorising about such functions, the ad-hoc tribunals have most often cited deterrence and retribution as justifying punishment in ICL.[12] The Security Council Resolutions establishing these tribunals also makes reference to both of these justifications as the key functions of the prosecutions.[13]

Deterrence is a controversial function of criminal law, particularly when it comes to mass atrocity. The theory of prevention relies on the theory of rational choice: an actor will weigh up the potential benefits of committing a crime against the probability of being caught and punished, and the extent of the punishment.[14] Where the possibility for such rational choice is removed or reduced due to the context of a norm and culture of mass violence which is driven by political or other ideologies, it becomes very difficult to assert that any kind of prosecutions actually deter individuals from taking part in these collective crimes.[15]

However the deterrence theory is still cited as one of the primary functions of ICL,[16] and is even mentioned in the Rome Statute preamble, where it is stated that the States Parties wish to 'put an end to impunity for the perpetrators of [the most serious] crimes and thus to contribute to the prevention of such crimes.'[17] If we take this function as a given, at the very least the notion of general prevention may logically lead to distinguishing those participants higher up in the decision making echelons of collective mass atrocity crimes, in an attempt to end impunity.[18]

From this perspective, there would be more utility in focusing upon punishing those who are in a position to causally affect the general norm of violence, and to make a freer rational choice, namely those who are at the higher levels of a system hierarchy, involved in the collective deliberative process at the 'programmatic

---

[12] Erdemović, Sentencing Judgment 1996, para 58; Delalić Trial Judgment 1998, para 1234; Krajisňik Trial Judgment 2006, para 1134; Danner 2001, p. 444; Heller 2011, p. 601. See also the reference to deterrence in the appeal decision convicting Duch at the ECCC, Duch, Appeal Judgment 2012, para 380.

[13] S/Res 827/1993 1993, p. 1204; S/Res 955/1994 1994, p. 1601.

[14] Cooter and Ulen 2000.

[15] Hinton 1998, p. 113.

[16] Erdemović, Sentencing Judgment 1996, para 58; Delalić Trial Judgment 1998, para 1234; Krajisňik Trial Judgment 2006, para 1134; Duch, Appeal Judgment 2012, para 357; Heller 2011, p. 601; Werle and Jessberger 2014, p. 35.

[17] Preamble, Statute of the ICC 1998; Heller 2011, p. 601.

[18] As expressed in the preamble, Statute of the ICC 1998; Heller 2011, p. 601.

level'.[19] Only at this level might there be any notion of deterrence when there is a threat of punishment.

When it comes to retribution, in order to redress the widespread, collective and systematic nature of these crimes, and the community norm of violence that had been created and which had caused the physical perpetrators to act the way they did, retribution would demand that the leaders receive a greater degree of blame—and the most severe punishments—than someone on the periphery. Precisely because we cannot allot responsibility for the collective to an individual based solely on associative guilt, it can only satisfy the need for retribution if those who truly represent the collective are punished in a representative way.

Yet this can be problematic, since as Hannah Arendt put it, mass atrocity demands retribution because it is 'radical evil', yet at the same time it seems impossible to punish the unforgivable.[20] One of the greatest challenges faced in applying retributive theories to ICL is the problem of putting 'radical evil on trial', as described by Argentinean professor Carlos Nino: it may be impossible to find a punishment that measures up against the crime sufficiently.[21] Nino argues that a retributive theory is focused primarily on a descriptive understanding of blame, namely the decisions, intentions and choices of an actor, regardless to what extent these are actually under the actor's control.[22]

H.L.A. Hart drew a useful distinction in 'Punishment and Responsibility' between general justifying aims on the one hand, and on the other hand the question 'to whom punishment may be applied'.[23] General justifying aims include utilitarian theories and reprobation; the application of the pain of punishment to an offender who has breached the law with moral guilt. Or as George Fletcher has put it, 'crime pollutes the moral order, and punishment serves to restore the law and the world as it should be.'[24] Given the nature of mass atrocity crimes, this tends to be the general justifying aim which is referred to most often.[25] However this is different from limiting principles in terms of determining who should be punished and with what distribution. The latter has to do with why we attach moral importance to the role played by an individual in the completion of a crime.[26]

Punishment that is incommensurate with the blameworthiness of conduct risks becoming unclear or even counter-productive in terms of communicating the reprobation attempted by the 'international community'; indeed the crimes may become lost in technical legalese.[27] Yet here again the distinction drawn by Hart may be use-

---

[19] Tallgren 2002, p. 572.

[20] Arendt 1958, p. 135.

[21] Nino 1996, p. 141.

[22] Nino 1996, p. 139.

[23] Hart 1970, p. 9.

[24] Fletcher 2007, p. 305.

[25] See e.g. Kupreškić Trial Judgment 2001, para 848; Krajisňik Trial Judgment 2006, para 1134; Drumbl 2005, p. 559; Fletcher 2007, p. 250; Osiel 2009, p. 147; Tallgren 2002, p. 565.

[26] Hart 1970, p. 9.

[27] Tallgren 2002, p. 583; Simpson 2007, p. 83.

ful: when it comes to a general justifying aim for ICL, reprobation seems the most likely. When it comes to retribution in distribution, it may be harder to seek commensurate punishment, but it would therefore be necessary to distinguish between those more blameworthy and those less blameworthy, even where every participant in the collective is in *some way* blameworthy. The duty to suffer punishment is grounded in the wrongdoing and its consequences, taking into account all circumstances, and not just the mental fault of acting intentionally.[28] This is where the reality of mass atrocity crimes and the norm of violence within which they take place is important to understand. If there is to be appropriate retribution in distribution, then the levels of blameworthiness should be distinguished in the telling of the facts, and in the qualification of a defendant's role in a collective crime.

As former chief prosecutor at the ICTY Richard Goldstone has written, leaders and those in superior positions in the chain of command are, owing to their positive governance obligations, more deserving of prosecution and weightier punishment for their involvement in mass atrocity.[29] Precisely because we cannot allot responsibility for the collective to an individual based solely on associative guilt, it can only satisfy the need for retribution if those who exert influence over the collective are punished in a representative way.

This last point relates also to the narrative and expressive functions of ICL, in terms of history-writing and truth-telling. While criminal trials are first and foremost the determination of the liability of an accused, when it comes to choosing which system of liability will be the foundation of this finding, the distinction between those who were instrumental in the authoritarian creation of the collective norm of violence and those who were on the periphery of the commission of specific crimes is of great importance to the narrative that arises, and to the sense of accountability for what occurred. Furthermore, the disclosure of a substantive truth through the trials—recording the facts and circumstances, and the role of those with the most influence over these circumstances, as accurately as possible—can encourage a collective consciousness and self-reflection regarding the norm of violence that arose.[30] Accurate history writing is a necessary aspect of fulfilling this narrative or expressive function.[31]

The greatest advantage of a normatively differentiated system is that there is a very strong expressive value in the final judgment.[32] Not only is there a legal re-telling of the moral facts such that it is clear who had the most authority, the most influence over the occurrence of mass atrocity, but there is also a difference in the qualification of all those convicted. In a crime-oriented or naturalistic approach, there is limited expressive value in the final qualification; everyone involved in a collective crime is deemed to have committed it. But under a normatively differentiated approach, a secondary participant whose degree of blameworthiness is lesser is not convicted as

[28] Fletcher 2000, p. 418.

[29] Goldstone 1995, p. 7.

[30] Nino 1996, p. 147.

[31] van der Wilt 2010, p. 55.

[32] van Sliedregt 2012, p. 73.

if she committed the crime, as if she 'pulled the trigger' when she did not, but rather is convicted of having contributed to the crime committed by the principal.

10.2.2 *Objectivity as a Doctrinal Choice: A Better Fit for the Context of ICL*

Although the majority of judges at the ICC has thus far opted for an objective approach, as evidenced in the lengthy didactic passages of the *Katanga* and *Lubanga* Confirmation of Charges decisions, the failure on the part of these judges has been in the justification for *why* this is the preferred approach. Judges Fulford and Van den Wyngaert are right in their dissenting opinions that it is unacceptable to simply import the German model without saying why this particular domestic solution is applicable in terms of domestic law being a direct source before the court, and without arguing further why it's more appropriate than a subjective appropriate for the context of ICL.[33]

In terms of the first criticism, with regards to domestic law as a source for ICL, it has been argued here that in fact domestic criminal law is the most important source for the development of the normative content of this field of international law. Precisely because the traditional, formal sources offer no clarity on criminal law doctrines, participants have no where else to turn. This has been the case right from the inception of the IMT at Nuremberg and continues to be the case. However this patchworking process of translation from the domestic to the international has often been arbitrary, or biased by the backgrounds of the individual participants asserting a particular interpretation. The preference for modes of liability based on the US notion of conspiracy liability on the part of participants hailing from the common law tradition is apparent in the tribunals, as is the preference for the German doctrine by participants familiar with this from their own backgrounds in the civil law tradition. As comparativist Alan Watson has taught, transplants take place based on the accessability and authority of the domestic systems from whence they are borrowed, and rarely based on any functional justification. This is what must change if the solutions offered at the international plane are to be more acceptable. As stated above, the task for international criminal lawyers is to continue the patchworking process, but in an explicit and reasoned manner.

The second criticism levelled by the dissenting judges at the ICC is also founded; that there should be justification of why the objective approach is more appropriate to the context of ICL. The fact that an objective approach has been the foundation of modes of liability developed at the domestic level as a specific policy response to organised atrocities should factor very heavily. The subjective approach and its correlative system of equal liability for all parties to a crime was an effective policy response to organised mafia and gang crime, but when there are State or State-like

[33] Separate Opinion of Judge Christine van den Wyngaert, Ngudjolo Trial Judgment 2012, para 17; Judge Fulford, Separate Opinion, Lubanga Trial Judgment 2012, para 10.

apparatuses being utilised by those with the power and influence to do so, the extent of the crimes being committed, and the extent of penetration into the societal norms are typically far greater. The objective approach, and specifically the test of control over the crime, was developed in response to the observation of the Eichmann trial, and the need to deal with similar atrocity prosecutions at a domestic level in Germany. For the same reasons it was applied judicially in response to domestic atrocities enacted by those who had usurped the State power in Argentina, Bolivia, Chile, Peru and Uruguay.[34]

Furthermore, the subjective approach in the common law tradition has developed over time to compensate for the fact even though the old Common Law terminology did in fact make a normative distinction between principals and accomplices in the second degree, the legal distinction has disappeared. Today there is a lack of any coherent doctrinal definition for the distinction.[35] In order to capture leaders of group criminality who were not physically present and did not fulfil the *actus reus* of the crimes, the collapse between principal and accomplice had to be compensated by some other factor. The subjective approach fulfils this function by extending liability based on the original agreement. But this creates a risk that the principle of culpability is ignored or even breached.

Mass atrocity can only become widespread and occur on a grand scale if a central authority overtly encourages it.[36] The subjectivist argument that an individual taking part in a collective with the intention to commit a crime increases the risk that further crimes will take place, and thus that liability should follow for all crimes committed, amounts to a risk-based assessment of liability which is unacceptable in the context of these kinds of crimes.

In collective crimes committed within a norm of violence, by groups of people against other groups of people, all the parties tend to act intentionally and maliciously. There is therefore no basis for distributing responsibility on the basis of *mens rea*.[37] A subjective theory does away with this problem by not distributing responsibility at all, but rather attributing all the acts committed by any member of the collective to all other parties to the collective. But given the gravity of mass atrocity crimes, what this does is assign liability beyond moral culpability in cases where an individual had no possible way of intervening in the acts of others, nor distancing herself from these acts.[38]

Such an all-encompassing extended liability which includes all crimes committed by any other member of the collective need not be the logical extent of a subjective approach. Indeed, the *Pinkerton* rule of conspiratorial liability in the US, upon which this is based, is unique among common law jurisdictions in how far it extends liability

---

[34] Causa 13 Trial 1985; Leaders of the Left Revolutionary Movement 1993; Condor Plan 2009; Case Against Leaders of the Shining Path 2006; Fujimori Extradition Request 2007; Fujimori 2009; Salazar and Urquiaga 2010, pp. 84–85.

[35] Ohlin 2014, p. 4.

[36] Semelin 2007, p. 167.

[37] Fletcher and Ohlin 2005, p. 550.

[38] Fletcher and Ohlin 2005, p. 550; Osiel 2005, p. 1772.

based upon an original agreement.[39] Yet at the same time there are no doctrinal limits on this expansion. In Canada there is a conscious effort to require the highest tests of intention as the proof of culpability when it comes to party liability, according to the constitutional protection of moral innocence, however this is specific to the Canadian Charter of Rights. When we consider how the subjective approach has been applied at the ad-hoc tribunals, particularly with respect to JCE, there have been no such judicially created limits, and since there are no inherent doctrinal limits, the results have been endlessly expansive.

In order to widen the net of liability, the subjective approach is more focused on the intention or agreement. This is not the same as the pure subjective theory that prevailed in Germany in the nineteenth century, which asked whether the accused *intended* the acts committed by others as her own, since this may be a near impossible evidentiary burden.[40] However the subjectivity evident in the common law tradition does have similar roots, in that it focuses on the original intention and leads to the attribution to the accused of all acts committed by all other parties to the crime as if they were his own. This amounts to a kind of vicarious liability, which is reflected in JCE in particular.

On the other hand the objective approach delineates rather than expands liability, by using contribution as the foremost test of liability, including to what extent the accused had dominion over crime. The important difference in this approach is that this leads to attribution of blame, but not of the acts themselves. Blame, or responsibility, is imputed based on the role played by an individual, and brings with it a limitation on the extent of liability for two reasons. The first is that where a person has not contributed significantly, or cannot be said to have dominion over the crime, he will not be considered equally responsible for the acts of others. The second, related limitation is in the qualification that follows from this: a person who plays a lesser role has not fulfilled the elements of the crime and is therefore not convicted for having committed the crime proper, but for *assisting* another person in the crime.[41]

John Gardner describes very clearly why there is a moral difference between a principal and an accomplice; an accomplice is indeed responsible for her own actions in contributing to a crime, however there is a causal difference between what her actions bring about and what the actions of the principal bring about.[42] An accomplice still has a causal relationship to the crime, but it is a difference of degree, since what she does is make a difference to the actions of the principal. She contributes to a killing but she is not the killer, causally speaking, and thus should not be held responsible as such.[43] Under an objective test of liability, the person who instigates or abets does not have the same causal impact on the crime as a principal perpetrator.[44]

---

[39] Pinkerton 1946.

[40] As decided and discussed in Staschynski 1962, para 134; Weigend 2011, p. 95.

[41] Welzel 1970, p. 165; Stratenwerth 2005, p. 416.

[42] Gardner 2007, p. 58.

[43] Gardner 2007, p. 64.

[44] Jescheck and Weigend 1996, p. 677; Hamdorf 2007, p. 210.

This normative distinction is embedded in the notion of rational agency: morally speaking there is a difference between what I should do 'simpliciter', or the morality of principalship, and what I should do by way of contribution to what you do, or the morality ofcomplicity.[45] The latter is still wrong if you commit a crime, but it is a secondary wrong. For this reason a principal is convicted for the commission of the crime proper, and an accomplice, in whatever form, carries less responsibility.[46] Because accomplice liability is derivative of the principal, the former is convicted for her role in assisting the principal or principals, but legally speaking she is not convicted for having committed the crime herself, since she did not fulfil the elements of the crime.[47] This difference in qualification is important. Given that ICL is moving towards its own characterisation of liability, we ought to insist that our legal doctrine is sophisticated enough to distinguish between different these levels of participation in crimes of mass atrocity, if only out of our commitment to conceptual clarity.[48]

With the policy preference for focusing on 'those most responsible' in mind, as well as the central goal of increasing the sense of accountability for massive human rights breaches, it would seem that the major problem with the subjective approach is that it could be considered to lead to overly broad results, which blur moral distinctions as to blameworthiness.[49] Given that the limits of moral blame are at the core of the test of 'guilt' in both the civil law tradition and increasingly in the common law tradition, it should be applied in ICL.

10.2.3 A Normative Theory of Culpability

At the core of any system of liability is the principle of culpability yet even this principle is defined somewhat differently in the two different legal traditions compared in this study.

On the one hand, a descriptive theory of culpability measures 'guilt' by the presence of the 'guilty mind', or *mens rea*, which could be seen as the equivalent of punishing for 'free standing' culpability, rather than for wrongdoing.[50] Although there is certainly a normative element, since someone who acts intentionally is considered to be more blameworthy than someone who acts recklessly, still the descriptive theory is focused upon the mental state and not upon any other circumstances;[51] the eviden-

---

45 Gardner 2007, p. 66.

46 This must be distinguished from the way in which a commander is held liable for an omission, a dereliction of duty, rather than for the crimes committed by subordinates. As was argued in Sect. 8.3, command responsibility is a stand-alone crime of omission, the grounds of which are the predicate crimes committed by others, but it is not a conviction for those crimes.

47 Bohlander 2009, p. 155.

48 Ohlin 2007b, p. 84.

49 Damaška 2008, pp. 351, 353.

50 Fletcher 2007, p. 318.

51 Jescheck 2004, p. 44.

tiary test of the mental state of the accused as an element of the crime becomes at once the moral test for blameworthiness. This is why in jurisdictions which operate on a descriptive theory of culpability, attempt is considered to be equally punishable as the actual crime, even though the harm is less, since in both cases the requisite mental state is present.[52]

On the other hand, a normative theory of culpability is the determination that an individual is justifiably to blame for their conduct. The 'principle of culpability', *nulla poena sin culpa*, can be defined as something more than the statement that there must be a 'guilty mind', or *mens rea* present according to the definition of the crime. Rather, under a normative theory of culpability, *'culpa'* is read to mean blame, which is entirely personal, measured in degrees according to the circumstances of the act. Thus the statement that there should be no punishment without blame has the corollary that the punishment should match the degree of blame: the measure of punishment must be based on a court's conviction that the defendant is personally reproachable for the crime he or she has committed.[53]

The advantage of the descriptive approach is often said to be that it deals with 'moral luck' in a satisfying way: if an offender shoots a victim with the intention to kill him, but misses and does no harm at all, this is considered merely moral luck.[54] Because of the inherent dangerousness of such a person and such an intent, it is considered to be on par with having successfully killed.[55] However this leans towards a character or psychological based theory,[56] based on the premise that it's possible to distinguish the 'good guys' from the 'bad guys', and to punish someone who is dangerous. In some domestic criminal law jurisdictions this may be more appealing, given that 'good guys' tend not to act in violation of the law very often. However in the context of mass atrocity, it's not just 'bad guys' committing the crimes. Good people commit extraordinarily heinous acts when their own kin has been attacked, or when they are living under a regime where violence has become the norm and a way to belong to the dominant group, perhaps even a way to survive.

For this reason, it is argued here that the criminal law should treat wrongdoers in ways that are appropriate to their wrongdoing,[57] and that therefore findings of guilt should be about (i) *blaming* a person (ii) for their wrongful actions (iii) in violation of the law.[58] In reverse order these factors can be described as follows: violation of the law has to do with the elements of the crime, and includes the requisite mental state; the wrongfulness of the action has to do with whether or not it can be justified; and blame is a subjective test, taking all the circumstances into account. Rather than

---

52 Manning and Sankoff 2009, p. 324; Roach 2012, p. 161. See e.g. Dynar 1997, para 67.

53 Jescheck 2004, p. 44.

54 Ashworth 2009, p. 447; Stewart 2012a, p. 180.

55 Ashworth 2009, p. 446.

56 Fletcher 2007, pp. 28, 33.

57 Duff 2008, p. 4.

58 See the discussions on these aspects in Fletcher 2007, p. 319; and Duff 2007, p. 16, where he distinguishes between responsibility and liability, the latter being avertable by an exculpatory answer.

focusing on the offender's dangerousness in terms of character or in terms of a particular mental state, all the circumstances are taken into account. The principle of legality, if properly construed, does not lead us to focus on the actor's intent, but rather on the question whether the actor's conduct meets the preannounced definition of a crime.[59] The word 'guilt' as a psychological or emotional state can then be replaced with 'responsibility', which depends on the context as a whole.[60] Is the kidnapper, who is clearly intentionally doing something wrong, more responsible than those who are creating and enforcing the policy that all those deemed to be subversive should be rooted out of society, especially given the context that if the kidnapper refuses or questions this policy, she will no doubt become the next victim? Perhaps she is not excused, but she is also not more responsible than the leaders of the system of enforcement, given the circumstances; these leaders, too, are principals.

But why should such a normative theory of culpability prevail in ICL? When compared to the domestic jurisdictions discussed in this study, the answer may become clearer. As described in Chap. 6, in the US a descriptive theory of culpability is predominant, and culpability is considered a part of the definition of liability: either the requisite mental state is present, in which case an accused can be found guilty, or it is not, in which case a finding of 'not guilty' must follow.[61] While a higher sentence may be attached to an intentional crime than a reckless one, this normativity does not extend to parties who are considered equally liable for the crimes in which they assisted but which were committed by others, even if their own mental state was merely knowledge or even recklessness. Canada, on the other hand, has moved from a descriptive to normative theory, as there has been increased concern for moral blame as a limit on liability. Especially with respect to high stigma crimes such as murder and war crimes, there have been constitutional limits placed on the extent of party liability, since it has been considered a 'principle of fundamental justice' that a person should not be held liable for a high stigma crime in which they did not intend to take part.[62] Since wrongdoing and blame are not assessed separately in the Canadian criminal law, this limit of protecting the morally innocent does not extend to all forms of party liability for all crimes, but it has been made explicit in the case law and literature that the higher the stigma of the crime, the more important such a limit is. The intention to take part in a particular crime should not lead to liability for further high stigma crimes committed by other parties.[63]

Similarly, as described in Chap. 7, in Germany an explicit and deliberate doctrinal move was made in the twentieth century from a descriptive to a normative theory of culpability, a model which was inherited in Argentina due to the influence of German

59 Fletcher 2000, p. 181.

60 Fletcher 2007, p. 320; Duff 2007, p. 23.

61 Fletcher 2000, p. 308; Kadish 1980, p. 10.

62 Creighton 1993; Roach 2012, p. 165; Stuart 2007, p. 219.

63 Vaillancourt 1987, para 652; De Sousa 1992; Creighton 1993; Gosset 1993; Finlay 1993; Naglik 1993.

doctrine.[64] In this model, proof must be given of wrongfulness, the elements of the crime (including the mental state) and blameworthiness as three separate definitional aspects of criminal liability, which is predicated on a normative theory of culpability. Degrees of blame are measured based upon the extent of the contribution made by an accused toward the fulfilment of the crime.

The move from a descriptive to a normative approach in three of the four domestic jurisdictions considered here has been an express doctrinal response to problems such as how to take into account actors who may commit a crime with the requisite intent, but under duress, or mental incapacity, neither of whom can be said to be 'responsible' for their acts.[65] It has also been an express policy response in Canada to the problem of extended constructed party liability when it comes to high stigma crimes such as murder or war crimes. Indeed, in the Nuremberg Principles, the ILC stated very clearly that the criminal responsibility of inferiors acting on superior orders must be determined according to 'the true test …whether moral choice was in fact possible'.[66]

These factors all play a role in ICL when we consider the specific circumstances under which mass atrocity takes place. Responsibility expressed as criminal liability is what justifies punishment, and by determining responsibility according to *degrees* of blameworthiness, under a normative theory of culpability, it becomes possible to deal with these factors in a more comprehensive way.

As a reflection of this, some would argue that 'no punishment without guilt', in the sense of blameworthiness, can also be read to mean that punishment must never exceed the measure of the actor's blameworthiness.[67] Nor must it fall short of the degree of blameworthiness; punishment that does not live up to the extent of culpability is not appropriate in terms of retribution and compensation towards society.[68] Thus a normative theory of culpability fulfils both an expressive and a retributive function of punishment.

A normative theory also has more to say about the specific nature of collective mass atrocity crimes as system crimes, and the deliberative decision making structures which contribute to their commission.

10.2.4 Deliberative Structure as a Basis for Distributing Liability

The first two criteria posited here for an appropriate system of liability in ICL, and possibly the most important, are that it must capture the collective nature of the

[64] Bohlander 2009, p 13; Krey 2002, p. 97; Vormbaum and Bohlander 2014, p. 132; Nino 1980, p. 418 and further; Bacigalupo 1994, p. 184.

[65] Fletcher 2000, p. 455; Vormbaum and Bohlander 2014, p. 132; Duff 2007, p. 16.

[66] ILC 1950, Principle IV.

[67] Roxin 2005, § 3.

[68] Krey 2002, p. 113; Jescheck and Weigend 1996, p. 82.

crimes and it must successfully identify an individual responsible within the group even when there is no strict chain of command between parties. With respect to the translation that criminal law must make from collective agency to individual agency, and collective guilt to individual blame, participants asserting different modes of liability on the international plane have attempted to remain faithful to the collective nature of the crimes at issue, to a greater or lesser degree of success.

Where a descriptive theory of culpability prevails, whenever an individual participates in a collective crime, she might say 'the group made me do it'. However there is still some individual agency possible, since it is possible to interpret our actions as our own. According to this approach, complicit and direct actors must be seen to be equally and jointly culpable for collective crimes, and any unintended consequences of the collective action that are foreseeable, including further or different criminal acts committed by other members of the collective, should be ascribed to the group and back again to all the rest of its individual members. The acts of one become the acts of all, just as a group of picnickers are collectively and individually responsible for ruining flowers as a result of their picnic.[69]

But the question may need to be posed differently when faced with a norm of violence.[70] Where the physical perpetrators have the requisite *mens rea*, but at the same time are acting according to the norm and perhaps out of self-preservation their second-order ability to make moral choices is reduced, as illustrated in the well-known Erdemović case, where a group of executioners threatened the life of a fellow soldier who did not wish to participate in the execution of thousands of men and boys at Srebrenica.[71] In these situations, the individual agency upon which criminal responsibility depends tends to diminish in direct proportion to the scale of mass violence.[72]

This is where a normative theory of culpability can aid in coming to a basis for qualifying and distributing liability. The starting point is that anyone involved in the collective crime is to some extent blameworthy, but the further question is posed to what *degree* an individual's contribution to the crime deserves blame. The extraordinary context of mass atrocity, where violence has become horrifically ordinary, requires this further investigation.

The notion of overlapping agency discussed in Sect. 2.4 offers a way to maintain a recognition of the collective nature of mass atrocity crimes while at the same time being able to discern which individuals are more or less responsible within the collective. While a group can act with a certain collective rationality, the individual still retains individual agency, even though there is a submission of some individual reason to the group.[73] In this way the group and the individual are overlapping agents.[74] This means that while the group can be said to be acting collectively, the individual

[69] Kutz 2000, pp. 147–155.

[70] Drumbl 2005; Tallgren 2002; Salazar and Urquiaga 2010, p. 7 and further.

[71] Erdemović, Sentencing Judgment 1996.

[72] Stewart 2012b, p. 1216.

[73] Ohlin 2007a, p. 181.

[74] Ohlin 2007a, p. 185.

can still be held responsible for his or her own actions, and possibly—though not always—for the actions of the group. The determining factor is the extent to which there was a possibility for the individual to be involved in the group deliberation, which led to the collective will. Those lower down in the organisation can be said to fade more into the background, because they had not surrendered their individual reason to the group deliberative process. They can be held responsible for their own actions, but are less likely to have had any influence over the actions of others, and thus their criminal liability will be limited with respect to the crimes committed by others. Those higher up have a different role in the deliberative structure of the collective, and therefore a higher degree of moral responsibility as individuals, and can be held responsible for the collective will and action.[75] A normative theory of culpability combined with this notion of overlapping agency allows for clarification of the collective rationality, while at the same time not all of the individual members are subsumed into the abstract of the collective.

Thus, the more horizontal the participation of individuals in the deliberative process of the collective, the more justification there is to consider all of the members equally culpable,[76] but where there is more of a discernible hierarchy, overlapping agency describes the dynamic of collective criminality in such a way that the focus is upon those who have more influence over the crimes. The extraordinary nature of mass atrocity crimes come about because a collective has begun to accept them as normal behaviour—but it takes some individuals with specific political aims and sufficient influence over others to convince them to give up their individual reason to the group and use violent means to fulfil these ends. The notion of overlapping agency is highly compelling as a guiding principle to help select a model for modes of liability appropriate to ICL.

A subjective approach, with which a descriptive theory of culpability is associated, doesn't take this deliberative process into account, and we are left with the Hitler-as-accomplice dilemma: is the instructing and orchestrating leader really equally as culpable in a normative sense as the driver of a truck transporting victims to a concentration camp? To find this would be to get 'the moral valences entirely wrong—almost backwards, in fact.'[77] An objective approach is more reflective of the systematic way in which mass atrocity is committed, and the distribution of responsibility among those in the collective.

The moral opprobrium for contributing to acts under the control of others is less than for perpetrating or controlling those acts. For reasons discussed next, this normative differentiation should be expressed in the conviction through the modes of liability.

[75] Ohlin 2007a, p. 197.

[76] Ohlin 2007a.

[77] Osiel 2009, p. 85.

10.3 Sentencing or Qualification: Should Modes of Liability Do the Work of Differentiation?

Some authors have insisted that the differences between a unitary and a differentiated model in ICL have become negligible over time, as they have begun to borrow elements from each other, and since there is not much practical importance in a differentiated system if there is no correlating mandatory differentiation in sentencing.[78] There are indeed systems which differentiate normatively in the qualification and conviction of each individual, and which may or may not apply a mandatory sentencing mitigation. The point, however, is that the conviction itself in these systems aims to express something quite different from the conviction in either a unitary or a descriptively differentiated system, and the sentencing is only one function of this expression.

As Damaška puts it:

> By including large numbers of individuals with different contributions to crime and different modes of culpability into a single stigmatizing category, the doctrine disregards differences that in many societies are morally relevant ...[and] ignores the effects of the indiscriminate stigma of conviction.[79]

Judge Fulford has argued that a 'plain language' reading of the Rome Statute would mean the modes of liability are simply alternative descriptions of complicity, in a functionally unitary sense.[80] Some would agree with him that a 'plain language' conviction should suffice, amounting to a binary finding of 'guilty' or 'not guilty', since it is unnecessary to shoulder modes of liability with such an expressive burden.[81] In his plea for a unitary outcome in convictions, Stewart argues that a differentiated approach:

> does not tell us nearly as much about the culpability of the accused, because [a] formalistic concept [such as] "aiding and abetting" varies so widely from jurisdiction to jurisdiction and, just as importantly, may spoil the identity of the crime.[82]

This argument fails on several points. To start with, precisely because many modes of liability vary from jurisdiction to jurisdiction it is necessary to continue to clarify the requirements under ICL, in order to move away from the confusions of translation that have taken place during the arbitrary processes of patchworking, and create a more consistent and predictable system of liability specific to the context of ICL. Secondly, the very notion that the identity of the crime is important lies at the core of a differentiated approach. To convict a person as if she has committed the crime proper

[78] See e.g. Judge Fulford, Separate Opinion Lubanga Trial Judgment 2012, para 9; Separate Opinion of Judge Christine van den Wyngaert, Ngudjolo Trial Judgment 2012, para 27; van Sliedregt 2012, pp. 70, 73; Stewart 2012a, p. 207.

[79] Damaška 2008, p. 353.

[80] Judge Fulford, Separate Opinion, Lubanga Trial Judgment 2012, para 15.

[81] Separate Opinion of Judge Christine van den Wyngaert, Ngudjolo Trial Judgment 2012, paras 11, 16; Stewart 2012a, p. 212.

[82] Stewart 2012a, p. 212.

when she was on the periphery, when her assistance was not essential, and when she aided the physical perpetrator but did not herself commit the crime, is precisely to 'spoil the identity of the crime', and of the accomplice. By differentiating in the conviction, and utilising modes of liability to carry out this expressive function, the identity of the crime is preserved, in the sense that it remains clear exactly what is required both in terms of the objective *actus reus* requirements and the subjective *mens rea* elements, both of which may differ for an accomplice.

Furthermore, a secondary party is not convicted for having committed the crime proper under a normatively differentiated system, but rather for assisting the physical perpetrator; once again, the aider and abettor is not the *genocidaire*, but an assistant to the *genocidaire*. This difference in qualification records the relationships between the parties in a more factually accurate way than simply convicting all parties for genocide.

To carry out the expressive function by way of specifying the modes of liability in the conviction is to fulfil an important function of a trial that is focused more on an enquiry into the objective truth. Although criminal prosecutions are not necessarily the ideal route to reconciliation or truth-telling, where the choice has been made to undertake prosecutions, every attempt must be made to have the outcome reflect the truth as much as possible.[83] A trial in the international system reflects this tradition more than it does a trial focused on conflict-resolution, with its umpire-like judge and two competing versions of the truth in the hands of two opposing parties.[84]

Judge Fulford argued in his Separate Opinion to the *Lubanga* judgment at the ICC that a normative distinction between modes of liability is only necessary in a system where sentencing is determined by the mode of liability under which an individual is convicted. He pointed out this is the case in the German criminal law system, from which the interpretations applied in the Pre-Trial and Trial Chambers at the ICC has been drawn, but that under the Rome Statute no such sentencing regime exists.[85] Indeed, as Thomas Weigend points out, the Rome Statute allows judges a great deal of discretion when it comes to sentencing, given the flexibility of factors that may be included according to Article 78 of the Rome Statute and the Rules of Evidence and Procedure.[86] Weigend argues there is therefore no guarantee that defining differentiated modes of liability will lead to consistent outcomes. However it would be instructive to judges if the conviction were to express this degree of responsibility in a manner that is consistent over time, by applying the modes of liability as more than simply alternatives from which to choose, but rather as a system of liability that is normatively differentiated. A number of authors, including Weigend himself, assert that a normative distinction at the stage of qualification is a useful tool of expressing the necessary distinction.[87]

---

[83] Sancinetti and Ferrante 1999, p. 57.

[84] Damaška 1984, p. 162; Safferling 2003, p. 267; van der Wilt 2010, p. 54.

[85] Judge Fulford Separate Opinion, Lubanga Trial Judgment 2012, para 11.

[86] Weigend 2011, p. 102; Statute of the ICC 1998, Article 78(1); Rules of Procedure and Evidence of the ICC 2002, Rule 145(1).

[87] Weigend 2011, p. 103; Damaška 2008, p. 353; Drumbl 2007, p. 173; Jain 2012, p. 197.

The fact that there is currently discrepancy in sentencing among different tribunals is particularly problematic, since despite the general policy trend of focusing upon those in leadership positions, there is neither any judicial policy nor any consistency in the sentencing with respect to reflecting the role of an individual in a crime. For example, since the ICTY Statute provides no rules for mitigation of punishment according to the role played by a convicted individual, all those involved in a collective crime can be punished equally *regardless* of whether there has been a determination of differentiated modes of liability. There has been no discernable pattern of sentencing throughout the case law of the ICTY, since it is decided entirely on a case by case basis, taking into account numerous factors.[88]

Similarly, in his oft-cited dissenting opinion to the *Gacumbitsi* appeal judgment at the ICTR, Judge Schomburg pointed out that there is a choice to be made between a doctrine of modes of liability and simply leaving it up to sentencing. Conversely, in *Krnojelac* the judges in ICTY Trial Chamber II insisted that no distinction between principal and secondary participation is necessary precisely because there is no distinction made for sentencing. Someone who orders or solicits, which are so-called secondary modes of liability, may be punished greater than a so-called 'principal'.[89]

With respect to aiding and abetting there is also no clarity as to wether this should lead to lower sentencing. For instance, in *Krstić* the Appeals Chamber asserted 'that "accomplice liability" denotes a secondary form of participation which stands in contrast to the responsibility of the direct or principal perpetrators',[90] yet at the same time the judges noted that the case law of that tribunal 'does not evidence a discerning pattern of …imposing sentences on subordinates that differs greatly from those imposed on their superiors'.[91] Still Krstić's sentence was reduced due to the conviction based on aiding and abetting.[92] By comparison, Charles Taylor was convicted before the SCSL for aiding and abetting crimes against humanity and was sentenced to 50 years, certainly a higher sentence than any convictions at the other tribunals.[93]

While one possible conclusion from this is that modes of liability play no part in the convictions at the ad-hoc tribunals,[94] it is clear that the discussion on modes of liability looms in all the judgments, particularly where there have been express debates on JCE, aiding and abetting and indirect perpetration.

As Ohlin has put it, 'if culpability were simply a function of sentencing we could dispense with much of the substantive elements of international crimes' and simply replace them with one single offence of Felony.[95] Of course, he says, we would never impose such a penal scheme because it 'violates a firmly held intuition that

---

[88] See the discussion of this fact in Krstić Appeal Judgment 2004, para 247.

[89] Krnojelac Appeal Judgment 2003, para 77.

[90] Krstić Trial Judgment 2001, paras 642, 643.

[91] Krstić Trial Judgment 2001, para 709.

[92] Krstić Appeal Judgment 2004, paras 266, 275.

[93] Taylor Sentencing Judgment 2012.

[94] Stewart 2012a, p. 211.

[95] Ohlin 2007b, p. 87.

guilt and innocence must be determined relative to the elements of each offence.'[96] Related to this is the requirement that it must be clear to the accused exactly what the charge is, of precisely what he is being accused. This would require clarity of language, and consistency over time as to how modes of liability are indicted and interpreted. In this sense, modes of liability operate as a first order ranking system; in the indictment a normatively differentiated system fulfils the requirement that the accused know his charges and can prepare his defence, and in the conviction a normatively differentiated system expresses the role played by the convicted person more accurately than a functionally unitary one. The expressive function is thus best fulfilled at the stage of conviction, in specific modes of liability, predicated on an objective approach, rather than only at sentencing.

The requirement that guilt and innocence must be determined relative to each individual offence further relates to the notion of fair labelling, which is also the subject of some debate.

10.4 Fair Labelling

The notion that 'the label applied to an offence ought fairly to represent the offender's wrongdoing' has gained significant currency both in domestic criminal law and in ICL.[97] Originally when Andrew Ashworth wrote about fair labelling he was describing the necessity of attaching the correct *crime* to the conduct, particularly where an offender may have only had the intention to commit a smaller crime, but where greater consequences may have been an unlucky outcome of the situation. One example would be where a person intends to physically harm another, but nothing more, and the victim dies as a result. Ashworth argued that although it may be possible to convict a person of the crime of homicide, and to leave the consideration of moral culpability to the judge in the sentencing, this would create a possibility of misunderstanding in the case of an appeal, since the conviction still describes a weightier wrongdoing that may be at odds with the actual standard of fault or *mens rea*. This same problem applies to modes of liability, since a person involved in a collective crime may not have had the requisite *mens rea* for the crimes committed by the collective, and may have contributed only enough for a lesser crime.

Labels tell the story of the offender's criminality, so when it comes to the roles played by various individuals in collective criminality, the principle of fair labelling aims to ensure that the label applicable to each offender accurately describes the wrongfulness and severity of his or her specific involvement.[98] Labels are therefore important not only to describe the offending behaviour to the general public, but also

---

[96] Ohlin 2007b, p. 87.

[97] Ashworth 1981, p. 53; Chalmers and Leverick 2008; Guilfoyle 2011; Williams 1983; Zawati 2014.

[98] Nersessian 2007, p. 255.

to differentiate between different forms of wrongdoing.[99] As argued above, to leave all the work of differentiation to the phase of sentencing would be to lose something very important both in the description that is recorded in the conviction, and in the expression of the wrongdoing in a public, communicative sense.[100] Fairness requires not only the sentencing but also the labelling of the offender to be proportionate to the wrongdoing.[101]

The labels applied through a normatively differentiated system of liability work as a short-hand for moral facts,[102] whereby those who share a greater degree of moral culpability, due to either their direct involvement (they 'pulled the trigger'), or their ability to influence and control the entire scenario of mass violence (the leaders and intellectual authors), receive a conviction for the actual crime, as principals. On the other hand, those whose involvement was a lesser degree, or whose direct involvement was less culpable, receive a conviction for having helped the principals in the crime, but not for having committed the crime themselves. The expressive function of fair labelling therefore has a component that fulfills the requirements of both fairness and justice. As Ashworth has put it:

> once the label is entered on the person's criminal record the passage of time will dim recollections of the precise nature of the offence and may result in the label being taken at face value. Both out of fairness to the individual and in order to ensure accuracy in our penal system, therefore, the designation of an offence should fairly represent the nature of the offender's criminality.[103]

This can be considered fair labelling both with respect to the individual convicted, and with respect to victims and other relevant audiences. It is fair to the victims in the sense that the truth-telling that is expressed in the convictions paints a picture as accurate as possible, and can better reflect the wrongs that victims have suffered.[104] It is fair to the convicted person who was accused of some secondary form of liability, since the conviction records her participation in a realistic manner and she is not convicted for the crime proper. It is also fair to all those accused if there is a clear system of liability developed based upon a strong doctrinal foundation, which is predictable and expressive.

Some would argue that a legal distinction based on complex theories of liability will amount to technical legalese, unintelligible to any lay person, and ineffective in communicating the level of culpability of an individual to most of the community of victims affected by the crimes at issue, particularly once we reach the complexity of indirect perpetration by means of an organisation.[105] While there may be some truth to this, the difference between this complexity and the absurdity of convicting a leader in the highest echelon such as Charles Taylor for aiding and abetting, a term

[99] Ashworth 2009, p. 78; Chalmers and Leverick 2008, p. 222; Tadros 2012, p. 70.

[100] Chalmers and Leverick 2008, p. 224; Ohlin et al. 2013, p. 726.

[101] Ashworth 2009, p. 78.

[102] Robinson 2012.

[103] Ashworth 1981, p. 56.

[104] Drumbl 2007, p. 174; Ohlin et al. 2013, p. 745; Tadros 2012, p. 68.

[105] Stewart 2012a, p. 212; Weigend 2011, p. 102.

with which most lay people will have a relatively accurate association, is important. As discussed in Sect. 7.4, the convictions of the Junta regime in Argentina, and the subsequent application of the theory underpinning the identification of leaders as principal indirect perpetrators in other Latin American countries, demonstrates this desire to express the facts towards victims and survivors.

Fair labelling impacts positively on other interested parties in society as well, since it helps to convey information publicly in an expressive sense, and it also enables criminal justice professionals to make fair and sensible decisions on conviction and sentencing.[106] It acts as a check on sentencing discretion, in the sense that the sentencer is not left to determine some facts which are critical to the sentence, since these are part of the actual conviction.[107]

In his plea for a unitary approach to liability, Stewart has complained that one of the problems with differentiation is that there are different mental elements, and even objective *actus reus* elements for secondary parties, without there being a justification for the divergence from the definition of a crime. This has proven to be problematic in the case of JCE in its most extended form, where convictions have been possible based on a subjective approach for individuals who lack the requisite mental element for a crime, most notably with respect to the specific intent requirement for genocide.[108] A truly unitary approach, with neither a normative nor a descriptive differentiation between physical perpetrators and accomplices, would indeed do away with such a discrepancy. However a truly unitary approach has never been favoured at the international level, and even a functionally unitary approach in domestic systems such as Canada justifies the discrepancy in mental states precisely because the expansion of liability to include those other than the physical perpetrator in fact requires a counter-balance by requiring *more* of the mental state of an accomplice.[109]

Moreover, it is worth emphasising once again that under a normatively differentiated system, a secondary party is not convicted for committing a crime in the same way as the perpetrator, but rather for assisting the perpetrator. The fact that a secondary participant is convicted with a different label, for assisting the principal and not for committing the crime proper, means that a different level of stigma and qualification applies, and so it is justified to base this on requirements that differ from those for the commission of the crime. As one author has put it, ‘other mental states related to their influencing or assisting can be deontologically justified, because they are only being blamed for influencing or assisting.’[110]

Given the primacy of the expressive function of ICL, which is linked to the narrative and history-recording functions, if the imperative is to prosecute those ‘most

---

[106] Chalmers and Leverick 2008, p. 234; Ashworth 2009, p. 79.

[107] Chalmers and Leverick 2008, p. 224.

[108] Brđanin Appeal Judgment 2007.

[109] See for instance Article 21(1) Criminal Code 1985, which requires that a party liable for aiding and abetting acted ‘for the purpose of’ assisting the crime, even if the predicate crime does not require purpose on the part of the physical perpetrator. See Steer 2014.

[110] Robinson 2012.

responsible', we must have a way of identifying who those are, and expressing this in any conviction that results from such prosecutions.[111]

Thus the most appropriate modes of liability in ICL can best be said to be expressed in a normatively differentiated system that is responsibility-based, due to the complex systems and hierarchies within which mass atrocity takes place. This is a contribution-oriented, top-down approach, rather than a crime-oriented, bottom-up approach.[112] A bottom-up approach starts with the physical perpetrators of the crimes, and attempts to construct a causal chain up through the ranks to the leaders, which was the approach at Nuremberg. Where there is a clear chain of command and an exceptional paper trail, such as the Nazi regime left behind, this approach may be satisfactory, however the longer the chain of command, the more difficult the links become to make, and the immediate actors simply may not know enough to be liable for the grander scale of the crimes.[113] A responsibility-based top-down approach rather begins with those who have the main responsibility in organised collectives, such as the directors, generals and political leaders, and the links are made down the ranks to the 'small fish'; if these 'small fish' are principal perpetrators, then the contribution of the leaders must also be considered that of principal as well.[114]

While this must be balanced against the risk of prosecuting a certain individual as exemplary, or as a scapegoat for the crimes of an entire society,[115] and while the difficulty is in ensuring that 'defendants be convicted for their own conduct and not merely for the violent trauma experienced by entire nations,'[116] at the same time the political nature of international trials cannot be denied. Part of their narrative and expressive function is that they are sometimes instrumental in delegitimising the previous regime in order to create a new rule of law,[117] and expressing the norms that impunity is unacceptable and responsibility will be enforced.

10.5 Who Would You Rather Dine With? The Mastermind or the Executioner

The question was posed in the beginning of this book: if you had to share the dinner table with one of them, who would you prefer to dine with, the mastermind or the

[111] Guilfoyle 2011, p. 257.

[112] Vogel 2002, p. 155.

[113] Vogel 2002, p. 155.

[114] Vogel 2002, p. 155.

[115] Fletcher 2002, p. 1543; Nino 1996, p. 15.

[116] Danner and Martinez 2005, p. 100.

[117] See S/Res 827/1993 1993, para 7: 'Believing that the establishment of an international tribunal and the prosecution of persons responsible for ...violations of international humanitarian law will contribute to ensuring that such violations are halted and effectively redressed.' See also Danner and Martinez 2005, p. 90.

executioner?[118] The underlying assertion of this entire study is that there is something more understandable and more telling of the fallibility of the human spirit in the actions of an executioner who takes part in a collective where heinous acts are not only expected of him, but they have become the norm. There is something more distasteful, more horrendous, more blameworthy about the 'armchair killer', the leader who is in a position to influence others such that they become her instruments, such that they become capable of atrocities which serve an ideology that the leader wishes served by her subordinates. When she is fully capable of intervening, of using her influence to minimise rather than maximise the atrocities acted out by others at her behest, there is something more disturbing about the evil expressed through her when she does the opposite, than the evil expressed through the subordinates. We are all capable of the same kind of evil as the executioner under the right (or wrong) circumstances. One would hope not all of us are capable of the other kind of evil. This author would therefore choose to dine with the executioner, with the human being who may have done the unforgivable, but not the unthinkable.

From the perspective of individual victims and their family members, one could imagine that sitting across the dinner table from the physical perpetrators would be more difficult, because of the proximity of the evil done at the hands of this actor. There is a desire to see the physical perpetrator punished in order to redress the harms done to them. However the claim that there should be a distinction between participants does not deny there is culpability on the part of these physical perpetrators, since the wrongfulness of their acts is undeniable, and since they act with the requisite *mens rea*. It does not even deny that these individuals should be punished for their actions. However due to the contextual norm of violence under which they acted, and their reduced rational agency and reduced ability to act otherwise, there is no reason to hold these actors more responsible than the leaders who did not physically commit the atrocities, but who orchestrated them and ensured that others would take such actions. They are at least equally responsible. However there is most certainly a difference between the physical perpetrators and the leaders on the one hand, and those contributing on the periphery of these horrendous acts on the other. The latter are less morally blameworthy. They deserve to be labelled as such and punished to a lesser degree than the physical and intellectual authors of the crimes.

In the search for the most appropriate system of liability, and the most appropriate specific modes applicable to leaders, there is no domestic model that wins out due to a superior approach. Rather a choice must be made, by selecting which approach is more appropriate to the specific context of mass atrocity crimes. There are multiple translations from domestic to international, from collective to individual agency, and from the acts of the physical perpetrators to the responsibility of the leaders, all of which must be made with greater sophistication and awareness of all the factors discussed here. International criminal lawyers must become better comparativists,

[118] Gideon Yaffe, Professor of Law, Professor of Philosophy, and Professor of Psychology at Yale University, posed this question during a presentation at Yale Law School. I am grateful to James Stewart for presenting the question to me during discussions at the University of British Columbia.

and must look beyond the lens of their own individual legal backgrounds, such that the body of ICL can be improved with respect to the principles of criminal law.

An understanding of the policy reasons behind an objective or a subjective approach is core to making this choice. It has been argued here that an objective approach is more appropriate because it looks beyond the agreement to take part in collective action, and delineates between parties based on their actual contribution. The test of 'control over the crime' identifies those who have influence over the result, either by contributing to the division of labour directly, or to the orchestration indirectly. An objective approach leads to a system of normative differentiation between principals and secondary parties, based on degrees of blameworthiness, leading to fair labelling and the desired norm expression.

Modes of liability are necessarily open norms, and have always been subject to judicial interpretation and clarification in the doctrine. Because ICL instruments have not provided enough detail to support such interpretation and clarification, individual participants whose role is that of law application and interpretation have had to turn to domestic criminal law notions for help. The task of these participants now is to undertake this process with due attention to the policy considerations behind each model, as well as to the context of mass atrocity, and with justification of the selection of models being drawn upon. In doing so, the translation of guilt can become more consistent. There must always be casuistic decisions and factual determinations, but where there is a central and well informed model of liability upon which to base these decisions, the outcome can find more support and agreement. The 'clash of legal cultures' that has been prevalent in the development of modes of liability to date can be minimised, and a more centralised model can emerge in ICL to hold the leaders accountable for their definitive role in humanity's worst expressions of collective evil.

References

Arendt H (1958) The human condition. University of Chicago Press, Chicago

Ashworth AJ (1981) The elasticity of mens rea. In: Tapper CFH (ed) Crime, proof and punishment: Essays in memory of Sir Rupert Cross. Butterworth, London, p 45

Ashworth AJ (2009) Principles of criminal law, 6th edn. Oxford University Press, Oxford

Bacigalupo E (1994) Manual de Derecho Penal: Parte General, segundo edn. Editorial Temis SA, Santa Fé

Bohlander M (2009) Principles of German criminal law. Hart Publishing, Oxford

Chalmers J, Leverick F (2008) Fair labelling in criminal law. Mod Law Rev 71(2):217–246

Cooter R, Ulen T (2000) Law and economics. Scott Foresman, Glenview, Illinois

Criminal Code (1985) Criminal Code of Canada

Damaška M (1984) The faces of justice and state authority: A comparative approach to the legal process. Yale University Press, New Haven

Damaška M (2008) What is the point of international criminal justice? Chicago-Kent Law Rev 83:329–364

Danner AM (2001) Constructing a hierarchy of crimes in international criminal law sentencing. Va Law Rev 87(3):415–501

Danner AM, Martinez JS (2005) Guilty associations: Joint criminal enterprise, command responsibility, and the development of international criminal law. Calif Law Rev pp 75–169
Drumbl MA (2005) Collective violence and individual punishment: The criminality of mass atrocity. Northwestern University Law Review, Winter p, p 539
Drumbl MA (2007) Atrocity, Punishment, and International Law. Cambridge University Press
Duff RA (2007) Answering for crime: Responsibility and liability in the criminal law. Hart Publishing
Duff RA (2008) Whose luck is it amyway? In: Clarkson C, Cunningham S (eds) Criminal liability for non-aggressive death. Ashgate, Aldershot
Fletcher GP (2000) Rethinking criminal law. Oxford University Press, USA
Fletcher GP (2002) The Storrs lectures: liberals and romantics at war; the problem of collective guilt. Yale Law J 111:1499–1573
Fletcher GP (2007) The grammar of criminal law: American, comparative and international, Vol One: Foundations. Oxford University Press, New York
Fletcher GP, Ohlin JD (2005) Reclaiming fundamental principles of criminal law in the Darfur case. JICJ 3:539–561
Fujimori Extradition Request (2007) Decision on Extradition Request Fujimori. www.latinreporters.com/perouFujimoriFalloExtradicionChile21092007.pdf
Gardner J (2007) Offences and defences: Selected essays in the philosophy of criminal law. Oxford University Press, Oxford
Goldstone R (1995) The International Tribunal for Former Yugoslavia: A case study in Security Council action. Duke J Comp Int Law 6:5
Guilfoyle D (2011) Responsibility for collective atrocities: Fair labelling and approaches to commission in international criminal law. Curr Leg Prob 64(1):255–286
Hamdorf K (2007) The concept of joint criminal enterprise and domestic modes of liability for parties to a crime: A comparison of German and English law. J Int Crim Justice 5:208
Hart HLA (1970) Punishment and responsibility: Essays in the philosophy of law, 2nd edn. Oxford University Press
Heller KJ (2011) The Rome Statute of the International Criminal Court. In: Heller KJ, Dubber MD (eds) The handbook of comparative criminal law. Stanford University Press, Stanford, pp 593–634
Hinton AL (1998) Why did you kill? The Cambodian genocide and the dark face of honour. J Asian Stud 57:93–122
ILC (1950) Principles of international law recognized in the Charter of the Nuremberg Tribunal and in the judgment of the Tribunal
Jain N (2012) The control theory of perpetration in international criminal law. Chic J Int Law 12:159
Jescheck HH (2004) The general principles of international criminal law set out in Nuremberg, as mirrored in the ICC Statute. J Int Crim Justice 2:38–55
Jescheck HH, Weigend T (1996) Lehrbuch des Starfrecht, Allgemeiner Teil, 5th edn. A.T., Berlin, Berlin
Kadish SH (1980) Why substantive criminal law? A dialogue. Clevel St Law Rev 29:1
Krey V (2002) Deutsches Strafrecht Allgemeiner Teil: Lehrbuch in Deutsch und Englisch. W. Kohlhammer, Stuttgart
Kutz C (2000) Complicity: Ethics and law for a collective age. Cambridge University Press, New York
Manning M, Sankoff P (2009) Criminal law, 4th edn. Lexis Nexis, Ontario, Canada
Nersessian D (2007) Comparative approaches to punishing hate: The intersection of genocide and crimes against humanity. Stanf J Int Law 43:221
Nino C (1980) Los Limites de la Responsibilidad Penal. Editorial Astrea
Nino C (1996) Radical evil on trial. Yale University Press, New Haven
Ohlin JD (2007a) Group think: The law of conspiracy and collective reason. J Crim Law Crim 98(1):147–206

Ohlin JD (2007b) Three conceptual problems with the doctrine of joint criminal enterprise. J Int Crim Justice 5:69–90
Ohlin JD (2014) Co-perpetration: German dogmatik or German invasion? In: Stahn C (ed) The law and practice of the International Criminal Court: A critical account of challenges and achievements. Oxford University Press
Ohlin JD, van Sliedregt E, Weigend T (2013) Assessing the control theory. Leiden J Int Law 26(3):725–746
Osiel M (2005) The banality of good: Aligning incentives against mass atrocity. Columbia Law Rev 105:1751–1862
Osiel M (2009) Making sense of mass atrocity. Cambridge University Press, Cambridge, UK
Roach K (2012) Criminal law, 5th edn. Irwin Law, Toronto
Robinson D (2012) Darryl Robinson comments on James Stewart's "end of modes of liability". http://opiniojuris.org/2012/03/21/ljil-robinson-comments-on-stewart/
Robinson D (2013) The two liberalisms of international criminal law. In: Stahn C, van den Herik L (eds) Future perspectives on international criminal law. Ashgate, Surrey, pp 307–322
Roxin C (2005) Täterschaft und Tatherrschaft, 8th edn. De Gruyter Recht, Berlin
Rules of Procedure and Evidence of the ICC (2002) Rules of Procedure and Evidence of the International Criminal Court
Safferling C (2003) Towards an international criminal procedure. Oxford University Press, Oxford
Salazar K, Urquiaga XM (2010) Digest of Latin American jurisprudence on international crimes. Due Process of Law Foundation, Washington
Sancinetti MA, Ferrante M (1999) El Derecho Penal en la Protección de los Derechos Humanos. Hammurabi, Buenos Aires
Semelin J (ed) (2007) Purify and destroy: the political uses of massacre and genocide. Hurst and Company, London
Simpson G (2007) Law, war, and crime. Law Oxford University Press, Oxford
S/Res 827/1993 (1993) On the Establishment of an International Tribunal and Adoption of the Statute of the Tribunal
S/Res 955/1994 (1994) On the Establishment of an International Tribunal and Adoption of the Statute of the Tribunal
Statute of the ICC (1998) Statute of the International Criminal Court
Steer C (2014) A comparative perspective on the 'purpose' requirement. http://jamesgstewart.com/a-comparative-perspective-on-the-purpose-requirement
Stewart J (2012a) The end of 'modes of liability' for international crimes. Leiden J Int Law 25(1):165
Stewart J (2012b) Overdetermined atrocities. J Int Crim Justice 10:1189–1218
Stratenwerth G (2005) Strafrecht, Allgemeiner Teil I: Der Straftat (Translated as: Derecho Penal, Parte General I: El hecho punible). Hammurabi, Buenos Aires
Stuart D (2007) Canadian criminal law: A treatise, 5th edn. Carswell, Toronto
Tadros V (2012) Fair labelling and social solidarity. In: Roberts JV, Zedner L (eds) Principles and Values in Criminal Law and Criminal Justice. Oxford University Press, Essays in honour of Andrew Ashworth, pp 67–80
Tallgren I (2002) The sensibility and sense of international criminal law. Eur J Int Law 13(3):561–595
van der Wilt H (2010) Why international criminal lawyers should read Mirjan Damaška. In: Stahn C, van den Herik L (eds) Future perspectives in international criminal justice. TMC Asser Press, The Hague
van Sliedregt E (2012) Individual criminal responsibility in international law. Oxford University Press, Oxford
Vogel J (2002) How to determine individual criminal responsibility. In systemic contexts: twelve models. Cahiers de Défense Social, pp 151–169
Vormbaum T, Bohlander M (eds) (2014) A modern history of German criminal law. Springer, Heidelberg

Weigend T (2011) Perpetration through an organization: The unexpected career of a German legal doctrine. J Int Crim Justice 9:91–111
Welzel H (1970) Das Deutsche Strafrecht (Translated as Derecho Penal Aleman: Parte General). Editorial Juridica de Chile, Santiago
Werle G, Jessberger F (2014) Principles of international criminal law, 3rd edn. Cambridge University Press
Williams G (1983) Convictions and fair labelling. Camb Law J 42(1):85–95
Zawati HM (2014) Fair labelling and the dilemma of prosecuting gender based crimes at the international criminal tribunals. Oxford University Press, Oxford

Cases

Brđanin Appeal Judgment (2007) The Prosecutor v. Brđanin and Talić, Appeal Judgment IT-99-36-A
Case Against Leaders of the Shining Path (2006) Manuel Rubén Abimael Guzmán Reynoso y otros, R.N. no. 5385-2006, Segunda Sala Penal Transitoria, Suprema Corte de Justicia
Causa 13 Trial (1985) Sentencía del juicia a las juntas, Cámera Nacional de Apelaciones en lo Criminal y Correccional Federal de Buenos Aires
Condor Plan (2009) José Nino Gavazzo Pereira y otros, Sentencia 036, Ficha 98-247/2006
Creighton (1993) R. v. Creighton 91, 3 SCR 3
De Sousa (1992) R. v. de Sousa, 2 SCR 944 (Supreme Court of Canada)
Delalić Trial Judgment (1998) The Prosecutor v. Delalić et al, Trial Judgment IT-96-21-T
Duch, Appeal Judgment (2012) Kaing Guek Eav alias Duch, Appeal Judgment 001/18-07-2007/ECCC/SC
Dynar (1997) The United States v. Dynar, 2 SCR 462 (Supreme Court of Canada)
Erdemović, Sentencing Judgment (1996) The Prosecutor v. Erdemović, Sentencing Judgment IT-96-22-T
Finlay (1993) R. v. Finlay 3 SCR 103 (Supreme Court of Canada)
Fujimori (2009) Exp. No. AV 19-2001
Gosset (1993) R. v. Gosset 3 SCR 76 (Supreme Court of Canada)
Krajisňik Trial Judgment (2006) The Prosecutor v. Momciilo Krajisňik, Trial Judgment IT-00-39
Krnojelac Appeal Judgment (2003) The Prosecutor v. Milorad Krnojelac, Appeal Judgment IT-97-25-A
Krstić Appeal Judgment (2004) The Prosecutor v. Radislav Krstić, Appeal Judgment IT-98-33-A
Krstić Trial Judgment (2001) The Prosecutor v. Radislav Krstić, Trial Judgment IT-98-33
Kupreškić Trial Judgment (2001) The Prosecutor v. Zoran Kupreškić, Trial Judgment IT-95-15/T
Leaders of the Left Revolutionary Movement (1993) Luis García Meza Tejada y otros (Corte Suprema de Justicia de la Nación, 23 April 1993)
Lubanga Trial Judgment (2012) Prosecutor v. Thomas Lubanga Dyilo ICC-01/01-01/06
Naglik (1993) R. v. Naglik 3 SCR 122 (Supreme Court of Canada)
Ngudjolo Trial Judgment (2012) Prosecutor v. Mathieu Ngudjolo Chui ICC-01/04-02/12
Pinkerton (1946) Pinkerton v. United States 328 US 640 (Supreme Court of the United States)
Staschynski (1962) StE 4/62
Taylor Sentencing Judgment (2012) The Prosecutor v. Charles Ghankay Taylor, Sentencing Judgment SCSL-03-01-T
Vaillancourt (1987) R. v. Vaillancourt 2 SCR 636 (Supreme Court of Canada)

Appendix A
Full Texts of Articles on Liability

All translations of the following legislation are the author's own, with the exception of the French translation of the Canadian Criminal Code, which is an official translation. The greatest care was taken to ensure the choice of English words represent the meanings given to terminology within each of the respective jurisdictions, however any mistakes made are the responsibility of the author.

A.1 Modes of Liability Under the Charter of the IMT at Nuremberg

Article 6.

The following acts, or any of them, are crimes coming within the jurisdiction of the Tribunal for which there shall be individual responsibility:

(a) Crimes Against Peace: namely, planning, preparation, initiation or waging of a war of aggression, or a war in violation of international treaties, agreements or assurances, or participation in a common plan or conspiracy for the accomplishment of any of the foregoing;

(b) War Crimes: namely, violations of the laws or customs of war. [...]

(c) Crimes Against Humanity [...]

Leaders, organizers, instigators and accomplices participating in the formulation or execution of a common plan or conspiracy to commit any of the foregoing crimes are responsible for all acts performed by any persons in execution of such plan.

Article 8.

The fact that the Defendant acted pursuant to order of his Government or of a superior shall not free him from responsibility, but may be considered in mitigation of punishment if the Tribunal determines that justice so requires.

C. Steer, *Translating Guilt*, International Criminal Justice Series 9,
DOI 10.1007/978-94-6265-171-5

A.2 Modes of Liability Under the Model Penal Code of the US (Not Applied in Most States of the US)

Section 2.06. Liability for Conduct of Another; Complicity.

(1) A person is guilty of an offense if it is committed by his own conduct or by the conduct of another person for which he is legally accountable, or both.

(2) A person is legally accountable for the conduct of another person when:
(a) acting with the kind of culpability that is sufficient for the commission of the offense, he causes an innocent or irresponsible person to engage in such conduct; or
(b) he is made accountable for the conduct of such other person by the Code or by the law defining the offense; or
(c) he is an accomplice of such other person in the commission of the offense.

(3) A person is an accomplice of another person in the commission of an offense if:
(a) with the purpose of promoting or facilitating the commission of the offense, he
(i) solicits such other person to commit it; or
(ii) aids or agrees or attempts to aid such other person in planning or committing it; or
(iii) having a legal duty to prevent the commission of the offense, fails to make proper effort so to do; or
(b) his conduct is expressly declared by law to establish his complicity.

(4) When causing a particular result is an element of an offense, an accomplice in the conduct causing such result is an accomplice in the commission of that offense, if he acts with the kind of culpability, if any, with respect to that result that is sufficient for the commission of the offense.

(5) A person who is legally incapable of committing a particular offense himself may be guilty thereof if it is committed by the conduct of another person for which he is legally accountable, unless such liability is inconsistent with the purpose of the provision establishing his incapacity.

(6) Unless otherwise provided by the Code or by the law defining the offense, a person is not an accomplice in an offense committed by another person if:
(a) he is a victim of that offense; or
(b) the offense is so defined that his conduct is inevitably incident to its commission; or

(c) he terminates his complicity prior to the commission of the offense and

(i) wholly deprives it of effectiveness in the commission of the offense; or
(ii) gives timely warning to the law enforcement authorities or otherwise makes proper effort to prevent the commission of the offense.

(7) An accomplice may be convicted on proof of the commission of the offense and of his complicity therein, though the person claimed to have committed the offense has not been prosecuted or convicted or has been convicted of a different offense or degree of offense or has an immunity to prosecution or conviction or has been acquitted.

A.3 Modes of Liability Under the Canadian Criminal Code

Parties to Offences

Article 21.
(1) Every one is a party to an offence who
(a) actually commits it;
(b) does or omits to do anything for the purpose of aiding any person to commit it; or
(c) abets any person in committing it.
(2) Where two or more persons form an intention in common to carry out an unlawful purpose and to assist each other therein and any one of them, in carrying out the common purpose, commits an offence, each of them who knew or ought to have known that the commission of the offence would be a probable consequence of carrying out the common purpose is a party to that offence.

Article 22.
(1) Where a person counsels another person to be a party to an offence and that other person is afterwards a party to that offence, the person who counselled is a party to that offence, notwithstanding that the offence was committed in a way different from that which was counselled.
(2) Every one who counsels another person to be a party to an offence is a party to every offence that the other commits in consequence of the counselling that the person who counselled knew or ought to have known was likely to be committed in consequence of the counselling.
(3) For the purposes of this Act, 'counsel' includes procure, solicit or incite.

Attempts—Conspiracies—Accessories

Article 463. Except where otherwise expressly provided by law, the following provisions apply in respect of persons who attempt to commit or are accessories after the fact to the commission of offences:

(a) every one who attempts to commit or is an accessory after the fact to the commission of an indictable offence for which, on conviction, an accused is liable to be sentenced to imprisonment for life is guilty of an indictable offence and liable to imprisonment for a term not exceeding 14 years;

(b) every one who attempts to commit or is an accessory after the fact to the commission of an indictable offence for which, on conviction, an accused is liable to imprisonment for 14 years or less is guilty of an indictable offence and liable to imprisonment for a term that is one-half of the longest term to which a person who is guilty of that offence is liable;...

Article 464. Except where otherwise expressly provided by law, the following provisions apply in respect of persons who counsel other persons to commit offences, namely

(a) every one who counsels another person to commit an indictable offence is, if the offence is not committed, guilty of an indictable offence and liable to the same punishment to which a person who attempts to commit that offence is liable ...

Participants Aux Infractions

Article 21.

(1) Participent à une infraction:

(a) quiconque la commet réellement;

(b) quiconque accomplit ou omet d'accomplir quelque chose en vue d'aider quelqu'un à la commettre;

(c) quiconque encourage quelqu'un à la commettre.

(2) Quand deux ou plusieurs personnes forment ensemble le projet de poursuivre une fin illégale et de s'y entraider et que l'une d'entre elles commet une infraction en réalisant cette fin commune, chacune d'elles qui savait ou devait savoir que la réalisation de l'intention commune aurait pour conséquence probable la perpétration de l'infraction, participe à cette infraction.

Article 22.

(1) Lorsqu'une personne conseille à une autre personne de participer à une infraction et que cette dernière y participe subséquemment, la personne qui a conseillé participe à cette infraction, même si l'infraction a été commise d'une manière différente de celle qui avait été conseillée.

(2) Quiconque conseille à une autre personne de participer à une infraction participe à chaque infraction que l'autre commet en conséquence du conseil et qui, d'après ce que savait ou aurait dû savoir celui qui a conseillé, était susceptible d'être commise en conséquence du conseil.

(3) Pour l'application de la présente loi, ≪conseiller≫ s'entend d'amener et d'inciter, et ≪conseil≫ s'entend de l'encouragement visant à amener ou à inciter.

Tentatives—Complots—Complices

Article 463. Sauf disposition expressément contraire de la loi, les dispositions suivantes s'appliquent à l'égard des personnes qui tentent de commettre des infractions ou sont complices, après le fait, de la perpétration d'infractions:

(a) quiconque tente de commettre un acte criminel pour lequel, sur déclaration de culpabilité, un accusé est passible de l'emprisonnement à perpétuité, ou est complice, après le fait, de la perpétration d'un tel acte criminel, est coupable d'un acte criminel passible d'un emprisonnement maximal de quatorze ans;

(b) quiconque tente de commettre un acte criminel pour lequel, sur déclaration de culpabilité, un accusé est passible d'un emprisonnement de quatorze ans ou moins, ou est complice, après le fait, de la perpétration d'un tel acte criminel, est coupable d'un acte criminel et passible d'un emprisonnement égal à la moitié de la durée de l'emprisonnement maximal encouru par une personne coupable de cet acte …

Article 464. Sauf disposition expressément contraire de la loi, les dispositions suivantes s'appliquent à l'égard des personnes qui conseillent à d'autres personnes de commettre des infractions:

(a) quiconque conseille à une autre personne de commettre un acte criminel est, si l'infraction n'est pas commise, coupable d'un acte criminel et passible de la même peine que celui qui tente de commettre cette infraction …

A.4 Modes of Liability Under the German *Strafgezetsbuch*

Section Two
Third Title
Principal (Authorship) and Participation

§25 Principals

(1) Any person who commits the offence himself or through another shall be liable as a principal.
(2) If more than one person commits the crime jointly, each shall be liable as a principal (joint principals).

§26 Abetting (or Inducing)

(1) Any person who intentionally induces another to intentionally commit an unlawful act shall be liable to be sentenced as if he were a principal.

§27 Aiding

(1) Any person who intentionally assists another in the intentional commission of an unlawful act shall be convicted and sentenced as an aider.
(2) The sentence for the aider shall be based on the penalty for a principal. It shall be mitigated pursuant to §49(1).

§29 Separate criminal liability of the accomplice

Each accomplice shall be liable according to the measure of his own guilt and irrespective of the guilt of the others.

§49 Special statutory mitigating circumstances

(1) When a sentencing mitigation is required or permitted according to this provision, the following mitigation shall apply:

1. In place of life imprisonment, the mitigated sentence shall not be less than 3 years.
2. In the case of a limited term of imprisonment, a sentence of at most three quarters of the maximum sentence applicable for the crime may be given. For a monetary fine, the same applies with respect to the maximum daily rate.

Zweiter Abschnitt
Dritter Titel
Täterschaft und Teilnahme

§25 Täter

(1) Als Täter wird bestraft, wer die Straftat selbst oder durch einen anderen begeht.
(2) Begehen mehrere die Straftat gemeinschaftlich, so wird jeder als Täter bestraft (Mittäter).

§26 Anstiftung

Als Anstifter wird gleich einem Täter bestraft, wer vorsätzlich einen anderen zu dessen vorsätzlich begangener rechtswidriger Tat bestimmt hat.

§27 Beihilfe

(1) Als Gehilfe wird bestraft, wer vorsätzlich einem anderen zu dessen vorsätzlich begangener rechtswidriger Tat Hilfe geleistet hat.
(2) Die Strafe für den Gehilfen richtet sich nach der Strafdrohung für den Täter. Sie ist nach §49 Abs. 1 zu mildern.

§29 Selbständige Strafbarkeit des Beteiligten

Jeder Beteiligte wird ohne Rücksicht auf die Schuld des anderen nach seiner Schuld bestraft.

§49 Besondere gesetzliche Milderungsgründe

(1) Ist eine Milderung nach dieser Vorschrift vorgeschrieben oder zugelassen, so gilt für die Milderung folgendes:

1. An die Stelle von lebenslanger Freiheitsstrafe tritt Freiheitsstrafe nicht unter drei Jahren.

2. Bei zeitiger Freiheitsstrafe darf höchstens auf drei Viertel des angedrohten Höchstmaßes erkannt werden. Bei Geldstrafe gilt dasselbe für die Höchstzahl der Tagessätze.

A.5 Modes of Liability Under the Argentine *Codigo Penal*

Title VII
Criminal Participation

Article 45—Those who take part in the execution of the crime or provide assistance or cooperation to the author or authors, without which it would not have been possible to the commit the crime, will receive the sentence established for the crime. Those who directly bring about that another person commit the crime will also receive the same punishment.

Article 46—Those who co-operate in any other way in the execution of the crime, and those who provide subsequent assistance in fulfilment of promises made prior to the crime, are punished with one third to one half of the sentence corresponding the crime. If the sentence is reclusion for life, then reclusion for 15–20 years will apply, and if the sentence is life imprisonment, a sentence of 10–15 years imprisonment will apply.

Titulo VII
Participacion Criminal

Articulo 45—Los que tomasen parte en la ejecución del hecho o prestasen al autor o autores un auxilio o cooperación sin los cuales no habría podido cometerse, tendrán la pena establecida para el delito. En la misma pena incurrirán los que hubiesen determinado directamente a otro a cometerlo.

Articulo 46—Los que cooperen de cualquier otro modo a la ejecución del hecho y los que presten una ayuda posterior cumpliendo promesas anteriores al mismo, serán reprimidos con la pena correspondiente al delito, disminuida de un tercio a la mitad. Si la pena fuere de reclusión perpetua, se aplicará reclusión de quince a veinte años y si fuere de prisión perpetua, se aplicará prisión de diez a quince años.

A.6 Modes of Liability Under the Rome Statute

Article 25

Individual Criminal Responsibility

1. The Court shall have jurisdiction over natural persons pursuant to this Statute.

2. A person who commits a crime within the jurisdiction of the Court shall be individually responsible and liable for punishment in accordance with this Statute.

3. In accordance with this Statute, a person shall be criminally responsible and liable for punishment for a crime within the jurisdiction of the Court if that person:

(a) Commits such a crime, whether as an individual, jointly with another or through another person, regardless of whether that other person is criminally responsible;

(b) Orders, solicits or induces the commission of such a crime which in fact occurs or is attempted;

(c) For the purpose of facilitating the commission of such a crime, aids, abets or otherwise assists in its commission or its attempted commission, including providing the means for its commission;

(d) In any other way contributes to the commission or attempted commission of such a crime by a group of persons acting with a common purpose. Such contribution shall be intentional and shall either:

(i) Be made with the aim of furthering the criminal activity or criminal purpose of the group, where such activity or purpose involves the commission of a crime within the jurisdiction of the Court; or

(ii) Be made in the knowledge of the intention of the group to commit the crime;

(e) In respect of the crime of genocide, directly and publicly incites others to commit genocide …

Index

C. Steer, *Translating Guilt*, International Criminal Justice Series 9,
DOI 10.1007/978-94-6265-171-5

MIX
Papier aus verantwortungsvollen Quellen
Paper from responsible sources
FSC® C105338

If you have any concerns about our products,
you can contact us on
ProductSafety@springernature.com

In case Publisher is established outside the EU,
the EU authorized representative is:
Springer Nature Customer Service Center GmbH
Europaplatz 3, 69115 Heidelberg, Germany

Printed by Libri Plureos GmbH
in Hamburg, Germany

Methods for Multilevel Analysis and Visualisation of Geographical Networks

METHODOS SERIES

VOLUME 11

This Book Series is devoted to examining and solving the major methodological problems social sciences are facing. Take for example the gap between empirical and theoretical research, the explanatory power of models, the relevance of multilevel analysis, the weakness of cumulative knowledge, the role of ordinary knowledge in the research process, or the place which should be reserved to "time, change and history" when explaining social facts. These problems are well known and yet they are seldom treated in depth in scientific literature because of their general nature.
So that these problems may be examined and solutions found, the series prompts and fosters the setting-up of international multidisciplinary research teams, and it is work by these teams that appears in the Book Series. The series can also host books produced by a single author which follow the same objectives. Proposals for manuscripts and plans for collective books will be carefully examined.
The epistemological scope of these methodological problems is obvious and resorting to Philosophy of Science becomes a necessity. The main objective of the Series remains however the methodological solutions that can be applied to the problems in hand. Therefore the books of the Series are closely connected to the research practices.

For further volumes:
http://www.springer.com/series/6279

Céline Rozenblat • Guy Melançon
Editors

Methods for Multilevel Analysis and Visualisation of Geographical Networks

Springer

Editors
Céline Rozenblat
Institute of Geography
University of Lausanne
Lausanne, Switzerland

Guy Melançon
LABRI
University of Bordeaux 1
Talence cedex, France

ISBN 978-94-007-6676-1 ISBN 978-94-007-6677-8 (eBook)
DOI 10.1007/978-94-007-6677-8
Springer Dordrecht Heidelberg New York London

Library of Congress Control Number: 2013940873

Printed on acid-free paper

Springer is part of Springer Science+Business Media (www.springer.com)

Preface

During the last decade, research on networks has developed rapidly in most scientific disciplines. Several factors explain this sudden interest, which has led to an exponential growth in the dedicated literature from the natural to the social sciences. Technical advancements have certainly facilitated this exceptional development. Although graph theory and analytic concepts about graph structure and dynamics did exist before the 1960s (Berge, 1958), the processing of large connectivity matrices remained limited for a long time by the insufficient power of computers. On the empirical side, relational data were scarce or were very expensive to construct. Indeed, the significant increase in the speed of computing that has occurred recently as well as the proliferation of new types of empirical relational data that have been made viable because of the Internet have boosted the creation of powerful tools for network analysis. In parallel, many important modern societal trends have emphasized the need for developing adapted concepts and theories to help understanding these trends, which include the globalization processes that mainly operate through networking relationships; the trend toward decentralized management, the trend toward more participative or cooperative organization than the classical hierarchies; and the unfolding of individual connections through increasing mobility and educational level of population accompanied by various communication tools, to the extent that social networks are now considered to be a major source of "big data".

Among all of the social sciences, geography is well advanced in analyzing and modeling networks. Physical networks, such as the hydrographic, transportation or infrastructure networks, are part of the structuring of the geographical space. The descriptions of these networks were formalized decades ago, for instance, by models such as Horton's laws or accessibility and connectivity indices (Garrison, Berry, Marble, Nystuen, & Morrill, 1959). Thereafter, exchanges of goods, people and information between places were considered to be a possible fundamental explanation of geographical diversity leading to a conception of geography as a science of "spatial interaction" (Ullmann, 1954). Theoretical and operational models of spatial interaction were already well established in the 1970s (Wilson, 1970). The gravity model representing trip distribution was even considered to be "the first law of geography" (Tobler, 1970).

However, until recently, there were few methods for fully exploiting the vision of geographical space as a relational space, defined according to the many possible configurations by the interactions between georeferenced entities. Introducing that vision is the most salient aspect of this book. This book develops an integrated set of theoretical interpretations and adapted methods for exploring a variety of networks. These original and reproducible methods were elaborate, due to a long and involved interdisciplinary collaboration, between a laboratory of data processing and a network of geography researchers. A geographer, Céline Rozenblat, and a computer scientist, Guy Melançon, together developed a theoretical conception of geographical networks produced by dynamic processes in complex systems. They also conceived and adapted methods and software for visualizing the specific configurations of relational spaces that are operating at different scales of analysis within these networks. These scholars both succeeded at stimulating interest and animating a group of scientists from Canada, France and Switzerland, who worked together since 2005 when the SPANGEO program was initiated.

A major achievement of this project is the continuous development of the TULIP software, which is dedicated to the visualization of large networks according to the measurements of centralities and proximities in a variety of ways. As the networks analyzed by the geographers connect located objects, they are often rather strongly structured by the constraints that the distance exerts on the practices of communication or displacement. Explaining their structure implies combining the identification of topological organization (for example, small world configuration) and measurement of metrics on weighted interaction flows (representing, for instance, group cohesion). Moreover, as urban centers are the places where social interaction organizes in networks on many scales, from daily commuting patterns to worldwide air transportation system, there was a specific need for integrating within the measurement procedure—the hierarchical structure of urban systems that guides the many patterns of socio-spatial interaction. In addition to the usual methods for detecting communities within the network through various clustering methods, the TULIP software allows for a multi-level analysis of connections that are considered both at the intra-urban and the inter-urban level. This component is especially useful, for instance, in applications to urban economy when exploring the diversity of linkages among firms or among scientific researchers inside—as well as between—cities. This type of tool opens the door to a more specific appraisal of the so-called agglomeration economies, which may in fact represent "network effects".

Thanks to the efforts of the SPANGEO interdisciplinary group, we warmly welcome the novel feasibility of approaching the geographical space as a relational space. In the contemporary expansion of methods for network analysis, too many workers are satisfied with implementing standard algorithms or measurements on data files that are collected quickly, which yields results that are trivial or are difficult to interpret. This type of data appraisal was not followed in this work. Instead, this work proposes a method that fully integrates geographical theory with topological analysis, providing relevant measurements of centralities and proximities inside a method of visualization. Historically, geographers have used the integrative power of eyesight to envision the peculiarities of landscapes and places, constructing maps for encoding the accumulated knowledge from this process. It is now time to integrate sophisticated visualization tools in the cognitive elaboration of models and theories of spatial interaction that are occurring within the complex geographical systems and, in the process, continuously reshaping them.

Paris, France
December 2012

Denise Pumain

References

Berge, C. (1958). *The theory of graph*. New-York: Dover Publications.

Garrison, W. L., Berry, B. J. L., Marble, D. F., Nystuen, J. D., & Morrill, R. L. (1959). *Studies of highway development and geographic change*. US: Greenwood Press.

Tobler, W. (1970). A computer movie simulating urban growth in the Detroit region. *Economic Geography, 46*(2), 234–240.

Ullmann, E. L. (1954). Geography as spatial interaction. In *Interregional linkages (Proceedings of the Western committee on regional economic analysis)* (pp. 63–71). Berkeley, CA: University of California.

Wilson, A. (1970). *Entropy in urban and regional modelling*. UK: Routledge.

Contents

Chapter 1
Introduction

Céline Rozenblat and Guy Melançon

Geography is primarily concerned with society in space. Places are locations that interact through many types of flows: people moving to a different town or commuting between home and the workplace to benefit from services and to socialise; airplanes and ships travelling varying distances through webs of corridors; goods being traded; and information and financial flows that govern the spatial division of labour and organise companies into a complex hierarchies of subsidiaries. The interaction of these flows with a territory's inherent constraints and dynamics reveals the individual and collective goals that motivate these flows. The heterogeneity of social space, which is produced by unequal resources and the spread of wealth, as well as asymmetrical spatial relations, creates a "structural duality", due to continual trade-offs between structure and dynamics (Giddens, 1984). Dicken, Kelly, Olds, and Yeung (2001) claim that social networks are simultaneously structural and relational: they are structural because they are composed of established networks, such as infrastructure networks or networks of forces, and they are relational because they link agents – institutions, objects and knowledge, as well as individuals – across a wide variety of domains (Latour, 2005; Monge & Contractor, 2003). Many network interactions involve exchanges of information and human energy-producing power (Castells, 2009; Raffestin, 1980) that intertwine, nest and overlap with each other (see Fig. 1.1); these exchanges may exhibit cooperation, competition or both in "coopetition" (Badaracco, 1991; Doz, Santos, & Williamson, 2001).

C. Rozenblat (✉)
Institute of Geography, University of Lausanne, GEOPOLIS Building,
1015 Lausanne, Switzerland
e-mail: Celine.Rozenblat@unil.ch

G. Melançon
CNRS UMR 5800 LaBRI, GRAVITE INRIA Sud-Ouest, Université Bordeaux 1,
33405 Talence Cedex, France
e-mail: Guy.Melancon@labri.fr

C. Rozenblat and G. Melançon (eds.), *Methods for Multilevel Analysis and Visualisation of Geographical Networks*, Methodos Series 11, DOI 10.1007/978-94-007-6677-8_1,

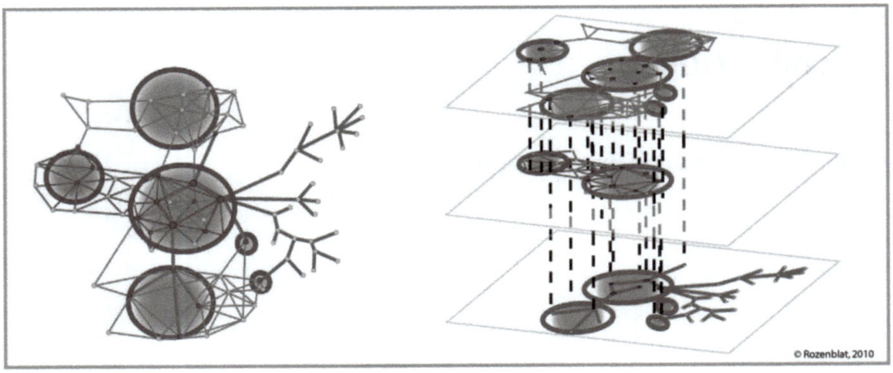

Fig. 1.1 The city as the connector of multiple networks

Thus, cities can be considered to be "connectors" that favour the emergence of new links and maximise interaction, and a city's wealth depends on the heterogeneity that this valuable combination creates (Allen, 1999; Castells, 1996; Jacobs, 1969). Each type of network develops its own particular rules, regulations, conventions and values (Castells, 2009; Storper & Salais, 1997; Uzzi, 1997).

From a geographical perspective, the assumption that transformations of places strongly depend on their components' internal and external relations with other places makes the positions of places in networks at different geographical scales crucial. Geographical interactions can be studied through networks that are represented as graphs. To date, most geographical network studies have been based either on "gravitation models" (Wilson, 1967) or general measures of graph connectivity or accessibility (Kansky, 1963). Graph theory – in particular, Watts' small-world networks model (Watts, 1999) – can contribute to this research area and enrich spatial analysis by complementing systems theory.

Through graph theory, network analysis has also been associated for a long time with sociology (Freeman, 2004). Graph visualization was employed much earlier than is typically reported. Freeman identified kinship trees (Morgan, 1851) and simple graphs of two-step marriage prohibitions (Macfarlane, 1883) as seminal graphical representations. Hobson's (1909) study of the interlocking directorates of the De Beers and Rand Mines industry in South Africa was the first example of two-mode data collection (Hobson, 1909).

Freeman (2004) noted that these pioneers were followed by a succession of twentieth-century schools that incorporated new issues and methods into the study of social networks – among them, Lund University's leading geographer, Hägerstrand.His approach to spatial diffusion (Hägerstrand, 1952), which assumed that the social network acted through space and time, highly influenced structural approaches to geography in the 1960s(Berry, 1964; Dacey, 1964; Garrison, 1960; Garrison, Berry, Marble, Nystuen, & Morrill, 1959; P. Gould & White, 1974; Tobler, 1965). However, Freeman criticised the lack of communication between social geographers and other social scientists that prevented the exchange and diffusion of ideas between these scientific communities. Similarly, many researchers in sociology, mathematics and physics, such as Rashevsky, Landau, Landahl, Lazerfeld, Merton, Moreno, and White in sociology and Erdős, Rényi, and Rapoport in mathematics, continued to develop the field of network analysis to evaluate network structure and the relative position of different network nodes. However, their structural network theories exerted little influence on research in social geography and spatial analysis, which focused investigations at the territorial scale (Chorley & Haggett, 1969).

Networks, which are organised through both direct links and indirect paths, induce topological proximities that simultaneously integrate spatial, social, cultural and organisational dimensions. Although the spatial systems approaches that emerged in the 1970s and the new paradigms of the 1990s have increasingly examined networks in which the "space of flows" interacted with the "space of places" (Castells, 1996), the framework guiding empirical studies of "networks in spatial systems" continues to lack relevant tools and concepts.

Network synergies based either on the similarities or complementarities between and within cities create specialised and relatively stable proximities between these entities (Powell, 1990). Interrelations between geographical levels of organisation also provide information regarding the equilibrium or disequilibrium of the territories that emerge at different geographical scales. From a geographical perspective that focuses on spatial interactions, the concepts and methods of the small-world model are particularly relevant because these interactions can be described and studied using large volume exchange and similarity matrices.

To the best of our knowledge, the field of geography has not employed this type of empirical model, which integrates a "network theory" approach with specialised methods. This perspective is based on Small-World Theory (Sect. 1.1 below), Scaling laws (Sect. 1.2), and multilevel approaches (Sect. 1.3), which lead to the multiscale (Sect. 1.4) and multidimensional (Sect. 1.5) analyses that are presented in this volume.

1.1 "Small-World" Networks

Networks develop privileged links between subset of nodes that subsequently are distinct from other nodes, while the network globally establishes a "small-world" web. This "small world" approach first appeared in the Hungarian writer Karinthy's "Chain-links" story, which described a fictional world in which all individuals were connected with each other through five acquaintances (Karinthy, 1929, as cited in Newman, Watts, & Barabasi, 2006). The Milgram experiment, which involved sending a document from a random panel of "starting persons" to a "target person", empirically evaluated the "six degrees of separation" concept (Milgram, 1967; Travers and Milgram, 1969) and was inspired by Pool and Kochen, who addressed this problem in a 1958 working paper that was not published until 1978 (Pool & Kochen, 1978, cited in Newman et al., 2006). The "small world" model was explicitly reintroduced by Watts in the 1990s (Watts, 1999; Watts & Strogatz, 1998). Pred (1974, 1977) identified the potential consequences of the structural properties of these networks for geographical places, particularly for cities, which accentuate "small world" structures in virtue of internal connectivity and interurban channels. The organisation of the overall society into a "space of networks and space of places" (Castells, 1996) create complex systems of "multiterritories" (Hassaert da Costa, 2004) with properties that exhibit multilevel, multiscale and multidimensional characteristics.

1.2 Scale-Free Networks

A multi-territory is a system of interconnected territories. However, every territory does not exhibit the same level of connection and integration within the system because the hierarchy is based on past uneven diffusions and new diffusions of innovations. Cities with established connections to multiple networks are more likely to attract new connections, as demonstrated by Batty's simulation (2006) of the formation of urban systems. Batty found that preferential attachment provided a better explanation for the formation of urban system hierarchies, which conform to Zipf's law (1949) rather than to Gibrat's "law of proportional effect" (Gibrat, 1931; Guerin Pace, 1995; Pumain, 1982). Airline networks illustrate this process: cities of secondary importance attempt to become airline hubs to more rapidly develop links to the principal worldwide network (Amiel, Melançon, & Rozenblat, 2005; Guimerà, Mossa, Turtschi, & Amaral, 2005).

The scale-free model describes the distribution of networks as a type of power law "because a power law is the only distribution that is the same whatever scale we look at it on" (Newman, 2005, p. 334). In the same paper, Newman noted that "'Zipf's law' and 'Pareto distribution' are effectively synonymous with 'power-law distribution'" (p. 327). Newman found that although the Yule "richer-get-richer" effect continued to explain power-law distributions (Simon, 1955; Willis & Yule, 1922; Yule, 1925), it also could reflect processes of self-organising systems and critical phenomena. The power law effect is estimated by two kinds of adjustments: the log-log distribution of the cumulative frequencies of a variable, which is commonly termed a "power law distribution", or the interaction of the log-distribution of a variable with the log-distribution of another variable describing the general size of the system, which is termed a "scaling law distribution", which characterises certain functions or specific phenomena based on cities' populations (Pumain, 2006).

Although physicists have proposed a single "universal" model, empirical studies reveal many different cases and interpretations (Newman, 2005). For example, three classes of "small-world" networks have been identified based on their adjustments of "power law distributions" (Amaral, Buldyrev, Havlin, Salinger, & Stanley, 1998), and "Apollonian networks" specifically describe networks as simultaneously scale-free, ultra-small-world, Euclidean, matching and space-filling (Andrade, Herrmann, Andrade, & Da Silva, 2005).

For urban systems more generally, the process implemented at the level of actors and organisations is primarily responsible for establishing the urban hierarchy. The hierarchy is not unitary but is composed of multiple subsystems that are simultaneously interwoven at different thematic and specialised levels of geographical space (Batty, 2006; Dicken, 2011; Rozenblat, 2007).

1.3 Multilevel Networks

Social networks are multilevel because different levels of aggregation allow the emergence of collective processes. Regardless of their spatial scope, networks are created, maintained and destroyed by individual actors or organisations that deploy assets and relationships to consolidate or improve their position in a space that is structurally identified and perceived at different spatiotemporal scales (Pumain, Paulus, Vacchiani-Marcuzzo, & Lobo, 2006). Due to the immediacy of access (within timespans of an hour), these networks rely on local territories – cities – from which they extract, exchange and expand human, material and informational resources to varying degrees (see Fig. 1.2).

As Pumain et al. (2006, p. 172) note:

> [N]ew properties emerge and characterize the city as a collective entity. Some of these new properties can be directly related to the intention of some institution, but most of the time they are the unexpected (and sometimes unwanted) result of collective interaction.

The city operates at an intermediate level in the organisation of societies, which Pumain (2006) termed the "meso level" (see Fig. 1.2). The intermediate level catalyses the social system daily on a local scale in a self-organising fashion (Batty 2005, 2006; Pumain, 1997; Pumain, Sanders, & Saint-Julien, 1989; Wilson, 2000). These places, which exhibit "immediacies of access" in interactions at the regional, national, continental or global level, due to actors' competitive or collaborative global strategies, comprise the total urban system that constitutes the macro level (Berry, 1964; Pred, 1977; Pumain et al., 1989; Sanders, 1992). This macro level provides feedback to the meso level that integrates the inner city's networks into larger interurban networks. However, the initial power belongs to individual micro level actors who create, maintain or destroy social and economic networks, although this process operates under macro and the meso-level constraints despite bottom-up feedback effects.

The "small-world" network approach identifies network components, termed "clusters", that are denser and more cohesive than the entire network (Barabasi, 2002; Barabasi & Albers, 1999; Newman et al. (2006); Watts, 1999; Watts & Strogatz, 1998). Clusters not only exhibit a high level of dependency between individual components but also a low level of "vulnerability" in networks that contain many redundancies (Burt, 2000, 2005). Thus, networks organise new topological "proximities" through direct links and more indirect pathways that are simultaneously spatial, social, cultural and organisational. Geography incorporates economic and

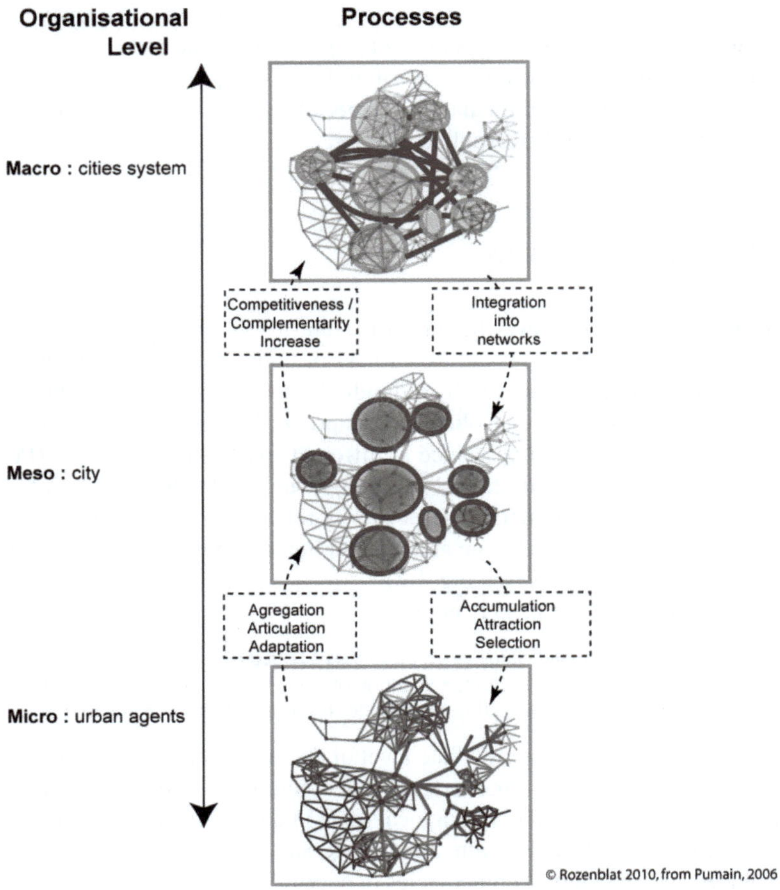

Fig. 1.2 The multilevel organisation of cities

social phenomena through the powerful and extremely relevant notion of level. Social relations naturally contribute to multilevel network dynamics. Although the view that the population dynamic simultaneously acts at the individual and group levels is plausible, formulating and validating a multilevel model is a painstaking endeavour. From the computer science perspective, the "level" concept provides strategies for downsizing data and potential methods for handling extremely large datasets. Modelling networks and flows as weighted graphs involves the use of graph theory, which transforms a multilevel network into nested subgraphs. Actors or entities interacting at a lower level establish higher-level entities that interact with each other, and flow exchanges at lower levels establish higher-level flows.

Incorporating cities into multilevel networks appears to have occurred not only in the field of transportation but also in the area of social and organisational networks (Dicken, 2011; Grabher, 2006). Although agent networks transcend the limitations of geographical scales (Castells, 2000), organisational levels of the "global commodity chain" are currently transforming from local to worldwide scales (Whitley, 1998).

1.4 Multiscale Networks

The multiscale analysis and visualisation of graphs enhances the study of networks. A careful visual analysis of the ways in which networks link similar actors provides insight into the reinforcement of social proximities, specialisations and segregations, as well as the extent to which transportation, communication, social and economic networks increase the anisotropy of space.

These networks thus produce a multiscale system. Although Ohmae (1990) claimed that multinational corporations develop within homogeneous and open territories, multiscale analysis reveals that these firms transversally weave together a system of potentially overlapping territories with their own rules and regulations. The structure of the "transnational field", which continues to be strongly influenced by the "international field", still depends on "the exchange rates, labor and fiscal regulation, those 'externalities' of which the company can take advantage, and the size and quality of the market" (Dollfus, 2001, pp. 104–105). From this perspective, the continental level increases its cohesion through the creation of free-trade zones that reinforce continental systems (Ohmae, 1995; Pomfret, 2007; Rozenblat, 2004; Rugman, 2001; Yeung, 2002), although it is not the only level from a multidimensional perspective.

Similarly, intense economic specialisation creates groups of cities that are increasingly interrelated – a tendency that is reinforced by "poles of excellence" or "poles of competitiveness" policies. The most striking example is financial specialisation, which has created a "global city" linking New York, London and Tokyo (Sassen, 1991). The financial capitals of the world link themselves to each other either directly or indirectly by forming specialised and/or geographical subgroups.

"Relay" cities provide "mandatory pathways" between these ensembles of highly interlinked cities. For example, national capitals host the national head-offices of the foreign subsidiaries of multinational corporations that, in turn, invest in businesses or organisations elsewhere in the host country. This situation is also observed for continental centres at a continental level (Alderson & Beckfield, 2004; R. Gould & Fernandez, 1989; Rozenblat, 2004; Rozenblat & Pumain, 1993, 2007; Rugman, 2001). Personnel meetings are often held at these "intermediate" head-offices. At the city level, capitals serve as "bridges" to other national cities; for example, capital cities provide preferred locations for representational offices (e.g., Brussels in Europe). Through this role of "display relay" to the country's other cities, proportionally more relevant relationships – upstream and downstream as well as national and international – are concentrated within the capital city. Serving as a "relay" or "bridge" provides the capital city with better access to the entire network, as well as increased control over information transfers (Burt, 2005). Additionally,

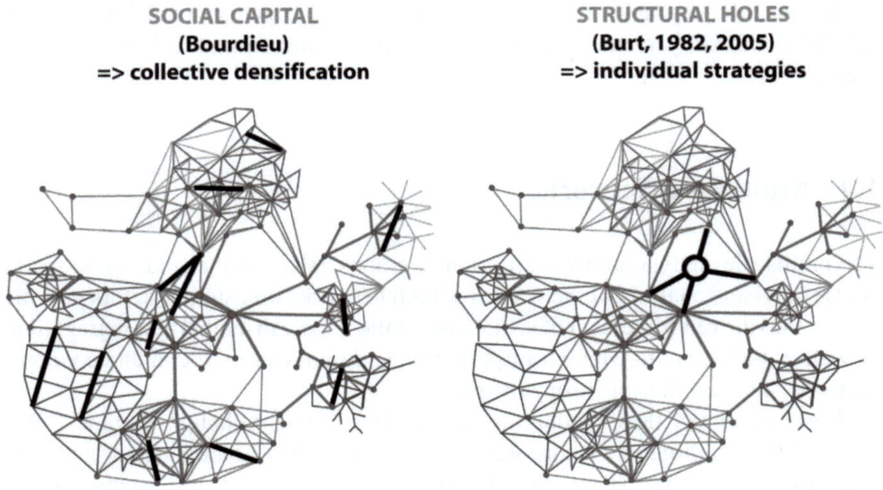

Fig. 1.3 Structural holes opposing social capital reproduction

the network positions of cities that share similar attributes of "dependency" with regard to "relay" cities display "equivalent" characteristics. Each "subgroup" maintains the hierarchy of cities' levels of centrality in the network, while other cities remain peripheral and are attached only indirectly to the network through their connection to participating cities. The recent reinforcement of the centrality of European national capitals reflects this process (Rozenblat, Bohan, & Benet, 2008; Rozenblat & Cicille, 2003).

Adopting sociologists' argument, we can claim that this cumulative process of concentration reinforces the social capital of large cities at the micro level (Walker, Kogut, & Shan, 1997; see Fig. 1.3).

Following Walker et al. (1997), we claim that Burt's "bridges" (1992, 2005) could be distinguished from places that "reproduce social capital" (Bourdieu, 1980) because they are places where nobody is located; this distinction focuses on the "structural holes" that individual actors might exploit to obtain a competitive advantage over other cities located in the same places. These places might generate network innovation by developing new kinds of spatial or social organisations that are disconnected, at least initially, from the dominant networks (Burt, 2005).

1.5 Multidimensional Networks

Innovative networks also appear at the crossing of different networks, which creates shared synergies. Differences between networks can be found in the endogenous network structure, as well as in the exogenous variables characterising the network. For example, the economic dimension of the network is composed of three levels, which Capello & Rietvelt (1998) have defined as the network itself (micro level), its economic relationships (the meso level) and the urban or regional organisation as a whole (the macro level). Thus multiple dimensions – each consisting of several levels – contribute to the development of networks. Any account of the spatial dimension of networks must consider these other dimensions. For example, in the "Global Commodity Chain", Gereffi (1996) identified four major dimensions that determine the distribution of functions and forces between different production units that are often distant from each other: the input-output economic structure of production, services and resources; the governance structure; the differentiated spatial field; and territorialised institutions. In a territorialised approach, the spatial and institutional dimensions are incorporated in the single dimension of the territory. This system posits three interdependent dimensions: corporate governance; the economic system in which it functions; and the territory that produces it (Dicken & Malmberg, 2001; Rozenblat, 2007). This three-dimensional system particularly applies at the urban system scale because the relational properties of amplitude, connectivity and diversity provide the variety, flexibility and stability that networked businesses require (Camagni, 1999).

1.6 Inherent Spatial Properties of Networks: The SPANGEO Project

Although increasing our knowledge of the properties of networks of cities is essential, these properties can be measured at the city level and must be assessed by analysing actor networks. The shift from the actor level to the city level must consider the processes that develop within each individual city because there are many processes that permit the emergence of urban properties in which "the whole becomes not only more than but very different from the sum of its parts" (Anderson (1972) cited in Lane (2006)).

The present volume, which seems similar to an earlier volume, "Methodology and epistemology of multi-level analysis" edited by Courgeau (2003), shifts the focus from the individual to the group and from individualism to holism (known here as the territory) because it focuses less on individual characteristics and more on the interactions of actors and institutions that create functional territories in which the structure of existing links constrains emerging links. Rather than basing explanations on external factors, the goal is to determine the extent to which network properties reflect spatial distributions and create local synergies at the meso level that are incorporated into global networks at the macro level where different geographical scales occur. Although the multilevel analysis developed by Courgeau proposes that *a priori* groups and levels explain differences between individuals, the goal of the present volume is to use the graphs' structure to identify empirically relevant groups and levels that explain dynamics. This initial exploration applies network analysis and visualisation to determine spatial path dependencies and the emergence of functional territories as clusters forming a whole multiterritoriality.

The convergence of the network multiterritoriality paradigm collaboratively formulated and manipulated by geographers and computer scientists produced the SPANGEO project (Spatial Networks in Geography) in 2005. Although the methods inspired by graph analysis and visualisation could improve the ability to comprehend large, complex networks and flows, geographers still must confirm the relevance of these methods through solid and informative conclusions based on the data. The robustness of the computing methodologies can only be assessed based on the knowledge that the proposed technology enables geographers to acquire. The entire project was developed by the Visual Analytics Loop (Thomas & Cook, 2006; van Wijk, 2005) in which geographers contributed to the visualisation design. Insights based on the data and its visualisation inspired geographers to adopt methods for exploring the phenomena underlying different data through an iterative loop that began on the first day and continued until the project's end. SPANGEO, which was supported by the ANR in France from 2005 to 2008, initially involved a core team that continuously expanded to include new teams and new people through support from European Research Group S4 (Spatial Simulations for Social Sciences) under the direction of Denise Pumain.

The increasing amount of available networks data presents a challenge to quantitative geography. Compared to other disciplines, most of the data are available through population surveys, although the networks data in international comparative studies is not always homogeneous. The computer sciences offer geographers the opportunity to fully exploit their data and confront complex issues that require new methodologies and strategies of analysis. Until recently, the primary tool for visualising and analysing geographical and social data was use visual cues to superimpose quantities over a world map (e.g. Bertin, 1973). Currently, it is possible to employ cybergeography and new visual and analytical paradigms that complement and improve the traditional visualisations using world maps. Because new visualisations generate new issues, the multilevel analysis and visualisation of geographical networks transforms the geographical approach to the definition of places and hierarchies (Pumain, 2006).

1.7 Outline of This Volume

The multilevel, multiscale and multidimensional aspects of spatial analysis provided a privileged terrain that geographers and computer scientists shared within the SPANGEO project. This volume assembles some of the contributions based on this collaborative project. Most of the papers represent collaborations between geographers and computer scientists. Many conferences and workshops created a common language and conceptual background to develop approaches and methods to address questions in geography. New issues also emerged from the SPANGEO project research.

In the first section, Céline Rozenblat and Guy Melançon – a computer scientist with 5 years of practice and training in geographical issues – describe the "multilevel and multitheoretical framework for geographical studies" that they developed. This paper provides the initial formalisation of the process of creating a coherent corpus for the Spangeo project. Incorporating the concepts of centrality and proximity developed in physics, information visualisation, sociology, and regional science, the two authors describe how the concepts might be adapted to fit the geographical context. Next, Alain L'Hostis describes how the concept of anisotropic space developed in geography – in particular, by Tobler (1963) – contributes to this approach. The vision of networks from the computer scientists' point of view is developed by Antoine Lambert, Romain Bourqui and David.

The methods developed by the SPANGEO team – in particular, the TULIP software[1] designed by the LaBRI team in Bordeaux – are presented in a second part with three papers. Guy Melançon describes the basic measures for identifying graph centralities and clustering, and Jean-François Gleyze presents specific approaches to the topological aggregation of graphs. Finally, Pierre-Yves Koenig explains the various methods for analysing and visualising networks and their inherent or exogenous properties.

The papers in a third part describe five applications of these methods, which are presented at the global scale, the European scale and the city scale. Cesar Ducruet describes sea networks, and Céline Rozenblat, Romain Bourqui and Guy Melançon describe air traffic networks. At the global scale, sea and air transportation networks exhibit different types of clustering based on internal network structures. These transport networks have the methodological advantage of being quasisymmetrical, unrestricted by physical limits, and global. The clustering methodologies of these applications operate at the global scale and at the macro level of urban systems. The third paper by Charles Bohan and Berengere Gautier presents a micro-level analysis of multinational firms' behaviour at the global scale and introduces space as exogenous variable that makes it possible to compare worldwide networks of corporations through graphs and DAGMaps. Marie-Noelle Comin describes research networks at the European scale that define classes of cities according to their specialisations, associations or type of research, alternating between the macro level of the urban system in which groups are based on topological proximities and the meso level of processes occurring within cities due to the diffusion of innovations and specialisations. Finally, Patrice Tissandier, Trung Tien Phan-Quang and Daniel Archambault couple topological and geographical proximities, describing the polycentrism that emerged in cities from 1975 to 1999 and describe commuting networks, which are explicitly positioned at the meso-level of the cities because they involve people's daily movement but also involve the macro-level because people increasingly fragment their work and home lives between distant locations by commuting weekly or tele-commuting.

---

[1] The TULIP Software is free and open access: http://tulip.labri.fr/

All these presentations describe the emergence of networks at different levels through diachronic analysis. To develop dynamic models of networks, it was necessary to explore the temporal and attribute structure of the different networks in depth. Thus, we wanted to share this collaborative work with readers to identify the issues in adapting these methods to different themes, scales and levels.

The successful cross-disciplinary work presented in this volume should promote further collaborative efforts. For the participants, this collaboration was a valuable opportunity that forced each author to explicitly define the "basic" concepts of his or her field for colleagues in other disciplines. The process required an open mind, as well as the ability to view the knowledge and principles of our fields critically. This volume is unique and original, not only because it provides information on the state of the art, but also because it represents a product of collaboration. We hope that the enthusiasm we share will persuade readers in the fields of geography, sociology, computer science and related areas to contribute to this fruitful cross-disciplinary endeavour.

References

Alderson, A. S., & Beckfield, J. (2004). Power and position in the world city system. *American Journal of Sociology, 109*(4), 811–851.

Allen, J. (1999). Cities of power and influence: Settled formations. In J. Allen, D. Massey, & M. Pryke (Eds.), *Unsettling cities* (pp. 182–227). London: Routledge.

Amaral, L. A. N., Buldyrev, S. V., Havlin, S., Salinger, M. A., & Stanley, H. E. (1998). Power law scaling for a system of interacting units with complex internal structure. *Physical Review Letters, 80*(7), 1385–1388.

Amiel, M., Melançon, G., & Rozenblat, C. (2005). Réseaux multi-niveaux: l'exemple des échanges aériens mondiaux. *Mappemonde, 79*(3), 20 p. Retrieved July 21, 2009, from http://mappemonde.mgm.fr/num7/index.html

Anderson, P. (1972). More is different: Broken symmetry and the hierarchical nature of science. *Science, 177*, 393–396.

Andrade, J. S., Herrmann, H. J., Andrade, R. F. S., & Da Silva, L. R. (2005). Apollonian networks: Simultaneously scale-free, small world, Euclidean, space filling, and with matching graphs. *Physical Review Letters, 94*(1), 018702.

Badaracco, J. (1991). *The knowledge link: How firms compete through strategic alliances*. Boston: Harvard Business School Press.

Barabasi, A.-L. (2002). *The new science of networks*. Cambridge, MA: Perseus.

Barabasi, A.-L., & Albers, R. (1999). Emergence of scaling in random networks. *Science, 286*, 509–512.

Batty, M. (2005). *Cities and complexity: Understanding cities with cellular automata, agent-based models, and fractals*. Cambridge, MA/London: MIT.

Batty, M. (2006). Hierarchy in cities and city systems. In D. Pumain (Ed.), *Hierarchy in natural and social sciences* (pp. 143–168). New York: Springer.

Berry, B. (1964). Cities as systems within systems of cities. In J. Friedmann, & W. Alonso (Eds.), *Regional development and planning*. Cambridge, MA: MIT Press (also in Paper of Regional Science Association, 1964, 13).

Bertin, J. (1973). *Sémiologie graphique: Les diagrammes-Les réseaux-Les cartes*. Paris, France: Gauthier-VillarsMouton & Cie.

Bourdieu, P. (1980). Le capital social: Notes provisoires. *Actes de la Recherche en Sciences Sociales, 3*, 2–3.
Burt, R. S. (1992). *Structural holes: The social structure of competition*. Cambridge, MA: Harvard University Press.
Burt, R. S. (2000). The network structure of social capital. In R. Sutton, & B. Staw (Eds.), *Research in organizational behavior*. Greenwich, CT: JAI Press.
Burt, R. S. (2005). *Brokerage and closure: An introduction of social capital*. New York: Oxford University Press.
Camagni, R. (1999). The city as a milieu: Applying GREMI's approach to urban evolution. *Revue d'Economie Régionale et Urbaine (RERU), 3*, 591–606.
Capello, R., & Rietvelt, P. (1998). The concept of network synergies in economic theory: Policy implications. In K. Button, P. Nijkamp, & H. Priemus (Eds.), *Transport networks in Europe: Concepts, analysis and policies* (pp. 57–83). Northampton, MA: Edward Elgar.
Castells, M. (1996). *The rise of the network society*. Oxford: Oxford University Press/Blackwell.
Castells, M. (2000). *The information age: Economy, society and culture*. Malden, MA: Blackwell.
Castells, M. (2009). *Communication power*. Oxford: Oxford University Press/Blackwell.
Chorley, R. J., & Haggett, P. (1969). *Network analysis in geography*. London: Edward Arnold.
Courgeau, D. (Ed.). (2003). *Methodology and epistemology of multilevel analysis. Approaches from different social sciences* (Vol. 2). Dordrecht: Kluwer Academic Publisher.
Dacey, M. F. (1964). A note of some number properties of a hexagonal hierarchical plane lattice. *Journal of the Regional Science Association, 5*, 63–67.
Dicken, P. (2011). *Global shift: Mapping the changing contours of the world economy*. New York: Guilford Press.
Dicken, P., Kelly, P. F., Olds, K., & Yeung, H. W. C. (2001). Chain networks, territories and scales: Toward a relational framework for analysing the global economy. *Global Networks, 1*(2), 89–112.
Dicken, P., & Malmberg, A. (2001). Firms in territories: A relational perspective. *Economic Geography, 77*(4), 345–363.
Dollfus, O. (2001). *La mondialisation*. Paris, France: Presses de science Po.
Doz, Y., Santos, J., & Williamson, P. (2001). *From global to metanational: How companies win in the knowledge economy*. Boston: Harvard Business School Press.
Freeman, L. C. (2004). *The development of social network analysis: A study in the sociology of science*. Vancouver, British Columbia, Canada: Empirical Press.
Garrison, W. L. (1960). Connectivity of the interstate highway system. *Papers and Proceedings of the Regional Science Association, 6*, 121–137.
Garrison, W. L., Berry, B. J. L., Marble, D. F., Nystuen, J. D., & Morrill, R. L. (1959). *Studies of highway development and geographic change*. New York: Greenwood Press.
Gereffi, G. (1996). Global commodity chains: New forms of coordination and control among nations and firms in international industries. *Competition and Change, 1*(4), 427–439.
Gibrat, R. (1931). *Les inégalités économiques*. Paris, France: SIRET.
Giddens, A. (1984). *The constitution of society. Outline of the theory of structuration*. Berkeley, CA: University of California Press.
Gould, P., & White, R. (1974). *Mental maps*. Harmondsworth: Penguin.
Gould, R., & Fernandez, R. (1989). Structures of mediation: A formal approach to brokerage in transaction networks. *Sociological Methodology, 19*, 89–126.
Grabher, G. (2006). Trading routes, bypasses, and risky intersections: Mapping travels of "networks" between economic sociology and economic geography. *Progress in Human Geography, 30*(2), 163–189.
Guerin Pace, F. (1995). Rank-size distribution and the process of urban growth. *Urban Studies, 32*(3), (551–562).
Guimerà, R., Mossa, S., Turtschi, A., & Amaral, L. A. N. (2005). The worldwide air transportation network: Anomalous centrality, community structure, and cities' global roles. *Proceedings of the National Academy of Sciences of the United States of America, 102*(22), 7794–7799.

Hassaert da Costa, R. (2004). De la déterritorialisation à la multiterritorialité. In S. Allemand, F.R. Ascher, & J. Lévy (dir.), *Les sens du mouvement* (pp. 69–79). Paris, Berlin: Institut pour la ville en mouvement.

Hägerstrand, T. (1952). *The propagation of innovation waves* (Vol. 4). Lund, Sweden: Gleerup.

Hobson, J. A. (1909). [1894] *The evolution of modern capitalism: a study of machine production*. New York: Walter Scott.

Jacobs, J. (1969). *The economy of cities*. New York: Random House.

Kansky, K. (1963). *Structure of transportation networks: Relationships between network geometry and regional characteristics* (Technical Rep. No. 84). Chicago: Department of Geography, University of Chicago.

Lane, D. (2006). Hierarchy, complexity, society. In D. Pumain (Ed.), *Hierarchy in natural and social sciences* (pp. 81–119). Dordrecht: Springer.

Latour, B. (2005). *Reassembling the social: An introduction to actor-network theory*. Oxford: Oxford University Press.

Macfarlane, A. (1883). Analysis of relationships of consanguinity and affinity. *Journal of the Royal Anthropological Institute of Great Britain and Ireland, 12*, 46–63.

Milgram, S. (1967). The small world problem. *Psychology Today, 1*, 61.

Monge, P., & Contractor, N. (2003). *Theories of communication networks*. Oxford: Oxford University Press.

Morgan, L. (1851). *League of the Ho-dé-no-sau-nee or Iroquois*. Rochester, NY: Sage.

Newman, M. (2005). Power laws, Pareto distributions and Zipf's law. *Contemporary Physics, 46*(5), 323–351.

Newman, M., Watts, D., & Barabasi, A.-L. (Eds.). (2006). *The structure and dynamics of networks*. Princeton, NJ: Princeton University Press.

Ohmae, K. (1990). *The borderless world*. New York: Harper & Row.

Ohmae, K. (1995). *The end of the nation state – How region states harness the prosperity of the global economy*. New York: Free Press.

Pumain, D., Paulus, F., Vacchiani-Marcuzzo, C., & Lobo, J. (2006). An evolutionary theory for interpreting urban scaling laws. *Cybergéo: Revue européenne de géographie, 343*, 20 p.

Pomfret, R. (2007). Is regionalism an increasing feature of the world economy. *The World Economy, 30*(6), 923–947.

Pool, I. d. S., & Kochen, M. (1978). Contacts and influence. *Social Networks, 1*, 1–48.

Powell, W. (1990). Neither market nor hierarchy: Network form of organization. *Research in Organizational Behaviour, 12*, 295–336.

Pred, A. (1974). *Major job-providing organizations and systems of cities*. Washington, DC: Association of American Geographers.

Pred, A. (1977). *City-systems in advanced economies*. London: Hutchinson University Library.

Pumain, D. (1982). *La dynamique des villes*. Paris, France: Economica.

Pumain, D. (1997). Pour une théorie évolutive des villes. *L'Espace Géographique, 2*, 119–134.

Pumain, D. (2006). Alternative explanations of hierarchical differentiation in urban systems. In D. Pumain (Ed.), *Hierarchy in natural and social sciences* (pp. 169–222). Dordrecht: Springer.

Pumain, D., Sanders, L., & Saint-Julien, T. (1989). *Villes et auto-organisation*. Paris, France: Economica.

Raffestin, C. (1980). *Pour une géographie du pouvoir*. Paris, France: Library Techniques.

Rozenblat, C. (2004). Intégration dans le commerce international: l'évidence du graphique triangulaire. *Mappemonde, 75*, 15 p.

Rozenblat, C. (2007). Villes et réseaux "petits-mondes". In A. Da Cunha & L. Matthey (Eds.), *La ville et l'urbain: des savoirs émergents* (pp. 81–105). Lausanne, Switzerland: Presses Polytechniques Romandes.

Rozenblat, C., Bohan, C., & Benet, G. (2008). Multinational corporate networks in Central European cities. In M.-C. Maurel (Ed.), Central European cities in globalization. CEFRES: Prague.

Rozenblat, C., & Cicille, P. (2003). *Les villes européennes: Analyse comparative/Die Stadte Europas. Ein vergleichende Analyse*. DATAR-La Documentation française Paris / Bündesamt für Bauwesen und Raumordnung. Bonn, 123 p.

Rozenblat, C., & Pumain, D. (1993). The location of multinational firms in the European urban system. *Urban studies, 10*, 1691–1709.

Rozenblat, C., & Pumain, D. (2007). Firm linkages, innovation and the evolution of urban systems. In P. J. Taylor, B. Derudder, P. Saey, F. Witlox (Eds.), *Cities in globalization. Practices, policies and theories* (pp. 130–156). London: Routledge.

Rugman, A. (2001). *The end of globalization: Why global strategy is a myth and how to profit from realities of regional markets*. New York: AMACOM.

Sanders, L. (1992). *Système de villes et synergétique*. Paris, France: Anthropos.

Sassen, S. (1991). *The global vity: New York – London – Tokyo*. Oxford: Princeton University Press.

Simon, H. A. (1955). On a class of skew distribution functions. *Biometrika, 42*(3–40), 425–440.

Storper, M., & Salais, R. (1997). *Worlds of production. The action frameworks of the economy*. Cambridge, MA: Harvard University Press.

Thomas, J., & Cook, K. (2006). A visual analytics agenda. *Computer Graphics and Applications, IEEE, 26*(1), 10–13.

Tobler, W. R. (1963). Geographic area and map projections. *Geographical Review, 53*(1), 59–78.

Tobler, W. R. (1965). Computation of the correspondence of geographical patterns. *Papers of the Regional Science Association, 15*(1), 131–139.

Travers, J., & Milgram, S. (1969). An experimental study of the small world problem. *Sociometry*, 32, 425–443.

Uzzi, B. (1997). Social structure and competition in interfirm networks: The paradox of embeddedness. *Administrative Science Quarterly, 42*, 35–67.

van Wijk, J. (2005). The value of visualization. In C. Silva, E. Groeller & H. Rushmeier (Eds.), *IEEE visualization conference (VIS 2005)* (pp. 79–86).

Walker, G., Kogut, B., & Shan, W. (1997). Social capital, structural holes and the formation of an industry network. *Organization Science*, *8*(2), 109–125.

Watts, D. (1999). *Small worlds*. Princeton, NJ: Princeton University Press.

Watts, D., & Strogatz, S. (1998). Collective dynamics of "small-world" networks. *Nature, 393*, 440–442.

Whitley, R. (1998). Internationalization and varieties of capitalism: The limited effects of cross-national coordination of the economic activities on the nature of business systems. *Review of International Political Economy, 5*, 445–381.

Willis, J., & Yule, G. (1922). Some statistics of evolution and geographical distribution in plants and animals, and their significance. *Nature, 109*, 177–179.

Wilson, A. (1967). A statistical theory of spatial distribution models. *Transportation Research, 1*, 253–269.

Wilson, A. (2000). *Complex spatial systems: The modelling foundations of urban and regional analysis*. Harlow: Prentice Hall Education.

Yeung, H. W. C. (2002). The limits to globalization theory: A geographic perspective on global economic change. *Economic Geography, 78*, 285–305.

Yule, G. (1925). A mathematical theory of evolution, based on the conclusions of Dr. J. C. Willis, F.R.S. *Philosophical Transactions of the Royal Society B: Biological Sciences 213*, 21–87.

Zipf, G. K. (1949). *Human behaviour and the principal of least effort*. Cambridge, MA: Adison-Wesley.

Part I
Concepts and Visualizations of Multilevel Spatial Networks

Chapter 2
A Small World Perspective on Urban Systems

Céline Rozenblat and Guy Melançon

2.1 Introduction

The increasing complexity of spatial organizations, in particular the organization of cities, challenges the study of spatial distributions. Places tend to be seen more and more as nodes in specialized networks, and their future is undoubtedly strongly dependent on their position in these networks (Castells, 1996). Geographical distances and accessibility always matter in spatial dynamics, but in parallel, other distances appear as relevant: social, economical, cultural and organizational distances can also interfere in spatial dynamics. As a consequence, geography, as a science of society in space, must work at defining and measuring how spatial accessibility dynamics influence other kinds of distances and how processes emerge from spatial or territorial dimensions. From this perspective, in a social-science division of work, geography could be defined as the science dedicated to the identification of spatial patterns and their implications at specific geographical levels of scale.

These geographical levels are themselves evolving according to systems of social processes. For example, globalization processes are developing as transport and communication speeds are increasing (and as costs are decreasing) but are also changing according to organizational transformations of society, deriving from actors themselves, like companies and subsidiaries, social groups and institutional arrangements. These types of interactions and their dynamics explain why

C. Rozenblat (✉)
Institute of Geography, University of Lausanne, GEOPOLIS Building,
1015 Lausanne, Switzerland
e-mail: Celine.Rozenblat@unil.ch

G. Melançon
CNRS UMR 5800 LaBRI, GRAVITE INRIA Sud-Ouest, Université Bordeaux 1,
33405 Talence Cedex, France
e-mail: Guy.Melancon@labri.fr

C. Rozenblat and G. Melançon (eds.), *Methods for Multilevel Analysis and Visualisation of Geographical Networks*, Methodos Series 11, DOI 10.1007/978-94-007-6677-8_2,

globalization has been developed at a worldwide scale for stock exchange, while "global value chains" of production are developing, most of the time, at a continental scale, leveraging from "regional" international trade arrangements (Dicken, Kelly, Olds, & Yeung, 2001; Doz, Santos, & Williamson, 2001).

The hypothesis we just stated suggests that all transformations strongly depend on the internal and external relations of network components, calling for a study of the relative positions of places in networks and at different geographical scales. Geographical interactions can be studied through multi-dimensional networks, which can be considered as graphs. To date, most geographical network studies were based either on gravity models (Wilson, 1967) or on general indexes of graph connectivity or accessibility (Kansky, 1963). Graph theory and more particularly Watts' theory of small-world networks (Watts, 1999) can feed this trend of research and enrich spatial analysis as a complement to systems theory. Networks indeed organize through direct links or by indirect paths, inducing topological proximities that simultaneously involve spatial, social, cultural or organizational dimensions. Network synergies, either based on the similarities or the complementarities between or inside cities, create specialized and more or less stable proximities between these entities (Powell, 1990). Interrelations between geographical scales of organization also allow an understanding of the equilibrium or disequilibrium of territories emerging at different geographical scales. The concepts and methods of the small-world theory are particularly relevant to geography, in which spatial interaction is mainstream and where interactions can be described and studied using large numbers of exchanges or similarity matrices. As far as we know, this type of empirical approach combining a conceptual approach of the "small-world theory" and dedicated tools has not been developed in geography.

Multi-level analyses and the visualization of graphs bring new questions. How do networks link similar actors, reinforcing their social proximities as well as spatial specializations and segregations? In this sense, how do networks (not only transport and communication networks but also social and economic networks) increase the anisotropy of space? How far are networks from random processes, instead creating densely connected groups of nodes separated from others, like "small-world" networks (Watts, 1999; Watts & Strogatz, 1998)? What are the properties of these networks, and what consequences can we infer for geographical places, especially for cities?

In this presentation, we will stress the advantages of analyzing and visualizing networks using their network properties (Sect. 2.2). Then, we will underline topological positions in a static state as well as from a dynamic perspective (Sect. 2.3) and transfer them onto properties of city systems (Sect. 2.4). The final section discusses levels of organizations, taking into account empirical findings on hierarchical clustering (Sect. 2.4).

2.2 Small World Theory and Scale-Free Networks

This chapter is devoted to the introduction of small-world theory and scale-free networks. Introducing these formal notions in a separate section will allow us to freely build the discussion around the examples studied by members of the SPANGEO project without having to pause for mathematical digressions. The next chapter goes into technical details and formally defines and compares the various network indices that we refer to throughout the book. Here, we aim to introduce the main concepts that we transpose from graph theory to geography.

Graph theory, developed especially by Erdös and his school since the 1950s, was based on the concept of a "random graph", making the strong assumption that all relationships are likely to occur with equal probability (Erdös & Rényi, 1959, 1960). Put simply, a random graph is one in which edges are drawn at random, playing heads or tails; that is, there are no regions in the graph where edges have a higher probability to appear, leading to uniformly distributed connections among nodes.

Since the late 1990s, many researchers have proposed alternatives to this model, showing the importance of "small worlds" in the organization of societies and their dynamics, where networks are not seen as homogeneous (Newman, 2000; Newman, Watts, & Barabàsi, 2006). Small worlds are built from groups of tightly connected communities, themselves connected through privileged nodes. The small-world phenomena, experimentally identified by Milgram (1967) in social networks, was observed in many other fields, giving this notion a hint of universality. In biology, for instance, networks of interacting proteins show a small-world character, where communities of proteins embody cell functions (Vespignani 2003). The internet is another well known and extensively studied example, beginning with Adamic (1999). Two main properties typify these networks:

- The average path length from one individual to another is small and compares to the average path length of random graphs (with the same number of nodes and edges);
- The networks have a strong propensity to create sub-groups (also called communities or clusters). This propensity is the strongest characteristic of small-world networks and is the result of several micro processes, like transitivity, homophily, and the range of ties.

In parallel to Watts and Strogatz's theory of small worlds, some authors observed and investigated another property of networks, referring to the networks as being scale-free. These networks organize into an order hierarchy with respect to the degree of nodes and evolve along a basic process. According to Barabási and Albert (1999) (see also Barabási, 2002), who popularized these "scale-free networks", this order hierarchy is due to the "preferential attachment" that characterizes the development of networks: new links preferentially link to nodes that already have a greater number of links, favoring the hierarchy. In reality, we can find early explanations of the Zipf law proposed by Simon (Simon, 1955), referring to the "Yule process" (Yule, 1925). This process thus induces the degree distribution of the nodes to follow a power law; that is, the frequency of nodes of degree k is approximately given by $p(k) \sim c \cdot exp(-k)$. In other words, a sharp inequality exists between strongly central entities and others who cling to these centralities (the interested reader is referred to Bornholdt & Schuster (2003) or Newman (2010), for instance).

The scale-freeness of spatial networks makes sense, as the forthcoming sections shall explain. However, this property makes their visual exploration and (automatic) analysis much harder. Most often, networks show both characteristics simultaneously. That is, networks do contain small communities of interacting entities, while the overall organization of the network is dominated by a scale-free process. The high-degree nodes thus make it hard to identify communities or bridges and hide the small-world structure. Hybrid approaches must be designed to reveal the structure and properties of these complex networks.

2.3 Visualizing Topological Proximities in Geographical Networks

Visualizations are common tools for geographers, who use cartography to stress geographical organizations and patterns. Nevertheless, because geographical distance is not the only factor constraining a system and its dynamics, a topological representation can be much more useful to highlight organizational structures.

A basic geographical example is the worldwide container-traffic networks, for which visualization allows a good overview of the overall structure (Fig. 2.1) (Ducruet, Rozenblat, & Zaidi, 2010). On one hand, the representation ignores the geographical coordinates of ports. On the other hand, the ports' location in the graph is determined by applying a layout based on topological proximities for maritime exchanges. Therefore, the geographical situation is preserved due to the continuum of movements from one port to another.

Fundamentally, two main sub-systems exist: Asia-Pacific and Europe-Atlantic. These sub-systems are connected through a limited number of nodes: the Panama and Suez canals. Removing those two global pivots would thus split the world system in two parts, although one may notice some other links (e.g., Magellan Strait and Cape of Good Hope), but those links remain very limited compared to the main trunk lines concentrated at the two canals. Also, port ranges appear, such as the Scandinavian range in the top left of the Fig. 2.1, close to American ranges.

Another example maintaining some geographical position in a topological network representation is the movement of commuters (Fig. 2.2). The example of Switzerland shows especially the linguistic proximities among municipalities with three visible isolated zones: French Swiss speaking at the left, Italian speaking at the right, and Swiss German in the center, cut into two parts between a system around Bern (the political federal capital) and another around Zürich (the economic capital).

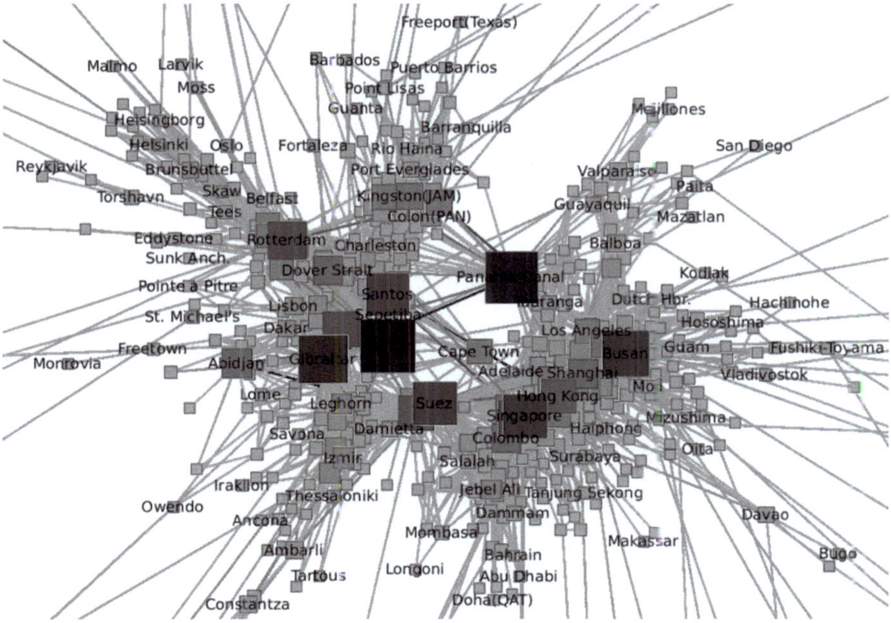

Fig. 2.1 Worldwide container traffic network 2006

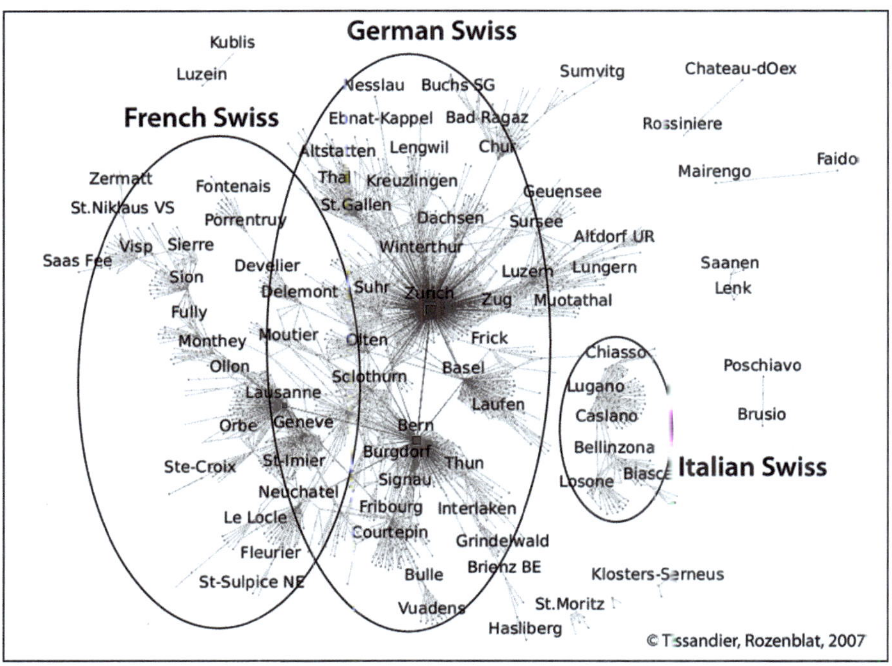

Fig. 2.2 Commuters between Swiss Municipalities in 2000

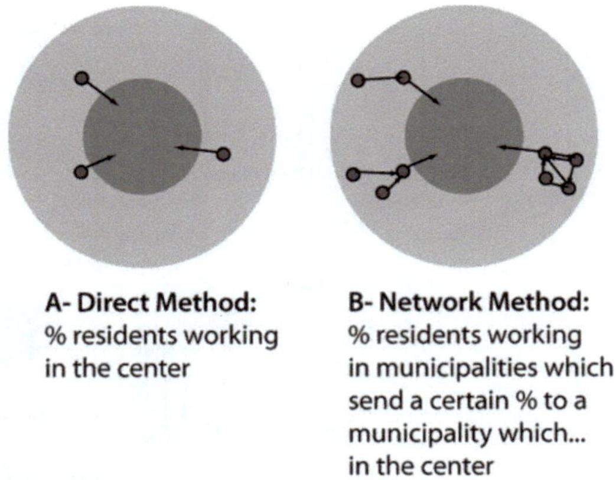

Fig. 2.3 Urban-area delineation using transitivity properties

2.4 Network Properties As Cities Systems Properties

The small-world property has a clear interpretation for territories in general and for urban systems in particular. The short average distance between entities can be seen as a consequence of redundant paths through different routes between city networks, strengthening direct and/or indirect interaction. This mutual interaction is mainly through "clusters", which can form inside cities or towns and that are heavily connected by groups at another level between cities.

2.4.1 Urban Systems As Small Worlds

Empirical evidence of these networks is founded on some basic properties reinforcing these strengths (Monge & Contractor, 2003), which, in transposition, could make sense for geographical network organizations:

- Transitivity: This elementary property was shown early by sociologists (Weimann, 1980): If an individual A is linked to B, and B is linked to C, then A is most likely linked to C.

  In geographical perspectives, we can transpose this property to interaction and interdependency effects. If a place A is closely related and interdependent on B, and B is linked to C then A is interdependent on C. For example, commuter exchanges between municipalities create these kinds of chain dependencies (Fig. 2.3). If a factory employing several thousands of people in A suddenly closes, it will affect the residential population of B that sends commuters to A, but in consequence, it also will affect C sending commuters to B. Sometimes, this transitive relation is materialized by real exchanges (like commuters), but even if there are not such direct materialized exchanges, interdependencies also occur through indirect effects. Therefore, some local chains create tangled networks, allowing very integrated local systems of territories (see Chap. 12).

At the inter-urban scale, interdependencies also make sense in mutual exchanges (symmetric or not), increasing proximities in terms of space (e.g., local or regional economical networks) or specialization (e.g., worldwide clusters linked through multinational firms' global value chains or through research centers). These new proximities stimulate the growth of exchanges, strengthening the closeness of some specific groups of cities or territories in spatial, economic, social or cultural proximities.

- Homophily: Networks between actors, territories or cities with similar attributes have a greater probability of occurring. These exogenous criteria actually interact with networks because exchanges could also transform places as they transform individuals with imitative behaviors and diffusions of many types of ideas, concepts and technologies.
- Range: In parallel to homophily, diversity of the elements of places and cities linked by networks seems essential to the reproduction and renewal of territorial systems. This diversity could be seen in terms of exogenous (for example, the diversity of activity sectors) and endogenous characteristics (the diversity of links to other places).

2.4.2 Urban Systems As Scale-Free Networks – Hierarchies of Individuals

There is evidence in urban systems that innovations spread preferentially from the largest cities, already a concentration of multiple networks, to the smallest ones (Batty, 2005; Pumain, 2006). At a micro level of individual behavior, these processes are largely due to the search for security rather than maximization, in a context of partial information in game theory (Luce & Raiffa, 1957, Chapter 3, Section 2). The "bounded rationality" concept mainly developed by Simon (1955, 1972, 1999) explains why, in a context of a random probability of meeting opportunities, individuals choose the first satisfactory opportunity rather than the best one. These trends lead to a strengthening of major cities, where opportunities are more diversified and where interactions are maximized, decreasing transaction costs (Coase, 1937). This process allows some specific properties of centralities, equivalence or structural equivalence for places or groups of places according to the positions of located individuals.

- Centralities: inequalities between places in terms of their relative position in networks are obviously dependent on accessibility or reachability into some set of places in a certain context and at a certain period (Bretagnolle, Pumain, & Rozenblat, 1999). Many kinds of indexes allow close analysis of these central

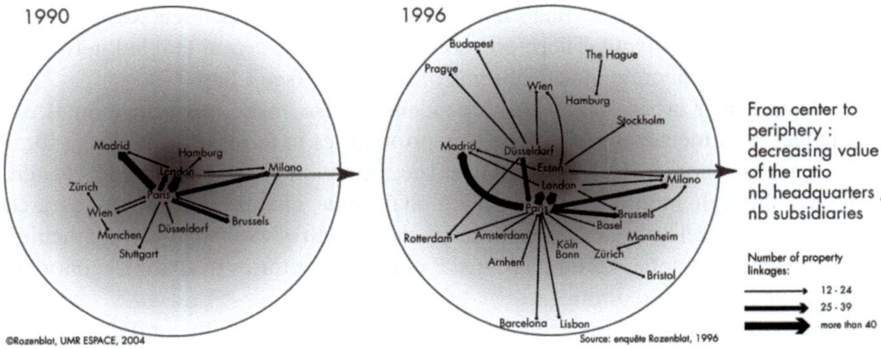

Fig. 2.4 Hierarchy of European cities according to their control of multinational firms from 1990 to 1996

positions, such as the node degree (number of links), reachability (distance to all other points), betweenness centrality (number of shortest paths passing through a point), or generalized Strahler index (Auber, Delest, & Chiricota, 2004; Delest & Auber, 2003), measuring how many trees can be built from a node; every measure can be weighted or not (Wasserman & Faust, 1994; Newman, 2003; Brandes & Erlebach, 2005). These indexes are all very useful to compare node positions in a graph or even the position of the same node in different graphs (multiplex graph).

- Dependencies and power: this characteristic is considered when social or economic networks are oriented either reflecting ownership between enterprises, decision-making, or hierarchy. The balance between incoming links (in-degree) and outgoing links (out-degree) can be a measure of such dependencies or power. For example, this method can be used to easily identify cities of power in multinational-firm networks according to the opposition between headquarters and subsidiaries (Fig. 2.4). Links are oriented and the centrality in the graph is calculated based on the share of the number of headquarters (representing the out-degree) relative to the number of subsidiaries (in-degree). Between 1990 and 1996, the graph has spread all over Europe, particularly areas like southern and eastern Europe. London, with a high share of headquarters/subsidiaries, is the most central, with many American headquarters controlling subsidiaries in the continent. However, Paris seems to be much more connected to other larger European cities.

  Also, one can identify dual positions in which some cities are only linked to one other. According to Nystuen and Dacey's methodology (1961), one can build a conceptual hierarchy with the highest link of each city going to a bigger city. All of these links create some "regional systems". Therefore, a hierarchy can also emerge from non-oriented links according to weighted dependencies. Figure 2.5 shows this construction applied to connections among European cities using complete graphs built inside each firm of a sample of 100 groups in 1996 (Fig. 2.5). The first link of the majority of the analyzed European cities is oriented to Paris, the most connected city according to this kind of graph of firms.

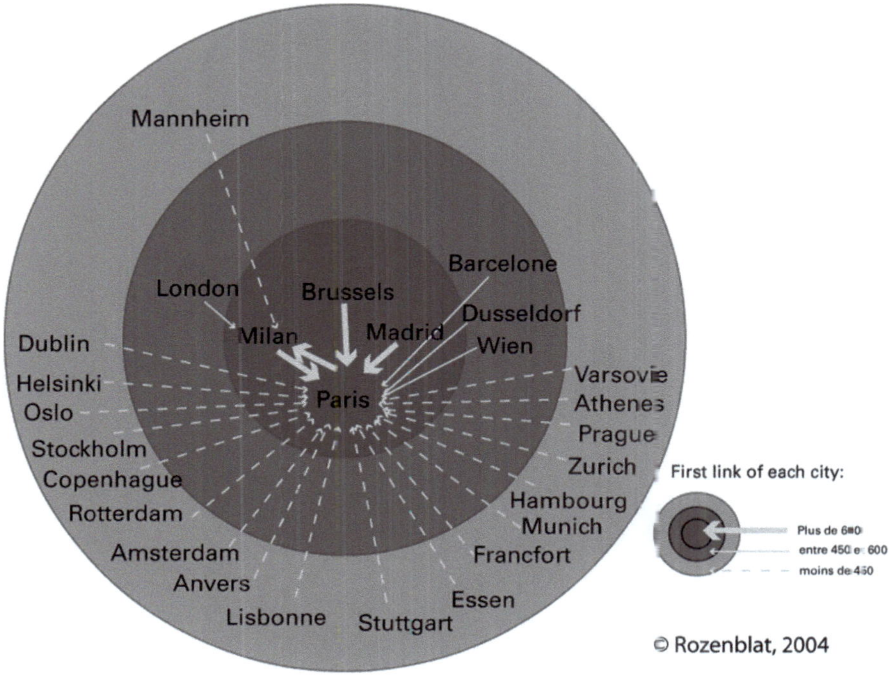

Fig. 2.5 Hierarchy of European cities according to their positions in multinational firms in 1996

- Equivalence and structural equivalence: organizational structure is viewed as a pattern of relations among positions. White, Boorman, and Breiger (1976) and Burt (2000) have developed positional theories of structural equivalence. According to Monge and Contractor (2003, p. 19), this theory "argues that people maintain attitudes, values, and beliefs consistent with their organizational positions irrespective of the amount of communication that they have with others in their organizational networks". From a geographical perspective, we could argue that places like cities are formed by tangled and overlapped social, economic and institutional networks that also put them in equivalent positions. For example, cities of the same country are equivalent with respect to their political national capital in many ways, from hierarchical administration to hierarchical economy and culture. In contrast, national capitals are in structural equivalence with their respective national urban systems, although some national networks are more hierarchical than others.

Bridge cities can be identified by confronting the weighted degree and betweenness centrality of each city at every level (Barrat, Barthélémy, & Vespignani, 2005; Guimerà, Mossa, Turtschi, & Amaral, 2005) In fact, degree and betweenness centrality are very correlated, although they have different natures: local (for degree) and global (for betweenness centrality). However, a low degree with a high betweenness centrality especially shows the role of bridges between clusters (Fig. 2.6 left), and the threshold to apply these measures must

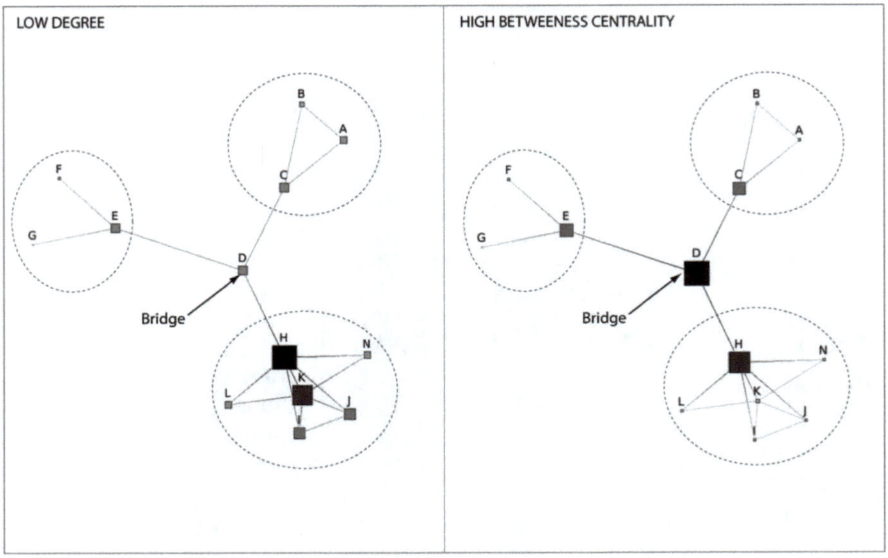

Fig. 2.6 Identification of bridges

be carefully selected to properly identify these bridges (Fig. 2.6 right). We can extend this approach to test all kinds of confrontations of local and global measures.

Individuals or organizations acting as bridges (or connectors) can override the "structural holes", strengthening the entire "social capital" but primarily their own capital (Burt, 2000, 2005). These bridges have a very strong strategic centrality even if they have a small number of links (Guimerà et al., 2005). A city can form a "bridge" between several cities when it is located at the interface between different groups or clusters or between different units (e.g., as a relay between a country's national cities and abroad or between a continent and the rest of the world).

In dynamic terms, such urban placements require high flexibility in networks, strong adaptability of cities and reactivity of the network actors. In contrast, a hierarchy that is too strong between nodes freezes the system but leaves individuals excluded from any network. The limitations of networks between what is integrated and not integrated are highly constitutive of the networks' operation and evolution (Barabási, 2003).

2.4.3 Hierarchies of Spatial Groups of Cities

If we assume that interaction implies interdependency, then groups of cities that exchange more together than with others are more dependent upon each other. Such cities constitute groups that can be defined in different ways (as explained below).

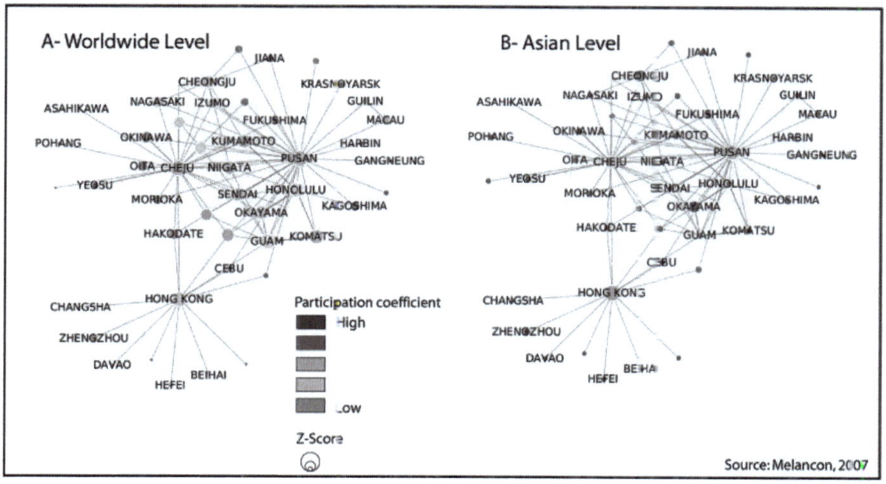

Fig. 2.7 Identification of hubs based on contribution indexes

Whatever method of classification is used, these groups imply some relative closure due in part to the property of transitivity, as previously discussed. In other cases, the network organizes into a star-shaped structure built around a central node that acts as a hub connecting the whole group to the outside world. This phenomenon can be tracked by measuring the relative density or "cohesion" of a group. As in classical classifications, some nodes (like cities) contribute more than other nodes to the cohesion of their groups, which may be seen as the "contribution" of each point to its group. Conversely, each city can also have links going out to the group, calling for a measure of the "participation" of a node in other groups in the network, as explored by (Guimerà et al., 2005). Contribution and participation indexes can be measured as individual shares in one node to one group or the opposite (details are discussed in the next chapter).

These indexes can help identify hubs in networks The role of hubs is often concentrated in or between particular communities. Guimerà et al. (2005) presented a z-score and "participation coefficient" of nodes that can be extended to take edge weights into account. The z-score of a node measures how many of the node's connections are devoted to its own cluster, while a node's participation coefficient measures how much its connections cover all of the other clusters in the network. Because we cluster the graph into a hierarchy of subgraphs, we can moreover measure these indices at every level and study how the indices vary down the hierarchy of clusters. Again, the full definition of these measures is delayed until the next chapter.

The figure below provides a good illustration (Fig. 2.7). The nodes have been sized according to their z-score (internal connections) and colored according to their participation coefficient (external connections). For example, Hong-Kong has a greater role at the worldwide level than at the Asian level.

Looking at node contribution and node participation as measures of interactions between nodes and groups and computing entropy help to provide a global view of the overall dynamics of a network. The significance of these entropy indexes could be interpreted as usual for entropy as signifying the degree of information contained in each group or information for each node (Lin, 1999; Shannon, 1948). With greater information included inside a node or group, the node or group's resilience and capability to renew are greater. Therefore, maximum entropy corresponds to a great variety of links for each node (balanced participation in many groups) or, at the level of clusters, various contributions of the nodes forming each group (and not only one in the case of star patterns). This variety contains a high disorder and could be interpreted as the system's high capability to maintain and renew itself (Bailey, 1990; Burt, 2005). This pattern illustrates for a particular node the "strength of weak ties" introduced by (Granovetter, 1973), in which a node has more opportunity for a large number of weak links by relying on differentiated clusters formed by larger links. In those clusters, the vicinities of economic links can develop processes of agglomeration economies or network economies. However, weak links between clusters are a source of diversity, which also produces a savings network, which allows cities both to operate and renew themselves.

It is the coupling between strong and weak links that allows the reproduction of the urban system and its transformations (Guimerà et al., 2005; Uzzi, 1997). We could identify such types of clusters between cities in different kinds of networks (e.g., transport, communication, migrations, multinational companies, etc.), and we could specify for each city the intensity of its membership in a cluster and assess to what extent these clusters correspond to territorial logic (of continents, countries or urban areas) or the logic of economic specialization (cities specialized in the same fields).

Besides diversity, specialization and "closure" are common patterns in networks, according to the general form of "scale-free" laws (see above). The slope index of the "scale-free function" can also constitute a norm used to measure high concentrations (Batty, 2005; Pumain, Paulus, Vacchiani-Marcuzzo, & Lobo, 2006). Then, the scale-free measures of the contributions inside each group could be adopted, but difficulty remains in defining these groups and their relative positions.

2.5 Conclusion

It is essential that our knowledge of all of the properties of these networks of cities be deepened through a multi-level perspective. Identifying, qualifying and measuring each level could better facilitate globalization because, at each level, there are many processes that allow the emergence of urban properties in which "the whole becomes not only more than but very different from the sum of its parts" (Anderson, 1972) (as cited in Lane, 2006).

These methods facilitate the definition and qualification of new properties of spatial resources based on networks, and new spatial groupings then lead to very new approaches in geography. We have to test the contributions of these methods to empirical studies and underline the new questions they bring out.

References

Adamic, L. A. (1999). The small world web. In *Third European conference on research and advanced technology for digital libraries* (Vol. 1696, pp. 443–452). Berlin: Springer.

Anderson, P. (1972). More is different: Broken symmetry and the hierarchical nature of science. *Science, 177*, 393–396.

Auber, D., Chiricota, Y., Jourdan, F., & Melançon, G. (2003). Multiscale navigation of small world networks. In *IEEE symposium on information visualisation* (pp. 75–81). Los Alamitos, CA: IEEE Computer Science Press.

Auber, D., Delest, M., & Chiricota, Y. (2004). Strahler based graph clustering using convolution. In *Eighth international conference on information visualisation (IV'04)* (pp. 44–51). Washington, DC: IEEE Computer Society.

Bailey, K. D. (1990). *Social entropy theory*. New York: State University of New York Press.

Barabási, A.-L. (2002). *The new science of networks*. Cambridge, MA: Perseus.

Barabási, A.-L. (2003). *Linked: How everything is connected to everything else and what it means for business, science, and everyday life*. New York: A Plume Book

Barabási, A.-L., & Albert, R. (1999). Emergence of scaling in random networks. *Science, 286*, 509–512.

Barrat, A., Barthélémy, M., & Vespignani, A. (2005). The effects of spatial constraints on the evolution of weighted complex networks. *Journal of Statistical Mechanics: Theory and Experiment, P05003*.

Batty, M. (2005). *Cities and complexity: Understanding cities with cellular automata, agent-based models and fractals*. Cambridge, MA: The MIT Press.

Bornholdt, S., & Schuster, G. (Eds.). (2003). *Handbook of graphs and networks: From the genome to the internet*. Cambridge, MA: Wiley-VCH.

Brandes, U., & Erlebach, T. (2005). *Handbook of network analysis: Methodological foundations*. Berlin: Springer.

Bretagnolle, A., Pumain, D., & Rozenblat, C. (1999). Space-time contraction and the dynamics of urban systems. *Cybergeo*.

Burt, R. (2000). The network structure of social capital. In R. Sutton & B. Staw (Eds.), *Research in organizational behavior*. Greenwich, CT: JAI Press.

Burt, R. (2005). *Brokerage and closure: An introduction of social capital*. New York: Oxford University Press.

Castells, M. (1996). *The rise of the network society*. Oxford: Oxford University Press/Blackwell.

Coase, R. (1937). The nature of the firm. *Economica, 4*, 386–405.

Delest, M., & Auber, D. (2003). A clustering algorithm for huge trees. *Advances in Applied Mathematics, 31*, 46–60.

Dicken, P., Kelly, P. F., Olds, K., & Yeung, W. C. H. (2001). Chain networks, territories and scales: Toward a relational framework for analysing the global economy. *Global Networks, 1*(2), 89–112.

Doz, Y., Santos, J., & Williamson, P. (2001). *From global to metanational: How companies win in the knowledge economy*. Boston: Harvard Business School Press.

Ducruet, C., Rozenblat, C., & Zaidi, F. (2010). Ports in multi-level maritime networks: Evidence from the Atlantic (1996–2006). *Journal of Transport Geography, 18*, 508–518.

Erdös, P., & Rényi, A. (1959). On random graphs. I. *Publicationes Mathematicae, 6*, 290–297.

Erdös, P., & Rényi, A. (1960). The evolution of random graphs. *Magyar Tud Akad Mat Kutató Int Kőzl, 5*, 17–61.

Granovetter, M. (1973). The strength of weak ties. *American Journal of Sociology, 78*(6), 1360–1380.

Guimerà, R., Mossa, S., Turtschi, A., & Amaral, L. A. N. (2005). The worldwide air transportation network: Anomalous centrality, community structure, and cities' global roles. *Proceedings of the national academy of sciences of the United States of America, 102*(22), 7794–7799.

Kansky, K. (1963). *Structure of transportation networks: Relationships between network geography and regional characteristics* (Technical Rep. Research Paper No. 84). Chicago: Department of Geography, University of Chicago.

Lane, D. (2006). Hierarchy, complexity, society. In D. Pumain (Ed.), *Hierarchy in natural and social sciences* (pp. 81–119). Dordrecht: Springer.

Lin, S. K. (1999). Diversity and entropy. *Entropy, 1*(1), 1–3.

Luce, R., & Raiffa, H. (1957). *Games and decisions: Introduction and critical survey*. New York: Wiley.

Milgram, S. (1967). The small world problem. *Psychology Today, 1*, (61).

Monge, P., & Contractor, N. (2003). *Theories of communication networks*. Cambridge: Oxford University Press.

Newman, M. (2000). Models of the amall world. *Journal Statistical Physics, 101*, 819–841.

Newman, M. (2003). The structure and function of complex networks. *SIAM Review, 45*, 167–256 (arXiv:cond-mat/0303516v1).

Newman, M. (2010). *Networks: An introduction*. Cambridge: Oxford University Press.

Newman, M., Watts, D., & Barabàsi, A.-L. (Eds.). (2006). *The structure and dynamics of networks*. Princeton, NJ: Princeton University Press.

Nystuen, J., & Dacey, M. (1961). A graph theory interpretation of nodal regions. *Regional Science Association, Papers and Proceedings, 7*, 29–42.

Pumain, D., Paulus, F., Vacchiani-Marcuzzo, C., & Lobo, J. (2006). An evolutionary theory for interpreting urban scaling laws. *Cybergéo: Revue européenne de géographie, 343*, 20 p.

Powell, W. (1990). Neither market nor hiearchy: Network forms of organization. *Research in Organizational Research, 12*, 295–336.

Pumain, D. (2006). Alternative explanations of hierarchical differentiation in urban systems. In D. Pumain (Ed.), *Hierarchy in natural and social sciences* (pp. 169–222). Dordrecht: Springer.

Shannon, C. E. (1948). A mathematical theory of communication. *The Bell System Technical Journal, 27*, 379–423, 623–656.

Simon, H. A. (1955). On a class of skew distribution functions. *Biometrika, 42*(3–4), 425–440.

Simon, H. A. (1972). Theory of bounded rationality. In C. B. Mc Guire & R. Radner (Eds.), *Decision and organization* (pp. 161–176). Amsterdam: North Holland Publishing Company.

Simon, H. A. (1999). Bounded rationality in social science: Today and tomorrow. *Mind and Society, 1*, 25–39.

Uzzi, B. (1997). Social structure and competition in interfirm networks: The paradox of embeddedness. *Administrative Science Quarterly, 42*, 35–67.

Vespignani, A. (2003). Evolution thinks modular. *Nature, 35*(2), 118–119.

Wasserman, S., & Faust, K. (1994). *Social network analysis*. Cambridge: Cambridge University Press.

Watts, D. (1999). *Small worlds*. Princeton, NJ: Princeton University Press.

Watts, D., & Strogatz, S. H. (1998). Collective dynamics of "small-world" networks. *Nature, 393*, 440–442.

Weimann, G. (1980) *Conversation networks as communication networks*. Abstract of Ph.D. dissertation, University of Haifa, Israel.

White, H., Boorman, S., & Breiger, R. (1976). Social structure from multiple networks. I. Blockmodels of roles and positions. *The American Journal of Sociology, 81*(4), 730–780.

Wilson, A. (1967). A statistical theory of spatial distribution models. *Transportation Research, 1*, 253–269.

Yule, G. (1925). A mathematical theory of evolution, based on the conclusions of Dr. J. C. Willis, F.R.S. *Philosophical Transactions of the Royal Society B: Biological Sciences, 213*, 21–87.

Chapter 3
Topological Clustering for Geographical Networks

Jean-François Gleyze

3.1 Introduction

In addition to transportation, energy, communications, city and human networks, the study of reticular structures in geography faces problems relative to the size and complexity of the structures. These problems are particularly daunting when considering networks at scales smaller than their usual scale of representation. In this case, the mass of components makes it difficult to understand the network and synthesize related information.

In addition to cartographic tools to generalize maps and make them easier to read at lower scales, tools have been developed to cluster network components and simplify their representation and assessment. These tools stem from graph theory and have been extended to a particular degree in the field of social networks. However, graph theory tools have not been widely applied in geography because spatial organizations cannot be described only by graph structures; they also require specific information to build relevant models.

Nevertheless, to provide useful guidelines to adapt these tools to issues facing geography, this chapter presents an overview of these clustering methods, with particular attention to the criteria of implementation. We illustrate these methods with examples of simple graphs and discuss the possible applications to the study of reticular structures in geography. This discussion reveals that these methods cannot be applied alone in most cases, but that the way they are constructed provides relevant criteria that can be used to develop suitable methods for the geographical context.

J.-F. Gleyze (✉)
Institut Géographique National – Laboratoire COGIT, Saint-Mandé, France
e-mail: jean-francois.gleyze@ign.fr

C. Rozenblat and G. Melançon (eds.), *Methods for Multilevel Analysis and Visualisation of Geographical Networks*, Methodos Series 11, DOI 10.1007/978-94-007-6677-8_3,

3.2 Definitions and Notations

Building clusters inside a graph requires the identification of structures that fulfill criteria related to graph theory. In that respect, the following paragraphs briefly present the graph theory concepts required to understand the clustering methods (for further information, see Berge, 1973; Bollobás, 1998).

3.2.1 Specific Notions of Graph Theory

A subgraph (or subgroup) corresponds to the union of a subset of nodes and of the edges linking those nodes.

A subgraph is called "maximal" in relation to a given property if it follows this property, and if every extension of this subgraph in which one or several outer nodes are added leads the graph to no longer follow this property.

Symmetrically, a subgraph is called "minimal" in relation to a given property if it follows this property and if every restriction of this subgraph in which one or several of its nodes are removed leads the graph to no longer follow this property.

A path between two nodes is a sequence of nodes and edges linking those nodes. The length of a path is given by the number of its edges. Two nodes are reachable from each other if there is at least one path between the nodes. The length of the shortest path(s) between two nodes corresponds to the distance between them (cf. Fig. 3.1).

Two nodes are said to be "connected" if there is at least one path between them. The graph is said to be "connected" if each pair of nodes is connected.

Two paths between i and j are independent if and only if they do not share any edges.

The number of independent paths between two nodes i and j corresponds to the minimum number of edges to be removed to disconnect i and j: this number is the connectivity index of the node pair $\{i, j\}$, and it provides information about the robustness of the paths between i and j on the graph (cf. Fig. 3.2).

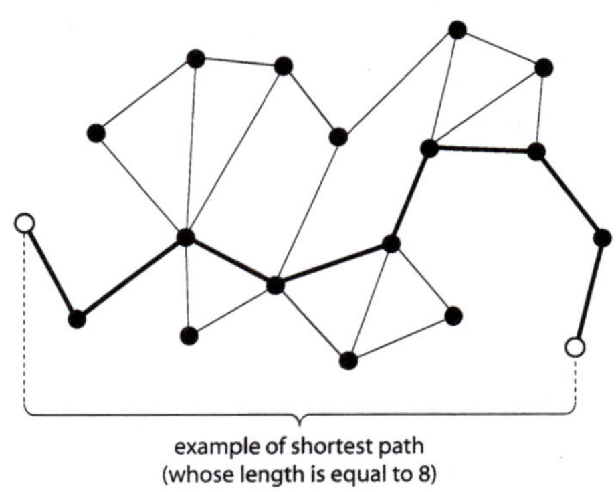

Fig. 3.1 Example of a shortest path on a graph

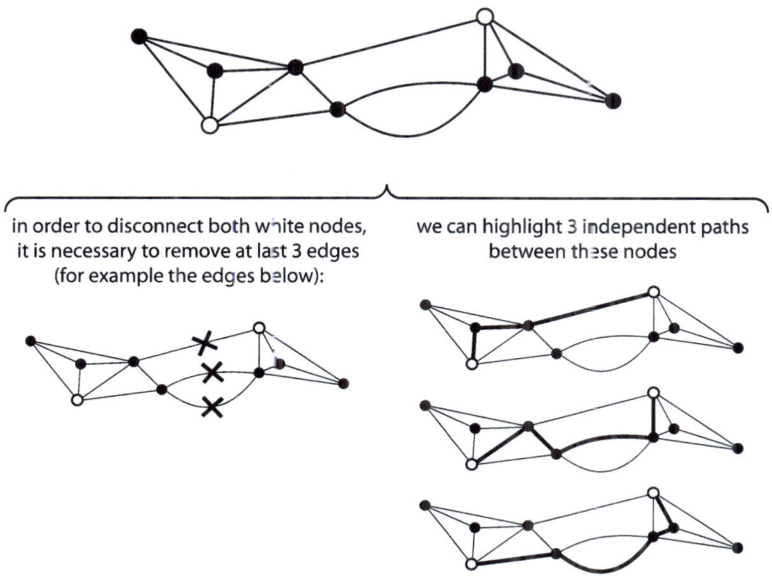

Fig. 3.2 The connectivity concept and its interpretation in terms of independent paths

By extension, the connectivity index of a graph corresponds to the minimum number of edges to be removed to disconnect this graph: as defined, the index is equal to the minimum of the connectivity indexes computed on all the node pairs of the graph.

The minimal number of edge sets or node sets that can be removed to disconnect the graph is called the number of "cut-sets".

In practice, the concepts from graph theory and their applications cannot be directly applied to geography. Studying a geographical network requires contextual adaptations that go beyond the strictly topological network description and integrate factors such as node location, node weighting, and edge values.

Among these factors, graph theory provides tools that take the direction and value of the edges into account: these extensions are important in a geographical context because they help to model the right way to pass through a section on the one hand (for example in road networks with one-way streets or expressways), and the cost of a section on the other hand (for example, the time required to pass through a section).

In the next paragraphs we present ways of integrating edge direction and value into network modeling and later into clustering methods.

3.2.2 Specific Features of Geographical Networks

Graph theory includes tools to take into account the possible edge directions (Berge, 1973; Bollobás, 1998). The suitable graphs correspond to "directed" graphs: inside

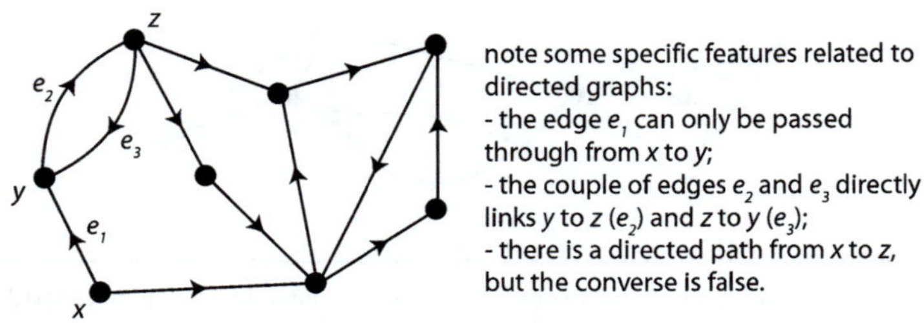

Fig. 3.3 Example of directed graph

these graphs, the edges are directed from an initial node to a final node. This modeling helps to represent asymmetric relationships in sociology ("I appreciate sb's company") and to consider the direction of a section of an infrastructure network (cf. Fig. 3.3). In this representation, the two-way sections have to be modeled by two edges opposite from each other.

Taking the possible impact of edge direction into account affects the meaning of the relationships between nodes. In this connection, edge directions may be integrated in several ways depending on the context of the study:

- a tie between two nodes x and y may be considered valid if either both edges $x \rightarrow y$ and $y \rightarrow x$ exist or if at least one of these edges exists;
- a connection between two nodes x and y may be assessed through both the shortest paths $x \rightarrow y$ and $y \rightarrow x$ or through the shortest path computed on the graph without taking edge direction into account.

In practice, the interpretation of edge direction depends on the type of network under study and on the meaning of the relationship between two nodes within a network. For example, within the context of risk, it is relevant to allow for two-way circulation on one-way sections because it may provide relief. This reasoning cannot be extended to air transport networks in which directed edges correspond to flights from one airport to another: for such graphs it is essential to take edge direction into account for path computation.

Graph theory is also a useful way to improve network modeling by taking the possible lengths or costs of sections into account (Berge, 1973; Bollobás, 1998). The operation consists of "valuating" edges by a scalar: from then on, the path lengths are no longer measured by the number of edges but by the total value of their edges (cf. Fig. 3.4).

As for edge orientation, edge values may be integrated in several ways:

- a direct link between two nodes x and y may depend on an additional condition related to the edge value (for example, the tie is considered valid if there is an edge linking both nodes and if its value is lower than a given threshold);
- in the same way, the path from x to y may be considered valid if its total length is below a given threshold.

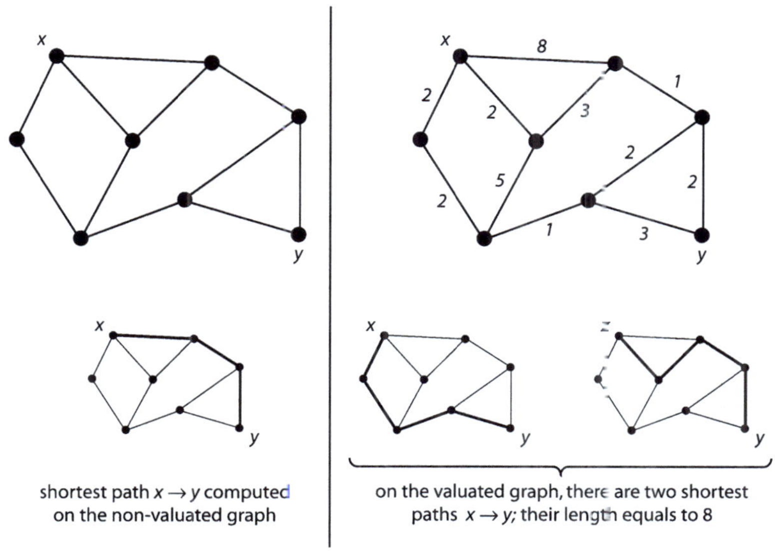

Fig. 3.4 Examples of a non-valuated graph (*left*) and a valuated graph (*right*) – impact of edge values on the computation of shortest paths

In practice, the interpretation of edge value depends on the meaning of the relationship between two nodes. Choosing a threshold (such that every edge with a greater value is considered unavailable) may be relevant for issues where the "metric" closeness is a main condition within the context of the study, such that nodes have to be topologically and geometrically well-connected.

Based on these notions and extensions of graph theory, we now propose to discuss the meaning of a clustering operation inside a graph and its possible expressions in terms of graph analysis.

3.3 Identifying Cohesive Subgroups Inside a Graph

Clustering operations inside a graph aim to gather strongly connected nodes according to criteria on the presence and nature of ties between those nodes.

Secondly, these operations may lead to the replacement of the identified groups with clustered nodes to produce a simplified representation of the graph (cf. Fig. 3.5). Nevertheless, this simplification represents a topic of research in that it also requires simplification of the edges and the component attributes (values, weightings, structural information such as paths, etc.).

Graph theory provides several mathematical tools to carry out node clustering within a graph. These tools have been widely formalized and developed in sociology

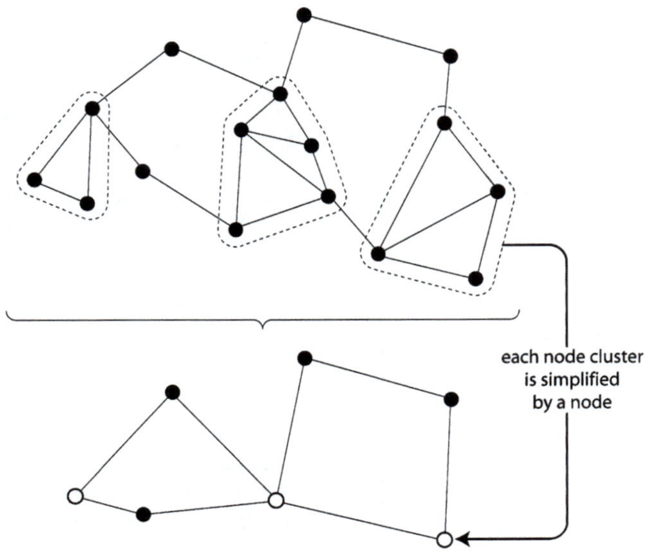

Fig. 3.5 Example of graph simplification by node clustering

because this research field needs graphs to model sets of people and the relationships between them. Within these "social graphs", the nodes and edges represent the individuals and the relationships that they maintain. These relationships may have different natures and concerns:

- individual relationships (e.g., "I have made sb's acquaintance") and group relationships ("I work in the same company as sb");
- similarity relationships ("I speak the same language as sb") and dissimilarity relationships ("I was not born in the same country as sb");
- symmetric relationships ("I am related to sb") and asymmetric relationships ("I appreciate sb's company");
- binary relationships ("I am indebted to sb") and quantitative (or "valuated") relationships ("I owe sb x euros").

In such graphs, a relationship is basically modeled by an edge between the nodes related to the individuals involved in the relationship. When considering group relationships, the edges are drawn between each pair of individuals stemming from the group.

However, some types of relationships demand more complex modeling:

- asymmetrical relationships require work with directed graphs, i.e., with graphs in which each edge connects an initial node to a final node;
- quantitative relationships need to value the edges.

These models need to adapt the methods of analysis by considering the directions and values of edges.

The clustering tools stemming from graph theory help to highlight the similarities between the individuals of a social graph, taking into account the relationships they maintain based on several criteria.

The different clustering methods presented in this chapter are based on the research work of Wasserman and Faust (1994) and Moody and White (2003), and we discuss those works from a geographical perspective.

3.3.1 *The Cohesion Concept*

The following clustering methods are based on the concept of "cohesion" between the "members", i.e., individuals, of the same subgroup.

The Petit Robert French dictionary defines cohesion as:

- the strength that brings together the parts of a material substance;
- the feature of a set whose parts are united, matched.

This definition is very close to the meaning of cohesion in graph theory as applied to sociology, as it defines subgroups of individuals who are strongly united by the relationships between them.

It is quite important to distinguish the cohesion concept from the adhesion concept related to structures centered on a node. Such structures are characterized by the convergence of many ties towards a "leader", node and therefore are strongly dependent on the status of this node. For this reason, "adhesive" structures are fragile because they can easily be affected by acting on this node.

On the contrary, "cohesive" structures are characterized by many independent ties that unite their members.[1] Groups with stronger cohesion are more difficult to split.

Finally, the group cohesion is affected by the arrangement of ties within the group. Groups with stronger inner ties exhibit a better ability to hold together.

3.3.2 *Qualitative Elements to Identify Cohesive Structures*

The structure of a group is best held together when all its members are linked in pairs. Such a structure corresponds to the "clique" notion of graph theory, but is rarely observed because the lack of a single tie is enough to "downgrade" this structure. However, the criterion of edge "exhaustiveness" specific to cliques can be relaxed and lead to more convincing clustering methods that remain operational.

---

[1] The contrast between cohesive and adhesive structures appears in the etymology of these words. On the one hand, the words cohesion, colleague, and companion refer to notions of equality or numerous connections; on the other hand, the words adhesion, adversary, and administrator refer to notions of centrality or reference Moody & White (2003).

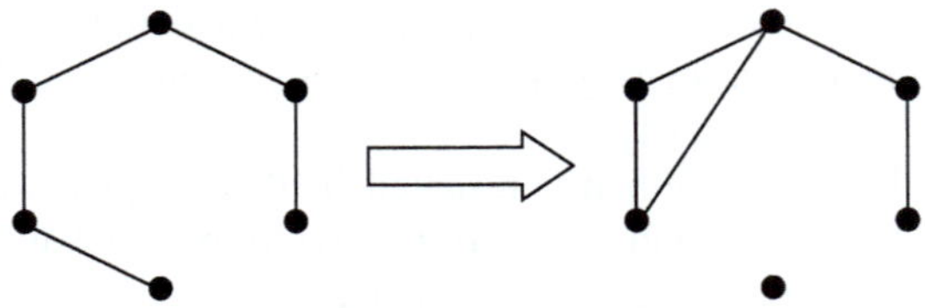

Fig. 3.6 The edge density is not sufficient to describe the cohesion of a group (From Moody and White (2003))

The nodes of a group may form a cohesive structure without pair linkages if there is a sufficient number of relationships symbolized by the edges: in this respect, cohesion is compared to the ease of reaching one node from another, i.e., to their "reachability" (this notion is connected to the notion of accessibility in geography).

Another criterion is the association between the group cohesion and the edge density (or volume) inside the group. Although this parallel at first seems acceptable, this characterization is actually inadequate: the linear structure and its slightly modified version in Fig. 3.6 show that the node reachability can differ widely for constant density.

Although the edge density does not constitute an exclusive indicator to characterize cohesion, the edge layout should be uniform to prevent the group from splitting. If the layout is not uniform, the subgroups with lower edge densities may be more vulnerable, as slight changes may split them more easily.

In this way, a third set of clustering methods demands that conditions regarding the node adjacency be fulfilled inside the subgroups: these criteria are based on the node "degree".

In terms of adjacency, it can be relevant to take into account not only the edges inside the group (these "intra-group" edges should have the highest possible density) but also the edges linking the group nodes with outer nodes (these "inter-group" edges should have the lowest possible density).

In a more formal way, it is possible to build subgroups, not by gathering nodes together, but by identifying the sets of edges that separate them from each other. To that end, it seems worthwhile to take advantage of the structural weakness of these sets with low densities. Then, these sets become small sets of edges whose removal splits the graph. This approach matches well with the whole area of graph theory devoted to the "cut-sets".

Based on these qualitative elements, we present the clustering methods and their possible applications in geography according to the following notions:

- the edge exhaustiveness;
- the closeness or reachability of nodes inside the subgroup;

- the degree of the subgroup nodes;
- the comparison between intra- and inter-group edges;
- the identification of cut-sets.

3.4 Methods for Identifying Subgroups Inside a Graph – Guidelines for Clustering Geographical Networks

3.4.1 Subgroups Built from the Edge-Exhaustiveness Criterion

The most direct clustering method consists of picking out the subgroups where the nodes are directly linked in pairs: this criterion corresponds to the clique notion of graph theory.

A clique is formally a maximal complete subgroup of at least three nodes. In other words, each node pair of a clique is materialized by an edge, and it is not possible to find an outer node that would be linked at each node of the clique (cf. Fig. 3.7).

By definition, the notion of cliques is restrictive for subgroup definition. The lack of a single tie between two nodes of a potential clique prevents clique formation. In practice, cliques are extremely rare inside wide networks, and they are generally very small: for that reason, this notion has to be extended to highlight larger but less cohesive structures.

3.4.2 Subgroups Built from the Reachability Criterion

The reachability notion is a useful extension to the clique notion. Reachability is particularly relevant when studying information networks: in this case, the issue often consists in identifying subgroups where information can be quickly transmitted between nodes. Therefore, the nodes do not have to be adjacent to each other, as they do in a clique (exhaustiveness criterion), but they do have to be close to each other in terms of shortest path length (reachability criterion).

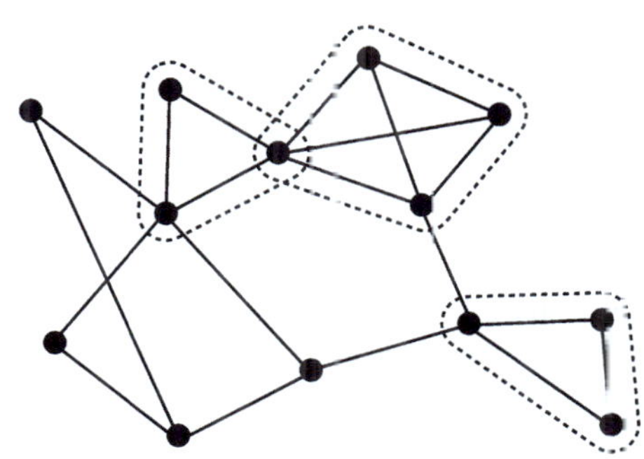

Fig. 3.7 Identification of cliques inside a graph (For this example, we note that the same node can be part of several cliques)

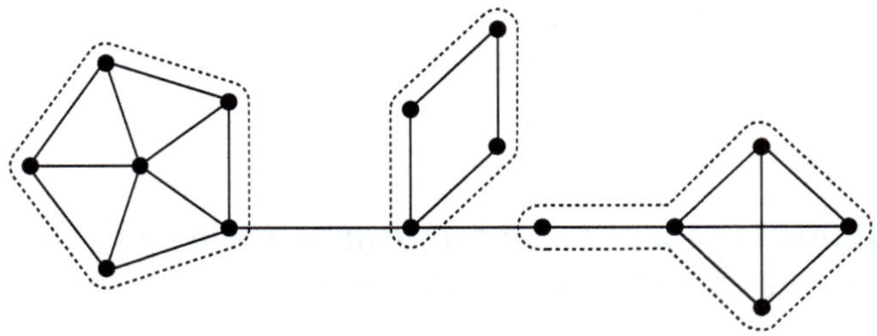

Fig. 3.8 Examples of 2-cliques

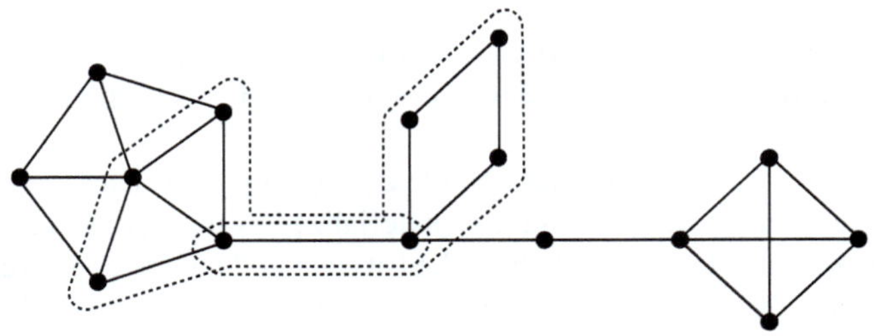

Fig. 3.9 The graph of the Fig. 3.8 also contains two 2-cliques that are difficult to identify at first sight

In this connection, the n-clique is defined as a maximal subgraph in which each pair of nodes is no more than n edges apart (the shortest path may pass through the outer edges). For this reason, it is impossible[2] to find an outer node that would be distant from each node of this subgraph with a distance less than or equal to n.

On the example of the Fig. 3.8, the spatial layout of edges and nodes shows three obvious 2-cliques.

However, these are not the only 2-cliques of this graph: in Fig. 3.9 we have identified two other 2-cliques, straddling two of the previously identified 2-cliques.

In this connection, reachability appears to be an extension of edge exhaustiveness insofar as the node closeness is under control (this criterion no longer concerns the ties between nodes, but rather the shortest path length).

For example, such a criterion is realistic and relevant in geography for accessibility issues. The need to cluster nodes based on the condition that they are "close enough" to each other amounts to selecting zones in which the shortest paths are shorter than a given threshold.

---

[2]As defined, a n-clique may have a diameter greater than n because the shortest paths between its nodes may pass through the outer edges of the n-clique. Nevertheless, there are variants of this clustering method that control the subgroup diameter. Subgroups extracted by such methods are called "n-clans" and "n-clubs" (for further information, see Moody & White, 2003; Wasserman & Faust, 1994).

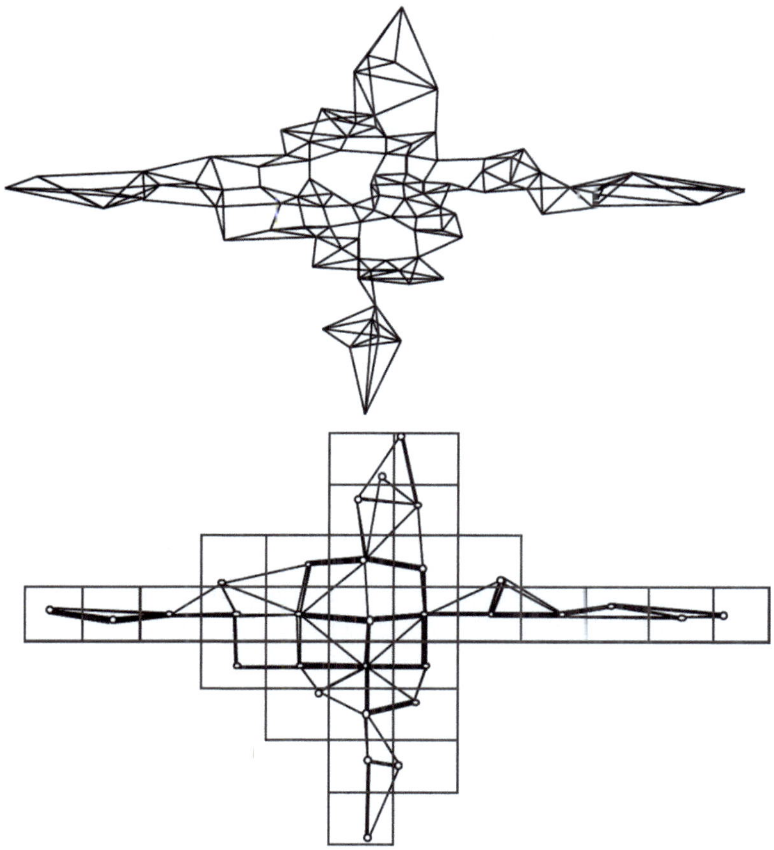

Fig. 3.10 A large social network and its simplification by a clustering method based on a reachability criterion (From Levine (2009))

Levine (2009) uses this principle to simplify a large social network with nodes that are widely spread out over the territory. According to the fact that n-cliques may overlap each other, this author proposes to cluster distinct sets of nodes that he selects inside the different cells of a grid (cf. Fig. 3.10).

In this application, the reachability is used to gather nodes that are close to each other, but some open-ended questions remain, such as the reachability of nodes inside clusters (although they are included in a cell, two nodes of the same cluster may be very far from each other if the shortest path between them passes through other cells) or the relevance of the grid in the clustering process.

From a mathematical viewpoint, the reachability criterion does not necessarily involve a high edge density inside the n-clique: actually, n-cliques may show star-shaped structures as in Fig. 3.9.

Fig. 3.11 Examples of 2-plexes

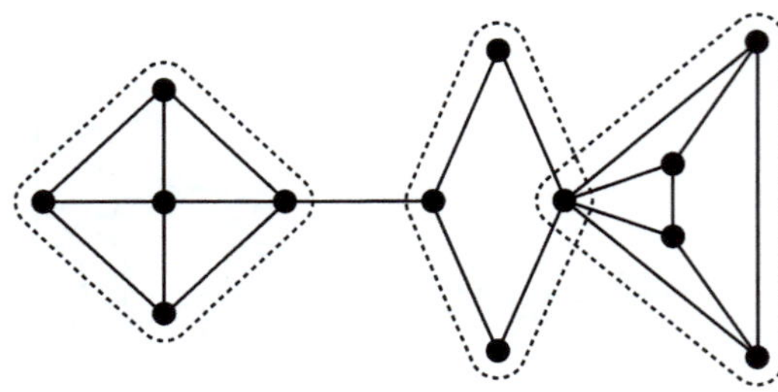

The following criterion helps to reject such structures by controlling the minimum number of ties between the subgroup nodes.

3.4.3 Subgroups Built from the Node Degree Criterion

In a subgroup built from node degree criteria, the nodes have to be adjacent to a "relatively high" number of other nodes.

In this respect, such subgroups are built according to the same rule as cliques (each node has to be directly linked to other nodes) but with weaker conditions (each node has to be directly linked to a minimum number of nodes).

This notion seems to be an interesting way to comprehend subgroup vulnerability: vulnerability occurs when the relationships between nodes are not sufficiently redundant, so that removing some target edges causes nodes disconnections. It turns out that *n*-cliques are not robust enough in the sense that their structure may be easily broken by the removal of a single node (a star-shaped structure is a 2-clique and is vulnerable to the removal of the central node).

In contrast, *k*-plexes are more robust subgroups because they are built so that each node has a "relatively high" adjacency number.

A *k*-plex is defined as a maximal subgroup where each node is adjacent to every other node of the subgroup except for *k* nodes at most.[3]

In that way, we have identified three 2-plexes on the graph of Fig. 3.11. In these 2-plexes of five nodes, the nodes are adjacent to 3 (or more) other nodes of the 2-plex. In other words, inside the subgroup built from the 2-plex nodes, the nodes have a degree greater than or equal to 3: the criterion specific to the k-plex is explicit here.

A clustering method based on a degree criterion has been used by Gleyze (2009, September 4–8) to identify groups of mutually supportive countries in the Eurovision Song Contest. In this study, Gleyze first determines which pairs of

[3]Subgroups with a similar criterion can also be built by considering the minimum number of adjacencies (rather than the maximum number of "non-adjacencies"), which we define as the *k*-core. A *k*-core is then a maximal subgroup of n nodes where each node is adjacent to at least *k* other nodes (Moody & White, 2003; Wasserman & Faust, 1994).

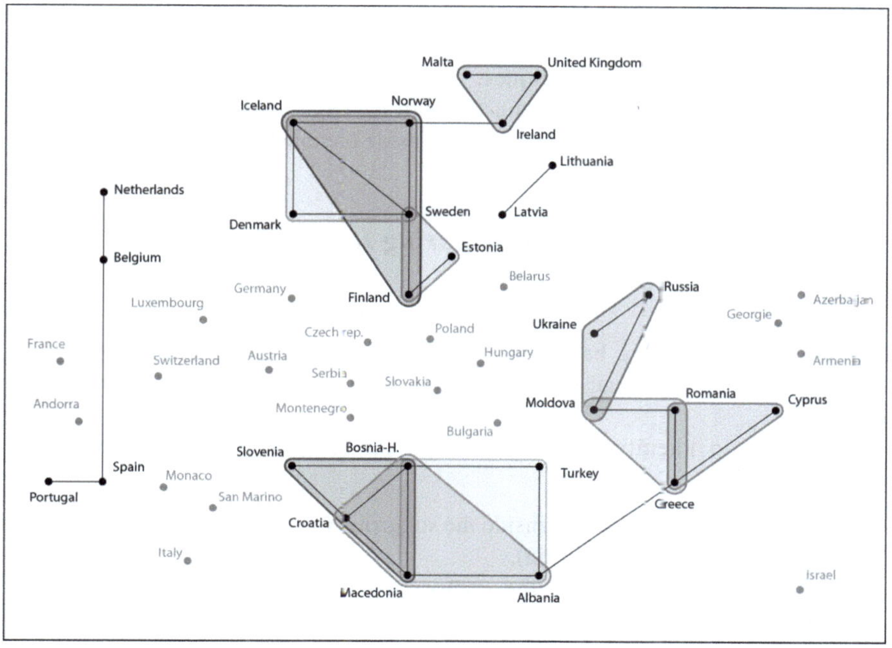

Fig. 3.12 Clusters of countries swapping over-votes between each other in the Eurovision Song Contest between 1993 and 2008 (From Gleyze (2009, September 4–8)) (Note: all the over-votes were taken into account in the cluster computation, but to make the figure more legible, only the reciprocal over-votes are represented here (the edges correspond to reciprocal over-votes between pairs of countries and the clusters are highlighted by *bold lines*))

countries voter performers show an abnormal tendency to produce a high number of votes. Then, he considers the country graph of such relations (called "over-votes"). In this graph, the number of cliques (4) seems to be low compared with the quantity and concentration of nodes (49) and over-votes (135). To detect larger structures, Gleyze suggests lightening the exhaustiveness criterion. In that respect, he defines a cluster as a set of nodes so that the over-vote density is not necessarily equal to 1, but must be higher than three quarters. In compensation, he demands that two degree criteria be verified:

- each country of a cluster must give and receive at least one over-vote from another country of the cluster,
- there must be at least one over-vote between each pair of countries in the cluster.

The Fig. 3.12 shows the country clusters highlighted by this method. While only four cliques were identified at first, these degree criteria allow for the detection of larger cohesive structures and their possible interactions.

Finally, the clustering methods based on the degree criterion guarantee the redundancy of relations and thus strong cohesion among subgroup nodes. In the mathematical definition, the parameter k corresponds to the tolerance (in terms of lack of ties) inside the subgroup in comparison to the maximal cohesion theoretically observed inside cliques. In the geographical application shown in Fig. 3.12, this criterion has been adjusted with the edge-density and two other rules relating to the adjacent edges of each node inside the cluster.

3.4.4 Subgroups Built by Comparing Intra- and Inter-Group Edges

The previous criteria are based on the properties of inner subgroup ties. Indeed, members in cohesive social network groups are closer to each other than to outer members.

Therefore, to identify a cohesive subgroup inside a graph, it is relevant to consider:

- not only the density of edges inside the subgroup (as we did with exhaustiveness, reachability and degree criteria),
- but also the intensities and the comparative frequencies of intra- and inter-group ties.

Lambda-sets are subgroups built from this second criterion. For such subgroups, cohesion is linked to the notion of connectivity: the identified subgroups must be difficult to disconnect by edge removal. Thus, lambda sets are maximal subgraphs for which it is harder to disconnect an inner node pair than a pair made up of an inner node and an outer node.

In other words, the connectivity index of inner node pairs must be greater than the same index computed for pairs composed of an inner node and an outer node (cf. Fig. 3.13).

The computational methods used to identify lambda-sets are difficult to implement. However, the criterion based on the comparison between inner and outer ties may be relevant for graphs whose clusters:

- may be made up of nodes that are not close to each other (actually, this criterion does not depend on degree or reachability criteria),
- and need to be disjoint (as they are defined, lambda-sets do not overlap each other unless they are included in each other).

Thus, Amiel, Melançon, and Rozenblat (2005) use a similar method to identify clusters of international airports, given that the airports of a given cluster have to be better connected to each other than to other airports. With this method, the authors aim to explain how air relations are organized and to highlight their structures at different scales (from global to local).

In practice, Amiel et al. build a graph of the air relations (on this graph, the nodes represent the airports and the edges correspond to air relations between airports). On this graph, they consider each edge and determine whether it can be part of a cluster.

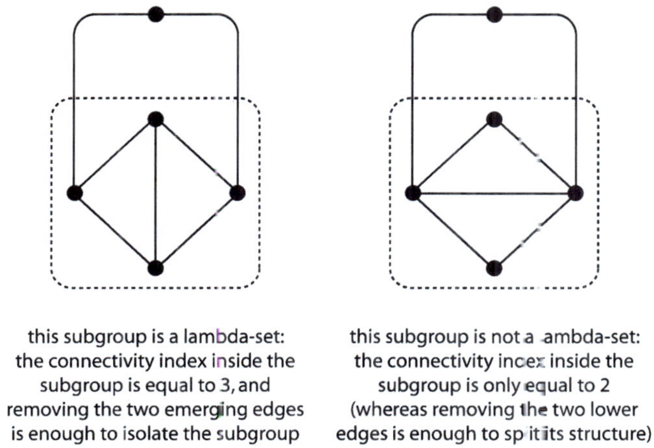

Fig. 3.13 Illustration of the lambda-set notion: the local weakening of a lambda-set (here by decreasing the degree of some nodes) can lead to a "downgrade" of the set

For this analysis, they compare the sets of nodes that are adjacent to each end of the edge. If these sets overlap each other enough (in comparison to a given threshold), then the edge is considered an inner-edge. Finally, the sets of inner edges delimit several clusters according to a criterion based on an inner/outer edge comparison.

Figure 3.14 shows the Tokyo cluster identified with this method (for a threshold equal to 1/2).

3.4.5 Subgroups Built from Cut-Sets

As previously mentioned, the notion of connectivity is linked with the minimal cut-set notion.

Moody and White (2003) propose to use these notions to identify cohesive subgroups inside a graph, not by comparing intra- and inter-group connectivity indexes, but by gradually splitting the graph around its cut-sets.

This "cohesive blocking" method is recursive: for a given graph, it aims to identify the minimal cut-set, pick out subgraphs on both sides of this cut-set, and then repeat this process on these subgraphs.

Figure 3.15 is 1-connected: the minimal connectivity index computed on its node pairs is equal to 1 (additionally, this graph can be disconnected by removing only one node).

By applying the "cohesive blocking" method, we identify two subgroups on both sides of this critical node. In the same way, these subgroups are split around their cut-sets. The new subgroups have connectivity indexes equal to 2 and 3.

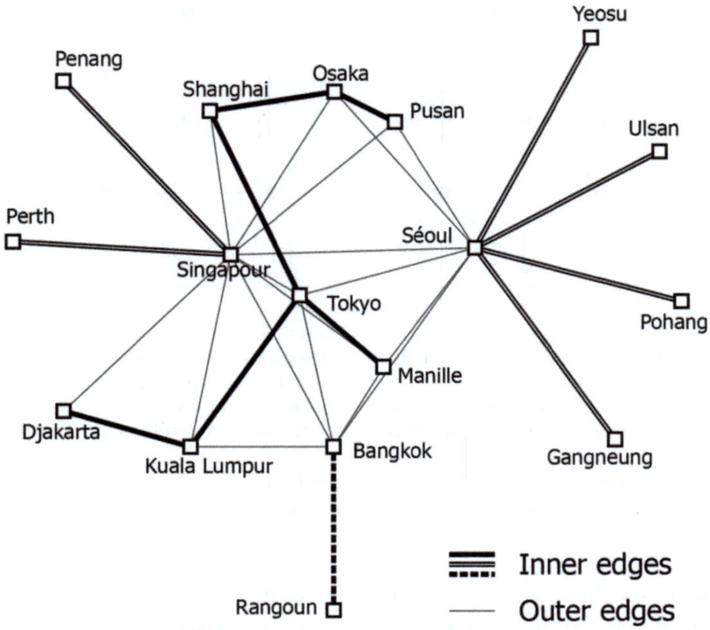

Fig. 3.14 Inner and outer edges identified around Tokyo on the air relations graph – the set of airports linked by bold ties represents the Tokyo cluster (From Amiel et al. (2005)) (Color figure online)

Fig. 3.15 Illustration of the "cohesive blocking" method (From Moody and White (2003))

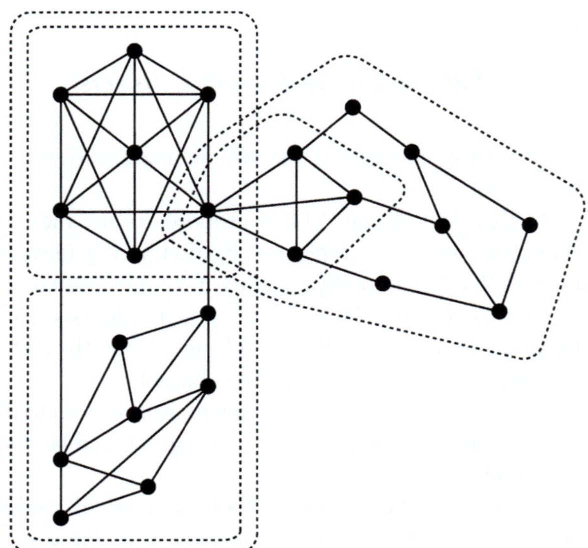

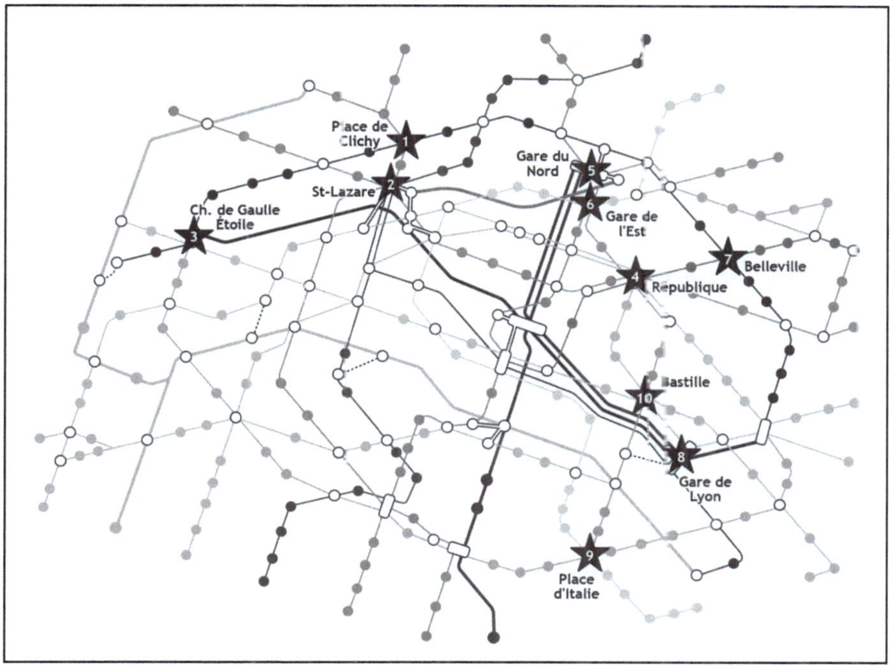

Fig. 3.16 Stations of the Parisian Metro network whose successive removal leads to the greatest reduction in mean accessibility (From Gleyze (2005))

This method finally helps to hierarchically segment the graph into increasingly cohesive subgroups.

The connectivity notion rarely appears in geographical studies as reticular structures are often robust: actually, transportation, energy and communications networks contain several redundant ties and are not considered to be systems that could be gradually disconnected.

There is an exception to this fact: when analyzing risks, it may be relevant to identify the most fragile points in the network. However, the edge density of geographical networks is so high that it is not possible to identify clusters by successively damaging such points. In fact, these networks are almost never disconnected, but some paths become so long that they could be considered as disconnected.

Nevertheless, Gleyze (2005) proposes to identify the node of a transportation network whose removal would lead to the greatest decrease in the mean accessibility inside the network. Then, he considers the network deprived of this node, identifies the node in this network whose damage would exert the greatest effect on the mean accessibility, removes this node, and so on.

Figure 3.16 shows the succession of the "most fragile" nodes of the Parisian Metro network.

Finally, the clustering methods based on cut-sets cannot be directly applied to geographical issues because of the high edge-density of geographical networks. However, a criterion based on network weakness may indicate the components that are likely to segment the network in the same way as cut-sets. For that reason, this criterion has to be considered as a segmentation method instead of a clustering method.

3.5 Conclusion

Graph theory provides clustering methods that help to simplify networks into cohesive subgroups, i.e., into subgroups in which nodes are close to each other and are difficult to disconnect. These methods depend on criteria whose implementation may lead to splits of various sizes and natures. This review is not exhaustive; nevertheless, it is representative of the main topological criteria that are commonly used to build cohesive subgroups (cf. Fig. 3.17).

These clustering methods fit with three main ways to comprehend the relationship between two nodes of a graph:

- according to the exhaustiveness and degree criteria, the relationship is assessed in view of the presence or absence of an edge between these nodes;
- according to the reachability criterion, the relationship is assessed in view of the connection quality, i.e., the length of the best path linking these nodes;
- finally, according to the inter-/intra-group and connectivity criteria, the relationship is assessed in view of the robustness of the connection, i.e., of the

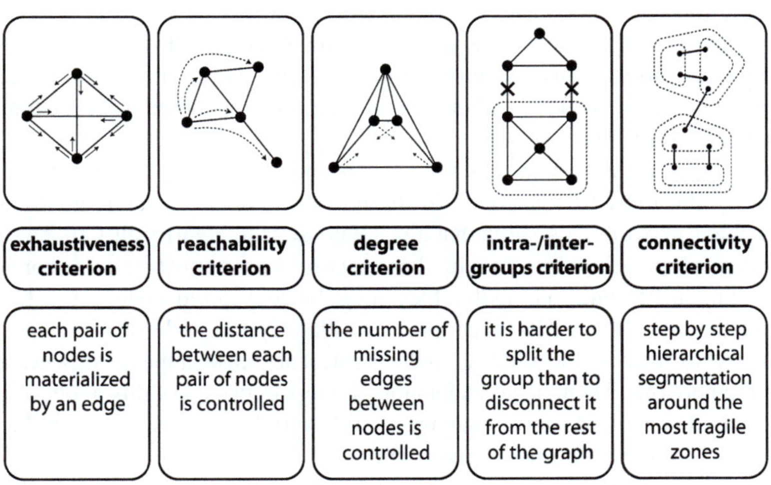

Fig. 3.17 Main topological clustering criteria

maintenance of at least one path between these nodes when some network components happen to be unavailable.

These criteria are exclusively topological because they are based on the presence of edges, the connection quality or the connection robustness between node pairs. Nevertheless, graph theory provides extensions that integrate the edge directions and values, and geography itself provides suitable models that help to take the specific features of infrastructure networks into account.

From then on, the clustering criteria become relevant within an application context because they provide a topological framework within which network simplifying methods can be contextually adapted. However, this approach leaves several issues unaddressed. These issues need to be considered based on the meaning of the topological clustering within the context of the geographical study: How can cluster quality be assessed? How can an index be computed for a simplified network? How can nodes belonging to several clusters be managed? How is a path on a network affected by its simplification? Is it possible to simplify a network that has already been simplified?

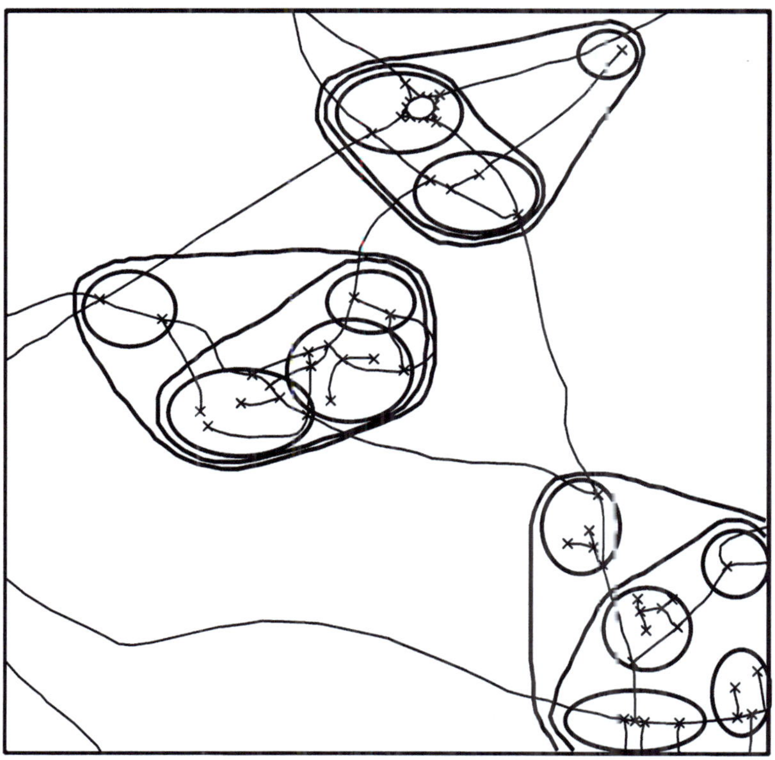

Fig. 3.18 Clustering on road networks for simplification of junctions from Mackechnie and Mackaness (1999)

These different ways of comprehending the relationship between two nodes finally influence the choice of the clustering criterion used to study a network. In that respect, the geographical examples presented in this chapter often adapt the clustering criterion to the context of the study.

Several criteria may be adapted and mixed to provide a relevant clustering method. For example, MacKechnie and Mackaness (1999) propose to simplify a cartographic representation of a network by sequential clustering of nodes that are closest to each other (cf. Fig. 3.18).

In this geographical application, Mackechnie and Mackaness finally use a similar criterion:

- with the definition of a n-clique,
- and with the hierarchical approach relating to cut-sets (except that components are progressively gathered together instead of being separated).

Ultimately, they build a method that is relevant for identifying possible spatial clusters according to the level of observation.

In addition, it is important to underline that these research directions are not exhaustive. Actually, they depend on the notion of cohesion, but within some contexts it may be relevant to consider other clustering criteria:

- the adhesion notion: adhesive structures are characterized by the convergence of ties towards a "leader" node and are representative of local hierarchies within the network;
- the similarity notion: the associated structures contain similar nodes considering the relationships they maintain with the rest of the network (Wasserman & Faust, 1994).

Finally, if the study of a geographical network is strongly dependent on the context, application of a clustering method requires the accurate identification and interpretation of the topological criteria that affect this operation.

References

Amiel, M., Melançon, G., & Rozenblat, C. (2005). Réseaux multi-niveaux: l'exemple des échanges aériens mondiaux. *Mappemonde, 79*(3). Retrieved July 21, 2009, from http://mappemonde.mgm.fr/num7/index.html

Berge, C. (1973). *Graphes* (3rd ed., 1983, 400 p.). Paris, France: Gauthier-Villars.

Bollobás, B. (1998). *Modern graph theory* (394 p.). New York: Springer.

Gleyze, J.-F. (2005). *La vulnérabilité structurelle des réseaux de transport* (848 p.). Unpublished doctoral dissertation, Université Paris VII, Paris, France.

Gleyze, J.-F. (2009, September 4–8). Friendship and neighbourhood in the eurovision song contest. In *Proceedings of ecqtg09, 17th European Colloquium on theoretical and quantitative geography*. Maynooth, Ireland.

Levine, J. (2009). *Large networks* (13 p.). Unpublished draft, Hanover, NH: Dartmouth University. Retrieved July 21, 2009, from http://www.dartmouth.edu/~jlevine/NSF%20%205.pdf

Mackechnie, G., & Mackaness, W. A. (1999). Detection and simplification of road junctions in automated map generalization. *GeoInformatica, 3*, 185–200.
Moody, J., & White, D. (2003). Social cohesion and embeddedness: A hierarchical conception of social groups. *American Sociological Review, 68*(1), 103–127.
Wasserman, S., & Faust, K. (1994). *Social network analysis – Methods and applications* (857 p.). Cambridge: Cambridge University Press.

Chapter 4
Theoretical Models of Time-Space: The Role of Transport Networks in the Shrinking and Shrivelling of Geographical Space

Alain L'Hostis

4.1 Introduction

Understanding the distance between places is a fundamental task for the geographer, while the representation of distances constitutes the primary function of cartography. This is why a time-distance representation is a critical tool for the contemporary geographer. Among the types of maps introduced to represent time-space, anamorphoses were supplemented with time-space relief cartography in the 1990s. In one aspect, globalization has been made possible primarily through a reduction in time-distances, which are allowable by high travel speeds, particularly through the development of air transportation. However, the metropolitanization process, which is seen as the urban counterpart of globalization, is deeply associated with the development of air connection platforms. Both phenomena are intrinsically linked with the formation of distance, specifically with time-distances.

The purpose of this chapter is to present cartographic models of time-space and discuss their theoretical meaning for geography. The first step is to show a set of solutions for the cartographical representations of distances, which are transformed by different means of transportation. In the second step, we will discuss the relations between these cartographical solutions and the theoretical geographical discourse on time-space, which is structured around classical theoretical models.

A. L'Hostis (✉)
Université Paris-Est, LVMT, IFSTTAR, 59666 Villeneuve D'ascq Cedex, France
e-mail: alain.lhostis@ifsttar.fr

C. Rozenblat and G. Melançon (eds.), *Methods for Multilevel Analysis and Visualisation of Geographical Networks*, Methodos Series 11, DOI 10.1007/978-94-007-6677-8_4,

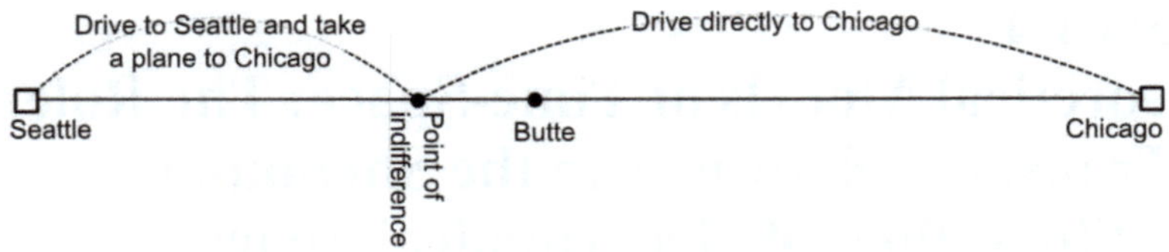

Fig. 4.1 The phenomenon of spatial inversion

4.1.1 Transport Networks Create Distance: A Cartographic Problem

To understand the present global space, it is necessary to emphasize the major role played by high travel speeds in the structuring of metropolitan spaces. These high travel speeds have initiated the development of communications on a global scale. However, high travel speeds also control a dramatic variety of places.

While this high-speed movement highlights the importance of global communication, the secondary networks and spaces remain present in the interstices. A key point in the formation of global time-distances is the fact that these secondary networks are obsolete, when compared to the main high-speed networks.

How can this complicated set of networks and relations in space be represented? To answer this question, a broader perspective of the representation of distances must be considered. In this field, Bunge has stated that two methods are available: "representing complicated distances on simple maps or representing simple distances on complicated maps" (Bunge, 1962).

In his classical example of the complicated relations in space generated by contemporary transportation, Bunge considers the movements from intermediary space to higher level cities with a trip from a location in Montana to the city of Chicago (Fig. 4.1). In this example, the use of the transport modes of cars and planes, implies that the shortest path in time-space takes a completely different shape beyond and after a particular point of space. Departing from Butte would mean driving directly to Chicago, while starting from a location that is closer to the Pacific Coast involves a trip by car to Seattle, followed by a flight. This phenomenon of spatial inversion that follows intuitive logic from a transportation perspective provokes a disturbance in the order of proximities. With regard to Cauvins's formalism (1984), although the Rocky Mountains are located between Seattle and Chicago in chorotaxic space, which is the normal geographic space, their position in functional space is different. In the transport space, Seattle stands between the Rocky Mountains and the Great Lakes.

The identification of the phenomenon of spatial inversion constitutes a key explanation of the research for new representations of time-space that would render this complicated set of distances. Allowing for the representation of distances that would be simpler to read would certainly complicate the map.

From this perspective, time-space anamorphic cartography is the first application in which one moves the locations to represent time-distances more effectively. An example is given by Shimizu, showing the contraction of Japan due to the development of the high-speed train networks between 1962 and 1992 (Shimizu, 1992). In the field of the representation of distances, anamorphosis aligns with the type of cartography defined by Bunge as simpler distances on a complicated map. The two elements of information on time-distances can be found from such a representation: the overall space contraction and the local deformations produced by high-speed lines. If the new transport networks had been characterized by homogeneity and anisotropy, the shape of the external borders of the country would have remained unchanged, and only size would have been reduced. All of the distortions from the normal and conventional shape of Japan indicate directions privileged by the shape of the networks. The literature on networks has abundantly expressed the idea that modern transport provokes heterogeneity in space (Castells, 1996; Dupuy, 1991; Graham & Marvin, 2001; Knowles, 2006).

However, this model is subject to limitations. The major criticism of the application of anamorphosis to the representation of distances is the fact that if two locations, such as two cities, are becoming closer due to a new transport link, this does not mean that the space between the cities is also growing in accessibility. Toll roads are examples of the "tunnel effect" of some infrastructures where the limited access points reduce the accessibility growth to a set of subspaces and are not distributed evenly along the line (F. Plassard, 1976). This phenomenon is even more pronounced in the case of the high-speed rail (Mathis, 2007; Murayama, 1994) and is one of the major characteristics of air transportation (Haggett, 2001). Providing an illustration of this limitation, the phenomenon of spatial inversion cannot be read from the anamorphic map because of the principle of the preservation of the order of proximities, which can be found in most methods developed in the literature (Clark, 1999; Kotoh, 2001; Shimizu, 1992; Spiekermann & Wegener, 1994).

Displacing the locations on the map is not the only way in which distances can be represented. The idea of drawing transport lines between places in such a way that different distances are shown was introduced in the 1980s (H. Plassard & Routhier, 1987; Tobler, 1997). In the example proposed by Tobler, the location of cities and network nodes remains unchanged, when compared to their normal cartographic position. The length of the roads between the nodes is displayed in the form of a spring: the intensity of the tension indicating the sinuousness of roads unevenly distributed in the mountainous area in western Colorado. In this model, one can obtain the information on the difficulty of linking two places by observing the visual length of the links. The notion of visual length was introduced (L'Hostis, 2003) to describe the capacity of a map reader in extracting the information on the length of a route from the analysis of the path's shape. A straight segment can be converted in kilometers through the direct use of the scale, while a sinuous curve indicates a longer road. This principle is used in the spring map to express the idea of privileged and handicapped directions (Fig. 4.2).

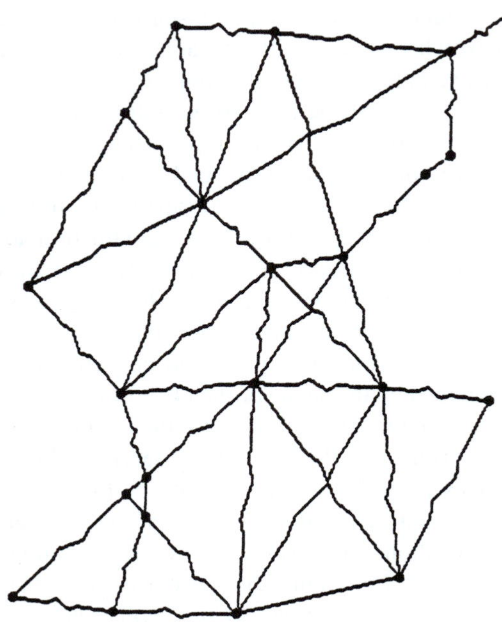

Fig. 4.2 Spring map of roads in western Colorado

The spring map model can indicate the shortest directions in space. A non-Euclidean representation by design, it presents the idea that the shortest paths often differ from a straight line. In this perspective, it constitutes a possible proposal to call for non-Euclidean geography (Golledge & Hubert, 1982; Müller, 1982). Recently, but in the same vein, a model has been formulated to introduce a three-dimensional surface that allows for the representation of different speeds in urban spaces (Hyman & Mayhew, 2004).

Sharing a principle of construction similar to that of the spring map, the time-space relief map was introduced in the 1990s (L'Hostis, 1996; L'Hostis, Mathis, & Polombo, 1993; Mathis, 1996). This type of representation preserves the location of places but exploits three-dimensional resources to draw the various speeds and the corresponding time-distances in a multimodal network.

4.1.2 Air and Road Modes As Major Inter- and Intra-Metropolitan Transport Systems

Globalization, along with metropolitanization as its urban counterpart, is made possible through the development of efficient, long- and short-haul transportation systems. If metropolises can be defined as urban entities that communicate on a global scale, then the air transportation mode constitutes the primary passenger transportation system associated with globalization (Sassen, 1991; Haggett, 2001).

In addition to the characteristics of metropolises and airport infrastructure, the number of flights and destinations available as well as the air distances are often used as indicators to position cities within the global economy (Grubesic & Zook, 2007). Nevertheless, the development of the air transportation mode from the twentieth century to now has not led to the replacement of slower transport systems. Each transport mode has developed within its own space of predominance, which contains fierce competition in the margins. On the scale of metropolitan space, the road system is considered the major mode of transport, even when regional specificities are stressed. The overall picture of mobility involves two distinct levels: the agglomeration or local level dominated by cars and the long-distance level dominated by air. For a complete analysis, this picture must be enriched marginally with the development of other transport systems, such as urban public transport and high-speed rail, which operate as a complement, rather than a substitution for cars and planes.

L'Hostis proposed considering the simultaneous high-speed and lower-speed transport modes on a continental scale in a single time-space relief cartography (L'Hostis, 2009). This representation is proposed from a different angle of view (Fig. 4.3). While the typical 30° was used in almost all previous time-space cartographies (L'Hostis, 1996; L'Hostis et al., 1993; Mathis, 1996), this map adopts a 45° angle of view. As the differential of speed for each subspace reaches a ratio greater than 6, this angle highlights the dichotomy of time-spaces, which occurs between the top level high-speed network and the common lower speed non-metropolitanized space.

In the USA, the air and road transport systems create a complicated time-space that the relief cartography helps to understand and describe.

After developing cartographic solutions to the problems of distance, we now interpret the theoretical background of this cartography, relating it to one of the most fundamental tasks of the geographer: understanding space and distance.

4.1.3 The Theoretical Models of Time-Space, from Shrinking to Shrivelling

In geography, there are several theoretical models elaborated to understand time-space. These models consider the movements of contraction or shrinking, convergence, divergence and crumpling or shrivelling. The rhetoric of world contraction can be considered a fundamental observation. Early references from the ancient Greeks reveal that they perceived the evolution of vessel techniques at the time-scale of the life of a human being and linked it to the reduction of distance between places in the Mediterranean Sea (Abler, Janelle, Philbrick, & Sommer, 1975; Braudel, 1979). In the nineteenth century, maps of now classical French geography depicted the contraction of the national territory with improvements in their land transport, while German cartographers mapped the improvements in maritime transport on a global scale, showing the reduction in travel times between Europe and the rest of

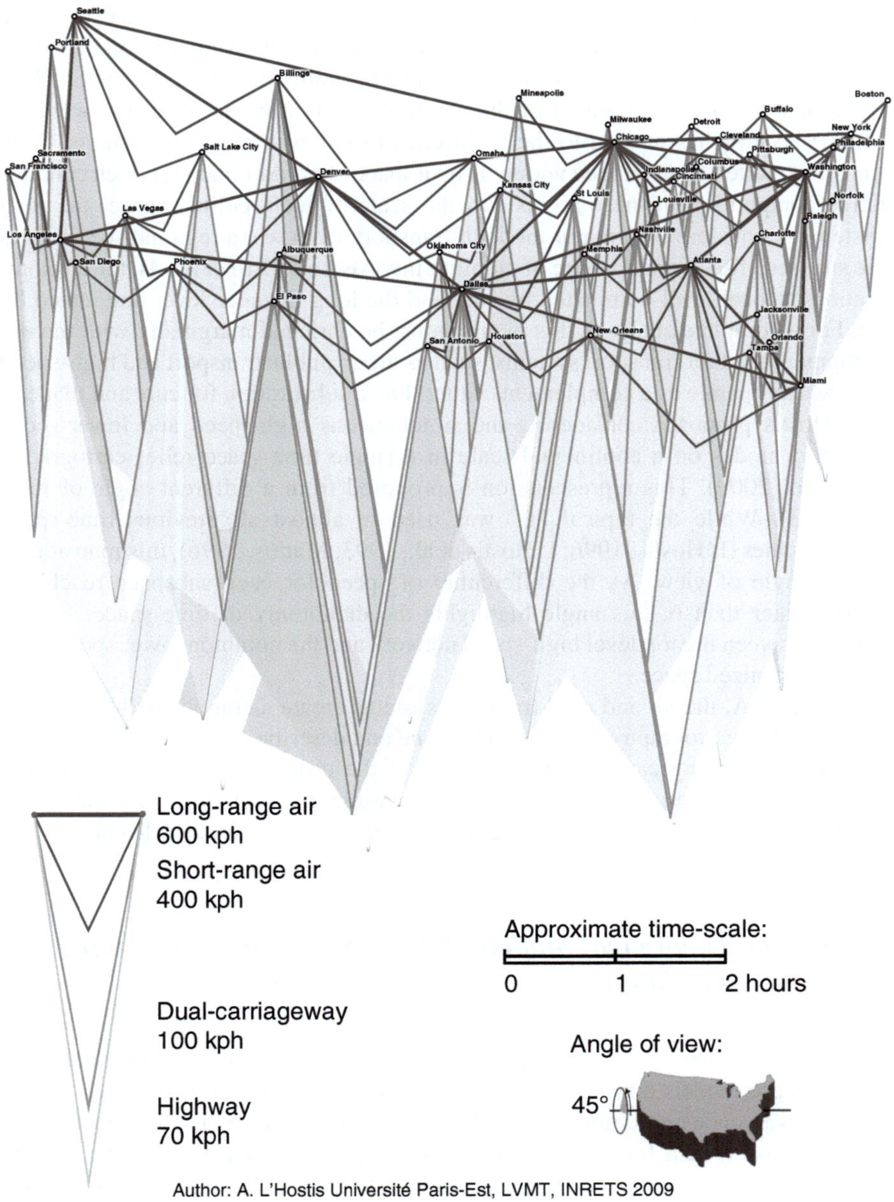

Fig. 4.3 High-speed by air versus lower speed by land: The shrivelling USA

the world (Brunet, 1987). Recently, in the domain of theoretical geography, Forer has developed the principle of time-space contraction (Forer, 1978).

Nevertheless, prominent criticisms of the contraction model have been made by Kirsch stating that "space is not "shrinking" but must rather be perpetually recast"

(Kirsch, 1995). According to Kirsch, the evolution of time-space cannot be seen as a simple contraction or shrinking process. Various disagreements with the uniform contraction idea can be found in several geographers' works on cartographical developments (Abler et al., 1975; Boggs, 1941; Haggett, 1990). More recently, Knowles uses the expression of a "shrunken but misshapen world" to describe the present time-space where contraction is all but uniform (Knowles, 2006). The abandonment of commercial supersonic flight, the development of congestion and the increasing concern about security in the air transportation system are the cause of contradictory movements implying both convergence and divergence in time-space.

The notion of time-space convergence was introduced (Janelle, 1968; Janelle & Gillespie, 2004) to show that larger cities benefited more than smaller cities from the contraction of time-space, which resulted from the development of faster modes of transportation. For Janelle, the modernization of transport systems is seen as a factor of concentration in urban agglomerations. Throughout history, increased transport speeds have benefited larger cities more than smaller settlements. According to Christaller's theory of urban hierarchy, large cities are more dispersed in space than smaller ones. Two large neighboring cities will be separated by a much longer distance, than any two smaller neighboring cities. Janelle demonstrated that the increase in speed produces a stronger effect on the time-space contraction of long distances than on short distances. The evolution of the transportation systems speed provides an advantage to the larger cities, rather than the smaller ones, which is in agreement with the literature on metropolitanization. For Knowles, the identification of privileged places in the process of space contraction is still an idea that must be demonstrated to challenge the misrepresentations of uniform time-space shrinkage (Knowles, 2006).

Recently, the literature on globalization has adopted the principle of fast communication and transportation on a global scale as its base (Sassen, 1991; Smith & Timberlake, 2002; Taylor, 2004). Global transport systems are solely responsible for making the present movement of globalization and the unprecedented concentration of population in major metropolises possible.

The initial discussion on the three-dimensional model presented here considered the terrestrial relief as the space of reference. The expressions of time-space peaks and time-space valleys referred to the relative position of the portions of space in the representation. In the time-space deformation, following the introduction of the high-speed train system within French national space (L'Hostis et al., 1993), the development of a new high-speed tangential axis was seen as a way to fill up the valleys created from the introduction of a differential in transport speeds, combined with a star shape in the network. These initial contributions emphasized the analogy with terrestrial relief.

Later, the idea of a time-space 'crumpling' was introduced to develop a metaphor (L'Hostis, 2000) to represent European space deformed by the high-speed rail network. The image developed is the deformation of a shrivelled sheet of paper, which is an idea of the three-dimensional treatment of a plane. The metaphor refers to a shortening of some distances while preserving the initial geographical surface. The metaphor of 'crumpling' opposes the two spaces of high and low speeds and suggests a three-dimensional geometric construction coherent with the concept behind the representation. It also proposes a rather negative image, with the crumpled surface observed as a degradation of the ideal unaltered flat plane. A positive perspective can be established when the crumpled shape, being more compact than the flat surface, is considered because it allows for shorter paths between locations. The crumpling metaphor gives an evocative image of the poor treatment reserved for interstitial spaces but is not as expressive of the global contraction of geographical space, given the increases in speed.

It is relevant to associate these reflections on the representation of geographic time-space with some recent developments in theoretical physics: The astrophysicist Luminet describes the shape of the universe as crumpled around the idea of multiple folding due to time-space deformation, according to the theory of general relativity (Luminet, 2001). Even while establishing relations between completely different fields should be made with caution (a use of common words is not sufficient to validate a comparison), it is apparent that both systems of representation are tightly linked to the identification of a maximum speed. Indeed, the speed of light is one of the three fundamental constants in physics. Astrophysicists have to work with the existence of a maximum speed, which is the speed of light on the universal scale, while geographers have to take into account the maximum speed of transport provided by the air transportation mode to understand terrestrial time-space. To better understand space, both reflections seem to require building complicated representations that divert radically from common Euclidean geometry.

In a different field, psycho-analysts have been studying crumpled spaces (Diener, 2008) to understand the links realized in the time-space of dreams, where the associations of ideas lead to direct relations between locations that can be very distant in time-space. In reference to the astrophysicist Luminet's work, a crumpled space is defined as "welded on itself by several points" and can establish these types of connections (Diener). The analogy goes much further: In the geographic space, fast transport systems directly link remote locations, advocating for the development of crumpled time-space cartography, while in the time-space of dreams, direct connections between remote locations constitutes an essential property. The movement of crumpling generates new connections and new proximities that reflect both the properties of geographical space and the time-space of dreams.

In the evolution of the discourse produced on the time-space relief, the following step explores a different type of metaphor based on the idea of shrivelling. The first use of the word must be credited to Tobler through his commentary on the L'Hostis-Mathis image, when he stated that "the world is shrivelling as it shrinks" (Tobler, 1999). We move from an image of the shape of an inanimate entity or artifact, such as the crumpling of a sheet of paper, to a principle of natural evolution of a living organism, such as the shrivelling of a fruit. This shrivelling expresses the idea of contraction with the deformation of an envelope; i.e., the volume decreases while the external envelope keeps its initial surface.

The strength of the metaphor lies in the combination and the linking of two complementary movements of both the contraction and deformation of the surface.

It is the reduction of the volume due to a loss of substance that provokes the deformation of the skin of the fruit. In geographic terms, it is then possible to explain the complicated shape of the map from global contraction due to the high speeds of transportation. The model generates forces of contraction along air routes that apply to high-speed nodes, such as metropolises. As L'Hostis proposed (2009), "high speeds and metropolitanization make the world shrivel as it shrinks" would then be a reformulation of Tobler's statement.

4.2 Conclusion

To conclude this chapter we will discuss a social dimension of time-space that can be understood from the analysis of time-space relief cartography.

4.2.1 A Social Space in Maps: "It Took Me One Hour to Get There"

Let's consider terminal T2 at the Roissy-Charles-de-Gaulle Airport near Paris. Several people inside the terminal may state "it took me one hour to get there". This includes the airline travelers at 600 km/h, the high-speed train travelers at an average of 250 km/h, as well as the airport employees that went to work by car through the congested Paris urban region at an average of 20 km/h and those who came by public transportation at 10 km/h from neighborhoods around the airport. Each person took 1 h to arrive there, but each moved at a different speed, with a profoundly differing kinesthetic experience (Hall, 1966).

The first person belongs to the global network; they occupy the Higher Metropolitan Jobs market, whose concentration is one of the most revealing markers of metropolitanization (Rozenblat & Cicille, 2003). They progress in the space of horizontal relations between metropolises, which is the space of globalization. Their time-space is the top level network space in the relief cartography.

The second person belongs to a multi-polarized urban space, which is made possible by the "High-Speed Train web" that links the French cities where those living there are "neighbours of one of the rare global cities" (Viard, 2008).

The third person belongs to the local or slow network. They live inside the time-space abysses located around the high-speed nodes. They may pass members of the two other categories, but they do not obtain access to high-speed transportation. Their time-space has a steep slope. They live in the folds of the crumpled time-space. Their time-space is nearly orthogonal to that of the metropolitan workers.

The time-space crumpling produces time-distances that coexist and renders the propagation of movement at differing speeds, with differing means of transportation.

Building a representation of contemporary time-space requires building a synthesis of these distinct spaces within these distinct time-spaces. This is the purpose of time-space relief cartography.

4.2.2 The Contribution of Cartography to the Models of Time-Space

In this chapter, we have discussed several models of cartographical representation for the problem of distances produced by different means of transportation. This is a problem of fundamental interest for the geographer. In this section, anamorphosis, spring maps and crumpled relief maps are discussed. In a second step we have analyzed the relations between these cartographical solutions and the general discourse on time-space in geography. Complex and sometimes contradictory phenomena occur in time-space, and each cartographical model is able to account for parts of these phenomena. Time-space relief cartography produces an image associated with metaphors, linking it to the crumpling of a sheet of paper and to the shrivelling of a fruit. This discussion leads to the astrophysicist's conceptions and to recent developments in psycho-analysis. Finally, consideration is made for the social dimension of time-space; thus, time-space relief cartography is seen as a representation of the encounter between socially distinct spaces in nodal points located inside metropolises.

References

Abler, R., Janelle, D. D., Philbrick, A., & Sommer, J. (1975). *Human geography in a shrinking world.* North Scituate, MA: Duxbury Press.

Boggs, S. W. (1941). Mapping the changing world: Suggested developments in maps. *Annals of the Associations of American Geographers, 31*(2), 119–128.

Braudel, F. (1979). *Civilisation matérielle, économie et capitalisme, Le Temps du monde*. Paris, France: Armand Colin.

Brunet, R. (1987). *La Carte mode d'emploi*. Paris, France: Fayard/RECLUS.

Bunge, W. (1962). *Theoretical geography*. Lund, Sweden: Gleerup.

Castells, M. (1996). *The rise of the network society*. Oxford, Blackwell.

Cauvin, C. (1984). Espaces cognitifs et transformations cartographiques. *Lettres et Sciences Humaines (Strasbourg), 304*.

Clark, G. (1999). Where is Stranraer now? Space-time convergence re-visited. *World Transport Policy & Practice, 5*(2), 11–17.

Diener, Y. (2008). Un Espace chiffonné est-il un espace lacanien? *Essaim, 2*(12), 9–15.

Dupuy, G. (1991). *L'Urbanisme des réseaux*. Paris, France: Armand Colin.

Forer, P. (1978). A place for plastic space. *Progress in Human Geography, 2*(2), 230–267.

Golledge, R. G., & Hubert, L. J. (1982). Some comments on non-Euclidean mental maps. *Environment and Planning A, 14*, 107–118.

Graham, S., & Marvin, S. (2001). *Splintering urbanism, networked infrastructures, technological mobilities and the urban condition*. London: Routledge.

Grubesic, T., & Zook, M. (2007). A ticket to ride: Evolving landscapes of air travel accessibility in the United States. *Journal of Transport Geography, 15*(6), 417–430.
Haggett, P. (1990). *The geographer's art*. Oxford: Basil Blackwell.
Haggett, P. (2001). *Geography, a global synthesis*. Harlow: Prentice Hall.
Hall, E. T. (1966). *The hidden dimension*. Garden City, NY: Anchor Books.
Hyman, G. M., & Mayhew, L. (2004). Advances in travel geometry and urban modelling. *GeoJournal, 59*, 191–207.
Janelle, D. G. (1968). Central place development in a time-space framework. *Professional Geographer, 20*, 5–10.
Janelle, D. G., & Gillespie, A. (2004). Space-time constructs for linking information and communication technologies with issues in sustainable transportation. *Transport Reviews, 24*(6), 665–677.
Kirsch, S. (1995). The incredible shrinking world? Technology and the production of space. *Environment and Planning D: Society and Space, 13*(5), 529–555.
Knowles, R. D. (2006). Transport shaping space: Differential collapse in time-space. *Journal of Transport Geography, 14*(6), 407–425.
Kotoh, H. (2001). New ideas of time maps which shows exact times partially or maintain the topological structure. In *Colloque Européen de Géographie Théorique et Quantitative, Saint-Valéry-en-Caux.*
L'Hostis, A. (1996). Transports et Aménagement du territoire: cartographie par images de synthèse d'une métrique réseau. *Mappemonde, 3*, 37–43.
L'Hostis, A. (2000). Multimodalité et intermodalité dans les transports. *Atlas de France: transport et énergie, 11*, 99–112.
L'Hostis, A. (2003). De l'espace contracté à l'espace chiffonné. Apports de l'animation à la cartographie en relief des distances-temps modifiées par les réseaux de transport rapides. *Revue Internationale de Géomatique, 13*(1), 69–80.
L'Hostis, A. (2009). The shrivelled USA: Representing time-space in the context of metropolitanization and the development high speeds. *Journal of Transport Geography, 17*(6), 433–439.
L'Hostis, A., Mathis, P., & Polombo, N. (1993). Les grandes vitesses. In A. Bonnafous, F. Plassard, & B. Vulin (Eds.), *Circuler demain* (pp. 129–142). Paris/La Tour d'Aigues: Datar/éditions de l'aube.
Luminet, J.-P. (2001). *L'Univers chiffonné*. Paris, France: Fayard.
Mathis, P. (1996). La Stratégie des réseaux de transport dans le grand Ouest. In Y. Morvan (Ed.), *L'Entreprise Atlantique* (pp. 97–111). Paris, France: Editions de l'Aube.
Mathis, P. (2007). *Graphs and networks: Multilevel modelling*. London: ISTE.
Müller, J.-C. (1982). Non-Euclidean geographic spaces: Mapping functional distances. *Geographical Analysis, 14*, 189–203.
Murayama, Y. (1994). The impact of railways on accessibility in the Japanese urban system. *Journal of Transport Geography, 2*(2), 87–100.
Plassard, F. (1976). *Les Autoroutes et le développement régional*. Lyon, France: Economica/Presses Universitaires de Lyon.
Plassard, H., & Routhier, J.-L. (1987). *Sémiologie graphique et évaluation*. Lyon, France: A.R.T.U.R.
Rozenblat, C., & Cicille, P. (2003). *Les Villes européennes, analyse comparative*. Paris, France: La Documentation française/DATAR.
Sassen, S. (1991). *The global city: New York – London – Tokyo*. Oxford: Princeton University Press.
Shimizu, E. (1992). Time-space mapping based on topological transformation of physical map. In *W.C.T.R. Sixième conférence mondiale sur la recherche dans les transports, Lyon.*
Smith, D., & Timberlake, M. (2002). Hierarchies of dominance among world cities: A network approach. In S. Sassen (Ed.), *Global networks, linked cities* (pp. 117–141). London: Routledge.
Spiekermann, K., & Wegener, M. (1994). The Shrinking continent: New time-space maps of Europe. *Environment and Planning B: Planning and Design, 21*, 653–673.

Taylor, P. (2004). *World city network, a global urban analysis*. Oxon: Routledge.
Tobler, W. R. (1997). *Visualizing the impact of transportation on spatial relations*. Western Regional Science Association meeting, Hawaii.
Tobler, W. R. (1999). *The world is shriveling as it shrinks*. ESRI user conference, San Diego.
Viard, J. (2008). *Eloge de la mobilité, essai sur le capital temps libre et la valeur travail*. Paris, France: Editions de l'Aube.

Part II
Tools for Networks Analysis

Chapter 5
Structural Analysis of Networks

Guy Melançon and Céline Rozenblat

5.1 Introduction

Discovering patterns in a network, understanding how information flows between the regions of the network, and being able to locate these regions are central issues in a number of scientific areas or application domains. Previous chapters encouraged the use of graph theoretical concepts to capture notions related to geography, such as *reticularity* (Chap. 5.3), *reachability*, *accessibility* (Chap. 5.4) and *cohesion* (Chaps. 5.2 and 5.3). Such notions actually apply to other fields and, in fact, graph theory now plays a key role in the study of complex systems Studying the structure of a graph can reveal deep insights about the phenomenon it models.

In this chapter, we adopt a more quantitative perspective and introduce a series of measures used to typify the roles of nodes in a network, ultimately allowing the location of regions of interest. Notions such as node reachability and group cohesion will be revisited, and node metrics that can help quantify these notions will be discussed.

G. Melançon (✉)
CNRS UMR 5800 LaBRI, GRAVITE INRIA Sud-Ouest, Université Bordeaux 1,
33405 Talence Cedex, France
e-mail: Guy.Melancon@labri.fr

C. Rozenblat
Institute of Geography, University of Lausanne, GEOPOLIS Building,
1015 Lausanne, Switzerland
e-mail: Celine.Rozenblat@unil.ch

C. Rozenblat and G. Melançon (eds.), *Methods for Multilevel Analysis and Visualisation of Geographical Networks*, Methodos Series 11, DOI 10.1007/978-94-007-6677-8_5,

5.2 Measuring Reachability

Roughly speaking, the most easily reachable entities in a network are those sitting at a minimum distance from all others. This informal definition calls for a precise definition of what is understood as the *distance* between any two nodes in a network. We first establish the notation that will be used throughout this chapter and the rest of the book.

A graph $G = (V,E)$ is a pair where V is the set of *nodes* in the graph and E is its set of *edges*. Formally speaking, the set E is defined as a subset of the Cartesian product $V \times V$; that is, an edge $e = (u,v)$ is an ordered pair of nodes in V. Note that this definition implicitly assigns a *direction* to the edges. The particular case where E always simultaneously contains the pairs (u,v) and (v,u) allows us to neglect the edge direction, as we can alternatively represent edges as unordered pairs $\{u,v\}$. This chapter will only be concerned with undirected graphs.

A *path* in a graph can be described as a sequence of nodes u_0, u_1, ..., u_k, where pairs $\{u_i,u_{i+1}\}$ each form an edge in E.[1] (A similar definition can be given for directed graphs.) The notion of paths between nodes is central to connectivity: a graph is connected if for any two nodes $u,v \in V$, there exists a path from u to v. Equivalently, a graph is connected is there is a node from which any other node can be reached.

The special case where u_0 coincides with u_k is called a *cycle*. Given that all nodes are distinct (except possibly for u_0 and u_k), we say that the path (cycle) is *elementary*. The *length* of a path is its number of edges. Thus, the *distance* between any two nodes in V is the length of the shortest path connecting them. The distance between u and v in G is denoted as $d_G(u,v)$; if a path between u and v does not exist, we set $d_G(u,v) = \infty$. Note that the latter case only happens when the graph is disconnected. Also note that the graph theoretical distance $d_G(\cdot,\cdot)$ is a distance in the pure mathematical sense: it is always positive and satisfies $d_G(u,u) = 0$, it is symmetric ($d_G(u,v) = d_G(v,u)$), and it obeys the triangular inequality ($d_G(u,v) \leq d_G(u,w) + d_G(w,v)$).

The neighbors of a node u are the nodes sitting at distance 1 from u. The neighborhood of u, denoted as $N_G(u)$, thus equals $N_G(u) = \{v \in V | d_G(u,v) = 1\}$, and we have $|N_G(u)| = d_G(u)$. The neighborhood can be extended to the set of distance-k neighbors denoted as $N_G^{(k)}(u)$, which is the set of nodes sitting at a distance of at most k from a node, namely $N_G^{(k)}(u) = \{v \in V | d_G(u,v) \leq k\}$.

[1] Strictly speaking, this requires that the graph does not have multiple edges, which is the case we consider here. Otherwise, paths should specify which edges are followed along the path.

5.2.1 Distances and Closeness

Although reachability essentially relies on the *existence* of paths between nodes, one can attempt to provide a measure of *how easily* a node $u \in V$ can be reached in G – with the idea that longer paths require more effort to be traversed. Thus, the reachability of a node can be expressed in terms of graph distances.

The metrics we are about to describe use the graph theoretical distance to infer the properties of nodes and ultimately indicate how/where information flows in the graph. The rationale underlying all these metrics is that nodes sitting close to most other nodes have the easiest access to information; here, ease is measured in terms of distances.

$$s_G(u) = \sum_{v \in V} d_G(u,v) \tag{5.1}$$

The metric in Eq. (5.1) was introduced by Harary (1959) as the *status* of a node (in a graph). Observe that this metric is integer valued and can take any positive value (and even $S_G(u) = \infty$ can occur when G is disconnected). Beauchamp (1965) (although often cited through Sabidussi, 1966) later suggested computing the inverse ratio, which is currently called the *closeness centrality* of a node:

$$C_G(u) = \frac{1}{\sum_{v \in V} d_G(u,v)}. \tag{5.2}$$

Note that $C_G(u) \in [0,1]$ and that $C_G(u) > 0$ when G is connected. Additionally, $C_G(u) \leq \frac{1}{d_G(u)}$; the equality $C_G(u) = 1$ can only occur in the very special case of a two node graph.

An alternate definition is the *eccentricity* of a node $e_G(u)$ (see Hage & Harary, 1995), with the idea that central nodes are those having minimal eccentricity because they sit relatively close to all other nodes in the graph:

$$e_G(u) = \max_{v \in V} d_G(u,v). \tag{5.3}$$

Note, however, that the eccentricity of nodes does not lie in a bounded interval, which makes it difficult to compare nodes of different graphs. *Node centrality* (see Hage & Harary, 1995) overcomes this problem by taking the inverse ratio $1/e_G(u)$. The *diameter* of a graph is defined as the maximum eccentricity:

$$diam(G) = \max_{u,v \in V} d_G(u,v) = \max_{u \in V} e_G(u). \tag{5.4}$$

The *average distance* in a graph is also often used as an indicator of the graph structure:

$$\frac{1}{\frac{|V|(|V|-1)}{2}} \sum_{u,v \in V} d_G(u,v). \tag{5.5}$$

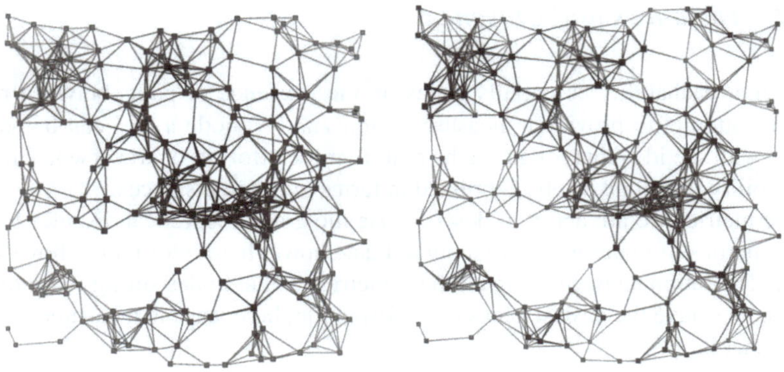

Fig. 5.1 The figure shows how values of the *(left)* Beauchamp Eq. (5.2) and *(right)* integration Eq. (5.6) distribute over nodes of a typical small world graph. Note how the Beauchamp reachability acts as a global measure, whereas the integration metric reveals local features. The node color changes from *light orange* to *purple* as the values increase. The node size also correlates with the node values (Color figure online)

Several authors have proposed variations to attempt to measure reachability; all of these variations use the distances between nodes as a basic component. Valente and Foreman (1998), for instance, define the *integration metric* as:

$$I_G(u) = \frac{\sum_{v \in V} \frac{1}{d_G(u,v)}}{|V| - 1}. \tag{5.6}$$

Just as with Beauchamp's reachability metric, the integration metric lies in the bounded interval $[0,1]$. At first glance, one may think that these metrics actually compare or even correlate with each other. Figure 5.1 shows that this is not the case and illustrates how Beauchamp's closeness centrality is a global measure, unlike integration, which acts as a local indicator of the degree to which a node "integrates" in its close-by neighborhood. It is true, however, that the closeness centrality correlates well with eccentricity.

Dangalchev (2006) recently proposed exponentially altering distances before computing the same sum $\sum_{v \in V} 2^{-d_G(u,v)}$. Surprisingly, Dangalchev's residual closeness strongly correlates with Beauchamps's closeness centrality; this correlation indicates that the geometric transform has a "globalization effect" on the distances. Note also that Dangalchev's centrality can be computed even when the graph is disconnected as $2^{-d_G(u,v)} = 0$ if $d_G(u,v) = \infty$.

5.2.2 Shortest Paths

Other authors designed measures that capture the degree to which a node is central in connecting the other elements of a network. This is somewhat different from

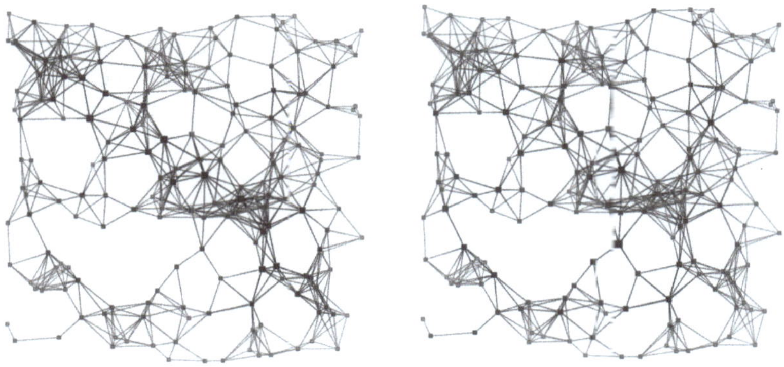

Fig. 5.2 The figure shows how values of the (*left*) stress centralities Eq. (5.7) and (*right*) betweenness Eq. (5.8) are distributed over the nodes of a typical small world graph. Note how both metrics compare well. The node color changes from *light orange* to *purple* as the values increase. The node size also correlates with the node values (Color figure online)

all distance-based metrics described in the preceding paragraph. In fact, distances between nodes do not fully describe *how* nodes are connected together and do not explicitly indicate whether actors have an essential role in distributing information through the network. Roughly speaking, focusing on the notion of *paths*, we aim to identify nodes that play an essential role in making information available to others, as opposed to the preceding paragraph, in which we tried to identify nodes that had easy access to information. That being said, distance-based and/or shortest path-based metrics can nevertheless be correlated, as we shall describe.

More precisely, let σ_{vw} denote the number of shortest paths connecting two nodes v and w. We set $\sigma_{vv} = 1$ by convention. This number may be greater than '1': imagine a cycle of size $2n$ for which this number is equal to 2 when nodes sit at distance n (that is, when they are one diameter away from each other). Let us also write $\sigma_{vw}(u)$ for the number of shortest paths connecting v and w *and* going through u (this number may be equal to 0). Shimbel (1953) defined the *stress centrality* of a node as:

$$S_G(u) = \sum_{v,w \in V} \sigma_{vw}(u). \tag{5.7}$$

Anthonisse (1971) and Freeman (1979) independently considered a ratio involving both $\sigma_{vw}(u)$ and σ_{vw}, which is now known as a node's *betweenness centrality*:

$$B_G(u) = \sum_{v,w \in V} \frac{\sigma_{vw}(u)}{\sigma_{vw}}. \tag{5.8}$$

Figure 5.2 shows both metrics mapped in terms of color and node size. As can be observed, these two variations correlate well, which can be confirmed, for instance, by examining the histogram of values.

A useful exercise is to compute distance-based and shortest paths metrics on the same graph and compare the results. As we have argued earlier, we can consider these metrics as belonging to different classes. Distance-based metrics highlight nodes having easy access to information, while shortest path metrics identify nodes playing a central role in distributing information. As such, both classes of metrics will indicate nodes involved in both roles: those sitting at the center will presumably play a role in distributing information, while efficient broadcaster nodes should not sit at the periphery. This exercise can be performed by visually comparing the colormaps in Figs. 5.1 and 5.2.

5.2.3 Group Cohesion

As emphasized above, singular nodes can embody specific roles in terms of reachability and/or the distribution of information through the network. As for the other nodes, although they do not have decisive impacts as individual entities, they might collectively contribute to the overall network structure. What is usually meant here is that the network organizes into tighter subgroups and that, in a sense, the whole network divides into "communities", a term coined by Newman (2001), see also Newman (2003, 2006), interacting with each other.

We seek to identify the *community structure* of a network: find the subgroups and observe how they interact. Being able to identify these communities is an issue that differs somewhat from the questions we addressed in previous sections. More precisely, the perimeter of a community cannot be deduced based on the role of a single node but rather from the overall properties of a subset of nodes. We shall see, however, how these two perspectives act dually and relate to each other.

Watts and Strogatz (1998) had introduced the *clustering index* of nodes in a graph as an index measuring the overall connectivity between the neighbors of a node u:

$$c_G(u) = \frac{e(N_G(u), N_G(u))}{\binom{|N_G(u)|}{2}} \tag{5.9}$$

where $e(A,B)$ is the number of edges connecting the nodes in set A to nodes in set B. The expression $\binom{k}{2}$ denotes the binomial coefficient, which incidentally computes the maximum number of edges that can exist in a graph with k nodes. That is, the index $c_G(u)$ roughly computes the probability that any two neighbors of u are connected. The *clustering index of a graph* then equals the average value of its nodes $c(G) = \frac{\sum_u c_G(u)}{|V|}$.

Until Watts and Strogatz started studying *small world networks*, the only available graph model was that defined by Erdős and Rényi (1959, 1960). The Erdős-Rényi model includes the following characteristics: a set of n nodes is given; edges are then added independently between node pairs following a random procedure (heads or tails, for instance). These graphs can be shown to have a rather

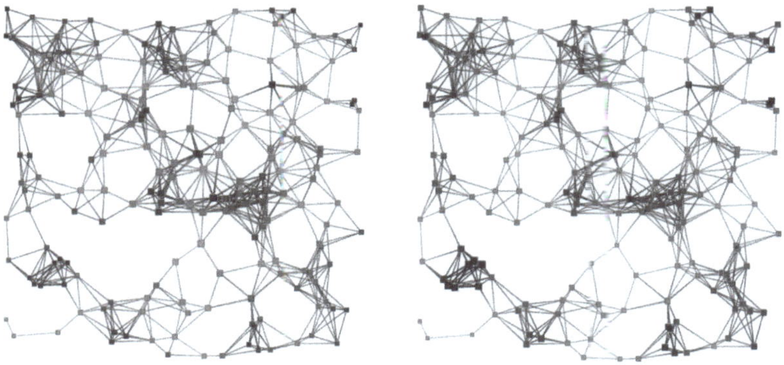

Fig. 5.3 The figure shows how values of the clustering index (*left*) and Jaccard index (*right*) are distributed over the nodes/edges of a typical small world graph. Note how both metrics act locally, as opposed to the metrics in Fig. 5.1 or Fig. 5.2. The node color changes from *light orange* to *purple* as the values increase. The node size also correlates with the node values (Color figure online)

low average distance (see Eq. 5.5) between nodes. However, nodes do not split into distinct communities because the connections are uniformly distributed among the nodes.[2] In other words, the clustering coefficient of an Erdős-Renyi graph can be expected to be somewhat small.

Inspired by the popular saying "It's a small world", Watts and Strogatz developed different graph models that showed characteristics similar to those of the real world networks observed by Milgram (1967). More precisely, a graph is said to be small world if its clustering coefficient is high while the average distance between nodes is low when compared to the values of random Erdős-Renyi graphs with the same number of nodes and edges.

The clustering index does just what it is expected to do, as it indicates just how well connected is the neighborhood of a node u; the clustering index can thus be presented as a *cohesion metric*. Another approach is to focus instead on edges and consider how many common neighbors that the end nodes of an edge $e = \{u, v\}$ have:

$$J_G(e) = \frac{|N_G(u) \cap N_G(v)|}{|N_G(u) \cup N_G(v)|} \tag{5.10}$$

The metric in Eq. (5.10) was first introduced by Jaccard (1901), who studied the similarity between floral species. However, the Jaccard index can be computed on sets and can thus be used to compare any type of attribute-based data (Fig. 5.3). Also note that, when applied to neighbor sets $N_G(u)$, the numerator actually computes the number of triads (cycle of length 3) in a graph G going through an edge e.

[2]This being said, a random Erdős-Rényi graph might not be necessarily connected. For more details, see Bollobás (1985).

This idea converges with that of Granovetter, in which triads are a key component in social cohesion (Granovetter, 1973). Other edge metrics that were proposed much later were designed to achieve similar goals (see Sallaberry & Melançon, 2008 for more details).

The Jaccard index has a useful interpretation in terms of group cohesion. Note that when an edge $e = \{u, v\}$ connects two distinct neighborhoods $N_G(u) \cap N_G(v) = \emptyset$, we have $J_G(e) = 0$. That is, the Jaccard index seems capable of locating regions where communities break apart. In the same line of reasoning, one may expect the clustering coefficient to help locate such transition areas in networks.

The two metrics we have described so far are *local*, meaning that the value assigned to a node or edge is computed by examining neighboring nodes and/or edges. Two other types of metrics can be defined to capture community structures. A well-known and widely used metric that has recently attracted the attention of a wide audience through the fertile work of Newman et al. (Clauset, Newman, & Moore, 2004; Girvan & Newman, 2002; Newman, 2001, 2006) is the betweenness centrality metric *extended to the edges* of a graph. Recall that the betweenness centrality Eq. (5.8) is defined in terms of the number of shortest paths going through a node (or edge). That is, an edge with a high betweenness centrality can be expected to act as a *bridge* between communities. Burt (2005) also highlights high centrality edges as bridges passing through structural holes in the network. Bridges have been exploited as a central component for breaking down a graph into communities (see, for instance, Amiel, Melançon, & Rozenblat, 2005; Auber, Chiricota, Jourdan, & Melançon, 2003; Clauset et al., 2004; Girvan & Newman, 2002).

5.3 Community Dynamics

Finding cohesive subgroups or communities in a network helps obtain a better understanding of how network entities interact. When performing a visual inspection or exploration of a network, one will typically compute communities and then build a simpler graph, the "quotient graph", from the community structure. More precisely, let $G = (V, E)$ be a graph and $\mathbf{C}$ be a partition $\mathbf{C} = (C_1, \ldots, C_k)$ where $C_i \subset V$ are distinct subsets ($C_i \cap C_j = \emptyset$ when $i \neq j$) covering $V = \cup_{i=1,\ldots,k} C_i$.

Then, we can form the quotient graph $G/\mathbf{C}$ with vertices $\{C_1, \ldots, C_k\}$, where edges connect C_i and C_j when there is at least one edge $e = \{u, v\}$ connecting two nodes $u \in C_i$ and $v \in C_j$. Fig. 5.4 illustrates this construction. Each box shows the subgraph contained in a cluster, which is shown as a metanode. Metanodes are connected according to the edges between the nodes they contain.

Providing an overview of all existing graph clustering is beyond the scope of this book. The interested reader should see Brandes, Gaertler, and Wagner (2007) and Schaeffer (2007), for instance. Here, we shall limit our discussion to a few popular clustering strategies that rely on network indices (see Amiel et al., 2005; Auber et al., 2003; Clauset et al., 2004; Girvan & Newman, 2002). As mentioned earlier, centrality indices can be used to identify the bottleneck edges in networks. Hence,

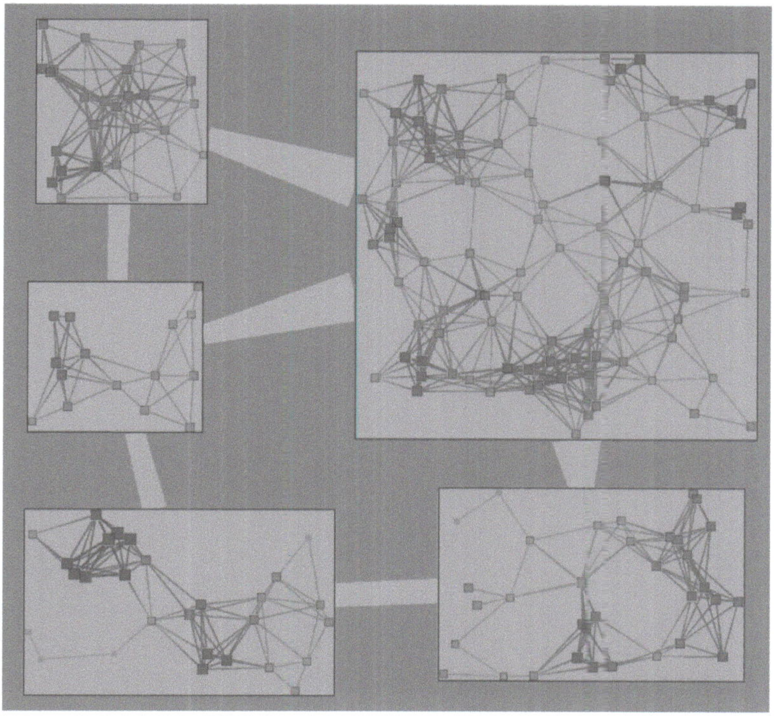

Fig. 5.4 A quotient graph consists of metanodes that themselves contain subgraphs. Links between metanodes consist of the edges connecting the nodes in the underlying subgraphs. This quotient graph is built from the graphs shown in the previous figures

bridges (edges with low centrality values) can be temporarily discarded, turning the graph into disconnected components that all appear as communities and defined clusters. The quotient graph can thus be formed where metanodes connect through bridges. Note that this procedure can be repeated at will on each community to provide a multilevel decomposition of the original graph, as suggested in Fig. 5.5.

The procedure described here proceeds in a top-down manner, further decomposing communities into sub-communities. Other algorithms proceed in a bottom-up manner and build super-communities from smaller subgroups.

A number of questions can be addressed once a graph has been decomposed into communities. Note that in the visual representation of a quotient graph (Figs. 5.4 and 5.5), all edges connecting members of two distinct communities are abstracted as a single higher-level edge. It is interesting, however, to study how clusters exchange through their members. To this end, Guimerá, Mossa, Turtschi, and Amaral (2005) introduced what they call the *participation coefficient* of a node. Before giving a formal definition for this coefficient, we need to introduce some notation. Recall that $d_G(v)$ denotes the degree of a node in a graph G (simply written as $d(v)$ when G is clear from context). Now, we denote as $d_{C_i}(v)$ the degree of a node in a cluster

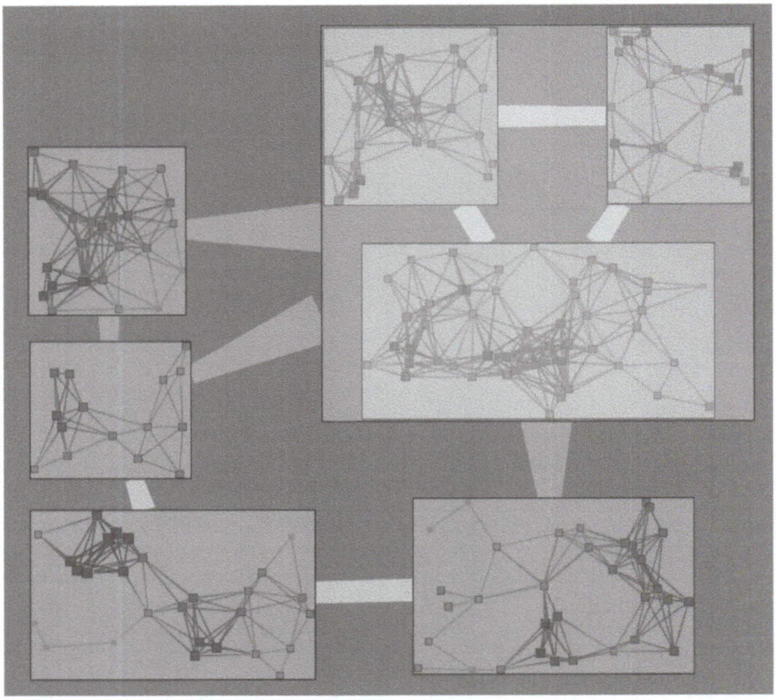

Fig. 5.5 A multilevel quotient graph is one where higher level clusters are themselves decomposed into sub-communities. The upper right cluster in Fig. 5.4 has been further decomposed into three sub-communities

C_i. That is, $d_{C_i}(v)$ equals the number of neighbors the node v has in cluster C_i, i.e., $d(v) = \sum_{i=1,\ldots,k} d_{C_i}(v)$. The participation coefficient of a node $v \in V$ is defined as:

$$p(v) = 1 - \sum_{C_i \in \mathbf{C}} \left(\frac{d_{C_i}(v)}{d_G(v)}\right)^2 \tag{5.11}$$

Note that $p(v) = 0$ exactly when v has all its neighbors in the same cluster, meaning that the participation of v in the overall community structure concentrates on a single subgroup, which is most likely the one it belongs to. Conversely, $p(v)$ is minimized when its neighbors are uniformly distributed over all clusters. The participation coefficient of a node compares with *Shannon's entropy* (Shannon, 1948):

$$H(v) = \sum_{C_i \in \mathbf{C}} \frac{d_{C_i}(v)}{d_G(v)} \log\left(\frac{d_{C_i}(v)}{d_G(v)}\right) \tag{5.12}$$

Note that both of these coefficients (and those defined below) allow for straightforward generalizations to weighted graphs. Similarly, Burt also introduced the *hierarchy index* of a node (see Burt, 2005). The participation coefficient, like Shannon's entropy and Burt's index, can be used to evaluate the nodes' participation in the overall network dynamics.

In Eq. (5.11), we may distinguish the term involving node v and consider the quantity

$$\frac{d_{C_i}(v)}{d_G(v)}$$

as the *representation* of node v within its cluster C_i. Similarly, we can define the *contribution* of a node v to its class as:

$$\frac{d_{C_i}(v)}{\sum_{u \in C_i} d_{C_i}(u)}$$

These indices have proven to be useful and complementary to Guimera's participation coefficient and z-score (Guimerà et al., 2005) when analyzing community dynamics.

5.4 Conclusion

This chapter surveyed a number of nodes and edges metrics, echoing the notions explored in the previous chapter, and implicitly classified these metrics into a taxonomy. Local metrics only rely on neighbor information and can usually be computed efficiently. Distance-based metrics typically require traversing the graph and computing all node distances. Centrality measures rely on the shortest paths between nodes.

All of these types of metrics can help identify densely connected subgroups by either assessing the local edge density, such as with the Jaccard index for edges Eq. (5.10), evaluating the node clustering index Eq. (5.9), or identifying the nodes or edges acting as bottleneck routes or points in the network. These bottleneck elements can then be used to split the network into subgroups and capture its community structure. The last metrics we described apply to clustered graphs and aim to measure the degree to which a node interacts with other clusters in the network. Examples showing how these metrics can be used to analyze networks shall be discussed at length in the forthcoming chapters.

References

Amiel, M., Melançon, G., & Rozenblat, C. (2005). Réseaux multi-niveaux l'exemple des échanges aériens mondiaux de passagers. *Mappemonde*, *79*, 12 p. Retrieved July 21, 2009, from http:// mappemonde.mgm.fr/num7/articles/art05302.html

Anthonisse, J. M. (1971). *The rush in a directed graph* (Technical Rep. No. BN 9/71). Amsterdam: Stichting Mathematisch Centrum.

Auber, D., Chiricota, Y., Jourdan, F., & Melançon, G. (2003). Multiscale navigation of small world networks. In *IEEE symposium on information visualisation* (pp. 75–81). Seattle, GA: IEEE Computer Science Press.

Beauchamp, M. A. (1965). An improved index of centrality. *Behavior Science, 10*(2), 161–163.

Bollobás, B. (1985). *Random graphs*. London: Academic Press.

Brandes, U., Gaertler, M., & Wagner, D. (2007). Engineering graph clustering: Models and experimental evaluation. *Journal of Experimental Algorithmics, 12*, 1.1.

Burt, R. S. (2005). *Brokerage and closure*. Oxford: Oxford University Press.

Clauset, A., Newman, M. E. J., & Moore, C. (2004). Finding community structure in very large networks. *Physics Reviews E, 70*, 066111.

Dangalchev, C. (2006). Residual closeness in networks. *Physica A, 365*, 556–564.

Erdös, P., & Rényi, A. (1959). On random graphs. I. *Publicationes Mathematicae Debrecen, 6*, 290–297.

Erdös, P., & Rényi, A. (1960). On the evolution of random graphs. *Publications of the Mathematical Institute of the Hungarian Academy of Sciences, 5*, 17–61.

Freeman, L. C. (1979). Centrality in social networks: Conceptual clarification. *Social Networks, 1*, 215–239.

Girvan, M., & Newman, M. E. J. (2002). Community structure in social and biological networks. *Proceedings of the National Academy Science USA, 99*, 7821–7826.

Granovetter, M. S. (1973). The strength of weak ties. *The American Journal of Sociology, 78*(6), 1360–1380.

Guimerà, R., Mossa, S., Turtschi, A., & Amaral, L. A. N. (2005). The worldwide air transportation network: Anomalous centrality, community structure, and cities' global roles. *Proceedings of the National Academy of Sciences of the United States of America, 102*(22), 7794–7799.

Hage, P., & Harary, F. (1995). Eccentricity and centrality in networks. *Social Networks, 17*, 57–63.

Harary, F. (1959). Status and contrastatus. *Sociometry, 22*, 23–43.

Jaccard, P. (1901). Distribution de la flore alpine dans la bassin de dranses et dans quelques regions voisines. *Bulletin del la Société Vaudoise des Sciences Naturelles, 37*, 241–272.

Milgram, S. (1967). The small world problem. *Psychology Today, 1*, 61.

Newman, M. E. J. (2001). The structure of scientific collaboration networks. *Proceedings of the National Academy of Science USA, 98*, 404–409.

Newman, M. E. J. (2003). The structure and function of complex networks. *SIAM Review, 45*, 167–256.

Newman, M. E. J. (2006). Modularity and community structure in networks. *Proceedings of the National Acadademy of Sciences USA, 103*, 8577–8582.

Sabidussi, G. (1966). The centrality index of a graph. *Psychometrika, 31*(4), 581–603.

Sallaberry, A., & Melançon, G. (2008). Edge metrics for visual graph analytics: A comparative study. In *Information visualization conference* (pp. 610–615). London: IEEE Computer Society.

Schaeffer, S. E. (2007). Graph clustering. *Computer Science Review, 1*, 27–64.

Shannon, C. E. (1948). A mathematical theory of communication. *The Bell System Technical Journal, 27*, 379–423, 623–656.

Shimbel, A. (1953). Structural parameters of communication networks. *Bulletin of Mathematical Biology, 15*(4), 501–507.

Valente, T. W., & Foreman, R. K. (1998). Integration and radiality: Measuring the extent of an individual's connectedness and reachability in a network. *Social Networks, 20*(1), 89–105.

Watts, D., & Strogatz, S. H. (1998). Collective dynamics of "small-world" networks. *Nature, 393*, 440–442.

Chapter 6
Graph Visualization For Geography

Antoine Lambert, Romain Bourqui, and David Auber

6.1 Introduction

Network analysis is usually performed with combinatorial measures such as those presented in Chap. 5 or traditional statistical tools such as principal component analysis. However, it is extremely difficult for some patterns, even with these powerful approaches, and the results of these methods can be difficult to interpret without an appropriate representation.

The purpose of this chapter is to present methods for the automatic drawing of networks, which is a complex field that can raise several intractable problems related to graph theory. However, for the purpose of geographic network analysis, there are two main approaches that are sufficiently general for use in this application. The first method is the so-called force-directed approach that uses a physical analogy to construct a model network where the Euclidean distances are very similar to the distances in the actual network. The second method creates an abstract network that reduces the visual complexity of the representation. Because the geographic positions of the network elements can provide crucial information, one specific approach relates the topology of the network to the geographical positions. Then, the flow map layout technique can be applied to emphasize high level patterns in that network.

The remainder of this chapter is structured as follows. Section 6.2 reviews some related work on general graph drawing methods and, in particular, force-directed approaches. In Sect. 6.3, we present the compound graph visualization technique. We next explain how node positions can be taken into account and high level patterns can be highlighted in Sect. 6.3.1. Finally, we draw our conclusions in Sect. 6.5.

A. Lambert (✉) • R. Bourqui • D. Auber
CNRS UMR 5800 LaBRI, Université Bordeaux 1 – GRAVITE INRIA Sud-Ouest,
33405 Talence Cedex, France
e-mail: Antoine.Lambert@labri.fr; Romain.Bourqui@labri.fr; David.Auber@labri.fr

C. Rozenblat and G. Melançon (eds.), *Methods for Multilevel Analysis and Visualisation of Geographical Networks*, Methodos Series 11, DOI 10.1007/978-94-007-6677-8_6,

6.2 General Graph Drawing Method

The most frequently used approach for laying out general graphs is the so-called force-directed method that uses particles and attractive forces to model the nodes and their edges in a network, respectively. The algorithm creates a physical system and simulates its evolution to a state of equilibrium. Figure 6.1 shows a representation of a physical system that models nodes with particles and their edges with springs.

That method has become increasingly popular in recent years because it presents some interesting advantages. First, the physical analogy is easy to understand for practitioners without a background in graph theory. Thus, it generates visualizations of graphs that can be shared and presented to large audiences. Second, the naive implementation of this algorithm is straightforward, and one can easily modify it and make customized versions to meet particular needs. Finally, this method generates drawings that are visually pleasant, structurally significant and usually better optimized than those obtained with other graph drawing algorithms.

When using this naïve approach (e.g., Eades, 1984; Frick, Ludwig, & Mehldau, 1995; Fruchterman & Reingold, 1991), system convergence is quite slow and this method is not useful for large datasets. Recent force-directed algorithms overcome this limitation by making a trade-off between time efficiency and aesthetic criteria (e.g., Gajer & Kobourov, 2002; Hachul & Junger, 2005; Lauther, 2007). In this section, we first present the main classical force-directed algorithms and then explain how to improve the time complexity of such algorithms.

6.2.1 Force Directed Drawing

In 1984, Eades introduced the first algorithm based on the use of a force system. His algorithm is initialized by placing the nodes at random positions. Then, at each round, each node is moved in the direction of the attractive (springs) and repulsive forces (nodes). After several rounds, the system automatically converges to equilibrium. The bottlenecks of this approach are the force computation complexity and, in some cases, the system can oscillate and never reach equilibrium. Several

Fig. 6.1 Example of a physical system built using particles and springs to represent the nodes and their edges, respectively. Particles create repulsive forces while springs create attractive forces between particles

improvements have been proposed. Even if these methods do not scale to large datasets, they guarantee the convergence of the system.

All force-based algorithms follow the same concept, and algorithm 1 summarizes the four main steps of these methods. In the following, we compare three of the most commonly used implementations for each step of the algorithm (Eades, 1984; Frick et al., 1995; Fruchterman & Reingold, 1991).

```
Data: A graph
Result: A force directed layout
Calculate Initial Position (CIP)
while not equilibrium do
    Compute Attractive/Repulsive Forces (CARF)
    Compute Displacement Vector (CDV)
    Move Nodes (MN)
end
```

Force-directed algorithms can be summarized in four steps. CIP initializes the system before running the simulation, CARF implements the interaction rules of the system, CDV manages convergence of the system and MN applies the result to the system.

Eades (1984):

To speed up the convergence of the system Eades (1984) proposed replacing the Hooke's forces or Hooke's law with his own force model where P_u is the coordinates of a node u, $\overrightarrow{p_u p_v}$ is the unit vector co-directional with $\overrightarrow{P_u P_v}$, $dist(u,v)$ is the norm of $\overrightarrow{P_u P_v}$, l is the ideal length of an edge (i.e., spring), and c_{att}, c_{rep} are two constants that define the strength of the spring. Attractive and repulsive forces are given by the following formulas:

$$f_{att}(u,v) = c_{att} \cdot \log\left(\frac{dist(u,v)}{length}\right) \cdot \overrightarrow{p_u p_v} \tag{6.1}$$

$$f_{rep}(u,v) = \frac{c_{rep}}{dist(u,v)^2} \cdot \overrightarrow{p_v p_u} \tag{6.2}$$

and the total force exerted on a node u is as follows:

$$f_{total}(u) = \sum_{(u,v)\in E} f_{att}(u,v) + \sum_{(u,v)\notin E} f_{rep}(u,v) \tag{6.3}$$

Note that the repulsive forces are computed between all pairs of non-neighbor nodes. Because the MN phase is performed in a synchronous way, i.e., all nodes are moved at the same time, Eades' method used a constant $\delta \in [0,1]$ to limit the movements of a nodes. The displacement vector is computed as follows:

$$disp(u) = \delta \cdot f_{total}(u) \tag{6.4}$$

Fruchterman and Reingold (1991):

The force system was then modified to reach equilibrium with fewer iterations. This is achieved by using a quadratic force model:

$$f_{att}(u,v) = \frac{dist(u,v)^2}{length} \cdot \overrightarrow{p_u p_v} \tag{6.5}$$

$$f_{rep}(u,v) = c_{rep} \cdot \frac{length^2}{dist(u,v)} \cdot \overrightarrow{p_v p_u} \tag{6.6}$$

and the total force exerted on node u is as follows:

$$f_{total}(u) = \sum_{(u,v) \in E} f_{att}(u,v) + \sum_{v \in V, v \neq u} f_{rep}(u,v) \tag{6.7}$$

Note that in Fruchterman and Reingold's algorithm, the repulsive forces are computed between all pair of nodes. To optimize the time computation, Fruchterman and Reingold use a grid that helps to approximate the repulsive forces and a decreasing function δ that bounds the movements of the nodes (during the CDV step):

$$disp(u) = \delta(iteration\_number) \cdot f_{total}(u) \tag{6.8}$$

where

$$\delta : [1..maxIterations] \to [0,1]$$
$$itNumber \to \delta(itNumber)$$

is a decreasing function. The extent to which the algorithm restricts the node movements increases with the number of times the algorithm has iterated, which improves the convergence of the algorithm.

GEM by Frick et al. (1995):

In this article, the authors give their own force model:

$$f_{att}(u,v) = \frac{dist(u,v)^2}{length \cdot \Phi(u)} \cdot \overrightarrow{p_u p_v} \tag{6.9}$$

where $\Phi(u) = 1 + \frac{\deg_G(u)}{2}$.

$$f_{rep}(u,v) = c_{rep} \cdot \frac{length^2}{dist(u,v)} \cdot \overrightarrow{p_v p_u} \tag{6.10}$$

and the total force exerted on node u is as follows:

$$f_{total}(u) = \sum_{(u,v)\in E} f_{att}(u,v) + \sum_{v\in V, v\neq u} f_{rep}(u,v) + f_{gravity}(u) - f_{random}(u) \quad (6.11)$$

The force-directed approach determines the node positions by minimizing the total energy of the system. However, the solution is usually sub-optimal and the algorithm finds a local minima. Frick et al. (1995) added a random force in an attempt to overcome that problem. Frick et al. also introduced a new displacement scheme (CDV phase) in which the δ function is defined for each node of the network:

$$\delta : V[1..maxIterations] \rightarrow [0,1]$$
$$(u, itNumber) \rightarrow \delta(u, itNumber)$$

Again, the more the algorithm has iterated, the more the nodes movements are restricted. This function also decreases when the node rotates and/or oscillates around a point, which allows the algorithm to converge faster.

6.2.2 *Fast Force Directed Drawing*

To reduce the computation times, particular nodes are extracted from the network. These nodes must be as "central" as possible because they will represent the whole graph. This process is repeated until the set of elected nodes is small enough to be quickly drawn, typically several dozen nodes. This results in a hierarchy of sets of elected nodes, from the highest set that contains only a few nodes to the lowest set that contains all of the nodes from the original network. Let $S_0, S_1, \ldots, S_k$ be these sets of elected nodes, then $S_0 \supset S_1 \supset \ldots \supset S_k$ and $S_0 = V$. Each set of elected nodes is then drawn in a top-down manner, i.e., from S_k to S_0, using a classical force-directed algorithm (see Sect. 6.2.1). To reduce the computation time, when drawing the set S_i, only nodes not contained in S_{i+1} are taken into account so that each node is embedded only once.

In the following section, we present a more detailed description of GRIP (Gajer & Kobourov, 2002; Gajer, Goodrich, & Kobourov, 2004), one of the most popular fast force-directed algorithm.

GRIP: Graph dRawing with Intelligent Placement (Gajer & Kobourov, 2002; Gajer et al., 2004):

The GRIP algorithm consists of two main steps: an extraction step and a drawing step.

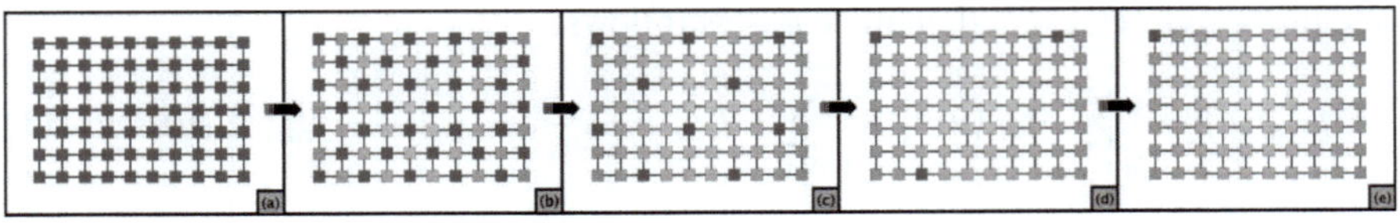

Fig. 6.2 (**a**) Example of a graph containing 70 nodes and 123 edges; (**b**) Maximal set S_1 of nodes at distance at least 2 is shown in *red*; (**c**) Maximal set S_2 of nodes at a distance at least $2^2 = 4$ is shown in *red*; (**d**) Maximal set S_3 of nodes at a distance at least $2^3 = 8$ is shown in *red*; (**e**) Maximal set S_4 of nodes at a distance at least $2^4 = 16$ is shown in *red* (Color figure online)

Extraction Step:

The first step of the GRIP algorithm computes a *Maximal Independent Set Filtration* (MISF) ν of the set of vertices V of G. The filtration, ν, is defined as follows: $\nu = S_0, S_1, \ldots, S_k$ where $S_0 \supset S_1 \supset \ldots \supset S_k \supset \emptyset$, $S_0 = V$ and $\forall i, 0 < i \leq k, S_i$ is a maximal subset of S_{i+1} such that for all pairs of vertices u, v of S_i, the graph distance between u and v is greater or equal to 2^i. The idea is to extract the nodes at a distance of at least 2 to represent the whole graph, then to extract the nodes at a distance of at least 4 to represent the nodes at a distance of 2, and so on until the set of elected nodes contains at most 3 nodes as shown in Fig. 6.2. This process makes it possible to compute sets of elected nodes that are well-distributed in the network.

Drawing Step:

During the second step of the GRIP algorithm (Gajer & Kobourov, 2002; Gajer et al., 2004), the process first places nodes of S_k and then each filtration set $S_{k-1}, S_{k-2}, \ldots, S_0$. For each level, the placement is split into two phases: an “intelligent” initial placement (i.e., nodes are placed closed to their final positions), and then a refinement phase that uses a force-directed algorithm (either the Kamada-Kawai (1989) or the Fruchterman-Reingold (1991) algorithm).

Intelligent Initial Placement:

For the smallest set S_k, the authors consider that S_k contains exactly 3 nodes. (If this is not the case, then the nodes of S_{k-1} are included in S_k). Let u_1, u_2 and u_3 be the three nodes of S_k, which are laid out on a triangle as follows:

$$dist_{\mathbb{R}}(u_1, u_2) = dist_G(u_1, u_2)$$
$$dist_{\mathbb{R}}(u_1, u_3) = dist_G(u_1, u_3)$$
$$dist_{\mathbb{R}}(u_2, u_3) = dist_G(u_2, u_3)$$

where $dist_{\mathbb{R}}(u, v)$ is the Euclidean distance between u and v and $dist_G(u, v)$ is the theoretical graph distance between them.

Table 6.1 Comparison of computation times based on our networks in seconds

	AT 2000	AT 2004	US migration	Daily migrations
# nodes / # edges	1,525 / 16,434	1,767 / 19,353	1,716 / 9,775	1,181 / 10,829
GEM (s)	251.1	384.9	291.5	105.8
GRIP (s)	1.5	2.2	1.1	1.3

For every other level i, $0 \leq i < k$, let u be a vertex of S_i, and v_1, v_2 and v_3 be the three closest and already placed neighbors of u. The authors lay out u at the barycenter of v_1, v_2 and v_3 that is evaluated from the graph distances between u and the three neighbors.

Refinement:

To refine the placement, the authors use force-directed algorithms. For each level i, $1 \leq i < k$, they use the well-known Kamada-Kawai algorithm (1989). This algorithm takes into account the graph distances in the computation of the forces. And, for the last filtering set S_0, computing all graph distances would take too much time and/or memory space, so they apply the Fruchtermann and Reingold algorithm (1991), which uses only Euclidean distances (see Sect. 6.2.1). To further improve the computation time, the forces applied to a vertex v of S_i include only the interactions between v and its closest neighbors in S_i.

6.2.3 *Results Comparison*

In this section, we compare two well-known force-directed approaches in terms of their computation time and drawing quality.

Table 6.1 shows the computation times obtained with GRIP (Gajer & Kobourov, 2002; Gajer et al., 2004) and GEM (Frick et al., 1995), a fast force-directed algorithm and a classical force-directed algorithm, respectively. These networks are 2000 and 2004 international air interconnections, US migrations and French daily migrations. Using GRIP improves the computation times by a factor of more than 100. This approach handles large networks (in this example, up to more than 1,500 nodes and 25,000 edges). The gap between the force-directed and fast force-directed approaches is even wider when drawing networks containing hundreds of thousands of elements.

Figure 6.3 shows drawings of 2004 international air interconnections and French daily migrations networks by GEM (Frick et al., 1995) and GRIP (Gajer & Kobourov, 2002; Gajer et al., 2004). When looking carefully at these results, one can see that the drawings obtained with GEM emphasize more information than

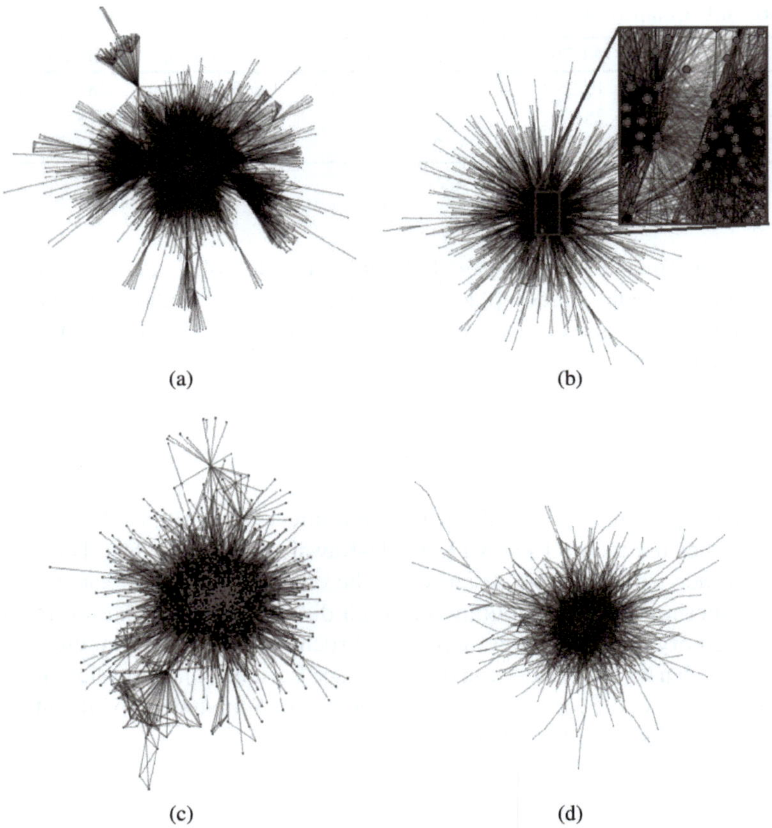

Fig. 6.3 Drawings obtained from 2004 international interconnections and US migration networks by: (**a, b**) GEM (Frick et al., 1995) and (**c, d**) GRIP (Gajer & Kobourov, 2002; Gajer et al., 2004)

those from GRIP. For instance, in Fig. 6.3b, one can see two large components that split the US migration networks into two regions while no such phenomenon is observed in Fig. 6.3d. Actually, these two regions correspond to groups of western and eastern states of the United States.

When dealing with reasonable size networks (several hundred or thousand elements), it seems that the classical force-directed approaches are preferable to the fast force-directed ones, because they offer better results in terms of aesthetic criteria and they emphasize more information. On the contrary, when dealing with large networks, one should use fast force-directed algorithms because they offer much better computation times.

6.3 Compound Representation

To find an algorithm that gives good results (in term of computation time, aesthetic criteria and information emphasized) for arbitrary networks is a very difficult problem. One of the most popular approaches to drawing such general graphs, namely the force-directed approach, which produces visually pleasant and structurally significant results (e.g., Frick et al., 1995; Fruchterman & Reingold, 1991; Gajer & Kobourov, 2002; Hachul & Junger, 2005; Lauther, 2007).

Even if fast graph layout algorithms can scale up to thousands of elements, two important issues still remain: cluttering and slow rendering. The clutter (node-node overlaps, node-edge overlaps and edge-edge crossings) increases with the size and complexity of the graph, resulting in unreadable visualizations. Additionally, displaying a large number of elements may lead to slow rendering and thereby hinder interactive exploration. Therefore, both cluttering and slow rendering make the analysis task harder for end-users and must be avoided.

An effective way to reduce the visual complexity of the representation, thereby reducing visual clutter and speeding up the rendering process, is to build an abstraction of the original graph. That abstraction is then displayed on the screen and the user can explore his/her data via a dedicated interactive system that allows the end-user to reduce/increase the level of detail.

6.3.1 Building an Abstraction

To build an abstraction, one usually applies clustering algorithms that group together similar elements of the network (e.g., Auber, Chiricota, Jourdan, & Melançon, 2003; Dongen, 2000; McSherry, 2004; Newman & Girvan, 2004; Schaeffer, 2005). After the first partition of the elements is computed, two options are offered: either refine each cluster by applying the clustering algorithm, or contract each cluster into a single node and apply the clustering algorithm to the resulting graph. If the first partition of a few clusters contains many elements, then the first option should be preferred. On the contrary, if the partition of many clusters contains only a few elements, then the second option will increase the level of abstraction. In Fig. 6.4a, one can see the result of an iterative decomposition of an illustrative network. The network has been first decomposed into three clusters (green, pink and yellow) and then the yellow cluster has been refined and a brown cluster has been found. In Fig. 6.4b, one can see the usual data structure that was used to represent that hierarchical partition of the nodes.

Once the hierarchical partition has been computed, the abstraction of the network (according to that partition) is then built by replacing each cluster of nodes with a single node (called a metanode), replacing all edges linking two of these subsets with a single edge (called a metaedge), and linking the corresponding metanodes. This operation reduces the number of displayed nodes and edges and, therefore,

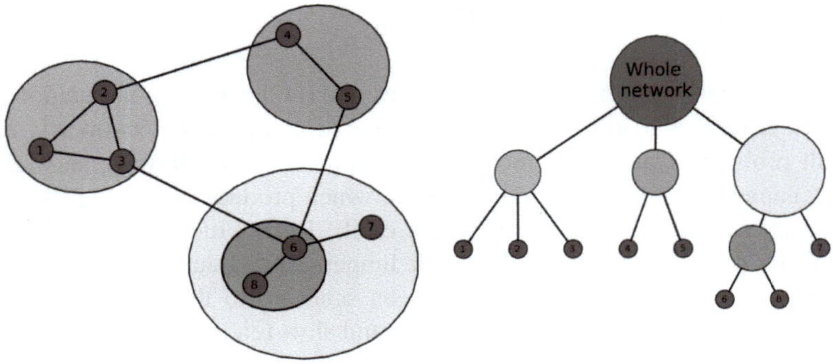

(a) (b)

Fig. 6.4 (**a**) Example of a network that has been iteratively decomposed into clusters and sub-clusters. Each of these clusters is shown in a particular color. (**b**) The corresponding data representation where each leaf of the tree represents a node of the network, and an internal node represents a cluster (Color figure online)

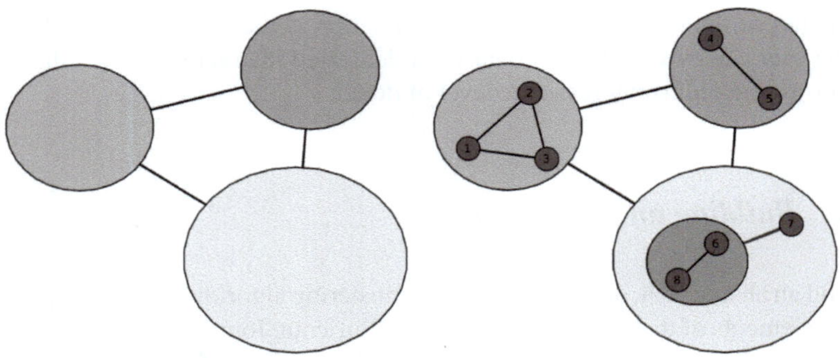

(a) (b)

Fig. 6.5 (**a**) Classical compound visualization of the highest level of abstraction of the network shown in Fig. 6.4a; (**b**) corresponding multiscale visualization

reduces the visual clutter and increases the rendering speed. This abstraction is usually called a compound graph, which is associated with the original graph and the partition (see Fig. 6.5). If repeated iteratively, this process allows the construction of abstractions of higher and higher levels.

6.3.2 Drawing the Abstraction

After the abstraction has been built, one must draw it in order to present a readable image to the end-user. To draw an abstraction, there exist two strategies called top-down and bottom-up. The top-down strategy consists of first drawing the abstraction at the highest level, and then drawing deeper and deeper clusters in the hierarchy. For instance, applying this strategy to the network of Fig. 6.4a one would first draw the compound graph of the highest level (containing the yellow, the green and the pink metanodes), then draw the compound graph contained in the yellow cluster (containing node "7" and the brown metanode) and finally draw the content of the brown cluster (containing nodes "6" and "8"). The main drawback of that method is that one cannot predict the size of each metanode when drawing a cluster. For instance, when drawing the yellow cluster, one does not know the size of the brown metanode. To overcome this problem, the bottom-up strategy first draws the deepest clusters and then draws clusters at lower and lower levels. In Fig. 6.4a, the brown cluster would be drawn before the yellow one, and the size of the brown metanode can therefore be determined before drawing the yellow cluster.

Some particular tools (Abello, van Ham, & Krishnan, 2006; Archambault, Munzner, & Auber, 2007, 2008) that focus on the visualization of very large graphs (even those graphs that are too large to fit in memory) support on-demand drawing to quickly give the user a first picture of the data. Only the highest level of abstraction is displayed, and on-demand clusters can be drawn in a top-down manner. However, because these tools use the top-down strategy, metanodes sizes in the representation do not necessarily correspond to the number of elements they represent.

6.3.3 Compound Visualization Methods

In this section, we present the two compound visualization methods that we call classical and multiscale compound visualization methods.

In the classical compound visualization, the metanodes appear opaque. Figure 6.5 shows an example of classical compound visualization corresponding to the highest level of abstraction of the network of Fig. 6.4. One of the first interactive classical compound visualization techniques, developed by Schaffer et al. (1996), shows the effectiveness of using graph clustering and compound visualization when exploring large networks. To enable exploration, the authors introduce a variable zoom method that consists of using a combination of a metanode expansion (or contraction) and a geometrical fish-eye. While this interaction (Schaffer et al., 1996) only allows the expansion of a single metanode, van Ham and van Wijk (2004) proposed a method to define the desired level of abstraction.

In contrast to classical compound visualization, multiscale compound visualization displays both the original graph and the whole hierarchical partition in a single view. This is done by filling the "interior" of each metanode (that represents a cluster) with a drawing of the highest level of abstraction (i.e., the compound graph of the sub-hierarchy). For instance, in Fig. 6.5b, the interior of the green metanode is filled with a drawing of the compound graph corresponding to the top level of the subtree rooted on the green node, here a graph with three vertices and edges. And the interior of the yellow metanode is filled with an abstraction of the subgraph it represents, here a compound graph containing the brown metanode and

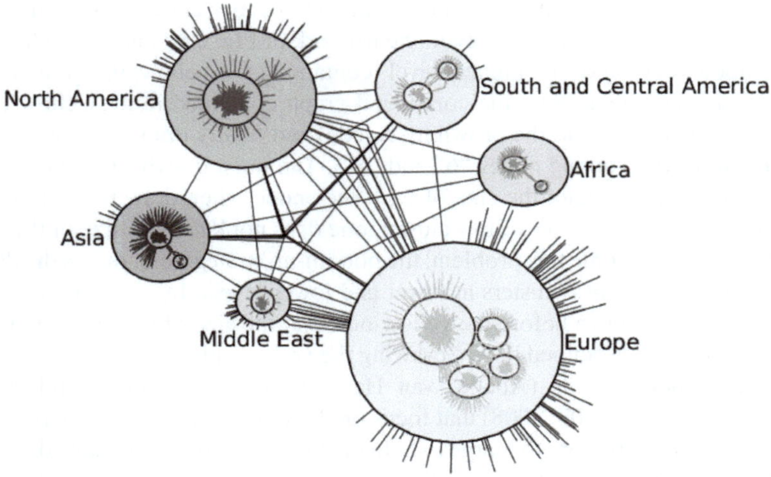

Fig. 6.6 Multiscale compound visualization of 2000 international air traffic network clustered according to continents and then to subnetwork density

node "7". The ability of this method to display both abstracted inter-cluster and intra-cluster relationships has been shown in numerous work (Auber et al., 2003; Balzer & Deussen, 2007; Eades & Feng, 1996; Huang & Eades, 1999).

6.3.4 Example Based the 2000 International Air Traffic Network

In this section, we present some results that we obtained by analyzing the international 2000 air traffic network.

Figure 6.6 shows a multiscale compound visualization of the 2000 international air traffic network. In this representation, the first partition has been computed according to geographical regions and then each cluster has been refined using a clustering algorithm. That view of the network allows one to visualize intra-region relationships while having an overview of inter-region ones.

However, the representation in Fig. 6.6 does not allow one to identify the central airports (or hubs) of the network. This is mainly due to the utilization of geographical positioning of the airports in the clustering process. Figure 6.7 shows another representation of the same network. Connections with fewer than 300,000 passengers have been filtered out to remove "noise" from the data. Indeed, connections having few passengers may lead to inconsistent clusters (usually, clustering algorithms are based on cluster density). When not using the geographical positioning of airports, one can see in Fig. 6.7 that clusters do not necessary correspond to geographical regions of the network. For instance, the main international (e.g., Paris,

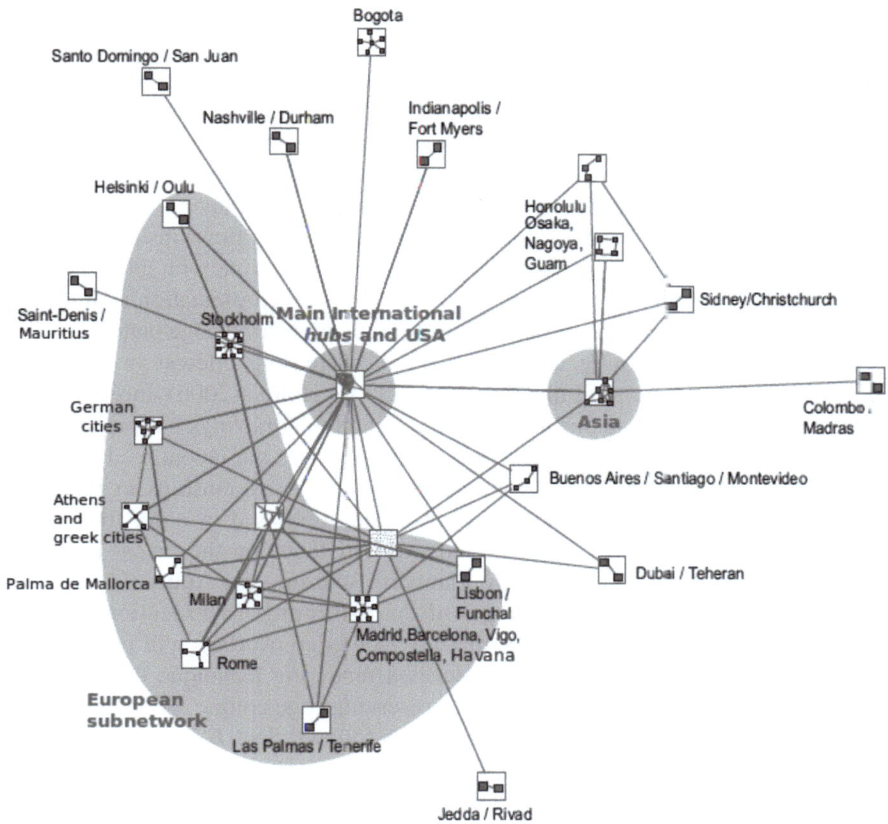

Fig. 6.7 Multiscale compound visualization of the 2000 international air traffic network. Connections with fewer than 300,000 passengers have been removed and the resulting graph clustered using a strength clustering algorithm (Auber et al., 2003) (Image courtesy Amiel, Rozenblat, and Melançon (2005))

London, Frankfurt) and North American airports have been clustered together (cf. Fig. 6.7), because they are highly connected and form one of the main hubs of the network. One can also see that Asian airports have been clustered together and that these airports are sparsely connected to the rest of the network (in Fig 6.7 the "Asia" metanode is linked only to the main hub and to several other clusters).

6.4 Flow Maps Representation Of Graphs

Graph drawing algorithms focus primarily on the computation of node positions and try to achieve various aesthetics in order to provide readable layouts. A major graph drawing aesthetic is edge crossing minimization (di Battista, Eades, Tamassia, &

Tollis, 1998). The effect of this aesthetic on human understanding was demonstrated in previous user studies (Purchase, 1998; Purchase, Cohen, & James, 1995). In some cases, however, graph layout algorithms cannot avoid producing edge cluttering due to high edge density or intrinsic connectivity. This is typical of force-directed layouts that are applied to real world dense graphs: edges connecting close neighbors in the drawing mix with long range edges impair readability or even introduce confusion. The situation is even worse when laying out data using geographical positions. In such cases, the task of identifying flows inside a network is challenging because displaying a large number of connections with lines results in visual clutter. Therefore, reducing the clutter in a graph representation is of utmost importance to the identification of relationships and high-level edge patterns.

Currently, edge bundling techniques are of increasing interest in the graph visualization community. Put under the spotlight by Holten (2006), this technique addresses the issue of reducing edge cluttering in graph drawings by routing edges into bundles. The resulting graph visualization improves the layout readability by uncovering high level edge patterns and emphasizing relationships in relational data. Several rendering techniques have also been proposed in order to improve this type of visualization. In this section, we will focus on the recent and intuitive bundling technique proposed by Lambert, Bourqui, and Auber (2010). This bundling algorithm has the advantages of outperforming the execution times of existing methods and can guarantee that no node-edge overlaps will occur in the drawing. Furthermore, this work introduces a rendering technique that makes it possible to perceive bundle density while preserving edge colors.

The remainder of this section is structured as follows: Sect. 6.4.1 reviews related work on edge bundling methods. In Sect. 6.4.2, we present the bundling algorithm of Lambert et al., while Sect. 6.4.3 refers to rendering techniques that can be applied to enhance edge bundling visualization.

6.4.1 Previous Work

Phan, Ling, Yeh, and Hanrahan (2005) present a flow map layout technique based on geometrical node clustering. Edges are routed along the hierarchy tree branches. This idea has also been used by Holten (2006) to enhance relationships in hierarchical (and relational) data. The main drawback of both methods is that the edges are routed by using a hierarchy tree that can be restrictive in the general case.

Gansner, Koren, and North (2005) give an improved circular layout algorithm where the edges are routed on either the outer face or the inner face of the circle. Edges routed inside the circle are bundled using an edge clustering algorithm that tries to optimize area utilization. Another edge clustering method is given by Cui, Zhou, Qu, Wong, and Li (2008). In this paper, they propose a geometric approach to create bundles of edges. They build a control mesh based on user interaction or a Delaunay triangulation. The mesh is then used to compute regions where edges should be merged. Then, a clustering algorithm based on the orientation of edges is used to merge the edges. A post processing step is applied to reduce the effect of "zigzag" edges.

Recently, Holten & Wijk (2009) introduced a force-directed heuristic to bundle edges and, therefore, to unclutter a representation of a graph where the node positions are fixed. In this heuristic, dummy nodes are inserted in order to split edges into segments. A similarity measure between edges is computed to determine which of the dummy nodes should interact. Dummy nodes of any two interacting edges are linked by inserting dummy edges. Bundles are obtained by running a force-directed algorithm while preserving the positions of the original nodes.

6.4.2 *Edge Bundling Method of Lambert et al.*

This technique uses edge routing to bundle edges. A grid is computed according to the node positions. This grid is then used to compute the shortest routes for each edge. In the same way that highways attract more drivers than smaller roads, frequently taken paths are used to bundle edges.

Grid Computation

To compute the shortest paths for each original edge, a grid graph is created on which the nodes of the original graph are connected. This grid is computed by discretizing the plane into cells using original node positions.

To obtain a multi-resolution grid graph, one can use a quad tree where the plane is decomposed in four parts until it contains at most one element. Such an approach is efficient in terms of computation time because its complexity is $O(|V| \cdot \log(|V|))$. On one hand, it generates a large grid, and on the other hand, large cells promote horizontal and vertical paths. Voronoi diagrams can also be used to generate the grid graph. In a Voronoi diagram, the cells are regions of the plane in which points of a cell are closer to the cell's site (here, the original nodes) than to any other site. Using classical Voronoi diagram does not guarantee that node edge overlaps will be prevented in the case of non-point size nodes. However, that problem can be easily addressed by using a constrained Voronoi diagram that takes into account the node size. This method generates a small grid graph that can be computed in $O(|V| \cdot \log(|V|))$, but it generates large cells for a sparse region. Due to the routing method used in the next step of the algorithm, these large cells will create large detours. Lambert et al. propose a hybrid algorithm based on both quad trees and Voronoi diagrams. In this algorithm, the quad tree cell sizes are parameterized to generate different levels of clutter reduction. Then, Voronoi diagrams are used to construct the final grid graph. Figure 6.8b shows the grid graph obtained with this hybrid approach. Because the quad tree adds $O(|V|)$ nodes, the $O(|V| \cdot \log(|V|))$ time complexity is preserved. Therefore, the resulting grid graph is of reasonable size.

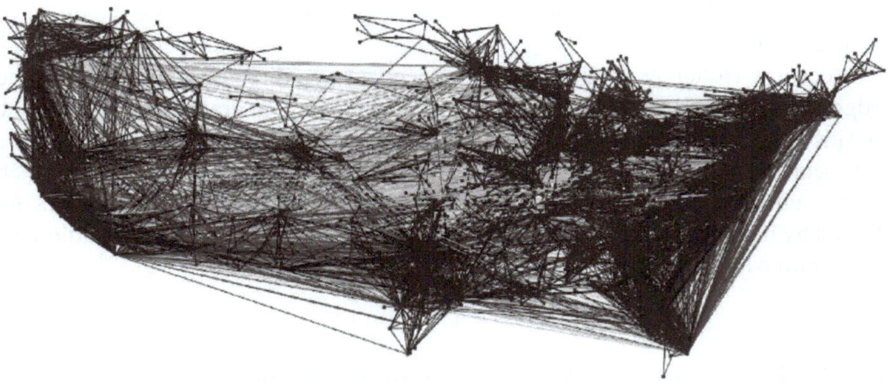

(a)

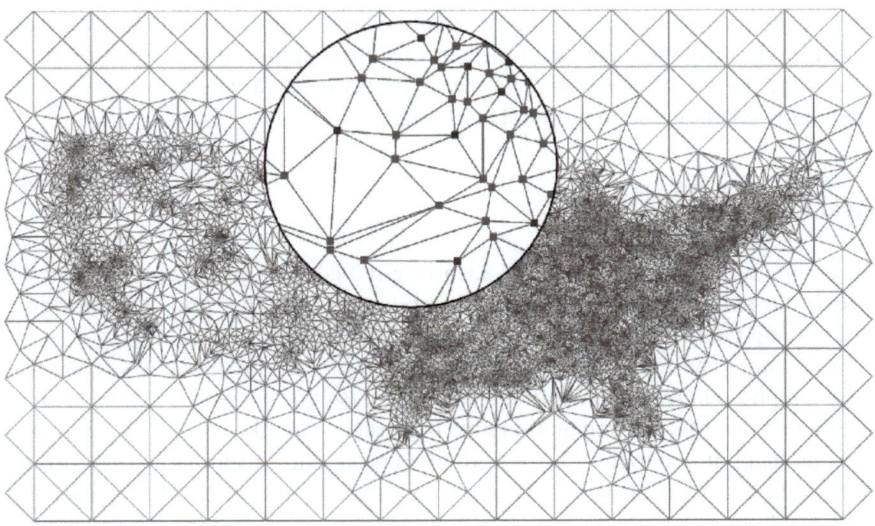

(b)

Fig. 6.8 (**a**) Original graph layout before the bundling process. This network represents migrations of workers between major cities in the USA. (**b**) The grid graph created to perform the edge rerouting. *Red nodes* are grid nodes and *blue nodes* are those from the original graph (Color figure online)

Edge Routing

The next step in the method consists of routing edges in the original graph onto the grid obtained in the previous step. A shortest path algorithm could be used directly to perform this operation. However, this method does not guarantee that edges follow the same path and, thus, it creates few bundles. To augment this bundling effect, the metaphor of roads and highways is used. Regular roads that are frequently used are

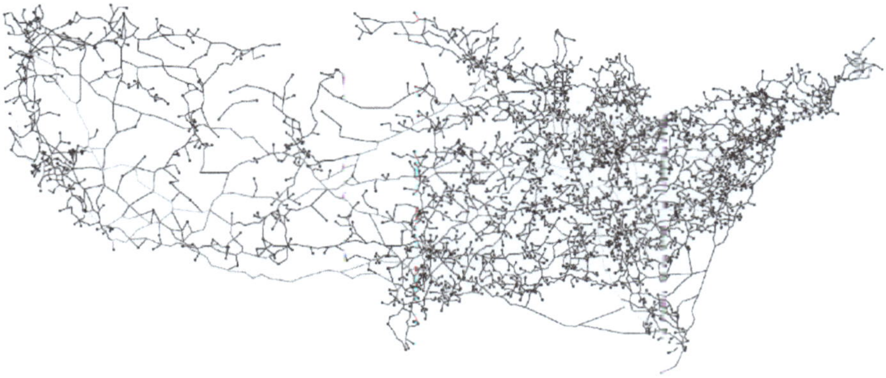

Fig. 6.9 The layout of the graph introduced in Fig. 6.8a after the edge bundling process

transformed into larger ones. This effect is reproduced by first computing all of the shortest paths between linked nodes on the original graph. Then, according to the number of shortest paths through an edge of the grid, the weights of the grid edges are adjusted. Reducing the weight of an edge is equivalent to transforming it into a highway because it is possible to go faster from one point to another. A shortest path for each edge of the original graph is then computed. This adjustment of the weights creates new bundles because the new distance matrix of our graph promotes frequently used edges. To compute the shortest paths, we use the well-known Dijkstra's algorithm (1971), leading to $O(|V_{grid}| \cdot |E_{grid}| + |V_{grid}|^2 \cdot \log(|E_{grid}|))$ time complexity. The result of this edge routing phase is presented in Fig. 6.9.

6.4.3 Enhancing Edge Bundled Graph Visualization

Edge bundled graphs leverage several issues with respect to rendering the graphs on a screen. To obtain an aesthetic drawing and to ease the retrieval of information from the visualization, some rendering methods and visual encodings have been designed specifically for this type of visualization technique. The following summarizes these techniques.

Smoothing Edges with Curves

The main feature common to every edge-bundled graph visualization is the drawing of edges as curves. Indeed, rendering graph edges as curves makes it easier to follow the edges and gives a more visually appealing graph drawing. In 2006, Holten renders bundled edges piecewise with cubic B-splines. By using this type of spline, which provides local control of the curve shape, one can produce distinct

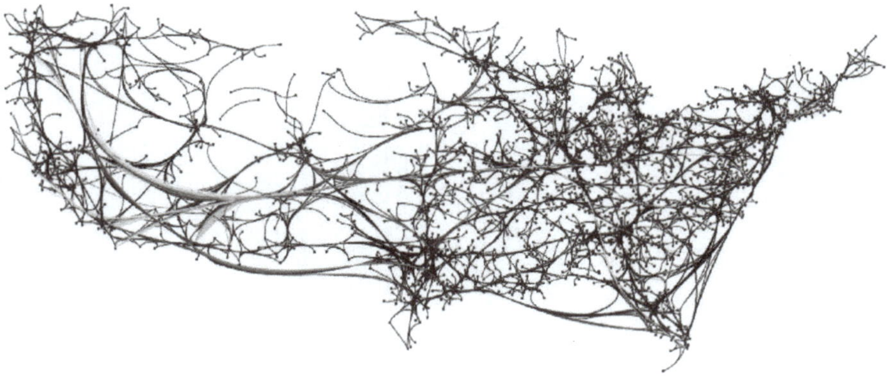

Fig. 6.10 Smoothing the edges of the graph from Fig. 6.9 with Bézier curves

and coherent bundles. Lambert et al. (2010) use this type of spline as well as others such as Bézier curves or Catmull-Rom splines. Another method used by Holten and Wijk (2009) and Cui et al. (2008) is to apply a smoothing technique to the edges that are drawn as polylines in order to morph them into curves. A rendering of the same graph in Fig. 6.9 with edges drawn as Bézier curves is introduced in Fig. 6.10 for visual comparison.

Coloring Edges

Another method to enhance edge-bundled graph visualization is to use edge colors and opacities to encode information. Edge colors are mapped to the directions of the original links (Cui et al., 2008). In a similar technique, edge direction is encoded by an interpolated color gradient running from a fixed color for the source to a fixed color for the target (Holten, 2006). Holten (2006) maps edge opacities to their length where long curves are more transparent than short ones, which prevents short curves from becoming obscured. Cui et al. (2008) use the opacity of each segment of the polyline representing an edge to map the density of the lines overlapping it.

Perceiving Bundle Density

After graph edges have been bundled, some of the bundles share successive bends. Consequently, several edge segments are merged into a single segment and the information about the number of edges contained in a bundle is not readily visible in the drawing. To distinguish strong bundles from weak ones, some techniques have been designed to visually enhance the bundle strength. The first rendering technique for estimating the quantity of merged edge segments is proposed by Holten & Wijk (2009). A GPU-based method is used to compute the amount of overdraw for each pixel in the graph visualization. This value is then used to map pixel colors to a user-defined gradient color scale after the minimum and maximum values of the overdraw have been computed. Lambert et al. (2010) introduce an "edge splatting" technique that allows the viewer to perceive bundle density while preserving edge colors. In a similar manner, the GraphSplatting technique introduced by Liere and Leeuw (2003), a splat field is computed to encode continuous variations in the density of the merged edges. To obtain the splat field, the first step is to compute the number of edges crossing each pixel in the drawing. A GPU-based method (Holten & Wijk, 2009) is used to perform that task. Then, the splat field is generated by convoluting the discrete density values (associated with each pixel) with a Gaussian kernel. The larger the kernel radius and the standard deviation, the more the splat field is smoothed. This splat field can be rendered on a screen in a various ways. After computing the minimum and maximum values of the splat field, one can perform a simple color mapping, for instance. To preserve edge colors, Lambert et al. use a *bump mapping* technique. Bump mapping is a computer graphics technique that allows a rendered surface to appear more realistic without modifying its geometry. Bump mapping adds a per-pixel shading that makes the surface appears bumpy, by changing the surface normals. The color and brightness of each pixel are then altered with respect to these normals by using an illumination algorithm. The final color of a pixel is computed from the light properties and a color map. This color map can correspond to a splat field color mapping or to the original edge colors. By mapping the splatting values to the heights, bundles with high density edges appear taller than others and visually emerge from the layout. Results of this rendering technique can be found in Fig. 6.11.

6.5 Conclusion

In this chapter, we presented the main methods for the automatic drawing of networks for geographic data visualization. Among these methods, one of the most popular approaches is the force-directed method, which produces visually pleasant and structurally significant results. In that method, nodes are considered as particles and edges represent attractions between these particles. To overcome the computational cost problem of such methods, fast algorithms have been designed to make a trade-off between computation time and aesthetic criteria.

We also presented the compound graph visualization that displays an abstraction of the network. This is achieved by first producing a partition of the nodes using a clustering algorithm. Then, each cluster is collapsed into a single node to build the abstraction. In the case of large and complex data, compound graph visualization allows the reduction of visual clutter in the representation while preserving the rendering speed and, therefore, supporting interactive exploration.

Finally, we reviewed the main flow map layout techniques. In this approach, node positions must be preserved because they can provide information. The main problem is that it thereby creates many edge crossings that clutter the representation.

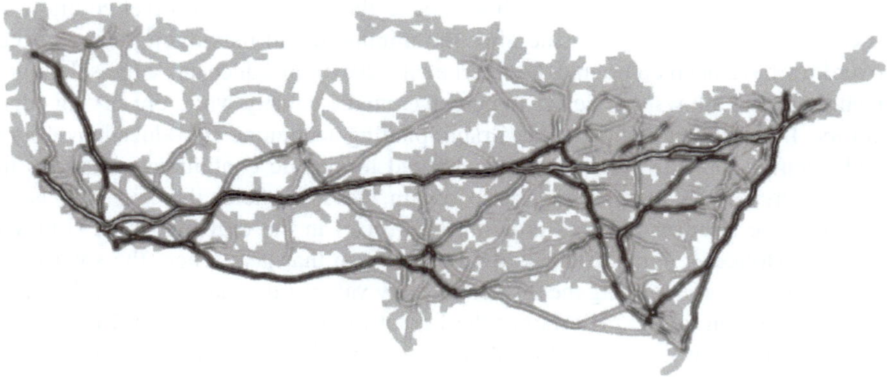

(a)

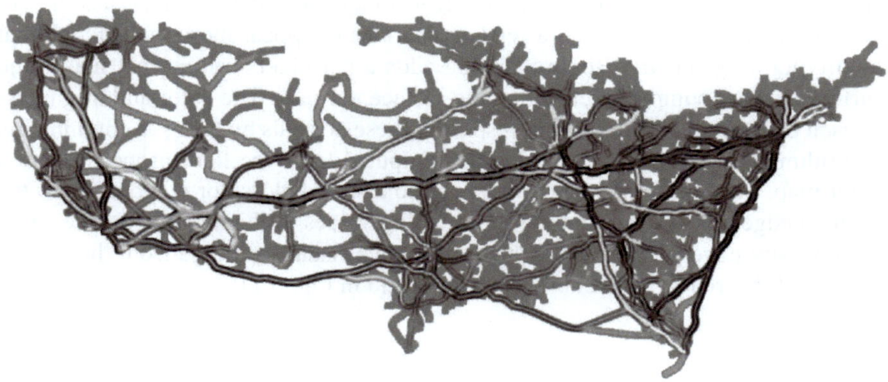

(b)

Fig. 6.11 Perceiving bundles densities with a bump mapping rendering technique. Edges are rendered as cubic B-splines. (**a**) A splat field color mapping is used as a color map. (**b**) Original edge colors are used as a color map (Color figure online)

To remove this clutter and to emphasize high-level edge patterns, the edges are routed and grouped into bundles. Finally, rendering techniques such as bump mapping are used to illustrate bundle densities.

References

Abello, J., van Ham, F., & Krishnan, N. (2006). Ask-graphview: A large scale graph visualization system. *IEEE Transactions on Visualization and Computer Graphics*, *12*(5), 669–676.

Amiel, M., Rozenblat, C., & Melançon, G. (2005). Réseaux muti-niveaux: l'exemple des échanges aériens mondiaux de passagers. *Mappemonde*, *79*(3). Retrieved July 21, 2009, from http://mappemonde.mgm.fr/num7/index.html

Archambault, D., Munzner, T., & Auber, D. (2007). Grouse: Feature-based, steerable graph hierarchy exploration. In K. Museth, T. Möller, & A. Ynnerman (Eds.), *Eurographics/IEEE-VGTC symposium on visualization* (pp. 67–74). Geneva, Switzerland: Eurographics Association.

Archambault, D., Munzner, T., & Auber, D. (2008). Grouseflocks: Steerable exploration of graph hierarchy space. *IEEE Transactions on Visualization and Computer Graphics, 14*(4), 900–913.

Auber, D., Chiricota, Y., Jourdan, F., & Melançon, G. (2003). Multiscale exploration of small world networks. In *IEEE symposium on information visualisation* (pp. 75–81). Seattle, GA: IEEE Computer Science Press.

Balzer, M., & Deussen, O. (2007). Level-of-detail visualization of clustered graph layouts. In *6th international Asia-Pacific symposium on visualization APVIS'07* (pp. 133–140). Sydney, NSW: IEEE.

Cui, W., Zhou, H., Qu, H., Wong, P. C , & Li, X. (2008). Geometry-based edge clustering for graph visualization. *IEEE Transactions on Visualization and Computer Graphics 14*(6), 1277–1284.

di Battista, G., Eades, P., Tamassia, R., & Tollis, I. G. (1998). *Graph drawing: Algorithms for the visualisation of graphs*. Englewood Cliffs, NJ: Prentice-Hall.

Dijkstra, E. (1971). *Ewd316: A short introduction to the art of programming*. Eindhoven, The Netherlands: Technische Hogeschool.

Dongen, S. V. (2000). *Graph clustering by flow simulation*. PhD thesis, University of Utrecht, The Netherlands.

Eades, P. (1984). A heuristic for graph drawing. *Congressus Numerantium, 42*, 149–160.

Eades, P., & Feng, Q.-W. (1996). Multilevel visualization of clustered graphs. In S. North (Ed.), *4th international symposium on graph drawing* (Vol. 1190, pp. 101–112). Berlin, Germany: Springer.

Frick, A., Ludwig, A., & Mehldau, H. (1995). A fast adaptive layout algorithm for undirected graphs. In *Symposium on graph drawing GD'94* (Vol. 894, pp. 389–403). London: Springer-Verlag.

Fruchterman, T., & Reingold, E. (1991). Graph drawing by force-directed placement. *Software—Practice & Experience, 21*, 1129–1164.

Gajer, P., Goodrich, M. T., & Kobourov, S. G. (2004). A multi-dimensional approach to force-directed layouts of large graphs. *Computational Geometry, 29*(1), 3–18.

Gajer, P., & Kobourov, S. G. (2002). Grip: Graph drawing with intelligent placement. *Journal of Graph Algorithms and Applications, 6*(3), 203–224.

Gansner, E. R., Koren, Y., & North, S. C. (2005). Topological fisheye views for visualizing large graphs. *IEEE Transactions on Visualization and Computer Graphics, 11*(4), 457–468.

Hachul, S., & Junger, M. (2005). Drawing large graphs with a potential field based multilevel algorithm. In *International symposium on graph drawing GD'04* (Vol. 3383, pp. 285–295). New York: Springer-Verlag.

Holten, D. (2006). Hierarchical edge bundles: Visualization of adjacency relations in hierarchical data. *IEEE transactions on visualization and computer graphics (Proceedings of Vis/InfoVis 2006), 12*(5), 741–748.

Holten, D., & Wijk, J. J. V. (2009). Force-directed edge bundling for graph visualization. *Computer Graphics Forum, 28*(3), 983–990.

Huang, M. L., & Eades, P. (1999). A fully animated interactive system for clustering and navigating huge graphs. In S. H. Whitesides (Ed.), *Symposium on graph drawing GD'98* (Vol. 1547, pp. 374–383). Montréal, Canada: Springer.

Kamada, T., & Kawai, S. (1989). An algorithm for drawing general undirected graphs. *Information Processing Letters, 31*, 7–15.

Lambert, A., Bourqui, R., & Auber, D. (2010). Winding roads: Routing edges into bundles. *Computer Graphics Forum, 29*(3), 853–862.

Lauther, U. (2007). Multipole-based force approximation revisited—A simple but fast implementation using a dynamized enclosing-circle-enhanced k-d-tree. In M. Kaufmann & D. Wagner (Eds.), *Graph drawing* (Vol. 4372, pp. 20–29). Berlin, Germany: Springer.

Liere, R. V., & Leeuw, W. D. (2003). Graphsplatting: Visualizing graphs as continuous fields. *IEEE Transactions on Visualization and Computer Graphics, 9*(2), 206–212.

McSherry, F. (2004). *Spectral methods for the data analysis*. PhD thesis, University of Washington, Seattle, WA.

Newman, M. E. J., & Girvan, M. (2004). Finding and evaluating community structure in networks. *Physics Reviews E, 69*, 026113.

Phan, D., Ling, X., Yeh, R., & Hanrahan, P. (2005). Flow map layout. In *IEEE symposium on information visualization* (pp. 219–224).

Purchase, H. (1998). Which aesthetic has the greatest effect on human understanding?. In *Symposium on graph drawing GD'97* (Vol. 1353, pp. 248–261). Rome, Italy: Springer-Verlag.

Purchase, H., Cohen, R. F., & James, M. (1995). Validating graph drawing aesthetics. In *Symposium graph drawing GD'95* (Vol. 1027, p. 435–446). Passau, Germany: Springer.

Schaeffer, S. E. (2005). Stochastic local clustering for massive graphs. In T. Ho, D. Cheung, & H. Liu (Eds.), *Advances in knowledge discovery and data mining* (Vol. 3518, pp. 413–424). Berlin, Germany: Springer.

Schaffer, D., Zuo, Z., Greenberg, S., Bartram, L., Dill, J., Dubs, S., et al. (1996). Navigating hierarchically clustered networks through fisheye and full-zoom methods. *ACM Transactions on Computer-Human Interaction, 3*(2), 162–188.

van Ham, F., & van Wijk, J. J. (2004). Interactive visualization of small world graphs. In T. Munzner & M. Ward (Eds.), *IEEE symposium on information visualisation* (pp. 199–206). Seattle, GA: IEEE Computer Science Press.

Chapter 7
Exploring Hierarchies Using the DAGMap

Pierre-Yves Koenig

7.1 Introduction

Hierarchical data appears naturally in many applications. In their simplest form, hierarchies are trees. Tree structures, such as classifications (for instance, geolocalization classifies continent, country, states or region), phylogenetic trees, and company organization charts, are but a few examples. In richer hierarchical organizations, cross links allow elements to belong to several non-distinct classes. Inheritance relations of classes in object oriented programming is a typical example. Classifying concepts extracted from documents, for instance, can sometimes require concepts to be "duplicated", depending on the possible interpretations. The "network" concept, for example, can refer to both social network analysis and low level computer hardware. Additionally, both concepts may be required to index a collection of documents. Such a classification naturally leads to the construction of a directed acyclic graph (a *DAG*).

In a DAG, nodes are ordered using *ancestor/child relationships* just as with trees, with the exception that nodes may have multiple parents. Nodes without ancestors are called *source* nodes, while those without any children are called *sinks*.

The case study we shall explore in this chapter is made up of companies linked to one another through subsidiary links, i.e., company c_1 is linked to company c_2 if c_2 is a subsidiary of c_1. Clearly, this graph structure is a DAG: subsidiaries themselves have subsidiaries and a subsidiary may be held by several "ancestor" companies. The dataset also collects attributes measuring how much of a subsidiary is held

P.-Y. Koenig (✉)
CNRS UMR 5506 LIRMM – Montpellier, INRIA Bordeaux Sud-Ouest, Montpellier, France

Valtech, Avenue Marcel Dassault 5, Toulouse 31500, France
e-mail: pykoenig@gmail.com

C. Rozenblat and G. Melançon (eds.), *Methods for Multilevel Analysis and Visualisation of Geographical Networks*, Methodos Series 11, DOI 10.1007/978-94-007-6677-8_7,

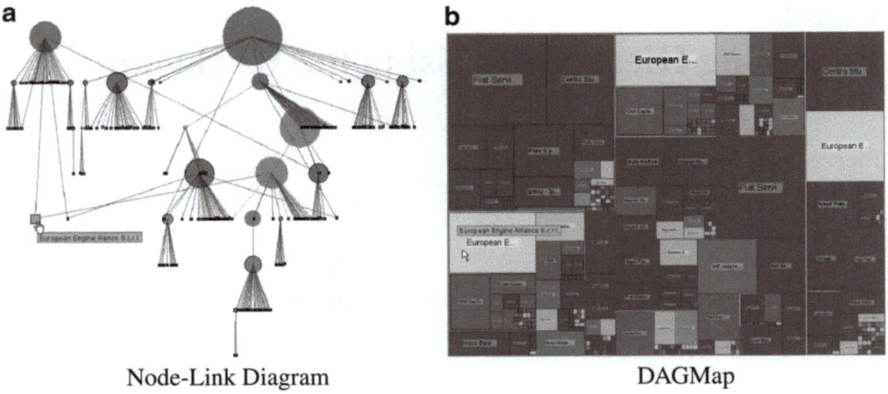

Fig. 7.1 (**a**) Traditional Sugiyama node-link layout. (**b**) Corresponding DAGMap view of a DAG. The node size and color correspond to node attributes such as the companies' assets and the location of headquarters (color figure online)

by each of its parent companies. Image (a) in Fig. 7.1a shows a part of a DAG describing links between companies and their subsidiaries. The node (and link) color and map size correspond to different attributes that can be computed from the DAG (see Sect. 7.3).

Natural questions emerge when studying data modeled by DAGs. The level of a node v in the DAG roughly corresponds to how "general" the node is. More precisely, the level can correspond to how much of the DAG the node spans, or how many nodes can be accessed going downwards from v. For companies, the node level corresponds to the degree to which a company controls its subsidiaries (and subsidiaries of its subsidiaries, etc.). This is even more the case when one takes into account edge or node attributes, such as a company's assets. In our case, the DAG structure also reveals the extent to which two (or more) companies' strategies or interests overlap over a set of subsidiaries. We have developed a technique that combines a node-link view with the DAG, explicitly showing links between companies and visually indicating where a company is located in the whole hierarchy. However, this type of view is not optimal for showing and comparing node attributes. We have thus designed a DAGMap view, which constructs a TreeMap (Shneiderman, 1992) from a DAG, in order to emphasize node attributes, such as the total capital of a company, making it easy to compare companies based on attributes. Color coding semantic attributes, such as the country (or region) for the company's headquarters, has been shown to be efficient in helping geographers study how companies build strategies to avoid taxes, compete or collaborate with others. For instance, the DAGMap view facilitates identifying small subsidiaries as "tax havens", while the node-link view helps clarify how small subsidiaries link to ancestor companies or other subsidiaries.

The remainder of the chapter is organized as follows. We first present related work, followed by a description of our case study, motivating the use of the combined DAGMap and node-link layout. We then provide details on how the DAGMap is computed from the DAG. Returning to our case study, we explain how the node-link and DAGMap views are linked through user interaction.

7.2 Related Work

7.2.1 Node Link Diagram

A node-link diagram is an intuitive representation of a hierarchical dataset. Each element of the dataset is represented as a node, and relationships between elements are represented as edges linking two nodes. Tree hierarchies have been widely explored, and different types of layout algorithms have been proposed. Figure 7.2 shows different representations of the same hierarchy using several algorithms.

These algorithms can be extended to a directed acyclic graph. When drawing a DAG, source nodes are often placed at the top and are said to have level 0. The level of a node is then set to the length of the longest path connecting the node to a source node. In this process, we make sure that all of the edges are directed downwards (see Fig. 7.1a).

Tree layout algorithms (see di Battista, Eades, Tamassia, & Tollis, 1998) draw internal nodes as well as leaf nodes, thus explicitly showing the relative positions of the nodes in the tree. These properties also remain the same for the layout of the DAGs (Eades & Sugiyama, 1990).

The quality and readability of these node-link representations are usually measured in terms of the ability of the representations to avoid edge crossings (see Gutwenger & Mutzel., 2004, for instance). Node-link diagrams are useful for representing DAGs but are less efficient when the representation involves semantic information in terms of node sizes and colors. Indeed, even when drawing DAGs with nodes of equal sizes, layers must be kept sufficiently apart to ensure that the diagram is readable.

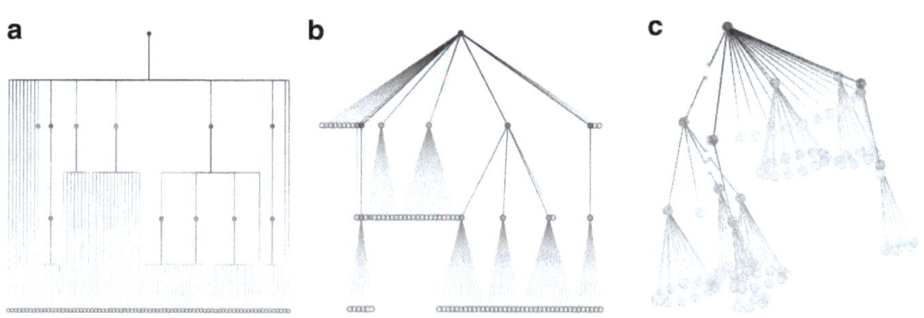

Fig. 7.2 Different layouts of the same hierarchy using different algorithms. (**a**) Dendogram (**b**) Walker (Reingold & Tilford, 1981; Walker, 1990). (**c**) Cone tree (Carriere & Kazman, 1995)

Additional space is obviously required when dealing with nodes of various sizes. Edge readability requires that layers be kept apart and that horizontal edges be avoided. Even with medium-sized DAGs, overlapping neighboring nodes can only be avoided at the price of a bad aspect ratio. Consequently, the identification of color or size patterns is difficult because nodes stemming from a common ancestor are placed in rows. Note that this is also true for the standard node-link representations of trees because in both cases, the layout is partly devoted to the encoding of the structure of the graph (dominance relations), leaving empty screen space between layers.

7.2.2 Space-Filling Approaches

Space-filling approaches (Shneiderman, 1992) were a solution to the visualization of trees with node attributes at the price of illustrating less structure. Treemaps are a visualization technique for presenting hierarchical information for two-dimensional displays (Shneiderman, 1992). TreeMaps follow a space-filling approach, mapping the leaf nodes of a tree onto contiguous areas in a plane, with several graphical cues reflecting attributes in the data. The area itself can be computed by taking the attributes into account. The internal nodes of the tree are only apparent through nesting and encode several attributes through node size or color.

This space-filling strategy radically differs from classical node-link representations for trees (see di Battista et al., 1998), where much care is taken with the relative node position, reflecting the structure of the hierarchy, as opposed to semantics in the data (leaf nodes).

The commercial success and adoption of TreeMaps in several domains of application corroborates the usability of TreeMaps. Improved interactivity (Chintalapani, Plaisant, & Shneiderman, 2004) and versatility (Vliegen, van Wijk, & van der Linden, 2006) often make TreeMaps an obvious choice for designing visualization systems for hierarchical data.

Many improvements on the original TreeMap representation have been suggested (Bruls, Huizing, & van Wijk, 2000; van Wijk & van de Wetering, 1999). In the original TreeMap algorithm, the display is cut into alternating directions parallel to the X-axis and Y-axis. The resulting drawings (see Fig. 7.3) seem to make it difficult for users to compare two cells in the TreeMap. The TreeMap cells end up being long, thin slices, which are difficult to read both in terms of shape and color.

In the "Squarified" TreeMap layout, the authors attempt to keep the ratio between width and height of the rectangle close to 1 (a square). Large square areas are easier to compare with each other (see Fig. 7.4a). Likewise, in Voronoï TreeMaps (Balzer & Deussen, 2005), cells are represented by polygons rather than by rectangles. Two polygons are easier to compare with each other (polygons are closer to circles and two circles are easier to compare with each other) (see Fig. 7.4c).

Other techniques focus on the perception of the TreeMap structure. In (van Wijk & van de Wetering, 1999), the authors draw cushions over cells to highlight

Fig. 7.3 Shneiderman TreeMap algorithm

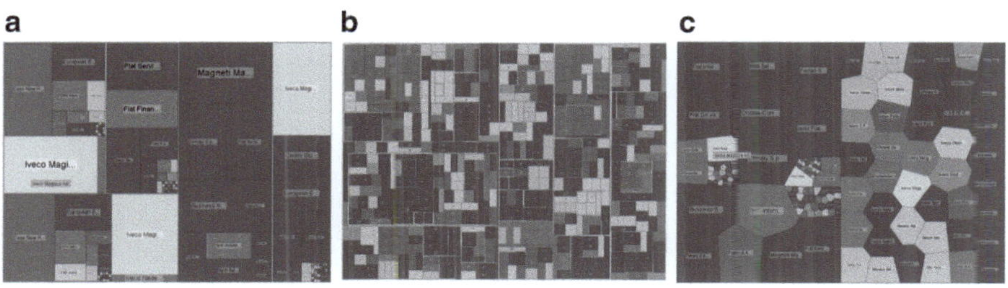

Fig. 7.4 Different layouts of the same hierarchy using different algorithms. (**a**) Squarified (Bruls et al., 2000). (**b**) Ordered (Bederson et al., 2002). (**c**) Voronoi TreeMap (Balzer & Deussen, 2005)

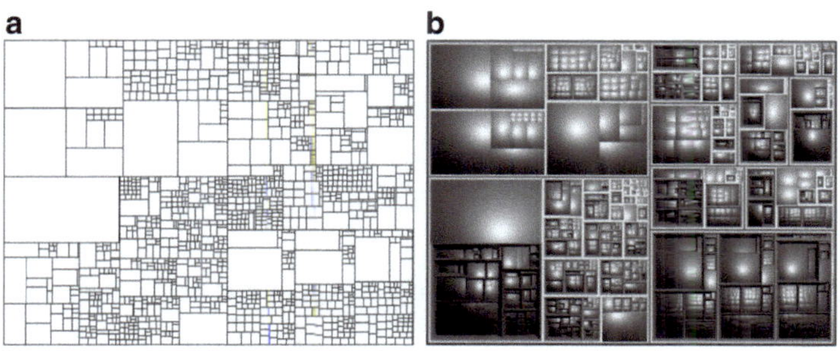

Fig. 7.5 Use of the gap and cushion to emphasize hierarchical structure. (**a**) The basic algorithm computes the plain underlying cell structure. (**b**) The TreeMap is augmented with a cushion effect to enhance the nesting structure

the hierarchy (see Fig. 7.5b). Gaps between cells can also be added. As shown in (Wattenberg & Fisher, 2003), gaps play a very important role in the human perception of levels. Adding gaps between cells in the TreeMap can better illustrate different hierarchical levels (see Fig. 7.6a). In (Lü & Fogarty, 2008), offsets illustrate the relationships between elements as an alternative to nesting (see Fig. 7.6b).

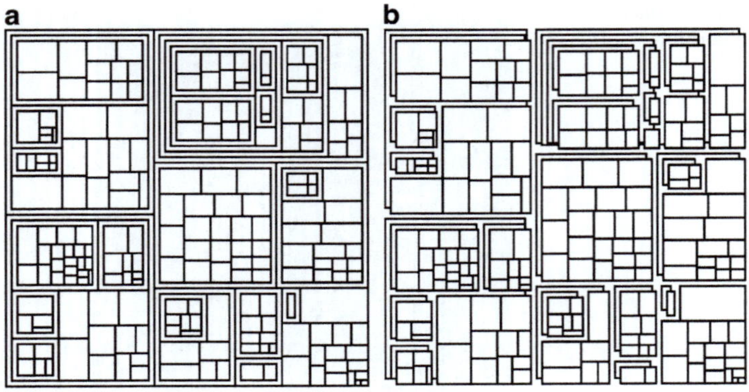

Fig. 7.6 Use of the gap and cushion to emphasize hierarchical structure. (**a**) Extra space is inserted between cells to make the nesting structure apparent. (**b**) The cushion effect is computed using a small shift of cells

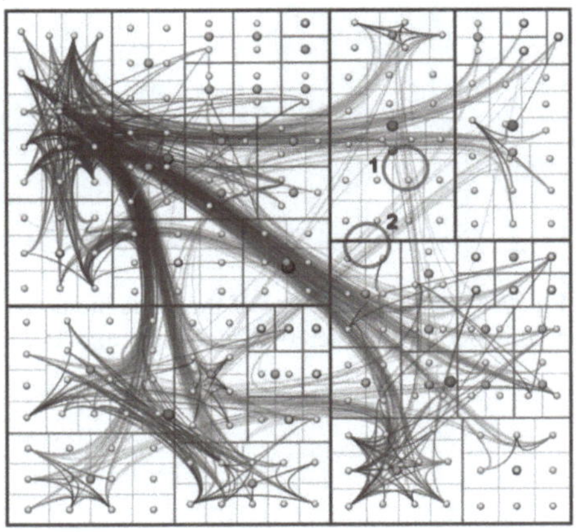

Fig. 7.7 Edge bundling

7.2.3 *Combined Views*

Fekete et al. have extended TreeMaps to general graphs (Fekete, Wang, Dang, Aris, & Plaisant, 2003) by extracting a spanning tree from the graph and then drawing edges over the TreeMap. Incidentally, Holten's edge bundling technique (Holten, 2006) can be used to improve the readability of this technique (Fig. 7.7).

We believe our approach is an improvement over previous techniques for visualizing hierarchical data (DAGs) for two reasons. First, in our case, edges do not connect two arbitrary cells but correspond to inheritance relationships that incorporate specific meaning. Users should refer to the Sugiyama layout to understand inheritance relationships and the level of companies before returning to the Treemap to view the attributes. Frequently, users switch between these two views to obtain a complete understanding of the system.

Linked highlighting enables structures in the Sugiyama layout to be easily identified at a glance. Consulting the Sugiyama layout orients the user in the DAG because spatial position is used to encode the node position and the level in the hierarchy. Simply drawing DAG edges on the treemap view obscures the level and the position of the structure in the hierarchy.

Second, extracting a spanning tree biases an arbitrary subtree of the DAG, randomly assigning the subtree more to one owner company than another owner company. However, a subtree can only be chosen by using information obtained from exploring and visualizing the datasets.

The DAGMap indeed lays out useful and complementary information on different parts of the screen while linking corresponding cells through user interaction. In a sense, unfolding the DAG along with our visualization system provides a framework for exploring any hypothesis.

7.3 Case Study

Part of this work was completed in collaboration with geographers in the National French Research Program SPANGEO[1]. Some data were obtained from the Orbis database maintained by the Bureau van Dijk office[2], and some of the data were manually collected. The full dataset encompasses approximately 140,000 companies, which emerge as subsidiaries of 597 main companies that cover numerous industrial sectors. Companies (and subsidiaries) are linked to one another by approximately 250,000 links. However, geographers are not interested in studying the full dataset but would rather focus on specific sectors of activity (NACE code). Alternatively, a part of the DAG that corresponds to selected major companies (source nodes) can be computed and explored separately. In this case study, we consider a DAG that spans the subsidiaries of a major European car manufacturer.

The visualization helps geographers determine how companies follow territorial logics or, in contrast, develop their activities based on other concerns. Strategies differ from one group to another. For example, Peugeot's organization is completely different from that of Renault. However, strategies also differ between sectors. Food companies tend to sell their products locally. The car industry is organized differently, producing parts in lower-cost countries, assembling the cars in another country, and shipping the vehicles to wherever markets exist. The picture becomes even more complex when companies expand their activities over several industrial sectors. Typically, a company will develop subsidiaries in the financial sector that manages financial flows generated by the overall activity of the company. Consequently, commercial and financial strategies intertwine over several industrial sectors.

---

[1]See the URL s4.parisgeo.cnrs.fr/spangeo/spangeo.htm

[2]See the URL www.bvdep.com/en/orbis.html

The combined DAGMap and node-link visualization provide an efficient exploratory device in order to develop an overview of these phenomena. Figure 7.1b shows the DAGMap view of subsidiaries in the car industry for the Fiat international group. Cells have been colored according to whether subsidiaries are in Europe (blue to dark blue is used to indicate that a company belongs to one of the 15 founding European countries), Asia (yellow), North America (orange), South America (green) or Africa (gray). This color coding was designed with the help of geographers. The geographers assigned a special color code for subsidiaries that are suspected tax havens (magenta). The DAGMap relies on a squarified treemap algorithm (Bruls et al., 2000), where the cell size corresponds to the amount of company assets.

The DAGMap does not simply encode a tree but actually encodes a DAG. This translates into being able to determine, in the treemap itself, if a subsidiary depends on a single ancestor company or is held by more than one of the competing companies. By clicking on a cell, the user can readily see whether the underlying element encompasses other cells in the treemap. This visual feedback provides a quick measure of the presence of a subsidiary in the network and aids with analyzing company strategies. This is the case for the subsidiary "Centro Studi sui Sistemi di Trasporto" (CSST for short), which depends on both "Iveco S.p.A." and "Fiat Auto Holdings B.V." Examining the node link diagram, we indeed see that CSST sits between these two companies. Moreover, we observe that "Fiat Auto Holdings B.V." does not hold its CSST subsidiary directly but rather through "Fiat Auto S.p.A.". We do not wish to draw any conclusions about the relationships between these companies and their subsidiaries. We rather wish to draw attention to the utility of this combined view, as was reported to us during our interviews with geographers.

7.4 DAGMaps: Extending TreeMaps to Directed Acyclic Graphs (DAGS)

Treemaps appear to be an excellent alternative to the traditional (node-link) layout for trees (Shneiderman, 1992), emphasizing the attributes of leaf nodes as opposed to the relative position of nodes in the tree. The DAGMap we describe here aims to do the same for DAGs.

We now explain how to adapt the basic TreeMap algorithm to address DAGs. Roughly speaking, the DAG is unfolded into a tree by duplicating nodes wherever necessary. Thus, a node v with multiple parents is duplicated as many times as necessary so that each parent has its own copy of v. In formal terms, let $G = (V, E)$ be a DAG. Given a node $v \in V$, let $F(v)$ denote the parents of v in G. In other words, $F(v) = \{u \in V | (u, v) \in E\}$. Now, assume that the subgraph H induced by the set of descendant nodes of $v \in V$ (nodes accessible by going downwards from v, including v itself) is a *tree*. For each parent $u \in F(v)$, we clone the subtree H and attach it below u. In doing so, the nodes $u \in F(v)$ now have distinct descendants (as far as v is concerned) (see Fig. 7.8). Performing this transformation

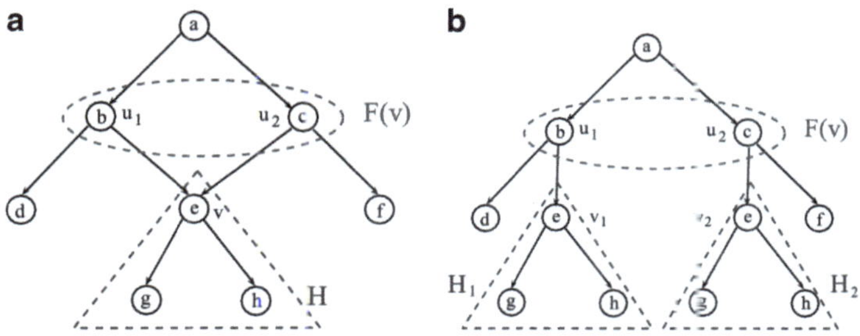

Fig. 7.8 A tree is obtained from a DAG by duplicating nodes (and labels) with multiple ancestors. (**a**) Two sibling nodes share a common subgraph. (**b**) The subgraph is duplicated to recover a tree structure

(denote $\mathscr{T}$) bottom-up from the sink nodes to the sources will eventually produce a tree $T = \mathscr{T}(G)$.

Conversely, let $T = (V, E)$ be a tree where the nodes have labels, that is, T comes equipped with the label $\lambda : V \to \mathscr{L}$. If two nodes $u, v \in V$ in T have equal labels $\lambda(u) = \lambda(v)$, then the subtrees $T_u = (V_u, E_u)$, $T_v = (V_v, E_v)$ extending from u and v are isomorphic (the subtrees have the same tree structure) and the corresponding nodes have equal labels. Thus, a bijective correspondence $\phi : V_u \to V_v$ satisfies $\lambda(x) = \lambda(\phi(x))$ for all $x \in V_u$. There is also a correspondence for preserving edges: (x, y) is an edge in E_u if and only if $(\phi(x), \phi(y))$ is an edge in E_v. We can then define on T an equivalence relation for nodes where $u \equiv v \iff \lambda(u) = \lambda(v)$. The quotient set $V/\equiv$ can then be equipped with a graph structure G, where the classes $[u]$, $[v]$ connect (in G) if two nodes $x \in [u], y \in [v]$ connect in T. As can be easily seen, the graph computed from T is a DAG and the previous unfolding process unfolds G back into T (see Fig. 7.8).

This bijective correspondence must be implemented in order to track user interaction on the DAGMap and to map the user interaction back to the original DAG (or over the Sugiyama layout when synchronizing views).

The TreeMap algorithm proceeds recursively, dividing the cell associated with a node into sub-cells for each of the child nodes, ordering the child nodes $u_1, \ldots, u_k$ according to their number of leaf nodes (in $T_{u_1}, \ldots, T_{u_k}$). This also holds true with the DAGMap: the number of sink nodes accessible from node u (in G) equals the number of leaf nodes in the subtree T_u (in $T = \mathscr{T}(G)$). Consequently, cells with greater area are placed in the top left part of the TreeMap because companies with many subsidiaries are expected to have greater assets.

7.5 Exploring the Hierarchy: Coordinating DAGMaps with the Classical Sugiyama Layout

We explain below that using a DAGMap alone is not sufficient to explore the entire hierarchy of companies and subsidiaries. The exploration focuses on the discovery

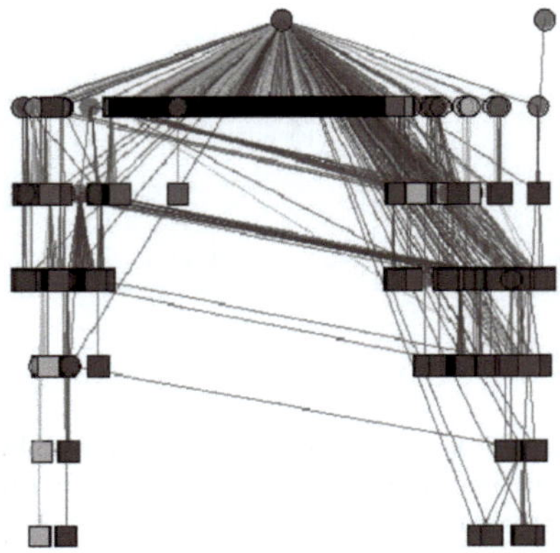

Fig. 7.9 The DAG shown here describes the links between Nestlé and its subsidiaries. Although the DAG is sparse, the node-link diagram obscures the distribution of subsidiaries over world regions and the ways in which higher level companies control lower level subsidiaries

of different commercial strategies and companies and their relationship to the territory. Note that the visual exploration only represents a first step toward a full explanation of such phenomena. Typically, users try to build hypotheses by combining standard (text-based) exploration with Google searches. Users subsequently return to the classical node-link visualization with questions in mind, iterating a sense-making loop (Thomas & Cook, 2006) by engaging in various intellectual postures (from wild guesses to strong and documented assertions).

The complexity of the analysis, however, is partly due to needing to quickly perceive the level and place of a node in the hierarchy, while simultaneously visualizing attributes such as a company's assets and locality (country/continent)[3]. Simply mapping a company's assets and territorial membership using node size and color in a traditional layout (image (a) in Fig. 7.1) proved to be inefficient because the node areas could not be easily compared.

The traditional layout displays at least twice as much visual information as the DAGMap (because ancestor nodes are mapped onto colored disks as well) or edge crossings and occlusions. Notice also that there is a lot of free space with a Sugiyama layout, as is the case with almost all node-link layouts. Consequently, displaying all of the nodes needed to report companies' assets requires much more space than using TreeMaps and argues against a visualization based solely on a Sugiyama layout.

Our experience with end-users clearly indicates the need to visualize the hierarchical structure directly on the DagMap. To this end, we enable the user to visualize how ancestor nodes cover cells in the DagMap, i.e., the user can actually define a "strip" that contains all of the elements at a given level in the tree (which come from the expanded DAG). In this way, the user can visualize the dependence of elements on ancestors.

---

[3]In fact, the original data are not exactly a DAG: there may indeed be links from a subsidiary controlling part of an ancestor company, thus introducing cycles into the network of links. However, this only happens exceptionally and cycles are considered marginal by our expert users, who actually discarded them.

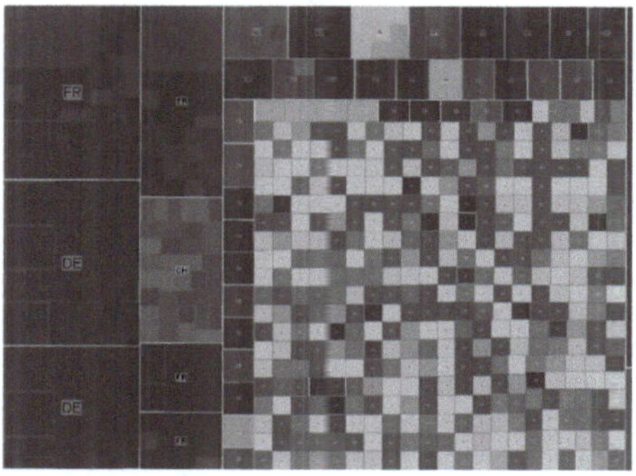

Fig. 7.10 Locating patterns in node attributes (*color/area*) is easier with the DagMap. By overlaying higher level cells on the DAGMap, hypotheses can be constructed on the potential management strategies of higher level companies (color figure online)

Even when the DAG is sparse, the node-link diagram in Fig. 7.9 cannot be used to analyze the distribution of subsidiaries over the regions of the world, or the ways in which higher level companies control lower level subsidiaries. The computation and interaction with the strip is designed to overcome this difficulty.

The strip is visualized using alpha values, by overlaying subsidiary cells by ancestor cells (Fig. 7.11). The mix of colors naturally indicates whether subsidiaries are controlled by mid-level companies residing in the same region of the world, or whether control is concentrated toward headquarters. When colors were mixed, indicating that subsidiaries from different parts of the world were all controlled through the same mother company, users suspected that the organization of companies relied on by-product logic (as opposed to territorial logic).

By varying the strip height, the user obtains information about the attributes of the ancestor cells and the size of their neighborhood. This functionality is helpful when combined with color variation. Color coding was shown to be useful with our test data, where the subsidiaries under the control of a suspected tax haven company were highlighted (purple colored cells), as shown in Fig. 7.10. In other situations, geographers could directly see that the South American and/or Asian subsidiaries occupying the lowest level of the hierarchy were controlled by European headquarters. Compare, for instance, the left side of the DagMap in Figs. 7.1b, 7.12, and 7.13.

7.6 Conclusion

In this chapter, we presented the DAGMap as an extension of TreeMaps to directed acyclic graphs or general hierarchies. The DAGMaps, together with the interactions we described, were designed with the help of geographers. Our case study showed that this combined view was particularly well-suited to our tasks. This was not only because the dataset was intrinsically encoded as a DAG but also due to the geographers' need for a visualization technique that astutely combines both the data attributes and hierarchical structure into a single view. The technique and the tool

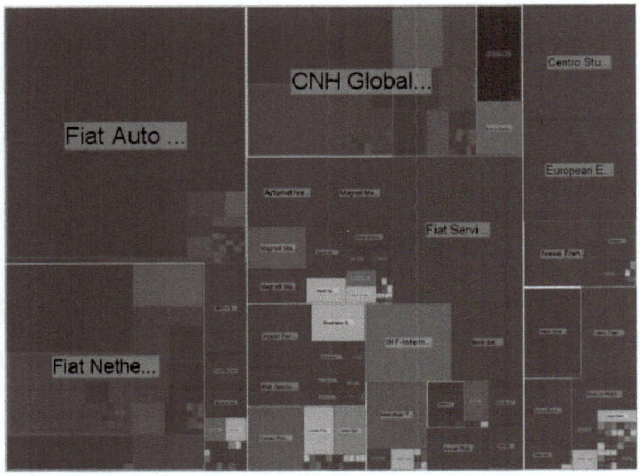

Fig. 7.11 The DAGMap shown here is obtained from that in Fig. 7.1b, using the "strip" (Fig. 7.12). The nesting of subsidiaries is revealed by visualizing the mother companies overlaying the subsidiaries. Additionally, the *color code* indicates that these subsidiaries are held by a mother company, which is suspected to be a tax haven (Fiat Netherland, *bottom left*; CNH Global, *center row*) (color figure online)

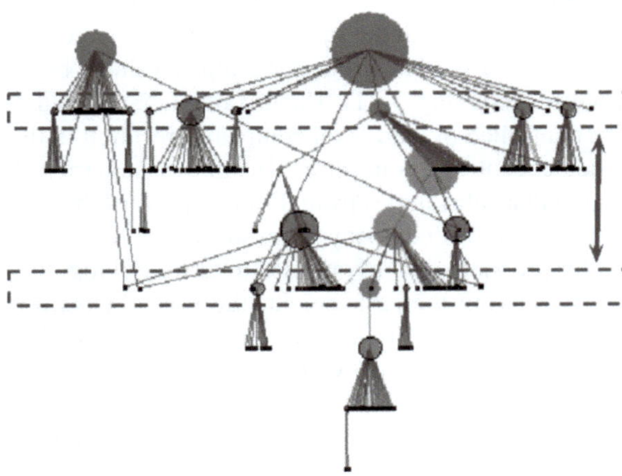

Fig. 7.12 The user can define a strip positioned at a certain level in the tree (the unfolded DAG) and can visualize the strip on the DAGMap with superimposed colored cells. Varying the height of the strip provides direct visual feedback on the DAGMap (color figure online)

were designed with expert geographers who were able to informally evaluate our prototype.

By exploring the Fiat dataset with the combined DAGMap-Sugiyama visualization, we were able to observe (and confirm) different strategies that have already been identified by Porter (Porter, 1986): these strategies have been developed on a continental level with Fiat, which most likely obeys a "car culture" differentiation. Additionally, (Fiat) Europe appears to traditionally control the hierarchy, while emergent South American countries occupy lower levels, thus confirming a classical center-periphery schema (Porter).

Further analysis is needed to confirm the following results of the DAGMap-Sugiyama visualization: the lower levels of the hierarchy obey a global model, which may be linked to the development of the "world car", explaining why

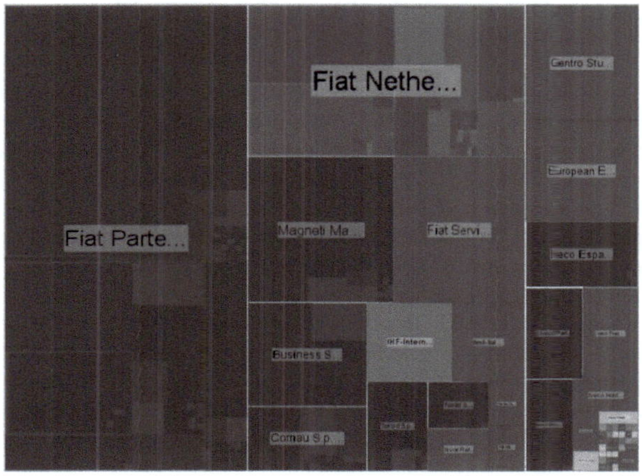

Fig. 7.13 The same DAGMap as in Fig. 7.1b, where the "strip" has a greater height. The *color code* indicates that the tax havens on the left of Fig. 7.11, themselves being subsidiaries of higher level companies, are actually held by mother companies that are all located in Europe. Thus, there may be a strategy on the part of European companies to control globalized subsidiaries through "tax haven" subsidiaries (color figure online)

subsidiaries from different world regions end up in the same DAGMap cell. Conversely, higher levels of the hierarchy represent the concentration of control on European soil, revealing a pyramidal logic in the entire structure: headquarters at the top of the pyramid, which relies on financial subsidiaries (2nd level) right above a market structured by cultural (continental) differences (3rd level, mainly involving European firms) and finally, an undifferentiated "world car" market at the bottom spread over emergent countries.

Different extensions of this technique have been proposed for dealing with larger hierarchies (Koenig, 2007). Using focus+context techniques, the user can explore the data using a specific interactor.

We also formally evaluated the combined DAGMap-Sugiyama technique versus the Sugiyama technique alone and the DAGMap alone. The evaluation showed that the DAGMap-Sugiyama visualization did not produce the best results but represented the best compromise between the number of interactions and the task completion time.

Our DAGMap visualization corresponds to a single view point showing low level details. Currently, there is no animated transition when the view point changes. In future work, an animated transition between two points of view in the DAGMap will be designed.

References

Balzer, M., & Deussen, O. (2005). Voronoi treemaps. In *IEEE symposium on information visualization InfoVis '05* (pp. 165–172). Washington, DC: IEEE Computer Society.

Bederson, B. B., Shneiderman, B., & Wattenberg, M. (2002). Ordered and quantum treemaps: Making effective use of 2D space to display hierarchies. *ACM Transactions on Graphics*, *21*(4), 833–854. doi: http://doi.acm.org/10.1145/571647.571649

Bruls, M., Huizing, K., & van Wijk, J. J. (2000). Squarified treemaps. In W. D. Leeuw, & R. V. Liere (Eds.), *Joint eurographics and IEEE TCVG symposium on visualization (data visualization '00)* (pp. 33–43). Amsterdam: Springer.

Carriere, J., & Kazman, R. (1995). Interacting with huge hierarchies: Beyond cone trees. In *Proceedings of IEEE information visualization '95* (pp. 74–81). Los Alamitos, CA: IEEE Computer Society Press.

Chintalapani, G., Plaisant, C., & Shneiderman, B. (2004). Extending the utility of treemaps with flexible hierarchy. In *Eighth international conference on information visualisation (IV'04)* (pp. 335–344). Los Alamitos, CA: IEEE Computer Society.

di Battista, G., Eades, P., Tamassia, R., & Tollis, I. G. (1998). *Graph drawing: Algorithms for the visualisation of graphs*. Upper Saddle River, NJ: Prentice Hall.

Eades, P., & Sugiyama, K. (1990). How to draw a directed graph. *Journal of Information Processing*, *13*(4), 424–437.

Fekete, J., Wang, D., Dang, N., Aris, A., & Plaisant, C. (2003). Overlaying graph links on treemaps. In *IEEE information visualization 2003 symposium poster compendium* (pp. 82–83). Los Alamitos, CA: IEEE Press.

Gutwenger, C., & Mutzel., P. (2004). An experimental study of crossing minimization heuristics. In B. Liotta (Ed.), *Graph drawing 2003* (Vol. 2912, pp. 13–24). New York: Springer.

Holten, D. (2006). Hierarchical edge bundles: Visualization of adjacency relations in hierarchical data. *IEEE Transactions on Visualization and Computer Graphics*, *12*(5), 741–748.

Koenig, P.-Y. (2007). Dagmap: un outil d'exploration adapté aux relations d'héritages multiples. In *Ihm '07: Proceedings of the 19th international conference of the association francophone d'interaction homme-machine* (pp. 237–240). New York: ACM. doi: http://doi.acm.org/10.1145/1541436.1541482

Lü, H., & Fogarty, J. (2008). Cascaded treemaps: examining the visibility and stability of structure in treemaps. In *Graphics interface 2008* (Vol. 322, pp. 259–266). Windsor, ON: Canadian Information Processing Society.

Porter, M. (1986). *L'avantage concurrentiel: comment devancer ses concurrents et maintenir son avance*. Paris: Intereditions.

Reingold, E. M., & Tilford, J. S. (1981). Tidier drawings of trees. *IEEE Transactions on Software Engineering*, *7*(2), 223–228. doi: http://dx.doi.org/10.1109/TSE.1981.234519

Shneiderman, B. (1992). Tree visualization with tree-maps: 2-d space-filling approach. *ACM Transactions on Graphics*, *11*(1), 92–99.

Thomas, J. J., & Cook, K. A. (2006). Illuminating the path: The research and development agenda for visual analytics. In *Illuminating the path: The research and development agenda for visual analytics* (pp. 33–68). Los Alamitos, CA: IEEE Computer Society.

van Wijk, J. J., & van de Wetering, H. (1999). Cushion treemaps: visualisation of hierarchical information. In G. Wills & D. Keim (Eds.), *Ieee symposium on information visualization (InfoVis '99)* (pp. 73–78). Los Alamitos, CA: IEEE CS Press.

Vliegen, R., van Wijk, J. J., & van der Linden, E.-J. (2006). Visualizing business data with generalized treemaps. *IEEE Transactions on Visualization and Computer Graphics*, *12*(5), 789–796.

Walker, J. Q., II. (1990). A node-positioning algorithm for general trees. *Software Practice & Experience*, *20*(7), 685–705. doi: http://dx.doi.org/10.1002/spe.4380200705

Wattenberg, M., & Fisher, D. (2003). A model of multi-scale perceptual organization in information graphics. In S. C. North, & T. Munzner (Eds.), *Ieee symposium on information visualization.* Seattle, WA: IEEE Computer Press.

Part III
Empirical Studies of Spatial Multilevel Networks

Chapter 8
Ports in a World Maritime System: A Multilevel Analysis

César Ducruet

8.1 Introduction

Contemporary sea transport carries approximately 90 % of the world trade volume. Despite this enormous importance, very little attention has been paid to the spatial organization of maritime networks throughout transport geography. Extensive research has been performed on air transport and other land-based networks; in the latter, urban centrality is the chief concern, whereas seaports are often considered aberrant cases and of peripheral interest because of a predominantly continental conception of urban and economic geography.

The aim of this chapter is to further understand the relative position of seaports within a world maritime system formed by the circulation of containerships. A global database of daily vessel movements allows for the study of the individual network attributes of seaports. Such a study will shed new light on the relationship between the individual network attributes of seaports and the classical indicators of throughput volume. In addition, this research applies clustering methodologies to the world graph to reveal the functional regions in which ports are embedded, while also evaluating the respective importance of geographical proximity and economical linkages.

The remainder of this chapter is organized as follows: The first section reviews the lack of network analysis on maritime networks throughout port geography and economics. It also provides some possible explanations for the lack of significant research in this area and some key directions for potential improvements. The second section introduces the data and the methodology together with some preliminary results of port rankings and a visualization of the world graph in 1996

C. Ducruet (✉)
Centre National de la Recherche Scientifique (CNRS), Université de Paris I Panthéon-Sorbonne, UMR 8504 Géographie-cités / Equipe P.A.R.I.S., rue du Four 13, Paris 75006, France
e-mail: ducruet@parisgeo.cnrs.fr; cdu@parisgeo.cnrs.fr

C. Rozenblat and G. Melançon (eds.), *Methods for Multilevel Analysis and Visualisation of Geographical Networks*, Methodos Series 11, DOI 10.1007/978-94-007-6677-8_8,

and 2006. Based on these preliminary results, the third section provides a multilevel analysis based on the application of strength clustering. The conclusion presents implications for policy decisions and some possibilities for further research.

8.2 The Network Analysis of Maritime Transport

8.2.1 A Lack of Network Analysis on Maritime Transport

Transport geographers have faced many obstacles in the study of maritime networks. One obstacle is contextual. Recent technological innovations in maritime transport that displace port functions from urban areas have eroded the importance of maritime transport in public representations. While these circumstances have likely reinforced the influence of the central place theory as stated by Bird (1970), they have paradoxically renewed scholars' interest in the transformation of sea transport itself, notably through its impact on port development (Slack, 1993). Shipping networks have become more footloose, partly due to shrinking transport costs and reduced trade barriers in general (Clark, Dollar, & Micco, 2004) but also due to the capacity of shipping lines to reorganize their networks globally with increased bargaining power and to integrate land transport through vertical integration (Robinson, 2002). However, the spatial complexity realized by such changes worldwide seems to have discouraged scholars from mapping the new organization of sea transport. Instead, port and maritime specialists have concentrated their efforts on the changes occurring within ports (Slack & Frémont, 2005), between ports of a given region (Ducruet, Roussin, & Jo, 2009), and across port hinterlands (Notteboom & Rodrigue, 2005), where various actors intervene in a territorial context that is more visible than across the oceans. In addition, containerization has permitted the integration of maritime transport with land-based logistics through the actions of giant companies (Robinson, 2002), while the land leg has remained the most expensive portion and has witnessed the most concentrated scholarly efforts. Therefore, the dynamics occurring across the maritime space between ports are not well known, although the current evolution may lead us toward an "ultimate system of maritime transportation [...] whereby every port node can theoretically be linked to every other port node" (Bird, 1984, p. 26).

A second obstacle in the current study of maritime networks is economic and technical. Detailed data on maritime traffic between ports are often difficult to obtain because of their rarity and their very high cost. Thus, most port and maritime specialists use aggregated measures of port throughput for comparative analysis (e.g., traffic volume, growth, and concentration) because these are simplified and easily accessible measures that indirectly analyze the spatial changes in sea transport such as the impact of port selection and the hub-and-spoke strategies of ocean carriers. Some researchers have used data from meteorological offices to obtain a snapshot of the worldwide location of sea-going vessels (Brocard, Joly, & Steck, 1995), while others have used manually encoded and computed paper-based sources to capture the sequences of port calls by vessel and by company (Joly, 1999). Recent research has used more easily accessible data derived from the annually published *Containerisation International Yearbooks*. These data sources provide information about the regular services of the world's main shipping lines, which permits the mapping of their geographical coverage by port or by region, as measured by the Weekly Containerized Transport Capacity (Frémont & Soppé, 2005). However, the lack of available software other than classic Geographical Information Systems (GIS) prevented any innovations in the visualization and analysis of large maritime networks.

In the context of continuous growth, increased spatial complexity, and the widening power of global alliances and shipping lines around the world, the utilization of new visualization and analytical tools becomes necessary. However, such an effort cannot ignore the specific issues in maritime transport with regard to the other transportation networks.

8.2.2 The Specificities of Maritime Transport

Despite the rapid success of air transport for commercial use, maritime transport has retained a very important role supporting, and even enhancing, globalization (Frémont, 2005). From the 1950s onward, containerization has facilitated the regional and global integration of transport and value chains (Robinson, 2002). Of course, this integration has been made possible in a context of lowering trade barriers and geopolitical stabilization after the decolonization of the then-Third World and the collapse of the socialist block, resulting in higher freedom of circulation, lower transport costs, and continuous growth. Even the Soviet Union increased its domestic share of maritime transport from 2 to 9 %, not only for geopolitical purposes (e.g., supporting distant brother countries such as Cuba) but also to enlarge its commercial power in response to Western imperialism (Vigarié, 1995).

These evolutions have provided shipping lines the liberty to invest heavily in new technologies, and the regular and dramatic increase in vessel sizes is proof of their success, as reflected in the enormous amount of literature on this subject in maritime and port studies. One key aspect of global transportation by sea is the evolution from trade support (demand-driven) to trade stimulation (offer-driven). This means that shipping lines have become proactive by providing efficient door-to-door services across oceans over ever-longer distances and also across continents through the integration of terminal and logistics operations, notwithstanding the role of shippers, forwarders, and intermodal operators in ensuring the space-time continuum of freight flows (Ducruet & Van Der Horst, 2009).

The implications for ports are enormous and varied. Ports are bound to waterside locations where physical conditions are increasingly important due to growing vessel sizes (e.g., 14–15 m of maximal quay depth are required for large containerships). In addition, the technological revolution in sea transport (i.e., containerization) has provoked a drastic selection of ports capable of planning new terminals equipped with modern and costly cargo-handling facilities (Slack, 1985). Given the aforementioned freedom of circulation, shipping networks have become increasingly footloose, selecting ports based on various factors such as centrality and intermediacy (Fleming & Hayuth, 1994). Centrality refers to the landward position of the port with regard to hinterlands, markets, and intermodal arrangements, while intermediacy corresponds to the seaward position, facilitating the implementation of hub-and-spoke strategies with other ports. Such dynamics have fostered competition among neighboring ports, resulting in traffic concentrations at a few load centers and hub ports and traffic dispersion due to diseconomies of scale and the preference of ocean carriers for brand new facilities created on greenfield sites (Slack, 1999).

Thus, our traditional conception of port-development processes has radically changed in the last few decades (Olivier & Slack, 2006). While ports and supply chains are "terminalized" by ocean carriers and terminal operators acquiring global portfolios (Slack, 2007), the factors contributing to port growth or decline seem to have shifted in the hands of the shipping lines. No longer is proximity to a market or a densely occupied hinterland sufficient to explain the distribution of traffic along a maritime range, although it is clear that despite a few exceptions for terminals built in the "desert" for transshipment purposes, most of the world's container traffic is concentrated within large urban agglomerations, and this trend is actually increasing (Ducruet, 2008b)[1]. For instance, Europe's largest container ports tend to be located as close as possible to the megalopolis ranging from London to Milan; the northern European range is a perfect example of high port concentration near Europe's core economic region (i.e., the Rhine). Nevertheless, the maritime component of port evolution has gained unprecedented importance. The focus of this chapter is to determine new ways of measuring and comparing this importance worldwide.

8.2.3 Possible Improvements

Two main approaches in the literature on ports and maritime transport are essential to an understanding of the positioning of ports in shipping networks.

Functions and performance of seaports

The first approach is concerned with the geographical functions of ports. Depending on the quality of their insertion into maritime networks, container ports can be

[1] Ducruet (2008a) calculated that the proportion of world container traffic at ports located in urban areas of over one million inhabitants has increased from 66 % in 1980 to 77 % in 2005. Of course, these increases are influenced by the inclusion of many hub ports for which transshipment traffic is counted twice, such as in Singapore, Hong Kong, and Busan.

defined as global pivots, load centers, regional ports and minor ports (Bichou & Gray, 2005; De Langen, Van der Lugt, & Eenhuizen, 2002). This typology is rarely verified empirically because of the complexity of liner networks. As explained by De Langen et al. (2002), four main factors contribute to this complexity: the continuous increase in traffic volume, the increase in the number of connected ports, the increase in vessel sizes, and the increase in the spatial freedom of ocean carriers. However, as the authors argue, "the role of different ports in maritime networks has not been documented, nor has a precise typology that would allow for such a precise exercise been proposed" (De Langen et al., p. 3).

An indirect analysis of ports' insertion into maritime networks through empirical case studies is more feasible. For instance, research has been performed on the distribution of seaborne connections of ports of Australia, the United Kingdom, and France on various scales (Bird, 1969; Britton, 1965; Von Schirach-Szmigiel, 1973). These studies have been complemented by analyses of the number of vessel calls as a proxy for maritime performance in the United States (Lago et al., 2001 cited in De Langen et al., 2002) and worldwide, combined with other port and urban indicators (Ducruet, 2008b). As mentioned above, relevant data sources and analytical tools to measure more complex realities based on large maritime networks are lacking.

In fact, the relative position of ports within a given maritime system is analyzed either theoretically, through the definitions of hub functions, centrality, and intermediacy (Fleming, 2000; Fleming & Hayuth, 1994) and the formulation of port-concentration models within a port system, or empirically through the analysis of traffic distribution and concentration (Ducruet et al., 2009). However, the latter analyses are based on aggregated individual measures (e.g., port throughputs) that hide the extent to which ports that handle similar traffic may, in fact, be very different in terms of seaborne connections and their positions in the networks in terms of vulnerability. One good example is Frémont's (2005) mapping of the global port network of Maersk, the world's main liner shipping company; however, it does not include references to the changes in local performance resulting from global insertion.

Port regions and port systems

Another approach attempts to position ports within a precise port region or maritime region. Despite the lack of clarity over the definition of such terms[2], these studies are similar in that they give more importance to the architecture of the maritime networks, where ports are only one aspect. An original insight inspired from

---

[2]Ducruet et al. (2009) differentiates between the maritime façade (coastal alignment of ports), the port region (the inland area smaller than the hinterland but wider than the port city where the port activities influence the economic structure), the port range (the coastal system of interdependent ports), the port network (the portfolio distribution of a given carrier), and the port system (interconnection of ports by shipping networks within a given area).

maritime history is proposed by Westerdahl (1996), who examines how maritime itineraries shape functional regions in Scandinavia and Europe in the early Middle Ages. In a similar vein but based on the application of network algorithms, Joly (1999) proposed a worldwide analysis of maritime linkages among main regions, resulting in the calculation of estimated port throughputs based on vessel movements. Other specific studies of liner networks cover specific regional areas such as the North Atlantic (Helmick, 1994), the Caribbean (McCalla, Slack, & Comtois, 2005; Veenstra, Mulder, & Sels, 2005; Wilmsmeier & Hoffmann, 2008), the Mediterranean (Cisic, Komadina, & Hlaca, 2007), and the relative position of North Korean ports in Northeast Asia (Ducruet, 2008a). While such studies clearly focus on port performance and maritime network design, notably visualizing the situation of ports in such networks, they do not have a multi-level approach, and they rarely relate network attributes with port performance[3].

Despite their fundamental legacy in clarifying our understanding of contemporary shipping and port development, two main weaknesses can be highlighted in such studies. First, the regional scale in which ports are studied is often defined arbitrarily; this argument is defended by Slack (1999) in his study on the evolution of containerization in the North Atlantic. Therefore, the regional areas within which different ports share privileged linkages are taken for granted and are not well-studied. Second, individual measures of port performance rarely include different levels of observation, from the local to the global, although it is recognized that contemporary ports are better compared through their ability to connect scales rather than through their traffic volume rankings (De Roo, 1994). It seems that those objectives – definition of regional areas and individual measures – strongly require an engagement as indicated by current research on social network analysis and small worlds.

8.3 Structure of the World Maritime System

8.3.1 Data Source and Methodology

Three main data sources exist for the analysis of maritime networks from a graph perspective. The previously mentioned *Containerisation International Yearbooks* offer an overview of the main service schedules of the world's shipping lines,

[3]This statement should not ignore that some authors point to specific situations in which a relationship is established between performance and network design. For instance, Lago et al. (2001) note that last ports of call attract more cargo on average because of transit time advantages; Notteboom (2006) specifies that upstream ports generate more cargo throughputs because ocean carriers compensate for the deviation between distance and time; Ducruet (2008a) finds that ports situated within larger urban regions often have a higher share of long-distance connections in their traffic. The main problem underscored in this chapter is the lack of systematic empirical verification, especially on a global scale.

Table 8.1 Overview of the maritime database

	1996	2006
No. ports	975	1,240
No. countries	157	173
No. vessels movements	176,439	390,740
No. vessels	1,759	3,973
No. operators	497	720
Total slot capacity (TEUs)	3,352,849	9,590,309
Total deadweight capacity (DWTs)	50,644,151	130,742,023
Share world fleet (% containerships)	91.52	98.33
Share world fleet (% TEUs)	92.15	97.91
Share world fleet (% DWTs)	6.23	12.54

Source: calculated by the author based on LMIU and UNCTAD

together with the fleet capacity of the companies by vessel. Although such information would permit the building of a world graph of inter-port linkages and has the advantage of a cheaper cost compared to other sources, it necessitates large amounts of manual data encoding. Another problem is the lack of coverage on local and regional services and the probable mismatch with effective ports of call.

Two other sources provide numeric information on effective vessel movements. First, the French ship broker Barry Rogliano Salles through its branch company AXS Marine maintains the Alphaliner database on container vessel movements. Second, the source used in this chapter is derived from Lloyd's Marine Intelligence Unit (LMIU), a service of Lloyd's, the world leader in maritime insurance and shipping information based in London. This data source is selected because of its wide coverage of the world fleet (98 % TEUs[4] in 2006), as seen in Table 8.1. The data also include ship operators' names and daily ports of call (previous, current, and next), thus allowing for many measurements by link and by port.

A fundamental reflection is necessary about the way vessel movements should be computed to address issues of port performance and port regions. Several aspects deserve careful attention:

- *Weight and frequency of linkages*: Many vessel movements of different capacities pass through the same links within a given period of time. Therefore, edges can be weighted according to the total traffic (the sum of all vessel capacities), the number of vessels, the number of calls, or by ratios such as the average vessel capacity, the weekly average number of calls, etc. In this study, we retain the total traffic realized by the overall circulation of vessels within 1 year in 1996 and 2006. The same applies to vertices (ports) that can be weighted and compared based on their total capacity circulated.

[4]Twenty-Foot Equivalent Unit (TEU): a normalized measure of container traffic and vessel capacity referring to the number of 20-foot container boxes. Vessel capacities can also be measured in deadweight tonnage (DWT) or "commercial capacity".

- *Canals, straits, strategic passages*: The original data contain several places that are not container ports or even seaports, such as canals (Panama, Suez), straits (Gibraltar, Dover, Dardanelles, Messina), channels (Yucatan), and other passage points at which the vessel reported a call (e.g., Tarifa and Cape Finisterre on the Iberian peninsula; Skaw in Denmark; Brixham in the UK). While such "nodes" are part of the effective movement of vessels, they do not account for port commercial operations, nor are they part of voluntarily selected logistics routes or transport chains. Because almost all pendulum services pass through the Panama and Suez canals, their centrality in the graph would go beyond those of the biggest commercial ports. Thus, they were omitted from the data[5].
- *Direct or indirect inter-port connections*: An analysis based on direct connections would imply a simplification of the reality of shipping networks. Although direct connections provide useful insights into the way ports are related to their close neighbors, liner networks are built upon a majority of indirect calls. Line-bundling and hub-and-spoke[6] are the most common services provided by ocean carriers willing to extend their influence across trading areas and continents. Thus, vessels are operated through rotating patterns that make an analysis based on indirect linkages more relevant. In addition, direct linkages may deprive some ports of their true foreland extent: a direct degree of 2 may hide an indirect degree of 30 if the port is connected to wide international logistics chains.

The methodology for building the graph consists of retrieving for each port all its direct and indirect connections through the circulation of each vessel during 1 year, regardless of the exact time of the connection. This enables Rotterdam to be connected with Tokyo, although in reality, those two ports are not directly connected by a single voyage; usually, pendulum services run for 3 weeks between Europe and Asia through the Suez Canal, including many intermediate calls. In the end, every vessel's circulation creates a complete graph in which all ports are fully interconnected. The resulting world graph is therefore a combination of all individual vessels' complete graphs.

---

[5]Other nodes that are commercial ports such as Istanbul (Turkey) and Port Said (Egypt) were also omitted because of their enormous number of calls compared to their actual traffic. Their proximity to important strategic passages has caused them to be reported by many vessels, although not every call represents a commercial operation at the terminals. Other cases include Brunsbuttel (Germany), a port at the mouth of the Elbe River, whose calls were attributed to Hamburg, the actual destination. Finally, some terminals have been merged in the data with their representative port, such as Port Botany and Sydney (Australia).

[6]The hub service is a combination of line-bundling and local services centralized upon one main transshipment center. A mother vessel calls at a transshipment hub where containers can be transferred to another mother vessel (i.e., interchange) or to a feeder vessel that carries out the rest of the local or regional service (Brocard et al., 1995). There are a few round-the-world (RTW) services where ships are bigger. Other services are pendulum services such as Europe-Asia and local services such as Rotterdam-United Kingdom. Line-bundling services often connect different global regions while hub-and-spoke services are more intra-regional.

Table 8.2 Topological characteristics of the world maritime network, 1996–2006

	Measure	Description	Calculation	Direct links 1996	Direct links 2006	Indirect links 1996	Indirect links 2006
e	Length	Number of edges	Sum	6,322	9,493	29,251	51,054
v	Population	Number of vertices	Sum	975	1,240	975	1,240
k	Cyclomatic number	Number of independent cycles	$e - v + p$	5,347	8,253	28,676	49,814
α	Alpha	Lattice degree	$k / \frac{v(v-1)}{2}$	0.011	0.010	0.011	0.010
β	Beta	Degree of graph complexity	$\frac{e}{v}$	6.484	7.655	30.001	41.172
γ	Gamma	Global connectivity	$e / \frac{v(v-1)}{2}$	0.013	0.012	0.061	0.066
C	Connectivity degree	Observed vs. optimal connectivity	$\frac{v(v-1)}{2} / e$	75.106	80.920	16.232	15.046

Source: calculated by the author based on LMIU and Joly (1999)

8.3.2 Topological Characteristics and Geographical Structure

The resulting graphs are characterized by a high complexity (Table 8.2). The high cyclomatic number, where p is the number of separated components, indicates a high connexity of the graph. The higher the connexity, the more accessible is a node from all other nodes. The connexity has increased between 1996 and 2006, and it is always higher for indirect links because of a higher density of inter-port relations. However, the lattice levels are relatively low, most likely due to the hierarchical nature of the network. The global connectivity is also low, but it is higher for indirect links, for the reasons cited above. In terms of complexity level, the high values for non-planar graphs indicate a very complex pattern. Notably, the growth rate of the number of ports connected is 27 %, but the maritime links grew 64 % over the same period, resulting in a denser network. This is most likely an effect of the factors cited by De Langen et al. (2002) on the current evolution of container networks. The observed connectivity (c) has increased for direct links but has decreased for indirect links, where it is significantly lower. All these facts indicate that the graph of indirect links is denser and more complex due to the richness of inter-port relations, taking into account the overall circulation of vessels instead of direct inter-port links, which tend to break the continuum of shipping.

The structure of the graph can also be analyzed based on the relationship between the number of ports and the number of connections (i.e., maritime degree). The results presented in Fig. 8.1 indicate that the organization of liner networks corresponds to a scale-free network that is defined by a power-law distribution. A few ports dominate the network by their high number of connections, while a majority of the other ports have only limited connections. Some exceptions in

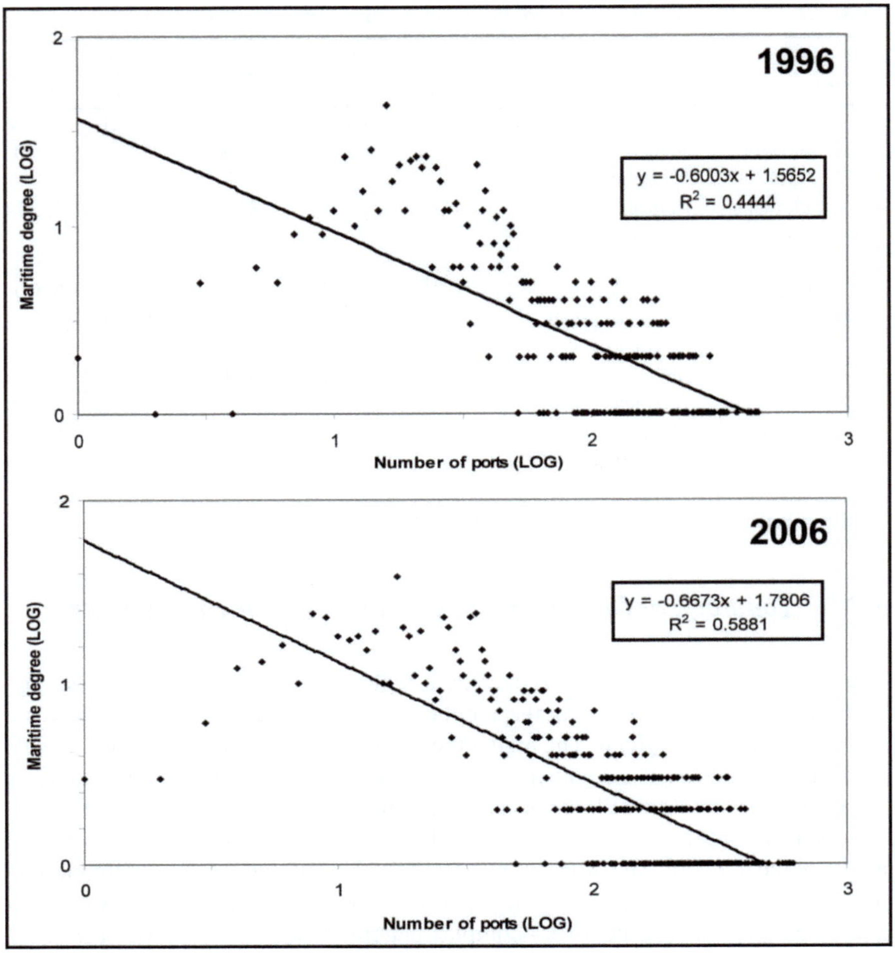

Fig. 8.1 Scale-free dimension of the world maritime network, 1996–2006 (Source: calculated by the author based on LMIU)

the figures – a few ports having the fewest connections – are explained by the methodology of data collection: only twelve ports are not very well-connected to the rest of the network in both years. In terms of evolution, we note that the slope of the line and the coefficient have increased from 1996 to 2006, highlighting a concentration of the network. This result most likely stems from the rationalization of carrier services, notably through the implementation of hub strategies throughout the world.

The spatial structure is observable through TULIP software based on the ports' betweenness centrality that is the sum by port of all their positions on their shortest paths. This measure is equivalent to a level of relative accessibility within the graph.

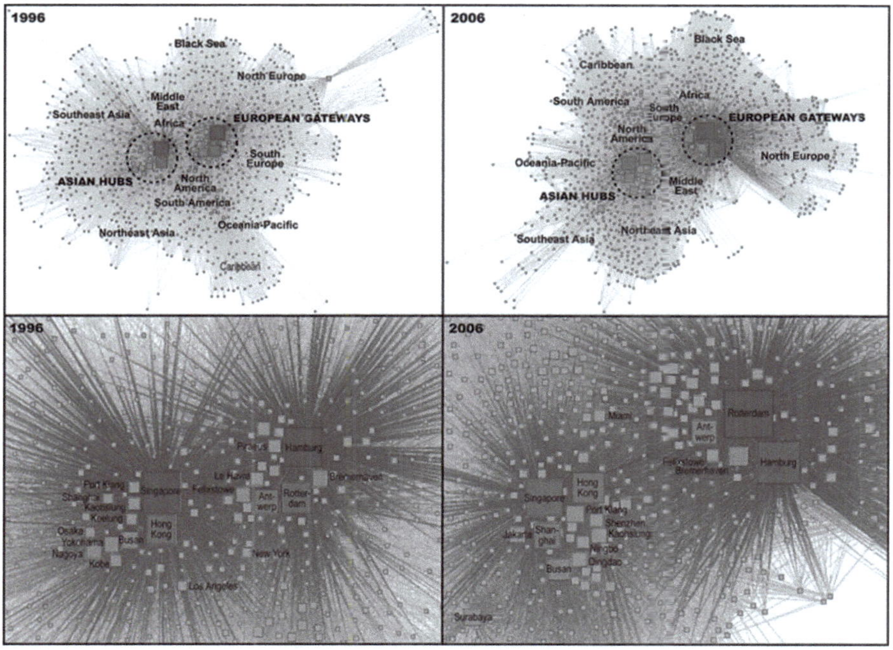

Fig. 8.2 Visualization of the world maritime system, 2006 (Source: realized by the author based on LMIU and TULIP)

Although it does not correspond to the daily preoccupations of port authorities or carriers, it helps to distinguish the positioning of ports in the overall pattern of circulations on various scales. A Gem-Frick layout is applied to situate the most central ports in the center and the least central ports in the periphery (Fig. 8.1).

The results confirm that very few ports dominate the hierarchy of centrality: "European gateways" (Rotterdam, Hamburg, Antwerp, and Bremerhaven as the northern range) and "Asian hubs" (Singapore, Hong Kong, Shanghai, and Busan as the Asian corridor). Despite the lower score of Asian hubs in general compared with the most central European ports (Rotterdam and Hamburg), we see many more Asian ports with a relatively high centrality: Port Klang (Malaysia), Jakarta and Surabaya (Indonesia), Kaohsiung (Taiwan), and Shenzhen, Ningbo and Qingdao (China). Thus, the measure of betweenness centrality seems to provide more advantage to hub ports upon which numerous smaller ports depend. Scandinavia and the British Isles tend to depend on Rotterdam and Hamburg, while Southeast and Northeast Asia tend to depend on their respective Asian hubs. In comparison, ports that usually rank high in traffic volume such as New York or Yokohama have a limited centrality. Outside the two cores of the system, only Miami scores high, most likely because of its strategic position as a hub between the Caribbean and the North Atlantic regions (Fig. 8.2).

The geographical dimension of the system is also apparent when examining the position of the ports and the region to which they belong. Despite the borderless crossing of multiple services and circulations allowed by indirect linkages, a strong spatial proximity characterizes the graph. The two large clusters are separated by Middle Eastern ports and Mediterranean ports (the Europe-Asia route), while other regions are logically distributed around their respective cores: the Black Sea, Scandinavia, and Africa for European gateways; Asia, Pacific and the Americas for Asian hubs. Although this global snapshot of the geographical structure does not account for a precise distinction among spatial and functional proximities, it demonstrates that the organization of the world maritime system is fundamentally geographical in scope. More advanced methods of clustering allow for a better investigation of possible hidden sub-structures in the global graph.

8.4 Multilevel Analysis of Port Performance

8.4.1 Comparison of Ports' Network Attributes

Two measures are compared with the performance of container ports (Table 8.3). The maritime degree is the total number of connections, as described above; here, the maritime degree is distinguished between direct and indirect numbers. The level of hub dependence corresponds to the traffic share (% TEUs) of the biggest connection in total port traffic. It highlights the extent to which a given port depends on another port for the distribution of its traffic: the more dependent a given port is, the more vulnerable it is in the system.

There is a close relationship between the direct degree and throughput in 2006: Singapore (210) and Hong Kong (193) have the largest throughput and the highest degree. Thus, larger ports are those that deploy a wider set of connections. This indicates that in general, network attributes closely follow (or explain) traffic hierarchies. Gateway or hub functions do not necessarily influence the results. However, specific locational factors may have an influence; for example, Guangzhou (Pearl River Delta) has a lower degree as explained by a combination of its upstream position – constraining the accessibility for large containerships – and its proximity to the major ports of Hong Kong and Shenzhen. The traffic and the degree are also very much interrelated in their evolution: their respective growth rates each have a coefficient of 0.83, with Shenzhen (China) at the top of the dynamic. We also see that ports in advanced economies tend to stabilize and rationalize (e.g., decreased degree) their position in the network, while ports in developing and emerging economies see their position improving quickly by catching-up.

Indirect connections show a somewhat more distinct pattern than that of direct connections. Singapore (591), Hong Kong (612) and Shanghai (569) have much fewer worldwide connections than Shenzhen (1,224). This measure is therefore better related with the actual foreland (i.e., commercial relations) of the port. Shenzhen is known as the world's factory; it has welcomed direct calls from global

Table 8.3 Throughput volumes and network attributes of the world's 30 main container ports

Rank in 2006	Port	Container throughput (000s TEU) 2006	Container throughput (000s TEU) 1996	Maritime degree – Indirect links 2006	Maritime degree – Indirect links 1996	Maritime degree – Direct links 2006	Maritime degree – Direct links 1996	Hub dependence (% TEU) – Indirect links 2006	Hub dependence (% TEU) – Indirect links 1996	Hub dependence (% TEU) – Direct links 2006	Hub dependence (% TEU) – Direct links 1996
1	Singapore	24,792	12,943	591	439	210	154	11.5	6.5	21.1	32.3
2	Hong Kong	23,230	13,460	612	407	193	132	8.0	5.7	15.6	30.5
3	Shanghai	21,710	1,930	569	235	185	63	9.9	7.4	18.3	33.5
4	Shenzhen	18,468	1,032	1,224	103	177	30	6.8	6.3	50.0	52.5
5	Busan	12,030	4,725	502	328	180	110	8.8	6.7	13.4	19.7
6	Kaohsiung	9,774	5,063	456	294	127	90	13.9	7.1	31.2	40.9
7	Rotterdam	9,690	4,935	622	433	145	137	5.4	5.1	24.3	20.3
8	Dubai	8,923	2,247	394	233	143	55	8.8	4.9	17.3	18.3
9	Hamburg	8,861	3,054	874	443	123	117	8.1	6.6	25.6	27.7
10	Los Angeles	8,469	2,682	287	241	56	63	8.3	6.7	25.6	33.4
11	Qingdao	7,702	810	445	142	100	28	8.6	6.6	22.6	30.1
12	Long Beach	7,290	3,007	317	–	46	–	10.5	–	27.1	–
13	Ningbo	7,068	–	434	70	101	14	11.0	8.7	32.0	34.4
14	Antwerp	7,018	2,653	548	416	96	101	5.0	4.8	32.3	18.6
15	Guangzhou	6,600	–	236	50	31	14	13.0	8.6	41.2	53.8
16	Port Klang	6,320	1,409	473	264	134	54	15.0	7.1	27.3	62.4
17	Xingang	5,900	800	357	151	54	42	12.6	6.9	31.9	28.3
18	New York	5,092	2,269	324	300	75	63	4.6	4.5	15.3	28.1
19	Tanjung Pelepas	4,770	–	340	–	74	–	8.6	–	16.1	–
20	Bremerhaven	4,450	1,543	433	324	92	57	10.1	6.2	22.4	34.9
21	Laem Chabang	4,123	820	313	166	60	25	14.2	7.2	29.9	34.1
22	Xiamen	4,019	400	392	86	74	20	15.0	7.0	35.2	36.9
23	Tokyo	3,665	2,311	330	208	74	51	9.3	7.3	18.7	14.9
24	Jawaharlal Nehru	3,298	423	343	149	50	20	10.3	4.4	16.7	23.5
25	Algeciras	3,245	1,307	370	247	104	62	4.5	5.0	28.7	40.5
26	Dalian	3,212	–	340	113	69	24	9.0	6.4	24.0	19.5
27	Yokohama	3,200	2,334	389	318	100	86	10.1	6.5	17.2	19.5
28	Colombo	3,079	1,356	381	258	63	51	9.3	5.8	20.3	40.2
29	Felixstowe	3,000	2,042	415	369	64	86	8.0	6.0	28.0	21.5
30	Jeddah	2,964	827	341	256	59	41	8.2	5.6	43.7	45.4

Source: Calculated by author based on LMIU and Containerisation International

shipping lines since 1998 (Wang, 1998). Here, gateway and hub functions create noticeable discrepancies along the hierarchy. For instance, European gateway ports such as Hamburg, Rotterdam, Antwerp, and Felixstowe all enjoy wider global connections than ports of equivalent throughput. Gateway ports are more likely to drain vast hinterlands and thus to connect with a variety of overseas markets. In comparison, hub ports have a simplified foreland because of narrowed hinterlands. The relationship between the respective growth rates of traffic and indirect degree remains quite significant with a coefficient of 0.74.

In terms of hub dependence, the direct vulnerability of some ports despite their large throughput is evident. The traffic share of one direct link reveals a higher vulnerability for Shenzhen (50 %), Kaohsiung (31 %), Guangzhou (41 %), Xingang (32 %), Laem Chabang (30 %), Xiamen (35 %), and Jeddah (43 %). This is because many Asian ports depend on another hub for accessing the rest of the system, although the case of Kaohsiung is highly political, with the obligation of connecting Hong Kong instead of mainland Chinese ports across the straits. Hub ports and gateway ports have a more even spread of traffic connections, as seen with New York (15 %) and Busan (13 %). European gateway ports have intermediate values of approximately 25 % due to the existence of port systems such as the North European range from Le Havre to Hamburg. Local strategies of port development also greatly impact the level of hub dependence. The important drop for Port Klang (−35 %) clearly reflects successful relief from the dominance of neighboring Singapore following Malaysian government support for gateway development. In comparison, Shenzhen and Guangzhou could not significantly decrease their dependence on the Hong Kong hub, nor could Ningbo reduce its dependence on Shanghai.

For indirect hub dependence, the highest values in 2006 are for Asian ports because of the unavoidable concentration of circulation at a neighboring hub such as Kaohsiung (14 %) with Hong Kong and Port Klang (15 %) and Laem Chabang (14 %) with Singapore. Continental gateways such as New York and Antwerp have the lowest hub dependence – approximately 5 % – for the reasons cited above, but some western interregional hubs such as Algeciras also have low hub dependence. Due to the comparatively limited importance of hinterlands in Asia, more interactions among Asian ports tend to increase their reciprocal interdependency and connectivity. Western ports function more as gateways linking the hinterland with the foreland, with less inter-port corridors. Thus, the indirect hub dependence can be seen as an indicator of logistics chain diversity through the importance of hinterland coverage.

8.4.2 The Clustering of Port Regions

In this section of the analysis, we use strength clustering to highlight possible "small worlds" in the graph. However, there are inherent limitations in using such methods because the maritime global network is a scale-free network and not a small-world network. Small worlds or "communities" – here, port regions – exist less in scale-free networks because of their very hierarchical dimensions: every node is in some way dominated by a "star" or "hub", making the formation of relatively independent groups of ports enjoying preferential relations difficult. We apply strength clustering on both the structure of the network itself (links without weight) and on the weighted graph (weight in TEUs, i.e., sum of circulated capacities) to verify the impact of traffic volume on small-world formation. Such an analysis will help to answer two sets of questions:

- *Multilevel maritime regions*: What are the tributary areas of the world's largest ports? Are they regionally specialized? What is the respective role of geographical proximity and carrier decisions to re-route their services based on economic factors (time and cost)? Are there resilient historical trade patterns still visible?

- *Multilevel maritime dynamics*: How do such strongly interconnected groups of ports evolve over time, given that carrier strategies and port competition profoundly modified the structure of shipping networks in the 1990s?

In 1996, there is some correspondence between the results from the weighted analysis and the non-weighted analysis. For instance, Hamburg appears as a key node in each analysis, although the two clusters are distinguished by the number of ports contained and by the location of those ports. In the non-weighted cluster, Hamburg and Rotterdam stand out as the two most central ports of a predominantly European cluster (Scandinavia, South Europe), while only Belawan and Jakarta (Indonesia) are included from outside Europe. In the weighted cluster, Hamburg also has a dominance of its European counterparts (Iberian Peninsula, North), of which many are similar to the non-weighted cluster, including Jakarta. Some important ports that were not included in the non-weighted cluster are Port Klang (Malaysia) and Keelung (Taiwan). Thus, it can be argued that weighted clusters underscore important corridors on a global level (Europe-Asia), while non-weighted clusters are better related to the neighboring architecture of a port's network (Europe) (Fig. 8.3).

Conversely, the geographical affinity of some central ports does not change much from the non-weighted to the weighted clusters. Amsterdam (with Ymuiden) is the center of a non-weighted cluster with a majority of Latin American ports and two Australian ports. It also appears in a weighted cluster alongside many Latin American ports, although in a less central position, due to the inclusion of large gateways such as Santos (Brazil) and Buenos Aires and some Oceania ports (e.g., Fremantle, Auckland). Therefore, taking into account the weight of the links does not necessarily disturb the pattern. However, some very central ports such as Nagoya (Japan) and Southampton (UK) that have a strong position in some non-weighted clusters are not part of any weighted cluster, most likely because of their lower weight in the network. In addition, some very central ports in some clusters have a relatively low rank in the usual port rankings: Dakar (Senegal), Damietta (Egypt), and Szczecin (Poland), while others are strategic places for carriers to develop gateway strategies, such as Hampton Roads in the United States (Starr, 1994).

In 2006, all clusters show a somewhat stronger geographical coherence than in 1996. For instance, the non-weighted cluster centered on Antwerp includes some major European gateways (e.g., Rotterdam, Hamburg) together with a series of Latin American ports, which highlights the importance of maintained transatlantic ties. This is also the case with the non-weighted cluster centered on La Guaira (Venezuela) linking Western Europe (Genoa, Liverpool) with North Africa (Morocco) and several Latin American ports. The two other non-weighted clusters also reveal interesting patterns. The cluster centered on the rapidly growing Black Sea port of Constantza (Romania) is mostly intraregional, as it concentrates mostly other North or South European ports. Conversely, the cluster centered on Ashdod

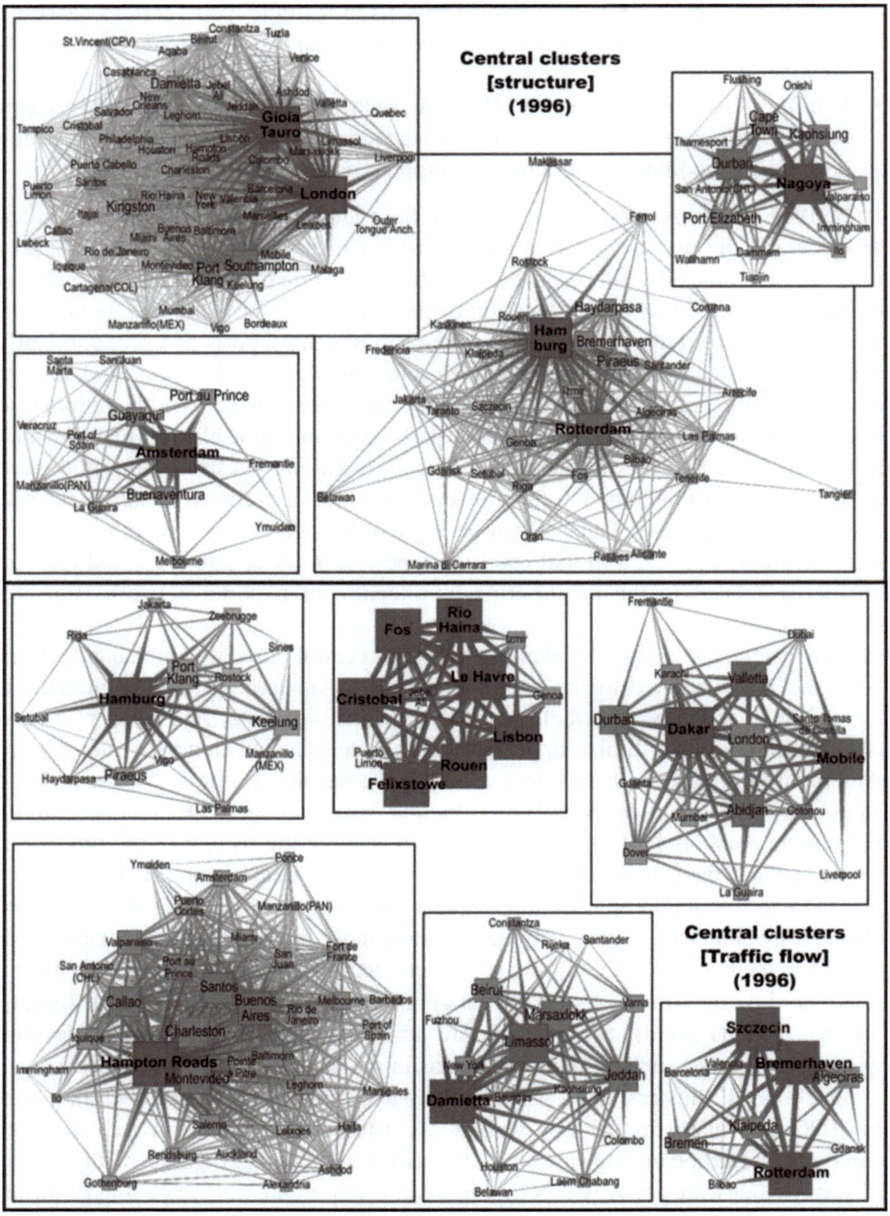

Fig. 8.3 Central clusters in the world maritime system, 1996 (Source: realized by the author based on LMIU and TULIP)

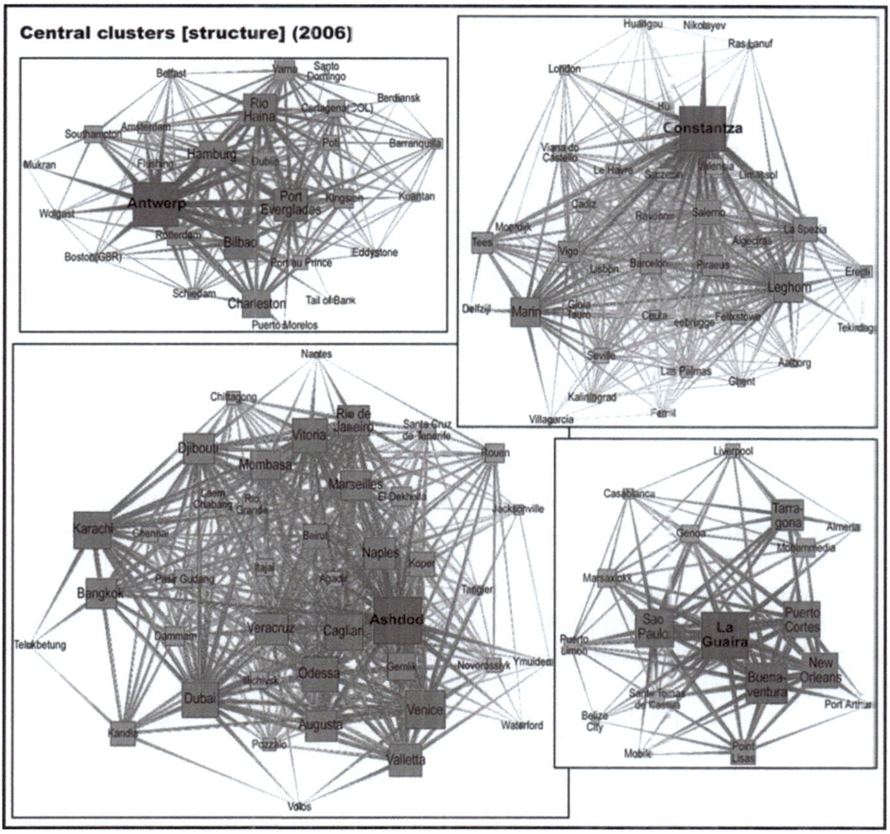

Fig. 8.4 Central clusters in the world maritime system, 2006 (continued) (Source: realized by the author based on LMIU and TULIP)

(Israel) is more diverse geographically, but one can notice a spatial continuity from Southeast Asia (Bangkok) to Brazil (Rio Grande, Itajai) across the Middle Eastern, Mediterranean, and North African ports. Only a few exceptions deviate from this corridor of ports (Fig. 8.4).

The very important role of the geographical and historical continuum becomes clear with weighted clusters. A Northeast Asian cluster centered on Yokohama interestingly connects some outer ports; a "Latin" cluster centered on Cartagena (Colombia) primarily gathers South European and Latin American ports; another cluster centered on Antwerp is predominantly European with some US Gulf Coast ports; and finally, a cluster centered on Dakar gathers a mix of African and Asian ports. The evolution between 1996 and 2006 for weighted clusters shows some permanency, with a majority of Iberian Peninsula / Latin American / Northeast Asia / European ports differently represented within a few central clusters.

8.5 Conclusion

This chapter has for the first time applied a multilevel analysis to maritime transport and, in particular, to liner-shipping networks formed by the daily circulation of container vessels on a global level. What can be learned from such an attempt? First, we confirm that liner shipping is characterized by high complexity, while the network properties of seaports interestingly are in line with the usual port hierarchies provided by economic intelligence and earlier academic studies on port performance. Second, the clustering methodology applied to the world graph of direct and indirect inter-port links, with or without traffic, provides a rather unusual combinations of ports that do not correspond to our vision of the world delineated by political and cultural criteria.

Thus, retrieving the inherent functional, spatial, economic – and perhaps accidental – logic of each cluster would necessitate enormous efforts of data checking from the raw dataset itself. What should port specialists understand from the results? Some clusters are not easy to "define" because they do not exactly overlap with the usually defined geographical areas or systems in port geography, such as port ranges and hub-port regions. For instance, Yokohama appears as the most central port in a Northeast Asian cluster, whereas all indications in the literature are that Busan (South Korea) has the strongest position in this region as a hub. This is perhaps where the methodology follows a somewhat different direction than the mainstream literature: hub ports (i.e., stars) where many links converge have a relatively lesser importance in the results because instead of forming – or belonging to – small worlds, they tend to create hierarchy and polarity in the system.

However, there are very positive lessons to learn from the application of such methods to liner shipping for port and maritime geography. Despite the "automatic" dimension of clustering, strong coherence is found in a number of clusters where either geographic or commercial/cultural logic dominates, as seen with transatlantic groupings or South-South groupings. Surprisingly, we did not find many clusters including the biggest ports of the Triade as altogether dominant poles of a global system (e.g., North America, Western Europe, and East Asia). This indicates that the world maritime system is regionally polarized but not globally polarized: some corridors link different regions through a complex mix of pendulum, round-the-world, and other line-bundling services, while intraregional services tend to be most concentrated on a few ports. This confirms that liner shipping does not exactly overlap with trade routes: the hub strategies of global ocean carriers have rerouted the dominant flows toward many hub ports in the Caribbean, the Mediterranean, and Asia as a whole, for the aforementioned reasons. Perhaps further research focused on finding a better match between the world economic system and the world maritime system would "delete" such hub ports from the map. Thus, the question is as follows: which ports are more hubs than others? Unfortunately, the answer or the methodology to answer is not yet available in the literature.

Further research should follow several key directions to improve the analysis. First, selecting only a category of vessels of a certain size would allow for the results to be refined: vessels over 2,000 TEUs often operate on longer-distance services (e.g., pendulum, round-the-world), while lighter vessels tend to operate on intraregional services (e.g., barging, feeder). The present analysis has mixed together different sizes of vessels, which in the end may have blurred the geographical logic behind the vessel movements. Second, because liner shipping is operated as a scale-free network – whether it is analyzed through direct or indirect linkages – it would be interesting to make use of other types of vessels carrying commodities such as bulks or general cargo, for which hub-and-spoke strategies do not apply. Therefore, bulk and general cargo shipping tend to overlap better with trade routes and commercial patterns. Third, as mentioned above, excluding hub ports from the analysis would permit a better appreciation of the trade continuum among ports of the world and possibly provide better results in terms of clustering and small-world analysis.

References

Bichou, K., & Gray, R. (2005). A critical review of conventional terminology for classifying seaports. *Transportation Research Part A: Policy and Practice*, *39*(1), 75–92.

Bird, J. (1969). Traffic flows to and from british seaports. *Geography*, *54*, 284–301.

Bird, J. (1970). Seaports are not aberrant cases. *Area*, *4*, 65–68.

Bird, J. (1984). Seaport development: Some questions of scale. In B. Hoyle & D. Hilling (Eds.), *Seaport systems and spatial change* (pp. 21–41). Chichester, UK: Wiley.

Britton, J. N. H. (1965). Coastwise external relations of the ports of victoria. *Australian Geographer*, *9*(5), 269–281.

Brocard, M., Joly, O., & Steck, B. (1995). Les réseaux de circulation maritime. *Mappemonde*, *37*(1), pp. 23–28. (http://mgm.fr/PUB/Mappemonde/M195/NAVIRES.pdf)

Cisic, D., Komadina, P., & Hlaca, B. (2007). Network analysis applied to mediterranean liner transport system. In *International association of maritime economists conference*. Athens, Greece, July 4–6.

Clark, X., Dollar, D., & Micco, A. (2004). Port efficiency, maritime transport costs, and bilateral trade. *Journal of Development Economics*, *75*(2), 417–450.

De Langen, P., Van der Lugt, L., & Eenhuizen, J. (2002). A stylised container port hierarchy: A theoretical and empirical exploration. In *International association of maritime economists conference*. Panama, November 13–15.

De Roo, P. (1994). Marseille: de l'aire portuaire à l'aire métropolitaine. In M. Collin (Ed.), *Ville et Port 18–20 siècles* (pp. 107–113). Paris: L'Harmattan.

Ducruet, C. (2008a). Hub dependence in constrained economies: The case of North Korea. *Maritime Policy & Management*, *35*(4), 377–394.

Ducruet, C. (2008b). World typology of port-city relationships. *Cybergeo*, *417*. (http://www.cybergeo.eu/index17332.html)

Ducruet, C., Roussin, S., & Jo, J.-C. (2009). Going west? Spatial polarization of the North Korean port system. *Journal of Transport Geography*, *17*(5), 357–368.

Ducruet, C., & Van Der Horst, M. R. (2009). Transport integration at European ports. *European Journal of Transport and Infrastructure Research*, *9*(2), 121–142.

Fleming, D. (2000). A geographical perspective of the transhipment function. *International Journal of Maritime Economics*, *2*(3), 163–176.

Fleming, D., & Hayuth, Y. (1994). Spatial characteristics of transportation hubs: Centrality and intermediacy. *Journal of Transport Geography*, *2*(1), 3–18.

Frémont, A. (2005). *Conteneurisation et mondialisation. Les logiques des armements de lignes régulières*. Habilitation à diriger des recherches. Paris: Université Paris I Sorbonne.

Frémont, A., & Soppé, M. (2005). Transport maritime conteneurisé et mondialisation. *Annales de Géographie*, *642*, 187–200.

Helmick, J. (1994). *Concentration and connectivity in the North Atlantic liner port network 1970–1990*. Miami: University of Miami.

Joly, O. (1999). *La Structuration des Réseaux de Circulation Maritime*. PhD thesis, Le Havre: Le Havre University.

McCalla, R., Slack, B., & Comtois, C. (2005). The caribbean basin: adjusting to global trends in containerization. *Maritime Policy & Management*, *32*(3), 245–261.

Notteboom, T. E. (2006). Traffic inequality in seaport systems revisited. *Journal of Transport Geography*, *14*(2), 95–108.

Notteboom, T. E., & Rodrigue, J.-P. (2005). Port regionalization: towards a new phase in port development. *Maritime Policy & Management*, *32*(3), 297–313.

Olivier, D., & Slack, B. (2006). Rethinking the port. *Environment and Planning A*, *38*(8), 1409–1427.

Robinson, R. (2002). Ports as elements in value-driven chain systems: the new paradigm. *Maritime Policy & Management*, *29*(3), 241–255.

Slack, B. (1985). Containerisation, inter-port competition and port selection. *Maritime Policy and Management, 12*, 293–303.

Slack, B. (1993). Pawns in the game: Ports in a global transportation system. *Growth and Change*, *24*(4), 579–588.

Slack, B. (1999). Satellite terminals: A local solution to hub congestion? *Journal of Transport Geography*, *7*(4), 241–246.

Slack, B. (2007). The terminalization of ports: An academic question? In *International Workshop on New Generation port cities and their insertion in global supply chains.* Hong Kong, December 12–14.

Slack, B., & Frémont, A. (2005). Transformation of port terminal operations: From the local to the global. *Transport Reviews*, *25*(1), 117–130.

Starr, J. T. (1994). The Mid-Atlantic load centre: Baltimore or Hampton roads? *Maritime Policy & Management*, *21*(3), 219–227.

Veenstra, A. W., Mulder, H. M., & Sels, R. A. (2005). Analysing container flows in the caribbean. *Journal of Transport Geography*, *13*(4), 295–305.

Vigarié, A. (1995). *La mer et la géostratégie des nations*. Paris: Economica & Institut de Stratégie Comparée.

Von Schirach-Szmigiel, C. (1973). Trading areas of the united kingdom ports. *Geografiska Annaler*, *55B*, 71–82.

Wang, J. J. (1998). A container load center with a developing hinterland: A case study of Hong Kong. *Journal of Transport Geography, 6*, 187–201.

Westerdahl, C. (1996). *Traditional zones of transport geography in relation to vessel types.* (Internet Document: http://www.abc.se/~pa/publ/transport.htm)

Wilmsmeier, G., & Hoffmann, J. (2008). Liner shipping connectivity and port infrastructure as determinants of freight rates in the caribbean. *Maritime Economics and Logistics*, *10*(1–2), 130–151.

Chapter 9
Comparing Multilevel Clustering Methods on Weighted Graphs: The Case of Worldwide Air Passenger Traffic 2000–2004

Céline Rozenblat, Guy Melançon, Romain Bourqui, and David Auber

9.1 Introduction

Worldwide deregulation of air transport began in 1978 and came to fruition in 2008, and the re-organization of flows and companies since 1978 is one of the main subjects of air transport network studies. Hubs and spokes are re-shaping the entire network in national and multinational contexts and shrinking international and national links. Thus, each city with specific activities, such as business, research or tourism, develops its own strategies to maximize accessibility from the relative position with the goal of strengthening its economic base. Diagnostics for cities increase the need to study the properties and dynamics of cities, which can be considered nodes in the entire network of flows.

The study of air traffic networks involves large flow matrices. A useful approach to understanding the structure of networks and the evolution of this structure is to search for the highest subgroup densities in a graph, assuming that the graph is neither random nor homogeneous (Barabasi & Albers, 1998; Watts & Strogatz, 1998). Then, clusters of the most connected city airports encapsulate the entire network structure underlying highly inter-connected groups of cities. Different approaches from this perspective were recently developed by physicists (Brandes, 2001; Newman, 2004; Newman & Girvan, 2004). In air transportation in particular, air traffic networks have been used by different network clustering methodologies as a case study (Guimerà, Mossa, Turtschi, & Amaral, 2005; Sales-Pardo et al.,

C. Rozenblat (✉)
Institute of Geography, University of Lausanne, GEOPOLIS Building,
1015 Lausanne, Switzerland
e-mail: Celine.Rozenblat@unil.ch

G. Melançon • R. Bourqui • D. Auber
CNRS UMR 5800 LaBRI, Université Bordeaux 1 - GRAVITE INRIA Sud-Ouest,
Cours de la Libération 351, Talence 33405, France
e-mail: melancon@labri.fr; bourqui@labri.fr; auber@labri.fr

C. Rozenblat and G. Melançon (eds.), *Methods for Multilevel Analysis and Visualisation of Geographical Networks*, Methodos Series 11, DOI 10.1007/978-94-007-6677-8_9,

2007) or to test the network properties (Colizza, Barrat, Barthelemy, & Vespignani, 2006). The procedure in these studies for identifying the advantages of a method that analyzes or clusters an air transportation network, in terms of static and dynamic properties, is not clear.

The aim of this paper is to compare different clustering methods for an air transport network to evaluate the ability of these methods to reveal hidden structures. Consequently, we first discuss the properties of the air traffic network (Sect. 9.2). Next, we present recent contributions to clustering approaches (Sect. 9.3). We then test and compare different methodologies using the same dataset (Sect. 9.4). The paper concludes with the results of the comparison of the methods used and describes future research directions.

9.2 Complex Logics of World Air Network Organization

The organizational characteristics of the world air network are perceived through its shape (i.e., the topology of the graph and the weight of the exchanges), which has been produced by the activity of economic actors and policies that have evolved largely over the past 10 years.

9.2.1 Principal Traffic of the World Air Network

Assuming that the strength of links between cities is reflected in annual passenger traffic, the network is obviously concentrated around a small number of dominant cities. The 49 connections of more than two million passengers per year (in 2000) connect only 41 cities (including 23 cities in the United States). Paradoxically, the connections that comprise the highest passenger traffic are the shortest ones (in terms of average distance). The most frequented connections are Pusan – Seoul (more than six million passengers), Fort Lauderdale – New York, Hong Kong – Taipei, Chicago – New York, New York – Orlando, Los Angeles – New York and Boston – New York (all having more than five million passengers). The highest ranked European connection, Dublin – London, is in 13th place (with four million passengers), followed by Barcelona – Madrid (three, nine million) in 15th place. On the whole, of the 49 connections of more than two million passengers, 32 connect cities in the United States, 10 connect cities in Europe, and 6 connect Asian cities. London – New York is the only major transcontinental connection in this group (ranked 12th with four, two million passengers).

Connections with more than one million passengers extend this world coverage slightly, but they increase the density of already well-served zones. Notice that no Latin American city appears in this major world subnetwork. The highest-ranked connection for this region is Buenos Aires – Santiago (850,000 passengers per year), followed by Buenos Aires – Sao Paulo (725,000 passengers per year).

9.2.2 *Effects of Deregulation on the Forms of Networks*

One could assume that this world concentration results from gravitating logics, where the richest and most populated cities of the planet interact more intensely, being closer to each other. However, the application of the gravitational model to air passenger traffic reveals the ineffectiveness of Euclidean space-time (Amiel, 2003) on the overall network structure. The current organization of world air traffic results from various deregulatory processes. Indeed, with the advent of deregulation in 1978 in the United States, which then spread across the world in 1993, plane routes no longer depend only on the exchange capacities of the places being connected or on the "technological" limits of the apparatuses.

Other processes contribute to the structure and organization of air exchanges:

- *economic competition between companies or partnerships within* "alliances", of which the most well-known are Sky TEAM, Star Alliance and One World. These alliances organize the division of air networks between various companies and offer more destinations to passengers. The economic logic of these alliances develops consumer loyalty through various programs providing specific advantages (such as guaranteed reservations, seat choice, and frequent-flyer programs);
- governorship within each company, which centers activities on air courses and higher added-value activity sectors. Another type of governorship is produced by company mergers, which create groups on an international scale that are able to maintain a strong position in a greater number of markets;
- development of airports and definition of hubs and spokes through agreements between companies and airports to concentrate decreased traffic or average traffic ranges to supply long-distance routes more regularly. The major factors in a company's choice of airport are dedicated piers, reserved departure gates, the number of takeoffs and increased landings.

Thus, the exchanges are actually organized in a reticular form where companies charge for airport stops, redistributing passengers according to partnership agreements between companies and between airports and companies. From any place in the world, one stops at different airports to connect to the dominant network, which in turn leads to stops at other airports to reach the desired destination. Thus, many airports function as successive relays between every city on the planet, and the great intercontinental exchanges simply reinforce the dominant network.

The network structure, which was hitherto subject to the will of the cities and countries, is founded on supervised company strategies. For the most part, companies are private, and their networks are organized to be as powerful and profitable as possible, even to the detriment of a territorial homogeneous service routes. Companies have fewer national loyalties (even if company's national history still appears in its current structure), so the routes planned by companies cross national borders and create a new transnational structuring of equivalent territories (i.e., those connected to the same focal point or hub). Thus, the territorial logics that prevailed prior to deregulation are becoming obsolete, and new territorialities are emerging.

The objective of our approach is to emphasize these "new reticular territories", which are defined by an inter-connected overhead grid that divides the world into various levels of road service through mandatory stops in some hubs. The star-like form of the networks, a consequence of deregulation of the air industry, supports the emergence of a system comprising multiple air platforms. The dynamic space of the center-periphery type is reinforced in these networks. With the emergence of secondary centers (Seoul, Fort Lauderdale) (Amiel, 2005), the great centers have often become more powerful (New York, London, Chicago, Paris).

9.2.3 "Small World" Form of the Air Network

Thus, the world air network has specific "small world" properties. It is necessary, for example, to take 15 different flights to go from Mount Pleasant (in the Falkland Islands) to Wasu (New Guinea): this is the largest short path in the world air network (Guimerà et al., 2005). We can now see how "small worlds" are created in this network. In fact, a high degree of connectivity exists between selected airports so that short paths can be found between airports, but there are also numerous peripheral airports that are very far from the entire network.

In particular, the different continents of the world are especially "coherent" in the sense that airport traffic can create some very strong intertwined systems of exchanges compared to sparser inter-continental links.

9.3 Clustering Methods for Air Transport Network

The methodologies applied to the air traffic network can be classified into three families:

(a) Q value methods maximizing internal densities within groups;
(b) methods based on the Betweenness centrality index;
(c) common neighbors methods, including Jaccard or strength methods.

These methods are described below in the context of non-weighted graphs (classical cases) and in the context of weighted graphs that we have developed.

9.3.1 The Q Value Method

The Q value method was developed by Newman (2004). Given a graph $G = (V,E)$, let $C = \{C_1, C_2, \ldots, C_p\}$ denote the partitioning of this graph into *clusters* $C_1, C_2, \ldots, C_p$. Also let $|V|, |E|$ denote the number of nodes and edges in G, $|E(Ci)|$ denote the number of edges with extremities in the cluster C_i, and $|E(Ci,*)|$ denote the number of edges with at least one extremity in the cluster C_i.

The *modularity measure* Q of the decomposition C is defined as $Q = \Sigma_{1\leq i\leq p} Q(C_i)$ where $Q(C_i) = \frac{|E(C_i)|}{|E|} - (\frac{|E(Ci,*)|}{|E|})^2$.

The weighting of the edges of a graph by a weight function $\omega : E \to \mathbb{R}$ can be accounted for by setting $Q(C_i) = \frac{\Sigma_{e\in E(C_i)}\,\omega(e)}{\Sigma_{e\in E}\,\omega(e)} - (\frac{\Sigma_{e\in E(Ci,*)}\,\omega(e)}{\Sigma_{e\in E}\,\omega(e)})^2$.

The modularity value $Q(C)$ is used to judge the relevance and efficiency of a decomposition in finding "natural clusters". In particular, when C' is obtained from a decomposition C by merging two clusters C_i, C_j, the modularity values $Q(C)$ and $Q(C')$ can be used to judge whether the decomposition C' is "better" than C (with respect to Q). Let $\Delta_{i,j}Q = Q(C') - Q(C)$ denote the increase in Q of this merging operation on clusters C_i and C_j (which may be negative).

Newman's method is agglomerative, i.e., we begin with a decomposition C such that clusters correspond to singleton sets containing a single node; larger clusters are then formed by aggregating smaller ones.

1. Begin with a decomposition $C = (C_1, C_2, \ldots, C_{|V|})$, where each cluster C_i contains a single node (and conversely, each node is included in a cluster).
2. Find the two clusters C_i, C_j with the largest $\Delta_{i,j}Q$ values; merge these two clusters.
3. Step 2 is repeated until a single cluster decomposition $C = (V)$ (containing all nodes) is reached.
4. Return to the decomposition produced from a cut in the aggregation tree that maximizes Q (i.e., the best cut in terms of Q).

Step 3 requires some explanation. The merging process may be described by a tree to denote which clusters are merged and how clusters are nested. Each cluster occupies a given level, depending on how far the cluster is from the root cluster $C = (V)$. A *cut* is obtained by selecting all the clusters at a given depth, for example, d. The tree structure is designed such that clusters of a cut induce a decomposition $C = (C_1, C_2, \ldots)$. Step 3 thus consists of selecting the cut with the maximum modularity among all possible cuts.

9.3.2 The Betweenness Centrality Method

The betweenness centrality method was developed by Newman and Girvan (2004) based on edge betweenness centrality (see Brandes, 2001).

Let $G = (V, E)$ be a weighted graph: for the example considered, an integer attribute is assigned to the edges that correspond to passenger traffic. We then compute the weight of an edge by taking the inverse of the number of passengers. Low weights correspond to a smaller "distance" between the nodes. The weights may then be used to compute weights for paths in the graph by summing the weights of the edges along a path. The (weighted) betweenness centrality metric of an edge e represents the number of weighted shortest paths containing this edge. A low $BC(e)$ value suggests that e is likely to be in a community, whereas a high $BC(e)$ value suggests that e is a mandatory passageway connecting distinct communities.

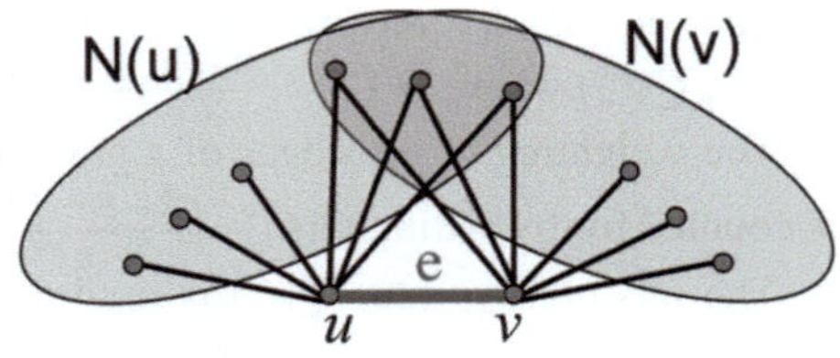

Fig. 9.1 Neighborhoods in the Jaccard method

In contrast to Newman's modularity method, the betweenness centrality method is *divisive*. The process starts with a decomposition containing a single cluster $C = (V)$ that is split into smaller clusters by removing edges before iterating on the newly created clusters.

1. Begin with a single cluster decomposition $C = (V)$.
   (a) Compute betweenness centrality on the edges.
   (b) Remove the edge with the maximum value. If edge removal disconnects a cluster into two smaller clusters, then update the decomposition.
2. Repeat steps 1(a) and 1(b) until there are no more edges to remove.
3. Return the decomposition induced by a cut in the aggregation tree that maximizes Q (i.e., the best cut in terms of Q).

This divisive process may be again encoded using a binary tree to indicate the nesting of the clusters. Additionally, the edge removal operation requires that the betweenness centrality of the remaining edges be updated.

9.3.3 *The Neighbors Method*

The Jaccard and strength methods were developed by Auber, Chiricota, Jourdan, and Melançon (2003) based on the extension of a metric designed by Jaccard (1901), which was later generalized to weighted networks (Amiel, Rozenblat, & Melançon, 2005).

The (weighted) "Jaccard metric" of an edge represents the (weighted) neighborhood similarity of the extremities of the edge. Assume that we are given a weight function on both nodes and edges, i.e., we are given the values $\omega(v)$ and $\omega(e)$ for all nodes $v \in V$ and edges $e \in E$ of a graph $G = (V, E)$. In our example, the weight of a node would typically be the passenger traffic going through that node.

Let $\omega(v)$ denote the weight of the node v (passenger traffic going through node v). Recall that $N(v)$ and $N(u)$ denote the neighborhoods of v and u, respectively, in G (see Fig. 9.1).

The weighted "Jaccard metric" of an edge $e = (u, v)$ is $jac(e) = \frac{\sum_{n \in N(u) \cap N(v)} \omega(n)}{\sum_{n \in N(u) \cup N(v)} \omega(n)}$. A high $jac(e)$ value suggests that e is likely to be in a community. The Jaccard method is also a divisive method that maximizes Q.

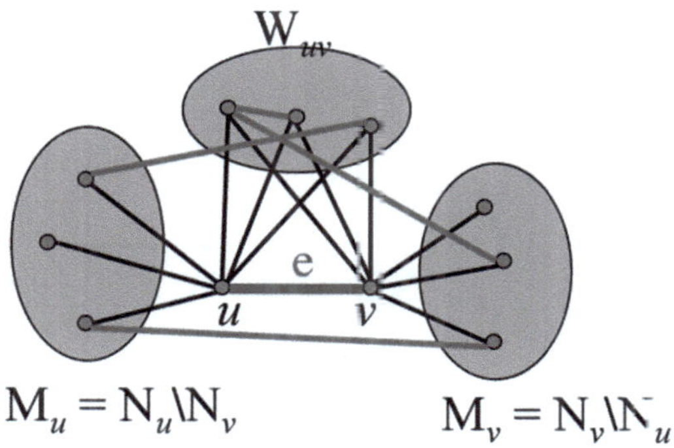

Fig. 9.2 Neighborhoods in the (Auber et al., 2003) strength clustering method

W_{uv} are the common neighbors of u and v, $M_u = N_u \setminus N_v$ are neighbors of u but not neighbors of v, and $M_v = N_v \setminus N_u$ are neighbors of v but not neighbors of u (Fig. 9.2).

For this operation, a ratio is calculated based on the number of cycles of length 3 involving an edge e:

$$\gamma_3(e) = \frac{|W_{uv}|}{|M_u| + |W_{uv}| + |M_v|}$$

A similar ratio can be defined for cycles of length 4:

$$\gamma_4(e) = s(M_u, W_{uv}) + s(M_v, W_{uv}) + s(M_u, M_v) + s(W_{uv}, W_{uv})$$

The strength index is then the sum of both indices $\Sigma(e) = \gamma_3(e) + \gamma_4(e)$.

Once again, the operation is a divisive method that maximizes Q. A quality measure is used to neglect edges with low strength values (Fig. 9.3).

This metric completes Burt's network constraint, which measures how much an actor is constrained by its direct neighborhood (Burt, 2000, 2005). In both cases, edges with low values are expected to act as bridges between tighter communities.

9.4 Comparison of Hierarchical Network Analyses

The three methods were compared by constructing datasets to describe the flow of air transport passengers between cities, based on data from ENAC (French National School of civil aviation). This data source adds national and airport sources to the classical OAG source (Official Airline Guide) for completeness. We also aggregated airports by city to reconstruct the actual traffic of each city.

Clustering using the Weighted Quality Measure for the international air-traffic network, whether non-valued (Guimerà et al., 2005) or valued, demonstrates that groups of cities are geographically organized by continents (Fig. 9.4).

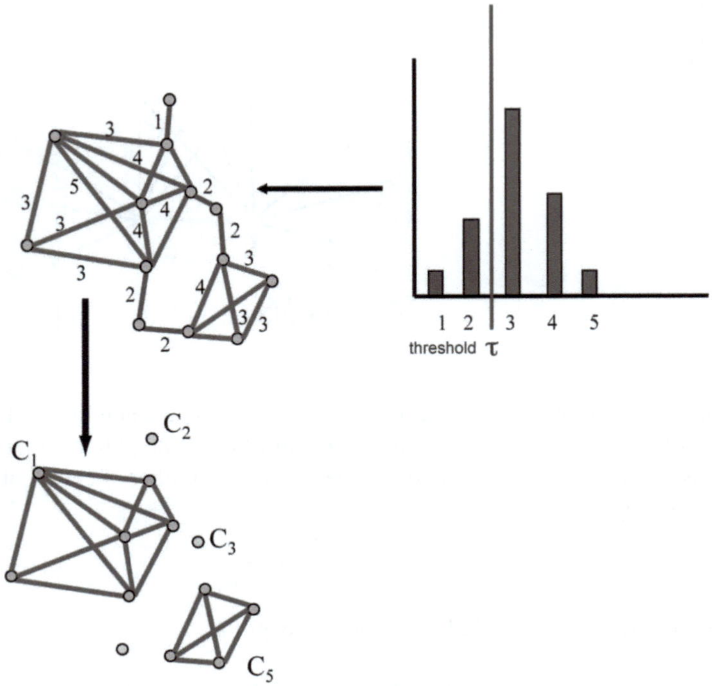

Fig. 9.3 The divisive approach of the (Auber et al., 2003) strength clustering method

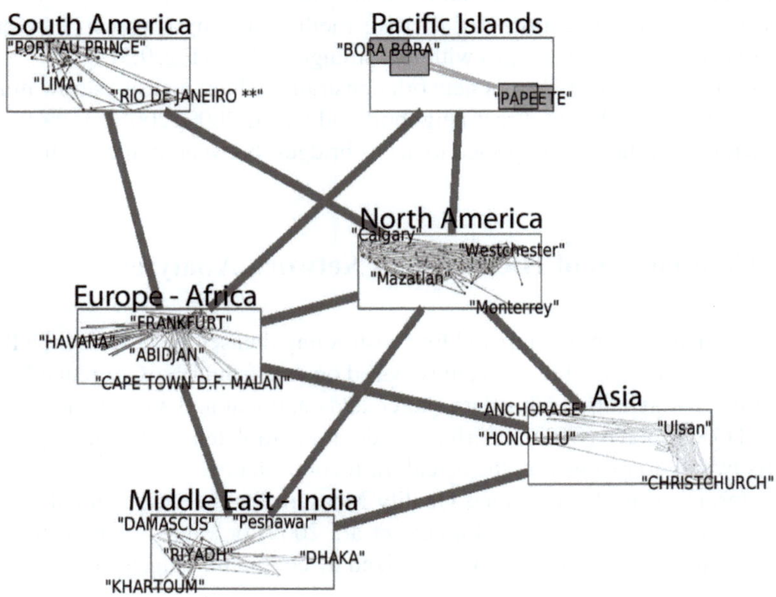

Fig. 9.4 Clustering of Worldwide air traffic using the Weighted Quality Measure Q

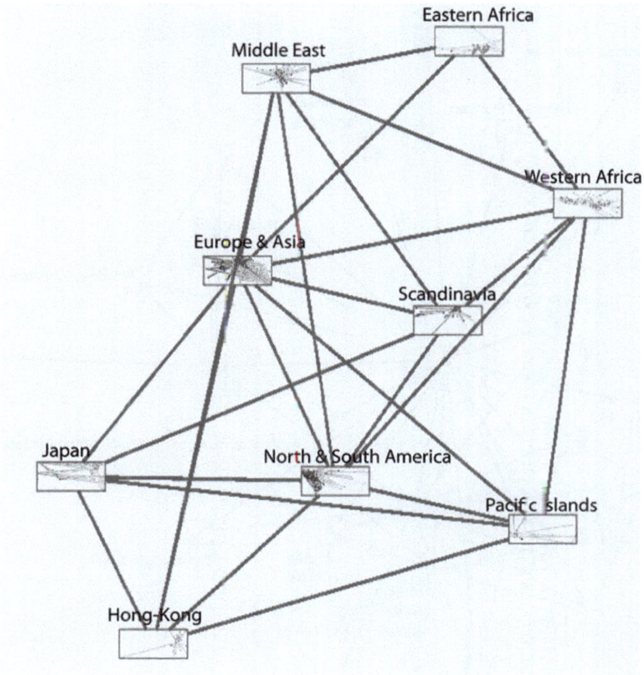

Fig. 9.5 Clustering of worldwide air traffic using betweenness centrality

Clustering using Betweenness Centrality also emphasizes the geographical proximity between cities but introduces three main changes:

1. Some continents are included in the same clusters, such as North and South America, Europe and Asia;
2. These continental groups also include non-continental cities. For example, Jerusalem and Aqaba are included in the Scandinavian group;
3. Other locally centered clusters emerge, such as Hong Kong or Japan (Fig. 9.5).

Clustering using the strength index was implemented via a multi-level approach, fragmenting each cluster into several denser clusters that were subsequently fragmented until the individual nodes were reached (Amiel et al., 2005). At the highest level of world air traffic (Fig. 9.6), the network is organized around a central component called the "principal worldwide hubs". A star topology spreads around this component, emphasizing the role of the intercontinental hubs. Small cores are strongly inter-connected at the same level as this central component. Among these cores, sub-networks with continental logic appear, as is the case with Europe and Asia. This general organization can be specified by descending the levels in the graph. Applying the decomposition algorithm to each component iteratively isolates the levels from the connected network, starting at the most general level and proceeding to the encased sub-graphs of levels 2, 3, etc.

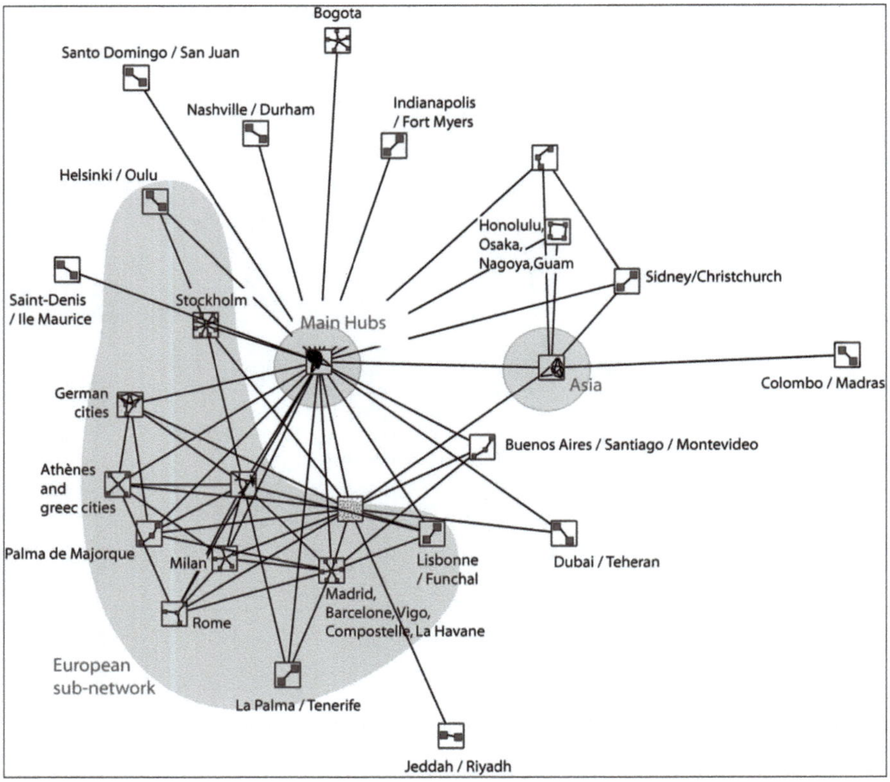

Fig. 9.6 Clustering of worldwide air traffic using the strength index

The cities forming the "principal worldwide hubs" group are American and European (Fig. 9.7). Examining the second, finer level of this area of the graph reveals European hubs such as Paris, London and Frankfurt, as well as the principal American hubs such as New York, Los Angeles, Houston and Miami. These American hubs are dissociated for two principal reasons:

- American hubs are the continental entrances for connections arriving from Europe;
- The majority of American hubs are chosen by large American companies, such as Delta Airlines in Miami or American Airlines in New York (Vowles, 2000).

An American sub-network is dissociated in the center, which can be specified at a third level. In fact, the American territorial extent creates a more complex air organization than does Europe. Indeed, air is the fastest and most effective means of transportation in the United States. This level mainly consists of American cities, but three European cities appear to be particularly connected to these cities: Amsterdam, Dublin and Manchester. This European presence can be explained by the bilateral agreements and alliances between companies, in particular Continental Airlines, which is present at all three airports.

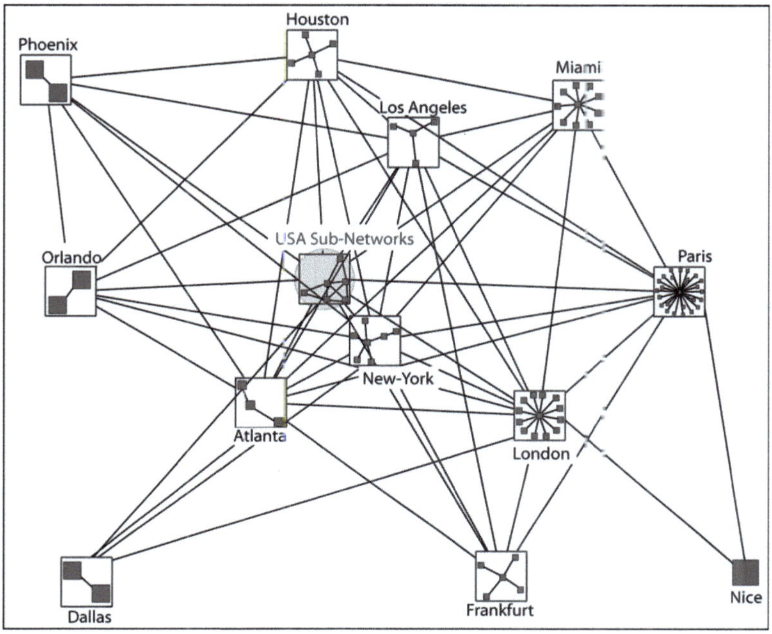

Fig. 9.7 Clustering of worldwide air traffic: focus on main hubs

The central component of the third level of the graph can also be divided into two lower levels, i.e., a fifth level is introduced (Fig. 9.7). This fifth level represents the local organization of the center and the western North American area. These are secondary areas, which are very dependent on the connections between them and are, at the same time, deeply "embedded" in the American hub system (levels 2 and 3) that is dominant at the world level (level 1). The strong connections between the secondary American cities and the world level arise from the characteristic organization of American airspace, which has been hierarchical, from very early on, in terms of three main hub categories:

- traditional "gateways" (Los Angeles, Chicago, Boston, and San Francisco);
- principal hubs (Philadelphia, Atlanta, Houston, and Dallas); and
- secondary hubs (Minneapolis, Denver, Phoenix, Cincinnati, St Louis, Tampa, and Las Vegas) (Connor, 2003; Weber & Williams, 2001).

This hierarchical organization integrates American secondary cities particularly well into the world air network.

The position of all these cities is thus at the core of air world traffic. The encapsulated shape of these graphs shows a highly integrated system, primarily on the part of the United States territory.

In comparison to the North American system, the Asian and European systems appear less integrated into the world system. However, while the European cities are

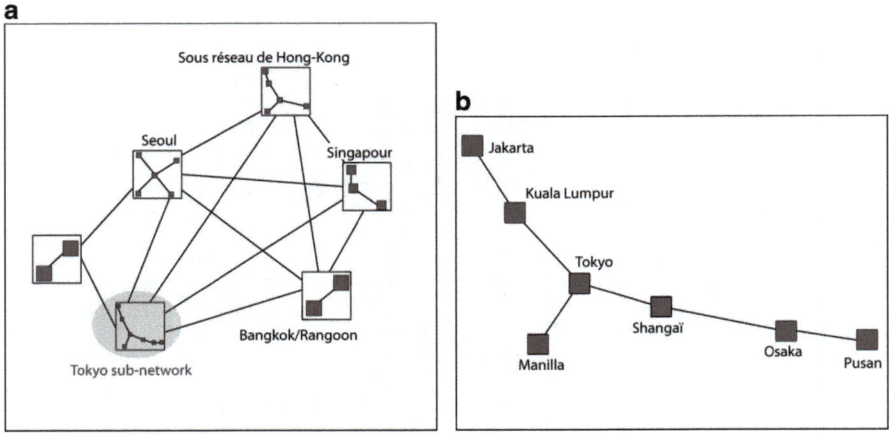

Fig. 9.8 Clustering of Worldwide air traffic: focus on Asian cluster

not well-integrated into the system, with the hubs displaying a burst structure in the form of various subgroups, the Asian system is very connected and forms a group with whole market share (Fig. 9.8).

The Asian system can be broken down at a second level (Fig. 9.8a). In particular, this system is organized around two poles, Hong Kong and Tokyo. Decomposition at a third level, around the Tokyo pole, shows an international system that is strongly integrated around the Japanese capital: the system spans Jakarta to Pusan while passing through Shanghai (Fig. 9.8b). The insularity of the Asian area partly justifies the strong integration of the regional network. Moreover, the potential and the demand for air transport in Asia involve increasingly strong competition between airports and cities (Park, 2003).

At the other end of the spectrum, the European system appears to be less well-integrated from Stockholm to Lisbon, undoubtedly because of the complex history of the airspace burst structure. Indeed, national companies, under pressure from the states, organized hubs for each large European capital. It was only with the deregulation of the European skies in 1993 that the management experience of the European air routes was transferred to private airlines (Burghouwt, Hakfoort, & Ritsema van Eck, 2003). Despite the efforts of the European Community to build a coherent airspace, great competition still exists between the principal European hubs, such as London, Paris, Frankfurt and Amsterdam (Graham, 1998).

Some cities, although strongly dependent on the major hubs or the two non-intermitting systems, have few connections between hubs. This is the case for cities in Latin America such as Bogota, Buenos Aires, and Santiago and for cities in the Middle East such as Dubai, Tehran, Jeddah, and Riyadh. There is intense traffic between the large South American capitals and the central network because the majority of the exchanges are with the international network, starting exclusively from the large capitals. These capitals have reproduced the traffic resulting from the neo-liberal policies of airspace openings, producing a star distribution network around the hub and spokes system (Lipovich, 2002).

9.5 Conclusion

The organization of the air transport network represented here reveals an interesting process of spatial integration. The networks are articulated around a central core, consisting of principal hubs where the arrangement of "reticular territories" distinguishes the three organizing zonal sets of the world economy: the United States, Europe and Asia. In the United States, structures are hierarchical, and the various territories are perfectly integrated with each other. Europe is an example of an airspace with very little organization and low levels of hierarchy: there is a multitude of airports and relations at the same level without a single organizing influence, as in the United States. One may assume that this phenomenon will attenuate over the next few years with the new European transportation policy (in particular, through the implementation of the White Paper on the "European transport policy by 2010: the hour of choices").

Attention should be given to the influence of Asian airspace. Indeed, Asia is the area currently showing the greatest growth in air transport. It will be necessary to closely follow the evolution of the Asian sector because one may assume that the principal worldwide hubs group will soon integrate with airports such as Tokyo, Hong Kong and Singapore.

Economies, markets and the internal reorganizations of airline companies are factors supplementing these tendencies. Power transfer from states to private companies is leaving airspace less divided and consequently more inclined toward competitiveness logics. Thus, private strategies have become the dominant organizing influence for the "reticular territories" of air transport. The only regulatory authorities in this system are the international air organizations (IATA, ICAO). With air transport being one of the most competitive sectors in the world, one must expect constant evolution of its networks and an increasingly complex organization of the reticular territories.

From a methodological point of view, the comparison of clustering methods can be classified into two types of final results:

- results that include a large number of geographically proximal areas but do not highlight the hubs in the network;
- results that highlight the denser parts of the networks and the hubs. These methods are interesting but represent the star shapes of the networks very poorly.

Repeating the clustering methods with different datasets shows that neither of these types of methods is satisfying in terms of the processes that need to be delineated: the methods do not operate on the same order or with similar weights. Thus, these types of exploratory methods are limited and could be supplemented in the future by more data-driven methods and the selection of different steps depending on the initial assumptions. These types of data-driven methods were first tested on commuter networks (see Chap. 11) and could be automated to test these methods on different datasets. Thus, the field of clustering is in progress and is open to many future research directions.

References

Amiel, M. (2003). *Le réseau aérien mondial en 2000: Étude empirique*. M.Sc. thesis, Université Montpellier 3.

Amiel, M. (2005). *La restructuration d'Air France, 1991–2001* (Tech. Rep. No. C7PQMR PO2005-01-X). Montreal, Canada: C.R.T., Université de Montréal.

Amiel, M., Rozenblat, C., & Melançon, G. (2005). Réseaux multi-niveaux: l'exemple des échanges aériens mondiaux. *Mappemonde, 79*(3). Retrieved July 21, 2009, from http://mappemonde.mgm.fr/num7/index.html

Auber, D., Chiricota, Y., Jourdan, F., & Melançon, G. (2003). Multiscale navigation of small world networks. In *Ieee symposium on information visualisation* (pp. 75–81). Los Alamitos, CA: IEEE Computer Science Press.

Brandes, U. (2001). A faster algorithm for betweenness centrality. *Journal of Mathematical Sociology, 25*(2), 163–177.

Burghouwt, G., Hakfoort, J., & Ritsema van Eck, J. (2003). The spatial configuration of airline networks in Europe. *Journal of Air Transport Management, 9*(5), 309–323.

Burt, R. (2000). The network structure of social capital. In R. Sutton, & B. Staw (Eds.), *Research in organizational behavior*. Greenwich, CT: JAI Press.

Burt, R. (2005). *Brokerage and closure: An introduction of social capital*. New York: Oxford University Press.

Colizza, V., Barrat, A., Barthelemy, M., & Vespignani, A. (2006). Prediction and predictability of global epidemics: The role of the airline transportation network. *Proceedings of the National Academy of Sciences of the United States of America, 103*, 2015.

Connor, K. O. (2003). Global air travel: Toward concentration or dispersal? *Journal of Transport Geography, 11*(2), 83–92.

Graham, B. (1998). Liberalization, regional economic development and the geography of demand for air transport in the European Union. *Journal of Transport Geography, 6*(2), 87–104.

Guimerà, R., Mossa, S., Turtschi, A., & Amaral, L. A. N. (2005). The worldwide air transportation network: Anomalous centrality, community structure, and cities' global roles. *Proceedings of the National Academy of Sciences of the United States of America, 102*(22), 7794–7799.

Jaccard, P. (1901). Distribution de la flore alpine dans la bassin de dranses et dans quelques régions voisines. *Bulletin de la Société Vaudoise des Sciences Naturelles, 37*, 241–272.

Lipovich, G. (2002). Cielos Cerrados en el Mercosur. La Integra(domina)ciòn Regional en el Transporte Aéreo. In *Cuarto Encuentro International Humboldt*. Puerto Iguazu, Argentina.

Newman, M. E. J. (2004). Fast algorithm for detecting community structure in networks. *Physical Review E, 69*, 066133.

Newman, M. E. J., & Girvan, M. (2004.). Finding and evaluating community structure in networks. *Physical Review E, 69*, 026113.

Park, Y. (2003). An analysis for the competitive strength of Asian major airports. *Journal of Air Transport Management, 9*(6), 353–360.

Sales-Pardo, M., Guimera, R., Moreira, A. A., & Amaral, L. (2007). Extracting the hierarchical organization of complex systems. *PNAS, 104*(39), 15224–15229.

Vowles, T. M. (2000). The geographic effects of us airline alliances. *Journal of Transport Geography, 8*(4), 277–285.

Watts, D., & Strogatz, S. (1998). Collective dynamics of "small-world" networks. *Nature, 393*, 440–442.

Weber, M., & Williams, G. (2001). Drivers of long-haul air transport route development. *Journal of Transport Geography, 9*(4), 243–254.

Chapter 10
Multilevel Analysis of Corporations Networks: A Comparison Between Agro-Food and Automobile Strategies for Urban Development

Charles Bohan and Bérengère Gautier

10.1 Introduction

The integration of territories is one of the major objectives of the convergence policies that Europe encourages in order to stabilize its borders and make them safer. In the Mediterranean region, as well as in Central and Eastern Europe, states use the structural funds that they provide for development to support policies that make the territories more attractive to multinational firms. By diffusing capital, information and expertise on a worldwide scale, firms became the principal driving force of development processes in the European Neighborhood Policies.

Most of the current studies on the economic integration of territories is based on the scale of national economies. However, in the context of the international division of labor and of the internationalization of the exchanges under the impulse of the NTIC, companies develop networks based on complementary and localized territorial competences, such as the local factors of attractivity and the faculty of networking production processes. Then, the networks of multinational firms seek locations that can enable them to achieve economies of scale and economies of networks (Doz, Santos, & Williamson, 2001; Rozenblat, 2004; Veltz, 2000). It is necessary to integrate this reticular dimension of the integration of cities in globalization. On the transnational scale, the firms stimulate economies of agglomeration (localization, urbanization and economies of scale) for cities and create original developments within each of them, thus reinforcing their position within the system

C. Bohan (✉)
IGD, University of Lausanne, Lausanne CH-1015, Switzerland
e-mail: charles.Albert-Bohan@unil.ch

B. Gautier
Université de Paris I Panthéon-Sorbonne, CNRS UMR 8504 Géographie-cités/Equipe P.A.R.I.S., rue du Four 13, Paris 75006, France
e-mail: berengere.gautier@parisgeo.cnrs.fr

C. Rozenblat and G. Melançon (eds.), *Methods for Multilevel Analysis and Visualisation of Geographical Networks*, Methodos Series 11, DOI 10.1007/978-94-007-6677-8_10,

of cities (Camagni, 1996; Rozenblat, 2004, 2010). The relationship between the networks of firms and the system of cities seems relevant for understanding this global/local articulation and the developments, which result from the interaction of intra-urban and inter-urban processes.

At the global scale, we distinguish two major types of organization strategies: a market-oriented organization (horizontal strategy) and a production-oriented organization (vertical strategy) (Porter, 1986). The market-oriented organization is based on a network of subsidiaries that are established within large areas of mass consumption and production and that adapt to the specific assets of local markets and of labor markets. This strategy is often planned on a continental scale and is observed in free trade areas. A production-oriented strategy exploits the specializations of various geographical areas in order to establish a different stage of its production in each area, based on its comparative advantages (Veltz, 2000). Such a strategy is responsible for the international division of labor. However, firms do not rigorously apply these two organization patterns, and a majority of large firms combine these two aspects, mixing strategies of production and geographical organization (Michalet, 1999; Rozenblat, 2004; Veltz, 1998).

Their strategies are organized in a three-dimensional system that is formed by the economic system to which each firm belongs, its governance or system by which the company is governed, and the territories in which it is embedded. The economic system corresponds to the external environment of the firm (Francfort, Osty, Sainsaulieu, & Uhalde, 1995). "This environment is based on more or less monopolistic competition, the technological projections and cycles of the products, imbalances between supply and demand (gone, price), of the fluctuations of the products value" (Rozenblat, 2004, p. 29). Within the economic system, the position of the invested firms and territories are constantly redefined. The governance of the firm is "the whole of the strategies and objectives of the companies that make it up, the human means and materials, which it implements, its organizational architecture, and its own managerial culture" (Rozenblat, 2004, p. 13). Taking an international perspective, the governance of the firm determines the relationships among its various establishments: it introduces hierarchical and functional relationships between territories. Within this system, the territory becomes a dynamic object, and it is therefore important to articulate the global and local dimensions in which each territory and each firm fits.

The economic literature on the agro-food and automobile sectors reveals that multinational companies in the agro-food sector rely more on market-oriented strategies, while multinational firms in the automotive industry primarily seek locations that minimize their production costs. The networks of multinational firms of the agro-food industry should therefore reinforce the existing urban hierarchies. Conversely, according to cost minimization strategies, the firms of the automobile sector develop their strategies on a worldwide scale. Such strategies should therefore be based on urban specializations that allow them to achieve important economies of networks by complementarities between locations. Thus, they play a larger role in the international division of labor.

To highlight the role of each economic sector in the integration of territories, we would like to demonstrate the role of each of these economic strategies in the articulation of the network of cities. In the first part of this paper, we will describe the governance of six multinational firms, starting from their individual network of filiations. From this individual firm network, we will collect measurements for each firm. These data enable us to compare individual networks of firms and to observe their strategies. In a second section, we will produce a visualization of the cities' networks based on the aggregation of the subsidiary networks of the firms. This stage enables us to observe their territorial strategies and to replace each city in the context of the global strategies of the firms. Finally, we compare our results for the automotive and agro-food sectors by comparing the cities' networks as formed by the main companies. Such graphs enable us to visualize and to compare the sector's strategies according to their inter- and intra-urban dimensions. This comparison will allow us to identify the intra-urban links that form the basis of the influence that each city has in the automobile and agro-food networks of cities.

10.2 Constraints of Economic Sector and Firm Governance

10.2.1 Two Economic Sectors

Up until now, international research on urban systems has been very difficult because of the lack of data relating to networks (Rozenblat & Pumain, 1993; Veltz, 2000). The ORBIS database (Bureau Van Dijk), which was created in 2003, however, now allows analysis of information on companies at a global scale (company names, location, level and nature of filiations, turnover, number of employees, etc.). Before explaining our procedure for analysis and visualization, we will provide an overview of the companies that we have selected to allow a comparative analysis. We have chosen to study 6 multinational firms in two different sectors: the agro-food industry and the motor industry.

10.2.1.1 Agro-Food Industry

The agro-food industry is a diversified economic sector that encompasses myriad sub-sectors (meats, dairy products, cereals, wine and spirits, milk, water, etc.), where small companies interact with large companies. Through the study of multinational firms in the agro-food industry, we focus on these "giant" enterprises. These enterprises are a product of the agro-industrial model, which has been growing since the beginning of the century, and is based on a model of mass consumption. The agro-industrial model of production is described as "specialized, concentrated, financialized and in the process of globalization" (Rastoin, 2000). However, food

Table 10.1 Selected firms in the agro-food industry

	City headquarter	Date of incorporation	Turnover billions US$	Number of employees
Nestlé SA	Vevey (Switzerland)	1866	69,709	250,000
Unilever Group	Rotterdam, London	1930	46,801	212,000
Kraft Foods	Chicago	1903	29,531	97,850

remains strongly anchored in cultural practices, which slows down the export of the model of Western food consumption. The ability of a firm to adapt to each mass consumption area constitutes an important factor of competitiveness for firms in a highly competitive sector. This sector is dominated by three multinational corporations (Table 10.1).

Nestlé S.A

Nestlé S.A, founded in 1866 in Switzerland, is the leading global agro-food company. With 250,000 employees throughout the world, distributed in 104 countries, the firm generates 70 billion dollars in sales each year. The firm acquired leadership in the sectors of water, ice cream, and baby food as a result of its innovation, which is the main competitive advantage of the firm. Nestlé has a strong policy of repurchasing local brands, allowing it to acquire local expertise and providing great adaptability in the international markets.

Unilever Group

Unilever was created in 1930 by an association between the Dutch margarine company UNI and the English soap manufacturer Lever Group Bothers Ltd. Since then, the group has had a bicephalous management, with Unilever PLC in the United Kingdom and Unilever NV in the Netherlands, and has diversified income due to its presence in both the hygiene and food markets. Consequently, Unilever profits from an important list of brands, which has allowed the company to establish in 106 countries and to become the second leader of the agro-food industry. The firm has acquired leadership in margarine, tea and olive oil.

Kraft Foods

In 1903, Kraft foods was just a creamery located in Chicago. The store opened its first cheese factory in 1914. The firm was then developed by a series of acquisitions and has acquired leadership in the sectors of coffee, chocolate, snacks and cookies.

Table 10.2 Selected firms in the motor industry

	City headquarter	Date of incorporation	Turnover millions US$	Number of employees
Toyota MC	Japan, Toyota city	1937	186,177	285,977
Peugeot SA	France, Paris	1890	83,193	211,700
Fiat S.p.A	Italy, Torino	1899	57,965	173,695

10.2.1.2 Motor Industry

On the economic scene, the motor industry is a tightening oligopoly, increasingly prone to fusions/acquisitions (F&A) with repurchases and alliances, and comprises only 14 main manufacturers (up to one billion sales per year, OICA, 2010). The industry is characterized by an extremely competitive climate at the international level, though there can be strong cooperation in local markets (Co-opetition model) where joint ventures are frequent (Toyota/Peugeot Citroen Automobile (TPCA), PSA/FIAT SevelNord, etc.).

We have chosen three multinational groups from the motor industry that have clearly differentiated strategies (Table 10.2). Toyota, PSA and FIAT have completely different modes of governance, based on their own histories and corporate cultures as well as their strategic choice of investments and localizations throughout their development.

Toyota Motor Corporation

Toyota Motor Corporation is now the world's leading car manufacturer, ahead of General Motors, in terms of sales (OICA, 2010). However, the group has only recently become the world's leading producer in terms of produced car units. Toyota is a very diversified group, whose activities range far beyond the sale of automobiles. Toyota produces automobile parts, components and accessories for its own needs and for other manufacturers. The strategy of Toyota is unique compared to its competitors in terms of production and locations. The group was first internationalized in the United States and then in China and other countries of Southeast Asia. Toyota has 52 assembly plants in 27 countries.

PSA Peugeot-Citroën

Peugeot-Citroen Group PSA is the second largest European car manufacturer. PSA is the result of the fusion of two French brands, which occurred in 1976. PSA's governance is characterized by its strong centralization in terms of locations and in terms of distribution of capacity to subsidiary companies. The group is engaged in the design, development, production and sale of private cars, commercial vehicles, scooters and motorcycles. The operations of the company include the bank PSA Finance, a transport and logistics group (Gefco) and Faurécia, which is the most important European equipment supplier. The market of PSA is primarily in Europe, but the firm has a strong presence in Latin America and operates in Africa, Asia Pacific and other parts of the world. PSA operates through five principal divisions: cars, automobile equipment (Faurécia), transport and logistics (Gefco), finance and other activities.

Fiat S.p.A

The Italian firm Fiat produces and sells commercial and agricultural vehicles, as well as construction equipment. The group also produces engines and transmissions, automobile components, metallurgical products and systems of production for its own needs and for sales to third parties. However, the group has some other interests such as press and communication. Fiat operates mainly in Europe and North America by means of an important network of subsidiary companies. Fiat has 12 operational sectors: Fiat Auto; agricultural and construction equipment (Box New Holland CNH); trucks (Iveco), component cars (Magneti Marrelli); systems of production (Comau); Ferrari; metallurgical products (Teskid); services (Business solutions); Maserati; editions and communication (Itedi); Fiat powertrain technologies and other companies. Fiat holds and uses 11 assembly plants (6 in Italy, 1 in each of Poland, Brazil, Argentina, India and Serbia) and participates in joint ventures with other car manufacturers.

Our analysis of the strategies of multinational firms largely focuses on the reticular field of firms' networks. Creating graphs makes it possible to compile a mass of information (labels) and allows a novel visualization of a city's place in the firm's network. Our analysis is comparative so that we can examine the way the various economic strategies articulate the territorial properties of the world's cities.

10.2.2 Comparison of Individual Firm's Networks

To connect the various components of a firm, we initially create graphs of individual corporate networks. The method consists of representing filiations of the companies, starting from their financial relationships. In a hierarchical way, these graphs are produced from the database ORBIS (Bureau Van Dijk) and are supplemented with information from firms' annual reports. These graphs enable us to observe the distribution of the financial relationships between each element of the multinational firm. Hereafter (Figs. 10.1 and 10.2), the nodes represent the companies and the edges denote the ownership relationships among them. We used the metric of betweenness centrality to highlight the centrality of each subsidiary in the governance of each group.

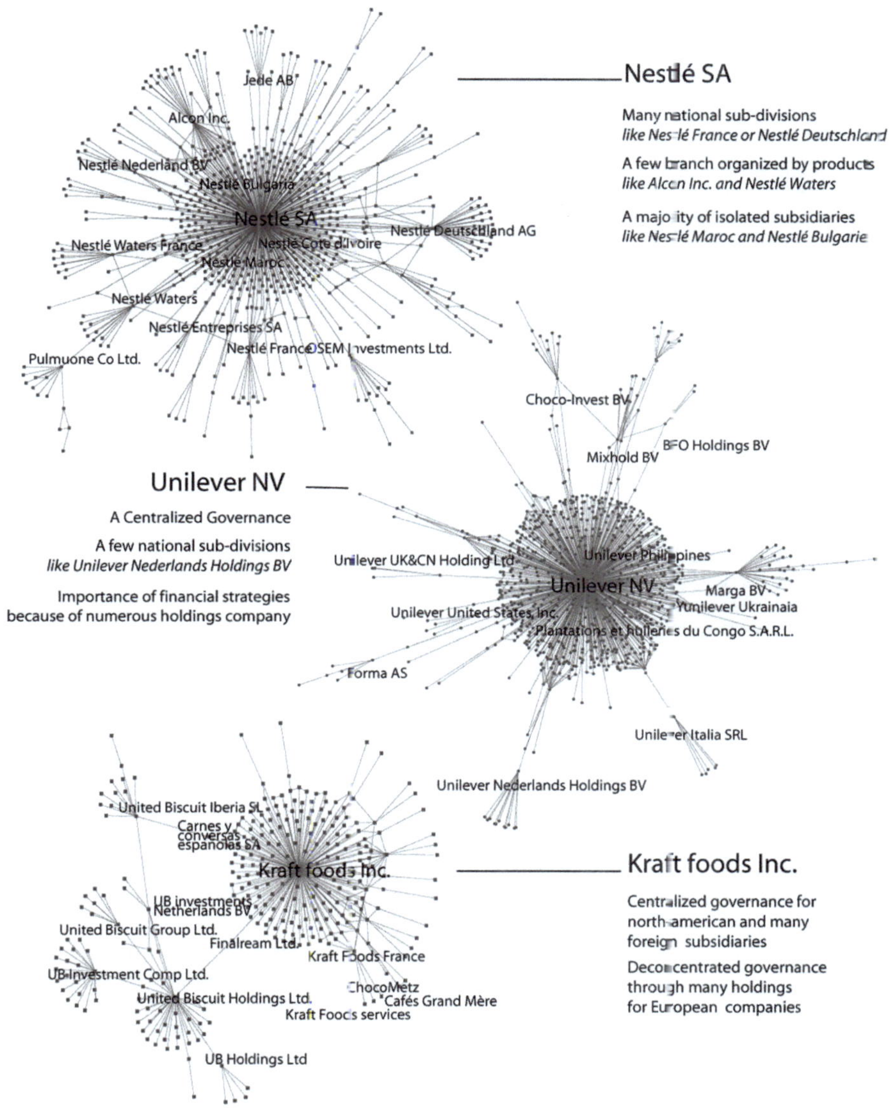

Fig. 10.1 Corporate governance in the agri-food sector (2005) (Source: ORBIS, 2005 – Gautier B. & SPANGEO @ IAMM, 2007)

10.2.2.1 Governance of Networked Companies

The network of a firm is a complex object, which deserves a graphic visualization tool to allow economists, managers, geographers or others to analyze its structure and evolution or to compare it to other networks. The governance of a firm varies

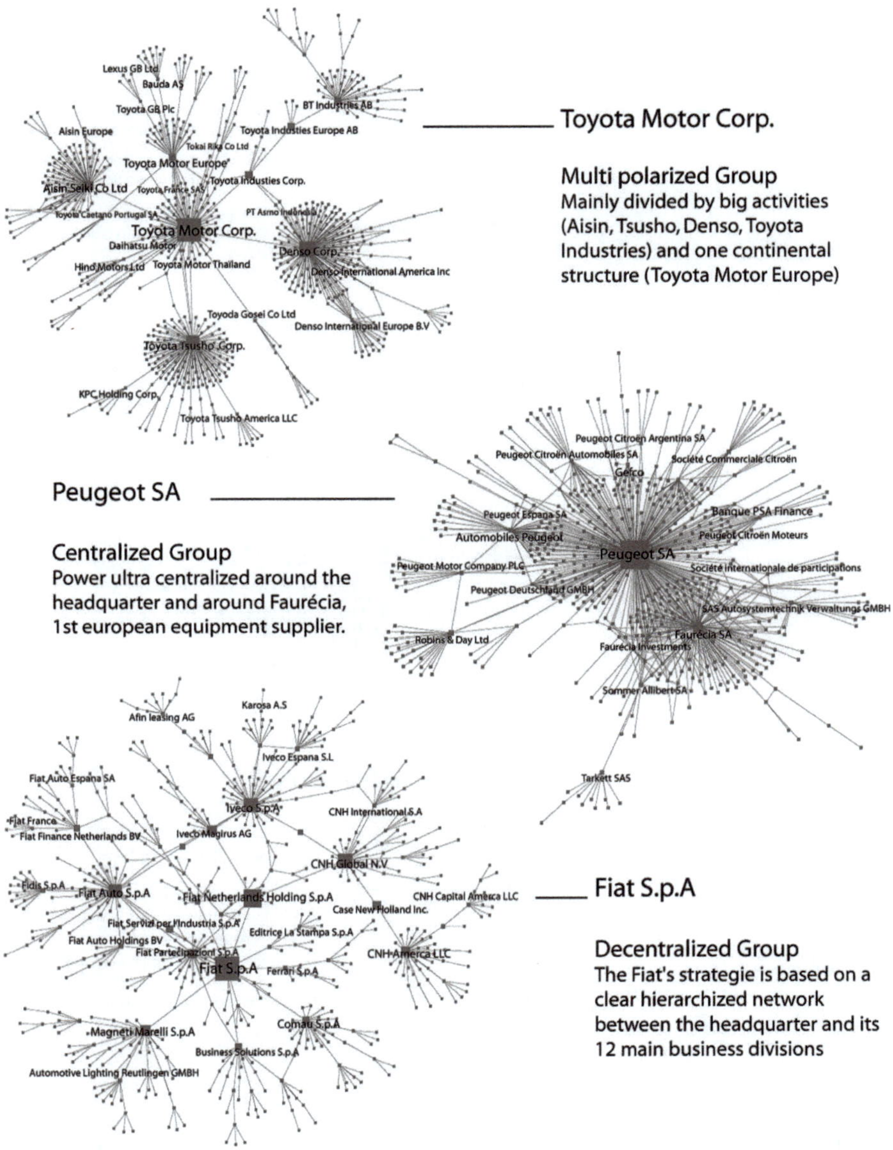

Fig. 10.2 Corporate governance in the automotive sector (2006) (Source: ORBIS, 2006 – Bohan C. @ Cefres, 2007)

according to its history, the corporate culture, its strategic decisions (such as acquisitions), the transfer of credits, and the establishment of subsidiary companies abroad, which modify the way it treats capacities inside its network on a hierarchical basis. Thus, the network of a multinational firm is an object in constant evolution, but we consider the actual shape as the legacy of past actions. The various forms that a network can take reveal the particular strategies according to its organizational structure, its localizations or the control of its capital and employees (Figs. 10.1 and 10.2).

The networks of the multinational food companies studied in this research show a strong centralization of power around the headquarters of Nestlé, Unilever and Kraft Foods (Figs. 10.1 and 10.3), where the headquarters owns an isolated subsidiary in each country. There are some branches organized within national spaces in countries that have a large market size. For example, Nestlé Deutschland AG and Nestlé France SA reflect market-oriented strategies with a strong national dimension. There is thus a strong constraint for national food network deployment, which can be interpreted to result from both a strong constraint related to the various markets in which these firms are located, with a strong cultural embeddedness of food consumption, and a constraint to food commodities that have a short life span, with a relatively low export capacity. All of these constraints on the agro-food sector hinder the internationalization of the groups. Thus, the networks of agrofood companies are highly organized according to geographical area. On the other hand, the motor industry networks present various structures. The power is distributed related to the governance organization and history of the firm. The three selected multinational companies are a good examples of different types of network structure. The Toyota group (Fig. 10.2) is mainly divided into five groups (Denso, Aisin Seiki, Toyota Motor Europe, Toyota Tsusho and BT industries). The headquarters holds five semi-autonomous groups that manage clearly delineated activities. The French group PSA is very centralized, except for its equipment group, Faurécia. The headquarters has quasi-direct control of its subsidiaries. In contrast, the Fiat group is very decentralized into twelve main divisions due to its governance structure and its diversification of activities (from agricultural machines to finance holdings). Even if these networks present different types of structure, the subsidiaries are regrouped around cluster activities. This type of structural organization is characteristic of a typical production-oriented governance.

10.2.2.2 *Betweeness Centrality*: Centrality of the Subsidiaries in the Network

The differences between the shapes of the networks can be evaluated by the distribution of the power inside each of them (measured by betweenness centrality) (Fig. 10.3). This metric was defined by Freeman in 1977 (Freeman, 1977, 1979). Betweenness centrality describes the control exerted by an individual on the interactions between the other actors of the network. The more often an individual is on a path that other individuals must use to communicate, the more central the network is Lazega (1998). To compare the distribution of power in the corporate networks, (we recorded) this metric for all of the subsidiary companies in each corporate network.

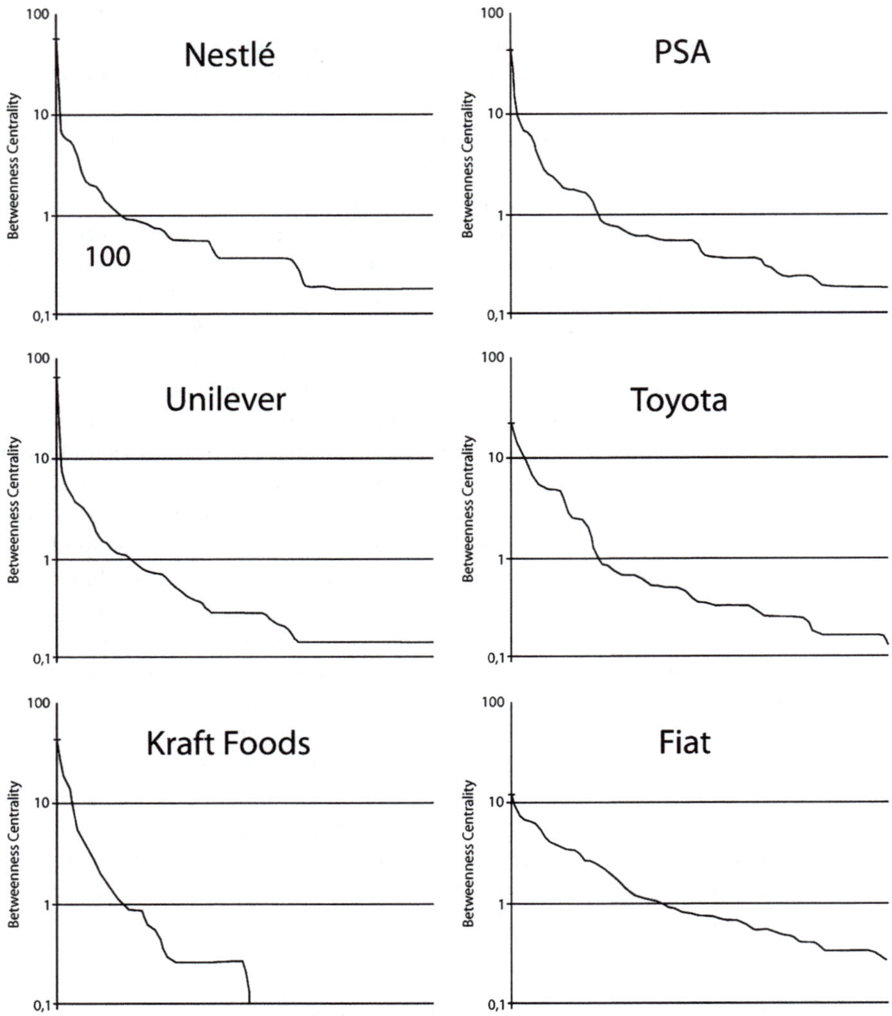

Fig. 10.3 Rank-size betweeness centrality distribution analysis (Source: ORBIS, 2006 – Gautier B., Bohan C. @ IAMM, Cefres, 2007)

These graphics show that the headquarters of the agro-food industry firms are much more centralized than the headquarters of the motor industry firms. Several models of governance are shown through the various graphics. While the agro-food industry presents similar modes of governance with a strong centralization, the strategies vary much more in the motor industry.

The three graphs of agro-food governance generally show similar characteristics, with a strong centralization of headquarters. At the heart of each network, we observe an important group of subsidiaries without any other relationships that bind it to the headquarters. The majority of the subsidiaries belonging to this star formation are in small, often-partitioned national markets (Morocco, Bulgaria, Congo, etc.). Nearby, we notice in each one of these graphs, sub-centers of control corresponding to the following:

- National branches, directing the activity of each subsidiary company in countries or small areas (e.g., Germany, Scandinavia or other countries and regional areas);
- More rarely, branches by product (e.g., Nestle waters, United Biscuit)

The three graphs of governance of the motor industry are different. Fiat and Toyota present strongly decentralized models (respectively, 11.75 and 22.62 % of betweenness centrality held by the headquarters of FIAT and Toyota). These models present branches organized by products at a global scale. These branches have strong connections between each other. On the other hand, PSA and its equipment supplier Faurécia have organized all of their global production around the Parisian headquarters, and hold 53.97 % of the total betweenness centrality of the graph.

Through this sectorial distinction of the governorship modes of the companies, we can distinguish two organizational models:

- An organizational model of agro-food companies, which articulate production-oriented strategies and market-oriented strategies inside each network. The strategies of localization for the subsidiaries mainly follow market dimensions of the territories. The general structure of each of these networks is relatively simple, with a concentration on the most important functions at the top and a weak specialization of the affiliated companies downstream. These mega-structures articulate affiliates involved in micro-strategies that are adapted to the markets, which can be drawn as vertical strategies (global geographical division structure (Dicken, 1992)).
- An organizational model of companies of the motor industry, where the strategies are articulated on a worldwide scale, according to branches organized by products. The networks have diversified morphologies because the strategies of the multinational corporations of the motor industry are complicated on a global scale. The complexity and the diversity of the organizational forms revealed by these graphs, according to a "federal" model (Veltz, 1998), attest to an organization by project, which is accompanied by a strong specialization of subsidiaries (Francfort et al., 1995) (global products division structure (Dicken, 1992)).

A part of the firms' governance is explained by the economic sector to which it belongs. As we can clearly see in these two examples, economic logic drives the governance of a firm depending on the sector. The multinational firms of the agro-food industry usually choose strategies of multi-localization. Indeed, they are constrained by strong market segmentation due to extremely diverse cultural practices. On the other hand, the companies of the automotive sector, compelled by a strong competitive climate at the global level, instead follow a "central logic of the globalization which consists in creating synergies in more and more widened geographical *scales*" (Veltz, 1998, p. 26). The activities within the subsidiary companies will be different: in the agro-food industry, diversified subsidiary companies are adapted to the same regional national markets, whereas in the car industry, subsidiary companies are directly specialized and integrated within a global project. Therefore, the competition between cities is not on the same spatial scale (it is on the regional or national scale in the agro-food industry and on the global scale in the automotive sector).

10.3 Networked Cities Through a Corporate Firm's Network

To ensure their competitiveness, the multinational firms studied in this research divide their activity according to localized territorial skills. These territorialized skills are the result of local networks of cooperation between private and public actors. The local scale allows close relationships between these actors (Storper & Venables, 2004). According to the classic definition of economies of agglomeration (Camagni, 1999; Duranton & Puga, 2004; Ellison & Glaeser, 1997; Henderson, 1988; Hoover, 1937, 1948; Jacobs, 1969; Marshall, 1920; Ohlin, 1933), these effects of proximity allow the companies to realize different kinds of economies:

- Economies of scale, allowing them to increase their productivity;
- Economies of urbanization, related to the collective use of generic services in the heart of the territories (e.g., grid systems, administration and unspecialized services);
- Economies of location, related to the collective use of specific agents (e.g., subcontractors, services, specialized infrastructures).

Economic actors such as multinationals will be based on these economies of agglomeration and will stimulate them, thus reinforcing the capacity and/or the specialization of the territory (Rozenblat, 2004). The company networks use these localized territorial properties and they are articulated in the heart of the network (Doz et al., 2001). The networks of multinationals comprise hierarchical and functional links between territories. Worldwide, the relations between territories follow two simultaneous and articulated plans:

- A hierarchical model that organizes the local market areas, as described by Christaller (1933), which relies more on networks of proximity, facilitated by close markets and regional hierarchies of cities.
- A model of specialization between specialized cities encouraged by long-range relationships defining "economies of archipelagos" (Veltz, 2000).

The approach of the firms' networks, aggregated by cities, highlights the spatial effects of the inherent strategies of the two sectors:

- For the agro-food industry, except for the cities that contain headquarters with broad functions, we observe that most large cities of the large economic regions contain a large number of subsidiaries: Paris and London for Nestlé, London and

Paris for Unilever and London and Amsterdam for Kraft Foods. Furthermore, we notice that the spatial organization of each of these firms creates "cliques" among cities belonging to cohesive economic systems. In every example, the cities belonging to the European Union form a distinct group where the interurban relations are denser than in the remainder of the network. This spatial organization is particularly visible within the group Nestlé (European origin) and to a lesser extent in the governance of the group Kraft Foods (American origin). Worldwide, we observe globally a respect for the urban hierarchy in very different areas of markets.

- On the other hand, the logics of production that we find throughout the automotive sector grant an important place for medium-sized cities: PSA is heavily localized in Coventry or Sarrebruck; Toyota, in the cities of Mjolby and Vantaa; and Fiat, in the cities of Wilmington (which is an international tax haven) and Ulm. More than agglomeration economies, the motor industry requires space and a labor pool.

In the automobile sector, the city to city relationships across national borders (which represent a strong constraint in the agro-food sector) increase the importance of long-range relations, such as Bratislava-Mexico City-Seoul at PSA; Sydney-Kariya-Mjolby-Brussels at Toyota or London-Shanghai-Zedelgem at FIAT. In this production-oriented strategy, relations between specialized cities cross national borders and integrate cities worldwide (Figs. 10.4 and 10.5).

We can thus define various spatial types of organizations:

- Logics of market within the agro-food framework, with relations contained within preset international frameworks. These spatial organizations respect the urban hierarchies and strengthen existing links between cities belonging to the same economic bloc.
- Logics of production of the automotive sector, where the specialized subsidiaries rely on localized territorial skills to strengthen or create relations between distant specialized cities.

According to the economic strategies adopted by the governances of the firms, we see that the firms look for different territorial skills and aim to utilize them in an original way within their network. Now the question is how these different networks affect the territories.

10.4 Competition Between Cities in Corporate Networks

By their choice of location, firms develop and stabilize their competitiveness. They find within territories the useful resources for their development, and through economies of agglomeration, they stimulate those resources. Moreover, economies of agglomeration result from cooperation between economic actors who unite infrastructure and information needs, guaranteeing their competitiveness.

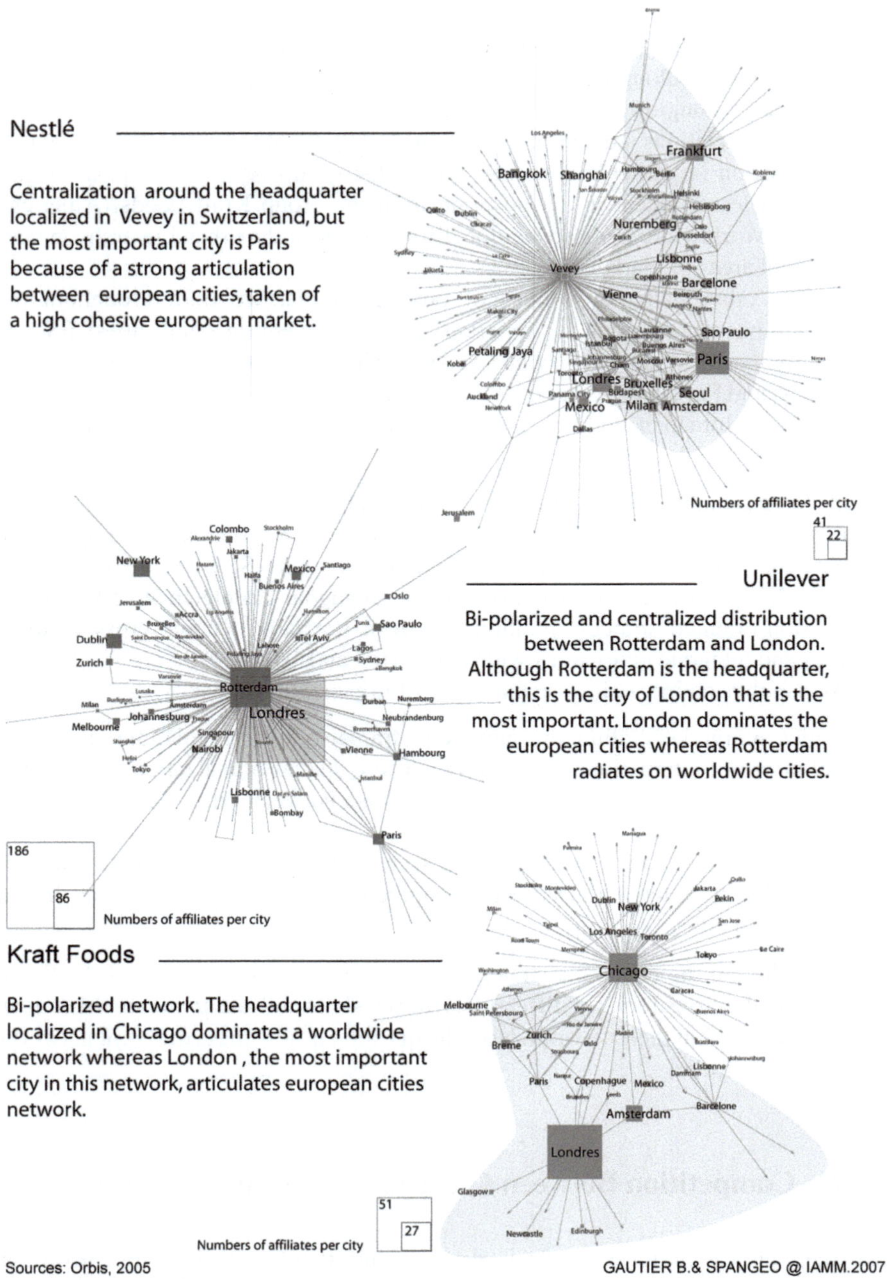

Fig. 10.4 Corporations and their urban networks – agri-food sector (Source: ORBIS, 2005 – Gautier B. & SPANGEO @ IAMM, 2007)

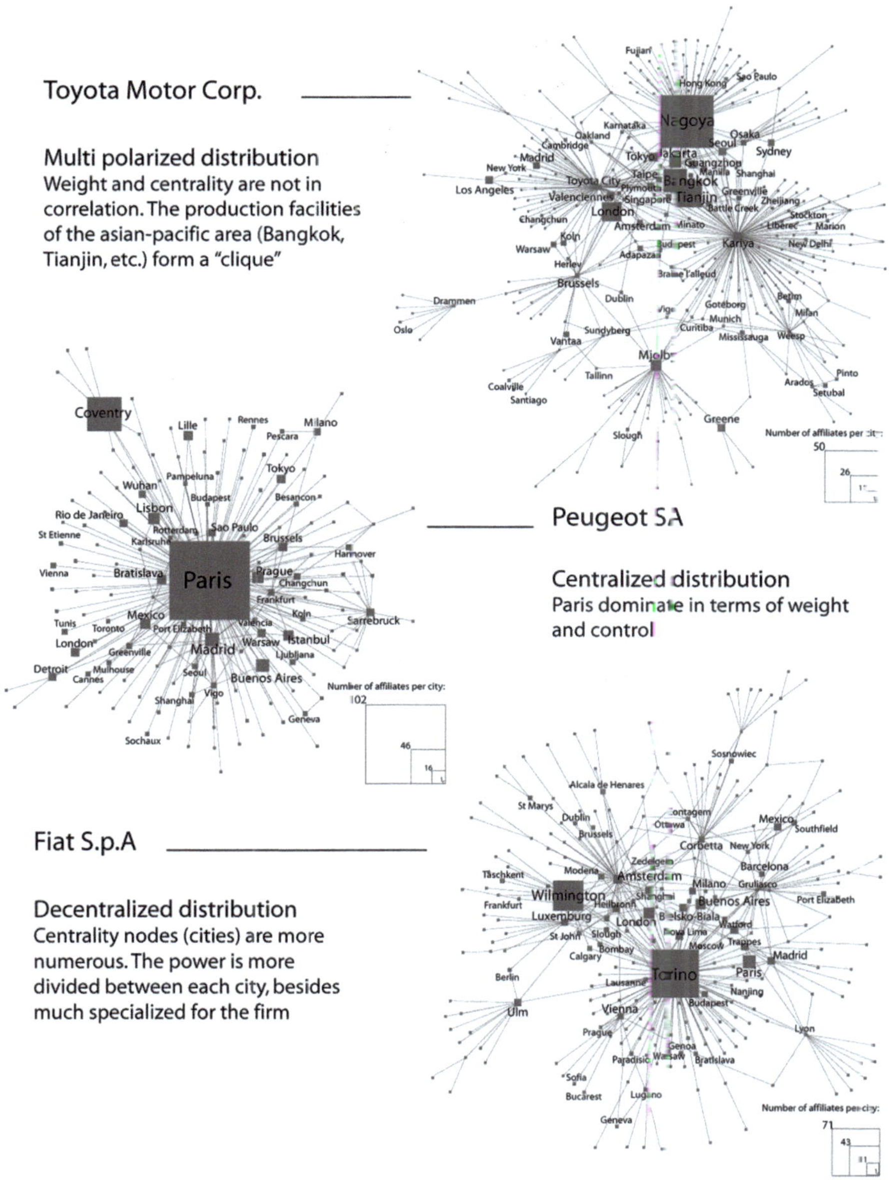

Fig. 10.5 Corporations and their urban networks – automotive sector (Source: ORBIS, 2005 – Bohan C. @ Cefres, 2007)

The territorial embeddedness of the firms assumes a paradoxical aspect: whereas those firms are concurrent on a worldwide scale, they cooperate directly (e.g., resource exchange or exchange of information) or indirectly (e.g., lobbying) within local area networks.

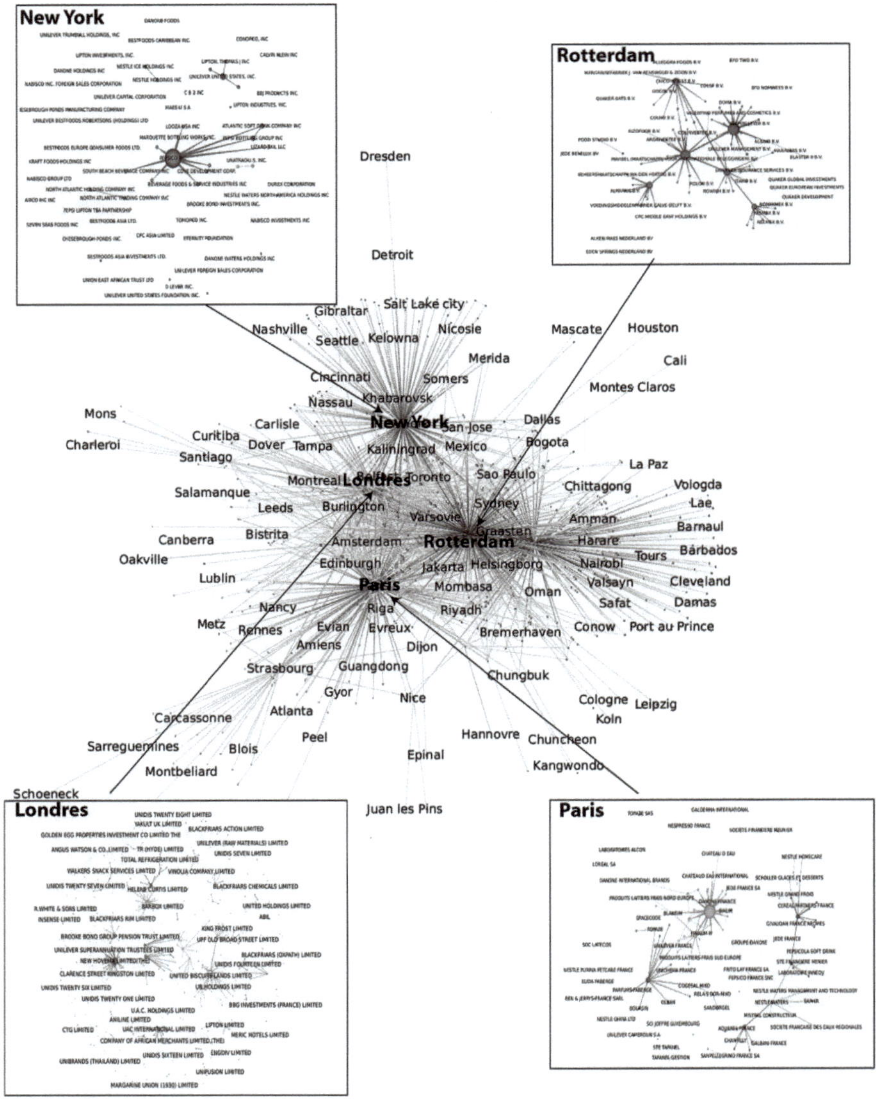

Fig. 10.6 Inter corporate competition through urban areas – urban network formed by Nestlé SA, Unilever, Kraft Foods, PepsiCo and Danone (Source: ORBIS, 2005 – Gautier B. & Rozenblat C. @ SPANGEO, 2007)

To observe this phenomenon, we aggregated all the networks of each economic sector (see Figs. 10.6 and 10.7). Combining the networks of several firms, we finally obtained a single network that represents inter-urban relationships. This kind of graph allows us to visualize not only relations between cities in one particular branch of an industry but also relations within these cities. In this way, every city of the

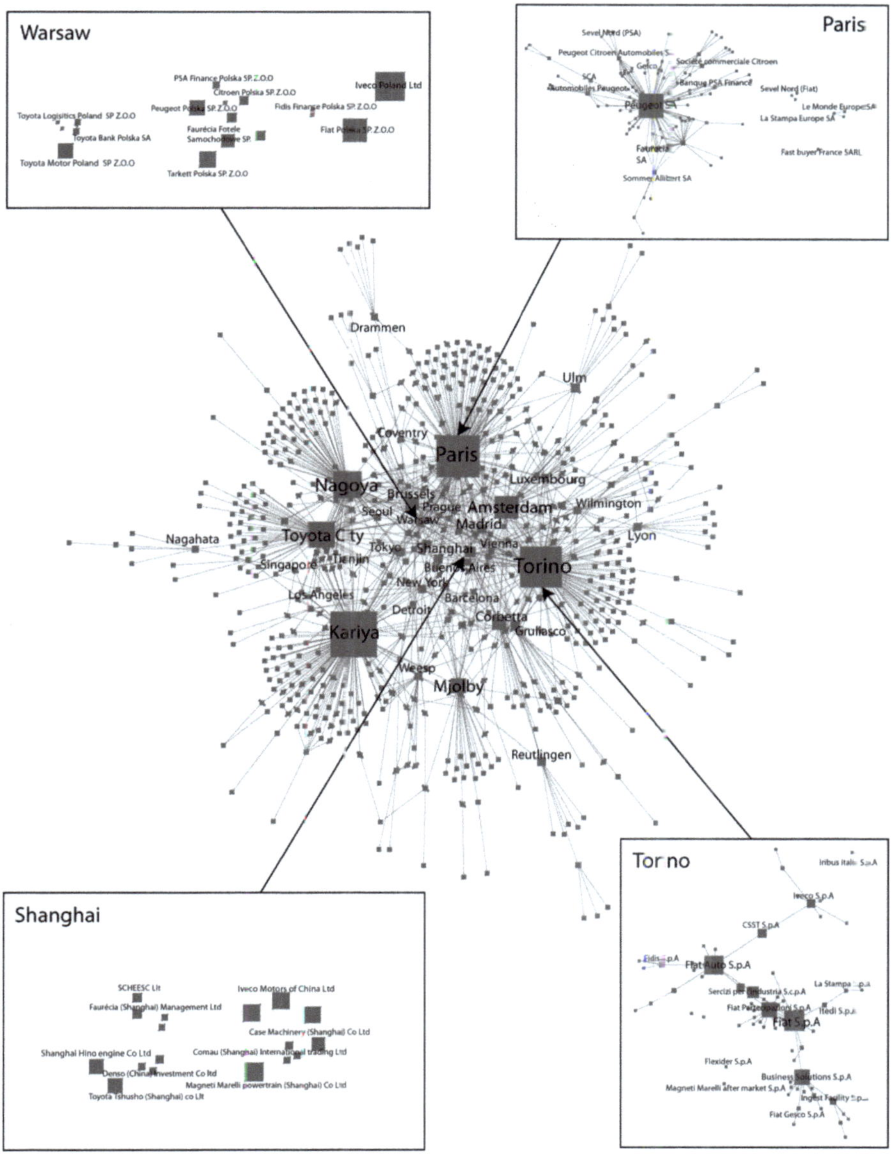

Fig. 10.7 Inter corporate competition through urban areas – urban network formed by Toyota MC, Peugeot SA and Fiat SpA (Source: ORBIS, 2006 – Bohan C. @ Cefres, 2007)

network becomes a cluster displaying every firm localized therein and represents an intra-urban network (Rozenblat, 2010). This operation reveals the main urban framework that firms use to localize their subsidiaries and where multinational companies locate, according to their needs, in terms of row, size and specializations (Pred, 1977; Veltz, 2000).

Table 10.3 Most important cities in each economic sector

Agro-food industry		Motor industry	
City	Nb afiliates	City	Nb afiliates
London	500	Paris	109
Paris	120	Torino	71
Rotterdam	95	Nagoya	50
Mexico city	92	Coventry	46
Edinburgh	81	Wilmington	43
New York	67	London	29
Dublin	56	Buenos Aires	28
Chicago	43	Tianjin	26
Amsterdam	41	Madrid	24
Vitoria	40	Bangkok	24

The inter-urban relations of the motor industry are made on a more global scale than those of the agro-food industry. We explain this pattern through the strong specialization of the territorial resources necessary for the competitiveness of the automotive companies and the close relations of the markets of the agro-food industry. In terms of weight (see Table 10.3), we notice a stronger polarization of the cities for the agro-food industry: London accumulates 500 subsidiaries, which is one-sixth of the total agro-food subsidiary sample used. The position of London in the strategic space of the agro-food sector is largely due to its European position: the city radiates to all of Europe for Kraft Foods, which uses the cultural and linguistic proximity between the United Kingdom and the United States. The city is also a pillar for the Unilever firm, which has a secondary headquarters there that influences all the European cities. Paris is the second city most heavily invested in by the subsidiary companies of the agro-food industry. Paris is the main city in Nestlé's network, because of the cultural proximity between French-speaking Switzerland and France. Paris has regional importance for the European agro-food networks. The weight of the European Union in the agro-food strategies is strong: six of the ten most heavily invested cities belong to the European Union. This result is due to the cohesion and the size of the European market (Figs. 10.8 and 10.9).

The first most heavily invested city in the network of the automotive sector is Paris, which contains 109 subsidiary companies, due to the presence of the PSA headquarters and its surroundings (Fig. 10.6), since PSA centralizes many functions at the top of the hierarchy. It is the same for Torino and Nagoya, respectively second and third most heavily invested cities of the classification. The fourth city is Coventry, a town in the United Kingdom of 305,000 inhabitants (UK census, 2002), which saw the establishment of Rover at the end of the nineteenth century. Since then, 130 industries connected to the automobile sector have settled nearby. This specialization is also visible in Shanghai and Warsaw, with the presence of the three selected firms (Fig. 10.6). Finally, this automobile classification reveals the strong proportion of cities belonging to emergent countries, such as Buenos Aires, Tianjin and Bangkok, which may result from strategies to minimize production costs.

Fig. 10.8 Main cities of the agro-food industry (Nestlé, Unilever, Kraft Foods, Pepsico et Danone) (Source: ORBIS BVD, 2006 – Bohan C. & Gautier B. @ IAMM, Cefres, 2007)

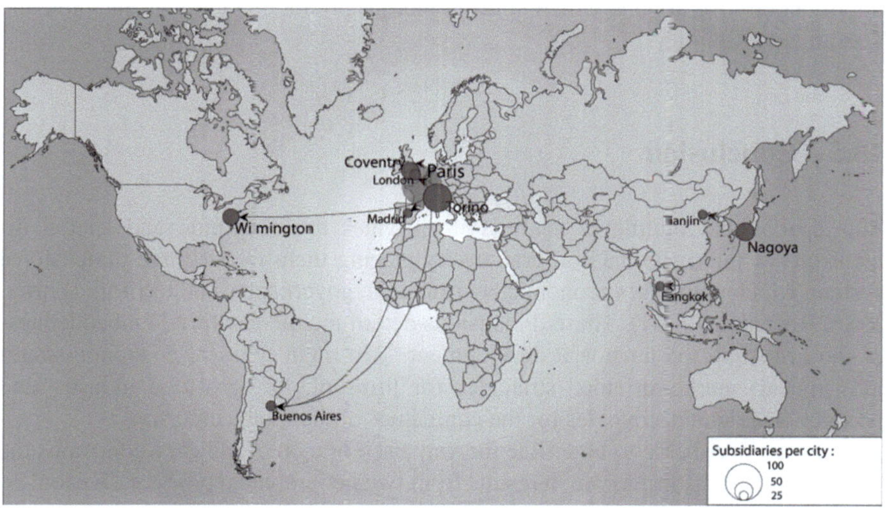

Fig. 10.9 Main cities of the motor industry (Toyota, PSA, Fiat) (Source: ORBIS BVD, 2006 – Bohan C. & Gautier B. @ IAMM, Cefres, 2007)

The heart of the network formed by the firms of the agro-food industry is marked by strong relations between cities benefiting from a powerful market and, more particularly, by the presence of European cities. This model follows a classic economic core/periphery model because it presupposes the existence of agro-food markets. The peripheries are not ignored by these strategies but remain marginal because of their weak cohesion networks. On the other hand, strong relations between distant cities form the heart of the automobile network. The heterogeneous spatial strategy of the motor industry makes it possible to combine the specific advantages of the territories through strategies to minimize costs, due to connections between distant cities. This economic strategy diffuses capital and expertise more widely globally and contributes to an international division of labor by creating competition among cities at a global scale.

Above, we discussed the table of the ten most important cities by sector (Table 10.3). On a global scale, the five largest cities for the agro-food industry are situated in Europe, with a disproportionate ratio in London because it is unambiguously superior for the administrators of this sector. A second pole is distinguishable in North America. New York presents the same advantages as London but on a North American platform. Chicago is the headquarters of Kraft Foods, and Mexico City controls activities for all the Latin America.

For the three firms of the automotive sector, Europe stands out because PSA and Fiat have their primary markets there. In the United States, Wilmington (situated in the tax haven of Delaware, which is between Washington DC and Baltimore) is the American pole of the activities of Fiat. On the Asian side, Nagoya is the most important city for Toyota, which holds numerous production units in Bangkok and Tianjin, near Beijing.

10.5 Conclusion

This spatial and multilevel analysis investigates the economic and spatial dimensions of multinational corporations by placing them in our three-dimensional system, which comprises economic environment, governance, and territory (Porter, 1986; Rozenblat, 2004). These dimensions determine the territorial embeddedness of the firms and the territorial development. Two main types of economic strategies appear: market-oriented strategies for firms of the agro-food industry and production-oriented strategies for the companies of the motor industry.

Thus, we were able to underline the way each branch of industry constrains the organization of the firms, which results from two geographical types of division:

- A geographical division where the localization and the connections between the subsidiaries of the agro-food industry respect the broad outlines of the regional markets;
- A division of labor based on globalization at the world level, placing specialized cities in competition for the companies of the automotive sector. The original aspect of our work is the scale of the inter-urban range. We explain how firms of the agro-food industry tend to gather in the most powerful cities, and how the companies of the automotive sector join in medium-sized and specialized cities.

All of this information allows us to generate some hypotheses on territorial development and to sketch the basis of a typology of the development of territories by the corporate networks. Although this information is incomplete, the typology based on the affiliated networks is able to include the financial links that bind subsidiaries to each other. We know that there are numerous other forms of exchange between subsidiaries (intra-firm business, information, decision-making power, etc.) as developed with the methodology of the global value chain (Gereffi, 2001). All these approaches must be developed on the local scale in order to understand the ways in which these local area networks influence a city at the global level. These analyses must also be conducted with territorial data.

The methodological approach of using networks of individual firms enabled us to cover two different geographical levels of cities, intra-urban and inter-urban scales. The complementarity of the processes occurring at and between these two levels has been analyzed, even if the explanatory processes are not yet demonstrated. Determining those explanatory processes is now a challenge opened by this analytical study.

References

Camagni, R. (1996). *Principes et modèles de l'économie urbaine*. Paris: Economica.

Camagni, R. (1999). The city as a milieu: Applying GREMI's approach to urban evolution. *Revue d'Economie Régionale et Urbaine (RERU)*, *3*, 591–606.

Christaller, W. (1933). *Die zentralen Orte in Süddeutschland*. Jena, Germany: Gustav Fischer.

Dicken, P. (1992). *Global Shift, the internationalization of economic activity*. New York: The Guilford Press.

Doz, Y., Santos, J., & Williamson, P. (2001). *From global to metanational: How companies win in the knowledge economy*. Boston: Harvard Business School Press.

Duranton, G., & Puga, D. (2004). Micro-foundations of urban agglomeration economies. In J. V. Henderson & J.-F. Thisse (Eds.), *Handbook of regional and urban economics* (Vol. 4, pp. 2063–2117). Amsterdam: Elsevier.

Ellison, G., & Glaeser, E. (1997). Geographic concentration in U.S. manufacturing industries: A dartboard approach. *Journal of Political Economy*, *105*, 889–927.

Francfort, I., Osty, F., Sainsaulieu, R., & Uhalde, M. (1995). *Les mondes sociaux de l'entreprise*. Paris: Desclée de Brouwer.

Freeman, L. C. (1977). A set of measures of centrality based upon betweenness. *Sociometry*, *40*, 35–41.

Freeman, L. C. (1979). Centrality in social networks: Conceptual clarifications. *Social Networks*, *1*, 215–239.

Gereffi, G. (2001). Beyond the producer-driven/buyer-driven dichotomy – the evolution of global value chains in the internet era. *IDS Bulletin*, *32*(3), 30–40.

Henderson, J. (1988). *Urban development: Theory, fact, and illusion*. New York: Oxford University Press.

Hoover, E. (1937). *Location theory and the shoe and leather industry*. Cambridge: Harvard University Press.

Hoover, E. (1948). *The location of economic activity*. New York: Mc Graw-Hill.

Jacobs, J. (1969). *The economy of cities*. New York: Vintage Books/Random House.

Lazega, E. (1998). *Réseaux sociaux et structures relationnelles*. Paris: PUF.

Marshall, A. (1920). *Principles of economics*. London: Mc Millan.
Michalet, C.-A. (1999). La séduction des nations ou comment attirer les investissements, Paris: Economica, 134 p.
Ohlin, B. (1933). *Inter-regional and international trade*. Cambridge: Harvard University.
OICA. (2010). World motor vehicle production by country and type, 1997–2009. Paris: Organisation Internationale des Constructeurs d'Automobile. http://oica.net/category/production-statistics/
Porter, M. (1986). *L'avantage concurrentiel: comment devancer ses concurrents et maintenir son avance*. Paris: InterEditions.
Pred, A. (1977). *City-systems in advanced economies*. London: Hutchinson University Library.
Rastoin, J.-L. (2000). Une brève histoire de l'industrie alimentaire. *Economie Rurale, (255–256)*, 61–71.
Rozenblat, C. (2004). *Tissus de villes: réseaux et systèmes urbains en Europe*. Habilitation à diriger des recherches, Université Montpellier 3.
Rozenblat, C. (2010). Opening the blackbox of agglomeration economies for measuring cities' competitiveness through international firm networks. *Urban Studies*, *47*(13), 2841–2865.
Rozenblat, C., & Pumain, D. (1993). The localition of multinational firms in the European urban system. *Urban Studies, 30*(10), 1691–1709.
Storper, M., & Venables, J. (2004). Buzz: Face-to-face contact and the urban economy. *Journal of Economic Geography*, *4*, 351–370.
Veltz, P. (1998). *Globalisation et territorialisation des groupes industriels: rapport de synthèse*. Paris, DATAR.
Veltz, P. (2000). *Mondialisation, villes et territoires: l'économie d'archipel* (3rd ed.). Paris: PUF.

Chapter 11
The Capture and Diffusion of Knowledge Spillovers: The Influence of the Position of Cities in a Network

Marie-Noëlle Comin

11.1 Introduction

Cities concentrate economic activities, information, power and people in both quantitative and qualitative ways. Cities are also nodes of complex, interconnected networks. In Europe, the establishment of a large supranational entity has resulted in the multiplication and intensification of the relationships between politically unified countries. The majority of these relationships pass through cities. Studying the complex links that underpin the European system of cities is critical to understanding the interdependencies in the system. Here, the interconnection of European national urban systems is studied by analyzing scientific and technological collaborative-research networks. Such networks make it possible to observe the flow of scientific knowledge within the European system of cities.

In Europe, as in the rest of the world, scientific and technological innovation has acquired increasing strategic importance in economic competition. Such innovation also plays a crucial role in the structure and dynamics of settlement systems: in the emerging "knowledge economy," the major dynamic feature characterizing the evolution of urban systems appears to be competition for the collection of knowledge and innovations. Knowledge spillovers are considered to be both an input and an output of the innovation process. Therefore, understanding how knowledge spillovers flow within European territories is critical to understanding innovation processes in Europe. Contrary to the widespread idea that innovation is indifferent to location, various empirical studies stress the importance of cities as nodes of accumulation, production and diffusion of scientific and technological knowledge (Bouinot, 2004; Castells, 2000; Lever, 1999; Simmie, 2001). Thus, analyzing

M.-N. Comin (✉)
Université de Paris I Panthéon-Sorbonne, CNRS UMR 8504 Géographie-cités/Equipe P.A.R.I.S., rue du Four 13, Paris 75006, France
e-mail: comin@parisgeo.cnrs.fr

C. Rozenblat and G. Melançon (eds.), *Methods for Multilevel Analysis and Visualisation of Geographical Networks*, Methodos Series 11, DOI 10.1007/978-94-007-6677-8_11,

scientific collaborative networks can yield critical insights regarding the wider geography of interactions among European cities.

The study of knowledge flows distinguishes two types of knowledge: codified knowledge and tacit knowledge. Codified knowledge is defined as knowledge that is stated in an explicit form. Explicit knowledge is relatively easily transferable and can be transmitted through information channels and infrastructures. Tacit knowledge is described by Michael Polanyi's famous quotation, "We can know more than we can tell". Hence, tacit knowledge is a non-linguistic and non-numerical form of knowledge that essentially arises from personal experience and skill. Tacit knowledge requires social interaction for transmission and is context specific. Thus, it is difficult to capture, codify and transfer tacit knowledge. Tacit knowledge is recognized as a central component of the "knowledge-based economy." As the key element for effective exploitation of innovative opportunities and abilities, tacit knowledge crucially influences the innovation process. Therefore, many studies focus on this form of knowledge. Generally, studies show that tacit knowledge can be transmitted via collaboration between research centers, training or personnel exchange.

From a geographical perspective, the mechanics of knowledge flows are unclear. Studies commonly state that codified knowledge does not require proximity to be shared and thus can flow easily across long distances. In contrast, tacit knowledge requires proximity to be shared; thus, it is difficult to exchange tacit knowledge over long distances. In reality, the geography of knowledge flows is far more complicated. In addition to geographical proximity, additional types of proximity make it possible to share tacit knowledge at long distances (Comin, 2009). Here, we focus on organizational proximity in analyzing scientific networks between European cities created through collaborative Research and Development (R&D) projects dedicated to converging technologies (Boschma, 2005). Converging technologies result from the merging of nanotechnology, biotechnology, information technology and cognitive science (NBIC). Such technologies are expected to drive the future innovation wave predicted to emerge by 2020 (Nordmann, 2004).

The position of cities in scientific networks highlights the geographic structure of European knowledge creation, which facilitates cities' economic competitiveness in the "knowledge economy". However, a large number of scientific exchanges with other cities does not by itself make a city well positioned in a scientific network. The criteria for judging cities' success also include their ability to ensure the interconnection of national urban systems at the European scale (Rozenblat, 1992, 2004).

The remainder of this paper is divided into three sections. First, the empirical data used to analyze European cities' position in scientific networks are presented. Next, the structure of the European scientific network of cities is described and compared to the structure of the European system of cities in terms of population size. In urban geography, the population of a city is a good indicator of its position in urban systems (Pumain, 1997). Finally, particular attention is paid to the cities that ensure the interconnection of national urban systems. These cities are essential in expanding the spatial dimension of knowledge flows and linking the 27th European national system of cities.

11.2 Sources: From Research Networks to Urban Networks

There are three principal types of indicators for studying scientific flows between cities: (i) patent documents, including references to previous patents (citation). According to Jaffe, Trajtenberg, and Henderson (1993), a patent citing earlier patents reveals knowledge flows between localized inventors. (ii) Co-authorship networks related to joint publication activities indicate close working relationships and potential knowledge flows between localized authors (Matthiessen, Winkel Schwarz, & Find, 2010). The main difficulty with these two indicators is collecting data related to the cities where the innovative actors are located. (iii) Finally, we consider scientific and technological collaborations between organizations engaged in innovation processes (research organizations including public research centers, universities and corporate research centers) within European-funded research and technology development projects (RTDs).

Data were extracted from the EC database CORDIS RTD-PROJECTS (Community Research and Development Information Service) and were drawn from the second to sixth European Framework Programs (FPs) for Research and Technological Development (the first FP is too incomplete to be used). The FPs are 4-year programs initiated in 1984 by the European Union to increase scientific collaboration between European countries.

The CORDIS database provides information about the evolution of European research support from 1986 to 2006. CORDIS data include the actual locations of the organizations involved (not simply the locations of their headquarters). Institutions may have several research centers located in different cities. Therefore, we chose to identify the precise location of each research center (laboratory) involved in NBIC-related projects. We then created urban networks by aggregating CORDIS data at the city level to measure the links that these networks create between cities (Besussi, 2006; Rota, 2008; Rozenblat & Cicille, 2003) These links represent bounds created by scientific collaborations between two cities involved in the same NBIC RTD project. Through these collaborations, cities produce, attract and diffuse knowledge spillovers that encompass the differential evolution of each city in the future innovation wave.

For urban aggregation, we considered cities to be functional urban areas (FUAs) defined in a comparable way throughout Europe (Comin, 2009). Each laboratory appearing in the networks was geographically located and then aggregated to the appropriate urban area.

In total, the urban database includes FUAs with 10,000 or more inhabitants and contains 9,300 research centers located in 800 functional urban areas within 117 countries. Among all functional urban areas in the database, 509 are located in UE 27. Two cities were considered to be connected if their institutions or enterprises had collaborated on at least one project. Here, we focus only on European cities. CORDIS data also make it possible to examine tacit-knowledge flows according to the approach of Candice Stevens (1997), who identified for the journal OECD Observer three main ways in which tacit knowledge can be exchanged: research and development cooperation between the public and private sectors, technology diffusion and personnel movements. We assume that personal contacts exist between people involved in the same European RTD projects because it seems likely that scientists choose collaborators with whom they are acquainted.

11.3 The Distribution of Knowledge Flows Within the European System of Cities

The resulting graph of European scientific-research collaboration networks (Fig. 11.1) has a large number (28,558) of links. How are these links distributed between cities?

First, we can study the whole properties of this network. The connectivity of the graph can be measured using three indices: (i) the density of links in the graph, or the number of links expressed as a proportion of the maximum possible number of links; (ii) the diameter of the graph, or the longest of the shortest paths between all pairs of vertices, which measures the maximal extent of the network; and (iii)

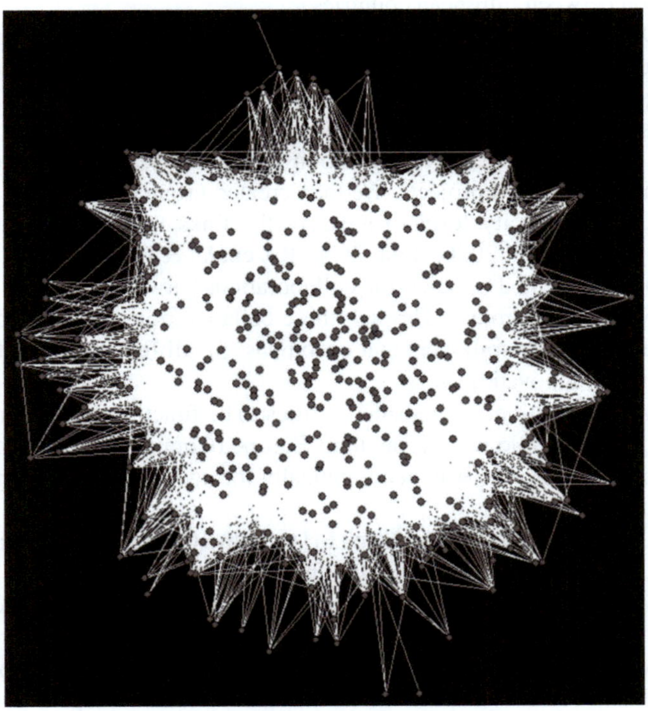

Fig. 11.1 Graph of Research Collaborations in NBIC among European cities, 1986–2006 (Made with Pajek software, Kamada-Kawai layout. Source: NBIC-Euro, Comin, 2009)

Table 11.1 Indices of connectivity

Density	6 %
Diameter	4
Average path length	2

Source: NBIC-Euro, Comin (2009)

the average path length of the graph, which is the average number of intermediate connections between all pairs of vertices.

These three indices (Table 11.1) show that the connectivity of the graph is high: although 22 % of all possible links are present, each city can reach all other cities in an average of two and a maximum of four steps. Thus, the scientific-collaboration network for converging technologies within European cities is strongly connected. These results underline the interdependence of European cities for R&D in converging technologies and provide promising information about the circulation of scientific knowledge between cities.

Nevertheless, knowledge flows are not evenly distributed among European cities. Rather, their patterns of accumulation are strongly affected by the pre-existing population-size structure of the European system of cities (Comin, 2009).

This phenomenon can be illustrated using the indices of centrality developed by L. C. Freeman (1977, 1979) to describe the relative importance of nodes within a network. Two examples can be shown. First, the non-valuated degree centrality of a city is the number of cities connected to it. This index measures the relational activity of a city. Second, the betweenness centrality of a city measures its potential intermediary role within a network: the more often a city occurs on the shortest paths between other cities within the network, the higher is its betweenness centrality. These two indices clearly distinguish the largest European cities from other cities. Indeed, cities of the European megalopolis have large degree centralities (Fig. 11.2). Table 11.2 shows that the 20 cities with the largest degree centralities are also the largest European cities, except for two university cities: Utrecht and Goteborg. The largest European cities, particularly capital cities, also have the largest betweenness centralities (Fig. 11.3).

The structure of the scientific-collaboration network dedicated to converging technologies reveals that European cities exchange knowledge through a classical hierarchical-diffusion pattern (Hägerstrand, 1952).

Moreover, our maps show that certain cities, such as Utrecht and Goteborg, are more central than expected based on their population size. While large cities concentrate the infrastructures that traditionally facilitate material and non-material flows, specialized cities also have important innovation and training capacities. Specialized cities involved in European-funded converging-technologies R&D vary in population size but are well connected to the network. Therefore, their hierarchical positions in the network are much higher than their rankings among European cities in terms of population size . For example, the empirical data analyzed here show that Utrecht, Goteborg and Cambridge are ranked 11th, 19th and 24th, respectively, in terms of degree and 84th, 92nd and 351st, respectively, in terms of population.

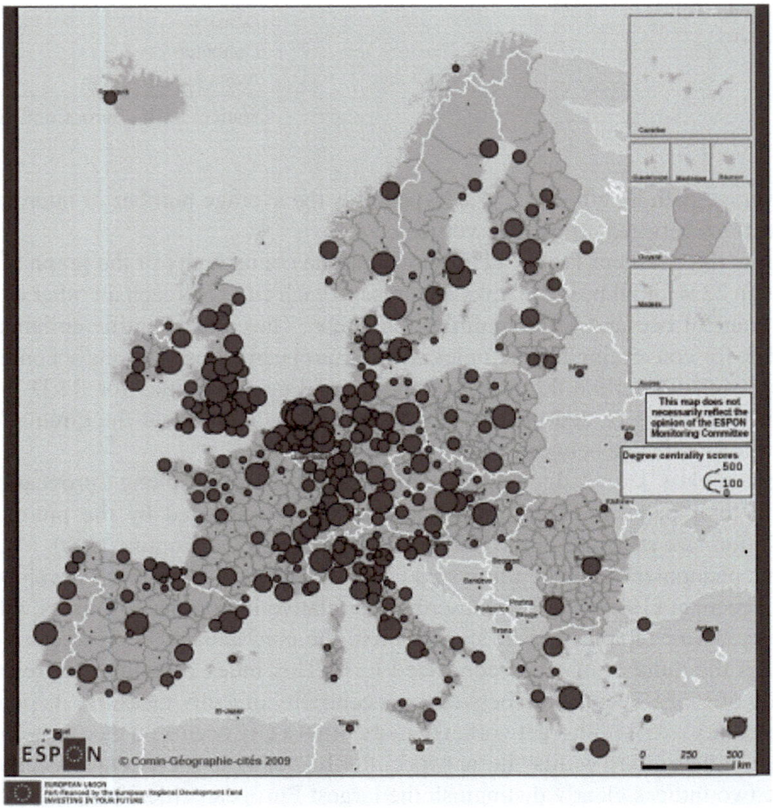

Fig. 11.2 Degree centrality (cumulative over the period 1986–2006)

Table 11.2 The 20 cities with the largest degree centralities (cumulative over the period 1986–2006)

Cities	Degrees	Rank	Cities	Degrees	Rank
Paris	416	1	Utrecht	351	11
London	407	2	Amsterdam	348	12
Athens	396	3	Barcelona	343	13
Madrid	385	4	Lisbon	342	14
Copenhagen	372	5	Munich	342	15
Milan	370	6	Berlin	340	16
Brussels	365	7	Stockholm	338	17
Helsinki	364	8	Stuttgart	337	18
Vienna	362	9	Goteborg	320	19
Rome	358	10	West-Midlands	318	20

Source: NBIC-Euro, Comin (2009)
Note: non-valuated degrees

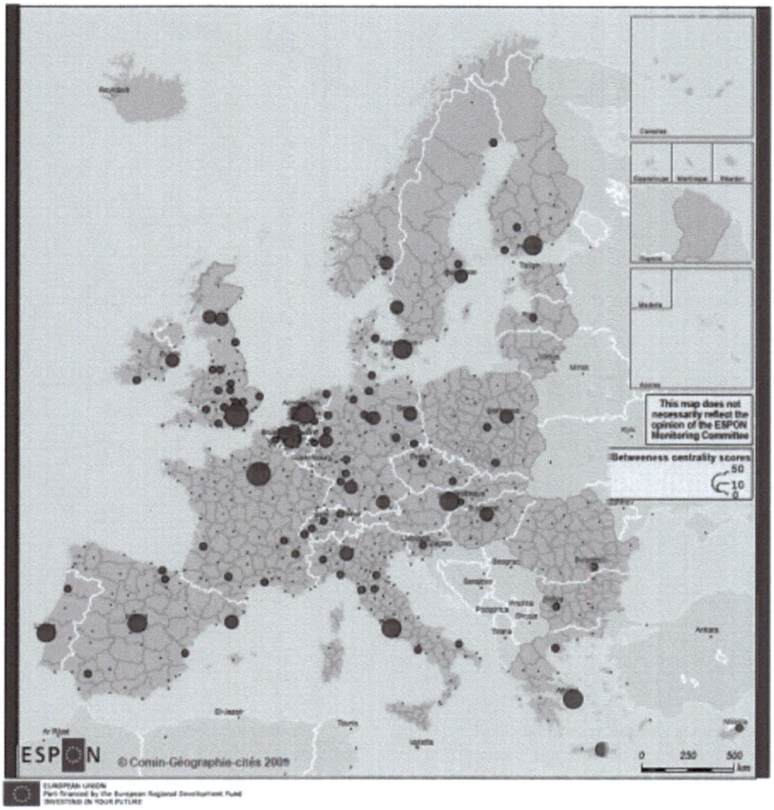

Fig. 11.3 Betweenness centrality (1986–2006)

The simplest way to validate this insight is to calculate the ratio between the number of scientific collaborations (or RTD projects) in a given city and the population of that city. A larger ratio indicates greater specialization of a city in European-funded converging-technologies R&D.

In Fig. 11.4, some small cities appear to have many RTD projects for their population size. These cities are considered to be specialized, as defined above. The most specialized cities are the most renowned university cities, such as Cambridge, Louvain, Oxford and Heidelberg (Table 11.3), because the location factors for R&D in converging technologies depend on the quality of the scientific infrastructure and the presence of a skilled population. These factors tend to favor the most renowned university cities.

As a whole, the structure of the European urban network dedicated to converging technologies is dominated partly by the largest European cities and partly by renowned university cities. We hypothesize that these "hub cities" are essential to the interconnection of national urban systems at the European scale. In the following

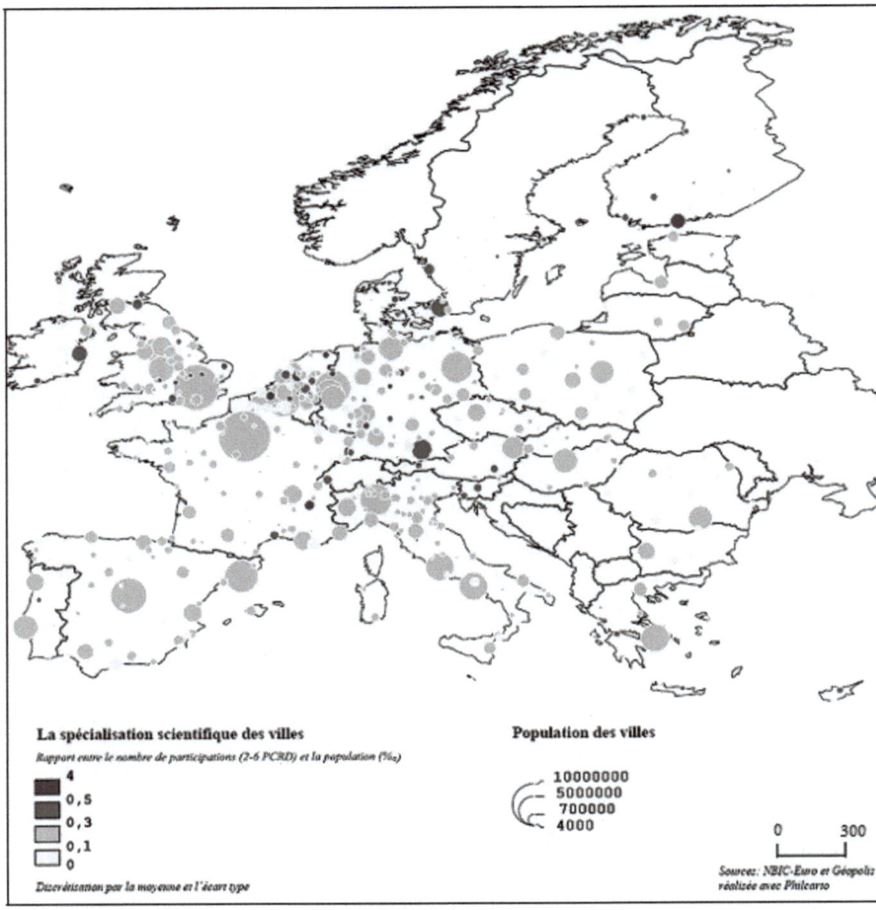

Fig. 11.4 Specialization of European cities in European-funded R&D dedicated to converging technologies

Table 11.3 The most specialized cities in European-funded R&D dedicated to converging technologies

Cities	Number of RTD projects per thousand inhabitants
Lulea	4.02
Cambridge	3.10
Aberystwyth	2.30
Louvain	1.52
Uppsala	1.46
Oxford	1.40
Heidelberg	1

Source: NBIC-Euro, Comin (2009)

section, we will test this hypothesis by distinguishing those cities that act as "relays," ensuring the articulation of scale for knowledge flows between European national urban systems. These cities are known as "relay-cities" (Rozenblat, 1992, 2004).

11.4 Interconnections of European National Urban Systems

To distinguish relay-cities from other European cities, we used the concept of "articulation nodes" from graph theory (Berge, 1958). In a connected graph, such as our scientific-collaboration graph of European cities, a node is an articulation node if the sub-graph obtained by removing it is no longer connected (if removal of the nodes creates disconnect components). Thus, an articulation node is a bottleneck in the graph.

In urban geography, articulation nodes can be interpreted as relay-cities that connect urban systems at different scales: national, European and global. Articulation nodes or relay-cities are crucial vectors for the transmission of knowledge spillovers within the network because they attract diversified knowledge and ensure its circulation among the networked European cities. Moreover, the position of such an intermediary confers a strategic role in controlling the diffusion of knowledge to other European cities: a relay-city may choose to diffuse knowledge among its networked cities or to appropriate knowledge for its own profit.

Our analysis shows that few cities are relay-cities (Fig. 11.5): only 98 cities are structurally positioned as articulation points in the network. Most of these relay-cities connect several components: Copenhagen connects 14 components; Paris connects 11; Athens, Utrecht and Brussels each connect 7; London connects 6; and Vienna, Berlin, Munich, Dublin and Leipzig each connect 5; …

We assume that a given relay-city tends to act as an interface at a particular scale rather than at all scales (national, European and global). To classify relay-cities according to the scale at which they act as an interface, we chose a factor-analysis approach, principal-components analysis (PCA). A PCA can reveal the latent structure of a set of variables by reducing the attribute space from a large number of variables to a smaller number of factors.

The rows of the data matrix used for PCA represented relay-cities, while the columns contained the number of cities with which each relay-city forms a bridge (i.e., an edge whose deletion increases the number of connected components). These cities were distributed in three modalities (national, European and global) according to their relative locations with respect to the relay-city concerned.

Figure 11.6 presents the first two factors or dimensions of the PCA, which explain 77 % of the total variance. Factor 1 (F1) clarifies the scale at which relay-cities act as interfaces, whereas factor 2 (F2) indicates cities' degree of specialization with respect to this scale of activity.

The majority of relay-cities tend to ensure interconnection at the European scale. However, most of them appear to be diversified, ensuring the interconnection of urban systems at multiple scales. Only Athens and Utrecht act principally as European

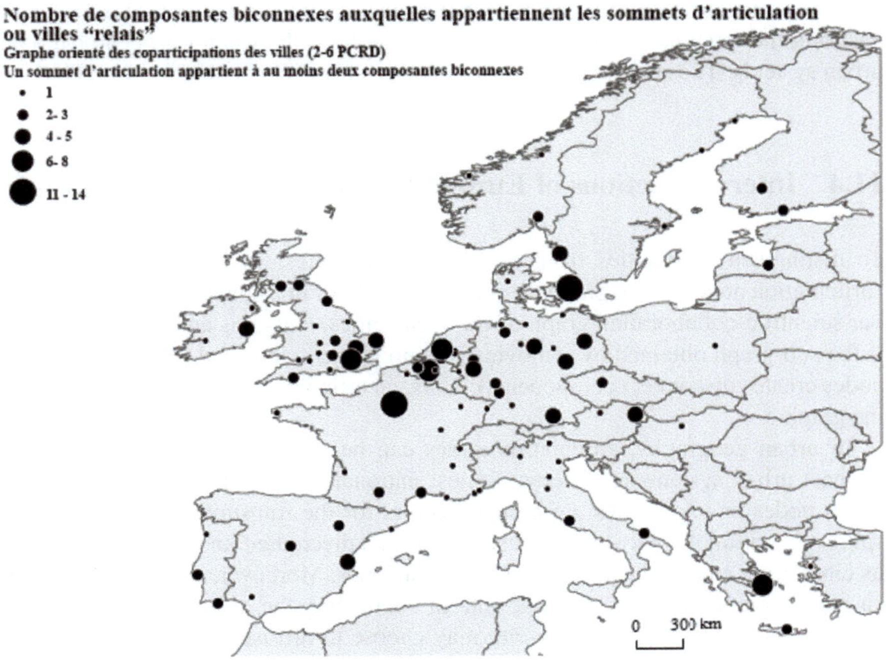

Fig. 11.5 Relay-cities identified as articulation nodes (1986–2006) (Source: NBIC-Euro, Comin, 2009)

interfaces. At the national level, relay-cities tend to be university cities and regional metropolises (e.g., Ålesund, Mytilene, Toulouse, Tampere, Bordeaux and Leipzig) or peripheral European capitals (Helsinki, Dublin, Heraklion and Budapest). Few relay-cities act at the global level; Copenhagen and Paris appear to be strongly specialized for this function, while other large international metropolises, such as Brussels and London, also tend to exercise this function.

11.5 Conclusion

In conclusion, this analysis of networks dedicated to R&D in converging technologies underlines the fact that patterns of accumulation in an urban network of knowledge flow are strongly driven by the pre-existing structure of the system of cities. Thus, the final impact of the knowledge-flow network on the pre-existing structure of the system in which it evolves is rather marginal. However, these results emphasize the trend for larger cities to capture knowledge spillovers, making it possible to predict the strength of the structural hierarchical disparities within the European urban network. This trend prefigures the reinforcement of metropolitan patterns

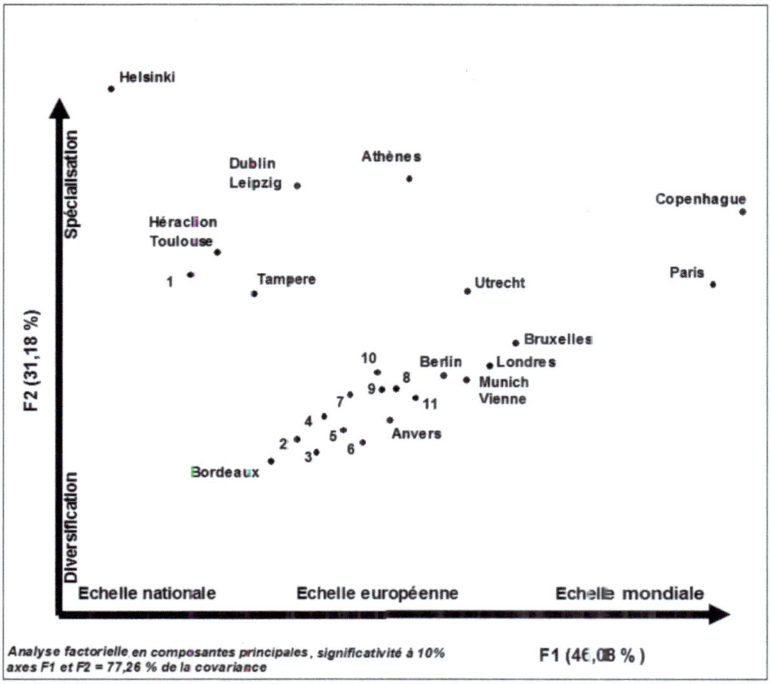

Fig. 11.6 Relay-cities tend to act as interfaces: scale of activity and degree of specialization (Source: NBIC-Euro, Comin, 2009)
1 Ålesund, Barcelone, Budapest, Mytilène, *2* Aachen, Bologne, Brest, Bristol, Cannes, Cheltenham, Florence, Genève, Haïfa, Hambourg, Linz, Louvain, Lulea, Marseille, Nijmegen, Umeå, Varsovie, *3* Arnhem, Belfast, Bergamo, Bolzano, Bremerhaven, Cardiff, Dijon, Grenoble, Hannover, Herning, Kingston, Lille, Limerick, Magdeburg, Manchester, Nancy, Nice, Padoue, Porto, Rostock, Seville, Southampton, Stockholm, Strasbourg, Stuttgart, Trondheim, West-Midlands (Birmingham), *4* Brème, Glasgow, Mannheim, Oxford, Riga, Tilburg, Mayence, *5* Edinburgh, Lisbonne, Madrid, *6* Prague, Essen, Torbay, *7* Bari, Faro, Gand, Leicester, Oslo, *8* Valence (Espagne), *9* Montpellier, Rome, Tyneside (Newcastle), Zaragoza, *10* Brunswick, Cambridge, Cologne, *11* Göteborg, Norwich

within the European urban system through the European-scale integration of cities for R&D activities in the most innovative economic sectors of the present time.

References

Berge, C. (1958). *The theory of graph*. New-York: Dover Publications.

Besussi, E. (2006). *Mapping European research networks* (Tech. Rep. Nos. Working Papers Series, Paper 103). CASA.

Boschma, R. (2005). Proximity and innovation: A critical assessment. *Regional Studies, 39*(1), 61–74.

Bouinot, J. (2004). *La ville intelligente*. Paris: LGDJ.

Castells, M. (2000). *The rise of the network society* (2nd ed.). Oxford: Blackwell.

Comin, M.-N. (2009). *Réseaux de villes et réseaux d'innovation en Europe: Structuration du système des villes européennes par les réseaux de recherche sur les technologies convergentes*. PhD thesis, University Paris 1, 722 p.

Freeman, L. C. (1977). A set of measures of centrality based upon betweenness. *Sociometry*, *40*, 35–41.

Freeman, L. C. (1979). Centrality in social networks: Conceptual clarifications. *Social Networks*, *1*, 215–239.

Hägerstrand, T. (1952). *The propagation of innovation waves* (Vol. 4). Lund: Gleerup.

Jaffe, A. B., Trajtenberg, M., & Henderson, R. (1993). Geographic localization of knowledge spillovers as evidenced by patent citations. *The Quarterly Journal of Economics*, *108*(3), 577–598.

Lever, W. F. (1999). Competitive cities in Europe. *Urban Studies*, *36*, 1029–1044.

Matthiessen, C., Winkel Schwarz, A., & Find, S. (2010). World cities of scientific knowledge: Systems, networks and potential dynamics. An analysis based on bibliometric indicators. *Urban Studies*, *47*, 1879.

Nordmann, A. R. (2004). *Converging technologies – shaping the future of European societies* (Vol. V; Tech. Rep.). European Commission, Luxembourg: Office for Official Publications of the European Communities.

Pumain, D. (1997). Pour une théorie évolutive des villes. *L'Espace géographique*, *2/97*, 119–134.

Rota, F. S. (2008). Cities as nodes of research networks in Europe. In N. Cattan (Ed.), *Cities and networks in Europe. A critical approach of polycentrism* (pp. 125–138). Montrouge, France/Esher, UK: John Libbey Eurotext.

Rozenblat, C. (1992). *Le réseau des entreprises multinationales dans le réseau des villes européennes*. PhD Thesis, Université Paris I.

Rozenblat, C. (2004). *Tissus de villes : réseaux et systèmes urbains en Europe*. Habilitation à diriger des recherches, Université Montpellier 3.

Rozenblat, C., & Cicille, P. (2003). *Les villes européennes: analyse comparative/Die Stadte Europas. Ein vergleichende Analyse*. DATAR – La Documentation française/Bündesamt für Bauwesen und Raumordnung.

Simmie, J. (2001). *Innovative cities*. London: Spon Press.

Stevens, C. (1997). Cartographier l'innovation. *L'observateur de l'OCDE, 207 – Août-Sept. 1997.*

Chapter 12
Defining Polycentric Urban Areas Through Commuting Cohesion in France

Patrice Tissandier, Trung Tien Phan Quang, and Daniel Archambault

12.1 Introduction

Recent urban processes show two principal ways in which cities are changing. First, at an international and national level, economic assets and power are concentrating in large cities, increasing their population. Second, at a local scale, population and employment is being redistributed, moving away from the urban center and towards the periphery. As a result, urban sprawl develops and polycentric structures materialize. This movement away from the traditional urban structures leads to difficulties in implementing urban management policies. Urban planning based on a halo territory and the actual reticular development of the cities are conflicting goals.

Several studies of daily commuters, all focused on Europe and North America, have been conducted based on methods introduced by Albert and Barrabási' (Patuelli, Reggiani, Gorman, Nijkamp, & Bade, 2007). We can cite, for example, case studies of the Netherlands (Schwanen, Dieleman, & Dijst, 2001, 2002), Germany (Patuelli et al., 2007), France (Aguilera & Mignot, 2004; Berroir, Mathian, & Saint-Julien, 2002; Berroir, Mathian, Saint-Julien, & Sanders, 2004), the United States (Cervero, 1996; Cervero & Wu, 1997; Giuliano & Small, 1991;

P. Tissandier (✉)
Pôle universitaire des Portes du Jura, Université de Franche-Comté – THEMA,
UFR STGI -Place Tharradin 4, Montbeliard 25211, France
e-mail: patrice.tissandier@univ-fcomte.fr

T.T.P. Quang
CNRS UMR 5800 LaBRI, Université Bordeaux 1 - GRAVITE INRIA Sud-Ouest,
Talence, France

D. Archambault
CNRS UMR 5800 LaBRI, Université Bordeaux 1 - GRAVITE INRIA Sud-Ouest,
Talence, France

Department of Computer Science, Swansea University, Swansea SA2 8PP, UK
e-mail: Daniel.Archambault@ucd.ie

C. Rozenblat and G. Melançon (eds.), *Methods for Multilevel Analysis and Visualisation of Geographical Networks*, Methodos Series 11, DOI 10.1007/978-94-007-6677-8_12,

Gordon & Richardson, 1996; Sultana, 2000), Canada (Bourne, 1989) or the comparative study on Stuttgart and Turin (Binder, Haag, & Rabino, 2003). According to these studies, the polycentric structure of major urban areas is now a self-evident truth. A polycentric structure can be defined by the existence of one or several sub-centers in addition to the city center. In our study, the commuters' flows illustrate the existence of these sub-centers, which are seen as a point of attraction for commuters. The aim of our work is to characterize these polycentric structures and to analyze their properties. Therefore, our approach, which is based on the visualization of the network of commuters using graphs and applying a measurement index, is quite different from the approaches used by existing studies.

The analysis of daily commuting patterns, represented as weighted, directed graphs, leads to new measurements and perspectives, potentially assisting in the process of urban planning. The flows of commuters are represented by the edges between places of residence and those of employment, seen as nodes of the graph. Three types of data are added as weights on the edges: the number of commuters that pass between the municipalities; the portion, in percent, of these commuters who are in the working population in their municipality of residence; and the same percentage of these commuters who are in the working population in the municipality of their employment.

In this study, we investigate the home to work commuting patterns among 36,000 French municipalities drawn from the census data for 1975, 1982, 1990 and 1999. For these 4 years, two scales of investigation are developed.

We need to highlight some inconsistencies in this data. First, we do not distinguish between effective and official residences, leading to incongruence between the graphs built on this data and the actual situation. A data cleaning performed at the national level and associated with a coloration of the graph based on its structural properties leads to the delineation of cities through the commuters' flows and is applied in the first stage of this paper.

Second, we show the impact that regional transportation had on the spread of the commuters' network around the French major cities. Comparing the commuters' graphs from four censuses and their cohesion, defined using the average strength index, we highlight the reticular extension of the cities' delineation, their spatial discontinuities and their overlap. At the same time, we show the development of urban sprawl. The number of municipalities connected, directly or indirectly, to a central municipality grows from 1975 to 1999. These municipalities are growing progressively more distant from the city center. However, each municipality is not necessarily either strongly or directly linked to primary city center. We see urban polycentric structures develop through subnetworks that function like subsystems. In this second section, we will apply different indexes to characterize these subcenters. We also explain a method for building a graph that combines the data from the four censuses with the aim of creating an evolving study of the commuters' network.

In the last stage of this paper, we investigate a new approach that considers the commuters' graphs to be scale-free networks. We explain this position by the structure of the commuters' graph: a few important municipalities import a large number of workers every day while the majority of municipalities import very few commuters. For this study, we return to the level of the entire French commuters' network. We apply a recursive algorithm to identify the polycentric structure of the urban areas. Finally, we present a large network evolution analysis.

12.2 Data Cleaning

The census data that is used to construct the commuters' network does not distinguish between the municipality of administrative residence and the municipality of actual residence (Talbot, 2001). Additionally, the respondent may not report a second place of residence in addition to their primary place of residence. These situations, as well as others, raise numerous problems in the visualization of the graphs that are inferred using this data. Several examples from the data acquired in the 1999 census illustrate the types of problems that arise from a lack of information and include the following:

- A total of 45 people commute over 1,000 km every day, essentially the distance between Corsica and the north of France.
- Over 400 people commute over 500 km every day.

To filter out these edges, which are most likely erroneous, most approaches filter out the links with low weight or the links that are distant geographically (Bonnefoy, Pumain, & Rozenblat, 1996; Talbot, 2001). These methods, applied on a French departmental or regional level, provide convincing results. However, the diversity of situations at the national level limits the applicability of these methods. Also, as with many graph problems in geography, we have the problem of selecting a threshold. We need to choose a threshold that depends on the geographic distance between the two municipalities or/and the number of commuters that make the trip on a daily basis.

We choose to apply a gravity model to help distinguish between the edges that represent practical daily commutes with those that, most likely, result from missing census information (Nijkamp, 1975; Pumain, 2001; Sen & Smith, 1995; Shen, 1999). A gravity model stipulates that the attraction between two elements is proportional to their mass P and inversely proportional to the distance d between them.

$$T_{ij} = k\frac{P_i P_j}{d_{ij}}$$

Because we would like to prevent reasonable links from being filtered, we replaced the weight of the municipalities of residence and employment with a specific attribute of the edges, namely the portion of the migrants in the total working population of the municipalities of residence and employment, as described in the introduction.

The gravitational filtering of the graph makes the data appear to be more reasonable. The excessively long commutes have disappeared, which were likely

Fig. 12.1 Graph of commuters for France in 1999 after data "cleaning"

average distances that were imputable to weekly commutes performed by working people who travel home only for the week-end. The reasonable commutes are preserved, as shown in (Fig. 12.1).

After filtering, we apply a coloring that is proportional to a metric value. The value is computed based on the structural properties of the graph. Frequently, we use the strength metric (Auber, Chiricota, Jourdan, & Melançon, 2003; Guimerà, Mossa, Turtschi, & Amaral, 2005) where the edges of low cohesion, or components with less local participation, are given a pale color and the components with high cohesion are given a darker color. These areas of the graph, defined as high cohesion sub-systems, are what we consider to be urban areas.

Our approach differs from the classical approach, which is applied in many countries (see, for example, the French Statistic Institute's definition), where it is assumed that at least *n* % of the active population work in the central locality or agglomeration area (Fig. 12.2a). The classical approach assumes centrifuge flows only occur from the peripheral municipalities to the city center, and it does not take into account any polycentric communities that may arise. Thus, our visualization method, applied to the network of commuters, provides a color to the nodes and edges in such a way that the polycentric sub-systems pop out visually. These sub-systems act like commuter satellites around a city center (Fig. 12.2b). This network approach is congruent with a polycentric model and allows for the cities to be directly or indirectly linked to their center (Fig. 12.2).

When the filtering stage is complete, we next work at the urban areas level. We illustrate the polycentric structure of the French urban areas through several examples and present the metrics to characterize this polycentrism.

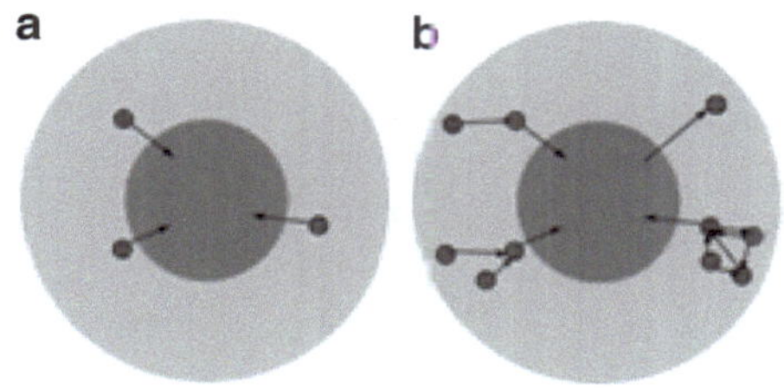

Fig. 12.2 Delineations of urban areas. (**a**) Classical delineation. (**b**) Network delineation

12.3 Characterization of Polycentrism in the French Urban Areas

A polycentric urban area organizes itself around several political, social and financial centers. The commuters' flows illustrate the existence of one or several employment sub-centers, which are seen as points of attraction. Consequently, the commuters' networks can help to illustrate polycentric structure by clarifying how the municipalities contribute to the urban area's polycentrism. In this section, we explain our methodology for defining and visualizing polycentrism through commuting cohesions.

A previous study (Rozenblat & Tissandier, 2007) shows the local polycentric structures but could not take the geographic positions of the cities into account. We fill this gap using the geographic coordinates of municipalities and, to deepen this study, we also choose to realize an evolving network analysis. The aim is, finally, to characterize the polycentric structure of French urban areas but also to underline the formation of sub-centers. So, this work is essentially a dynamic graph visualization problem. Therefore, at the urban area level and using a combined 4 years of census data, we visualize the formation of a polycentric structure by using the cohesions of the cities or by using the weighted connections in terms of the number of commuters between the municipalities.

12.3.1 Metrics Calculated

To illustrate our method, we provide many example networks around cities such as Lyon, Marseille, Lille, Bordeaux and Toulouse. Paris is not presented in this section because the network of commuters around it is very dense. We were not able to render images that were readable using our visualization system, which demonstrates a limitation of our approach.

12.3.1.1 Betweenness Centrality

There are many centrality indices that are based on the set of the shortest paths in a graph. The shortest path betweenness centrality can be viewed as a metric

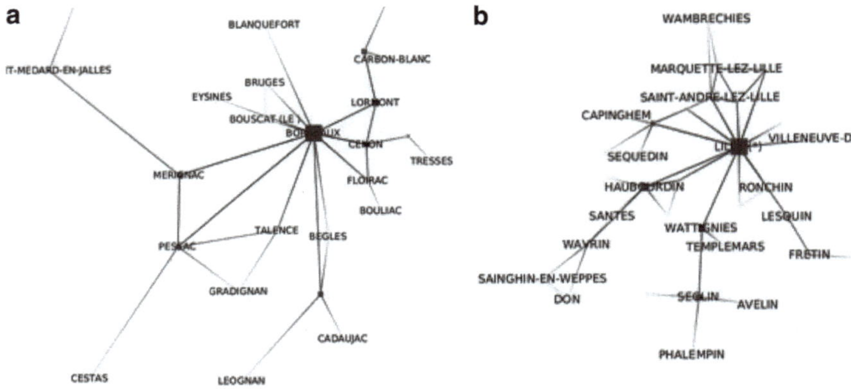

Fig. 12.3 Betweenness centrality index around Bordeaux (**a**) and Lille (**b**), year 1975

to quantify the centrality of a vertex or an edge in a graph. The vertices that appear on many shortest paths have higher betweenness than those that do not. The betweenness centrality (Brandes, 2001) of a vertex is as follows:

$$C_B(v) = \sum_{s \neq v \neq t \in V} \frac{\sigma_{st}(v)}{\sigma_{st}}$$

The value of σ_{st} is the number of shortest paths in the graph between s and t, and $\sigma_{st}(v)$ is the number of shortest paths between s and t that pass through the vertex v. If the value of the betweenness centrality is high, the node is a bridge between components in the graph. Otherwise, the node is part of a densely connected cluster. In the study of the daily network of commuters, the nodes with high betweenness centrality are municipalities that directly receive workers from many other municipalities. Consequently, these nodes are points of convergence for commuters because they are employment centers and major cities in France.

We give edges a color that is based on the following metric: the components with high betweenness centrality are darker colored while the components with low betweenness centrality are lighter colored. The examples of Bordeaux and Lille are developed next (Figs. 12.3, 12.4, 12.5, and 12.6).

12.3.1.2 Strength value

We calculate the strength value (Auber et al., 2003) for each node and edge in the graph. Their average of the strength values enable us to know the cohesion in some sub regions, which in this case are the urban areas. In the following figures, the strength index has been applied in the major French urban areas such as Marseille and Lyon, providing us with a visualization of the high cohesion zones that are a sign of subcenters (Figs. 12.7, 12.8, 12.9, and 12.10).

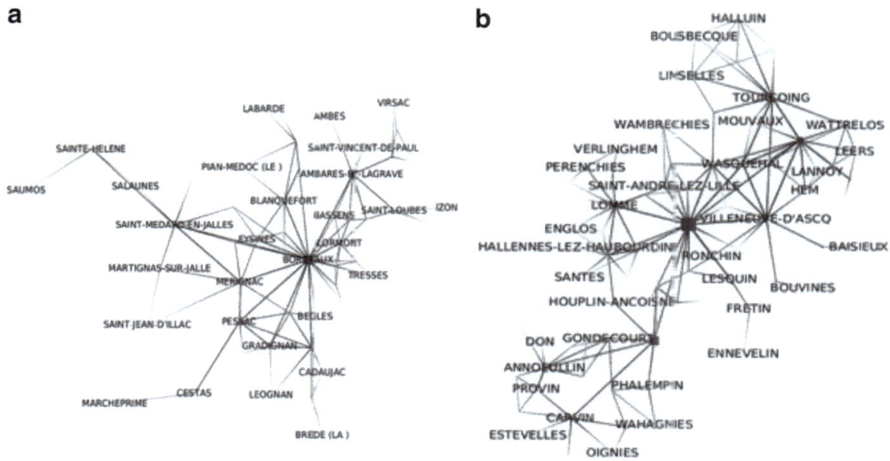

Fig. 12.4 Betweenness centrality index around Bordeaux (**a**) and Lille (**b**), year 1982

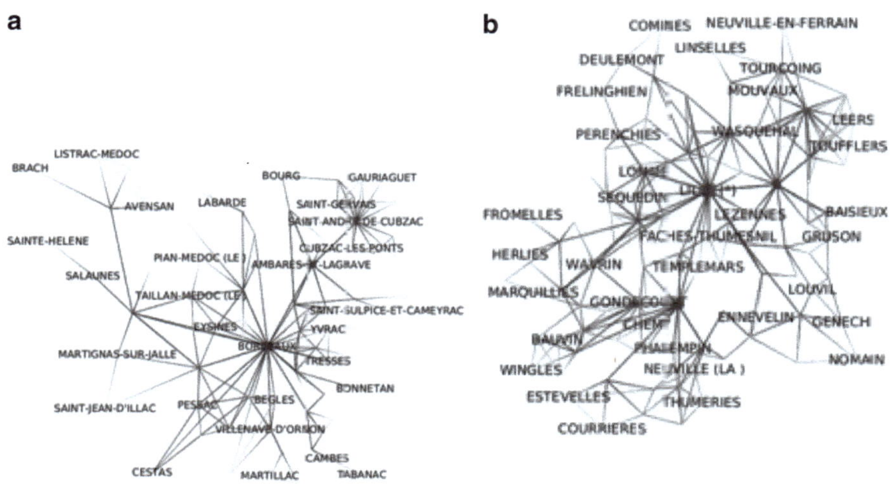

Fig. 12.5 Betweenness centrality index around Bordeaux (**a**) and Lille (**b**), year 1990

In these figures, the strength metric has been applied to the graphs. The color of each edge indicates the connectivity of each region. The areas drawn darker are tightly connected while the areas in pale are loosely connected.

12.3.1.3 Participation Coefficient

The participation coefficient measures the closure of the subnetworks as well as the intensity of the sub-centralities according to the contribution and participation indexes (Melançon et al., 2008). These indexes gauge the contribution of each

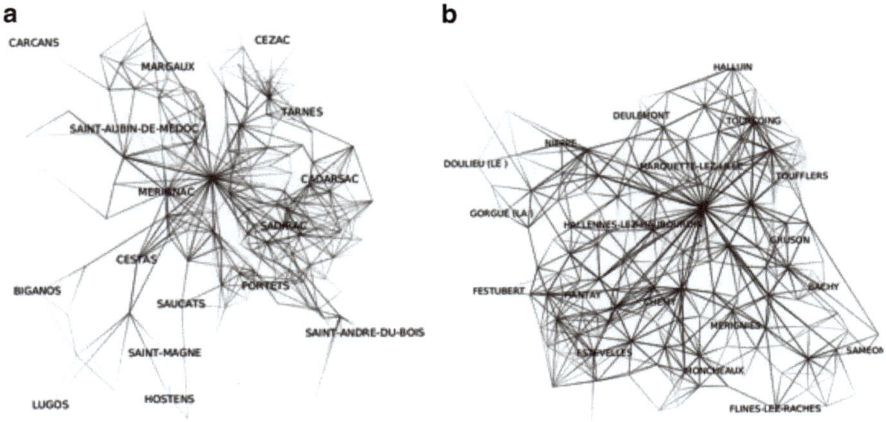

Fig. 12.6 Betweenness centrality index around Bordeaux (a) and Lille (b), year 1999

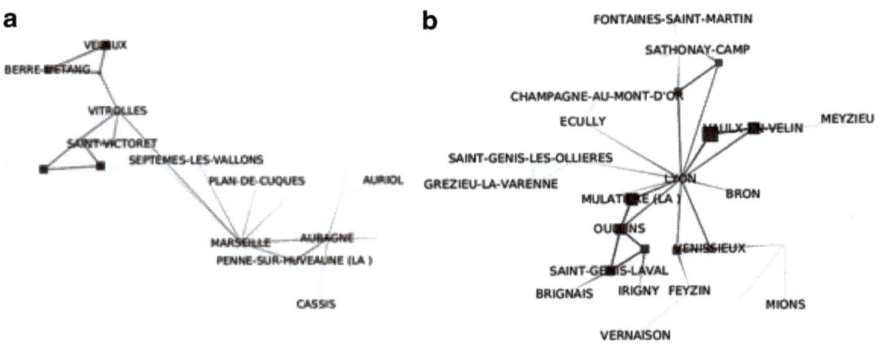

Fig. 12.7 Strength index around Marseille (a) and Lyon (b), year 1975

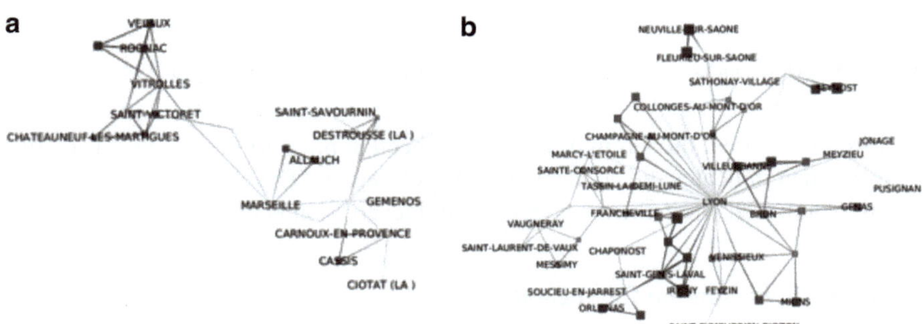

Fig. 12.8 Strength index around Marseille (a) and Lyon (b), year 1982

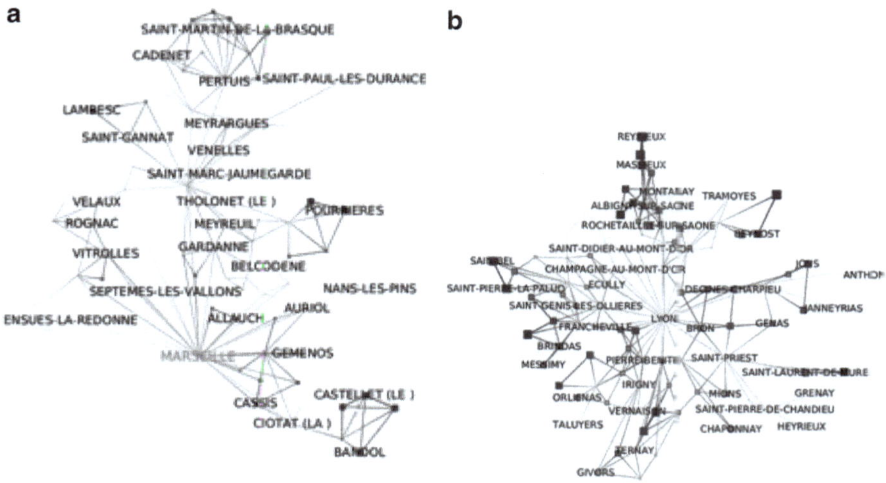

Fig. 12.9 Strength index around Marseille (**a**) and Lyon (**b**), year 1990

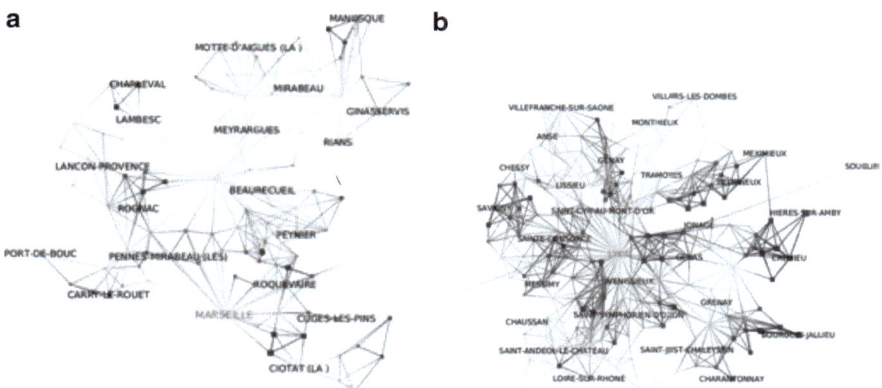

Fig. 12.10 Strength index around Marseille (**a**) and Lyon (**b**), year 1999

municipality to the cohesion of a subnetwork by measuring the city's representation in the subnetwork or the number of links that it shares with the cluster.

The participation coefficient can be computed on a directed graph using the incoming or outgoing degrees. The outgoing participation coefficient of a node v can be defined by the following:

$$p^{+}(v) = 1 - \sum_{w \in N_G^{+}(v)} \frac{\omega(v,w)}{d_G^{+}(v)}$$

where $N_G^{+}(v)$ are the outgoing nodes from v, $\omega(v,w)$ is the weight of the edge (v,w), and $d_G^{+}(v)$ is the sum of the weights for the outgoing edges of v. The incoming participation coefficient is similarly defined. The outgoing participation coefficient

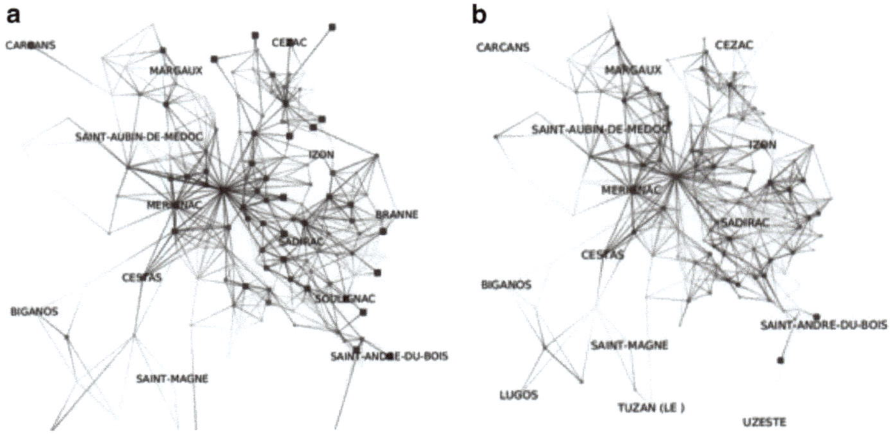

Fig. 12.11 (**a**) The incoming participation coefficient around Bordeaux in 1999. (**b**) The outgoing participation coefficient around Bordeaux in 1999

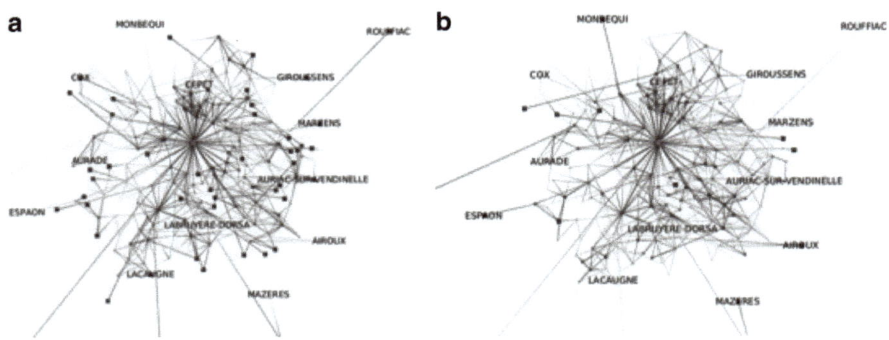

Fig. 12.12 (**a**) The incoming participation coefficient around Toulouse in 1999. (**b**) The outgoing participation coefficient around Toulouse in 1999

should reveal how much a municipality participates in the export of workers, which is another way to understand how much it acts as a residential area that feeds other regions with workers. Conversely, the incoming participation coefficient should reveal how significantly several municipalities act as an attractor, bringing workers into their local employment structure (Figs. 12.11, 12.12, and 12.13).

The above figures, which show smaller cities that contribute to local employment or that send their workers out to urban areas around major cities like Bordeaux, Toulouse, and Lyon have color/size matrices based on the incoming/outgoing participation coefficient.

These examples of major French urban areas illustrate the interdependent processes of urban sprawl and polycentrism while the different indexes applied during this stage lead to the characterization of these phenomena. Urban sprawl is undeniably present in all of the graphs that we show: the number of municipalities

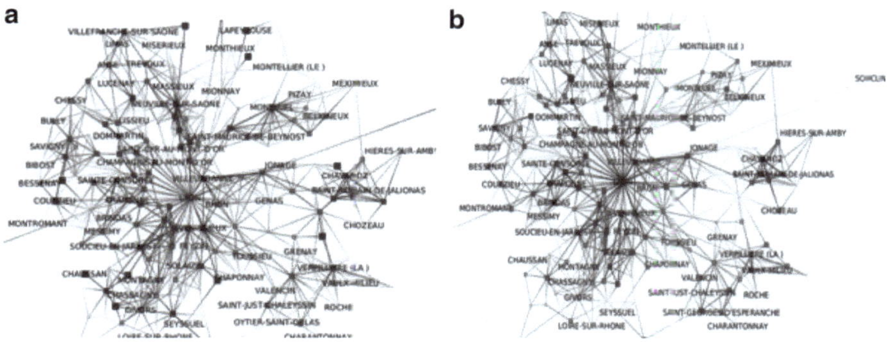

Fig. 12.13 (**a**) The incoming participation coefficient around Lyon in 1999. (**b**) The outgoing participation coefficient around Lyon in 1999

that are connected directly or indirectly with the city center grows during all of these periods. Moreover, these municipalities become more and more geographically distant from the city center over time.

The betweenness centrality index permits the identification of municipalities that polarize the territory and catch workers from many surrounding municipalities. This index illustrates the impact of the major French cities on the commuters' network.

The polycentric structure of the urban areas and its extension is also clearly identifiable through the application of the strength index. The increasing number of subcenters is evident, and their place and weight in the commuters' network also grows. The number of municipalities sending commuters to these subcenters becomes significant and increases the subcenter connectivity. More importantly, the municipalities situated in these subcenters become more tightly connected.

Finally, by applying the participation coefficient, we can characterize two types of municipalities: those who send workers out, defined as residential areas, and those who receive workers, defined as employment centers.

In summary, the commuters network can be characterized by a star topology, highlighted using the betweenness centrality index, with small world properties, illustrated by the strength metric. In the last section of this paper, we investigate this characterization further.

12.3.2 *Creating an Egocentric Network of an Urban City On-Demand*

In this work, we extracted egocentric networks from the complete data for all four census years to create a local view around a city of interest. The egocentric network of a city in graph G is defined as the subgraph of G induced by the nodes of the graph that are situated near, in a topological sense, the city of interest. The extraction is useful for finding local properties associated with a particular city. We had defined some terms that will be used when creating these egocentric networks.

Given **commuter data** for year D_i (this data includes the number of commuters between a pair of cities and the name and geographic position of each city), a commuter network $G_i(U_i, V_i)$ is defined where

- A node U_i is a city. Cities are mapped to their geographic coordinates. As a consequence, the relative position of the nodes reproduces the geography.
- An edge V_i connects two towns if they exchange commuters.

Given the set of commuter networks over all 4 years (G_1, G_2, G_3, G_4), the union of these graphs, $G(U,V)$, is defined as

$$U = \bigcup_i U_i \qquad V = \bigcup_i V_i$$

Therefore, G is also a network.

If C is the commuter data for 1 year, $C = Table(t_i, t_j, n)$ where t_i is the town of residence, t_j is the town of the workplace, and n is the number of commuters that pass between the towns. Let us denote $linkij$ as the weighted edge between t_i and t_j.

Path(t, length) is a ordered sequence of cities starting at t such that a link exists between the adjacent towns in the sequence, and the sequence is a given length.

Given a city, a number of paths of the distance length exist in the commuter data C for a given year. This data can be filtered where D = Filter(C) with ti in Path(t,length) or tj in Path(t,length). Next, we format all 4 years of our data, using the Tulip[7] framework, into a single file with all of the attributes available: the number of commuters for each year on all edges (a weight of 0 is entered if the edge did not exist for a given year); the geographic positions for each city; and the city names. The city names will be used as an identifier, and, as such, we must ensure that the names do not vary from census to census.

Building a visualization inside of the urban areas as delineated in the first stage can be achieved using a series of Tulip plugins (Auber, 2003). During this study, we choose to work on the major cities of France such as Bordeaux or Lyon. To begin, we work on the union of all 4 years between 1975 and 1999. Therefore, each edge in this graph has a collection of four weights.

All of the nodes that lie on a path of length less than or equal to three are included. A path length of at most three was chosen because the graph is densely connected. All of the cities of distance three or smaller from the query city are included. We could have used the weighted distance from the city center, but that remains for future work. The nodes and the edges of this graph were stocked in a vector for efficiency reasons. This city-centric view of the data can help the geographers to reason about the polycentric structures around a given city.

The next step is to filter the data, at several scales, around the regional and departmental levels surrounding the capitals. In the process of creating a graph on demand by combining all of the edges, we considered that the value of the edge could be positive (if this edge exists in 1 year of data) or negative (if there is no edge). So, imagining that we just need to have a graph in which the edges exist at all times, we could consider the possible values of the edge. We have suggested that it is better to approach the analysis as an extraction from the original network in which the urban area is easier to visualize.

12.4 Commuters Networks as a Scale-Free Network

Previously, we observed that the commuters' network has both a star topology and small world properties. In this section, we explain an approach that exploits these two qualities directly.

A network is said to be scale-free if the distribution of node degrees in the graph follows a power law. More formally, the probability distribution of node degrees, $p(k)$, has the following form:

$$p(k) \sim k^{-\gamma}$$

where k is the node degree and γ is the power law exponent. The principal characteristic of a scale-free network is that a few nodes have many connections while the majority of nodes have very few connections. The commuter network is, for the most part, scale free in nature. A few large cities import a number of workers every day while the vast majority of cities import very few workers. However, not all areas of the graph are scale free and the exponent y still needs to be determined. The areas of the graph that do not have a scale-free structure often correspond to places where polycentric activity occurs in the network.

Figure 12.14a shows the commuter network where the links have been already filtered by the number of commuters. The nodes in this graph are cities and directed edges exist if the cities exchange commuters. In the plot, the y-axis is the node degree and the x-axis is the node index. There are a few nodes with a very high degree, however, most nodes have low degree.

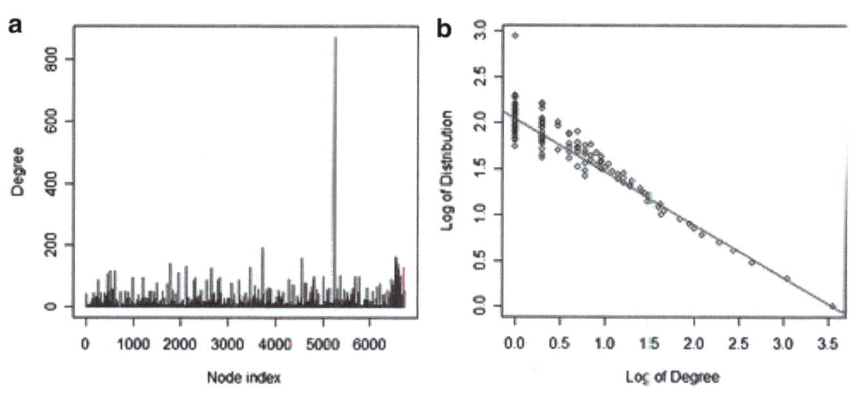

Fig. 12.14 (**a**) Degree distribution in network. (**b**) log-log plot as a straight line in migration network

Fig. 12.15 Migration network as scale free

Figure 12.14b shows a plot of the log of the degree versus the log of the frequency of that degree in the graph. If the degree distribution follows a power law, the data should fit a straight line with a slope equal to the exponent y. As shown in the figure, the commuter network appears to fit the power law distribution of degrees.

The high degree nodes in the commuter network often correspond to the major cities in France. Where there exists a central node with a very high degree, we try to detect these major cities and their star topology; the cities around it are often linked to it directly.

We now present a recursive algorithm that helps to illustrate the polycentrism in the commuter networks of France. The algorithm starts by checking to see if the passed graph is scale free. If the graph is scale free, we then decompose the graph into sub graphs using the nodes of highest degree in the network. If the graph is not scale free, the algorithm uses the strength metric to color the edges of the graph. The densely connected areas of the graph are highlighted in dark. Quite often, these areas correspond to the regions of the graph that exhibit polycentric behavior (Figs. 12.15 and 12.16).

12.4.1 Visualizing Polycentric Structures Using the Strength Metric

Although the commuter network for the most part is scale free, at the lower levels of the multi-level decomposition, the graph can exhibit small world properties. Because polycentric phenomena are more densely connected, they are more likely to have small world properties. The strength metric is used to quantify the participation of an edge in a cluster, and therefore, we can use the metric to assess whether or not an edge participates in a strongly connected cluster.

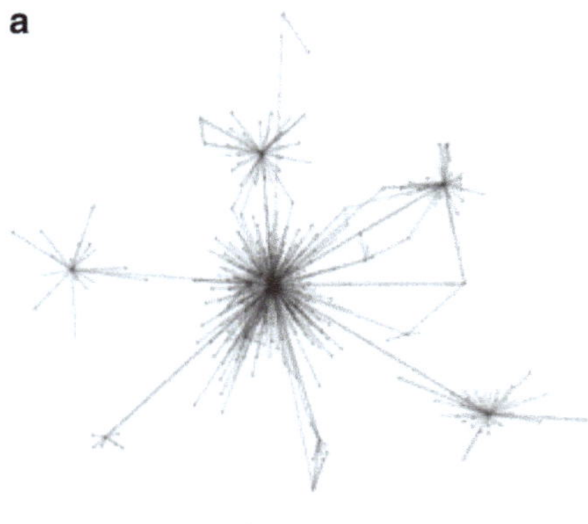

Fig. 12.16 (**a**) Star topology in the scale-free. (**b**) Cohesion on groups in sub-component

12.4.2 *Recursive Algorithm for Illustrating Polycentrism in a Network*

Step 1 (a) Find the cities that attract the most workers on a daily basis, or the powerful cities in the network. We determine these cities based on their weighted in-degree.

(b) Remove all powerful cities from the network.

(c) For each powerful city, place it back in the network with all of its edges and extract the connected component in which it is contained.

(d) Check whether the connected component has scale free topology. If yes, return to step 1 and recursively decompose this connected component. If not, go to step 2 to illustrate polycentric structure.

Step 2 To illustrate polycentric phenomena in the commuter network, we calculate the strength metric on each edge to determine the areas of high connectivity. Then, using this metric, we color the edges to illustrate the areas of high connectivity. Following figures illustrate polycentric structures around the major cities using darker color. The remaining areas of the graph are colored in pale (Figs. 12.17 and 12.18).

In this example, we illustrate the polycentric structure around "Pays de Montbéliard" in the department of Doubs (Northeast of France). Figure 12.18 shows the subcenter formed around the municipality of Audincourt.

12.5 Conclusion

In this chapter, we analyzed the data from four French national surveys of commuting networks. Because daily commuting patterns provide evidence for polycentric structures situated around urban areas, we have presented visual explorations

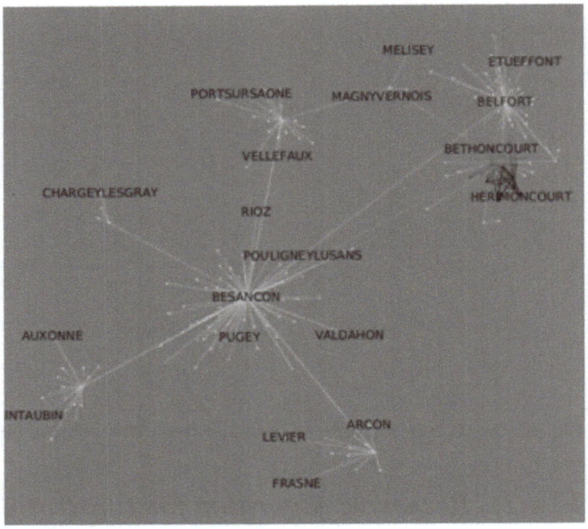

Fig. 12.17 Scale free structure of (star topology) and the structure of the dense sub-graph (zone colored *blue* by applying the strength – density cohesion zone) (color figure online)

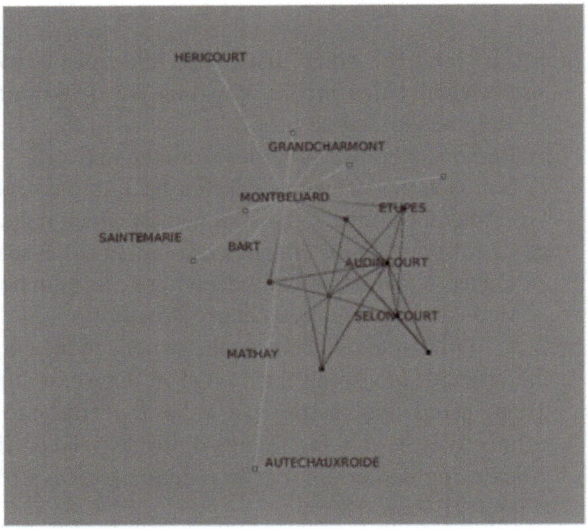

Fig. 12.18 Level of density cohesion zone – image of a polycentric event

of these phenomena. By filtering the data based on a gravity model in the first stage, we visualize the graph based on an accepted definition of polycentrism. Through metrics applied to the examples around Lyon, Marseille, Lille, Toulouse and Bordeaux, we then analyze the topological structure of the graph. The betweenness centrality allows us to locate the major French cities. Their power to attract workers from the surrounding municipalities is seen through the network growth around them. The strength index highlights the polycentric structure of the urban areas. In the examples of Marseille and Lyon, which evolve over time, the polycentric structure not only develops through the creation of new subcenters, but also through additional connectivity at two scales both from inside the municipalities forming the subcenter and from external municipalities. The coefficient of participation identifies where a community is situated in the sending/receiving commuter patterns of the network, characterizing areas as having either a residential or an employment dominant orientation. We also show how to construct a graph that may lead to a way to study network evolution over time. This work helps to demonstrate the impact of urban planning on the residential strategies of the active population inside urban areas.

Finally, based on the results of the second section, we describe a method to decompose the commuter network into its primarily scale-free parts while also highlighting some of its small world structure at a local scale. Our preliminary results demonstrate that a scale-free/small world approach to commuter network analysis is interesting but further investigation is needed.

References

Aguilera, A., & Mignot, D. (2004). Urban sprawl, polycentrism and commuting. A comparison of seven french urban areas. *Urban Public Economic Review*, *1*, 93–113.

Auber, D. (2003). Tulip – a huge graph visualization framework. In P. Mutzel, & M. Jünger (Eds.), *Graph drawing software* (pp. 105–126). Berlin/New York: Springer.

Auber, D., Chiricota, Y., Jourdan, F., & Melançon, G. (2003). Multiscale navigation of small world networks. In *Ieee symposium on information visualisation* (pp. 75–81). Los Alamitos, CA: IEEE Computer Society.

Berroir, S., Mathian, H., & Saint-Julien, T. (2002). Déplacements domicile-travail: vers le polycentrisme. *INSEE Ile-de-France Regards*, *54*, 5–8.

Berroir, S., Mathian, H., Saint-Julien, T., & Sanders, L. (2004). *Mobilités et polarisations: vers des métropoles polycentriques. Le cas des métropoles francilienne et méditerranéenne* (Etude dans le cadre du programme de recherche Mobilité et territoires urbains). UMR 8504 Géographie-cités.

Binder, J., Haag, G., & Rabino, G. (2003). Analysis and modelling of commuters flows: Application to the regions of Stuttgart and Turin. *Jahrbuch der Regionalwissenschaften*, *23*, 117–139.

Bonnefoy, J., Pumain, D., & Rozenblat, C. (1996). Théorie des graphes et interactions non gravitaires. In Bocquet-Appel, D. Courgeau, & D. Pumain (Eds.),*Analyse spatiale des données biodémographiques* (pp. 171–186). John Libbey/Ined Collection, 367p.

Bourne, L. (1989). Are new urban forms emerging? Empirical tests for Canadian urban areas. *The Canadian Geographer*, *33*(4), 312–328.

Brandes, U. (2001). A faster algorithm for betweenness centrality. *Journal of Mathematical Sociology*, *25*(2), 163–177.

Cervero, R. (1996). Mixed land-uses and commuting: evidence from the American housing survey. *Transportation research A*, *30*(5), 361–377.

Cervero, R., & Wu, K. (1997). Polycentrism, commuting and residential location in the San Francisco Bay area. *Environment and Planning A*, *29*, 865–886.

Giuliano, G., & Small, A. (1991). Subcenters in the Los Angeles region. *Regional Science and Urban Economics*, *2*(21), 163–182.

Gordon, P., & Richardson, H. (1996). Beyond policentricity. The dispersed metropolis, Los Angeles, 1970–1990. *Journal of The American Planning Association*, *62*(3), 289–295.

Guimerà, R., Mossa, S., Turtschi, A., & Amaral, L. A. N. (2005). The worldwide air transportation network: Anomalous centrality, community structure, and cities' global roles. *Proceedings of the National Academy of Sciences of the United States of America, 102*(22), 7794–7799.

Melançon, G., Trung Tien, P. Q., Tissandier, P., & Chiricota, Y. (2008). Visual exploration of (French) commuter networks. *Geovisualization of dynamics, movement, and change AGILE'08 satellile workshop*, Spain.

Nijkamp, P. (1975). Reflections on gravity and entropy models. *Regional Science and Urban Economics, 5*, 203–225.

Patuelli, R., Reggiani, A., Gorman, S. P., Nijkamp, P., & Bade, F. J. (2007). Network analysis of commuting flows: A comparative static approach to German data. *Network and Spatial Economics, 7*(4), 315–331.

Pumain, D. (2001). *Les interactions spatiales : flux et changements dans l'espace géographique*. Paris: A. Colin.

Rozenblat, C., & Tissandier, P. (2007). Commuters graphs and cities' polycentric cohesion. In *European Colloquium of Geographical and Quantitative Geography, Montreux - Switzerland.*

Schwanen, T., Dieleman, F. M., & Dijst, M. (2001). Travel behavior in Dutch monocentric and polycentric urban systems. *Journal of Transport Geography, 9*, 173–186.

Schwanen, T., Dieleman, F. M., & Dijst, M. (2002). The impact of metropolitan structure on commute behavior in the Netherlands, a multilevel approach. In *42nd ERSA Congress, Dortmund, 27-31 August.*

Sen, A., & Smith, T. (1995). *Gravity models of spatial interaction behavior.* Heidelberg/New York: Springer.

Shen, G. (1999). Estimating nodal attractions with exogenous spatial interaction and impedance data using the gravity model. *Papers in Regional Science, 78*, 213–220.

Sultana, S. (2000). Some effects of employment centers on commuting times in the Atlanta metropolitan area, 1990. *Southeastern Geographer, XXXXXI*(2), 225–233.

Talbot, J. (2001). Les déplacements domicile travail. De plus en plus d'actifs travaillent loin de chez eux. *Insee Première, 767, April 2001.*

Conclusion

Céline Rozenblat and Guy Melançon

This volume explored applications of networks to multilevel, multi-scale and multidimensional aspects of spatial analysis, discussed "complex network" theories and introduced the use of networks to represent the structural aspects of spatial phenomena. With the aid of visual analytics tools, empirical studies of large network samples examined hypotheses regarding the factors influencing the patterns of interactions represented by graphs.

The set of papers included in this volume not only described new analysis and visualisation techniques but also introduced new concepts relevant to geography and more widely for social sciences. Many different networks composed of individual entities were investigated: maritime networks linking ships, knowledge diffusion networks (e.g., scientific networks linking authors), mobility networks (e.g., commuter and airplane passenger traffic), and control networks (e.g., multinational firms' ownership of subsidiaries). In the studies presented in these chapters, networks were aggregated based on territorial attributes to group individual entities into new entities. In these new networks, the interacting entities were places or territorial units, and the individual entities of the original network generated the attributes of the newly formed entities. For example, aggregating companies based on their geographic location produced a set of cities that inherited attributes from the companies located in those cities (Chap. 10).

Using the processes designed to represent society and individuals to represent space has many implications for the analysis and representation of the results. This representation influences the meaning of the geographical interactions observed

C. Rozenblat (✉)
IGD – University of Lausanne, Lausanne CH-1015, Switzerland
e-mail: Celine.Rozenblat@unil.ch

G. Melançon
CNRS UMR 5800 LaBRI, Université Bordeaux 1 – GRAVITE INRIA Sud-Ouest, Cours de la Libération 351, Talence 33405, France
e-mail: melancon@labri.fr

C. Rozenblat and G. Melançon (eds.), *Methods for Multilevel Analysis and Visualisation of Geographical Networks*, Methodos Series 11, DOI 10.1007/978-94-007-6677-8,

in the social sciences and modifies views regarding the use of the individual networks analysed by sociologists (see Sect. 1 below). This representation transforms geographical space and introduces a new type of visualisation (see Sect. 2) to identify new resources in spatial networks (see Sect. 3). The transposition also identifies the empirical levels of hierarchies (Sect. 4) and cohesive spaces (Sect. 5), which enables interactions between these levels to be investigated (Sect. 6). Many perspectives emerge (Sect. 7) that relate these theoretical and empirical results to other approaches to complexity that employ dynamic models based on individual, as well as global, properties of systems of interactions.

1 Changing the Meaning of Social Networks Through Spatial Interaction

Places and their social components are shaped and transformed by internal and external networks, and conversely, geographical space is a major factor in the transformation of social networks and social entities (Adams, Faust, & Lovasi, 2012). Geography contributes to the understanding of social organisations by determining the impact of geographical and territorial distances on social, economical, cultural and organisational distances. Several contributors to this volume revealed the critical role of geography in identifying the relevant criteria for clustering these networks and the spatial cohesion that reinforces the interactions underlying cohesive territories. Although the concept of "functional territories" in geography (Gottman, 1961; Perroux, 1964) emerged 50 years ago, methods for identifying and understanding the evolution and transformation of these territories have only recently become available.

Just as networks are no longer viewed as random or homogeneous, territorial structures are shaped on spaces that are neither random nor homogeneous. Several issues were addressed – how space and social networks influence each other, the role of distance in linking similar individuals and reinforcing social proximities, and the extent to which networks increase the anisotropy of space. The examples presented in this volume demonstrated that there were two kinds of networks:

- Networks in which space was the major factor in the organisation, such as the slow transportation and daily exchanges of commuters;
- Networks that involved the contribution of other, nonspatial proximities, such as communication, collaborative research, and the global division and coordination of labour; these networks primarily support globalisation and multiscale development.

The second type of network transformed spaces and territories to create new "reticular territories" in which societies' physical space and organisation depended on economic or social distances that made distant spaces closer and neighbour spaces more distant. Examples of the phenomenon provided in this volume were air transportation networks (Chap. 9), and collaboration between researchers (Chap. 11). These two examples illustrated that certain distant cities were highly connected, while other closer cities were less connected. Thus, space and territories are more influenced by the density of networks compared to physical distance.

However, unlike such domains as sociology, the concepts and applications of graph theory could not be directly applied to geography. Studying geographical networks requires conceptual adaptations that extend beyond a strict topological description that focuses on the orientation of the links, weights, positions and attributes of nodes (Sect. 3.4.3). Gleyze argued that the concept of "reachability", which corresponds to the geographical concept of "accessibility", is not necessarily revealed by high density in the graph. Moreover, when networks are aggregated by geographical proximities, non central territories inherit centrality properties, dependencies and powers from institutions as firms, although these properties may not have existed prior to aggregation.

2 Transformation of Space By Networks

Networks create new territory properties by transforming the proximities of spaces. Networks reduce the distance of paths along links that are used to locate individuals based on their relative positions in the network. In Chap. 6, Lambert, Bourqui and Auber detailed general methods of network visualisation for graph layouts based on attractive forces. The process of moving individuals with more interactions closer to each other transforms geographical space into topological space. Network topology may sometimes coincide with geographical distance. However, networks often distort space, and cohesive territories or communities are not based on geographical proximity. The topological proximity of Malta and UK in the Eurovision network provides a typical example (e.g., Gleyze's example of votes for Eurovision songs in Sect. 3.4.3).

In Chap. 4, Alain L'Hostis considered the time-distance relationship and described the spatial inversion phenomenon and the time-space contraction created by transportation networks' selection of a limited number of places. This inversion is difficult to represent by maps, which preserve the order of spatial proximities because they adopt Euclidean principles for representing distance in two dimensions. Abandoning Euclidean space and the rule that the shortest path is represented as a straight line, Alain L'Hostis introduced the "shrivelling" of maps that create new spatial terrain through "time-space" peaks and valleys. This process is similar to the approach adopted in astrophysics, where the space-time of the universe is deformed by general relativity in crumpled spaces (Luminet, 2001). Applied to terrestrial transport, this "crumpling" generates new connections and new proximities that allow new properties of geographical space to emerge. Tunnel effects that reflect the nonaccessibility of the space between two nodes eliminate part of the space between two linked nodes, and geographical distances are no longer symmetrical between places. Space is transformed by conditional relative distance, as well as the new dominating factors of speed and distance evaluation, which creates new centres and peripheries that redistribute the concentration of wealth and resources.

3 New Resources Throughout Networks

Network analysis provides a new perspective for spatial analysis and shifts the focus from how many resources a location possesses to the extent to which resources can be accessed from that location. Thus, privileged places are nodes that access many resources through direct or indirect links and take advantage of other nodes to reach resources they do not possess. These intermediary nodes play new roles in the local or global network because they provide paths for accessing resources. This strategic role implicitly introduces weaknesses into the network as a whole. The network is fragile because reachability depends on these relay nodes and their elimination influences the overall accessibility of the network. Although the nodes themselves are not fragile, their position in the network and the possibility that they might be eliminated makes those nodes fragility sites for the network. In Sect. 3.4.3, Gleyze noted that vulnerability occurs when the relations between nodes are not sufficiently redundant because removing edges causes node disconnection and the fragmentation of clusters. In relation to the issue of vulnerability, Gleyze distinguished between *n*-cliques, *k*-plex and *k*-core methods.

Just as there are fragile nodes, there are also fragile edges. The network becomes a new resource because links become more important than nodes. This feature is important for transportation networks, where power in the network is more and more concentrated in enterprises that manage shipping lines (see Chap. 8) or in primary air traffic lines (see Chap. 9).

Conversely, some links or nodes are highly resilient. Robust economic and financial edges such as the links between London and New York dominate the entire network and connect economies in different domains. The mutual support provided by these different networks increases their strength. Recent physical simulations demonstrate that similar networks are more robust than different networks to the multiple effects of failure (Buldyrev, Parshani, Paul, Stanley, & Havlin, 2010; Parshani, Rozenblat, Ietri, Ducruet, & Havlin, 2010).

4 Hierarchies

Hierarchies naturally appear in various contexts, and spatial geography is no exception (Pumain, 2006). The term itself may refer to an order hierarchy or a structural hierarchy. Hierarchies may be based on organisational structure (e.g., NACE codes classify industrial activity into distinct sectors and subsectors) or may emerge from territorial considerations (e.g., continents are divided into countries that are subdivided into regions and towns). Hierarchies can also emerge from functional dependence, due to the asymmetry of social or economic relations. The various studies presented in this volume refer to hierarchies in different scenarios.

The interactive visualisation DAGMap referred to two different hierarchies, with one governing how subsidiaries organise and the other governing how they distribute over world regions. The two hierarchies were merged into a single visual structure, a TreeMap displaying the organisational hierarchy and colour-coding world regions. Examining how the hierarchy of subsidiaries was globally distributed supported the analysis of patterns reflecting the potential impacts of globalisation. Specific interactions were designed to allow the analyst to navigate through the different levels of the organisational hierarchy and identify colour-coded patterns (see Figs. 7.11 and 7.13).

In Chap. 6, the edge bundling method produced a hierarchy that integrated physical distances and graph topology. Although the goal of the method was to improve the readability of a map, edge bundles also reduced edge cluttering in graphs by routing edges along Bézier curves. Similar to creeks and smaller rivers that flow into a larger stream, the edges were partly merged. The original graph connections were combined into higher-level edges that described overall flows on the map, which created a new hierarchy in the graph by aggregating closely edged segments. This hierarchy of bundles of routes provided an original and novel method for visualising a hierarchy in space.

The order hierarchies present in networks are often used to compare or reveal node roles, as mentioned above. Node degree remains an essential node statistic that captures the overall structure of a network and often reveals the graph's scale-free structure. Although not systematically or explicitly stated in the various studies presented in this volume, most of the networks studied were scale-free. This property is not surprising because places are also hierarchised based on economic dependency. For example, the Christaller model (1933), which is the root of the settlement system, is hierarchised by access to services as well as economic and political functions. Further theoretical research into geographical organisations with different types and scales of hierarchies would be worthwhile (Castells, 1996) and would detect the specific scales that optimally represent a phenomenon's spatial organisation.

5 Detecting Critical Structural "Levels" in Global Organisations

In summary, the methodology adopted in the SPANGEO project consisted in comparing and contrasting the network topology of the regional partitioning of nodes (places). The network's modular and multilevel organisation was based on identifying key nodes or edges that decomposed the network into smaller, denser components. Various connectivity statistics were used to identify the minimal number of edges to eliminate to disconnect the graph. In Chap. 6, Lambert, Bourqui and Auber presented geographically nested and socially nested network dichotomies. The geographical organisation identified the intercontinental links that dominated the air passenger traffic network. Globalisation facilitated the emergence of clusters at the centre of the worldwide network, which emerged as essential links between continents. Because the analysis was driven by the network topology, we were able to detect which parts of the network remained in continental organisations and which parts transcended them.

The present volume revealed how each node of a network participates and contributes to the formation of clusters (Chap. 5). Because a node can participate in more than one cluster, the concepts of "participation" and "contribution" inspired from factor analysis were used to evaluate how each node participated in specific clusters and the extent to which participation in a specific cluster contributed to its internal cohesion. This approach identified several issues:

- The extent to which these procedures could be used to evaluate polycentric structures, such as the commuters described in Chap. 12;
- The extent to which these measures might be described at multiple levels;
- How to introduce fuzzy measures in cluster membership.

Various methods to define clusters were developed by the authors of this volume: DAGMaps, strength clustering, Lambda-sets, and multistep methodologies. Comparisons of different methods of clustering developed for air passenger traffic (in Chap. 9) and commuting (in Chap. 12) did not identify a universal method for detecting the communities comprising different patterns and indicated that clustering is a strategic approach that must be adapted to each specific problem.

Detecting clusters simplified large and complex networks by abstracting cohesive groups and aggregating these groups into higher-level entities to produce smaller, simpler networks. Addressing the issues raised by clustering graphs, Gleyze distinguished two concepts (Chap. 3). The concept of "cohesion" corresponds to the currently popular notion of modularity that is often embodied and measured by edge density. However, other types of clusters that do not qualify as dense subgroups might also be of interest, such as star-shape clusters with the central concept of "reachability". The same network may exhibit different patterns, depending on the level or scale at which it is defined. This heterogeneity of patterns requires new methods to detect the different levels of networks' organisation.

6 Interactions Between Levels

Examining interrelations between organisations at different geographical scales contributes to the understanding of the equilibrium and disequilibrium of territories (Chap. 2). Many social phenomena developed at one level have their root in other levels, and these levels interact. This interaction is illustrated by the subsidiarity of firms, which is produced both by local processes in cities and by their global position (Chap. 10). In Chap. 11, Comin examined relay cities between scales based on the assumption that most components in scientific collaboration networks were intranational and that certain cities linked these components to the structure as a whole. Interactions between levels influenced actors' ability to address each of these levels, and geographical aggregation was important for decision makers. Network patterns made it possible to identify individuals and their interactions, although further research is needed to determine routines and patterns fully.

7 Perspectives

This volume provided an initial exploration of the application of network analysis and visualisation to geography and introduced novel perspectives and new issues to the field. Future work should further this approach through theoretical accounts that link the description of networks to specific processes. For example, in which contexts and at what scales do network patterns, such as stars or clusters, emerge and remain robust and resilient? How are clusters and levels of organisations modified by the evolution of the system as a whole? To what extent does geographical distance contribute to the development of different networks at different scales? Adapting the methods presented in this volume to make them accessible to geographers will allow them to address these issues and many others.

It is worthwhile to review the origin of the project that brought the authors in this volume together. Many problems that were addressed in the present volume had never been fully visualised, and the case studies presented here often involved data sets that had never been fully explored because the data were too extensive to be spatially mapped. The interactive visualisation of graphs provided a means to analyse the entire data set based on the visual analysis and inspection of salient features.

One consequence of this approach was the realisation that physical distances were not central to the analysis, which was based on topological properties of the data sets. This pragmatic approach contrasted physical distances with "social" distances that were measured through the intensity of interactions between entities. This idea is a fundamental principle of network science that parallels similar concepts in the social sciences (Martin, 2009), and this perspective allowed computer scientists to contribute to the data analysis. The present volume documents the value of this type of idea exchange.

The graph visualisation methodology also exploited the central paradigm of "scale", which generated many debates on network hierarchies and the "small world" phenomenon. The constraints imposed by the manipulation of large datasets on standard visualisation displays drove downsizing approaches that transformed large graph structures into hierarchies of smaller subgraphs. Because hierarchies occur as a structural principle in nature (Pumain, 2006), many computational approaches have been successfully applied to these datasets to render the entire data set accessible; in addition, the downsizing process implicitly addressed the concept of data scale. Thus, the computational approaches were relevant to geographers and to social scientists more generally because the concept of scale is central to these fields. Thus, computational approaches based on mathematical features of the data sets provided a paradigm through which both geographers and computer scientists could construct hypotheses. The papers in this volume report the findings and results of SPANGEO project members that exhibit how concepts established through collaboration could generate insightful discoveries.

This new approach to theories of territorial and spatial organisation should strengthen the integration of network analysis into general spatial analysis through the transformation of space by networks and the new resources provided by network hierarchies and levels. To improve understanding of the common and specific properties of spatial networks in empirical studies, network simulations based on small theoretical sets of nodes and subsequently applied to empirical situations should be developed. Exchanges between theoretical studies, empirical studies and simulations should promote progress in this new field.

References

Adams, J., Faust, K., & Lovasi, G. (Eds.). (2012). Capturing context: integrating spatial and social network analyses. *Social Networks, 34*(1).

Buldyrev, S.-V., Parshani, R., Paul G., Stanley, H.-E., & Havlin S. (2010). Catastrophic cascade of failures in interdependent networks. *Nature*, *464*, 1025–1028.

Castells, M. (1996). *The rise of the network society*. Oxford: Blackwell.

Christaller, W. (1933). *Die zentralen Orte in Süddeutschland*. Jena, Germany: Gustav Fischer.

Gottman, J. (1961). *Megalopolis. The urbanized north-eastern seaboard of the United States*. New York: Twentieth Century Fund.

Luminet, J.-P. (2001). *L'Univers chiffonné*. Paris: Fayard.

Martin, J. L. (2009). *Social structures*. Princeton, NJ: Princeton University Press.

Parshani, R., Rozenblat, C., Ietri, D., Ducruet, C., & Havlin, S. (2010). Inter-similarity between coupled networks. *EPL (Europhysics Letters), 92*(6), 68002.

Perroux, F. (1964). Economic space: Theory and applications. In J. Friedmann, & W. Alonso (Eds.), *Regional development and planning* (pp. 21–36). Cambridge, MA: M.I.T. Press.

Pumain, D. (Ed.). (2006). *Hierarchy in natural and social sciences*. Dordrecht, the Netherlands: Springer.

Authors

C. Rozenblat and G. Melançon (eds.), *Methods for Multilevel Analysis and Visualisation of Geographical Networks*, Methodos Series 11, DOI 10.1007/978-94-007-6677-8,

N

O

P

R

S

T

Terms

C. Rozenblat and G. Melançon (eds.), *Methods for Multilevel Analysis and Visualisation of Geographical Networks*, Methodos Series 11, DOI 10.1007/978-94-007-6677-8,

MIX
Papier aus verantwortungsvollen Quellen
Paper from responsible sources
FSC® C105338

If you have any concerns about our products,
you can contact us on
ProductSafety@springernature.com

In case Publisher is established outside the EU,
the EU authorized representative is:
Springer Nature Customer Service Center GmbH
Europaplatz 3, 69115 Heidelberg, Germany

Printed by Libri Plureos GmbH
in Hamburg, Germany